Agrarian Reform as Unfinished Business

The Selected Papers of Wolf Ladejinsky

Agrarian Reform as Unfinished Business

Louis J. Walinsky, Editor

Published for the World Bank · Oxford University Press

Oxford University Press
NEW YORK OXFORD LONDON GLASGOW
TORONTO MELBOURNE WELLINGTON CAPE TOWN
IBADAN NAIROBI DAR ES SALAAM LUSAKA ADDIS ABABA
KUALA LUMPUR SINGAPORE JAKARTA HONG KONG TOKYO
DELHI BOMBAY CALCUTTA MADRAS KARACHI

The views and interpretations in this book are those of the author and editor and should not be attributed to the World Bank, to its affiliated organizations, or to any individual acting in their behalf.

Library of Congress Cataloging in Publication Data

Ladejinsky, Wolf Isaac
Agrarian reform as unfinished business: the selected papers of Wolf Ladejinsky.

Bibliography: p. 580
Includes index.
1. Land reform—Asia—Addresses, essays, lectures.
I. Walinsky, Louis Joseph, 1908– II. Title.
HD843.2.L3 1977 333.1'095 77-24254
ISBN 0-19-920095-5
ISBN 0-19-920098-X pbk.

FOREWORD

THE READER IS ABOUT TO EMBARK on a voyage which spans decades of agriculture history in a score of developing countries. The perspective offered should be valuable to anyone concerned with the unfinished business of alleviating rural poverty. These studies in agrarian policy and land reform also provide a fascinating glimpse into the life and work of an extraordinary man. They have been edited with love and care. The collection is a splendid one, and I would not detain the reader at all if it were not appropriate to say something about why this book came into being under the auspices of the World Bank.

For this is not the kind of publication which the World Bank normally sponsors. It does not conform to accepted canons of modern economic research. Nor does it derive entirely from experience with development projects of the Bank. It is unencumbered by equations and quite free of economic jargon. Its beautifully crafted language has an unusual, almost archaic flavor.

But then, Wolf Ladejinsky was an unlikely member of the World Bank staff. Some may have thought of him as old-fashioned. Yet he was worrying about the rural poor and doing something to help them in Japan and elsewhere long before they became uppermost in the preoccupations of development economists.

He joined the Bank some years after the normal retirement age. Hence, the bulk of the papers presented here were produced when Wolf Ladejinsky was working for institutions other than the World Bank. In any event, his was a fiercely independent mind, unfettered by institutional and conventional constraints. He spoke for peasants everywhere with urgency and realism. Indeed, it was because of his fresh approach to the development problems of traditional rural societies that I recommended that he join the agricultural team led by Sir John Crawford which eventually put forward major agriculture policy recommendations to the government of India and the World Bank in 1966, on the eve of what came to be known as the Green Revolution. Wolf Ladejinsky made a notable contribution to the work of this mission, and I subsequently asked him to remain in the service of the Bank. He did so until his death on July 3, 1975.

He toiled hard until the very end, keeping tabs on the huge agricultural and rural sector of India, with occasional and fruitful forays in neighboring countries. He was a tireless traveler, a voracious reader, and a patient listener. He knew how to get farmers and petty officials to talk and tell all. His reports were based on firsthand observation. They had an impact because the author was known as a man passionately dedicated to ferreting out and reporting the truth.

He rarely reported good news and never succumbed to the self-adulation prevalent among even the most enlightened of bureaucracies. As a result, he made enemies as well as friends. But his stature grew and, on the whole, given his obvious sincerity and his abiding intolerance of avoidable poverty, he was able to speak and write without giving offense where others would have remained silent.

Because of his selfless dedication to the realities of the development business, Wolf Ladejinsky never took the time to collect his writings in order to reach a wide audience. Yet the concerns eloquently expressed in this volume have become central to the task of development administrators everywhere and are arousing growing interest in the academic community and among the public at large. Given the new orientation toward rural development of World Bank lending, it is appropriate and timely for the World Bank to disseminate the insights and views—however controversial—of Wolf Ladejinsky, one of the few truly original development thinkers and practitioners of our times.

I. P. M. CARGILL
Vice President, Finance
The World Bank

ACKNOWLEDGMENTS

THANKS ARE DUE TO MANY PERSONS for invaluable assistance in locating and making available Ladejinsky's writings, for providing information about him and his work, for suggestions as to the selection of papers for inclusion, for comments on the introduction, and for editorial and other assistance. This appreciation is no less warm for the brevity with which it is expressed.

Robert Picciotto collected the papers prepared under the World Bank's auspices. Charles Olsen of the Joint World Bank–International Monetary Fund Library searched and provided copies of published papers. Eldon Jones, Ford Foundation archivist, searched and provided copies of papers and letters from the foundation's archives. Similar assistance was provided by T. H. Shen for the Joint (U.S. and China) Commission on Rural Reconstruction; Gladys Baker, Gill Picquette, and Penny Cate for the U.S. Department of Agriculture; Barbara Ennis and Betty Donovan for the U.S. State Department; and Helen F. Ulibarri for the U.S. National Archives. Marsha Swiss and Mr. and Mrs. Peyton Kerr, Ladejinsky's executors, granted free access to his personal papers and full discretion as to their use.

Colleagues, associates, and friends of Wolf Ladejinsky who provided enlightening information about the man and his work at various stages of his career included Moishe Borotkin, David Hopper, Abe Weisblat, Nathan Koffsky, Martin Abel, Shigaharu Takahashi, Horst Eschenberg, William Gilmartin, Robert Fearey, Ferdinand Kuhn and Delia Kuhn, General E. G. Lansdale, and Raymond Moyer.

Professor Abel, Sir John Crawford, Mr. Gilmartin, John Mellor, and Arun Shourie offered suggestions with respect to the selection of papers for inclusion. Anne Hamilton, Roberta Leary, and Messrs. Abel, Gilmartin, Picciotto, Koffsky, Enrique Lerdau, Ashok Mitra, Gregory Votaw, and Robert Nathan offered valuable editorial comments and suggestions on the Introduction. Mr. Mitra generously granted permission to cite in full his published tribute to Ladejinsky.

Roberta Leary provided helpful editorial assistance and efficient organization of index materials; Marlene Houben and Martha Sanchez Gavito provided invaluable secretarial assistance; Rachel C. Anderson and Jane Carroll provided technical editing, Mrs. Anderson prepared the index, and Brian J. Svikhart supervised the production of the book; John Adler, Vittoria Winterton, and Jivat Thadani were most helpful with respect to administrative and logistic arrangements.

Anne Hamilton and Messrs. Adler, Gilmartin, Picciotto, Votaw, and Yudelman of the World Bank took a special and continuing interest in the project from its inception and lent it their support. I. P. M. Cargill, who initiated and took chief responsibility for the project, is due special thanks for his interest, encouragement, and support throughout. I should record here that the World Bank imposed no constraints other than space on the selection or presentation of these papers. As the editor I have had a free hand and take sole responsibility for any errors of taste or judgment which may be found.

LOUIS J. WALINSKY

TABLE OF CONTENTS

FOREWORD v

ACKNOWLEDGMENTS vii

INTRODUCTION 3

I. THE WASHINGTON YEARS, 1935–1945 23

1. *Collectivization of Agriculture in the Soviet Union, March 1934 (two excepts)* 24
2. *Soviet State Grain Farms, October 1938 (conclusion)* 28
3. *Agricultural Problems of India, August 1939 (selection)* 30
4. *Japan's Agricultural Crisis, August 1939* 39
5. *Chosen's Agriculture and Its Problems, February 1940 (selection)* 49
6. *Agriculture of the Netherlands Indies, September 1940 (excerpt)* 58
7. *Agricultural Policies of British Malaya, April 1941* 61

II. THE TOKYO YEARS, 1945–1954 67

8. *Farm Tenancy in Japan, June 1947* 68
9. *Trial Balance in Japan, October 1948 (excerpt)* 93
10. *The Rent Reduction Program in Taiwan: Field Observations, September 1949* 95
11. *Land Commissions in Japan, September 1949* 109
12. *Field Trip in Szechwan: Tenant Conditions and Rent Reduction Program, October 1949* 113
13. *Too Late to Save Asia? July 1950* 130
14. *Rural Reconstruction under the China Aid Act, August 1950* 136
15. *Observations on Rural Conditions in Taiwan, June 1951 (summary and conclusions)* 142
16. *From a Landlord to a Land Reformer, June 1951* 148
17. *The Plow Outbids the Sword in Asia, June 1951* 151
18. *Field Observations in the Punjab, October 1952* 154
19. *Land Reform Observations in Madras, November 1952* 162
20. *Land Reform Observations in Kashmir, November 1952* 178
21. *Comments on Land Reform in India, December 1952* 189
22. *A Comment on the Report of the Ford Foundation Conference on Land Tenure, March 1953* 198

23. *The Status of the Land Reform Program in India, August 1954 (conclusions)* 201
24. *Advancing Human Welfare, November 1954* 204

III. The Vietnam Years, 1955–1961 215

25. *Field Trip Observations in Central Vietnam, April 1955* 217
26. *Field Trip in Southern Vietnam, April 1955* 229
27. *A Visit with President Ngo Dinh Diem, June 1955* 239
28. *South Vietnam Revisited, July 1955* 243
29. *U.S. Aid for Land Reform in Vietnam, July 1956* 268
30. *Making the Pending Land Redistribution Program More Practicable, October 1956* 271
31. *Toward a More Effective U.S. Aid Program in Vietnam, November 1956* 275
32. *Agrarian Revolution in Asia, April 1957* 279
33. *Agrarian Revolution in Japan, October 1959* 280
34. *Self-description/Appraisal, June 1960* 289
35. *Exploration of Job in Nepal, November 1960* 290
36. *Nepal's Five-year Plan, November 1960* 294
37. *Land Reform in Indonesia, January 1961* 297
38. *Agrarian Reform in the Republic of Vietnam, 1961* 299
39. *Corporate Farm Management for Japan? June 1961* 312

IV. The Ford Foundation Years, 1961–1964 315

40. *Tenurial Conditions in Nepal, February 1962* 316
41. *India after the China Border Clash, November 1962* 324
42. *Visit to the Philippines, January 1963* 325
43. *Agrarian Reform in Nepal, March 1963* 332
44. *Tenurial Conditions and the Package Program in India, 1963 (summary and sample recommendations)* 336
45. *Land Reform in Indonesia, February 1964* 340
46. *The Industrialization Bias in Economic Development, February 1964* 352
47. *Land Reform, July 1964* 354

V. The World Bank Years, 1964–1975 367

48. *Agrarian Reform in India, October 1965* 369
49. *The New Agricultural Strategy and Institutional Factors, June 1968* 405
50. *Punjab Field Trip, April 1969* 432
51. *Bihar Field Trip, July 1969* 442
52. *Agriculture—Some Broader Considerations, April 1970* 462
53. *A Note on Agriculture Reforms, August 1970 (summary)* 474
54. *Field Trip to Eastern Uttar Pradesh, May 1971* 475
55. *Agriculture—Problems and Needs, May 1971 (selection)* 481
56. *Refugees in West Bengal, August 1971* 487
57. *Agriculture—Issues and Programs, May 1972 (selection)* 493
58. *Drought in Maharashtra: Not in a Hundred Years, January 1973* 514
59. *The Rural Scene, May 1973 (selection)* 532
60. *Agrarian Reform à la Punjab, September 1973* 542

61. *Agrarian Reform in the Philippines, October 1974* 550
62. *Food Shortage in West Bengal—Crisis or Chronic? December 1974* 557

APPENDIXES 567

A. The Atcheson-Fearey Memorandum, October 26, 1945 569
B. SCAP Directive 411 on Rural Land Reform, December 9, 1945 579
Chronological Bibliography of Wolf Ladejinsky 580
Depository Libraries for the Ladejinsky Papers 586

INDEX 589

Agrarian Reform as Unfinished Business

INTRODUCTION

FROM THE END OF 1945 WHEN HE was posted to Japan to assist General Douglas MacArthur in planning the postwar land reform there, Wolf Ladejinsky spent the last thirty years of his life almost entirely in Asia. These three decades were devoted to the cause of agrarian reform on behalf of the hundreds of millions of submarginal farmers, tenants, sharecroppers, and landless laborers who had become his human as well as his professional concern, and whose cause he had made his own.

Another, broader concern was at least equally compelling. Having grown to young manhood in Czarist Russia in the midst of the impoverishment, alienation, indignity, and persecution that characterized life in the restricted Jewish community (*shtetl*), and having experienced at first hand the brutalities that accompanied the Bolshevik revolution, Ladejinsky had a passionate dedication to democracy. This had been intensified by the holocaust Hitler visited upon European Jewry during World War II. Ladejinsky thus committed himself entirely to the survival and flourishing of democracy in the newly independent countries of Asia. Such an outcome, he was convinced, depended on satisfying the basic needs and yearnings of impoverished rural Asians for a bit of land they could call their own, or at least for security of tenure and a tolerable rent on the land they cultivated for absentee and exploitative landowners. Only in this way could they escape from the grinding poverty and personal indignity they were increasingly unwilling passively to accept. Aware of how powerfully Lenin's promise of "land to the tiller" had influenced the Russian peasantry to accept and support the revolution, all too soon to be dispossessed again in favor of collective and state farms, Ladejinsky developed a profound sense of the *political* role of the land. He recognized the importance of ownership, tenurial rights, and the distribution of the land's rewards in determining whether democracy would indeed survive in Asia or whether its rural masses would succumb to the Communists' promise.

Ladejinsky's major contribution to the highly successful land reform in Japan (1946–48) brought him almost instant renown and led to requests for his advice and assistance in many other lands.[1] He made a significant contribution to the equally successful land reform in

1. Ironically, this recognition became even more pronounced following the notoriety which Ladejinsky received late in 1954. When responsibility for the work of agricultural attachés abroad was turned back to the Department of Agriculture by the State Department, U.S. Secretary of Agriculture Benson refused to re-employ Ladejinsky on the ground that he was a "national security risk." The outpouring of support for Ladejinsky which then ensued from the press, professional associates, the Japanese government, and farm organizations and individuals in the United States soon forced a retraction and apology from Secretary Benson and a revision in U.S. national security procedures. It also made Ladejinsky and the general nature of his overseas work known to millions, both at home and abroad.

Taiwan in the early postwar years, encouraged and provided guidance to agrarian reform efforts in India in the early 1950s, was a prime mover in less well-known land reform successes in South Vietnam in the late 1950s, and encouraged and provided guidance to incipient agrarian reform stirrings in Nepal, Indonesia, and the Philippines in the early 1960s. Most of his last ten years were spent fighting, against enormous odds, to channel ineffectual agrarian reform efforts in India into more constructive and practical channels and trying to generate the political *will* without which meaningful reforms could not be implemented. If, in consequence, he came to be regarded by many as "Mr. Land Reform," this was in a sense a misnomer. For humanity was his deeper cause and, for him, human welfare and dignity in Asia could be promoted only under democracy, not under Communism, which he saw as the only likely alternative.

Antecedents

Wolf Isaac Ladejinsky was born in 1899 in Ekaterinopol, a small town or hamlet in the Russian Ukraine. His father was engaged in flour milling and the timber trade—a man of relative means in a predominantly Jewish community in the Pale of Settlement to which Jews were by law restricted. Precluded from farming and most urban occupations, Russian Jews were almost uniformly poor or on the verge of destitution. Demeaned and persecuted at best, and subjected at worst to periodic pogroms, they were able to endure only by nurturing a close communality and an intensely religious inner life. The more intellectual and spiritual among them escaped into religious studies. The relatively few who achieved a degree of material success, as did Ladejinsky's father, enjoyed some prestige in the community (witness Tevya's song, "If I Were a Rich Man," in *Fiddler on the Roof*); but genuine reverence and honor were accorded only to learning and the learned—in the shtetl, invariably rabbis and biblical scholars. This tradition led Ladejinsky to scholarship. One of the very few of his local peers who completed the Gymnasium (the secular secondary school), he also attended in his younger years the elementary *cheder* for Hebrew and religious instruction.

Little is known of how the Ladejinsky family fared after their properties were expropriated by the revolution. According to one report, his brother was killed in the civil war. In 1921, two years after completing his studies, Ladejinsky walked out of the Soviet Union into Romania. He worked briefly in a flour mill and as a baker's apprentice before getting a job in Bucharest with the Hebrew Immigrant Aid Society. But the United States—the "Golden Land" which, from the 1880s to World War I, had drawn perhaps a third of all East European Jewry as immigrants—was almost surely his destination from the time he left home. The following year, 1922, with a group of Jewish orphans who had been entrusted to his care for the journey, Ladejinsky entered the United States. A variety of odd jobs sustained him while he learned enough English to enter Columbia University in 1926 and then while he earned his degree, which was granted in 1928. He sold newspapers at a stand on Sixth Avenue and 50th Street in New York City while he continued his graduate work in economics and history into the early 1930s. His first published paper, the classic "Collectivization of Agriculture in the Soviet Union," apparently was intended to have been his doctoral dissertation. In the early years of the Great Depression, however, Ladejinsky withdrew from the university and accepted a post with the Department of Agriculture offered by Rexford Tugwell, one of his professors, who had been called to Washington by the newly elected President Franklin D. Roosevelt. This job launched a richly productive professional career that was to span four decades, terminating only with Ladejinsky's death while serving as a member of the World Bank's resident mission in New Delhi.

Work Career

Ladejinsky's first eleven years in the U.S. government's service (1935–45) were spent in Washington with the Office of Foreign Agricultural Relations of the Department of Agriculture. Here his work was essentially scholarly. Working from source materials and internal documents, he completed for publication in the department's official journal some twenty-odd studies of agriculture, mostly in Asian countries. Although many of these were quite conventional (such as "Agriculture in Manchuria—Possibilities for Expansion," "The Japanese Cotton-Textile Industry and American Cotton," "Japan's Food Self-Sufficiency," and "Thailand's Agricultural Economy"), others made important contributions to the crystallization of Ladejinsky's interests, ideas, and values and to the shaping of his later career.

His 1937 study of "Farm Tenancy and Japanese Agriculture" was his first major venture into tenurial problems outside the Soviet Union. Together with his 1939 *Foreign Affairs* article, "Agrarian Unrest in Japan," it established him as the Department of Agriculture's expert on Japanese agriculture. When it later became Allied policy to "democratize" a defeated Japan, it was thus only natural that Ladejinsky would be called upon to assist in planning a thoroughgoing land reform in that country. His 1938 study of "Soviet State Farms" strengthened his already strong antipathies to forced collectivization, to the imposition of rigid central plans, policies, and administration on highly disparate agricultural production situations, and to measures which either violated the independence and integrity of the individual cultivator or smothered his incentive to produce. In 1939 Ladejinsky's substantial study of "Agricultural Problems in India" provided a solid base for his future work in that country and introduced him to tenurial and other problems in India's complex agricultural economy. For the first time he faced the severity of India's population problem and the limitations that the land-man ratio would necessarily impose on any future efforts at urgently needed agrarian reforms. Wartime studies of agricultural policy in colonial situations, such as "Agricultural Policies in British Malaya" and "Manchurian Agriculture under Japanese Control," confirmed his convictions that peasant cultivators could be constructively led, given an appreciation of their interests, but that they could not be manipulated or successfully controlled to act in ways contrary to their perceived interests. By the time Ladejinsky arrived in Tokyo to participate in the Japanese land reform effort, he had served his apprenticeship. Already in his middle years, he was a man who had experienced and studied much, reflected deeply, and developed powerful convictions. He was a man strong enough to impress General MacArthur.

Ladejinsky never wrote about his own role in the Japanese land reform, although many have credited him with having been its chief architect. While the State Department was still undecided on the matter, General MacArthur decided to proceed with land reform on the basis of a memorandum presented to him by his political adviser on October 26, 1945 (see Appendix A). This memorandum was the outcome of intensive consultations with Wolf Ladejinsky in Washington.[2] Agricultural specialists attached to the occupation headquarters

2. Robert Fearey, who prepared the memorandum, is the authority for this statement.

thereupon got to work, concentrating largely on the Fearey memorandum and Ladejinsky's earlier studies. Ladejinsky himself was called to Japan in December. Meanwhile the key SCAP[3] Directive 411, issued by order of General MacArthur December 9, instructed the Japanese government to submit by March 15, 1946, a program for rural land reform and established basic guidelines for such a program (see Appendix B). After almost a year of intensive study, planning, and negotiations back and forth with the Japanese, legislation acceptable to the occupation authorities was enacted. During this period, according to the official historian, Ladejinsky "knew more as a student about the general topic of land reform in its historical, political and economic aspects than anyone else. His tremendous enthusiasm was a continuous energizing factor in its accomplishment. Ladejinsky was the brilliant, indefatigable salesman of ideas about land tenure. His enthusiasm brooked no obstacles. Not even the sacred precincts of high military rank prevented his carrying the gospel directly to the fountainhead of all authority."[4] The fountainhead, of course, was General MacArthur. But Ladejinsky was also getting out into the countryside, in what was to become the hallmark of his working method, to obtain first-hand information on the attitudes of tenants and landlords toward the land reform then under consideration. These field observations are in part recorded in his "Landlord versus Tenant in Japan" and are reflected in the more substantial "Farm Tenancy in Japan," which updated and went beyond his earlier studies and obviously did much to shape the final land reform program.

For Ladejinsky himself, these field trips in Japan and his face-to-face contacts and communication with hundreds of Japanese peasants resulted in another kind of shaping—an internal one. When intellectuals from the shtetl background encountered modernization in a new setting, they typically sublimated their traditionally intense inner religious life by turning to Zionism, socialism, or trade unionism (with, to be sure, a heavily socialist tinge). Ladejinsky's scholarly vocation, had he remained in Washington, might well have found complementary expression in such a way. (He did, indeed, develop very strong feelings about Israel, to which, in his will, he left the beautiful Oriental works of art he had lovingly collected over the years.) But the peasants, their wives, and children whom he encountered in the countryside he did not see as primarily Japanese. He saw them as human beings in desperate need of help to achieve a bit of the security and dignity, and the prospect of a better life for their children, to which all human beings were entitled. His humanity, his sense of social justice, were deeply aroused. Henceforth Ladejinsky's religious inheritance would require no other expression.

Papers presented in this publication serve to block out, broadly, subsequent phases in Ladejinsky's career. He was borrowed in 1949 to assist the Joint (U.S. and China) Commission on Rural Reconstruction in its agrarian reform efforts in China and Formosa. From 1950 through 1954, in addition to his normal duties as agricultural attaché in Tokyo, he reviewed the effects of the Japanese land reform, assisted again in Formosa, and, at the urgent request of the ambassador to India, Chester Bowles, examined at first hand tenurial conditions and problems in Kashmir, Punjab, and Madras (in 1952) and the general status of the land reform program in India (1954). His final year in U.S. government service was spent as land reform advisor with the aid mission in Saigon. From 1956 to 1961 he continued his work

3. Supreme Commander for the Allied Powers.

4. Lawrence I. Hewes, Jr., *Japan—Land and Men* (Ames, Iowa: Iowa State College Press, 1955). The official history was Hewes's "Japanese Land Reform Program," Report no. 127, General Headquarters, Supreme Commander for the Allied Powers.

in Vietnam as personal advisor to President Diem. For the next three years he served the Ford Foundation as a kind of roving regional consultant, advising on the foundation's work in Nepal, India, Indonesia, and the Philippines. In the latter part of 1964 he began his consulting work for the World Bank. After participating in a major World Bank study of India's prospects for economic development and in other missions to Mexico and Iran, Ladejinsky was posted to India early in 1967 as a member of the Bank's resident mission in New Delhi. There he finished out his work with an increasing stoicism. While India was his primary responsibility and concern during this last period, he was also called upon to assist ad hoc with the Bank's work in Iran, the Philippines, and Sri Lanka.

Writings

Over the years, Ladejinsky's career was productive of a considerable body of written work. Most of it, of course, was done for the governments and development institutions for which he worked. His Department of Agriculture studies, most of them pre-1945, were all published in official journals (two appeared in *Foreign Crops and Markets*, the rest in *Foreign Agriculture*). Other U.S. government papers done in India and Vietnam appear here for the first time. In fact, the 1955 Vietnam papers have been declassified only by request for the present purpose. The Ford Foundation papers similarly have until now reposed in the foundation's archives. The World Bank papers have always been restricted in circulation and appear publicly for the first time. For these reasons Ladejinsky's postwar written work has been available only to the extent that it has appeared in public print or was distributed, on occasion, to a handful of close friends and professional colleagues. The best known, perhaps, are the seven articles published in *Foreign Affairs* over a thirty-year period and another six which appeared, beginning in 1969, in the *Economic and Political Weekly* (Bombay). Other articles appeared from time to time in more popular magazines such as *The Reporter*, *Country Gentleman*, and the *Saturday Evening Post*. But other than the *Foreign Affairs* articles, most of the published work, including many of the *Foreign Agriculture* articles, are not easily accessible. Ladejinsky's reputation as a development practitioner therefore rests more on the intimate knowledge of the relatively few government and institutional policymakers and officials with and for whom he worked and of a small circle of professional colleagues with whom he shared his experiences than it does on a wide familiarity with his writings. The present publication is intended to remedy this situation.

An intensive search for the purposes of this volume turned up some 142 studies, surveys, reports, articles, memoranda, conference statements, and a few substantive letters comprising approximately one million words. Although a number of significant items may not have been garnered in this process—especially internal memoranda and private letters of value—the papers collected and listed in the Chronological Bibliography almost surely incorporate all Ladejinsky's important work and are representative of all the countries, problems, and ideas with which he concerned himself throughout his career. As might be expected, Japan, Vietnam, and especially India—the countries where he spent the most time and did his most important work—account for a substantial portion of his total writings. Tenurial conditions and agrarian reform are recurrent themes in different countries, conditions, and times. But the papers deal in depth with much more: agricultural and development strategy, institutional barriers (credit, extension, administration, and so on), small farmer programs, drought in Maharashtra and refugees in Bengal, the Green Revolution and government procurement procedures and marketing, human welfare and Chinese communes, President Diem's views and the seriousness of Nepal's development intentions. Truly, this lode is rich with highly concentrated ore waiting to be mined by policymakers, development practitioners, rural development specialists, scholars and students, and also by all those genuinely interested in the human and operational aspects of economic development.

Those who read Ladejinsky here for the first time will find nothing that is academic or theoretical. Every piece Ladejinsky ever wrote is based on specific facts and situations addressed to *living* and urgent problems involving the burdens and agonies of millions of impoverished peasants. His writings were designed either to stimulate the required central government attention, necessary programs, and policy action, or to suggest to the World Bank and the Ford Foundation, his only nongovernmental employers, how they could most usefully assist one country or another. His published postwar articles were written to mobilize an élite or broader public support for necessary government actions. To repeat, then: this man was no academic, no theorizer; he was a *doer*. His written work, no less than his work in the field, was operational.

The pieces presented here have been selected, first, for their relevance to the agricultural, developmental, and human conditions in Aia today and the enlightenment they offer policymakers and development practitioners in addressing and improving that condition. A second criterion has been the contribution made to the understanding of agricultural and economic development. Third, an attempt was made to present papers that would be truly representative of the man and his work, the countries, problems, and ideas with which he was concerned, and the values he cherished. A fourth criterion, applicable in the case of papers such as those on Vietnam, has been their uniqueness and historical value. Fifth, previously unpublished papers have been given some slight preference, although not at the cost of quality, relevance, or comprehensiveness.

Limitations of space and the fact that Ladejinsky dealt with many topics more than once have made it necessary to curtail many excellent pieces. Sometimes he revisited at a later time a problem previously dealt with; sometimes he added a new dimension or depth, or underscored that a problem previously flagged was still there, naggingly awaiting action; sometimes he revised for a more popular periodical an earlier piece written for an institution or professional journal. Particularly in a number of the India papers over the last decade of his life, Ladejinsky repeatedly reviewed the evolving problems (and programs) of the small submarginal farmer, of land reform, of the Green Revolution, of such institutional constraints as cooperative credit, extension, and administration, and so on. The criterion of accessibility has also been invoked. All but two fragments of the classic "Collectivization of Agriculture in the Soviet Union" has reluctantly been excluded because of its length (close to 90 printed pages)—but only because the *Political Science Quarterly* in which it appeared in 1934 should be available in any good library. In several cases it has been possible to use Ladejinsky's own summary of, or a significant excerpt from, a paper. But no paper with a solid claim to inclusion on the grounds already described has consciously been omitted unless it is reasonably accessible in published form.

The Ladejinsky Thesis

I have dealt thus far with Ladejinsky's professional career, his antecedents, his motivations and values, and the subject matter of his written work. It is time to turn to the substance of his thinking—to the Ladejinsky "message" or thesis, to his work method and style, and to the qualities of the man himself.

Ladejinsky never wrote a book. Neither did he contemplate, even toward the end of his career, summing up in book form what he had learned or the message he wished to leave behind. Like the shoemaker who heeded the admonition "Stick to thy last," Ladejinsky was too humble a man to undertake—as have so many of vastly lesser experience, insight, and wisdom—organizing and presenting his views in a framework broader than that of the specific problems to which he habitually addressed himself. Thus, even among his articles, there are no such titles as "Toward a Strategy for Agricultural Development," "The Role of Agrarian Reform in Agricultural and Economic Development," "Agrarian Reform and Political and Social Stability," "The Possibilities and Limitations of Foreign Aid," or "The Role of Asia in a Postwar Democratic World." Yet Ladejinsky had deep insights and strong convictions about these and other large questions too. Although he never chose to isolate and develop them, his views about them nevertheless emerge, often in fragmentary form, in the course of his treatment of much more specific subjects. Like exploding flashbulbs, they illuminate the larger background of the more immediate and limited scenes on which he had been focusing.

In only one or two significant exceptions to this generality did Ladejinsky present his broader views in something more than a paragraph or two. In a private letter to Kenneth Iverson of the Ford Foundation late in 1954 Ladejinsky attempted to formulate, in very broad terms, his understandings of the problem facing those in the West who wished to advance the cause of human welfare in the poor countries. He dealt with the competition with communism in this regard; the kinds of assistance programs that might prove most effective, especially in relation to land tenure; and the qualities most essential and desirable in technical assistance personnel (II–24).[5] Another exception is a paper, "Land Reform," prepared for delivery to an MIT Conference on Productivity and Innovation in Underdeveloped Countries in 1964 (IV–47). Here Ladejinsky uncharacteristically dealt with the land reform problem in general terms rather than in the particular. He examined the scope and significance of land reform, its meaning and content, its politics, its relation to productivity and political stability, its prospects and limitations. These two items should not be missed even by the highly selective reader.

It should be useful to set forth here, albeit only briefly, what Ladejinsky was too humble to develop at length himself. Here, then, are what strike me as the chief elements in what might be called "the central Ladejinsky thesis," set in the larger perspective in which his views were framed.

1. In a world in which mankind's best hopes for human welfare under freedom are dependent on the fruition of the democratic dream, to which Communism represents both the major threat and likely alternative, the role of Asia is crucial. ". . . what happens in Asia in coming decades is as decisive for the future shape of the world as anything that happens in Western Europe." "From the point of view of the West and the preservation of its most cherished values, it is of the utmost importance whether, in their current efforts to modernize, the underdeveloped countries will lean towards the West, adapting its technology and political ideas to suit their special needs or, instead, accept the Communist promises and eventually the Communist system. It is therefore no exaggeration to say that the existence of the Western, democratic world will depend upon the choice, free or accidental, of the underdeveloped countries between following in the Communist path or proceeding with Western aid" (II–24).

5. References are to part and chapter numbers of this book as listed in the table of contents. The examples cited in this introduction are illustrative rather than exhaustive.

2. The welfare of the Asian peoples will play a definitive role in this outcome. "Every Asian, however illiterate, consciously or unconsciously aspires to a measure of this welfare which consists of better living conditions, better health, better social status or greater equality of status, better government, greater participation in local or national affairs, and a host of other values which spell out the ideas of human welfare and of 'the dignity of man'. . . The survival of the underdeveloped countries along Western lines dictates the application of these objectives . . ." (II–24).

3. Agricultural progress is basic and essential to economic development and welfare, whether in Asia or in other agrarian and developing societies (II–24, IV–43), not only because agriculture contributes directly to production but also because it provides a growing market for the domestic manufacturing industries which are also essential to economic growth and employment (CB–100).[6] Yet developing countries generally failed to recognize this through the 1960s, and recognition of agriculture's central role is still far from general. Ladejinsky was far ahead of his time when he perceived in the early 1950s that the developing countries were making a critical error in overemphasizing industrial development as the high road to economic growth, at the cost of agriculture and overall economic development (II–24; III–29, 38; IV–46). "More immediately to the point [as an obstacle of agricultural progress] is . . . the heavy bias in favor of industrialization in the typical underdeveloped, agrarian country, as against setting its agrarian house in order. This approach, favored by planners and economists, both Western and Asian, has tended to retard the much-needed agricultural effort. This criticism is not leveled against industrialization as such; its usefulness is all too obvious. What is at issue is *the overemphasis on industrialization in relation to agriculture . . .* In largely agrarian Asia this approach is an economic and political fallacy which, if persisted in, is bound to lead—has already led—to serious stresses and strains affecting the progress of developing countries" (IV–46).

4. Where the land-man ratio permits, redistribution leading to widespread land ownership for the great mass of cultivators is the best way to provide the incentives necessary to the agricultural investment, modernization, increased output, and higher levels of living and welfare that connote agricultural progress. Where the land-man ratio will permit such redistribution only in limited degree, the most practicable and therefore constructive solution to the problem of equity and incentives lies in achieving a truly secure land tenure for tenants and sharecroppers, combined with tolerable ceiling rents (IV–47). The price and terms of repayment for redistributed land must not burden excessively the new owner (II–18, IV–47). Where land redistribution programs enable absentee landlords to resume sizable farms for their own cultivation, "The net result is a new type of tenant . . . the 'evicted tenant'" (II–24). The satisfactory implementation of land redistribution programs requires active participation and execution by local or village land commissions, at least half of whose members should be tenant farmers (II–10, 11). Tenant tenure should be protected by written contracts with renewal safeguards. In the absence of secure tenure, effective rent ceilings are impossible to administer (II–24, V–51). Effective rent ceilings reduce the capitalized value of land and make it easier for tenants and sharecroppers to buy land of their own in the marketplace (II–12).

5. Widely distributed land ownership and secure low-rent tenancies do not alone suffice to make or ensure the success of an agrarian reform. Also needed are adequate and secure

6. This and a few following CB references are to papers listed in the Chronological Bibliography but not reproduced in this book.

water supplies and effective institutional arrangements for essential inputs, the credit necessary to obtain them, and the extension services needed to guide small farm operators in their efficient use. "In conditions of rural poverty even redistribution of the land will not suffice unless it is accompanied by the necessary means to work and improve the land" (IV–47).

6. This is not to say that agrarian reform is a panacea that provides an assured solution to all the problems of the countryside. Especially "Where the pressure of population on limited land resources is great, agrarian reform . . . is not a final solution . . . Rather it does away with the worst features of a system that has outlived its usefulness economically, socially, and politically" (IV–47). And "the issue in India is not one of *solving* the rural problem but of palliatives capable of wiping out the worst features which condemn the farmers to a below-subsistence level of existence" (II–23).

7. Because of the peasant's desperate need and hunger for land, and because exposure to the winds of change have aroused him ("an overworked and overexploited common man who for centuries was inertly miserable is now alertly miserable"), basic agrarian reform is inevitable and its character essentially revolutionary (II–24). ". . . despite the opposition of the landlords or of governments dominated by them, the transfer of ownership is inevitable; what is in question is the pace—quicker in a country where the government is bent on implementing a reform, slower where a government has neither the strength nor the desire to activate the issue . . . the reform movement is a revolutionary one even though landlords' heads do not roll and noblemen's nests are not set afire" (II–22). Land reform "involves a drastic redistribution of property and income at the expense of the landlords. It becomes a revolutionary measure when it passes property, political power, and social status from one group in the society to another . . . A reform worthy of its name is supposed to strengthen the principle of private property where it was weakest, at the base of the social pyramid . . . As the landlord loses much of his affluence, he loses much of his influence" (IV–47).

8. The key to who makes agrarian reform, and to what determines whether an attempted reform will be successful, is political. "Technical expertise in preparing and administering the necessary legislation is indispensable, but experts do not make reforms. Politicians, and only politicians, make good or poor reforms or do not make them at all. They control the political climate, which determines the will or lack of will to proceed with the task; the specific measures with which the reform is or is not endowed; the care or lack of care with which the enabling legislation is formulated; the preparation or lack of preparation of the pertinent and administrative services; the presence or absence of technical services with their bearing upon the success or failure of the reform; and, most important, the drive or lack of drive behind the enforcement of the provisions of the law" (IV–47; see also V–51).

9. Government then—its desire and its *will*—is the key to agrarian reform. "The built-in landlord opposition, abetted by public servants, can be dealt with successfully if the political leadership is bent on carrying out its goals. This is especially important because the peasantry has not developed a popular political movement of its own capable of effectively representing and advocating its own cause" (IV–47). But what if governments fail to act, or act only in a halfhearted way, so as to vitiate the announced purposes of an "intended" reform? "Government authority, after all, usually is controlled by the very forces likely to be adversely affected by progressive economic and social development" (CB–47). In such a case, "There is ample proof that sooner or later the dispossessed will take the law into their own hands, to the utter destruction of the governments and classes who failed to grant them peacefully what they [will] otherwise try to acquire through violence . . ." Experience has thus demonstrated that "the foundations of the social structure stand or fall in the countryside and

that the peasant and his interests and aspirations must be placed 'in the center of the piece.'" There is ample lesson and warning in postwar history "of the shape of things to come when the economic and social aspirations of the peasant are sacrificed for the sake of an outdated status quo" (III–32). ". . . somebody else will take over this long-overdue task, and much more will be at stake than a new rearrangement of income distribution and status in the countryside" (IV–47).

10. Complementary to this "or else" thesis is the conclusion that, where circumstances fairly cry out for agrarian reform, effective governmental action to achieve it is essential to economic, social, and political stability in the countryside and hence to the survival of democracy itself (III–32, IV–43, CB–100). ". . . farm reform can become a powerful political instrument. The native governments friendly to us would be more likely to win popular support, and popular support in Asia is 'peasant support or nothing.' An owner cultivator or a reasonably satisfied tenant would acquire a stake in society. He would guard that society against extremism. Private property would be strengthened where it has been weakest, at the huge base of the social pyramid . . . any effort to ease the peasant's burden . . . lays the foundation for a middle-of-the-road, stable rural society . . . 'Land and liberty' has ever been the ideal of all peasants" (II–13).

The Ladejinsky thesis speaks for itself and calls for little comment here. Attention may, however, be called to one element in his thesis about which Ladejinsky himself, late in his career, became shaky. And some questions need to be asked about another because events may seem to have negated it.

Until close to the end of his life, with his observation of Communist policies in the Japanese, Vietnamese, and Indian countrysides confirming his earlier observations and experience in the Soviet Union and China, Ladejinsky insisted that if basic agrarian reforms were not carried out from the top down, voluntarily, by supposedly democratic governments while they had the opportunity to do so, change would inevitably come about by violent action, from the bottom up, inspired by Communist propaganda. But he began to experience serious doubts about this as he puzzled over the passivity of the peasantry in Maharashtra during the devastating drought of 1972–73. Raising the question "Will peasants rebel?" Ladejinsky found it "a source of wonderment that the afflicted 'take it' without overt protests, let alone without resorting to violence . . . one is mystified . . . This observer couldn't help but cogitate about the causes explaining it . . ." (V–58). A few months later, addressing a World Bank seminar on land reform, this doubt had become a conclusion. "If we are to wait until the peasantry of India—or, for that matter, a number of other Asian countries—decide to take the law into their own hands and fight for an out and out radical agrarian revolution, I think we would have to wait a long, long time." Given the obvious lack of the essential *will* on the part of the Indian government, Ladejinsky for once was at a loss. "I really don't know what one does at this point." And then, with wry humor, "I suppose the only person who probably knows what the eventual solution may be is the Indian astrologer; I leave it to him at this point to provide the final answer" (CB–127).

What invites question is the complex of judgments at the heart of the Ladejinsky thesis that "the foundations of the social structure stand or fall in the countryside," that the peasant must be placed "in the center of the piece," and that the economic, social, and political stability essential to the future of democracy in Asia rest on finding solutions to the socially cancerous problems of the countryside. Are not these judgments contrary to the Asian experience of recent years? Has it not been the wretched conditions and insistent demands of a rapidly growing and highly politicized new urban proletariat which has dominated political decisionmaking or frightened weak governments into indecision and impotence? Has it not been the frustrated dissatisfaction of this urban proletariat, rather than the "alert misery" of the peasantry, which has eroded the political base of weak parliamentary governments and opened the political gates to military and other authoritarian governments?

So it would seem. But do these unhappy historical developments of recent years really negate the core of the Ladejinsky thesis, or do they serve rather to confirm it? What was it that sent the teeming millions of impoverished and landless cultivators streaming into the cities where they could not find employment, housing, nor schooling and health care for their children nor escape from misery? Was it not the failure to address the fundamental problems of agrarian reform in the countryside, and the misguided overemphasis on industrialization at the expense of agriculture, which increasingly exacerbated inequalities of income, unbalanced development, and stunted economic growth? Has the central problem not been and does it not remain the one Ladejinsky posed? It was with a piercingly prescient simplicity that he wrote in 1954: "Four-fifths of the people who populate the underdeveloped areas are peasants. Agriculture, not industry, is the pivot of their lives in all its principal manifestations. Industry has made but a small dent in the character of Asia, notwithstanding the industrialization of Japan, the oil gushers of the Middle East, the tin mines of Malay and Siam, and the jute and cotton mills of India. The factory may bring material advancement to the Asians some day, but that day is in the future. The heart of the problem of Asia today lies in the countryside. It is on the farm where solutions must be sought and found" (II–24).

Foreign Assistance

Ladejinsky firmly believed that economic development was basically a job which, for the most part, the poor countries had to do for themselves, making fundamental policy choices which only they could make. His views on foreign aid were therefore not incorporated into the core of his thinking just presented. The vital importance he attributed to the role of Western aid was made evident from the very outset, however, and was indeed fundamental to his thought. For this reason, and as a sensitive, wise, and highly experienced practitioner of development assistance over the years, he came to a number of highly insightful and useful ideas about it, which will be examined here. Ladejinsky's view of the critical choice confronting the poor countries between the Western and Communist systems provides a useful point of departure for this examination. The following quotations from an article written in 1950 develop this view.

"Now the forces that keep the peasant within well-defined bounds are breaking down under rising agrarian discontent. The peasantry is at last in motion. The Communists have exploited this fact and placed it in the center of Asiatic politics . . . Lenin . . . visualized the final crucial battle as a conflict between a Communist East and a capitalistic West . . . To win Chinese and Indian support, Stalin developed a program for those countries consisting of three stages: a struggle against foreign imperialism, an agrarian revolution under the leadership of the Communist Party, and finally a proletarian dictatorship. The key step was to be the wooing of the peasants . . . The only way to thwart Communist designs on Asia is to preclude such revolutionary outbursts through timely reforms, peacefully, before the peasants take the law into their own hands and set the countryside ablaze" (II–13).

On the nature and strategy of assistance, he had this to say about the Western role: "Whatever we may contribute to Asia's advancement and stability—be it in the form of dollars, of technical guidance, of organizational advice, or of military assistance—our policy and all our diplomatic competence and tact should be actively and sympathetically guided by the knowledge that the foundations of the social structure stand or fall in the countryside . . . We must make an effort to persuade the more conservative Asian groups that rural reform is essential to their own preservation as well as in the interest of the peasantry . . . From ambassadors and ministers to foreign service clerks, we must begin to feel and act in terms of the common man, and in Asia he is the peasant. We must make a special effort to seek out and encourage in every way possible the native liberal groups who might otherwise be lost to our common cause" (II–13).

Ladejinsky went to China and Taiwan in 1949 to assist the Joint Commission on Rural Reconstruction in a pre-Point IV foreign assistance program which impressed him greatly and which he thought might well serve as a model for future assistance efforts. One unusual aspect he admired was that the joint commission was itself a cooperative effort, with three Chinese and two American members. The chairman was Chinese while both American members were men who had lived and worked professionally in China for many years and who shared fully the aspirations of the farmers they sought to serve. The commission adopted "a bold rural program" with ten constituent elements that gave first priority to land reform. This was no accident. "It was . . . clear to the commission that the chief beneficiaries of the technical projects would be the landlords and not the majority of the farmers—the tenants—unless tenancy conditions were improved." And Ladejinsky described approvingly the basic principles and criteria the commission selected as a basis for its activities. It ruled out costly projects with modern equipment unsuited to small-scale Chinese agriculture. It chose not to set up new enterprises that would compete with existing ones, but chose rather to assist in enlarging and improving the latter. It shunned projects for which the farmers themselves had not expressed a strong preference. It concentrated on projects that would help the people to help themselves and that would benefit the great majority of farmers rather than just a few. To develop indigenous leadership it made special efforts to draw the educated youth of the country into the work of rural reconstruction. And, most important of all, it insisted as a condition of its aid that "physical reconstruction must be accompanied with social justice." In assessing the work of the commission, Ladejinsky sagely observed that "the experience of the commission shows that, unless it can be usefully absorbed, money alone cannot solve problems. And," he generalized, "this is a point worth remembering in all rural aid efforts in underdeveloped countries" (II–14).

Two letters to Ford Foundation officials in 1953 and 1955 are addressed to the question of how the foundation could most usefully be of assistance in Asian countries, but Ladejinsky's advice also has a more general application. The foundation's Report on the Conference on Land Tenure (in which he had himself participated by invitation some months earlier) was, he said, not realistically geared to the actual situation and problems of the countries the foundation sought to aid. Its stated principles of land reform policy, he pointed out, "constitute something akin to a perfectionist scheme, upon the fulfillment of which no land reform waited in the past or will wait in the future." Moreover, useful assistance in land reform does not lie in advising governments "how to formulate policies, administer, execute, and evaluate them." We know too little for this, he insisted, while "Every country in Asia has an articulate group with knowledge of the country's agrarian problems and ways of solving them. The difficulty is that often they choose to ignore them." It follows that "To be of real aid we could do no better than try to persuade this group that it is perhaps later than they think. This is where under propitious circumstances we could, with subtlety and experience, render useful service." A low profile, caution, and humility are other requisites (II–22).

The second letter, already cited in an earlier connection, is much broader in scope. Because backwardness in underdeveloped countries is not only economic but general, assistance programs "must not be limited to what is commonly known as technical or economic assistance." Moreover, with a sober prescience, he foresaw and stressed that "it is not at all certain that achievements in technical [or economic] assistance will automatically or necessarily insure a greater sharing of economic welfare . . . of political power, free public schools, the emergence of representative governments, and other developments which denote progress in democracy" (II–24).

Since it is the political climate for reform which is crucial, the strategy of assistance should aim at reaching those "groups whose support could make the difference between determined action . . . or none at all." In India Ladejinsky saw the universities as the key medium for such an approach. But he cautioned that outsiders seeking to advance the essential cause of land reform must be sensitive to the fact that it is a highly controversial issue. Landlords are bound to resist it, and "Foreign 'interference' may add fuel to the already burning issue by antagonizing not only the conservatives but also the pro-reform supersensitive nationalists" who are only too ready to fight any infringement on their national sovereignty (II–24). Two decades later Ladejinsky cautioned the World Bank to the same effect. Since agrarian reform was essentially a political decision, no government, large or small, could be prodded into such a decision if it were not itself politically ripe for it (CB–127, 141).

These views about foreign aid are, not unnaturally, reflected in the qualities Ladejinsky sought in technical assistance personnel. "In the light of the goals our aid should pursue, he [the field-worker] must have much more than technical competence . . . [He] must be able to understand the position the people of a given country are in, to grasp something of their attitudes, their feelings, their state of mind, their view of events and of the world . . . He must see what it is . . . that these people bring with them onto the world stage at the present time." In a former colonial dependency he must appreciate that "the economic legacy is often poverty . . . The psychological legacy is fear, suspicion, and hostility" stemming from "all that resulted in the imposition of enforced inferiority. The political legacy is authoritarianism, even when it is embellished with modern democratic forms . . . to a large degree the success or failure of his [the field-worker's] labor will depend upon the extent to which he understands and accepts these fundamentals" (II–24). Given these qualities as well as tact, a genuine humility, a gift for personal relations, and a sharing of purposes, said Ladejinsky, the field-worker will gain the *acceptance* which is so essential if his ideas are to reach their mark. "Advice from the outside has its place and will be accepted if based on knowledge and profferred in the spirit which induces Asians and Westerners to work together for a common goal" (II–24, III–28). (Not a bad—if unintended—description of Ladejinsky himself!) These views were reflected, not long after, in his assessment of the U.S. aid program in Vietnam (III–31).

Some twenty years later development institutions and the development profession are coming to a belated recognition of some of Ladejinsky's homely truths. Recognition that "the heart of the problem lies in the countryside" became central to the World Bank's development assistance policy in the 1970s, and much else that he had to say has become part of the conventional wisdom, if not yet the conventional practice. The Western world has learned to its sorrow that economic growth does not necessarily ensure human welfare, nor an increasing measure of social justice, nor even the continued adherence to, let alone the strengthening of, democratic institutions. Not until the 1970s did the appreciation develop that "overemphasis on industrialization in relation to agriculture . . . is an economic and political fallacy which, if persisted in, is bound to lead . . . to serious stresses and strains affecting the progress of the developing countries." This early insight alone should suffice to make Ladejinsky stand out like a giant in the field of economic development.

Work Method and Style

What imprints on Ladejinsky's writings their unmistakable stamp of reality and authenticity is the fact that they either reported on his personal field observations or derived directly from them. This is not to say that he failed to devour the relevant literature, to pore over the statistics, or to garner the wisdom and experience of other scholars, observers, and field-workers, domestic or foreign, living or dead. All these things he also did, with a prodigious energy and a burning intellectual curiosity, steeping himself at the same time in the culture of the place so that he could see from the inside as well as from the objective perspective he never lost. But he never made these things his own, whether by adoption or rejection, until he had tested them by his own direct observations. As Ashok Mitra observed (see below), "Ladejinsky did not just write reports, he lived through them." It is the inescapable realization that Ladejinsky had thought and felt deeply about what he had himself seen and heard, that he had "lived" his reports, that gives his papers the great immediacy, cogency, and force which characterize them.

Just as agrarian reform was central to Ladejinsky's concerns, so direct field observation was central to his work method and style. Typically he would set out, accompanied only by an interpreter and without a preset plan, and interview everyone he encountered by chance or sought out, who might be involved in one way or another in the problem at hand: tenant farmers, sharecroppers, agricultural laborers, landlords and sublandlords, district and local officials, representatives of cooperative banks and marketing agencies—in short, everyone who could possibly provide an insight into the prevailing condition, their interests in it, and their reactions to it. This took him not only into the fields, marketplaces, and offices but also into the hutments where he could observe how the people lived, what food was in the larder, and how the wives and children fared. If he stopped by the roadside to talk to a farmer or laborer, inevitably others drifted over to observe and listen and then to join in the conversation. For it was always plain to them that this foreign stranger was no alien being, questioning them from a plane other than their own. They sensed at once that he spoke with them as one man to another, that he shared with them the human condition, as well as their interests and concerns; they in turn were glad of the opportunity to share their interests and concerns with him. Over the years Ladejinsky must have thus visited and spoken intimately with many hundreds of peasants in every country in which he worked. It is doubtful whether any other man—agricultural economist, development practitioner, or, indeed, government official—had made so close and direct an acquaintance with the peasant cultivators of Asia or had come to know so well the peasant condition. It was quite natural, therefore, that Ladejinsky's work was never theoretical but invariably rooted in fact and specific situations and problems. His work was always set in a perspective that embraced both the immediate as well as the historical and broader background. It sought out causes, objectives, obstacles, and accomplishments and assessed all of these in one large, professionally disciplined and yet very human view.

One other aspect of Ladejinsky's working method or style is particularly noteworthy. In country after country in which he worked he managed, after conducting his surveys and reaching his conclusions, to gain an audience and the opportunity personally to present his findings and conclusions to the head of government. That this was not customary but rather highly unique any experienced development practitioner or advisor knows, ruefully, only too well; neither, of course, was it accidental. It happened for the most part by Ladejinsky's design. Although his reputation often preceded him and helped, just how he managed to do this is a political, diplomatic, and operational art that can neither be dissected nor emulated by any formula.

Ladejinsky would have been less than human if he had not been somewhat flattered and gratified by such opportunities. He could scarcely have avoided wonderment, as he approached such audiences, that he, so recently a poor immigrant, was about to converse with the high and mighty. But such access was essential to his purposes. He knew very well that any significant progress toward agrarian reform in any country required a political decision at the highest level of authority. Ministers of agriculture or planning or finance might be persuaded and might even relay the message to the head of government, but they were scarcely likely to convey it as cogently or persuasively as Ladejinsky himself, if indeed they could be relied upon to carry the message at all. This is the why if not the how of Ladejinsky's audiences, frequently repeated, with General MacArthur in Japan, Premier Chen Cheng in Taiwan, President Diem in Vietnam, Prime Minister Koirala and King Mahendra in Nepal, Prime Minister Hoveida in Iran, President Marcos in the Philippines, and President McNamara at the World Bank. In his single audience with Madame Gandhi, he once confided, he failed to "reach" her, and despite his long residence in India he did not try again. And he knew better than to try in the case of one other national leader. Describing a visit to Indonesia early in 1961, where he had met at some length with Minister Sadjarwo, he wrote: "Whether it is really worthwhile investing time—assuming that I had some to spare—is another matter. My skepticism stems from the knowledge that, in such a highly personalized government . . . the number one, Mr. Sukarno, must be influenced if real progress is to be made. Regretfully, I am not sure that I could reach him, and, even if reached, I am not sure that I could find the key to a man whose attitude toward affairs of state is a great puzzle and wonderment to me and countless others" (III–37).

That he was a superb reporter as well as analyst of the agrarian scene is evident in many of the papers presented, whether in Japan (CB–27), Szechwan (II–12), Vietnam (III–26), Punjab (V–50), Bihar (V–51), Eastern United Provinces (V–54), Maharashtra (V–58), or elsewhere. Nor was this superb reporting limited to the agrarian condition, as demonstrated by his description of the conditions in West Bengal as refugees fled from the horrors in Bangladesh (V–56), by his interview with President Diem (III–27), by his impressions of India after the China war (IV–41), or by his assessment of Nepal's development intentions (CB–86).

At least two observations should be made about Ladejinsky's writing style. The first and more obvious is that, especially in his published work, which he labored to polish, he had a beautiful command of the English language, a command all the more impressive because his mother tongues were Yiddish and Russian, and, like Joseph Conrad, he learned English painfully only after he had reached maturity. This command was enriched by his gift for the pithy phrase, the apt quotation, and the lively anecdote.[7] The second is that Ladejinsky always treated his subject in a full, thoroughly rounded out—almost prolix—fashion. I take this to

7. For example, in his "Traditional Agriculture and the Ejido" (CB–100) he tells of the farmer who, when asked about the extension service, replied, "Ah, you mean the fellow who drives by but never stops here."

be a reflection of both his seriousness and his pride. If a subject was important enough to engage Ladejinsky's attention, he would treat it thoroughly, with the seriousness it deserved and with no concession to the possibility that the reader might take it somewhat less seriously than he. He can sometimes almost be heard saying, "If this matter was serious enough to warrant my attention and examination, and you wish to learn what I have learned and come to think about it, you will have to give it your equally serious attention." In his case, this point of view is entirely justified.

The Man

Ladejinsky was one of those rare individuals of whom it can literally be said that the man was his work and the work, the man. A good deal of the essential Ladejinsky must already have emerged from what has been said in the preceding pages. Those who read all his papers will be impressed with many more aspects or qualities of the man than have already been stated or implied: the judicious evenhandedness shown in his assessment of colonial policies in the Netherlands Indies and in British Malaya (I–6, 7); the fair treatment for landlords called for in Taiwan (II–10); his insightful political acumen (everywhere, but see especially III–28, 31, 35; CB–86); his unerring eye for the essential; his great persuasiveness (IV–40, CB–101); his dogged persistence in getting his work done, despite the official discouragement or obstacles he sometimes encountered (II–21, CB–123, 141); and the humility with which he assessed his own role and achievements (II–16, 24; CB–49). These are only some of his personal qualities. Left untouched are others which would more appropriately be treated in a personal memoir—his capacity for friendship, his passion for good craftsmanship, his sense of humor, the stoic fortitude with which he endured ill health and disappointment, and his love of beauty in music, literature, and art. But at his innermost core and in the end perhaps most important were zeal, and faith, and love. It was of these that his long-time friend and colleague in the quest for genuine agrarian reform in India, Ashok Mitra, wrote in a touching tribute:[8]

> Wolf Ladejinsky was much more than a desiccated agrarian expert. Such experts are legion; without them, foundations set up by multibillionaires would face a problem of non-utilization of resources. They come, write their report, go, and soon forget the country for which they had written it, their mental horizon is always colored a cynical grey. Ladejinsky did not just write reports, he lived through them. He did not need, in his ripe old age, the trifle of the World Bank salary to keep coming to this country, base himself in an impersonalized hotel room in New Delhi, and restlessly wander across the Indian countryside to comprehend the essence of agrarian truth. It was bizarre, yet emotionally a moving experience, to see this near-octogenarian, who had lost the sight of practically both his eyes, who was in such a precarious state of health that he could hardly assimilate any food, still determined to catch the plane, land on the airstrip in a distant town, and get into the jeep or station wagon, or take the ferry across the river, so as to reach some remote village where the *bataidar* would be able to tell him a little more about the mystique of the local land system, or of the local wage rates. This was no run-of-the-mill technical expert, this was a zealot. And the zeal came from a deep love for people, whatever their civilization or the pigmentation of their skin.

8. *Economic and Political Weekly* (Bombay), July 19, 1975.

Yet Wolf Ladejinsky was no ideologue. He would not have minded being described as an old-world liberal, who takes it as his mission to analyze and state the truth, whatever its hue. Thus he could be of considerable service to General MacArthur in Japan in 1946; the Shah of Iran too called for him, and the State Department, in the late 1950s—before it lost its head completely over Vietnam—also thought that he could render some good to the authorities in Saigon with his advice and counsel. Ladejinsky did not stint on his advice, but he did not stint on telling the truth either. Thus, in the Indian milieu, he could say things—and be heard with respect—which the establishment would not be prepared to hear from others, things about Bihar's medieval feudalism, about the Green Revolution stifled three-quarters the way in Punjab and Haryana, about the agony of Bengal's deprived sharecroppers and landless labor, about the real nature of famine relief in Maharashtra. Land reform, he could say in his quaint American-heavily-tinged-with-Russian accent, was nine-tenths political will, and where there is a will there is a way, and not just legislation.

He was no ideologue—one cannot in any case afford to be one once one accepts World Bank sponsorship—but he was not afraid of ideas either.

And he knew how to bestow affection. Never very demonstrative, he would still make his little gestures, and there could be no mistaking the depth of his goodwill. An admirer of this journal, he would, every now and then, drop one an appreciative note about some altogether insignificant piece one might have written. It calls for a special genre of faith in humanity to assert that, whatever the circumstances, howsoever unmitigatedly unfavorable the objective conditions, in the short run, a people—any people—are capable of lifting themselves through a revolutionary upsurge. Wolf Ladejinsky held such faith, and he tried hard to convert others to it. As one mourns for him, one does not just mourn for a great American, one also mourns for a great romantic, who had wizened with the years, but who refused to forsake either hope or love.

The Ladejinsky Contribution

What was the Ladejinsky contribution? What did his forty years of dedicated work accomplish? This is not the place or time for such an assessment. His contribution will have to be evaluated eventually by his impact on agrarian reforms accomplished, stirred up, shunned neglected, or frustrated; on economic planners and development workers in the developing countries and the institutions with which they are associated; on the programs of development institutions like the World Bank and the Ford Foundation; on the thinking of academics and research workers in the field and on their institutions; and on the new generation of university graduates in both Western and developing countries who were inspired by his thinking, whether it reached them directly or indirectly. However his contribution might be assessed at this time, there can be no doubt that his impact is still growing. "The ideas of economists and philosophers," Keynes said, "are more powerful than is commonly understood. Indeed, the world is ruled by little else . . . I am sure that the power of vested interests is vastly exaggerated compared with the gradual encroachment of ideas."

It will not, however, be inappropriate to record here a few notes and suggestions concerning his contribution. Partial answers have already been placed on the record by persons in the best possible position to evaluate his role in particular situations. Concerning Ladejinsky's contribution in Japan, General MacArthur wrote (to the U.S. secretary of agriculture on January 15, 1947): "Mr. Ladejinsky performed outstanding service in connection with the initiation of a comprehensive Land Reform Program by the Japanese Government, thereby aiding in the attainment of important objectives of the Occupation. I wish to express my appreciation to you and the Department for having made his services available." And in a letter to Ladejinsky dated April 26, 1952, inscribed on an impressive silver plaque, Minister of Agriculture Kozen Hirokawa on behalf of the Japanese government wrote in part: "Our country owes you [a] heavy debt for your share in the successful accomplishment of the Land Reform Program. The world knows that this is the most significant reform which Japan accomplished under the Occupation, and I know that the reform was carried out with the closest cooperation and in the most friendly atmosphere between yourself and the personnel of our Ministry and I am convinced that the effect of the Land Reform will remain forever and so will your name in Japanese agriculture."

With respect to his contribution to agrarian reform in Taiwan, there is the testimony of Premier Chen Cheng, who cabled Ladejinsky in Tokyo on March 6, 1951: "I was glad to learn . . . of your willingness to come to Taiwan again. We are grateful for your invaluable assistance rendered to our country when you were last here [in 1949]. Now we are planning to carry out a limitation of land holdings program in which your advice is greatly needed. We have requested General MacArthur to approve lending your services to our Government and he has been good enough to give his approval. Please notify us of the date of your departure. You will be coming as my honored guest." Acknowledging the report "Observations on Rural Conditions in Taiwan" (II–15) which resulted, the premier wrote, on July 19, 1951, "We are grateful for your deep concern with Free China and the enthusiasm in presenting and analyzing the actual rural conditions in Taiwan. Please be advised that your valuable suggestions have been referred to [the] Economic-Financial Committee of [the] Executive Yuan for consideration to put them into effect." This relationship and appreciation continued. More than a year and a half later, on February 28, 1953, the premier wrote Ladejinsky again, inviting his views on the most recent land reform steps taken by his government. "It has been some time since I last heard from you. It is perhaps not necessary for me to say that you have always been in my thoughts . . . I am happy to report that at the beginning of this year we have put into execution the third and final step of land reform . . . As you have given invaluable assistance in the past . . . I enclose herewith a copy of the English translation of the *Land to the Tiller Act* . . . I shall be very grateful if you will be good enough to give me your comments . . ."

These testimonials by the highest authorities associated with the land reforms in Japan and Taiwan take on body and flesh when they are read in the light of Ladejinsky's own backward looks and assessments of the achievements of those reforms. For Japan the relevant papers are II–9, III–33, and CB–12, 49; for Taiwan, II–14 and CB–38. We have no comparable testimonial from President Diem of Vietnam; but the Vietnam papers presented here are perhaps testimony enough, especially the two backward looking and evaluative papers, III–38 and CB–75.

Ladejinsky's contributions to agrarian reform in India, where he strove mightily and long, will be more difficult to assess, even by those of his professional Indian colleagues who subscribed to the same values and goals and who are in the best position to attempt it. The papers included in this publication provide abundant evidence of Ladejinsky's efforts and the lines along which they were directed. As far back as the early 1950s he established close and cordial working relations with the Planning Commission in India. The commission subsequently formulated central guidelines for agrarian reform by the state governments (which have sole authority under the Indian constitution to legislate in this field) along lines which coincided with Ladejinsky's views. For this early period there is also the testimony of the then U.S. Ambassador Chester Bowles, at whose urgent request Ladejinsky was borrowed from his post in Tokyo in the latter part of 1952 to study and advise on projected plans for land reform in India. (This visit resulted, among other things, in land reform observations in the Punjab, Madras, and Kashmir, recorded as II–18, 19, 20.) Upon Ladejinsky's return to Tokyo, Ambassador Bowles wrote him on February 5, 1953: "It was wonderful to have you here. I am convinced that your work and your contacts have clarified Indian thinking on the whole problem of land tenure and speeded up the process of doing something about it." To Ambassador Robert Murphy in Tokyo, who had reluctantly loaned Ladejinsky for this purpose, Bowles wrote: "I can't tell you how much it meant to us to have Wolf Ladejinsky here over the last several months . . . There was no clear, current picture of the situation when Wolf got here and no really good evaluation of the actions that had been taken over the last several years by the central and state governments. Wolf's clear insight and unfailing energy gave us a good picture in a brief time of what had been done and, still more important, what needs to be done. I think he communicated his own sense of urgency about this to many of the state officials but particularly to the people concerned with the Five Year Plan and other officials here in the central government."

But over the years following, most state governments in India enacted land reform legislation that was either feeble or pockmarked with such loopholes as too high ceilings on acreage retention and resumption—and this only after landlords had been afforded ample opportunity nominally to "divide" their land among relatives, so as to gain additional escape from the application of "reform" legislation. The same state governments also failed to implement vigorously the weak legislation they did enact. Despite Ladejinsky's valiant efforts in later years to stimulate significant action on security of tenure and rent ceilings, where substantial gains were and remain possible, the government revived instead the chimera of new landholding ceilings and redistribution legislation, where only very limited gains were possible and still fewer actually made. (See especially on this aspect papers V–59 and CB–122.)

Ladejinsky's impact was felt, however, in the sense of concern and even guilt he introduced into the national conscience through his own writings and those of India's intellectual and academic elite to whom he provided inspiration and example. The basic questions he raised became part of a continuous public debate in academic, research, and government circles and in the press. National commissions were appointed, conferences organized, new programs undertaken, and new priorities established in large part because of the logic and moral force of Ladejinsky's insistence that the problems of India's agrarian poor were central and had to be addressed. The battery of rural "social justice" programs initiated in the early 1970s—such as those for small and marginal farmers and landless laborers and crash employment programs—owe their genesis to the ferment Ladejinsky did so much to create.

Another important outcome of his residence in India on behalf of the World Bank was undoubtedly his impact on the Bank itself. The World Bank Sector Policy Paper, "Land Reform," May 1975, notes that: "The position of the World Bank in regard to land reform has changed over the past decade . . . In the early years of the Bank's operations the focus was on providing adequate infrastructure for increasing agricultural production. In the early 1960s the approach to agricultural development was widened to include the provision of rural credit and on-farm inputs. Problems of tenure were seen to have an indirect bearing on production . . . By the end of the 1960s, however, concern was growing about distribution of income in the rural areas and the relationship between land distribution and income distribution." An internal Bank paper affirmed that "It is clear that agricultural development cannot do all it might to improve rural life if the distribution of land ownership is highly skewed." Since that time the Bank's concern has been reflected in both its technical assistance and lending policies.

In a large organization such as the World Bank many people are involved in the accommodation of policy to experience, changing conditions, and new insights into needs, goals, and optimum strategies. But there can be no doubt that Ladejinsky's work and views were a major influence in the evolution of these constructive changes in World Bank policy. Although the Bank recognizes, as did Ladejinsky, that its "potential for using the Bank's influence to press or even force the issue of structural reform on member countries is severely circumscribed," it does propose to "give overt priority in lending to those countries and projects which meet land reform criteria" and states that it will "not lend for projects if tenurial arrangements are so bad that they frustrate the achievement of the Bank's objectives." How much impact this new policy will have remains to be seen. Ladejinsky's contribution to it may prove to have been one of his crowning achievements.

Even these very preliminary notes suggest that Ladejinsky accomplished more than enough during his lifetime to have justified a real sense of pride in his achievements. But he was far more concerned with what he had failed to do, or get done, than with what he had helped to accomplish. And indeed one comes away from Ladejinsky's papers with the pronounced sense that the business of agrarian reform is largely unfinished and that the major part of the job to which he devoted his life remains to be done. Only in retrospect, however, do these opportunities lost or neglected seem distressing; for the future the opportunity is great. The publication of these papers, it is hoped, will bring about a greater awareness, in many countries where it is needed, that this essential, unfinished business still awaits constructive social action. To the developing countries, newly insistent on the creation of "a new international economic order," these papers issue a sober reminder that a necessary component of the social justice they seek can be achieved only through internal agrarian reforms, and that the sovereign power to initiate and carry through such reforms rests solely with themselves. This reminder is at the same time, therefore, also a challenge.

LOUIS J. WALINSKY

I. THE WASHINGTON YEARS, 1935–1945

DURING THESE ESSENTIALLY preparatory Washington years, Ladejinsky published some twenty-six articles—all but four in official journals of the U.S. Department of Agriculture.* Two were concerned with the food situation in Asia. The remainder were country studies: seven of Japan; five of the Soviet Union; two each of India, British Malaya, and Manchuria; and one each of Korea, Chosen (Formosa), Sri Lanka (Ceylon), Thailand, the Netherlands Indies (Indonesia), Australia, and the South Pacific islands. With the exception of the last two, all were to provide a useful preparation and much illumination for subsequent phases of Ladejinsky's work career.

Of the five Soviet Union pieces, three are outstanding. From the first of these, "Collectivization of Agriculture in the Soviet Union" (which because of its eighty-eight printed pages in length appeared in two parts in the March and June 1934 issues of the *Political Science Quarterly*), only two brief excerpts are presented here. The first of these, from the opening pages of the article, stresses the *political* role of the peasantry—a view which was greatly to influence Ladejinsky's subsequent thinking and work. The second deals with the Soviet leaders' perception of the organic interdependence between agriculture and industry in the country's economic development—another basic idea which was profoundly to influence Ladejinsky's thinking. The two remaining significant Soviet pieces—"Soviet State Farms," which because of its length was again published in two consecutive issues of the *Political Science Quarterly* (March and June 1938), and "Soviet State Grain Farms," which appeared in the *Journal of Farm Economics* in October 1938—substantially overlap one another in their subject matter. Only the concluding section from the shorter and later of the two is presented here.

Of the seven Japanese articles written during the Washington years, those dealing with the cotton textile and silk industries, food self-sufficiency, and agriculture in general are of no special interest, although they undoubtedly provided useful background for Ladejinsky's later work. And although the three remaining papers are all important, only one of them appears here. Ladejinsky's 1937 paper, "Farm Tenancy and Japanese Agriculture," which did so much to establish his reputation as an expert and proponent of land reform in Japan, was subsequently made redundant by his more thorough and updated "Farm Tenancy in Japan," which he wrote in 1946 while assisting in the actual planning of the reform. The

* See the Chronological Bibliography.

latter paper appears in Part II of this book. Of the two remaining papers, both significant—"Agrarian Unrest in Japan," which appeared in *Foreign Affairs* in January 1939, and "Japan's Agricultural Crisis," which appeared in the *Journal of Farm Economics* in August 1939—the first is a more popular, the second a more professional, treatment of essentially the same material. I have selected the latter for presentation here.

"Agricultural Problems of India," a fairly substantial yet long-forgotten 1939 article, has also been selected for partial presentation. A quite comprehensive survey of India's basic rainfall, backward production techniques, population pressure, fragmentation of holdings, land tenure, indebtedness, and other agricultural problems (as these appeared almost forty years ago), it obviously provided Ladejinsky with insights which germinated over the years and enabled him to approach his fieldwork in India in the early 1950s with a solid foundation and broad perspective. It also impressed him with the serious implications of the population problem for both agricultural and overall economic development long before this became a widely recognized concern.

"Chosen's Agriculture and Its Problems" (February 1940) and "Manchurian Agriculture Under Japanese Control" (August 1941) describe an insensitive colonial authority at work. The latter paper is especially interesting in its analysis of the attempt—and failure—of the Japanese authorities to force agricultural production and crop delivery by Manchu farmers to official agencies at uneconomically low prices—a lesson that was not lost on Ladejinsky when he observed the government of India's attempts at centralized wheat procurement more than thirty years later. "Agricultural Policies of British Malaya" (March 1941) and "Agriculture of the Netherlands Indies" (September 1940) show, in contrast, relatively enlightened and benevolent colonial administrations at work—a not unuseful reminder these days that all pre-independence prewar colonialism was not unmitigatedly baleful in its intentions and effects. I have included the short Malaya piece in full, a substantial part of the Korea paper, and Ladejinsky's own summary and conclusions from the Indonesia piece. The Manchuria article unfortunately did not lend itself to curtailment and has been omitted in its entirety.

The nature of these selected pieces provides the subtitle to Part I's "The Washington Years"—namely, "Early Studies of Agriculture Under Dictatorship and Semifeudal and Colonial Administration."

1. *Collectivization of Agriculture in the Soviet Union*

This, the first published of Ladejinsky's written works, was a fruit of his graduate studies at Columbia University and was probably to have been his dissertation. Concerning this and his later "Soviet State Farms" published in the same journal, he later wrote "They were the first studies of this kind in English, still used by students concerned with the origins of the Soviet economy." Ladejinsky's experience in the early days of the Soviet revolution and these studies made an indelible impression on him. The two brief excerpts presented here, for which I have provided the subtitles, help to explain why.

They are reprinted with permission from *Political Science Quarterly*; the article appeared in two parts in the March and June issues of 1934.

On the Political Role of the Peasantry

As early as 1906, Lenin, fully recognizing in the agrarian events of 1904–05 the harbingers of a social revolution, nevertheless saw fit to admonish his followers in these words: "We are supporting the peasant movement to the last, but we must remember that this is not the class which is capable of bringing about or will bring about a socialist revolution."[1] This attitude toward the peasantry flows directly from what Lenin called a "truism" known to every Marxian, that the "leading social forces in every capitalistic society are the proletariat and the bourgeoisie, while all the other social groups who occupy an intermediate position... inevitably gravitate in the direction of the first or second major group."[2] In the light of the position of the peasantry during the French Revolution, in the revolutionary movement of the nineteenth century in Western Europe, and in the first decade of the twentieth in Russia, Lenin held that "all the attempts of the petite-bourgeoisie in general, and the peasantry in particular, to assert their power and direct economic and political policies along their own lines, ended in defeat."[3] Soon after the outbreak of the February Revolution of 1917 he warned his followers that a union of the peasantry with the bourgeoisie might take place. Hence, "the proletarian party at present must not place any hopes on the community of interests of the proletariat and the peasantry. We are striving to win the peasantry over to our side; the peasantry, however, is more or less consciously on the side of the capitalists."[4]

Despite this critical attitude, Lenin was aware of the fact that the October Revolution would not have triumphed but for the support of the poor and middle peasants. He realized the importance of an alliance between the workers and the majority of the peasants, since the role of the latter group in a still overwhelmingly agrarian Russia was not less significant than that of the proletariat. If we may consider the October Revolution of 1917 as the resultant of several compounded forces of which the peasantry is a leading one, then, to a large extent as the peasantry shifts, so does the direction of the Revolution. In the crucial year of the Soviet Republic, in 1921, he insisted that "only an agreement with the peasantry can save the social revolution in Russia until the revolution takes place in other countries."[5] As the prospects of an early communist outbreak abroad faded away, an understanding with the peasants became the central point of Lenin's policies. When Lenin, through the promulgation of the New Economic Policy, spoke of establishing a closer link (*smychka*) between the workers and the peasants, it was more than the expression of a temporary policy; it was the premise upon which rested the possible solution of the peasant problem and, by the same token, the development and preservation of the Soviet state.

The practical expression of this all-important link is the socialization of agriculture. This proceeds along two lines: the organization of peasant collective farms and the organization of state farms. The present inquiry is concerned with the former, which the Communists view at this moment as the most important way of solving the peasant problem.

On the Relationship between Soviet Industry and Agriculture

The Soviet government viewed the situation in the village with great misgivings. It realized that it could not "eliminate the kulaks as a class through taxation and other restrictive measures, while leaving in their hands the means of production and the right of making free use of the land."[6] The Soviet government was compelled to acquiesce in a state of affairs where the kulaks played an important role in the village, for any other policy was bound to court "sure failure, strengthen the position of kulakism and

1. N. Lenin, *Pcresmotr Agrarnoi Programmy Rabochei Partii {A Reexamination of the Agrarian Program of the Workers Party}* (St. Petersburg [Leningrad], 1906), p. 27.
2. N. Lenin, *Sobranie Sochinenii {Collected Works}* (Moscow, 2d ed), vol. 23, p. 290.
3. Ibid., vol. 26, p. 290.
4. Ibid., vol. 20, p. 245.
5. Ibid., vol. 22, p. 238.
6. J. Stalin, *Voprosy Leninizma* [Problems of Leninism] (Moscow, 1931), p. 463.

leave . . . [the State] without grain."[7] The rich peasants of the Russian village had taken advantage of their strategic position as producers, and by the end of the decade they were a force sufficiently large to warrant serious attention.

The middle-class peasants, the *seredniaki*, who, in 1926, composed 56.4 percent of the total peasant population,[8] became the central figures in the village. The old agricultural system revolved about them. The aim of the Communists was to cultivate in them the element of the toiler at the expense of the seller and to draw them gradually into the ranks of the builders of socialism. But at this very time the sellers' proclivities asserted themselves with greater force than ever before, due largely to the government's policy of maintaining a high price level for manufactured goods and a corresponding low price for agricultural products. The so-called "scissors" problem, which has characterized Soviet economics for many years now, found its expression in this inequitable price disparity between manufactured goods and the products of the village. It is true that the "scissors" contributed much toward the rehabilitation of Soviet industry, but in view of the fact that this industrial growth was achieved largely at the expense of the village, it accentuated the already existing schism between the interests of the city and those of the village. All efforts on the part of the Soviet government to effect a close working union with the great mass of the middle peasantry were, for the time being, unavailing. In the quiet but grim economic struggle with the city, this most important section of the village identified its economic and political interests with the kulaks rather than with the proletarian state. Under these circumstances, eleven years after the October Revolution, there was ample justification for Lenin's statement made in 1921 to the effect that the village, after the Revolution, became more petit-bourgeois and as such represented "an independent class which after the crushing defeat and expulsion of the landed nobility and the capitalists, remains as the only class capable of challenging the political supremacy of the proletariat."[9]

The industrial progress attained by 1928 had fundamentally changed the relative position of industry and the prevailing system of agriculture. Although agriculture in the Soviet Union during its best years, 1925–1927, showed a year-to-year increase in its aggregate commodity production, its rate of progress was slower than that of industry. Whereas in 1926–1927 Soviet industry exceeded the prewar production level, the total agricultural production for the same year was only 88.1 percent of that of 1913.[10] The agricultural production in 1926–1927 was 30 percent greater than that of 1923–1924, but the percentage of marketable grain was even slightly less than in the earlier year.[11] This seemingly paradoxical situation is probably traceable to the refusal of the peasants to part with their produce at low prices. The generally slow progress in agriculture was traced by the Soviet government not so much to the dissatisfaction of the peasants with the government's price policy as to the general breakup of the large estates and some of the large peasant farms into smaller and smaller units where economic farming was impossible. Even if the price relationship had not adversely affected the development of agriculture, the existing system of small peasant farms producing 11 bushels of wheat and 4.5 bushels of barley as compared with 17 bushels of wheat and 9 bushels of barley per acre formerly produced on the large estates was bound to slow up the development of agriculture and to curtail the amount of marketable grain.[12]

The existing situation cannot be properly evaluated apart from its relationship to the vast scheme of industrialization of the country. If the existing tendencies in agriculture were allowed to proceed unchecked, the program of industrialization would be jeopardized. The fulfillment of the program depended upon impor-

7. Ibid., p. 458.

8. I. A. Iakovlev, *K Voprosu o Sotsialisticheskoi Rekonstruktsii Selskogo Khoziaistva [On the Problem of Socialist Reconstruction of Agriculture]* (Leningrad, 1928), p. 1. Iakovlev defines a middle peasant as one who cultivates from two to eight "desiatins" of land and owns from one to two head of work animals.

9. N. Lenin, op. cit., vol. 22, p. 289.

10. I. A. Iakovlev, op. cit., p. 357.

11. Ibid., p. 352.

12. Ibid., p. 353.

tation of machinery and of a great variety of raw materials from abroad. Soviet Russia, however, had not sufficient gold and was unable to procure long-term credits abroad to finance her purchases. The only way out was to finance imports with the proceeds from exports, primarily the export of agricultural products which had always played an important role in Russia's foreign trade, accounting for 73.7 percent of the gross Russian export in 1913.[13] In 1924 Soviet industry was started on the road toward rehabilitation because the Soviet Union was thus able to pay for the imported machines.[14]

Three years later, confronted with a vastly greater and more significant industrial plan, the Fifteenth Party Congress expressed its dissatisfaction with the agricultural situation in the following resolution: "The backwardness [of the old system of agriculture] hampers . . . the export of agricultural products which is the basis of import operations necessary for a more rapid industrialization of the country, and the further development of agriculture itself."[15] In other words, agriculture was still looked upon as the main item upon which the import plan had to be based. But in 1927 the relative importance of agricultural products in the total mass of exports declined from 73.7 to 57 percent, then to 46 percent in 1928.[16] The year 1928, when the five-year plan had assumed concrete form, ended with a 20 percent decrease in agricultural exports.[17] This decline of Soviet Russia's main item of export, along with the growing grain shortage in the cities and industrial centers, threatened to undermine the main pillars upon which the plan rested.

Immediately prior to the inauguration of the vast program of industrialization, it became obvious that the prevailing system of small-scale agriculture failed to meet the demands made upon it. With the system left intact, the situation was bound to become worse with the application of the five-year plan. From the point of view of the existing economic relationship, between the city and the village at the end of the decade 1918–1928, the former was moving toward a socialist form of industrial organization; the latter, toward a capitalistic form. Industry and agriculture did not constitute a single national economic whole. Notwithstanding the fact that the land was owned by the state, the city with its large centralized and partly socialized industry had its antithesis in the small and extremely backward individualistic system of peasant farming. From the Communist point of view, this was an economic anachronism which "would some day end in a complete breakdown of the entire national economy"[18] if it were not corrected. Only a system of large-scale mechanized agriculture could avert such an outcome. This was to be realized not through the introduction into the village of capitalistic large-scale enterprise but, on the contrary, by the eradication of all its vestiges. And the Communist Party decided that the surest way of achieving this end was the further consolidation of the small farms into large ones equipped with modern machinery. Therein lay the guarantee that the problem of "either back to capitalism or forward to socialism"[19] would be solved in favor of the latter alternative.

13. *Statisticheskii Spravochnik {Statistical Handbook of the U.S.S.R.}* (Moscow, 1932), p. 392.

14. L. L. Popov, "Osnovy Perspektivnogo Plana Razvitiia Selskogoi Lesnogo Khoziaistva" [The Basis of the Tentative Plan for the Development of Agricultural and Forest Economy], *Planovoe Khoziaistvo {Planned Economy}* (Moscow, August 1925), p. 19.

15. *Pravda* (December 21, 1927).

16. *Statisticheskii Spravochnik {Statistical Handbook of the U.S.S.R.}* (Moscow, 1929), p. 719.

17. Ibid.

18. Stalin, p. 444.

19. Ibid., p. 445.

2. *Soviet State Grain Farms*

Ladejinsky wrote two papers dealing with Soviet state farms. The first, much longer and more comprehensive, appeared in the *Political Science Quarterly* in two parts in March and June of 1938. The concluding section of the second and more circumscribed paper, which deals with the state grain farms only, is given here. Together with his first paper on the collectivization of Soviet agriculture, these papers provide important early background for Ladejinsky's anticollectivist convictions. Two other early conclusions are: that the problems of large-scale agriculture, contrary to Soviet expectations, are less amenable to solution than are those of large-scale industry; and that large-scale agriculture does not necessarily equate with greater efficiency and lower costs. The paper also has a certain contemporary, if background, interest in view of severe grain shortages in the Soviet Union in recent years and its consequent need to import substantial tonnages from the United States, Canada, and other Western sources.

This article appeared in *Foreign Agriculture*, a monthly publication of the U. S. Department of Agriculture, in October 1938.

IN SUMMARIZING THE DEVELOPMENT of state grain farms, it is well to recall that they were expected to achieve the following aims: increase the country's agricultural output, serve as "model" agricultural centers, and prove to the peasants that large-scale agriculture is vastly superior to their small-scale individual farming. Did the farms achieve these aims? To answer this question is to point out the role played by these farms in the Soviet agricultural economy.

The state grain farms unquestionably added to the country's agricultural resources. During the period 1931–1935, the total grain area of the Soviet Union was increased by an annual average of 9 million acres. Considering the great efforts exerted by the government to increase the country's agricultural output, the extension of the sown area under grain farms must be viewed as an achievement. On the other hand, it must be kept in mind that the grain output, in consequence of low yields, was not large enough to compensate for the huge investments that accompanied the rapid acreage expansion.

Since 1935 the work of state grain farms has shown considerable improvement, but not to the extent of raising them to the position of "cultural agronomic centers." If the proper application of agricultural science is the real test of a model agricultural undertaking, then the state grain farms have failed to pass it, judging by the following statement:

> It is no exaggeration to say that in the entire system of the Commissariat of State Farms there is not a single grain farm where agricultural science was raised to a proper level. By the proper level we have in mind essentially elementary conditions: sowings on fallowed and fall plowed land; good quality of the fallow and fall plowing; timeliness of sowings and utilization of a rational crop rotation system. These elementary measures are being applied in all the better organized . . . collective farms but they are not being made use of by the greater number of state farms.[1]

In view of the above, state grain farms were unable to convince the peasants of the alleged

1. V. Feigin, "Concerning the Profitability of State Grain Farms," *Socialist Reconstruction of Agriculture*, No. 12 (1935), p. 65.

benefits of large-scale farming. They had, therefore, little to do with the shift of the Russian peasants from their small, individualistic to a large, collectivistic system of agriculture. During the years 1929–1933 when the greater number of collective farms was organized, the work of the state grain farms was notoriously poor. If the latter had really been models of a reconstructed agricultural economy, collectivization of agriculture in the Soviet Union would have proceeded at a much slower pace.

The state grain farms were set up under difficult conditions. Everything that goes into the making of a modern large-scale agricultural undertaking had to be built from the ground up. This would not have been an easy task even if the necessary prerequisites had been at hand —such as good land, experienced management, skilled labor, and ample time in which to organize the farms. But during the period under consideration, many of these elements were completely lacking or available only to a limited degree. Furthermore, in organizing the state grain farms, the Soviet leaders proceeded on a number of theoretical assumptions that proved unworkable.

One of these theories was that large-scale industry and large-scale agriculture follow the same patterns of development. Without going into a discussion of whether or not this holds true under any or all circumstances, suffice it to say that in the Soviet Union, and in the period under consideration, it was not substantiated. It was revealed that, in both fields, given good management, proper equipment, and skillful workers, it is easier to manage and control the work of a large industrial plant than of a large farm; that, because of such an uncontrollable factor as climate, for instance, the planning programs of state grain farms must be less rigid and more flexible than those of industrial undertakings; that narrow specialization of state grain farms retards, rather than advances, their development; and that, notwithstanding the claims made by Soviet economists, the notion that "the larger the farm, the smaller the cost of production" cannot be supported as a general proposition.

Finally, it is obvious now that state grain farms are not mere adjuncts to tractors and combines, although the importance of the latter should not be underestimated. The validity of the following statement made by Professor Tulaikov in 1931 has been fully demonstrated:

> I understand very well that the utilization of the tractor is an exceedingly important aspect of the work to which we have to pay particular attention. It must be stressed, however, that we are not building grain farms in order to utilize tractors, but are utilizing tractors in order to receive a certain output of grain. . . . I believe that the farm as such has the right to exist and is not a mere adjunct to the tractor. Granting this premise, agricultural science is entitled to more consideration.[2]

Soviet authorities are aware now that, while modern agricultural machines can make the application of agricultural science more effective, they cannot take its place.

The state grain farm has always been viewed by the government as the true socialist type of agricultural enterprise and more in consonance with the Soviet economy as a whole than is the collective farm. But the government's policy of decreasing the land area of the state grain farms and adding it to the collectives indicates that the latter are the more efficient producers. It is possible that at some future time the state grain farms will become efficient producers, but this will be achieved not because of any inherent advantages peculiar to these farms but rather by a combination of all those factors that are essential to good farming under any economic system.

2. *The Organization of Large Socialist Grain Farms* (Moscow, 1931), p. 90.

3. *Agricultural Problems of India*

This 1939 study of India's agricultural problems is the first piece of work Ladejinsky ever did on India. It was part of the intellectual equipment Ladejinsky brought with him when he arrived in India for the first time in 1950 for discussions with the Planning Commission. Curiously, the paper was not uncovered by searches made at the U.S. Department of Agriculture. This first encounter in what was to be his long and arduous courtship with India was turned up, by chance, while rummaging among his personal papers. The more descriptive first half of the paper is omitted here.

This article appeared in *Foreign Agriculture* in August 1939.

Some Approaches to a Solution

THE AGRICULTURAL CONDITIONS described [in the section omitted] are not merely temporary depression phenomena but the depression aggravated the forces that have been undermining the country's agricultural economy for many years past. Relatively few measures have been advanced, however, with a view to easing the agricultural difficulties engendered by the depression. Despite the catastrophic fall of agricultural prices, hardly anything has been undertaken to raise them or stabilize them at definite levels. The devaluation of the rupee after the pound sterling went off gold had very little effect on the revival of prices. It was felt that the "natural" operation of economic forces should not be tampered with and that sooner or later the depression would work itself out. Certain provincial governments of India have helped the peasants by remitting some of the land revenue, by cancelling certain dues, and by forcing reduction of rent. On the whole, however, these measures were not sufficient to offset the additional decline in income created by the agricultural crisis of the last decade.

Although there is little in India of the kind of agricultural relief prevalent now in a great many countries, efforts have been made from time to time, and especially during the thirties, to ease some of the burdens from which the peasants suffered even before the depression. Since farm indebtedness and tenancy are perhaps the two greatest obstacles to India's rural progress, it is pertinent to inquire first into the attempts to deal with these problems.

Easing the debt burden

In regard to the problem of debt reduction, there must be distinguished two types of remedial action—that by the government through legislation and that through agricultural cooperatives. The latter will be discussed in the section dealing with the Indian cooperative movement as a whole.

The government has long realized the need of checking the activities of the usurious moneylender, upon whom the peasants chiefly rely for credit. The Deccan peasant riots of 1875, an important feature of which was an attack on the moneylenders, led to the passage of the Deccan Agriculturists' Relief Act of 1879. The act provided that, in suits by or against farmers, the court was permitted to examine the history of the debt and makes it obligatory for creditors to furnish a written account of the money actually due. Other provisions required creditors to furnish accounts and grant receipts, and stipulate that mortgages of farmers must be in writing. The court, in taking account of past transactions, could modify the contract by reducing an oppressive rate of interest and by preventing the sale of land unless specifically pledged. In the hope that litigation would decrease, the statute of limitations, which since 1859 had been for eight years only, was extended for suits against agriculturists to a period of twelve years if the suit was based on a registered deed and to a period of eight years if it was not. Minor amendments were made in the course of time, such as the giving of wider powers to the courts in order to determine the nature of the transaction independently of the provisions of the law regarding documentary evidence. This act is now in operation throughout the province of Bombay.

Years of application revealed that the act was ineffective as a means of safeguarding the interests of the cultivators. An investigation of the Famine Commission of 1900 found that "the Act had done but little good and that there was positive room for holding that the transfer of property both by sale and mortgage had become more frequent in the districts to which it applied."[1] Several years later the same commission summarized its conclusion on the same subject in the following words: "it is a contest of dishonesty, in which that side is likely to gain the upper hand which is prepared to go farthest in perjury and in the production of false evidence. . . . Hence it is that an Act whose main object was to put the relations between the agriculturists and moneylenders on a better footing, is actually having an opposite effect."[2]

Another measure, just as ineffective, was the Usurious Loans Act of 1918. An important feature of the act "is that the court, once seized of a case, may, of its own motion, re-open old transactions and enquire into the equity of the terms. . . . Where the debt is unsecured, the debtor can draw the creditor into court and, therefore, into the sphere of this Act, by the simple expedient of refusing to renew his loan."[3] It was hoped that this measure would lead to the reduction of interest as well as fix a maximum interest rate, but according to the Commission "the Act is practically a dead letter in all provinces." The Punjab Regulation of Accounts Bill of 1930 provided that all moneylenders must use regular account books and furnish each debtor every eight months with a legible statement of accounts signed by them or their agents. There must be indicated not only the amount outstanding but also all loan transactions entered into during the preceding eight months. If no accounts are kept, the court may in any suit disallow, either wholly or in part, the interest found due and also the cost of the suit. But like so much other legislation in this field, its provisions did not give adequate relief to debtors. The illiteracy of the peasants and sometimes of the moneylenders, the unwieldiness of court procedure, the expense of litigations, and the lack of sufficient alternative sources of credit contributed greatly to what the commission described as the "extraordinary difficulty of attaining the objects in view by means of legislation."

Despite the failure of the measures enacted prior to 1930, the worsening of agricultural conditions since then stimulated the promulgation of a number of very important laws intended to ease the burden of the debtors. What is being done in the central provinces and to a more limited extent in the province of Madras and in the native state of Bhavnagar, is characteristic of the general trend of legislation in other sections of the country.

In 1933 the government of the central provinces promulgated a Debt Conciliation Act, the main feature of which was the setting up of conciliation boards for debt adjustment. Having examined the transaction between the creditor and debtor, the board renders its decision as to the sum to be repaid, as well as the way of making the payment. The award is not unconditionally binding. If a creditor refuses to agree to terms considered by the board to be reasonable, the board makes clear who the obdurate party is through a special certificate. Should the creditor then resort to court, he would be penalized by not getting the costs of the suit and by having the interest charges from the date of the certificate reduced to 6 percent. If a debtor defaults in payments as agreed, the amount may be recovered in the same manner as arrears of land revenue on the application of the creditor. Failure to recover such arrears invalidates the agreement.

The work of the conciliation boards in the central provinces met with considerable success. Within a period of four years, 46,537 cases involving an amount of 56.8 million rupees (about U.S. $21,000,000) were settled. The

1. *Report of the Royal Commission on Agriculture in India*, p. 437.
2. Ibid., p. 438.
3. Ibid.

total amount was reduced to 29.3 million rupees (about U.S. $11,000,000), or a remission of 48 percent of the debt.[4] The government of the central provinces promulgated also the Usurious Loans Amendment Act of 1934, with a view to keeping interest rates within certain bounds; namely, 12 percent on secured and 18 percent on unsecured loans.

The most drastic attempt to scale down past debts in India is embodied in the Madras Agriculturists' Relief Act of 1938. A relatively small group of cultivators is brought within the scope of the act, but its main provisions indicate that the government of the province was approaching the problem in earnest. Thus, debts incurred before October 1, 1932, were reduced in the following manner: "(1) All interest outstanding on the 1st October 1937, in favour of any creditor of an agriculturist . . . shall be deemed to be discharged, and only the principal or such portion thereof as may be outstanding shall be deemed to be the amount repayable by the agriculturist on that date; (2) where an agriculturist has paid to any creditor twice the amount of the principal whether by way of principal or interest or both, such debt including the principal shall be deemed to be wholly discharged."[5]

Perhaps even more significant was the manner in which the native state of Bhavnagar attacked the debt problem. The plan called for the outright liquidation of the peasants' debts through the financial assistance of the government. The total debt having been determined, a settlement was arranged whereby the sum to be paid by an individual debtor was not larger than three times the annual land revenue assessment. The scheme was financed by the government, which advanced the necessary sum for the redemption of the individual debts, the advance being repayable in installments by the addition of 4 percent to the annual land tax. The total of all debts, of which the nominal amount according to the books of the moneylenders was 8.6 million rupees, was settled by a payment of 2 million rupees, or what was, in effect, a sum equal to the payment of a year's interest. The cultivators and the government revenue benefited from this change. The slate having been wiped clean of debts, payments of land revenue proceeded without any difficulty and in many cases even arrears were paid off. Moreover, debt litigation, of which there had previously been a great deal, almost disappeared.

It is to be noted, of course, that the sum involved was a small one and that there is no indication that such a scheme of debt redemption by the state and provincial governments is finding wide application. But even if all debts were wiped out now by government decrees, the problem would remain unsolved unless the factors making for indebtedness were also eliminated. So long as the peasant operates on a deficit budget, he will run into debt. In this connection the Bombay Banking Enquiry Committee made the following observation a decade ago: "Our examination of the problem of debt leads us to conclude that its incidence would not be so onerous and its growth not so serious, were the margin between income and expenditure higher than what it is today . . . even if the incidence of interest is reduced considerably, the success of any scheme for liquidation of debt would be problematical, should the agricultural conditions remain as they are now."[6] The statement is as true now as it was ten years ago. The lasting solution, therefore, is in turning India's largely deficit agricultural economy into a profitable economy.

Improving status of tenants

With respect to farm tenancy, the chief aim on the part of the various governments of India has not been to legislate it out of existence; rather the purpose has been to secure for the tenant greater certainty of tenure and to safeguard him against eviction, unfair rent, and illegal exactions. Practically nothing has been done along these lines in the past decade, but the legislation enacted in former years indicates the nature of the remedies designed for the improvement of the lot of the tenants.

4. P. J. Thomas, "Debt Relief in Central Provinces," *The Indian Co-operative Review* (July–September 1937), p. 394.

5. Law quoted in the editorial notes of *The Indian Co-operative Review* (January–March 1938), p. 4.

6. Quoted by Ambalal D. Patel in *Indian Agricultural Economics* (1937), p. 256.

The first experiment in tenancy legislation was made in 1859 when the Bengal Tenancy Act was promulgated. Its significance lay in the provision that tenants-at-will who had been cultivating the same land for twelve years continuously were to become occupancy tenants. This meant life tenure in the land and protection from arbitrary enhancement of rent. The act failed, however, to achieve its objective because the landlords saw to it that no tenant rented the same piece of land for a continuous twelve-year period. The new Tenancy Act of 1885 corrected this situation. Accordingly, a tenant could acquire occupancy rights by renting any land in the same village for twelve years. An occupancy tenant could transfer his right of life tenure in the land, subject to a special payment to the landlord. A landlord might ask for a higher rental only once in fifteen years, and the increase was restricted to one-eighth of the rent previously paid. The increase could be effected either through agreement between landlord and tenant or through court action. A civil court might increase the rent only on certain specified grounds, the most important of which were as follows: (a) When the rate paid was lower than that paid for similar land in the neighborhood; (b) when there had been a rise in the average local prices of staple foodcrops since the rent was last fixed; and (c) when the productive powers of the land had been increased by improvements effected at the landlord's expense. A certain degree of protection was extended also to nonoccupancy tenants; they could be evicted only by court order and their rents be raised only once in any five-year period. These provisions are found now in most of the tenancy acts of India.

The tenancy legislation of Bengal served as a basis for similar legislation in other parts of the country, although measures taken in some provinces were more comprehensive. The Tenancy Act (1920) of the central provinces brought about a very material change in the relations between the landlords and tenants, mainly to the benefit of the latter. It did away with the division of tenants into occupancy and nonoccupancy; under the provisions of the act every tenant, regardless of the length of time he held the land, was given the permanent right of occupation.

Developments in the province of Agra are of interest. Until 1928, the majority of tenants could be evicted or their rent raised at ten-year intervals. This brought about so much litigation that the number of evictions in 1923 reached a high of 157,000, and in 1924 a total of 620,000 suits were instituted. The existing tenancy act was amended in 1928 in one very important respect: it granted life tenure in the land to every tenant-at-will (nonoccupancy) and after his death the right of the heir to hold the land for another five years. Moreover, whereas formerly the landlord had a right to ask for a higher rent every ten years, the period has now been extended to twenty. The increase in rent is not determined by the landlord but by a special officer whose duty it is to propose fair and equitable rates, "based on genuine, adequate, and stable rents which are paid by the substantial tenants who depend for their livelihood on the produce of their holdings and can be paid without hardship over a series of years, due regard being had to movements in prices and rents and the letting value of land."[7]

The nature of the tenancy legislation just described is generally characteristic of the work done in this field throughout India. On the whole, it has provided tenants with greater security of tenure and has limited the arbitrariness with which landlords have raised rents. But considering the conditions under which tenants still cultivate the land, the inadequacy of the measures is all too obvious. Rents are still high and are getting higher because of the steadily rising competition for the land; illegal exactions are paid by the tenants as in days gone by, and so are premiums or lump sums over and above the rent when a transfer of occupancy rights is effected; and, even when life tenancy is granted, eviction is by no means uncommon because the same laws vested the landlords with extensive powers against the nonpayment of rents. Under the Agra Tenancy Act, litigation and evictions declined sharply, but in 1934–35 there were nearly 84,000 cases of eviction. Moreover, there are millions of so-called undertenants whose interests, except in the central provinces,

7. S. N. A. Jafri, *History and Status of Landlords and Tenants in the United Provinces* (Allahabad, The Pioneer Press, 1931), p. 138.

are not protected by the existing tenancy legislation. Many an occupancy tenant sublets the land to the actual cultivator of the soil (undertenant), thereby converting him "into a person without any adequate protection safeguarding to him the fruits of his toil."[8] There are also more than 30 million land laborers with a status inferior to that of the least protected undertenants. Finally, the measures intended for the protection of the tenants do not strike at the roots of the grave situation created by the constant rise in the number of landless peasants, the transfer of land to moneylenders, and the long array of middlemen who profit from the complexities of the present land system and widen the cleavage between the main rent receivers and the actual tillers of the land.

The problem of land tenure in India is admittedly an extremely complicated one and it does not lend itself to ready solution. The various tenancy acts have not materially affected the existing relationships which are responsible for many of the ills of the peasantry. There can be no visible improvement, however, in the Indian countryside unless the basic economic and legal conditions are changed in favor of those who actually cultivate the land. It is with this view in mind that Professor Mukerjee came to the following conclusions: "Some . . . revision of the old relations is necessary in India to ensure the peasant being economically as well as legally set free. The present deterioration of the position of the tenant forebodes an agrarian revolution; and unless the situation is handled boldly and sagaciously, it will end in disaster."[9]

Cooperation

While the palliatives dealing with indebtedness and land tenure have little helped the mass of the Indian peasantry, the cooperative movement is being looked upon as a real source of rural progress. The often referred-to report of the Royal Commission on Agriculture noted that "cooperative principles can be used in overcoming most of the obstacles to progress in rural communities," and that those of India were not exceptions. Having subjected many aspects of Indian agriculture to a thorough examination in looking for a remedy, the commission concluded that "if cooperation fails, there will fail the best hope of rural India."

The cooperative movement was officially launched in 1904, and its sole objective was to break the vicious credit system of the moneylenders by providing loans to cultivators at reasonable rates. In 1935 and 1936 India had a total of 94,000 agricultural cooperatives with over 3 million members and a working capital of 323 million rupees (U.S. $121 million). With the exception of Bengal, Punjab, and Madras provinces, the movement has so far reached only a small part of the rural population, and it has been estimated that in the Punjab only about 15 percent of the loans are provided by the credit societies. This indicates that the moneylenders still hold sway even in provinces where the cooperative movement is strongest. On the other hand, it is not denied that where cooperatives have taken root they have provided a fair amount of credit at rates ranging from 9 to 15 percent. The rates are high, but they are considerably lower than those charged by moneylenders. Where the cooperative movement is strongly entrenched, there has been a general lowering of the rate of interest charged by the moneylenders and their grip on the peasant has been loosened.

Certain achievements have been registered by the societies in the way of controlling credit and restricting unproductive debts. The moneylender seldom inquires into the purpose of the loan when the borrower's credit is good, protecting himself against loss by a high rate of interest. The aim of the credit society, however, is to make sure that the larger part of the loan will be utilized productively. Owing to custom, some nonproductive expenditure has to be accepted as necessary (in the sense of being socially useful or unavoidable, even if not economically beneficial); and it is better for members to borrow from the society than from the village moneylender even for such purposes. But in all such cases the loans are granted sparingly and under close supervision. The success of this policy is illustrated by the fact that from 60 to 70 percent of the loans advanced by the cooperatives are utilized for productive purposes.

8. Radhakamal, Mukerjee, *Land Problems of India* (London, 1933), p. 182.
9. Ibid., p. 197.

The influence exerted by the cooperative movement on the Indian village cannot be measured alone in terms of relieving peasants from the burden of usury. Although the credit society is still the cornerstone of the whole movement, within its province now fall numerous noncredit activities, all of which tend to promote the general welfare of the village community. It has long been felt that, if the great number of cultivators are to be won over to use of better seeds; to improved methods of cultivation; to better care of cattle; to adoption of precautions against human, animal, and plant disease; and to more efficient marketing, it must be through the agency of their own cooperative societies. In view of the widespread illiteracy among the peasantry, no official organization could possibly do it by pamphlets or written instruction, and no experts could reach the multitude through any medium other than cooperatives.

A mere enumeration of the principal types of noncredit societies brings out the vast scope of the work in which they are engaged with varying degrees of success. There are purchasing and sales societies as well as industrial, irrigation promotion, insurance, housing, better farming, better living, consolidation of holdings, and a number of other societies with objects ranging from education promotion to arbitration. The functions of the better farming and better living societies show to what extent rural reconstruction has become a part of the cooperative movement. The aims of the better farming societies include measures to promote better agricultural methods; to supply good seed and improved agricultural implements; to demonstrate the more recent improvements in agriculture and to render practical assistance in their application; and to create funds for loans to members for purchase of good implements, manure, and seed. The better living societies lay greater emphasis on reforms of social customs. They make efforts to improve the physical, moral, and spiritual conditions of the members; to inculcate habits of thrift; to teach and practice rules of hygiene; and to eradicate illiteracy among members. When the human and economic wastage of which an Indian village is capable is considered, it is evident that the successful carrying out of the above aims would contribute much toward the improvement of the social and economic status of the peasantry.

A significant example of what cooperatives can achieve is found in the consolidation of scattered holdings. Since fragmentation is in many parts of India one of the most important factors tending to block agricultural improvements, various attempts have been made to cope with the problem. The task is a difficult one; the fact that some cultivate their own land whereas others rent a part or all of it to tenants and that there are rights of mortgages to be considered and, above all, that there are great differences in the quality of the land set for consolidation produces an element of confusion in the discussion of the evils of subdivision and fragmentation and of the proposed remedies.

On the whole, action through legislation has been found wanting; a fair degree of compulsion and a revision of the country's basic laws of inheritance appear necessary if far-reaching results are to be secured. Measures of the type required are not contemplated, as evidenced by the observation of the Royal Commission: "The main policy of any government embarking upon a campaign for the consolidation of holdings must be to achieve progress by education." Educational action through the cooperatives in the Punjab, however, has produced some striking results. Special cooperative societies for consolidation are organized. Every member must agree to give and take possession of lands reallotted in accordance with any arrangement approved by two-thirds of the membership. The same provision holds good in case of future partitions or rearrangement of the consolidated land; and they must agree also to submit to arbitration any dispute that may arise in connection with the society's business. Consolidation is carried out in two ways: by mutual exchange of fields to bring the scattered fields of each owner into blocks as near to one another as possible, and by treating all land offered for consolidation as common land and dividing it into suitable compact blocks.

In the period 1921–1937, the cooperatives succeeded in consolidating in the Punjab an area of nearly 800,000 acres. But many millions of acres are still so fragmented that the application of improved methods of cultivation is virtually precluded. The point is, however, that the cooperatives have shown that the evil of fragmentation can be attacked successfully. In one case 35,000 scattered strips of land were consolidated into 4,500 blocks, while in another instance 50,000 acres were rearranged so that the average size of block of the 7,951 owners was increased from one-half to nearly 3 acres.[10] As a result of this process, the participants do not receive less land than they held before; on the contrary, some land may be added by eliminating the strips. And no one is asked to agree to the repartition until he has seen his new holding marked out on the ground. Now that the advantages of consolidation have been illustrated in a practical manner, the wish to do away with fragmentation is growing apace. The area consolidated in the Punjab during the period 1921–1925 averaged 8,000 acres yearly; this figure rose to 56,000 acres in 1934 and to 120,000 acres in 1937.

The history of the cooperative movement in the thirty-five years of its existence has not been one of continuous progress. There have been numerous failures and disappointments. As late as 1938 the codirector of the Department of Agriculture of the United Provinces pointed out that "the Co-operative Movement in India has accumulated so much frozen capital as a result of non-recovery of debts and the liquidation of societies on a large scale that the present period may be regarded as critical in the history of the movement."[11] It should be noted that the cooperatives had been suffering from non-recovery of loans even before the depression of recent years. The cure is in the educational efforts of cooperation to effect some change in the peasants listless and fatalistic outlook upon life, thereby undermining their adherence to customs and traditions in social and agricultural matters. Under conditions prevailing throughout the Indian countryside, the task is an extremely difficult one. But the beginning has been made, even if on a small scale. Outstanding examples of successful application of cooperation can be found in practically every province of the country. It is doubtful, of course, whether the cooperatives will prove a panacea for all the ills of the Indian village community. The fact is that the majority of those who cultivate the land lack the resources that should enable them to take advantage of the cooperative movement. The tens of millions of tenants and farm laborers who do not own land are not in a position to offer security for loans and, therefore, can get no help from the credit societies. But as cooperation becomes more and more an integral part of a comprehensive agricultural policy, millions of peasants may be helped to a better life.

10. Eleanor M. Hough, *The Co-operative Movement in India* (London, 1932), p. 185.

11. C. Maya Das, "The Co-operative Movement and Agricultural Development in the United Provinces," *The Indian Co-operative Review* (July–September 1938), p. 354.

Agricultural research and extension work

Cooperation, tenancy, and debt relief measures have been discussed at some length because, if effective, they are important elements in any scheme to render farming more profitable. Yet they are not the only remedies. Other recommendations and suggestions bearing even more directly upon a higher farm income are found in the *Report of the Royal Commission on Agriculture*. Some of them are pertinent to the discussion and must therefore be touched upon here.

The commission recognized that, while agricultural research was the foundation of rural progress, in India such work was only in its infancy. Left without the stimulus of a central organization, research carried on in the various provinces in a haphazard manner was not yielding the expected results—hence the recommendation and eventual organization of an Imperial Council of Agriculture with the principal function "to promote, guide, and coordinate agricultural research throughout India." At the present time the council is financing about ninety research projects covering crops, plant and animal diseases, cattle breeding, and marketing of agricultural products, as well as studies in soils and fertilizers. According to the International Institute of Agriculture, India now has 22 agricultural institutes and laboratories concerned with the improvement of crop production, about 300 experimental and demonstration farms, and a teaching and research staff of 800 officers and assistants; nearly 2,000 officials are engaged in introducing into general agricultural practice the successful results of research. The benefits derived by farmers from dissemination of useful knowledge cannot be overestimated.

Agricultural improvements evolved by departments of agriculture are of little use unless they are incorporated into the farm practice of the cultivators. Since so many of the cultivators are illiterate, great reliance is placed upon "ocular demonstration." That was considered to be the best method of convincing the cultivators of the advantages of agricultural improvement. The commission attached so much importance to the value of demonstration and propaganda that it advised the establishment of a new office of deputy director of agriculture, attached to that of the director of agriculture of each province, for the purpose of organizing and systematizing propaganda work. He was also to watch the various schemes of propaganda in operation, note their results, and suggest ways of making them more effective. In 1935 some 1,500 agricultural shows were held in various parts of the country and 86,000 demonstrations were given on the cultivators' own land.

It has been noted that the prosperity of India's agriculture is intimately connected with the quality of its livestock and that the latter is poor. The remedy, as seen by the Royal Commission, is better feeding and breeding. Increase in the existing grazing area is not possible. It is necessary, therefore, to increase the productivity of the land already growing grass by regulating grazing on a rotation system, cutting and storing dry grass, making silage, and cultivating fodder crops. In regard to the improvement of cattle by breeding, attempts have been made to get purebred and improved types of the best cattle available in India. The Royal Commission determined that this program was not to be weakened by efforts to produce a "dual purpose" animal suitable both for draught and milking. Another necessary link in the process of improving the quality of the cattle is the control of the contagious diseases prevalent throughout India. For the sake of greater effectiveness and to insure a uniform procedure in dealing with this problem, the commission recommended the promulgation of a Contagious Diseases of Animals Act. It meant the establishment in India for the first time of a veterinary service on a large scale and on a sound footing.

Expansion of irrigated areas

Of all the factors that limit both the extension of land under cultivation and the yield per acre in India, the supply of water is the most important one. The country's rainfall problem is one of unequal distribution in different parts of the country; of irregular distribution through seasons; and, very often, of total lack of rain. The remedy is irrigation and yet more irrigation. Large areas of wasteland have thereby become productive, creating additional wealth. From about 10.5 million acres in 1879 the area annually irrigated by state works alone has now risen to about 31 million acres. The government and the cultivators cooperate very closely in this work. The headworks of the canal; the main line, and the branches are all constructed and maintained by the government; but the field channels or water courses, by means of which the water is finally conveyed onto the fields, are usually constructed and invariably maintained by the cultivators themselves.

The commission's comprehensive study of irrigation conditions and needs has helped to stimulate the work in that field. On the commission's recommendation, a Central Board of Irrigation was established, which, among other things, aimed to coordinate research in irrigation matters throughout India and to disseminate the results achieved; to bring about the fullest collaboration between agricultural and irrigation departments; to secure an exchange of information and experience in irrigation between the various provinces; and, above all, to formulate schemes for extending and developing irrigation projects wherever they may be required. There are at present about 300 irrigation schemes in operation in British India alone, of which 70 are major works. The net effect of the constantly expanding artificial irrigation facilities is that millions of people no longer live, as they have continuously for centuries, in the shadow of periodically recurring crop failures caused by drought.

Reduction of illiteracy

Much of the evil that pervades the Indian countryside is generally attributed to illiteracy. "This curse," as one writer noted, "acts as a leaven to start the vicious circle of debt-disease-

and-death."[12] The commission recognized the immensity of the task and noted "that the only hope of substantial progress lies in the mobilization of all the available forces, both public and private, in a determined attack upon illiteracy."

In the belief that if a mother is literate a very strong influence will be brought to bear on keeping her children at school until literacy is assured, the first recommendation concerns the spread of literacy among women of India. But perhaps the most important remedy suggested by the commission is the introduction of a compulsory system of primary schools as the best means to overcome the unwillingness of parents to send their children to school. Inefficient teaching is to be corrected by better training of teachers who were to be selected, wherever possible, from rural districts. In addition, the commission also made detailed recommendations concerning secondary schools, agricultural colleges, and universities—all with a view to making their graduates fit to take the lead in the movement for the elimination of illiteracy and for the general uplift of the rural classes.

Industrialization

Finally, it must be noted that rural India cannot solve all its many problems by tilling the soil, no matter how efficiently. Villages are overcrowded. It is not certain how large the surplus farm population is but, if the view of India's agricultural experts that the lowest limit of an economic holding in that country is 15 acres is accepted, "in British India alone . . . as much as 44 percent (or 36 million) of the total workers on land were superfluous."[13] In order to find even a partial outlet for these millions, India must be industrialized. Efforts in this direction are being made, and the development of a modern textile industry, employing over 400,000 workers, is an outstanding feature of the industrial revolution in India. But to be really effective, that is, to be able to absorb millions of workers from the overcrowded villages, India must industrialize on a gigantic scale and at a very rapid pace. It is patent, however, that, because of lack of capital, relative shortage of mineral resources, limited domestic market due to poverty, and difficulty of acquiring foreign markets for manufactured goods, no such scheme can now be put into operation. This does not preclude, however, the possibility of further industrial expansion, for, to date, it is not commensurate with the size of the country or its population and natural resources. With India's return to its former status (roughly until the beginning of the nineteenth century) as both a manufacturing and an agricultural country, a portion of the rural population would be drawn off into industry. It would not bring about the desired balance between rural and urban population, but it would afford some relief to the hard-pressed villages struggling for mere subsistence.

Conclusion

The discussion in the preceding pages has indicated the nature of India's agricultural difficulties and the suggested remedies. Some advances have been made in regulating the relations between landlord and tenant, in scaling down indebtedness, and in adopting better farm practices. These, however, touched only the fringe of all those elements that make for the depressed state of the country's agriculture. Now, as a decade ago, the following statement of the Royal Commission holds good: "To a very great extent the cultivator in India labors not for profit nor for a net return but for subsistence. The crowding of the people on the land, the lack of alternative means of securing a living, the difficulty of finding any avenue of escape and the early age at which a man is burdened with dependents combine to force the cultivator to grow food wherever he can and on whatever terms he can."

When one weighs the problems against the remedies, it becomes clear that the immediate future cannot bring forth any substantial relief. At best, the spread of education and cooperation; the adoption of basic principles of agricultural science; and the all-important changes in social, religious, and political customs and traditions are slow processes. Illiteracy, primi-

12. Matu Ran Ahlawat, "Economic Future of the Indian Cultivator," *Indian Journal of Economics* (October 1938), p. 329.

13. B. T. Ranadive, *Population Problem of India* (London, 1930), p. 191.

tive equipment, poor cultivation of the soil, low yields, low standard of living, unsatisfactory landlord-tenant relations, indebtedness, diseases, and poverty—all these react upon one another and are inextricably intertwined. This militates against piecemeal reforms. It is impossible, for example, to solve the indebtedness problem without providing for conditions that would make recourse to unproductive loans unnecessary; nor is it possible to introduce improved methods of cultivation without first reducing the ignorance and illiteracy so prevalent among the peasants. By the same token, a higher standard of living of the peasantry depends not only upon improved yields but in no small degree also upon basic changes in the iniquitous land tenure system.

There is nothing inherent in the makeup of an Indian peasant that prevents him from assimilating knowledge or from realizing that "better farming" spells "better living." With the active assistance of various agencies, both official and private, the economic status of the peasants will probably be improved. The question is being raised, however, whether, in the long run, the material advancement would not be nullified by the growing pressure of the population on the land. It is the considered opinion of the Royal Commission that "no lasting improvement in the standard of living of the great mass of the population can possibly be attained if every enhancement in the purchasing power of the cultivator is followed by a proportionate increase in the population."

An area estimated at about 100 million acres is still available for cultivation, but as against this there is the uninterrupted increase in population so characteristic of India and the continued division of the land into smaller and smaller units. Thus the problem "of many men on little land" and all the adverse corollary problems that follow from it are likely to be perpetuated unless the birth rate declines or a new outlet is found for the surplus population. But these in turn raise a host of baffling new questions to which a ready answer cannot be given.

4. *Japan's Agricultural Crisis*

This is the fourth of Ladejinsky's published articles on Japan. The first, "Farm Tenancy and Japanese Agriculture," published in *Foreign Agriculture* in September 1937, contributed in major degree, together with the present article, to the reputation which made it inevitable that Ladejinsky would be called upon after the war to play a key role in the Japanese land reform. That article is omitted here only because its substance is built into his postwar "Farm Tenancy in Japan," presented later in this volume. His second ("The Japanese Cotton-Textile Industry and American Cotton") is omitted because it is conventional in both subject and treatment as well as lacking in current interest. The third, "Agrarian Unrest in Japan," published in *Foreign Affairs* in January 1939, is essentially a more popular treatment of the article presented here; it is therefore also omitted.

Ladejinsky deals here not only with farm tenancy but also with the high rents, interest charges, taxes, and other factors contributing to Japan's agricultural crisis and examines the inadequacy of government measures taken to deal with them. Interestingly, he draws attention to the Japanese military's special interest in agrarian problems.

This article appeared in the *Journal of Farm Economics* in August 1939 (vol. 21, pp. 614–31).

TODAY, AS IN THE PAST, agriculture remains the backbone upon which the economic life of Japan rests. The unprecedented progress of industrialization in Japan in recent years has tended to obscure this fact. In reality, however, the importance of agriculture in modern Japan is still very great. The rural population still constitutes approximately 45 percent of the total; agriculture continues to be practically the sole provider of the nation's food, while with respect to invested capital and value of net output, agriculture is the country's leading industry.

The place of agriculture in Japan's national economy does not bespeak either rural prosperity or rural progress. On the contrary, it is commonly agreed now that it has been suffering from an ever-deepening crisis at a time when Japanese industry and trade have been expanding and prospering. The feudal concept that a farmer should maintain a low standard of living has continued to pervade Japanese agriculture to this day. It helped to keep the village in a state of contentment as long as agriculture could satisfy the frugal needs of the farmers. In recent years, however, Japan's agricultural economy has not been able to provide even this minimum for a great number of farmers.

This is attributed to numerous burdens that Japanese farmers must shoulder and which operate to their detriment even in "normal" times. Chief among these are the country's very limited arable area and the extremely small farm units; inequitable land distribution; widespread tenancy system; sharp price fluctuations; heavy tax load; huge indebtedness and exorbitant interest rates, and, in consequence of these factors, the very low standard of living of the mass of Japanese farmers.

The Problems

The characteristic feature of Japan's agriculture that has given rise to a host of difficult problems is the small area under cultivation in relation to the rapidly growing population. The utilization of the land is highly developed and "not only the hillsides, but in some places even the mountain summits,"[1] are made to yield crops. The great efforts to extend the area under cultivation brought about an increase from 11 million acres in 1880 to not quite 15 million acres in 1936, or 16 percent of Japan's total land area. The extension of land under crops in Japan was achieved mainly during the first three decades of the period under consideration; since then the arable area remained practically stationary. The topography of the country being generally steep and mountainous, there is at present only a small undeveloped area which is suitable for cultivation.

The fact that Japan's 5,595,000 farm families cultivate less than 15 million acres indicates that the size of an individual holding is extremely small. If the land were evenly distributed, each household would cultivate 2.5 acres. In reality, great numbers of Japanese farmers cultivate much less than 2.5 acres. [See table.]

Thirty-four percent of all the households cultivate less than 1.2 acres each; 34.2 percent from 1.2 to 2.4 acres; 21.5 percent from 2.4 to 4.8 acres, and only 9.5 percent cultivate more than 4.8 acres. The first-mentioned group, comprising 1,896,000 families, cultivate an estimated total of 1,237,000 acres, while on the opposite end of the scale one finds 78,000 families who cultivate 1,482,000 acres of land. These figures reveal how small the holdings are and the extent to which a great deal of the cultivated land is found in comparatively few hands.

Japan's limited crop area precludes the concentration of land in great holdings. Official statistics record only 3,547 landowners with over 125 acres each, the average size of these holdings being 300 acres, and 46,000 with an area of from 25 to 124 acres, or an average of 57 acres. Notwithstanding the scarcity of large properties, there is a striking inequality in the distribution of land ownership in Japan. Data of the Ministry of Agriculture shed little light on the exact distribution of land ownership; it was estimated unofficially, however, that while 50 percent of all the farm households own only 9 percent of all the land, 7.5 percent of the households own 50 percent of the land.[2] The corollary of this is that the holding cultivated by a Japanese farmer does not correspond with the amount of land owned by the same farmer.

1. Shiroshi Nasu, "Population and Food Supply in Japan," *Problems of Pacific*, proceedings of Second Conference Institute of Pacific Relations, Honolulu, 1927, p. 346.

2. Hidetoshi Isobe, "Labor Conditions in Japanese Agriculture," *Bulletin of the Utsanomiya Agricultural College*, Section B, vol. 2, no. 1 (1937), p. 6, table 2 (B).

FARM HOUSEHOLDS ACCORDING TO SIZE OF CULTIVATED AREA (1936) AND TOTAL AREA CULTIVATED BY EACH GROUP (1933)

	Actual number		*Percentage of total*	
Arable land per family (acres)	*Number of families (thousands)*	*Area cultivated*[a] *(thousand acres)*	*Number of families (percent)*	*Area cultivated (percent)*
Less than 1.2	1,896	1,237	34.0	8.4
1.2 to 2.4	1,914	3,724	34.2	25.4
2.4 to 4.9	1,262	4,817	21.5	33.0
4.9 to 7.4	321	2,055	5.7	14.0
7.4 to 12.2	127	1,333	2.4	9.1
More than 12.2	78	1,482	1.4	10.1
Total	5,598	14,648	100.0	100.0

a. Estimated by Hidetoshi Isobe, "Labor Conditions in Japanese Agriculture," *Bulletin of the Utsanomiya Agricultural College*, vol. 2, no. 1 (1937), sec. B, p. 5.

Source: Government of Japan, Ministry of Agriculture and Forestry, *Statistical Abstract* (1936–37), pp. 1–3.

Many farmers own no land at all and many more own very little of it (2,556,630 farm households own less than 1.2 acres each). Under the circumstances they are compelled to rent land from those in whose hands it is largely concentrated. This has brought in its wake the development of tenancy in Japan on a large scale. At present, tenancy in Japan has reached a point where 54 percent of the irrigated rice land and 40 percent of the nonirrigated land are cultivated by tenants and part tenants. On the whole, the two groups cultivate 48 percent, or 7,200,000 acres of Japan's total arable land. Of the 5,597,000 farm households enumerated in 1936, almost 31 percent were composed of independent farm owners, 27 percent of tenants who owned no land at all, and 42 percent of part tenants and part owners.

In the United States "tenants still move with some freedom up the agricultural ladder,"[3] but in Japan tenants are in no position to do so. The scarcity of land and lack of alternative occupations bind the tenant to the land—regardless of the high rentals. For rice land, rent is paid in kind, while for other land it is usually paid in cash. The system of payment in kind is an added burden shouldered by the tenants. "It is all the heavier, too," a Japanese writer noted, "because of the arrangement whereby the tenant farmer pays his landlord so many bushels of rice per tan (0.2 of an acre), and this amount does not change much irrespective of whether the crop is large or small. In other words, the landowners are assured of a certain stabilized quantity of 'harvest' regardless of the yield."[4]

A survey of 9,134 villages by the Japanese Department of Agriculture showed that in 70 percent of the cases the rental from a single crop field constituted more than 50 percent of the crop; from a two-crop field the rent is around 60 percent of the crop. It may be noted also that, whereas the landlord pays only the land tax, the tenant has to pay a number of other assessments and dues; buy his own very expensive artificial fertilizer; and provide the farm house, farm buildings, implements, and seed. Under the circumstances a tenant's net share of the crop is considerably smaller than the above-mentioned figures.

The long-established landlord-tenant relationship continued till about the end of World

3. Government of Japan, "Report of the President's Committee on Farm Tenancy" (February 1937), p. 5.

4. Magohachiro Kimura, *Japan's Agrarian Problems* (Foreign Affairs Association of Japan, December 1937), p. 12.

War I without causing much open conflict, although the tenants were never satisfied with their economic status. But the growing agricultural distress and the spread of ideas opposed to old-established concepts brought about a decisive change in the attitude of the tenants toward the landlords. Since 1917 the number of recorded disputes increased from 85 to nearly 6,000 annually. The causes underlying the disputes are numerous, but the following two became outstanding: during the twenties, excessive rents; more recently, about 60 percent of the disputes were due to attempts on the part of the landlords to evict the tenants from the land.

The economic welfare of the Japanese farmers is closely bound up with two crops, rice and cocoons. The traditional practice of the rapidly growing Japanese population to depend largely on rice as the main article of diet caused more than half of the arable land to be devoted to rice cultivation. Because of the extension of acreage under rice and even greater rise in yield per unit of land, during the last sixty years total production of rice doubled, mainly from 155 million bushels to over 300 million bushels. In the past decade, however, the yield has remained practically stationary. Hence the suggestion that "Japan has just about reached the point where the returns on her agricultural efforts will begin to decrease irrespective of any added endeavors she may make to increase her production."[5]

The price of rice in Japan is characterized by sharp fluctuations which alternately work havoc with the producers and consumers. The fluctuations are due to a variety of factors: crop conditions with a consequent shortage or surplus of rice, the fact that Japanese rice is consumed at home almost exclusively and cannot be exported abroad at a profit, and that the consumed rice must be Japanese rice or that from Korea or Formosa. But whatever the causes, for nearly two decades the Japanese government has been confronted with the problem of how to eliminate sharp price fluctuations in a manner that would reconcile the interests of both producers and consumers and allay the discontent engendered by it.

5. Shiroshi Nasu, op. cit., p. 343.

The silk cocoon crop is only second in importance to the rice crop. In fact, for a third of Japanese farmers it is the principal source of cash income. Whether the latter is large or small depends entirely upon the price of silk in the United States, where more than 90 percent of Japan's raw silk exports is sold. The price is determined mainly by business conditions in the United States. This explains the precipitous drop in price from a record high of 10.6 yen per kwan (8.3 lbs.) at the close of the wartime boom in the United States to a low of 5.7 yen a year later. Chiefly for the same reason, and partly on account of increased competition of rayon, the price of cocoons declined from 7.5 yen in 1929 to 2.5 yen in 1934, which price was only two-thirds of the cost of production.

The burden of debt shouldered by the Japanese farmers is a basic factor underlying Japan's agricultural plight. Prior to World War I the total debt was estimated at less than a thousand million dollars. Since then, the increased disparity between income and expenditure in consequence of the distressed conditions in the countryside led to a rapidly mounting indebtedness, estimated at 6,000 million yen, or slightly over 1,000 yen per household. The net effect is that "Their [farmers'] excessive debts are the most important of the causes which compel the present proprietors to dispose of their own land and reduce them to the position of proletarian tenants or wage laborers, whereas otherwise they would remain as the sound middle class of the farming population."[6]

Most of the indebtedness is carried at high interest rates. A survey prepared by the Japanese Department of Agriculture on the amount of loans to farmers by various financial organs shows that the Hypothec and Agricultural and Industrial Banks, established primarily to facilitate agrarian financing, accounted only for 14 percent of the total farm loans. On the other hand, private banks and moneylenders accounted for 13 percent and 56 percent, respectively.[7]

6. Yoshinosuke Yagi, "The Problem of Farm Debt Adjustment," *Kyoto University Economic Review* (July 1937), p. 61.

7. Setsuo Uenoda, "Farm Problems in Japan Dates Back to the Early Meiji Era," *Trans-Pacific* (March 8, 1934).

As to interest rates charged, it was officially estimated that in 1932 nearly 43 percent of the debt carried a rate ranging from 7 to 10 percent; 51 percent of the loan at a rate from 10 to 15 percent; and the remaining at a rate of interest above 15 percent. In reality, however, the actual rates are considerably above the official estimates. According to one observer "of his [farmer's] total loans, 57 percent have been advanced by private lenders at a nominal rate of 12 percent and a real rate said to be between 20 and 30 percent."[8]

Japanese farmers have always been subjected to a heavy taxation burden. In feudal times this was epitomized in the saying: "The peasant is like a sesamum seed; the more you squeeze, the more you get." In modern times the process of exacting taxes from the farmers has been regularized and given legal form, but it remains true, nevertheless, that in Japan's economic scheme the farmers are still the main "beasts of burden." For this reason "the all-around progress from the feudal regime to the modern capitalistic system was achieved at the expense of farmers. The land tax, customs tax, consumption tax and local taxes were levied, whether intentionally or otherwise, as a means of facilitating this policy."[9] This attitude toward the farmers has persisted to this day, despite Japan's rapid and very successful industrial development.

Conditions described in the foregoing paragraphs had been gradually undermining Japan's agricultural economy for many years prior to the depression, but the full significance of the fundamentally adverse factors was not revealed, chiefly because of the relatively high prices of the two staple products—rice and silk.

The first sign of a downward trend was noted in 1926; by 1929 the price of rice had declined 30 percent and that of cocoons 37 percent, with a similar contraction noticeable in other products. In 1929 the total value of agricultural output declined to 3,500 million yen as against an annual average of 4,600 million yen during the period 1919–1928. But the real slump came in 1930 and 1931, particularly during the latter year, when the agricultural output was valued at only 2,200 million yen, or less than 50 percent of the 1919–1928 average. The price disparity between manufactured and agricultural products aggravated the conditions of the farmers; the decline of commodity prices purchased by the farmers was less marked; and, as in the case of the all-important fertilizers, prices were actually rising. The rise in prices proceeded at a very slow pace, but it became more pronounced from 1935 onwards. The estimated 1938 value of agricultural products is some 300 million yen greater than that of 1929. But here, too, one must note that the rise in prices lagged considerably behind those of manufactured articles, thereby lowering the purchasing power of the rural community. Despite this adverse factor, farm prices recovered sufficiently to mitigate to a certain degree the immediate problems given rise by the depression, having at the same time unsolved the fundamental difficulties which handicap Japan's agricultural economy.

The question to consider now is what the Japanese agricultural situation means to the farmer in terms of earning a livelihood. An investigation conducted in 1927, a relatively prosperous year, of the income of 132 farmers who cultivated their own land, revealed the following: the total income of a farmer averaged 1,350 yen as against an expenditure of 1,315 yen, or a surplus of 35 yen. The reliance upon subsidiary occupations as a means of making both ends meet is a basic pillar of Japan's agriculture. A survey on farm incomes for the period 1913–1934 shows how widespread this practice is: "With the single exception of 1913, agricultural receipts fell short of meeting household expenditures, i.e., cost of living, and forced the farmers to fall back upon nonagricultural income to make up the deficit."[10] The proportion of the income from subsidiary occupations to the total farm income ranged from 23 to 31 percent.

The conclusion one is likely to draw is that, for the majority of Japanese farmers, agriculture as such is not on a paying basis. The Bureau of Statistics of the Japanese government was well aware of that when it stated that "the average part-tenant as well as tenant farmer can count

8. *The Times*, London (July 5, 1932).
9. Magohachiro Kimura, op. cit., p. 7.
10. Hidetoshi Isobe, op. cit., p. 66.

upon a surplus only when the area of farm under his management is larger than 4.5 acres. The very fact that by far the largest number of Japanese farmers till an area of less than 2.4 acres is clear evidence of the difficulty of Japanese agriculture as a business."[11] Under the circumstances the question raised by Dr. Nasu "whether or not such high population supporting power [of the land] is partly due to a sacrifice paid by peasants in the form of a low standard of living,"[12] may be answered in the affirmative.

The Remedies

Prior to 1930 Japan's agricultural problem failed to attract serious attention, one of the important reasons being the farmers' inability to express their grievances in an articulate manner and failure to secure the support of the country's dominant political groups. But when, among other causes, that of the farmer also was taken up by the military, the agrarian problem, to quote ex-Premier Saito, immediately "caught fire." In the attempted assassination of Premier Hamaguchi on November 14, 1930, resulting in his death a few months later, rural discontent played no small part. The motives behind the subsequent bewildering series of assassinations of Japan's leading political and industrial figures were tinged by a combination of military-agrarian interests.[13] Considering the important role of the army in the nation's affairs, particularly after the occupation of Manchuria in 1931, it becomes clear why, in 1932 and after, every succeeding government of Japan became eager to placate the farmers.

In the attempts to stabilize the price of rice, government policies were directed toward maintaining prices at a level that would reconcile the producer-consumer differences. The origin of the policy goes back to postwar days, but its basic principles were definitely formulated in a law enacted on March 29, 1933. The government was thereby authorized to fix each year a minimum and maximum price at which it stood ready to buy or sell rice in order to maintain the market price between the two fixed levels. The minimum was to be based on the cost of production and the maximum on the cost of living. To carry out the chief provisions of the law, the government established a special rice fund of 800 million yen, which could be increased to a total of 1,150 million yen.

This law, which proved to be a financial burden on the government, was modified by the Autonomous Rice Control Law of 1936. In accordance with the provisions of the latter, the farmers were to be organized in local and federal rice control associations with a view to storing the surplus of their rice. The Department of Agriculture was authorized to grant special bounties to cover storage expenses and to make loans at a low rate of interest against the stored rice. The surplus was to be held until such time as the current market price should rise above the minimum official price.

Efforts to assist the farmers through price control were only partly successful, despite the more than 200 million yen loss sustained by the Japanese government in the course of the operation of the rice laws. It may be argued, of course, that without government aid prices would have declined below prevailing levels; but, from the rice producers' point of view, on many occasions prices were not sufficient to cover costs of rice production, much less a margin of profit.

The minimum price for rice now guaranteed to the producers may not satisfy them, yet they are assured of a certain income. A similar situation does not exist in the case of cocoons. Until very recently government attempts to aid cocoon raisers consisted chiefly of measures tending to improve quality and reduce costs of cocoon production. In 1935 and 1936 sales of cocoons were placed under government supervision with the stipulation that the quality of the product offered for sale must meet with official approval.

On occasions, as in 1930 and 1931, farmers engaged in sericulture have secured loans from the government amounting to 120 million yen.

11. Shiroshi Nasu, op. cit., p. 198.
12. Ibid., p. 196.
13. See S. D. Kennedy, "The Reactionary Movement of 1932," *Contemporary Japan* (March 1933). Minister of Finance Inouye was murdered on February 9, 1932; the managing director of the vast Mitsui interests, Dan, on March 5, 1932; and Premier of Japan Inoukai on May 15, 1932.

While intended as a relief measure, the government utilized these loans as a means for controlling the cocoon crop. But the reduction in output was ineffective as a method raising the price of cocoons because of a lagging demand for silk. Since prices of cocoons are governed by prices of raw silk, the Japanese government then promulgated numerous measures to increase silk prices. They included such devices as loans, subsidies, government purchases of surplus silk, and restrictions of silk output and silk sales. Yet, judging by price movements, all direct and indirect measures to raise cocoon and silk prices failed in their aim and the prosperity of Japanese sericulturists still depends upon the rising curve of economic activities in the United States rather than upon any other factor.

Aside from the attempts to aid agriculture through various schemes of price stabilization, the Japanese government concerned itself also with the problem of lowering cost of production. The main emphasis was upon cheaper supplies of artificial fertilizers. Because of the monopolistic character of the industry, fertilizer prices were rigidly maintained even during the years of severest depression, although prices of agricultural products were at record low levels. The complaints of farmers against this price disparity became so vociferous that on May 18, 1936, a law was enacted to correct the situation. The significance of the measure lies in the fact that the association of fertilizer manufacturers created by this law cannot make arrangements for controlling production, set up sales policies, or fix the price of the product without the approval of the government. Whether the Japanese farmers will be able, ultimately, to secure fertilizers at reduced prices remains to be seen; meanwhile, it should be noted that wholesale prices of fertilizers have been higher since the enactment of the law. The Sino-Japanese War increased the demands upon the industry, thereby stimulating a further rise in prices. It is not unlikely, of course, that prices would have been still higher but for the control measure enacted by the government.

Farm relief in Japan has come to be closely identified with easing of the debt burden. As the crisis grew in intensity and the demand for aid became insistent, in 1933 the Japanese government promulgated the Farm Debt Adjustment Act. It provided for the formation of farm associations with functions "to mediate between the creditor and debtor and make plans for redeeming the debts within 20 years by easing the conditions of debts through such methods as extending the period of redemption, establishing an annual installment plan, reducing the rate of interest, or scaling down the amount of the principal or of the accrued interest."[14] Within a period of five years the government planned to extend to the associations loans at a low interest rate, amounting to 200 million yen.

It may be inferred that the settlement of a relatively small portion of the farm debt was intended. Essentially, the act called on the farmers to solve the problem by their own means, as may be noted from the following: "Resources for the redemption of debts should be supplied by the surplus [farm] income . . . and also by disposing of property which is not needed for the efficient following of the household occupation."[15] Most of the Japanese farmers, however, were heavily in debt precisely because of a lack of surplus income. Thus the chief means whereby the government hoped to scale down the farm debts were reduced to nothing else than "making a more active use of the tradition of neighbourly fellowship and mutual aid."[16]

The basic features of the Debt Adjustment Act offered no solution for the debt problem. The hope that by virtue of the act farm indebtedness would be scaled down considerably did not materialize. Between August 1, 1933, and June 30, 1936, a period of nearly three years, the loans extended to the associations by the government amounted to only 15,440,000 yen, or 7.7 percent of the 200 million yen to be distributed in the course of five years. On March 30, 1936, debts actually adjusted amounted to 157 million yen, 26 percent of the minimum or 15 percent of the maximum sum planned and only 2.6 percent of the total estimated farm debt.

Japan's land tenure system has been the

14. Yoshinosuke Yagi, op. cit., p. 65.
15. Ibid.
16. Ibid.

fundamental cause of the widespread discontent in the countryside. How to remove it has been a question agitating the Japanese government for a number of years. Two measures have been adopted toward this end. One of them is the Conciliation of Tenancy Disputes Bill enacted in 1924. It provides, first, for a more simplified court procedure and, second, for arbitration by a committee composed of a judge and several citizens. If the parties in dispute express their satisfaction with the committee's decision, it becomes effective and is transferred to the court for a formal approval. More than half of the tenancy disputes are settled in this manner.

The second measure aims to do away with tenancy as an institution by assisting the tenants to become independent farm owners. A program was worked out in 1926 providing for the expenditure of 468 million yen in the forthcoming twenty-five years. This fund was not adequate to carry on land purchases on a large scale, for at prevailing prices a total of 287,000 acres could be acquired, or only 4 percent of the entire arable area rented by tenants.[17] To speed up the process, in 1932 a more ambitious plan was proposed in the form of a Peasants' Proprietors' Agricultural Land Bill. It provided for the purchase of 1,543,000 acres over a period of thirty-five years and at a total cost of 2.8 billion yen. The tenants were expected to repay this sum in annual installments.

The bill provoked a great deal of criticism on the ground that it was "a device invented to enable landowners to dispose of land that was a burden to them."[18] It failed to pass and there remains in operation only the plan inaugurated in 1926. In eight years, 1926–1933, nearly 120,000 tenants were assisted in the purchase of 126,000 acres, or less than 2 percent of the entire volume of rented land. The conclusion of a Japanese student of the country's land tenure system is that "the present plan is thus quite ineffectual as a measure for solving the land problem."[19]

In addition to the outlined plans, a number of unsuccessful attempts have been made to enact a law that would stabilize landlord-tenant relations. In view of Japanese tenure relations conditions, any tenancy law involves a downward revision of rents and an acknowledgment of the tenants' right to the land. Landlords see in this an infringement on their rights and they fight bitterly, and so far successfully, against any change.

Considerable importance has been attached in the past six years to the problem of settling Japanese farmers in Manchuria. According to Japan's secretary of agriculture, "Japanese emigration to that vast land means, *inter alia*, the extension of the rural community which forms the backbone of the Japanese race, a way out of the difficulties arising from the intensive system of cultivation in Japan Proper and the exploitation of natural resources in that latter country."[20] More concretely, Manchuria was to take the place of a safety valve, as it were, for relieving the pressure on land in Japan.

For the time being the accomplishments bear little relation to the outlined plan; nor do they augur well for the widely entertained idea in Japan that Manchuria is becoming an important outlet for the Japanese surplus farmers. During the years 1932–1937, a total of only 2,785 families, subsidized by the government to the extent of 1,000 yen per family, found their way to Manchuria. But the Japanese government refused to be dismayed by this result and in 1937 drew up a vastly more ambitious program. The latest move calls for the resettlement into Manchuria of a million farm households of 5 million people, in the course of the following twenty years.

The brighter side of Japan's agricultural economy is in the widespread development of the cooperative movement. It has been sponsored and financially aided by the Japanese government. The savings effected through the reduction of the great number of intermediate

17. S. Kawada, "The Establishment and Maintenance of Peasant Farms," *Kyoto University Economic Review* (July 1928), p. 77.

18. Y. Yagi, "The Current Land Problem and the Establishment of Peasant Proprietorship," *Kyoto University Economic Review* (December 1936), p. 73.

19. Ibid.

20. Yoriyasu Arima, "Japan's Agricultural Administration," *Contemporary Japan* (September 1937), p. 181.

merchants, for instance, is one of the factors stressed in favor of the movement. Farmers represent 70 percent of the total membership of about 6 million. The main functions of the Japanese cooperative societies are confined to providing credit; making sales and purchases; and the joint utilization of warehouses, milling establishments, farm machinery, and workshops. Almost half of the cooperatives combine all the mentioned activities.

The movement has been playing an increasingly important role in Japan's farm economy. Within the period 1920–1934, sales of rice through cooperatives increased ninefold. The same is true of wheat and to a lesser extent of cocoons. In the latter year, 28 percent of all the marketed rice, 27 percent of the wheat, and 12 percent of the cocoons were sold through the cooperatives. Considerable progress has been registered by the cooperative purchasing societies. Their main efforts are concentrated in the purchase of fertilizers, and in 1934 they handled one-third of the total consumption.

The cooperative credit societies extend loans for productive purposes such as purchases of land and agricultural machinery. Until recently about 70 percent of the loans were granted without security, but on the basis of a minute examination of the moral, financial, and political standing of the borrower. Of late there has been a reversal of this policy; the tendency is to grant loans only against some tangible form of property. The rate of interest is rather high, ranging as it does from 8 to 12 percent a year. There is little difference between these rates and those charged by commercial banks, but in certain cases the former are even higher. They generally do not relieve the rural population from the necessity of securing loans from private individuals at usurious rates. This is especially true in the case of the poorer peasants, because "in practice . . . credit societies are said to be made up of middle- and upper-class membership."[21] In the light of these facts, it becomes apparent why rural indebtedness to the cooperative credit societies comprises only from 10 to 15 percent of the total farm debt.

The cooperative movement has, on the whole, benefited Japan's agricultural economy, but the benefits are not evenly distributed. The more prosperous sections have profited most. Present plans for increasing the importance of the cooperative movement in the Japanese village might not be without its positive effects on the well-being of the least prosperous group. But the widely held view, particularly in official circles, that the very solution of Japanese agricultural problems is closely tied up with the cooperative movement is exaggerated; it can alleviate certain burdens but is in no position to remove them altogether.

The war waged by Japan against China is no boon for Japanese agriculture. The demand for farm products has been stimulated; but many farmers, having no surpluses to throw on the market, have not been able to reap the benefits of higher prices. Moreover, it is stated that the movement in prices of agricultural products have not kept pace with the rise in prices of articles which are essential to the farmers; while quotations of agricultural products have recently advanced, they are far below those of articles commonly purchased by the rural population.[22] In addition, the war deprived many farms of their principal workers, absorbed considerable numbers of livestock, and utilized so much fertilizer that the farmers have been having difficulty in obtaining a sufficient volume for their own use.

Various measures have been adopted since the outbreak of the Sino-Japanese War to assist the rural sections of the country. Chief among them are the Farm Adjustment Law and the Temporary Farm Debts Liquidation Law, both of which were enacted on April 2, 1938.

The first one is related to the creation and maintenance of new farm proprietors, the adjustment of tenant-landlord relations, and the disposition of farms owned by farmers who have been conscripted into the army. Perhaps the outstanding provision of the law is the strengthening of the position of the tenant with respect to leases. The landlord cannot evict a tenant unless sufficient reason, as indicated in the act, is given. The effectiveness or lack of effectiveness of the measure cannot be determined at this juncture; it should be stressed in

21. Ibid., p. 123.

22. Mitsubishi Economic Research Bureau, "Monthly Circular" (June 1938), p. 8.

this connection that, on the whole, the provisions of the act are rather vague and that the landlords still have an important voice in rural matters, particularly where their own interests are involved.

The Debt Liquidation Law does not extend to the farming population as a whole but only to "bereaved families and families of wounded soldiers living in the countryside."[23] Under the provision of the law, the Temporary Prefectural Committee for the Liquidation of Debts mediates between the debtor and creditor concerning the reduction of the principal, rate of interest, or period of redemption. The funds required to effect debt adjustment are to be supplied by a number of government banks, but in all such cases "the amount of this special financing by those banks shall be within the estimated value of the real property given in mortgage."[24] The liability of the government for any losses incurred in the process of this financing is limited to 120 million yen.[25] This latest attempt to deal with rural indebtedness differs little from that of 1933; the farmers are asked once again to ease the debt burden through the employment of their own meager resources. This method was not successful in the past and it holds little promise for the present or the future.

Conclusion

The urgent need of putting Japan's agricultural economy in order has been recognized now by those who shape the country's development. The former Minister of War, General Araki, stated that "if we should succeed in solving the agrarian problem, it would be easier to solve the remaining serious social problems."[26] But despite such expressions of grave concern, the measures enacted to aid agriculture were not sufficient to accomplish the task at hand.

To be sure, the task is a tremendous one and, under the present circumstances, some of its aspects hardly lend themselves to a solution. Assuming that Japan's farming area could be expanded by 3.5 million acres[27] and that it would be economically advisable to do so, the total crop area would reach 18.5 million acres or an average of 0.7 acres per capita. According to Japanese students of the problem, this would be 0.5 acre short of the minimum requirement. Only exceedingly rapid industrialization or large-scale overseas migration could achieve this minimum standard. It is reasonably certain, however, that for years to come neither development is likely to take place and the fundamental problem created by a great pressure of population on a very limited land area will undoubtedly continue to hamper the country's rural progress.

Notwithstanding the restrictions imposed by the problem of "many men on little land," the plight of the Japanese farmer can be alleviated to some extent by scaling down indebtedness, lowering interest rates, improving credit facilities, levying more equitable taxation, and by making a serious attempt to eradicate the evils of the existing tenancy system. All these things cannot be achieved by self-help, spiritual regeneration, and the oft-repeated statement that the farmers are the mainstay and backbone of the nation. Improvement of rural conditions can hardly be attained now without subsidizing the subsidizers—as the Japanese farmers may be truly called.

This is a line of action that requires large financial outlays which the farmers are manifestly in no position to supply. Moreover, the ever-mounting expenditures in consequence of Japan's war against China leaves little, if anything, for rural reconstruction. It is likely, however, that at some future date the very policy of expansion which involves the use of the army may give the demand of the farmers a friendlier reception. To the army, "the agricultural population constitutes Japan's first line of defense"; consequently, that line must be strengthened. Since stop-gap measures have proved inadequate, a thorough treatment of

23. Ibid., p. 18.

24. Mitsubishi Economic Research Bureau, "Monthly Circular," Farm Debts Redemption Fund Law (September 1932), art. 3.

25. Ibid., arts. 5, 6, and 8.

26. *Contemporary Japan* [Collection of essays in Russian] (Moscow, 1935), p. 73.

27. Shiroshi Nasu, "Population and Food Supply in Japan," *Problems of Pacific*, proceedings of Second Conference Institute of Pacific Relations (Honolulu, 1927), p. 343.

the country's number one problem may yet become the order of the day. Meanwhile, all one can do is to register the fact that, although talk about agricultural reform has become a favorite pastime of the leading political parties of the country, the farmers are still struggling against heavy odds which they cannot overcome single-handed.

5. *Chosen's Agriculture and Its Problems*

Although Ladejinsky was apparently called upon during the early postwar years to advise on land reform in Korea, this prewar article is the only piece of his own on that country I have been able to find. Farm tenancy plays an important role in the problems examined; so do the self-serving policies of the colonial master, Japan. Here, early on in Ladejinsky's career, he shows appreciation of the fact that economic growth does not necessarily equate with welfare. He notes "on the one hand, considerable expansion of production, and, on the other hand, the worsening of the economic conditions of the masses of Korean farmers. The benefits of enlarged acreage and augmented production seem to have eluded the very people whose efforts made them possible." One wonders, after reading, whether conditions for the poor farmer in this colonial setting were much worse than those experienced in some free Asian countries years after independence. The first and more descriptive part of the paper is omitted.

This article appeared in *Foreign Agriculture* in February 1940.

The Problem of Land Tenure

Small-scale farming

THE TOTAL CROP AREA of Chosen is distributed among 3 million households, or an average of about 4 acres per household. This is considerably larger than the average of 2.5 acres per farm family in Japan, yet it is clear that in Chosen, as well as in Japan, agriculture is on a very small scale. The scarcity of new arable land in conjunction with a growing farm population is tending to reduce the individual holdings still further. Thus, according to a Korean student of the problem, "land is so scarce that almost no farmer can cultivate as much land as he wishes to, and yet the acreage of arable land per farm household is declining."[1]

In the course of the years 1919–1930 the acreage of arable land per farm household declined by 10 percent.[2] Chosen, too, although to a somewhat lesser extent than Japan proper, is faced with the problem of many people on little land—and not rich land at that. It may be noted also that, whereas in Japan the proportion of the total farm income represented by income from subsidiary occupations ranged from 23 to 31 percent, in Chosen this item is of little significance because not more than 10 percent of the farmers derive any income from nonagricultural occupations.

Unequal distribution of land

The problem stated above is aggravated by the unequal distribution of the available land among the Korean farmers. The theoretical average of 4 acres per farm family has little relation to the amount of land actually culti-

1. Hoon K. Lee, *Land Utilization and Rural Economy in Korea* (Chicago: University of Chicago Press, 1936), p. 109.

2. Ibid., p. 120.

vated or owned by individual farmers. According to an official report of the Government-General of Chosen, published in 1938, fully 63 percent of all the farm households cultivate less than 2.4 acres each; and more than a third of this group cultivate less than 1.2 acres. Farmers working from 2.4 to 4.9 acres made up 21 percent of the total; and all others cultivate 5 acres or more. Only 457 households cultivate more than 50 acres each. These figures suggest that the amount of land cultivated by the great majority of the farmers is often not large enough to provide the farmer with the bare necessities. This becomes even more apparent upon the examination of the question of land ownership in Chosen.

The amount of land cultivated by a farmer often gives little indication of the amount of land owned by him. This is particularly true of Korean farmers. A study, prepared a decade ago,[3] of the Korean farming population (2,800,000 households) by classes sheds revealing light on this subject. Thus, 104,000 landlords, who comprised less than 4 percent of the total number of the households, owned 54.5 percent of all the arable land and two-thirds of the rice fields. Next in importance was the group of owner cultivators, consisting of 510,000 households—or 18 percent of the total—each one owning somewhat less than 5 acres per family. The balance of the land was distributed among the owner tenants—or 32 percent of all the farm families—who owned so little land that they were forced to enlarge their crop area by leasing some land from the landlords. In many cases the land owned by these farmers was so heavily mortgaged that the ownership was a purely nominal one. At the bottom of the scale were the tenants—or 46 percent of all farm householders—who owned no land at all and were renting an average of 3 acres per family. More recently, tenants have accounted for nearly half of the total. If the owner tenants were added to this group, then four out of every five Korean farmers would be tenants or semi-tenants. This situation has few counterparts in the world.

At the very bottom of the agricultural ladder is an estimated total of 256,000 so-called fire-field families who neither own nor lease any land. They are squatters living on tracts of ground in the mountains which they have cleared for cultivation by burning off the forest or brush. Ordinarily such land can be cultivated for only a few seasons.

Tenancy increasing

Tenancy in Chosen is a centuries-old institution; what is new and disturbing about it is its uninterrupted growth during the past twenty-five years. This came about chiefly through the loss of ownership in land by the part owners and part tenants and, to a smaller extent, through some loss of land by the cultivator owners (see table). Within the years 1914–1938, the total number of farmers increased 11 percent but tenants increased 66 percent, or six times as rapidly. During the same period the proportion of tenants to the total number of farm households increased from 35 to 53 percent. [See table.] This development, coupled

3. Ransford S. Miller, "The Farmers and Farm Lands of Chosen (Korea) in 1928, report by American Consul General at Keijo.

FARMING HOUSEHOLDS IN CHOSEN ACCORDING TO LANDHOLDING STATUS, 1914–1938

Year	*Landlords*	*Owner-farmers*	*Part owners–part tenants*	*Full tenants*	*Total*
1914	46,754	569,917	1,065,705	911,261	2,593,637
1924	102,183	525,689	934,208	1,142,192	2,704,272
1930	104,004	504,009	890,291	1,334,139	2,832,443
1938	—	543,481	814,293	1,511,424	2,869,198

Sources: Data for the years 1914, 1924, and 1930 were taken from *Agrarian Problem and Peasant Movement* (Moscow: International Agrarian Institute, 1937), vol. 4, p. 42. Data for 1938 were taken from U. Alexis Johnson, "Farming Households, Holdings, Ownership and Tenant Status in Chosen," (Keijo, Chosen, 1939).

with the concentration of land in the hands of the landlords and the terms under which the land is rented out, spells pauperization for the great majority of Korean farmers.

A number of factors are responsible for the loss of ownership in land sustained by an increasing number of farmers. There is rise in living expenses without a corresponding increase in income because of the low productivity of the soil and the primitive methods of cultivation. The fall of prices of agricultural products results in large deficits which lead to heavy indebtedness. Farmers are often in no position to discharge these debts except through the sale of their land. Another factor is the Japanese acquisition of large tracts of arable land since the annexation of Chosen.

Rise in land ownership by Japanese

Exact data concerning the amount of land owned by the Japanese are not available,[4] and the unofficial estimates vary. They are all agreed, however, that a relatively small number of Japanese have succeeded in acquiring a large share of the land. Official data show that at the end of 1927 the total number of Japanese households in Chosen was 10,300, comprising a population of 44,000. By 1936 the latter figure had been reduced to 35,000.

According to one estimate, at the end of 1930 the Japanese owned approximately 1,500,000 acres, or "about 11 percent of the total taxable land area in Korea."[5] On the other hand, "various careful estimates of fair-minded non-government Japanese and Koreans," a student of rural Korea wrote, "place the proportion of land owned, actually or virtually, by the Japanese at anywhere from 12 to 20 percent. In some counties in the south, Japanese ownership, based on tax records, is said to extend over half of the land. Thus in one county, Ikson, in South Keisho Province, an investigation by a Korean landlord and educator is reported to have shown 32 percent of the assessed property valuation in the hands of 120,000 Koreans and 68 percent in the hands of 8,000 Japanese. Since the great part of Japanese-owned land is in the south, it is probably fair to conclude that in this section about one-fourth of the land has passed out of Korean hands."[6]

One feature of Japanese land ownership is that the individual holdings of the Japanese are many times larger than those of the Koreans. The average size of such a holding is difficult to ascertain because of lack of official data and the conflicting nature of private estimates. In addition to the Japanese who are engaged in the actual farming of the land, there are also numerous Japanese and Japanese corporations who own and control what may be called in Chosen large landed estates, worked by tenants. In 1929 there were 538 such individuals or entities which owned 409,684 and controlled 65,858 acres, a total of 475,542 acres, or an average of 884 acres per owner.

The effect of Japanese infiltration on the fortunes of Korean farmers has been largely negative. Chosen has no free land to spare; the process of land accumulation in the hands of the Japanese (or Korean landlords) is mainly caused by the dispossessing of native farmers from their land. There have been many cases of replacement of entire Korean village communities by Japanese settlers.

An official survey of colonies made up of "free" or "protected" Japanese settlers revealed that "many of them have replaced Korean villages by driving out Korean peasant farmers. In 1930 there were sixty-one such colonies where the village is entirely Japanese in the fullest sense."[7] The Korean farmers have been objecting, for instance, to the organization of irrigation societies sponsored by the Government-General of Chosen. This attitude stems from a variety of reasons, "but one of the most important factors is that as soon as an Irrigation Association forms in a locality, the petty Korean landowners are not able to hold their land which, on the contrary, falls into the hands of large Japanese landowners and capitalists. This speedy accumulation of land necessarily

4. Latest official figures relate to 1927 and are incomplete.

5. Hoon K. Lee, op. cit., p. 148.

6. Edmund de Schweinitz Brunner, "Rural Korea," report of the International Missionary Council Jerusalem meeting, 1928, pp. 105–06.

7. Hoon K. Lee, op. cit., p. 288.

puts the Korean down into the tenant or landless classes."[8]

Conditions of tenancy severe

Considering the great predominance of tenant farming in Chosen, it may well be stated that the fortunes of the country's agriculture are closely bound up with the economic status of the tenants. It is pertinent, therefore, to inquire into the terms upon which they rent and work the landlords' land.

Korean tenants pay their rents both in cash and in kind. The latter, however, is the usual method since 85 to 90 percent of all the tenant households discharge their obligations in this manner. Rentals in kind fall into one of the following three types: payments of a certain quantity of produce per unit of land, regardless of the size of the crop; payment on the basis of the crop just harvested or on sample threshing; and payment of one-half of the crop—whatever its size. As to the amount of rent actually collected, the maximum may reach as high as 80 percent and the minimum no lower than 20 percent of the crop. For the country as a whole, rentals constitute from 50 to 55 percent of the total yield.

The tenant's net share is considerably below that of the landlord; all expenses connected with the cultivation of the soil such as human and animal labor, seed and fertilizers, as well as taxes are supplied by the tenant. There are instances when the landlords, too, pay taxes and carry other charges. The Oriental Development Company, a Japanese large-scale land enterprise, and other Japanese landlords follow that practice, but to compensate themselves "they . . . impose somewhat heavier rent upon their tenants."[9] To this must be added occasional gifts to the landlord and extra charges and services, illegal to be sure, shouldered by the tenant in order to maintain the goodwill of the landlord. It explains why a careful investigation of tenancy conditions in a village in the south of Korea "showed that the renter's actual *net* share was seventeen percent—a condition by no means exceptional in that part of the country."[10]

Both the tenant and the land he cultivates are at yet another disadvantage. In most cases the tenant's lease is for one year only, the result being that the tenant turnover at the end of any year is very considerable. In some provinces, especially in the more fertile southern regions, the replacements have been estimated at about one-third of the total. Where the tenants remain longer on the land and succeed in raising larger crops, the landlords are more likely than not to raise the rent accordingly. Under the circumstances, the tenants cannot be expected to make land improvements even if they had the means of doing so; all their efforts are bent on a maximum exploitation of the land in the course of their brief tenure. The onerous terms under which the tenants till the land are the consequence of scarcity of land and lack of alternative occupations for a growing rural population.

Try as hard as he may, the surplus of a tenant's crop after fixed annual expenses are paid is too small for the maintenance of his family until the next harvest. The lot of the Japanese tenants is notoriously bad, but a Japanese authority wrote that "Corean [*sic*] tenants are poorer than Japanese tenants and their economic condition is much worse than that of tenants in Japan Proper. One is indeed greatly amazed by the low and crude economy of those poor agricultural people in Corea."[11]

The Korean tenants were never satisfied with their lot, but until the early twenties customs and traditions regulating the landlord–tenant relations were sufficient to prevent open conflict and insure relative peace in the village. In the past two decades, however, the life and work of the Korean tenants have become increasingly conducive to discontent. The growing agricultural distress has brought about a sharp change in the attitude of the tenants toward the landlords. The number of disputes has increased by leaps and bounds, namely, from 15 in 1920 to 6,886 in the first half of 1935.[12]

8. Ibid., p. 148.
9. S. Kawada, "Tenant Systems in Japan and Corea," *Kyoto University Economic Review* (July 1926), p. 70.
10. E. Brunner, op. cit., p. 25.
11. S. Kawada, op. cit., p. 58.
12. *Tikhii Oksan*, [in Russian] no. 3–4 (1937), p. 139, note 5.

The causes underlying them were numerous but the principal ones, in order of importance, were termination of leases, excessive rents, and attempts to raise rents still higher.

Attempted remedies

Faced with the growing bitterness in the relation between landlords and tenants, the Government-General of Chosen was compelled to take official notice of the situation. It did so by the enactment of the Chosen Tenants Arbitration Ordinance (December 10, 1932) and at a later date (April 10, 1934) by the promulgation of the Organic Regulations of Chosen Prefectural, District and Island Tenant Committees.

The two laws aimed at providing a system for the arbitration of disputes between tenants and landlords. They enable the tenant as well as the landlord to request arbitration of a dispute at the local court. Refusal of one of the disputants to appear before the court without proper reasons is punishable with a fine of not more than 50 yen. The decision of the court is binding unless the parties concerned protest the decision in writing within a specified time limit. In addition, permanent regional tenant committees are created with the power to hear and arbitrate disputes over tenant rents and other matters of tenancy.

These measures help to settle disputes but they do not touch the causes underlying the conflicts and discontent. The first attempt along these lines was the enactment of the Chosen Agricultural Lands Ordinance on April 11, 1934. The basic features of this measure are expressed in two provisions: First, the term of lease to a tenant shall be three years instead of the usual one year, and seven years in the case of perennial crops such as ginseng and mulberry; second, restrictions are placed upon the arbitrary actions of the managers of tenant lands. It provides also that agreements in which a tenant waives certain of his rights are illegal (article 6); that renewals of tenant leases shall be for a similar term as the original leases; and that a tenant may propose a reduction or remission of the rental in case of crop failure (article 16).

The measure hardly enables the tenant to improve his economic status to any marked degree, but it undoubtedly has given him a greater degree of security through a more permanent employment. Judging by the official report of the Government-General of Chosen, which shows an increased number of disputes presented for court settlement, one gathers that the tenants have taken advantage of the enacted measures. According to the report, 732 cases were presented for arbitration in 1933 and 1,707 in 1934. The enactment of the Agricultural Lands Ordinance, which became effective in October 1934, led to a steep rise in the number of disputes brought for settlement; they jumped to 7,444 in 1935 and to 9,370 in 1936. In the latter year only 386 cases were due to complaints lodged by landlords; in 8,984 instances the complainants were tenants. The latter evidently found the arbitration system quite effective as a means of settling some of their differences with the landlords.

Aside from these measures, the Government-General of Chosen has been engaged in feeble attempts to cure the tenancy problem by assisting tenants to become farmer owners. A ten-year program was launched in 1932, aiming to create 2,000 farmer owners yearly. Every tenant selected for this purpose receives a government loan of 1,000 yen at 4.3 percent interest to be repaid in twenty-five yearly installments. The effectiveness of the scheme is questionable for two reasons. First, under the prevailing price of land in Chosen, a tenant could not acquire more than 1.2 acres. The income derived from the cultivation of so small a holding is not sufficient to provide the farmer with his meager needs at any time, let alone when the land is encumbered with debt. Second, the annual increase in the number of farmer owners created under the plan would be equal to not more than 8 percent of the yearly rise in tenant numbers. To be effective, therefore, the government plan would have to be on a much greater scale.

Other Problems

Low income

It is evident from the above that the income of the majority of Korean farmers is a very meager one indeed. The information on farm income in Chosen is not complete, but data available show what the Korean agricultural situation means to the farmers in terms of earning a livelihood.

An investigation of an agricultural region of the South Keisho Province carried out in 1922 revealed that only 30 percent of the cultivator owners made a profit at the close of the agricultural year, while the remaining 70 percent broke about even. Only 4 percent of the part tenants and 3 percent of the full tenants closed the year with a profit; 96 percent of the former and 97 percent of the latter wound up with a loss.[13] Two more recent surveys (1931) of two provinces summarized in the table below show approximately the same results. All farmers closed the year with a deficit except the part owners of one province who had an average profit of 17 yen. The average income of all types of farms, however, compared with the average outgo of the same farms was short by 44 yen in one case and by 14 in the other. [See table.]

High indebtedness, exorbitant interest rates

The only way in which a Korean farmer can cover the disparity between income and expenditure is by contracting a debt; hence, approximately 75 percent of all the farmers are in debt. Estimates of their total indebtedness vary. Before the agricultural depression of the thirties had set in, the debt, according to an official report, amounted to not less than 500,000,000 yen (about U.S.$144,000,000). No figures are available for the subsequent years, but it is safe to assume that indebtedness rose considerably during the years of the depression.

The average debt per household ranges from about 170 to over 200 yen. It is considerably smaller than the 1,000 yen per farm family in Japan; however, it is by no means small when one considers that the majority of Korean farmers operate on what might be called a deficit basis.

The burden of the debt is made heavier by the exceedingly high rate of interest at which it is carried. Incomplete information covering loans totaling 54 million yen extended to its members by the rural credit societies shows that 14 percent of all the loans was at 15 percent, while 40 percent carried an interest rate of over 30 percent a year. Another investigation revealed that the minimum rate for personal loans was 7 percent and the highest 70 percent; the respective figures for mortgage credits are 7 and 40 percent. The average for all types of loans is about 30 percent a year. The poorest section of Korean rural population pays the highest rates because they have little or no property to offer as security.

Having contracted a debt at such exorbitant charges, many a farmer would find it difficult to extricate himself from the debt entanglement even if the money were utilized for productive purposes. Actually, a great many of

13. E. Brunner, op. cit., p. 27.

INCOME AND OUTGO OF FARMERS IN TWO PROVINCES OF CHOSEN, 1931
(Yen)

	Total income according to investigation by		*Total outgo according to investigation by*		*Profit (+) or loss (—) according to*	
Class	*Agricultural Society of Chosen*	*Chosen Bureau of Agriculture and Forestry*	*Agricultural Society of Chosen*	*Chosen Bureau of Agriculture and Forestry*	*Agricultural Society of Chosen*	*Chosen Bureau of Agriculture and Forestry*
Owner-farmers	679	479.5	701	507.2	−22	−27.7
Part owners	392	484.6	473	468.0	−81	+16.6
Tenants	297	397.8	327	428.0	−30	−30.2
Average	456	453.9	500	467.7	−44	−13.8

Source: Hoon K. Lee, *Land Utilization and Rural Economy in Korea* (Chicago: University of Chicago Press, 1936), p. 272.

these loans are unproductive, being largely devoted to expenditures other than agricultural. For many Korean farmers indebtedness, therefore, spells "loss of land, discouragement, tenancy, greater debts, conditions approaching serfdom, then utter despair and barren stolidity."[14] The Government-General of Chosen has undertaken readjustment of farmers' debts by providing them with funds at low interest. "The number thus assisted," an official report stated, "increased six fold and the amount of funds seven fold over the figures previous to this movement [self-help movement inaugurated in 1932]."[15] The report does not reveal, however, the size of the special fund, the interest rates at which loans are made, or the number of farmers who actually benefited by it.

Falling prices

The fall of prices in the late twenties and early thirties added to the heavy burdens shouldered by the Korean farmers. As a Japanese colony, Chosen reacts immediately to the economic conditions prevailing in Japan. This is particularly true in the matter of agricultural prices. The price of Korean rice, for instance, is determined by the Osaka market quotations. There is this difference to be observed, however: low rice prices in Japan cause a still greater decline in Korean prices because of the urgency with which the farmers must sell their crop at the earliest possible date. The same conditions apply to Korean cocoons.

When prices of agricultural products in Japan began their downward trend in 1926, prices of Korea followed suit. By the middle of 1930 prices of all Korean grain crops had declined 20 percent and those of polished rice 28 percent; by October 1932 the respective figures were 39 and 43 percent. The total value of the principal agricultural products declined from 709 million yen in 1928 to 494 million yen in 1931, a reduction of 30.3 percent. That these developments played havoc with Korean agricultural economy is generally conceded.

14. Setsuo Uenoda, "Korean Rural Five Year Plan for Farms Will Likely Succeed," *The Trans-Pacific* (May 3, 1934).

15. *Annual Report on Administration of Chosen 1933–34*, p. 196.

Selling cheap, buying dear

An additional factor aggravating the agricultural situation was the price disparity between manufactured and agricultural products. The general commodity price level declined during the depression but to a smaller extent than prices of agricultural products. The price of such indispensable farm items as chemical fertilizers not only failed to decline but registered a new high. Taking 1925 as a base, the index number of the price of this product was 119 in 1930 and 122 in 1934.

The situation may be well summed up in the following words: "Grains and rice are produced and sold by farmers, while most of the general commodities are produced in Japanese factories and sold to farmers. Accordingly, the farmers in Korea have been in a disadvantageous position. They have been paying more and receiving less; their conditions of life are becoming harder."[16]

Farm relief

The plight of the Korean farmers goes back to predepression years; the late twenties and early thirties served only to accentuate a notoriously bad situation. Prior to 1932 the Government-General of Chosen took little or no notice of the farmers' difficulties, but by 1932 it became evident even to the powers-that-be that something (other than encouragement to produce larger crops in which Japan was particularly interested) had to be given to the farmers in order to relieve their distress. It is of interest in this connection to note the causes, as seen in official quarters, underlying the difficulties.

The Government-General of Chosen stated that "this miserable condition of affairs was due partly to the unconscious indifference of the farmers themselves and largely to the absence of governmental economic and educational provisions, as well as to the defective social organization, environment and lack of guidance."[17] In fact, it insists throughout that the farmers, more than any other element, must shoulder the blame for whatever ills have befallen them.

16. Hoon K. Lee, op. cit., p. 266.

17. *Annual Report on Administration of Chosen 1933–34*, p. 191.

This is so because, according to the official version, "in more recent years the farmers, carried away by the rush of material civilization, have lost any idea of self reliance and have forgotten the real character and true pride of farming communities, in joining the ill-advised pursuit of 'money economy,' deluded by the current idea of capitalism, worship of all powerful cash, and the supremacy of city life. Thus they have urged on their impoverishment."[18]

These views played a decisive role in the remedial measures adopted by the Korean authorities. Assistance through work relief projects was considered, and expenditures for public works programs increased from 7 million yen in 1931–32 to 13 million yen in 1936–37. The budget figure for 1937–38 shows a high of 32 million yen, declining to 19 million yen in 1938–39. The exact sum spent in the past two years has not been indicated. But in the main the emphasis was upon a cure through the farmers' own efforts. "To rescue the rural villages definitely," the official report continues, "and to see the farmers emerge with vigorous energy there remains the sole means—the Self-Help Movement—by which the farmers are urged to plan and work out their own salvation. Believing this an infallible and popular plan for the regeneration of Chosen, the Government-General, since 1932, has been encouraging and guiding the farmers in its practice."[19]

An attempt to clarify the nature of this very vague programmatic statement was made in 1933 when the Government-General issued instructions to the provincial officers "for the practical guidance and operation of the 'Self-Help' Plan."[20] From these one learns that:

> 1. Guidance should stress the mental awakening and self reliance of the farmers, in preference to urging them to material progress. . . .
> 2. Every year in each "Yu" and "Men" [small administrative units] one or more villages should be selected in which the living conditions of each family should be investigated and guidance given toward a new practical plan of family life, material as well as mental, covering a period of five years.
> 3. The intent of this plan will be (a) to meet the usual shortage of food and to rescue the farmers from "spring famine," (b) to maintain a balance between the annual cash income and disbursements, (c) to readjust and repay the harrowing debts.[21]

But how could all this be accomplished in the light of the meager resources of the Korean farmers? The answer was that "a government subsidy should be granted dependent on the progress of the mental awakening of the farmers and the development of their new life plans."[22]

According to the same official source, the actual working of the self-help plan brought considerable benefits to the farmers. Examples were cited showing that, whereas, at the beginning of 1933 out of 55,522 families selected from 1,988 villages, 31,581 suffered from food shortage; at the end of the year the number of such families was reduced by 6,939. Before the movement came into being, 43,329 out of the 55,522 families averaged a debt of 115 yen per farm, but in the course of the year they were able to reduce the individual debt by 25 yen. Also, the rate of tax collection increased by 2 percent, and "the Savings deposited in the Local Credit Associations increased by 20 percent in Shares and by 17 percent in Cash, while the arrears of payments decreased by 20 percent. . . . The Postal Savings increased by 10 percent in number of depositors and by 22 percent in amount."[23]

The actual increase both in savings and in deposits is not indicated. But whatever the size of the sum or the number of families relieved from the usual "spring famine," it is not clear to what extent the policy of self-help was responsible for these results. Even the so-called practical, detailed instructions on how to promote the self-help movement are only vague generalizations against which results can hardly be checked. The better prices that prevailed in

18. Ibid., pp. 191–92.
19. Ibid., p. 193.
20. Ibid., p. 194.
21. Ibid.
22. Ibid.
23. Ibid., p. 195.

1933 (and in the subsequent years) as compared with those of the immediately preceding years might have been the direct cause of certain improvements.

Furthermore, the official claims deserve careful scrutiny because of the tendency to see beneficial results where few are in evidence. From the official point of view, for instance, a cause for gratification is that "the consumption of Rubber Shoes decreased by 5.9 percent."[24] Decreased consumption of this type of commodity does not mean that the Korean farmers shifted to better quality leather shoes; on the contrary, it indicates a shift to the poorest quality shoes—straw shoes produced by the farmers themselves. This is characteristic of a type of self-sufficiency that feeds on a lower rather than a higher standard of living.

The kind of assistance exemplified by the self-help movement proceeds from the assumption that, basically, there is little the matter with Chosen's agriculture. "The future of these communities," we are assured, "should not be regarded with pessimism. Agriculture is favoured with good soil, good climate, and abundant labour. With the study of land productivity and the adaptation of farming methods, the yield can readily be doubled."[25] This statement is charged with an undue degree of optimism. The plentiful supply of labor is synonymous with a surplus farm population, finding no outlet for profitable employment; the soil is considered inferior to that of Japan, the land reserve fit for cultivation is small indeed, and there is no evidence that the yield could "readily be doubled."

Conclusion

Reviewing Chosen's agricultural development since the country's occupation by Japan, one observes, on the one hand, considerable expansion of production and, on the other hand, the worsening of the economic conditions of the masses of Korean farmers. The benefits of enlarged acreage and augmented production seem to have eluded the very people whose efforts made them possible.

The salient features of the seemingly paradoxical situation in Chosen are as follows: A cash income ranging from 50 to 100 yen per family; an insufficient food supply that spells hunger for considerable groups of farmers, coincidental with large exports of food products to Japan; indebtedness at usurious rates by which four-fifths of the farmers are harassed; and ever-increasing decline in land ownership and consequent swelling of the ranks of tenants. All this was well epitomized by a Japanese writer who stated that "the lot of the Korean farmer is as miserable as ever it was, a fact that explains why the Government-General, despite all its efforts on behalf of Korea, is not unqualifiedly popular."[26]

In Chosen as well as in a number of other Oriental countries, the inherent problem of a growing population pressing upon a limited arable acreage is responsible in a large measure for many of the difficulties mentioned above. They are accentuated, however, by Japanese economic policies relating to Chosen. Japan succeeded in turning Chosen into an ample source of raw materials, primarily foodstuffs, for which the former was greatly in need. In return, Japan supplied Chosen with manufactured products. This is a familiar colonial policy motivated mainly by the specific needs of the "mother country," in this case Japan, rather than by those of the colony.

The more recent very ambitious cotton production plan is yet another illustration of this policy. It is quite possible that the cotton program will benefit the farmers, but it must be noted that the expansion of acreage under cotton springs from Japan's desire to lessen the dependence of its textile industry upon foreign cotton. If this development should bring any advantage to the Korean farmers, it will be purely incidental to the main aim.

In the strenuous efforts made by the Government-General of Chosen to increase agricultural output, the immediate problems affecting the lives of the farmers were lost sight of until it became evident "to all those interested that it

24. Ibid.
25. Ibid., p. 192.

26. Yoichi Muto, "New Economic Trends in Korea," *Contemporary Japan* (September 1936), p. 210.

is of vital necessity to retrieve the rural communities from entire collapse."[27] The measures to combat this situation were not commensurate with its gravity. The attempts to scale down indebtedness, to arrest the growth in the number of landless farmers, to prevent the concentration of land in fewer hands, and to create a strong group of owner cultivators were quite ineffective.

These are admittedly difficult problems to solve, but their solution is not brought nearer when, as Korean authorities maintain, the real cure lies in spiritual regeneration, self-reliance, mental awakening, and the like. At best such measures could alleviate but slightly the plight of Korean farmers. Even the Japanese farmers, who have the qualities that the Koreans supposedly lack, failed to escape a goodly share of the ailments besetting the Korean farmers. The farmers of Japan were in no position to improve their economic status without considerable financial aid from the state; to a greater degree the same holds true of the Korean farmers.

To give the land back to the farmers or to arrest the process of growing tenancy, the government would have to render them financial assistance on a large scale, as well as introduce basic remedial legislation. Such measures, together with a greater emphasis on industrialization of the country, toward which a substantial beginning has already been made, might help to reduce considerably the distress in the Korean countryside.

27. *Annual Report on Administration of Chosen 1933–34*, p. 192.

6. *Agriculture of the Netherlands Indies*

This is a quite exhaustive—some 63 printed pages long—and interesting study of a benevolent colonialism, untypically protective of native-owned, small-scale farms and production. It is far too long for inclusion here in full, but Ladejinsky's own short summary and conclusions are presented. They convey the flavor of the piece and suggest the substantial background knowledge Ladejinsky brought with him when he made three work visits to Indonesia from 1961 to 1964. It may be noted that this study was published within months after the outbreak of World War II in Europe. Although the United States was not to become directly involved until more than a year later, this and other agricultural studies of the period—of British Malaya, Manchuria, Thailand, Australia, the South Pacific Islands, the food supply in Asia—were probably of considerable interest to the U.S. State and Defense Departments in their own strategic studies and planning.

This article appeared in *Foreign Agriculture* in September 1940.

SOME OUTSTANDING FEATURES of the economic development of the Netherlands Indies may be recapitulated here. The wide choice of climate and rich volcanic soil permit the cultivation not only of home food crops but also of a great variety of important tropical industrial products. The place of the Netherlands Indies among other producers of such commodities is important: in the case of three products it enjoys almost a monopoly. Thus the islands furnish 90 percent of the total world supply of cinchona, 30 percent of pepper, and 75 percent of kapok; they also supply a third of the world's rubber, a third of copra, nearly 20 percent of tea, 25 percent of palm oil, and more than 20 percent of sisal. In addition, they furnish sizeable quantities of sugar, coffee, and numerous other commodities. The islands are also important sources of oil and tin.

The agricultural economy of the islands is

characterized by its dual character of large plantations operated by Europeans and small 1- or 2-acre native farms. The plantations represent large capital investments and thorough utilization of agricultural science. These two elements, together with the fertile soil, abundant labor supply, and growing demand for tropical products, have enabled the enterprising Dutch to turn the Netherlands Indies into a country of typical plantation agriculture.

Unlike many other Asiatic countries, the land reserves of the Netherlands Indies are not exhausted; however, a sharp distinction in this connection must be drawn between Java and the outer provinces. Java is one of the most densely populated countries in the world, and the possibilities of increasing the crop acreage of the island are practically at an end. Expansion of output can come only from increase in yield; it is generally agreed that in this respect a great deal may be accomplished.

The outer provinces, on the other hand, are sparsely populated and the land under cultivation represents only a fraction of that available. To be sure, much of the soil is less fertile than that of Java, and irrigation is a prerequisite for the growing of food crops; but with considerable investment of capital, large tracts of land can be cleared and cultivated. This is actually being effected at the present, though on a small scale, through the government-sponsored and -directed colonization of southern Sumatra, Borneo, and Celebes, where large tracts of land are placed at the disposal of settlers from the most densely populated sections of central Java.

Colonial possessions are often referred to as the "white man's burden": for the Netherlands, however, this burden has not been difficult to carry. In the process it has become one of the world's leading colonial powers—not so much because 8.5 million Dutch rule a colonial empire of 65 million persons but because the Netherlands Indies, largely through development of its economic resources by the Dutch, has become one of the world's richest and best-paying colonial possessions.

At the turn of the century when Dutch investments were estimated at less than a million guilders, it was noted that "the welfare of the propertied classes in the Netherlands is very closely related to the retention of our colonies in East India."[1] Since then Dutch investment has almost trebled, and with it has grown the economic dependence upon the Indies of nearly every section of the population of the Netherlands. The number of Dutch whose economic welfare is directly or indirectly dependent on the colonies has been variously estimated at from 400,000 to more than four times that number.[2] The yearly profits from the colony have been estimated at about U.S.$160 million.[3]

On the whole it is clear that the East Indies place the Netherlands among the principal so-called "have" powers, as contrasted with the "have not" powers. Normally—that is, prior to the 1930s—such a position was beneficial to all concerned. Since the cost of production in the Netherlands Indies is relatively low, their exports have always been available at fair prices and profits have been distributed in varying degrees among the mother country, the colony itself, and the consumers. But in a period of a rising European and Asiatic totalitarianism, the very existence of a rich but poorly armed colonial power such as the Netherlands becomes gravely endangered. Their militarily strong neighbors, who depend on imports of the commodities with which the Netherlands Indies is so richly endowed, are not content with access to the market on equal terms with other countries.

Having noted the importance of the Netherlands Indies to the Dutch in terms of material well-being, the question may be considered of the effects of Dutch rule on the economic welfare of the natives. The low standard of living of the native population of Java has already been discussed. It is questionable, however, whether the responsibility rests solely with the Dutch. Less emphasis on the colony as an exporter of raw products and an importer of finished goods, or a greater measure of industrialization, would have added to the income of the natives: hence, the present policy of industrialization as a means of counteracting some of the ill effects of the economic crisis.

1. Amry Vandenbosch, *The Dutch East Indies* (Grand Rapids, Michigan, Wm. B. Eardmans, 1933), p. 209, quoting Van Deventer.

2. Ibid., and Arthur S. Keller, "Netherlands India as a Paying Proposition," *Far Eastern Survey* (January 17, 1940).

3. Amry Vandenbosch, op. cit., p. 209.

The failure to act in this direction before the depression must be laid to the government.

At the same time the importance of industrialization of the Netherlands Indies must not be exaggerated. Considering the natural resources of the Archipelago, it may be stated that industrialization could hardly affect its agricultural character. The natives of Java and, for that matter, of the other islands live by agriculture and are likely to continue to do so. The standard of living of the Javanese would be higher but for the fact that the island presents a problem of extremely dense population. This, perhaps, is at the root of the difficulty. On those islands in which this condition does not prevail, as in the outer provinces, the natives have facilities for bettering their economic status.

One of the cardinal policies of the Dutch colonial government has been that the native food supply must be as ample as possible. Because of the ever-increasing pressure on the land, per capita consumption of rice has declined over several years by 15 percent. Yet even in the worst years of depression the natives knew nothing of hunger. One cannot but contrast the situation with that in Chosen, under the Japanese, where huge numbers of farmers are subject to "spring hunger" year after year.

The outstanding achievement of the Dutch colonial administration is that it has prevented the natives from bartering away their land. Hardly any other colonial power has insisted that this, the natives' only capital, must be preserved for them; hardly any other colonial power has succeeded so well in carrying out this principle. Considering the eagerness with which European entrepreneurs and Chinese and Arabian moneylenders wished to obtain fertile native land, the Dutch achievement in this respect cannot be overestimated.

It may be argued that the European plantations occupy land that might have been cultivated by the natives; however, the acreage of state domain rented out to the plantations is too small, especially in Java, to make any appreciable difference in the size of the native individual holdings. When, as in the case of the sugar and tobacco plantations, the land is rented from the natives, they retain the use of it for cultivation of food crops and receive a fair rental set by the government. Finally, the absence of the tenancy system in the islands is a phenomenon for which Dutch colonial land legislation is largely responsible.

The economic development of the Netherlands Indies is largely synonymous with the increasing volume of export crops in which native agriculture has been playing an increasingly important part. The Dutch colonial government directly and the plantations indirectly have stimulated the cultivation of export crops by natives. The phenomenally rapid expansion of native rubber production to a point where it exceeds that of the plantations illustrates this point. While in the 1890s the natives contributed 10 percent of the total volume of export crops, this percentage has since risen to over 40 percent. It may be stated, then, that the natives, particularly of the outer provinces, are not only taking an active part and profiting in the economic development of the islands but may before long outstrip the plantations as a source of exports of tropical commodities.

Before the depression of the early 1930s exports from the Netherlands were valued at over U.S. $600 million; at the height of the crisis they declined to approximately one-third of that sum. The once-flourishing plantation industry was hard hit, since foreign markets could no longer absorb the output of the Netherlands Indies at prices insuring the producers a margin of profit. The laissez-faire system of free trade and uncontrolled private business enterprise gave way to an extensively regulated capitalistic economy. With decreased dependence on foreign markets, greater attention had to be devoted to the internal market and the development of industries based on domestic consumption. In a word, the new policy has been one of making the islands into a more self-sufficient economic entity. For the time being, this has helped to prevent a collapse of the economic structure of the Netherlands Indies, but it may be questioned whether it could restore the islands to their former prosperous state.

7. *Agricultural Policies of British Malaya*

This is a brief and even-handed, if somewhat sketchy, treatment of agricultural policies in Malaya under the British colonial administration. It touches on policies regarding land tenure and use, the balance between plantation and small-scale native agriculture, labor and immigration, research, and cooperative credit.

This article appeared in *Foreign Agriculture* in April 1941.

Land Tenure

THE PRINCIPLE THAT ALL LAND in British Malaya is owned by the sultans of the various states has been carried over, with little change, from ancient times. Even in the pre-British era, however, a farmer could acquire permanent tenure of the land he cleared and tilled along with the right to alienate and bequeath it, though his tenure was dependent on uninterrupted cultivation, payment to the sultan of part of the crop, and the contribution of labor. For, in a sparsely populated country where clearing the jungle was a difficult task, "Whatever form the subordination of the people to the chief might take in other matters, a certain security in land was essential if the labour of clearing was to be undertaken and the people were to be induced to cultivate at all."[1]

The coming of the British and later the demand for land caused by the rapid development of the rubber industry necessitated a revision of the land tenure system. Now, as in the past, the cultivators are assured of permanent tenure in addition to heritable, transferable rights subject to some conditions. But forced labor services have been abolished, and crop sharing has been replaced by cash rental. All public land in Malaya is still the property of the ruler, but a new and very important element has been introduced: the land is disposed of not directly by the ruler but by the British Resident in his behalf. Thus "The Resident is empowered to alienate land, and, with the approval of the Chief Secretary, to impose special conditions. The Resident may delegate certain defined powers to Collectors of Land Revenue in his state in connection with the alienation and occupation of State land."[2] The function is a significant one when it is considered how eager the British administration was to expand the rubber acreage.

The land tenure system of British Malaya is now regulated in accordance with the Land Code of 1926. This code was enacted for the Federated Malay States, but with minor modifications its provisions apply throughout the country. All agricultural land is divided into country land exceeding 10 acres and country land of 10 acres or less in order to draw a line of demarcation between the small-scale native agriculture and the large-scale plantation agriculture.

Title to lots of not more than 10 acres can be established through entry in the parish registry or through grant or lease. The first manner is the one under which small holders usually establish title to the land. Title to land exceeding 10 acres may be acquired only through grant or lease. Parish registry conveys absolute title in perpetuity, subject, however, not only to payment of rentals but also to a number of conditions laid down by the land code and the grant itself. The lease, usually for ninety-nine years, sets forth the terms decided upon in each case by British authorities. Title

1. H. Martin Laske, "Studies in Tropical Land Tenure," *Tropical Agriculture* (October 1932), p. 320.

2. *Handbook to British Malaya* (1935), p. 105.

under the land code conveys surface rights, the state retaining all rights in minerals below the surface.

The first charge on all land, irrespective of title, is an annual quit rent; the amount paid ranges from 60 cents to $4.00, depending on the quality of the land. Rents are subject to periodic revisions at intervals of not less than thirty years. In addition, a special per-acre payment must be made when the land is acquired. In the case of small holders, failure to pay rent or cultivate the land ". . . in a proper manner according to the methods of good husbandry" or to commence cultivation within a specified period is cause of forfeiture of part or all of the land. Similar conditions apply to the holdings exceeding 10 acres. In this case failure to commence cultivation within twelve months of the date upon which title has been established or to cultivate a quarter of the land during the first five years, half within ten years, and the remainder in the course of the subsequent three years constitutes default. Furthermore, the administration has a right to specify the crops that may or may not be grown on a given type of land.

The British land policy in Malaya stems from the desire to "open" and develop the country as rapidly as possible. The chief features of the policy are not considered burdensome, particularly as they affect the plantation industry. The provisions that certain areas must be cultivated within a given time were established in order to prevent land speculation. The net result is that it is very difficult to hold land and not to cultivate it. In general the provisions laid down before and after the enactment of the Land Code of 1926 have been liberally interpreted. For all practical purposes the grants as well as the leases are as good as outright sales. The transfer of rights from one person to another is also relatively simple. Thus the plantation industry is assured of permanence and thereby made attractive to investors.

The British land policy in Malaya aimed not only to encourage the development of the plantation industry but also to evolve a fair balance between the plantation and the small-scale native agriculture. This meant that the land requirements of the native population were to be satisfied. The administration has carried out its intentions with respect to the plantations, the acreage of which now exceeds that owned by the natives. There is no certainty that the expansion of the estates has not encroached upon the natives' only capital, the land. It is true that the relative scarcity of population in Malaya has not yet created a serious land shortage as in Java. It would seem, however, that a land problem does exist. As far back as 1913 the Federated Malay States passed the Malay Reservation Act, whereby the federation was empowered to designate any land as a Malay reservation. Such land cannot be sold to a non-Malay. Similar legislation has been enacted in more recent years. Nevertheless, all those willing to engage in large-scale agriculture in British Malaya have had no difficulty in acquiring all the land they have desired.

Labor

The development of plantation agriculture in British Malaya has depended on foreign labor. It has been almost impossible to lure the Malays from their own easy-going agricultural pursuits into the more arduous tasks of daily wage earners on the estates; hence the reliance on imported labor from India and China and, to a much smaller extent, from Java. The bulk of the labor force is composed of Indians. In 1938, out of 296,000 workers employed on the rubber plantations, the Indians constituted 70 percent, the Chinese 21 percent, the Malayans 5.5 percent, and the Japanese and others the remaining 3.5 percent of the entire labor force.

Before the depression of the 1930s, the government of British Malaya actively encouraged the importation of labor. Many Indians were recruited for work on the plantations and assisted to emigrate to Malaya, as distinguished from those who paid their own expenses. The process was closely supervised by the Indian Immigration Committee, set up to regulate the flow of labor in a manner defined by the Malayan government. Part of the function of this body was to manage the Indian Immigration Fund, built up from compulsory assessments on the employers of Indian labor to cover transportation expenses of the emigrants from their homes to the place of employment and for repatriation to India as well as subsistence and medical attention for those en route to Malaya.

In 1930, in consequence of the severe economic crisis, Malayan authorities suspended all emigration assistance. The system was reestablished in 1934, but the number to be assisted was greatly restricted. On the other hand, no restrictions were placed on those who paid their own passage.

Living and working conditions in Malaya are still poor but labor is subject to a considerable degree of protective legislation which covers general working conditions, wages, housing, and health. Pressure by the government of India has been partly responsible for the enactment of such legislation.

Labor on the plantations in British Malaya is free and does not operate under an indenture system. The enforcement of labor contracts through imprisonment was abolished for Indian labor in 1910 and for other labor not long thereafter. Indebtedness to the employer is not a cause for compelling workers to continue in employment; a worker is not liable to criminal prosecution for leaving his job without notice; finally, every worker is authorized to leave employment on a month's notice without subjecting himself to civil action.

One of the functions of the Indian Immigration Committee is to prescribe the standard rates of wages for practically all forms of work in which Indian immigrants are employed. But since the rubber industry chiefly employs such labor, wages there set the standard for all other industries. Wage rates are fixed by law in key areas, and experience has shown that these in turn effectively determine the wages in other districts. Accessibility of labor and sanitary conditions are the decisive factors in fixing the differences between the wages paid, not merely in different parts of the country but between different estates within the same district. The actual wage rates for men during 1936–1938 ranged from 23 to 29 cents per day and for women from 18 to 23 cents. The legal maximum number of hours per day is nine, but a rubber tapper usually works not more than five to six hours, and the number of working days per month guaranteed by the labor statutes is twenty-four.

In addition to wages, plantation labor is provided with housing, medical, and educational facilities. The larger the plantation, particularly if it is owned by Europeans, the better are the houses, medical attention, and the schools. On the whole, however, considerable progress is being made in these fields throughout the plantation industry, notably in the school system. The labor laws of Malaya empower the Controller of Labor to order the establishment of a school on plantations where ten or more children between the ages of 7 and 14 live. In 1938 the total number of schools on the plantations was 754 as compared with 565 in 1935.

A feature that has assumed much importance in the past decade is the allotment of land for the plantation labor force. At an early stage of plantation development one-sixteenth of an acre, to be utilized as a garden plot, was set aside by each plantation for workers with families. Before the slump of the 1930s, there was little economic pressure to make use of such facilities. Furthermore, "To many planters, and to nearly all visiting agents, a green shoot appearing above the soil which was not a rubber seedling was likely to be lallang (wild grass) and therefore something to be destroyed rather than encouraged."[3] This attitude on the part of the planters and agricultural laborers came to an end with the depression of 1929–1933. Many workers employed for years on the plantations refused to return to India despite the prevailing low wages and sometimes were given considerably larger areas for food crops than were specified by law.

To the Malayan administration, the increased use of garden plots remains a means of supplementing the earnings of the plantation laborers, but to the agent of the government of India, who is spokesman for Indian labor in Malaya, it is an opening wedge for permanent settlement on the land and a means of enabling the immigrant workers to become full-fledged citizens of Malaya. "It is the settlers themselves," he writes, "who by their own exertions open and develop the land, and it must be clearly understood that they have as permanent a stake in the welfare of the country as any mine or estate owner. Their title, therefore, must be permanent and unassailable, and in

3. Government of Malaya, *Annual Report of the Labour Department for the Year 1938* (Kuala Lumpur, 1939), p. 49.

any future political development which may take place they must receive recognition as permanent independent inhabitants of Malay."[4]

Agricultural Science

The chief causes underlying the decline in the crops that formerly dominated Malayan agriculture were plant diseases and poor methods of cultivation. Mindful of these experiences, the British administration realized that successful agricultural development in Malaya must go hand in hand with thorough application of sound techniques. The establishment of the Botanic Gardens in Singapore and in Penang was the first step in that direction. They were long the main sources for dissemination of technical agricultural knowledge. Their work in introducing rubber growing is especially noteworthy. When the Department of Agriculture was created in 1905, research was extended into every aspect of plant and soil. Numerous experiment stations were opened, some of them dealing with specific crops such as rice. The investigatory work of the experiment stations in Malaya has been supplemented by demonstration work. Agricultural education has also been provided by a school of agriculture, established in 1931, which devotes its attention particularly to the training of personnel to serve the needs of native agriculture.

The chief emphasis of the scientific work was on the development of rubber. From the outset, all efforts were directed toward the spread of scientific cultivation and the elimination of haphazard and destructive methods such as had been practiced in the Amazon Valley. For some time the work in this field was carried on by the department of agriculture and four other organizations, but it became increasingly evident that the problems of the rubber industry could be dealt with better by one institution devoted entirely to their solution. As a result, the Rubber Research Institute of Malaya was organized in 1926.

The two technical problems of fundamental importance to the industry at that time were crop improvement, through bud grafting from high-yielding trees, and soil management. The institute, amply financed by a special export tax on rubber, concerned itself with these problems as they presented themselves on large- as well as small-scale rubber holdings. In fact, the institute dealt with all scientific problems in relation to production, including investigations on soils, improvement in planting material, tapping systems, diseases and pests, the production of various types of raw rubber suitable for manufacturing purposes, and the marketing of latex of good quality. All of this is expected to bring about ". . . lower costs of production in the industry and at the same time should enable the industry to maintain itself in the face of increasing competition from outside and to safeguard its future prosperity."[5]

Cooperatives

The efforts of the British administration were concentrated on the promotion of the plantation industry, but certain deplorable conditions of native agriculture could not be overlooked. Farm indebtedness was one of those conditions. Although official data do not reveal how large the aggregate debt burden was, ". . . measures were needed to enable the small holder to free himself from the immense load of debt with which he is normally overwhelmed."[6] The farmers were deeply involved with moneylenders who not only charged exorbitant interest rates but also purchased their crops at prices much below those prevailing on the market.

In order to alleviate the natives' condition, a Cooperative Societies Department was established in 1921. The chief objectives of the institution were two: to improve agricultural finances through short-, medium-, and long-term credits and to improve marketing of agricultural products. However, these attempts to free the farmers from the moneylenders have failed. The government has provided the credit

4. "Immigrant Labour in British Malaya," *International Labour Review* (July 1940), p. 73.

5. G. Bryce, "Malaya Rubber Institute," *Tropical Agriculture* (September 1928), p. 215.

6. D. H. Grist, *An Outline of Malayan Agriculture* (Kuala Lumpur, 1936), p. 33.

cooperatives with constitutions but not with initial funds to carry on their work. The chief emphasis has been on teaching the farmers the virtues of thrift in the belief that the savings thus accumulated will enable them to establish their own credit institutions. The only difficulty has been that the economy of the Malayan farmer hardly permits surpluses. He has been forced, therefore, to continue to rely on the Indian and Chinese pawnbrokers and moneylenders as his chief source of credit, whether short- or long-term. The failure of the cooperative credit scheme is well-illustrated by the fact that in 1934 the Federated Malay States and Straits Settlements had a total of $50,886. Loans granted during the year by these bodies amounted to but $3,956, and their reserve funds totaled $19,233.[7]

Entirely different is the treatment accorded the large planters. They are in a more favorable position because they are considered the most dynamic factor in the country's agriculture. For this reason "The large private estate can usually obtain long-term credit . . . at a moderate rate of interest. The Federated Malay States Government has provided long-term credit . . . in the shape of a revolving fund of $4,000,000 called the 'Planter's Loans Fund' which is available for such purposes as may directly assist in the general development of the Federated Malay States."[8]

Nor have the government efforts to improve the lot of the natives by cooperative marketing met with success. Attempts have been made to organize the native rubber growers into cooperative groups for joint sales of the product, thus eliminating the Chinese middleman; but "The great fall in the price of rubber brought these societies to a stop, and at the end of 1934 all such societies were dormant or were under liquidation."[9]

Conclusion

In summary it may be stated that the British policy of land administration in Malaya was the key to the rapid development of large-scale agriculture by Europeans and to a lesser extent Chinese. The government, however, was not altogether oblivious to the land requirements of the natives. Though the British were not so determined to preserve the land for the natives as were the Dutch in the Netherlands Indies, an important protection measure was brought about by the establishment of the Malaya reservations, where land may be sold only to Malays. There is evidence that the reservations were aimed chiefly against encroachment by the Chinese rather than by the British land interests. The Malaya Reservations Committee noted in its unpublished report of 1931:

> We do not hold that the protection of a backward peasantry is the sole or the chief object of the policy of preservation. The policy is territorial, and whatever the competitive capacity of the Malay may be he cannot as a race compete with the far more populous peoples of the other races (Chinese and Indians) who are attracted to Malaya. It is a question of numbers. If the future of the Malay is to be assured, he must have room for expansion, and that requires land to be reserved.[10]

Whatever the underlying cause, however, the reservation's regulations, if properly enforced, are a guarantee that not all of the land of Malaya will be bartered away to the British or Chinese; at least part of the reserves of the states will pass into the hands of the natives.

7. *Handbook to British Malaya*, p. 153.
8. D. H. Grist, op. cit., p. 40.
9. Ibid., p. 153.
10. Rupert Emerson, *Malaysia: A Study in Direct and Indirect Rule* (New York, 1937), p. 479.

II. THE TOKYO YEARS, 1945–1954

ALTHOUGH THE NINE YEARS from December 1945 through 1954 were three times interrupted by tours of duty with the U.S. Department of Agriculture in Washington, for a total of some thirty months, it seems appropriate to describe the thirty-three papers Ladejinsky produced during this period under the heading "The Tokyo Years."*

Ladejinsky returned to Washington in January 1947, a few months after the Japanese land reform legislation was enacted. He was recalled to Tokyo in November of that year at a time when the land transfer program was moving into high gear; he remained there until September 1948, by which time a substantial part of the transfer program had been completed. His next return to Washington lasted until August 1949 when he was detailed to assist the Joint (U.S. and China) Commission on Rural Reconstruction on the mainland of China (Szechwan) and Taiwan. His final return to Washington in February 1950 lasted only until September. In October 1950 he was posted to Tokyo once more, this time as agricultural attaché.

The pattern of Ladejinsky's papers from 1946 through 1954 reflects these changing assignments. In the first group are papers which deal with the basis for the land reform in Japan, its promise, and its early results. Next come papers relating to rent control and land reform in Taiwan and mainland China. After his return to Tokyo as agricultural attaché, additional papers and despatches review the Japanese land reform in its broader social and political as well as economic effects and evaluate the U.S. contribution to it. But the many demands for Ladejinsky's advisory assistance in other countries during this attaché period are reflected in another visit to Taiwan and the resulting 1951 paper, in two papers emanating from his invited participation in the Ford Foundation Conference on Land Tenure in New York late in 1952, and in four papers on land reform which resulted from "on loan" ad hoc work in India in 1952 and 1954. Between the China-Taiwan papers and the second group of Japanese papers come two popular articles assessing the political problem and prospect for Asia as a whole. The last piece of writing produced during these years is a letter to the Ford Foundation, evidently part of a continuing contact with friends there which goes back at least as far as his participation in the 1952 conference, which contributes

* See the Chronological Bibliography.

importantly to a knowledge of Ladejinsky's broad views on human welfare, development, and external assistance in that process.

Of the total of thirty-three papers produced during these "Tokyo Years," eighteen dealt with Japan, six with China-Taiwan, four with India, two each with land tenure and the outlook for Asia, and one with what may broadly be called human welfare. Of these, seventeen are presented here—three of the Japanese pieces, three also of the China-Taiwan pieces, and all of the others. All but three of the seventeen are presented in full, chronologically, as they occurred in Ladejinsky's work experience and perspective.

8. *Farm Tenancy in Japan*

Although dated June 25, 1947, I have established that, like the field studies reported on in "Landlord versus Tenant in Japan" this study was also completed in 1946, well before the enactment of the land reform legislation in October of that year and in ample time to help shape the final legislation. It is therefore a piece of prime historical value as well as of prime importance in Ladejinsky's career and in the assessment of his contribution. I had given some thought whether to include here this longer and updated article or the earlier "Farm Tenancy and Japanese Agriculture" of 1937 which helped to establish his reputation as an expert on farm tenancy in Japan and brought about his later involvement in the reform itself. The great difficulty encountered in obtaining a copy of this paper—the National Archives could not help, the Library of Congress reported that its only copy was missing, and one was finally found only when some of Ladejinsky's private papers were located—tipped the balance in favor of its selection here. The twenty-two tables appended to the report have been omitted.

This study was published as Report no. 79 by Supreme Commander for the Allied Powers (SCAP), General Headquarters, Natural Resources Section, June 25, 1947.

Summary

AGRICULTURE, JAPAN'S LEADING INDUSTRY, provides total or partial occupation for 47 percent of the population and supplies 85 percent of the country's food. Important as it is, for many years before the war the majority of Japanese farmers could not operate profitably. The economic insecurity of most Japanese farmers is not a depression phenomenon. It is the result of the pressure of a large farm population on a small arable area and of the spread of land tenancy. Because of the inequitable land tenure system, 28 percent of the farmers own no land and must rent the land they cultivate; 40 percent own so little land that they are compelled to increase their cultivated holding by renting land. Thus, 68 percent of all Japanese farmers are tenants or part tenants, and only 32 percent are peasant proprietors. Tenancy practices in Japan place the tenant in a disadvantageous position. Rent is exorbitant, tenure is precarious, and income from farming is so low that it cannot satisfy even his frugal needs. Hence, the discontent and social unrest that characterizes the Japanese village.

The Japanese government has recognized that the tenancy system needs improvement. Yet since 1922 all attempts toward betterment have been half-hearted and have failed. Concessions on the part of landowners and large state expenditures were called for but did not take place. The tenants were not in a position to effect desired changes, and farm tenancy in

Japan remained a major social and economic problem. Abolition of all tenancy in Japan is neither feasible nor desirable. But the economic status of the tenants can be improved by aiding them to become farm owners and by revising and improving tenancy practices. The Japanese government should carry out such a program through legislative and financial measures.

To achieve greater diffusion of land ownership among tenants, a certain amount of land should be purchased from the landlords at a reasonable price for resale to the tenants. Resale terms should be designed to strengthen the new owners in their enterprises. Repayment terms of government loans should not be burdensome. The development of land ownership will still leave a considerable tenanted area. It is important, therefore, that improved tenancy practices attend the program of ownership among tenants. These include: written long-term agreements with fair rentals payable in cash, compensation for improvements in case of lease termination, and cheap credit facilities.

Improved tenancy practices and a well-devised and executed program of land ownership undoubtedly will raise the standard of living of tenants. But it will not solve the basic economic problem—the scarcity of arable land. Even the most thorough agrarian reform cannot add to the arable land. So long as the average farm holding remains at 2.4 acres or the farm population cannot find alternative occupations, Japanese agriculture will remain a problem.

Time did not permit thorough auditing of statistical data collected through official Japanese government sources.

Introduction

For centuries agriculture has been the backbone of the economic life of Japan. The industrial and commercial progress of the country obscured this fact. Between 1919 and 1941 when Japanese industry and foreign trade made their most notable advances, rural Japan presented a picture of poverty, distress, and social unrest. This is attributed to the fact that agriculture in Japan does not pay.

Many factors have contributed to this. First, a great farm population cultivates a very limited area which cannot be increased appreciably no matter how hard the farmers try. This results in small uneconomic farm units. Inequitable land ownership has given rise to widespread tenancy and absentee ownership. Governmental policies which favored development of industry rather than agriculture imposed a disproportionate tax load on farmers. It resulted in neglect of their economic plight. Sharp price fluctuations to which the principal agricultural products[1] were subjected, high interest rates on farm loans, and unfavorable disparity between prices of industrial and agricultural products added to the condition.

The uneconomic state of Japanese agriculture affects all classes cultivating the land. But the tenant farmers, who till nearly 50 percent of the cultivated land and pay an exorbitant rental, carry the heaviest burden. In the two decades before 1941 serious conflicts between landlord and tenant took the form of violent demonstrations, strikes, injunctions, arrests, and growth of farm tenant unions through which an organized struggle against landlords was waged. Much of the social unrest in Japan may be attributed to the tenancy system.

For years tenancy overshadowed other agricultural problems facing the Japanese government. Solution of subsidiary problems largely depended on basic changes in the land tenure system. In the words of a Japanese writer, tenancy "has epitomized everything reactionary and retrogressive in Japan since the Meiji revolution."[2]

The nature of the land tenure system of Japan, official attempts to change it, and suggestions to remedy the situation are treated in this report.

Physical and Population Factors in Japanese Agriculture

In topography and soil, Japan is one of the poorer agricultural countries. Roughly 75 percent of its area has slopes too steep and soils too thin for cropping.

This preponderance of highlands sets definite limits to the cultivated area. By the time of the

1. Rice and silkworm cocoons.
2. Seiyei Wakukawa, "The Japanese Farm Tenancy System," unpublished manuscript.

Meiji Restoration[3] most of the level land had already been farmed. Farmers turned to the remaining lowlands, mountain slopes, and hillsides. Between 1880 and 1939 the cultivated area increased from 11,000,000 to 15,000,000 acres, or 36 percent. Table 1 [omitted] shows, however, that practically no expansion of acreage has taken place since 1921 when Japan attained about 15,000,000 cultivated acres, largest on record.

Between 1940 and 1945 a sharp decline in the cultivated acreage, from 14,894,000 acres in 1940 to 13,633,000 acres in 1944, or 8.5 percent, was noted. The most drastic single reduction of 532,000 acres took place in 1941. New installations of a military or semimilitary character (airports, training and testing grounds, industrial establishments) resulted in loss of considerable acreage. Shortage of manpower in the rural districts, brought about by army drafts and high industrial employment, was another factor. Cultivated acreage reached the lowest figure since 1900.

The Japanese government cannot be charged with neglecting to seek expansion of its cultivated acreage. Reclamation programs have not been successful because of rugged topography. In the past twenty-five years not more than 15,000,000 acres, or less than 16 percent of the total estimated land area of Japan, have been cultivated. Compared with other countries, the ratio is small. Before World War II, Italy cultivated 41 percent, Germany 40 percent, France 39 percent, and Great Britain 22 percent of the total land area. In the continental United States, where much arable land has not been put to use, 18 percent of the total land is under cultivation. The acreage distribution by prefecture (table 2 [omitted]) shows that Hokkaido with 15 percent of the total cultivated acreage has by far the largest rate of cultivation. Distribution among the other prefectures ranges from a high of 4.0 percent (Niigata) to a low of 0.7 percent (Nara).

The soils of Japan vary owing to the different materials from which they are formed. Most of them are inferior in natural fertility. High production has been obtained only when the needs of each particular soil situation have been added. Nearly all Japanese soils are extremely deficient in nitrogen. Large amounts of lime, nitrogen, and phosphorus are usually necessary to obtain good yields.

A small cultivated area is a relative term. Its real meaning becomes apparent only when related to the rural population it must support. When this test is applied to Japan it appears that arable land is an even commodity.

Industry and commerce in Japan made great strides since 1880. But the remaining farm population was sufficiently large to permit one to speak of Japan's main agrarian problem as one of "many people on little land."

When Japan came into contact with the Occident in 1853, almost 80 percent of the people were farmers. In 1940 farmers constituted 40 percent of the total (table 3 [omitted]). This sharply reduced ratio was not due to decrease in the nonfarm population. The number of agricultural families reached a high of 5,642,500 in 1932 as against 5,518,000 in 1886. Most of the subsequent reduction in numbers (144,000 between 1932 and 1940) was regained. In 1944 Japanese farm families numbered 5,624,116, almost the level of the peak years of 1931 and 1932. It is concluded by Japanese industrialists that the number of people absorbed from the rural districts was not great enough to relieve the pressure on the land.

The scarcity of cultivated land in Japan is evidenced by the number of farm families as related to the cultivated area (tables 1, 2, and 3 [omitted]). If the land were evenly distributed, each household would cultivate 2.7 acres in 1939[4] and 2.4 acres in 1944. A cross section of Japan shows a gradual increase in the size of farm units from south to north but with considerable variations in individual prefectures (table 4 [omitted]). In southwestern Japan, the farm units are usually under two acres in size with some of the holdings along the Inland Sea averaging as low as 1.5 acres. In

3. In 1868 when Emperor Meiji ascended the throne to begin the "Enlightened Era," ending in 1912.

4. Double and multiple cropping so characteristic of Japanese agriculture raises the acreage actually harvested by one-third. On this basis the average land harvested per family would be 3.6 acres.

northern Honshu this figure is increased to over 3 acres, while in Hokkaido the average is 10 acres. Whatever the variations (excluding Hokkaido), the pressure of so many people on a small cultivated area can result only in very small holdings.

In the majority of cases, actual size of cultivated holdings is much smaller than the average of 2.7 acres. In 1941 about 34 percent of the families cultivated less than 1.2 acres each (table 5 [omitted]), or an average of 0.7 acres per family; 30 percent cultivated less than 2.5 acres per family. The farmers who cultivated from 7.4 acres (3 cho) to 12.2 acres (5 cho) number only 119,000 or 2.2 percent of all farm families; of these, 35 percent are in Hokkaido. Farmers who cultivated more than 12.2 acres total 70,000, of whom 89 percent are in Hokkaido. Small holdings often are broken into widely scattered plots, causing the majority of Japanese farmers to work on small plots which resemble gardens rather than fields.

Unequal Distribution of Land Ownership and Tenancy

The holding cultivated by a Japanese farmer has little relation to the amount of land owned by the same farmer. An analysis of land ownership in Japan bears this out (table 6 [omitted]). Data on the problem show that 50 percent of all farm owners possess only 18 percent of the land; 75 percent of all the farmers own not more than 34 percent of the land; and 97 percent of the farmers own 71 percent of the land. The remaining three percent constitute the real core of landlordism in Japan. This minority owns 29 percent of the cultivated land. It accounts for 48 percent of all the land worked by tenants and part tenants.

Japanese landlords are not a small district group of rent collectors. As a class they are less homogenous than any similar group in the Occident. Although not large, Japanese land ownership includes a number of owners with a great deal of land. The scale of operation of the average Japanese farmer is so small and the demand for land is so great that one may become a rent collector through ownership of an acre of land or less. Hence, in Japan "even landlordism is on a molecular scale."[5]

This is supported by data on the estimated acreage rented out by each owner group (table 7 [omitted]). Even the group of "less than 1.2 acres" rents out nearly a fourth of the land it owns. The same is true of the owners in the category of 1.2 to 2.4 acres. The remaining five categories show a sharp increase in the ratio of leased land to that of owned, depending upon the size of the owned holdings. In the "122 acre and over" category the ratio reaches a high of 98 percent. A prefectural breakdown (table 8 [omitted]), based on data collected in 1946, indicates that the composition of the Japanese landlords has not changed in the past decade.

Typical examples of molecular landlordism are the most parasitic group living off the tenants. Their rented holdings are so small that even the rack renting they practice is not sufficient to provide them with a sufficient income. They work as "salary-earners, merchants, officials, moneylenders or, as a supplement, hotel and brothel keepers."[6]

The grand landlords of Japan are those who own 122 acres (50 cho) or more. In 1940 they numbered 2,941 as against 3,410 in 1935. Recent statistics are inadequate on the variation in size of holdings owned by this group, amount of land they rent out, and land they cultivate. The only available data cover the year 1924 (table 9 [omitted]). But in view of the slow changes in the social structure of rural society in Japan, it may be assumed that they are representative of the large landowners of the more recent years.

The holdings of 89 percent of these landlords range from 166 to 323 acres. Only 25 such landlords out of 3,176 averaged 4,589 acres each. Hokkaido alone accounted for 10 of the biggest holdings. One-fifth of the owners in this group owned one-fourth of the land. In general, this prefecture is the principal center of concentration of land ownership.

An average of 11 acres was cultivated by each of the 1,454 cultivating owners of the 3,176 premier landlords of Japan recorded in 1924. This figure increased to 42 acres in Hok-

5. S. Wakukawa, "Japanese Farm Tenancy."
6. Ibid.

kaido and decreased to 5.5 acres elsewhere in Japan. Fifty-two percent of all landlords did not work any of their land, while the other 48 percent cultivated only 1.7 percent of the owned land. Since all of these large holdings are broken into small rented units, the number of tenants per landlord is rather large. Those landlords with 122 to 245 acres averaged 148 tenants each, while the 25 largest owners averaged 971 tenants each.

Extent of Tenancy

Because of the unequal distribution of land ownership, many farmers in Japan own no land at all and must cultivate that belonging to others. Some own so little that in order to improve their economic position they must rent additional land. This accounts for the three groups into which the Japanese land tenure system is divided: owners, part owners, and tenants. In 1944 the first category constituted 32.0 percent of the farm families; the second group accounted for 39.8 percent; and the third group, the tenants, who are completely divorced from ownership of the land they cultivate, constituted 28.3 percent (table 10 [omitted]). Tenants and part tenants represent 68 percent of all farm families.

In the thirty years before World War II, the social structure of the Japanese village showed little change. The number of owners declined slightly despite weak government attempts to reduce tenancy. This met with no success until 1937 when the category of part owners–part tenants showed an insignificant increase. A writer on Japanese agriculture noted that the structure of the Japanese village "has ossified, so to speak. Now one sees a picture slightly different from 30 years ago."[7]

The years 1939–44 considerably changed this picture when judged by the slow shifts in the past three decades. Farm owners and tenants have gained 100,000 and 122,000 families, respectively. The middle group, that of part owner–part tenant has declined by 90,000 families.

These shifts, despite the increase in the number of tenants, point to improvement rather than deterioration in the economic status of the farmers. In the past, the rise of tenancy was an index of further impoverishment of the independent owners or part owners–part tenants or both. Evidence does not show that in 1939–44 the latter categories helped to swell the ranks of the tenants. The large increase in tenants (8 percent more than in 1939) was essentially a wartime phenomenon, which helped the village economy. Many new war industries were established in rural districts, and the workers and their families were eager to grow their own food. Conditions were propitious since owners were unable to cultivate all of their land owing to the severe shortage of manpower and draftpower, and a number of prewar tenants were voluntarily returning some of their rented land. Increased tenancy kept that land in production, at the same time providing the owners with additional income they would have otherwise lost. Since part of the reduced acreage between 1940–44 was caused by the inability of farmers to keep all their land in production, it may be assumed that the increase in tenant numbers kept the reduction of cultivated acreage from being greater.

The increase in owners (6 percent over that of 1939) may be explained by the relatively prosperous war years as well as increased demand and high prices for agricultural products. Some farmers were able to move up the ladder of the Japanese land tenure system, a normally difficult process. Since it is easier for a part owner–part tenant to take advantage of improved economic conditions than it is for a tenant, it is possible that the augmented number of owners came mainly from the ranks of the part owner–part tenant group.

The prevalence of tenancy in Japan is evident from the amount of land tilled by tenants. Official data for the years immediately following the Meiji restoration are not available. Estimates indicate that in 1872 approximately 30 percent of all cultivated land was under tenancy. This figure rose to 37 percent in 1883; 39 percent in 1887; and 40 percent in 1892. For rice fields alone, the figure was 45 percent in that year. In the last decade of the nineteenth century, the pattern of tenancy was well established. In the intervening years, tenancy has

7. Andrew J. Grajdanzev, *Statistics of Japanese Agriculture* (New York: Institute of Pacific Relations, 1941), p. 9.

continued to increase, although slowly, and in recent years the tenants and part tenants cultivated 46 percent of all the land. The backbone of Japanese agriculture is the all-important rice crop raised on paddy fields, and it is there that tenants predominate. In 1943 they cultivated 53 percent of the paddy and only 37 percent of the dry fields (table 11 [omitted]).

Tenancy is prevalent in all agricultural areas of Japan. But the extent and pattern of tenant occupancy differs in each prefecture. The ratio of tenants to the total number of farm households is greatest in Osaka (table 12 [omitted]). Hokkaido is a close second. In only three prefectures is the proportion of tenants to all other farm families less than 20 percent.

The prefectural distribution of tenated acreage sheds further light on the land tenure system of Japan (table 13 [omitted]). The prefectures of Miyagi, Kagawa, and Akita show the heaviest concentration. Next in importance and except for the extreme northern part of Honshu (Aomori prefecture) is the region facing west to the Japan Sea, including the prefectures of Chiba, Tochigi, and Kumamoto. The proportion of land cultivated by tenants to the total cultivated area is smallest in southern Kyushu (and Okinawa), parts of southeastern Honshu, and the greater part of southeastern Shikoku.

The term "tenanted acreage" covers land cultivated by a farmer who owns practically no land (pure tenant) as well as land rented by a farmer who owns a considerable part of the holding he cultivates (part tenant or part owner). The pure tenant group, with which this report is mainly concerned, cultivates approximately 40 percent of all the rented land, averaging two acres per family. Table 14 [omitted] shows that the size of a tenant holding is much greater in Hokkaido than in the rest of Japan, and that in the majority of cases the actual size of a tenant holding is smaller than the average of 2 acres.

In Hokkaido the average size of a cultivated holding is 10.2 acres as compared with 2.1 acres elsewhere in Japan. This is reflected in the size of farms cultivated by tenants. In Hokkaido 57 percent of tenant-cultivated land is in holdings of more than 12.2 acres and 2.6 percent in holdings of less than 2.4 acres. Figures for the rest of Japan are 0.3 and 43.7 percent, respectively. Hokkaido, therefore, has relatively large-scale tenant farming, whereas the rest of Japan represents small-scale tenant farming. The predominance of uplands in Hokkaido enables the tenant family to cultivate, in Japanese terms, a fairly large holding. The lower reproductivity of the uplands as compared to paddy fields is another factor that compels the tenants of Hokkaido to cultivate a larger acreage.

But even with Hokkaido included, Japan is perhaps the only country where a tenant farm is so small. Thus 50 percent of all the tenants rent less than 1.2 acres each, or an average of 0.5 of an acre, and another 27 percent rent between 1.2 and 2.4 acres, or an average of 1.7 acres. Seventy-seven percent of the tenants rent and cultivate even less than the average of 2 acres per family. This explains the poverty of Japanese tenants.

Factors in the Rise of Tenancy

Farm tenancy in Japan began in the eighth century AD. It was firmly established during the Tokugawa Shogunate (feudal regime)[8] and assumed its present form with the passing of feudalism and the advent of the Meiji era. To escape from the oppression of local authorities, farmers would place themselves and their land under the protection of feudal lords. Some farmers lost title to the land in the process and in effect became cultivators of the soil at the pleasure of the feudal barons. As the landholdings of the lords continued to increase legally or illegally, the outright renting of land became a common practice.

Under the Tokugawa regime, all land was theoretically controlled by the Shogun. Its sale and purchase was prohibited. But toward the end of the feudal era various devices were invented to elude this restriction. Many farmers mortgaged their holdings to raise funds to satisfy the lords' demands and their own needs. Failure to repay resulted in foreclosures. Land passed, in fact although not in theory, into the

8. The term "kosaku" meaning land tenancy or tenant, came into usage during the Tokugawa regime. The Tokugawa dynasty was founded in 1603 and remained in power until 1868.

hands either of the rising merchant class, large landholders, or of the more prosperous farmers. Those who lost the land became tenant farmers.

After the abolition of feudalism in 1868, the clear title of peasants to the land was legally recognized. In cases where farmers mortgaged their land and agreed to pay rent to the mortgage holders in return for permanent use of the land, the mortgagor received title to the land. In such instances the farmers became ordinary tenants without privileges and rights held under their former arrangement of permanent tenancy. At the same time, "the erstwhile warrior-rulers were compensated for loss of their feudal privileges with government lands, cash, and subsidies amounting to ¥379 million, an enormous sum for the time. This outlay was assessed on the landowners in the form of a new land tax—not a few ex-feudal lords invested their bonds in land. Their hold on the peasantry thus shifted from feudal overlordship to modern landlordship. Some of them, by questionable means, succeeded in having their feudal holdings recognized as private property by the new regime."[9]

The change from a feudal to a money economy also helped to swell the tenant ranks. The land tax and all other taxes were paid in cash and not in kind as heretofore. Rapid adjustment from one type of economy to another worked hardship on many independent farmers. The necessity of obtaining cash to meet heavy taxation led many into indebtedness, frequently resulting in the loss of land. Moreover, the big landowners and well-to-do farmers who owned more land than they could cultivate themselves found it more profitable to rent it to tenants than to use hired labor. Moreover, the Japanese urban middle class for many years has invested its savings in land, which provided a safe and assured income.

Conditions of Tenancy

Tenancy as an agrarian institution is not an evil. Its existence in virtually every country shows that it has a sound economic basis. A just system of tenancy "is a partnership enterprise between the landlord, who supplies capital in the form of land and sometimes in the form of equipment, livestock, fertilizer, and a certain amount of management and the tenant who supplies labor and usually all, or most, of the operating expenses as well as most of the management of the farm."[10] Where such fair, joint enterprises develop, tenants may be economically better off as tenants than owners. This holds true in many regions in the Western countries. In Great Britain, for example, many tenants prefer to retain their status when given a chance to become owners.

Tenancy becomes an undesirable economic and social institution when certain evils develop. The Japanese land tenure system has been a "partnership" heavily weighted in favor of the landlord. It resulted from the heavy demand for a limited supply of land, the central problem of Japanese agriculture. The tenant has been willing to bind himself to the land regardless of the exactions. He has not wished to abandon the leased land since the opportunities of finding employment in other occupations are poor. For this reason in the decade before World War II the tenants' struggle against the landlord was not so much for a reduction of rent, exorbitant though it was, as for the right to cultivate the land. It was a struggle for security of tenure.

The right of the tenant to remain on the land is so vital that the question of long-term tenure and leasing arrangements in general are of paramount importance. Many Japanese tenants are not certain about the duration of their tenure. Leases for a long period are rare, although a "permanent" tenancy practice exists. This is a carry-over from feudal times when tenure was secure and generally continued undisturbed over long periods of years. Even in modern times this system insures a tenant's tenure for twenty to fifty years. However, a leasing arrangement extends to a very limited number of tenants. A survey (1921–36) revealed that less than 3 percent of tenants, cultivating about 1 percent of all the tenanted land, came under the "permanent" system.

It is not uncommon to find a tenant family

9. Andrew J. Grajdanzev, *Statistics of Japanese Agriculture*, pp. 13–14.

10. Lossing Buck, "Farm Tenancy in China," unpublished manuscript, p. 2.

working the same land for several generations. But most agreements are for from three to five years. In fruit and mulberry gardens, the term of agreement is usually from ten to fifteen years. Often agreements do not specify any fixed period. The farmer is allowed to work the land continuously, provided he has the goodwill of the landlord and uncomplainingly furnishes the exactions imposed upon him.

Agreements are both oral and written with the former predominating. In the northern prefectures (Hokkaido, Aomori, Yamagata, Akita, Fukushima, and Ibaraki) where the biggest landholdings are concentrated, from 70 to 80 percent of the agreements are written. These are initiated by the landlords and are drawn in their favor (annex 1). A typical written agreement is signed by the tenant and one or two guarantors. The latter must pay the rent in case the tenant fails to do so. With the possible exception of two or three prefectures, most agreements are oral (table 15 [omitted]). A Japanese economist states: "These (written) contracts for the first time included a time limit, usually extending over several years. But one must remember that this form of bargaining has been far from universal, being practiced only by a limited group of tenants, and the majority are still renting land by the old method, that is, they are at the mercy of their landlords."[11]

Whether written or oral, landlord-tenant leasing arrangements do not afford the tenant much protection. This is true particularly in the important matter of cultivation right. Even though in many cases agreements are for periods ranging from three to five years, they usually do not prevent the landlord from getting the land back whenever he wishes. Some agreements permit him to recover the land before its expiration, with no compensation to the tenant for the inconvenience (annexes 1A and 1C). Other agreements provide for compensation in cases where the landlord wishes to break the agreement (annex 1B). In both cases the tenant loses the right to cultivate the land. The tenant, too, has a right to abrogate an agreement. But in practice this privilege is exercised solely by the landlord.

An official statement of the Japanese Ministry of Agriculture and Forestry summarizes the nature of the one-sided partnership embodied in the landlord-tenant lease. "The land reverts to the landlord," it declares, "but with partial compensation if the field has standing grain, without compensation if the field has no standing grain, for the labor the tenant has put in, seeds, and fertilizer which has been used. In case the tenant wishes to return the field to the landlord, he gets no compensation at all."[12] This statement was made nearly a quarter of a century ago. Yet basically nothing has changed.

The right to cultivate the land has continued to be the chief concern of the tenant. Field investigations in 1946 revealed the same concern about the landlord's ability to abrogate an agreement and the bitter conflicts that such actions brought about. The landlords now are less prone to indulge in such acts. But this change in attitude is caused chiefly by the foreknowledge of events which are likely to establish a new rural pattern for future years.

Insecurity of tenure is only one phase of the inferior bargaining position of the tenant. The other phase which bears directly upon his low economic status concerns the rent he pays for the privilege of working the land and the manner in which it is paid.

Sharecropping, prevalent in the United States, is practiced very little in Japan. The "flexible tenant system" is adopted only in a few regions that are frequently subject to natural calamities. Under this system, rent is fixed every year after the condition of the crop is examined. The most widely used system of payment is fixed quantity of produce or cash per unit of land, agreed upon in advance (table 16 [omitted]). For the most part, the amount of the rent is based on yields in good years. This places the full burden of the loss from a poor harvest upon the tenant unless he can obtain a reduction in rent. According to a Japanese writer, the burden of rent shouldered by a tenant "is all the heavier, too, because of the

11. Shiroshi Nasu, *Aspects of Japanese Agriculture* (New York: Institute of Pacific Relations, 1941), p. 19.

12. S. Kawada, "Tenant System in Japan and Korea," *Kyoto University Economic Review* (1926), p. 40.

arrangement whereby the tenant farmer pays his landlord so many bushels of rice per tan.[13] This amount does not change much, irrespective of whether the crop is large or small. In other words, the landowners are assured of a certain stabilized quantity of harvest regardless of the yield."[14] But even when reductions are granted, the bargaining process involved in determining the amount or percentage of reduction causes many disputes between landlords and tenants.

Rents exacted are high, especially when compared with rents paid in Western countries. Rent for a single-crop rice field is approximately 50 percent of the crop. Rents on upland fields range from 30 to 40 percent, depending on cash payment or payment in kind, respectively. Rents vary in prefectures. They are highest in Kagoshima, Kochi, Shimane, Okayama, Nagano, and Hiroshima, with payments from 56 to 58 percent of the yield from a single-crop field. Rents are lowest (32 to 40 percent) in Hokkaido, Tokyo, and Okinawa.

In terms of the amount of rice, the average rental for a single crop paddy field was 1.03 koku[15] per tan, according to an investigation by the Japan Hypothec Bank in March 1943. Studies by the Ministry of Agriculture covering 1941–43 show that the average rental was 0.91 koku of rice for one-crop fields and 1.113 koku in cases where the fields were double cropped. On the whole, the average rent amounted to one koku of unhulled rice per tan, which is approximately one-half of the rice-yield per tan.

Occasionally the amount of rent collected by the landlord is greater than that specified in the agreement. This is the outgrowth of an old custom by which the tenant turns over an extra amount of rice in addition to the specified quantity. The practice known under various names such as "added rice," "mouth rice," and "spilled rice" is still found in some regions. This system of surcharges is additional evidence of the powerful position of the Japanese landlords. The cost to the tenants, where the custom still survives, has been estimated to be from 5 percent to 25 percent of the original rent.

Rents for rice lands are almost always paid in rice. In the less important upland fields, cash rents predominate. Cash rents are paid on less than 1 percent of the rice field acreage surveyed (table 15 [omitted]). Rent in kind converted into money equivalent accounts for 12 percent, and 87 percent is straight payment in kind. On the other hand, rents for 57 percent of the upland fields are paid in cash and only 29 percent in kind. Payment in kind for rice lands works hardship on the tenants because they are unable to take advantage of a favorable market situation. The landlords are the real beneficiaries.

The fact that until 1945 the principal payments of rent were made in kind is another tenant handicap. Rents in kind have declined since 1921, the year when they were exceptionally high (table 17 [omitted]). This decline, however, did not exceed 10 percent, even in the early 1930s when the agricultural depression was rather severe. The monetary value of rent in kind has actually risen. Rice prices have been subjected to severe fluctuations. But on the whole the trend has been upward since 1900. Even when prices declined sharply in the early 1930s, much of the loss of the landlords was absorbed by special state rice subsidies.

Rising prices have been a boon for the Japanese landlords who sell 85 percent of their crop, largely acquired in payment of rent in kind. Many of the tenants, although they produce fully one-half of the rice crop, are nevertheless compelled to purchase additional quantities of rice for their own consumption. A sample study on the degree of rice self-sufficiency showed that 60 percent of the tenants did not need to buy rice. The remaining 40 percent had to acquire varying quantities of rice before harvesting the new crop.

When the tenant does sell rice, it is usually immediately after harvest when prices are lowest. He is in no position to benefit from the higher prices later in the season. Because of the little influence exerted by tenants in the rice market, they were unable to take advantage of governmental price policies of the 1920s and 1930s, presumably designed to relieve general rural distress. "If, therefore, the agricultural price policy is to attain the desired end," a

13. One tan—0.245 acre.
14. Magohachiro Kimura, *Japan's Agrarian Problems* (Foreign Affairs Association of Japan, December 1937), p. 12.
15. One koku—5.119 bushels.

Japanese writer noted, "it is imperative that the farm tenancy system should be properly revised. In other words, insofar as the agricultural price policy aims at stabilization of agricultural life, the solution of the farm tenancy problem and, accordingly, the settlement of the land problem is required, as a matter of course, for its proper execution."[16] One might add that, in any scheme for improving the Japanese land tenure system, the substitution of a cash-rent system for payments in kind would undoubtedly have a salutary effect upon the economic status of the tenant.

The high rental the tenant must pay is not the only charge that he must bear in running the farm. The landlord pays the greater part of the various taxes and assessments, but the tenant is not free from them. He is responsible for the village and house taxes. But more important is the arrangement whereby all the tenant receives from the landlord is the land. The tenant must provide his own house, farm buildings, implements, commercial fertilizers, and seeds. He does not receive compensation from the landlord for expenditures involved in procuring these items. The tenant share of the crop must take care of them. Even if the tenant were concerned only with the purchase of fertilizer, his proportion of the crop would be much smaller than that of the landlord. The share of the crop from which the tenant draws sustenance frequently represents less than 30 percent of the crop.

Before World War II, the Japanese tenant was an exploited poor farmer. The exploitation was in accord with the feudal adage that "farmers should neither live nor die." In eagerness to retain the tenant status for the lack of alternative occupations, he has subjected himself to terms which even in good seasons keep him at the bottom of the economic and social scheme of a village. These terms are: (a) Uncertainty of tenure. Duration is subject to the whim of the landlord, since agreements, written or oral, are so drawn up that they offer the tenant little protection. (b) The tenant pays the landlord at least one-half of the crop and at the same time shoulders the major costs of bringing in the crop. The tenant is a mere tool of production.

What tenancy in Japan means in terms of economic well-being can be obtained from an examination of farm family budget studies. Such a study was made of the budgets of thirty-five tenants by the Japanese Ministry of Agriculture in 1920. The number of cases studied was small, yet it throws revealing light on the standard of living of a tenant. It shows that a tenant cultivating 3.7 acres (far above the average tenant holding) ended the year with a net deficit of 44 yen.[17] The deficit would have been greater but for the fact that the tenant spent nothing for the education of children[18] nor social activities and made no allowance for interest on debts.

An investigation into the living conditions of 208 tenants during September 1926–August 1927 revealed that the average income of a tenant farmer from all sources was not sufficient to cover living expenses. The tenant's yearly income from agriculture alone amounted to ¥567. His expenses for the corresponding period were ¥975, making a deficit of ¥408. Even the additional income derived from other sources was ¥25 short of meeting the deficit.[19]

A somewhat similar situation is shown by the family budget investigations covering 1932, 1935, and 1936–38. The year 1936 may be taken as representative. While farm conditions then were not as good as those of 1937 and 1938, they were much better than those of 1932 and 1935. A total of 84 owners, 103 part owners, and 86 tenants were investigated. The owners averaged a total net income of ¥1,030 as against total household expenditures of ¥841, or a balance of ¥189. For the part owners and tenants, the respective balances were ¥161 and ¥101.[20] Here, too, subsidiary

16. "The Establishment of Peasant Proprietorship," *Kyoto University Economic Review* (December 1936), pp. 61–62.

17. Shuichi Harada, *Labor Conditions in Japan* (New York, 1928), p. 84.

18. This compares with ¥427 and ¥29, the average spent for education by landlords and farm proprietors, respectively.

19. Shiroshi Nasu, *Land Utilization in Japan* (Tokyo: Institute of Pacific Relations, 1929), p. 198.

20. From table 17 [omitted]. Some results of family budget investigations in Andrew J. Grajdanzev, *Statistics of Japanese Agriculture.*

occupations were important; otherwise the tenants would have shown a deficit of ¥49 while the owners and part owners would have enjoyed surpluses of only ¥59 and ¥31, respectively.

The most recent investigation (March 1941–February 1942) by the Ministry of Agriculture and Forestry reveals a similar situation. The forty-three tenants covered averaged nearly 4 acres each as against a countrywide (excluding Hokkaido) average of 1.5 acres. Their gross agricultural income was larger than in previous years, but so were expenditures, particularly for food and clothing. The tenants averaged a net income of ¥202, made possible only by an income from subsidiary occupations averaging ¥238.

Subsidiary occupations of the farmer often turn a threatened deficit into a surplus. A survey of the income of farmers from all sources for 1913–34 revealed that, with the exception of 1913, income from their own farming each year fell short of meeting household expenditures. This forced the farmers to seek nonagricultural income to make up the deficit. Income from subsidiary occupations ranged from 21 to 31 percent of the total.[21] Income of the tenant from agricultural pursuits being lowest among the other farm categories, the tenant maintained his lowly position only by devoting more time to subsidiary occupations than did the peasant proprietor or part owner. The nature of the subsidiary occupations and the number of days the three categories of farmers devote to them is shown in tables 18 and 19 [omitted].

Household expenditures of a tenant or farmer, after income from all sources was added, accounted for his low standard of living. In 1936 such expenditures by a tenant amounted to ¥676, or ¥1.8 per family per day. As these households average six persons, the actual expenditure per person was ¥0.3. The meager income from agriculture and subsidiary occupations permitted the tenant to till the landlord's land only because of such small expenses.

The low net income of a tenant is immediately responsible for the basic feature of the Japanese tenancy system that a tenant remains in that category unless he has opportunity to change occupation. He has difficulty in becoming a part owner. It is almost impossible for him to achieve the status of independent farm proprietor.

In 1941 the equity of the farm proprietor in his establishment was worth an estimated ¥11,000, that of the part-owner ¥7,000. Even the tenant required a capital of ¥3,000. If a tenant, clear of all debt, employed his annual savings of ¥200 in addition to the money already invested and wished to become a part owner or a peasant proprietor, it would take him twenty and forty years, respectively, to achieve that status. Exceptions would be few. The normal financial position of the tenant reveals a tenant cannot rise above his status through his own efforts. It explains the failure of the limited Japanese government efforts to reduce tenancy before World War II.

During 1941–45, the position of many Japanese tenants improved. The basic equal division of the crop remained. But the tenant benefited through the general increase of agricultural prices. The price of rice, the most important crop, rose from ¥43 to ¥300 per koku between 1940–1946. Cost of production has also increased, but not sufficiently to eliminate the advantages the tenants received from the price increase. The tenants also benefited from a change in the rent collecting system introduced in wartime. The tenant delivers his rent in kind, not to the landlord but to a government agency where it is credited to the account of the landlord. For each koku so delivered, the tenant received ¥245 as against ¥55 paid to the landlord. The tenant's rent is still one koku, or 50 percent of the crop. But in terms of money received by the landlord, the rent represents only 9 percent of one crop. Since April 1946 all rents in Japan are payable in cash. The tenant's rent is ¥75 per tan instead of the usual one koku of rice per tan, the price of rice remaining ¥300 per koku. On this basis rent amounts to 12.5 percent of the crop.

Where land is double cropped, rent is considerably lower. If the wheat crop from the same piece of land is 1.4 koku, its money value is ¥327. The total value of the crops (2 koku of rice and 1.4 koku of wheat) will amount to ¥927. In that case the tenant's rental of ¥75

21. Hidetoshi Isobi, *Labor Conditions in Japanese Agriculture* (Utsunomiya Agricultural College, 1937), p. 66.

represents only 8 percent of the crops. Through such rent policies the burden of the tenant has been considerably lightened.

In wartime it was easier for a tenant to ask and receive better treatment from the landlord. This resulted from the shortage of farm labor and the increase in nonagricultural occupations. Some tenants were voluntarily returning rented land as shown by data collected on a field trip in Saitama prefecture. A survey[22] of 292 villages revealed that, between July 1941 and June 1943, 3,125 tenants returned 1,682 acres of paddy and upland fields to the landlords. The figures for the next two years were 3,919 tenants and 1,870 acres and 5,988 tenants and 3,657 acres, respectively. The land involved was large considering the average holding of a tenant. But the action itself was most important.

For the first time in years, the tenant found himself in an improved bargaining position. Those who took advantage of the development benefited. The tenant also gained to some extent from the rise in the agricultural prices, legal and illegal (particularly from the latter), as the war neared its end. Debts of long standing were discharged and savings were begun.

Before World War II, farm indebtedness in Japan was estimated at about 5,000 to 6,000 million yen. The war years witnessed the liquidation of the greater part of the debt. A survey of twenty-four prefectures in January 1946 revealed that about 47 percent of the surveyed farm households are in debt, the average for these families (peasant proprietors, part owners, and tenants) being less than ¥800. Only 38 percent of tenants investigated were in debt, the average per tenant family being ¥400. On the basis of this survey, total farm indebtedness may be estimated at about 1,800 million yen. The tenant share in this total is ¥241 million.

If these estimates are valid, the farm debt of Japan is only one-third of the prewar debt. It may be contended that this reduced indebtedness is more apparent than real and does not constitute a real burden. This fact was revealed by personal inquiries in a number of villages. It is supported by the increase in farm deposits from 2,300 million yen in 1940 to 16,200 million yen in 1945. This sum was deposited solely with the Agricultural Association (Nogyokai). Unascertained sums deposited with other institutions add to the total. It is obvious, therefore, that most farmers could liquidate their debts without outside aid if they wished to do so or if their creditors were insisting on immediate repayment. Many of them do not. They prefer to wait for economic stabilization before settlement.

The large deposit increase was not caused by cash accumulation of a particular group of farmers. The average deposit of a peasant proprietor was ¥2,200, that of a part owner ¥2,300. Even the tenants averaged ¥1,300. Tenants alone have deposited about ¥2 million with the agricultural associations. Compared with their estimated indebtedness of ¥241 million, the tenants' debt burden is virtually nonexistent.

The tenants' improved economic position indicated by indebtedness against deposits is temporary. It is a war phenomenon and does not come from changes in conditions of tenancy that might indicate a lasting improvement. The transitory character of the tenants' economic position is apparent. He cannot exchange the accumulated cash for tangible assets. Above all, he cannot now exchange it for land. Little land is for sale. The tenant's cash assets are only large enough to enable him to buy a tan or two of paddy land at the official price (¥750 per tan).

The tenant is again feeling the insecurity of tenure. Agriculture is the only segment of the Japanese economy that survived defeat in good condition. This adversely affects the tenants' hold on the land. Many small owners, whose only connection with the land in the past was in collecting rent, now are eager to cultivate some of the rented land. They hope to insure themselves of a greater food supply and to circumvent the Land Reform Bill of December 18, 1945, which intends to substitute rental in cash for that in kind. Many absentee owners, part of whose income was derived from now nonexisting industrial or commercial pursuits, have returned to the village willing to cultivate some of the land leased to tenants. In Saitama prefecture alone a survey of 136 villages revealed that 1,830 landlords claimed

22. Conducted by the Agricultural Bureau of the Prefectural office.

some of their rented land. Demobilized soldiers add to the competition for land. The bargaining position enjoyed by the tenant in wartime is gone. He finds himself in a position more difficult than in prewar days.

Tenant Unrest and Its Causes

The burdens shouldered by the tenant have never been accepted by him as a matter of course. At times he reacted violently. This is proven by the agrarian uprisings in the first two decades of the Meiji restoration.[23] The issues concerned all of the peasantry. They included government monetary policy, usurious loans, new land and taxation measures, and military conscription. Other issues also have affected the tenants including security of tenure, reduced rents, and rent payment in cash instead of kind.

The revolts were crushed, and the landlord-tenant relationship continued until about the end of World War I without causing much open conflict. A slight improvement in the position of tenants plus the conservatism and ingrained feudal custom of obedience to landlords helped reduce friction to a minimum. During the last twenty-five years the relationship changed. Growing industrial unrest, strengthening of the industrial labor movement in Japan, and the spread of ideas opposed to established concepts made landlord-tenant relations similar to strained conditions between industrial workers and employers.

The increasing dissatisfaction manifested itself in the growing number of disputes between landlords and tenants, so serious in character that they have assumed the proportion of a national problem. Tenant and landlord unions developed. The number and the seriousness of disputes increased rapidly.

Basic data on the number of disputes, people and land involved, and the causes of the disputes are presented in tables 20 and 21 [omitted]. The number of disputes in 1917 was 85. They rose to 1,680 in 1921 and more than 2,000 in 1930. They reached a peak of nearly 7,000 in 1935.[24] They then declined to 2,424 in 1943 because of wartime improvement in the economic condition of tenants and because of wartime appeals to patriotism and unity. Every dispute involved an average of four or more landlords and twenty or more tenants. A tenant often rents land from a number of landlords, all of whom may be involved in a dispute.

Until the late 1920s, tenant disputes were confined largely to a few prefectures such as Saitama, Kanagawa, Gifu, Aichi, Osaka, Hyogo, and Okayama, which is 40 percent tenanted acreage. As agricultural conditions worsened, militant farm organizations grew. Every prefecture except Okinawa became disrupted by disputes. The geographical distribution of the disputes reflected the extent of agricultural distress in each region. This fact explains the shift of a greater number of disputes from the rice to the predominantly silk areas when the price of cocoons declined and, later, to famine stricken northwest areas.

Underlying causes of the disputes are many. Two are outstanding: (a) excessive rent and (b) the attempt on the part of the landlords to terminate, or their refusal to renew, leases. Until the late 1920s reduction in rent was the dominating issue. Aided by their farm organization and rapidly growing unions, tenants demanded rent reduction not only in time of poor yields but irrespective of the size of the yield. Whereas in the past disagreements occurred after the harvest, in the late 1920s disputes would arise both before the sowing and after the harvest.

The excessive rent issue gradually moved into the background as a new dispute centered about efforts of the landlords to wrest the land from a given group or individual tenants and to dispose of it as they saw fit. This was an open attack on the tenants' most cherished right: security of tenure.

23. E. Herbert Norman, *Soldier and Peasant in Japan: The Origins of Conscription* (New York: Institute of Pacific Relations, 1943), p. 41, states, "In the standard collection of documents illustrating agrarian unrest in the early Meiji period, there are mentioned altogether 260 agrarian revolts and 14 uprisings in the larger towns; in the first ten years of the Meiji rule the same source mentions 185 peasant revolts."

24. The figures on disputes represent only those in the open. Actual discontent must have been greater than the official data indicated.

Japanese statistics on tenant disputes reflect this change very clearly. In 1921 two-thirds of the disputes were caused by excessive rents. Not a single dispute was recorded on the grounds of eviction. In 1930 eviction disputes accounted for 40 percent, while excessive rent disputes contributed 22 percent of the total. In 1937 the respective figures were 58 and 18 percent. Even in 1943 with the number of disputes greatly diminished, 1,000 of 2,424 cases were eviction cases. Only 580 were excessive rent cases (table 21 [omitted]).

Two chief motives are listed in the attempts of the landlords to evict individual tenants: (1) the desire of the landlord to cultivate the land himself; (2) "the desire to rent the land to more easily contented tenants."[25] Change in ownership, expiration of original agreements, and failure to pay the rent were other reasons for eviction disputes.

The landlord-tenant struggle gradually centered on the right of the tenant to cultivate the land. The Japanese courts ruled that, since the land belonged to the landlord, he was free to do with it as he pleased. The court injunction became a weapon for eviction. The tenants protested by refusing to pay rent and taxes, by withdrawing their children from public schools and setting up special tenant schools, and by sabotage and violence.

The Tenant Union Movement

In the quiescent days of landlord-tenant relationships, a conflict was dealt with on an individual basis. In a rent dispute the tenant was at a disadvantage. Any reduction depended more upon his final inability to pay than upon any equality of bargaining power. World War I marked the turning point in the development of tenant unions and the beginning of open conflict between tenant and landlord. Both developments arose because of the plight of the tenants and the growing desire for a remedy. The tenant unions became one of the principal means of effecting improvements.

As disputes increased and issues sharpened, tenants realized more fully the significance of collective bargaining. That the tenants were quick to see the economic advantage of concerted action is illustrated by the rapid growth of the tenant unions. They increased from 130 in 1917 to 681 in 1921 and to 4,582 in 1927 with a record membership of 365,000 (table 22 [omitted]), representing 25 percent of all tenant farmers. The movement was most vigorous in Niigata, Yamanashi, Hyogo, Okayama, Kagawa, and Gumma prefectures and weakest in Hokkaido, the northernmost prefectures of Honshu, and the prefectures of Kyushu. The Japan Farmers' Union (Nippon Nomin Kumiai), organized in 1922, was the first national body of farmers largely created by tenant unions. Political activity to achieve economic aims received early recognition. In 1925 the Farmer-Labor Party (Nomin Rodoto) was formed. Although suppressed within three hours of its birth, it carried on in various disguises. With the help of the Japan Farmers' Union, the Farmer-Labor Party succeeded in electing many candidates to prefectural assemblies where the tenant movement was strongest.

Official repressions and internal dissensions resulted in numerous splits and regroupings. In the main, the tenant unions during their greatest development were made up of three groups which represented politically the right wing, the center, and the left wing. Despite political differences, they agreed on such fundamentals as (a) legal recognition of the permanent right of the tenant to the land, (b) substitution of money rent for rent in kind, (c) reduction in rents, (d) development of cooperative buying and selling, (e) spread of education in the villages, and (f) development of political activity to influence local government and secure representation in prefectural and national governments.

Union membership was maintained through the late 1920s, although not at the peak of 1927. A decline began in 1934. By 1941 the tenant union movement, around which the militant part of the tenants rallied, was greatly depleted. During World War II few tenant leaders were in evidence. Many unions dissolved "voluntarily" in the name of national unity. Others were suppressed in the process of eliminating groups and individuals suspected of "dangerous thoughts." Remaining farm orga-

25. Shiroshi Nasu, *Aspects of Japanese Agriculture*, p. 20.

nizations were (a) the few local tenant unions; (b) the government-supported Imperial Agricultural Association (Teikoku Nokai); (c) the ultrareactionary, officially organized Farmers' Patriotic Association; and (d) the Agricultural Co-operative Association (Nogyokai). These organizations aimed to enlist the support of tenants and small farmers for the government's agricultural program.

Other factors contributed to the tenant union decline in the 1930s. The agricultural depression was perhaps the most important. Like the industrial depression, it was not conducive to the spread of unionism. The growing nationalist movement and opposition to unions by certain groups of tenants on the ground of the unions' radical tendencies was a factor. Furthermore, the frequent success of the unions in reducing rents or preventing evictions resulted in a loss of members. With the successful termination of his own fight against the landlord, the individual tenant saw no further need for the union.

Tenant union activities did not go unchallenged. The landlords with the cooperation of district authorities, organized joint landlord-tenant unions. They came to be known as "harmony" unions, or company unions. They aimed to benefit the membership through increased output, better buying and selling practices, and greater cooperation between landlord and tenant. They sought to improve the status of the tenant in an indirect way and, in turn, benefit the landlord. The regular unions were dealing with the problem more directly. They were concerned with the immediate needs and interests of tenants.

The "harmony" unions were organized with moderate aims and without radical leaning and were not subject to the changes of the tenant organized unions. They still had a membership of 208,000 in 1943 as against the peak of 279,000 in 1933, when tenant unions had almost disappeared. The war thinned out these and by 1943 it brought about a concentration of 81 percent of the membership in two prefectures, Yamagata and Fukushima.

The real struggle against the tenant unions and their demands was carried on by the Japanese Landowners' Association. They conceded the seriousness of the tenancy problem but, having no desire to solve it at their own expense, preferred the status quo of prewar days. Their attitude was expressed in the following statement issued in 1926 when the tenant unions were spreading rapidly.[26]

> The present condition of distress in agricultural villages cannot be regarded with indifference. We are suffering from heavy taxation. Our farms are desolated. Tenancy disputes are aggravated year by year. A catastrophe is close upon us. Is it not time to awaken and do something?
>
> Remembering the splendid tradition of our nation, with sovereign subjects forming one whole, and reflecting on the glorious history of our national development in the past, let us emphasize the harmonious relations between capital and labor, and especially cultivate peace between landowners and tenant farmers and thus contribute to the development of our agricultural villages. What sort of devils are they who furiously strike fire bells when there are no fires and incite to a class struggle, provoking animosity against landowners by exciting tenant farmers? If these malicious designs go unrestricted, what will become of our national existence? We are determined, therefore, to cooperate with those who hold the same ideas, to arouse public opinion, and to establish a more suitable national policy.

The statement does not admit that the tenants have just grievances, nor is there a suggestion on how to cope with the problem. Ten years later, with conditions worse than in 1926, the landlords exhibited the same "standpatism." In response to government inquiries concerning tenancy, the landlords recommended "the stabilization of relations between landowners and farmers by legislation regulating farming right." These recommendations were vague and noncommittal. They continued in the same view in subsequent years and throughout the war. Any government measures to improve the tenancy situation were made in the face of landlord opposition.

26. *Industrial and Labor Information* (Geneva: International Labour Office, September 27, 1929), pp. 492–93.

Government Tenant Policies

The growing disputes resulted in the Conciliation of Tenancy Disputes Act of 1924. The act sought prevention of disputes and arbitration of existing conflicts. A farm tenancy officer was assigned to each prefectural government to settle landlord-tenant difficulties before they became serious.

The act stressed arbitration, not prevention. It provided for the simple handling of disputes in court, avoiding the complicated procedure of civil law, for arbitration by a committee composed of a judge and several citizens. Either party or both parties jointly could apply for arbitration. Where parties to the dispute were numerous, they could appoint a representative or the court could order the appointment of such representation.

The court itself could settle a dispute or refer it to an arbitration committee. The court was bound to appoint the committee if requested by the disputants. This became the usual method. If the parties in dispute expressed satisfaction with the committee's decision, it became effective and was transferred to the court for formal approval. The court was bound to approve the decision unless it was "extraordinarily unjust." With court approval, the decision acquired the force of a court ruling.

The arbitration of disputes has been one of the really successful legislative actions to deal with the tenancy problem. While many settlements are reached by direct negotiations between the two parties, more than 60 percent of the disputes are settled in accordance with provisions of the act. A settlement may call for (a) reduction in rent, (b) payment of rent by a certain date, (c) dismissal of court proceedings upon payment of the rent, (d) staying of eviction, (e) written or oral apology from tenants to landlord or vice versa, and (f) division between landlord and tenants of expenses incurred in the conflict.

Such arbitration helped settle disputes. But it did not cure underlying causes of conflicts and discontent. Such problems require delving deeply into the economic and social conditions of the Japanese village. The government made only feeble attempts to do so. It preferred to promote an increase in agricultural production. Experimental stations were established. Improvements in fertilization, seed selection, and development of new varieties produced results. A Japanese economist described the program as "a chemical and botanical revolution" fostered by the government. With the government's concern for technical improvement of farming, little attention was paid to the farmer's welfare. Then in the 1930s the government became aware that something was amiss in rural Japan and that remedial policies were in order. But it dealt mostly with depression phenomena such as the low price of agricultural products. It avoided disturbing the established rural structure. Measures proposed or enacted to improve the lot of the tenants fall into two groups: (a) those concerning existing tenancy practices and (b) those concerning the raising of tenants to the status of peasant proprietors.

Measures to Improve Tenancy Practices

Growing landlord-tenant conflicts caused the government to organize a Commission on Tenancy to make proposals and recommendations. In 1929 the government's Commission on Social Policy recommended an agricultural tenancy bill similar to proposals made in 1926 by the Commission on Tenancy. The bill was introduced in the Diet in 1931, but it failed to overcome the opposition of the landlords and was never passed. The proposed law was constructive and would have aided the tenants. The following is a summary of the proposed legislation:

1. To guarantee tenants the benefits of short-term improvements, the minimum duration of a lease was to be five years. Unless six months' notice was given, a tenant was to continue to work the land even after termination of the lease. A tenant could not be removed from the land during an eviction appeal to the arbitration board. As much as a year's notice was required when tenancies were concluded for an indefinite period. Such notice was to be given after the harvesting of the principal crop and before sowing began.
2. Upon termination of a lease, the landlord was to buy the standing crop, remunerate

the tenant for farm improvements made with the landlord's consent, and pay him an indemnity for surrendering the right of tenancy to the land. The indemnity was not to exceed one year's rent.

3. Tenancies were also to be terminated by the landlord when no rent had been paid for a year or when a sum amounting to a year's rent had not been paid in two years or when any portion of the rent had not been paid in three years. Two months' notice demanding payment was to be given. If nonpayment was deliberate, two months' notice might terminate the tenancy. Against such demand for payment or such notice, the tenant might appeal to the arbitration board while tenancy continued. If the landlord contemplated the sale of land, he was required to give a month's notice to the tenant and allow him the right of preemption.
4. The tenant might claim a temporary reduction in rent on the ground of poor harvest. If a request for rent reduction was received, the landlord was required to inspect the crops fifteen days before harvesting began. If either party objected to this method, inspection was to be made by a public official or in a manner determined by a tenancy committee. The bill was tabled in the House of Peers.

In 1937 a new bill sought to curb the arbitrary manner in which landlords terminated leases. It was enacted on April 2, 1938, as the Agricultural Land Adjustment Act. But as approved it did not provide the tenants with needed relief and protection. Landlords could be compelled to negotiate with tenants for the sale of the land but could not be requested to sell the land. A landlord could not evict a tenant unless sufficient reason was given, but the broad interpretation of "sufficient reason" did not strengthen the position of the tenant with respect to leases.

No legislation had yet been enacted in Japan to redress the balance so heavily weighted in favor of the landlords. Because of the character of Japanese tenancy practices, any law regulating landlord-tenant relations had to be based on a downward revision of rents, payment of rent in cash rather than kind, and greater security of tenure. Landlords saw in such provisions an infringement of their rights as owners. They fought bitterly, and successfully, against any basic change.

Measures to Reduce Tenancy

The direct way of abolishing tenancy is to raise tenants to the status of land proprietors. The Japanese government's attempts along this line began on a very small scale in 1922. In the four years that followed, government loans to tenants for this purpose amounted to ¥14 million with interest ranging from 3 to 5 percent. A loan was not to exceed ¥600 and was to be redeemed in annual installments over a period of fifteen to twenty years. The area of land to be purchased by tenants was to average one-half acre.

Results were not impressive. The total amount of land purchased by tenants during 1922–25 was estimated at 17,000 acres. The program was expanded in 1926. A total of ¥468 million was to be advanced during the next twenty-five years at a 3.5 percent interest rate. Annual repayments were expected not to exceed the usual amount of rent. The plan provided for compulsory collection of the annual payments. In the event of nonpayment, the land would revert to the "rural cash office," a government agency in charge of financing the scheme. A Japanese economist declared, "provision exists for an entire or partial exemption of annual payments and for extension of the period of redemption if the lands have become desolate or the amount of crops is decreased through unavoidable causes."[27] It was thought, therefore, that nonpayment cases would be few.

Even this enlarged fund was not adequate to carry on land purchases on a large scale. At prevailing land prices, a total of only 287,000 acres could be acquired, or less than 5 percent of the entire acreage rented by tenants. Furthermore, the plan contained no provision compelling landlords to part with some of their land. A Japanese writer remarked, "The present organization for creating and maintaining peasant-

27. S. Kawada, "The Establishment and Maintenance of Peasant Farms," *Kyoto University Economic Review* (July 1, 1923), p. 81.

proprietors is very small in scale, and the methods adopted for its execution are very lukewarm."[28] This mild program had to overcome severe opposition on the ground that interest rates would tend to undermine the money market and that government sponsorship of this plan would involve it in serious disputes with tenants.

The idea of enabling tenants to become independent farmers was sound. But the process would take centuries under such a program. The government was aware of this. With agrarian unrest growing, another plan on a larger scale was proposed in 1932. This was embodied in a peasant proprietor's agricultural land bill, which did not pass. The bill provided for a bond issue in payment for the purchased land to be repaid by tenants in whose behalf land was acquired. The amount of bonds issued in any one year could not exceed ¥80 million or a total of ¥2,800 million in the course of the thirty-five–year period during which the scheme was to operate. The plan intended to convert 44,000 acres of rented land into proprietors' farms each year. It sought to turn over 1,543,000 acres to tenants in thirty-five years in addition to the 287,000 acres to be purchased under the twenty-five–year program. A total of 1,830,000 acres or 30 percent of the land farmed by tenants was expected to become their own property.

The bill provoked a great deal of criticism on the ground that it lacked effective provisions to fix the purchase price of agricultural land, that it carried no provision for compulsory sale of arable land, that it should have been accompanied or preceded by effective regulation of tenancy practices, and that it was a device to enable landowners to dispose of land that was a burden to them.[29]

The only plan in operation through 1936 was the one inaugurated in 1926. The government assisted 175,000 tenants in the purchase of 250,000 acres of land through the extension of ¥148 million in loans. Hokkaido was the principal center of this activity. During the later years the prefecture accounted for more than 50 percent of all the purchased land (table 23 [omitted]).

Considering the small beginnings of 1922–25, the annual average purchase of 23,000 acres subsequently would indicate considerable progress. But the progress was more apparent than real. A century would have been needed to acquire the land rented by full tenants alone and more than two centuries to establish farm ownership on all tenanted land. The 175,000 tenants who supposedly became owners or part owners are reflected to a small degree in Japanese statistics of the principal categories of farmers. Between 1926 and 1936, part owners increased from 2,314,000 to 2,349,000. But the number of owner cultivators remained unchanged. More significant, the number of tenants increased from 1,509,000 to 1,518,000. During 1931–36 when the amount of land purchased was largest and the greatest number of tenants aided, the owner cultivators declined by 25,000 and part owners by 35,000. Tenants increased by 23,000.

Official data showing that 175,000 tenants (1926–36) were aided in becoming peasant proprietors cannot be accepted at face value. Tenant movement into higher categories cannot be determined with certainty. It must have been counterbalanced by a greater movement of peasant proprietors into the tenancy class. Policies to convert tenants into owner cultivators are futile if the latter are permitted to lose their land and become tenants.

The inadequacy of the 1926 program was apparent to those who supported the tenants. The farmers were aided by the dominant political parties. When the military gave support, the agrarian problem, to quote ex-Premier Saito, immediately "caught fire." Rural discontent was partly responsible for the attempted assassination of Premier Hamaguchi on November 14, 1930, resulting in his death a few months later. The motives behind the subsequent bewildering series of assassinations of Japan's leading political figures[30] were tinged by a combination of military and agrarian interests.

28. Yoshinosuke Yagi, "The Current Land Problem and the Establishment of Peasant Proprietorship," *Kyoto University Economic Review* (December 1936), p. 72.

29. Ibid., p. 73.

30. Inouye, Minister of Finance, was murdered on February 9, 1932; Dan, managing director of the vast Mitsui interest, on March 5, 1932; and Premier Inoukai on May 15, 1932.

Agrarian reform became the watchword of every important group in the 1930s. The army wanted contented and peaceful villages. Close ties bound the army and agriculture and, as General Araki stated, "the agricultural population constitutes Japan's first line of defense." The manufacturing interests realized that they were suffering from the sharp decline in the farmers' purchasing power. They approved of agricultural reform because they wanted to dispel the belief that industrial progress had been achieved at the expense of agriculture. The professional class also welcomed reform. The industrial workers did, too, hoping that an improved economic status for farmers would reduce the flow of cheap labor to the cities.

Numerous programs were prepared. A twenty-five–year plan was inaugurated in 1937 to assist tenants in becoming peasant proprietors. The program was less ambitious than might have been expected. It was on a smaller scale than the contemplated 1932 plan in terms of the ¥1,000 million to be spent and the 1,021,000 acres to be purchased. Loans to farmers were made at an interest rate of 3.2 percent to be repaid in installments not exceeding current rates of rent. This program, like the two preceding, evaded the fundamental evils of the Japanese land system.

The statistical results are presented in table 23 [omitted]. From 1937 through 1943 less land was purchased and a smaller number of tenants were affected than in the seven years preceding. An unprecedented spurt of activity occurred during 1944. The tenants assisted in land purchase cannot be accounted for by changes in the number of owner cultivators, part owners, and tenants. The farmers' accumulation of cash and the difficulty many owners experienced to keep the land in cultivation undoubtedly aided the last minute increase in acquired acreage. In war as in peace, tenancy was a principal social and economic problem of Japanese agriculture.

The fact that the Japanese government failed to deal constructively with the problem does not mean that tenancy practices cannot be improved. The government emphasized such intangibles as self-help and spiritual regeneration and forgot badly needed subsidies and fair land prices. "The most important need of the village is not a state subsidy but initiative on the part of the farmer for economic rehabilitation through his own efforts," a former minister of finance said. But tenants could not raise their level by themselves. Drastic concessions by land owners and large state expenditures were needed. These would have caused a change in agrarian relations and a reduction in the huge military outlays of the 1930s. Unwilling to risk either, the government's efforts to aid tenants were of little practical significance.

Factors Underlying Land Reform

It remained for a defeated Japan to enact the Land Reform Law. The provisions of the law, their application, and their effects on Japan's farm economy will be the subject of another report. Many problems were considered in developing the program.

The land shortage makes the complete eradication of tenancy unlikely. The tenant's position could be improved by (a) assisting him to become owner of the land he tills and (b) improving the tenant status without altering existing titles. Only sustained action in these directions is likely to result in a more adequate share of agricultural income for the people who till the land.

Land Ownership among Tenants

The value of a reform that would institute peasant proprietorship cannot be denied. The fundamental cause of landlord-tenant disputes is the dependency of two sets of people on one piece of land and the unequal returns they share. Once tenants are converted into owners, the conflicts of interests would disappear. The national economy and social policy would be improved by peasant land ownership in place of pauperized tenancy. Ownership is the biggest asset in the development and maintenance of the productive power of the land. The Japanese are aware of the benefits of peasant proprietorship as indicated by the following statement: "Natural attachment of the newly established peasant-proprietors for their own land has become considerably stronger. They now manure more plentifully and improve their land more carefully than before and spare no effort to increase its productive power and improve the quality of its products. In many cases production has actually increased."[31] A defeated Japan is sorely in need of maximum soil productivity. The acquisition of land by most Japanese tenants therefore is most desirable. The real problems are: Who should make the acquisition, how much land should be acquired, and under what terms should a tenant obtain title to the land.

Normally the economic position of the Japanese farmer does not permit him to buy land. The state must purchase land for resale to the farmer or provide credit to permit the tenant to buy. The price of land must not be high, and it must be fixed by the national government. If the tenant and not the state were to deal with the landlord, he might be required to pay the fixed price plus whatever the landlord asked. Competition for land is too keen to expect tenant or landlord to abide by the official price. Such disregard would nullify the main purpose of land reform. The national government, therefore, must acquire the land from the landlords.

Moreover, individual negotiations between tenant and landlord in Japanese villages would delay completion of the task beyond the desirable period. Landlord obstructionism by delaying tactics would be the rule. Only central government action, setting purchase date and price, would insure transfer in good time. The national government, through central agencies such as the Ministry of Agriculture and Forestry and the Ministry of Finance, should assume the task of acquiring the land in behalf of the tenants. Local agencies could help distribute the land among tenants. Limited to this role, unsympathetic agricultural associations would find it hard to hinder the reform.

Landlords are unwilling to part with the land except at their own price. This makes compulsory sale of land necessary for an effective reform program. Failure of Japanese efforts since 1922 can be traced to the lack of condemnation provisions where landowners did not wish to sell. Compulsory sale has been adopted by other countries. It was introduced in Great Britain to speed tenancy reform.

The amount of land acquired from the landlords for resale to tenants depends largely upon the extent of the program. If complete abolition of tenancy is desired, it is necessary to purchase 2,600,000 acres leased by tenants who own no land and 3,800,000 acres leased by part owners and part tenants, or a total of 6,400,000 acres.

Compulsory sale of every acre of tenanted land is neither feasible nor desirable. Many landlords own small holdings rented out to tenants. They would request the right to cultivate some of their land to become peasant proprietors. There is no valid reason why such requests should not be granted. Moreover, purchase of every acre of tenanted land would create a tenantless Japan. This is not desirable for economic and social reasons. A certain amount of tenancy, with tenant rights well protected, is bound to persist as a form of land tenure. Legal prohibition of all tenancy, therefore, would not be in the interests of a flexible social structure in rural Japan.

The amount of land to be purchased for resale to tenants must be determined on the assumption that the furthering of land ownership among tenants would not preclude cultivation of some land by tenants. It is difficult to determine precisely how much land might be required in each case. If the size of individual farms and the total amount of farmland an individual landowner may own is limited to 3 cho (7.5 acres), approximately four-fifths or over 5 million acres would be available for government purchase resale. Over one million acres would remain for those remaining as tenants.

Price is the crucial element in government purchase of land. It could bring failure to an otherwise well-planned program. A Japanese writer noted that a successful plan to make owners out of tenants must be based on a reasonable price for peasant farmland.[32] High prices have been the main reason for the slow

31. Yoshinosuke Yagi, op. cit., p. 78; quotation from the Ministry of Agriculture and Forestry report, July 1935, on the results of the establishment of peasant proprietorship.

32. S. Kawada, "The Establishment and Maintenance of Peasant Farms," p. 95.

development of peasant proprietorship in Japan and other countries. Since 1922 the price of land in establishing peasant proprietorship was determined by free contract. The only legal limitation was that "the purchasing price of land should not exceed the standard price[33] and the usual price of the locality in which the land is situated."[34] The authorities aimed to prevent a further rise in prices. Even if they succeeded, and tenants bought land in accordance with official stipulation, the price was too high.

Two related factors are involved. First, the "standard price" and the "usual price of locality" are based on prevailing exorbitant tenant rents. "As the present farm rent is generally regarded as too high, the standard price worked out in this way cannot but be unduly high."[35] Second, keen competition for land is most important in determining high rent and the high land prices in Japan. This results in a high market price ("the usual price of locality") considerably above the productive value (capitalized earning power) of the land.[36]

The adverse effect of high land prices on a tenant-to-owner program was understood and criticized in Japanese circles outside the government. They maintained that "no success would be possible under such circumstances—that the plan was created to help landowners to get rid of undesirable lands, and to maintain the tenant rent and land value."[37] If a land reform program was to benefit by past mistakes of the old ones, the critics pointed out, the government should not fix prices on the basis of prevailing rents. The price of land should be determined in accordance with its productive value, they suggested. The critics asked that landowners be compelled to sell land on the basis of reduced rentals. The market price of land should be eliminated if the reform is to accomplish its purpose. Resale of land to tenants should include (a) small annual payments, (b) loans at low interest rates, (c) long period of repayment, and (d) state imposition of restriction upon utilization and disposition of the acquired land pending the full redemption of the loan.

Terms of sales should place the land securely in the hands of the tenants. If the sales plan required an appreciable initial payment or a short period of amortization, the new owners would have only a tenuous hold on the land. The new owner who feared foreclosure would have an instability of tenure similar to that of tenants. The owner who paid too high a proportion of his income to meet payments on his newly acquired land would be in the same position as a tenant who paid too high a proportion of his farm income as rent.

The 1926 program illustrated this fact. The annual payment on the principal was not to exceed the annual rent prior to the land purchase. The new small holder had to supplement this usually heavy charge by paying interest on the loan (at a low rate to be sure) and a land tax and other imposts not required of a tenant. Land ownership under such conditions failed to improve the low economic status of the farmer.

Fixed annual payments have helped nullify efforts of farmers to achieve permanent ownership. They have thereby increased tenancy rather than diminished it. A land ownership plan should provide flexible payments to allow for variations in crop yields and in prices received for agricultural products.

In countries where land reform programs have been carried out, conditions for payment

33. Determined by finding the capital value of the land by dividing the farm rent minus the land tax, land surtax, and similar local rates by a certain specified interest rate.

34. Ministry of Agriculture and Forestry, May 21, 1926, section 3, ordinance 10.

35. Yoshinosuke Yagi, op. cit., p. 80.

36. Shiroshi Nasu, in *Aspects of Japanese Agriculture*, pp. 129–30, states, "The actual price of land is usually higher than the rationalized land value calculated from the capitalization of profit. Among the reasons may be mentioned the social prestige attached to land ownership and the instinctive desire of the farmer to possess more land. The largest factor in determining the value of land is, however, the keen competition for acquisition of use of land, especially in a densely populated country like Japan. Thus the tenant farmer sometimes does not hesitate to propose an exceptionally high land rent at the expense of impairing his earnings from labor and other disbursements. Hence the comparatively high net revenue of the land owner. The equivalent of the capitalization of this revenue becomes the minimum price of the land. Usually, however, the actual price is considerably higher."

37. S. Kawada, "The Establishment and Maintenance of Peasant Farms," p. 98.

vary. They usually include long terms of amortization, large government loans at low interest rates, and deferment of payments to enable the new owner to build up a reserve and strengthen him in his new enterprise. The Japanese government considered the factors of successful foreign experience and are applying them with modifications in the new agrarian reform program.

Once tenants receive the purchase contract from the government, a rise of land values might induce them to pay off the loan and sell all or part of the land at a profit. Nonfarmers would take ownership. Provision should be made against this danger in any land reform plan. The Japanese government realizes that, to prevent such an occurrence, final payment should not be receivable until the end of the originally fixed amortization period, or the right to dispose of the property before the end of the amortization period should be prohibited. Exceptions to this rule will be at the discretion of the government.

Other legal limitations on the rights of possession of a farm acquired through the assistance of the state are necessary if the new owner is to be firmly established on the land. The new owner should carry on a type of farming that will maintain the fertility of the soil. He should not sublet or sell any part of the holding nor transfer his holdings freely by will to anyone but a natural heir. If he attempted to do so, the state should have the right to resume possession, buying the property at its original cost and reimbursing the occupant for any improvements he made. Or the owner could be given the right to dispose of his equity subject to state approval of the incoming purchaser.

The former tenant should not privately mortgage the holding until he has repaid his debt to the state. His credit should be restricted so that his holding will not be seized to recover a personal debt. Except for such restrictions, the new owner should enjoy all rights and privileges of landowners free from all obligations to the state.

Improvement of Tenancy Practices

Despite any attempt to spread ownership, a great deal of land in Japan will be still operated by tenants. Some tenancy is bound to develop even if all tenants and part tenants of Japan became owners. The factors leading to it have been summarized by Lossing Buck as follows:

> The development of business and industrial opportunities causes many owners to rent their farms and move to the city. A certain amount of renting results inevitably from changes to other business, retirement, illness, or death. Such renting seems desirable because it greatly aids individuals in these transitional periods. A man cannot always dispose of his land as soon as he changes from farming to another business. Division of land between brothers may make it desirable for one brother to rent his share to another brother. Heirs of farm land, again, may retain their land for sentimental reasons such as attachment to an old homestead and to the community and so prefer to rent rather than to sell. Farmers may be so limited in ability that they cannot completely manage a farm of their own.

Other reasons may be cited in support of renting land as against the idea of limiting land cultivation to owner cultivators only. A farmer might find himself with an insufficient amount of family labor to properly cultivate all the land he owns while another farmer would have excess of labor in terms of land he actually tills. The renting of an additional piece of land would meet the economic needs of the two. Some farmers may prefer not to invest in a title but utilize their capital for the purpose of cultivating a much larger acreage by renting rather than buying land. Perhaps the most telling factor is that given favorable tenancy conditions, many Japanese farmers would prefer tenancy to ownership.

Agrarian reform also should consider revision of the centuries-old tenancy practices. Even the best tenancy legislation may be ineffective unless the bargaining power of the tenant is improved. Bargaining equality is difficult to achieve so long as demand for land is greater than supply.

The tenants' economic position could be improved by these provisions: (a) long-term written leases, terminable by either party only after due notice is given as stipulated in the agreement; (b) fair rentals payable in cash; (c) payment for inconvenience or loss sustained by the other party when lease is terminated without cause; (d) limited tenant payments during emergencies such as serious crop failures or sudden fall of prices; (e) landlord and tenant sharing some of the expenses of raising crops (fertilizer, for instance); (f) removable improvements made by the tenant should be removed by him at the termination of the lease (compensation should be paid for improvements which the tenant cannot remove, such as soil and irrigation improvements); (g) measures leading to the increase of soil productivity; and (h) cheap credit, price stabilization, and improved marketing facilities.[38]

These elements of sound tenancy practices were practically unknown in Japan. Most rental contracts or leases have been oral. Terms are often vague and incomplete, leaving many important items dependent upon the uncertainties of memory or later discussions. A written lease is needed with terms completely stated and clearly understood by both tenant and landlord before signing. A long-term written lease, which provides for equitable rental and gives the tenant security and stability on the land he farms, can improve his economic status. Improved management also is important. A lease that gives the tenant stable and secure rights to the land for a definite period of time will encourage him to plan for the best development of the farm.

Measures taken by the Japanese government since 1945 have caused a drastic reduction in rents. But there is no assurance that the present provision will continue when the food crisis is past. High rentals have been an important factor in causing the difficult landlord-tenant relationship and the poverty of the tenant. The rent absorbed too much of the farm income for the tenant to operate economically.

The rental was based upon the strong bargaining position of the landlord growing out of intense competition for the relatively small cultivated area. Any improvement in tenancy practices called for a drastic revision in rents.

Equitable rent is difficult to determine in Japan. Probably the best method of determining a fair rental is to base it upon (a) the productive value of the land and (b) contributions made by each party in the way of supplies, equipment, taxes, and services. By listing investments and farm operating expenses and by assigning proper values to each, the relative importance of the contributions of each can be fairly well estimated. The division of income can then be calculated. Rents would probably have to be reduced about one-third or one-half, depending upon individual cases.

The benefits to tenants from such a reduction are evident. Prospect of a much-reduced income will force many small landowners to work their own land or hire labor. This will displace a number of tenants who have no alternative occupations. Rent reduction will also cause a decline in land value, a salutory effect. Nonfarmers will not be tempted to invest money in land at a time of declining land prices and uncertain returns. Land purchase by tenants might be facilitated thereby.

From the tenant's point of view, cash rent is a welcome departure from payment in kind. The majority of Japanese tenants prefer payments of a fixed cash rental per unit of paddy field as is done in payment for upland fields. With rent reduced, the tenant could bear the risk of price and production changes. If the tenant entered into a sliding-scale contract with the landlord, the two would share equitably in the benefits of rising farm prices. Equity also would prevail when prices fell to low levels. This plan provides a method by which cash rentals are based upon prices received by the tenant for his produce.

A tenant farmer whose lease contains "reform" provisions would be more inclined to buy equipment, keep buildings in good condition, manure more plentifully, and otherwise work his land more carefully than before. He would spare no effort to increase its productive power and improve the quality of its products. Experience of foreign countries, notably that of Britain, shows that many a farmer would rather rent than buy land if he is given a good lease. Equity in the land could be acquired without owning it, leaving capital free for oper-

38. These principles, incorporated in many programs for improved tenancy practices, are summed up in: Report of the Presidents' Committee, "Farm Tenancy," (Washington, February 1937).

ating expenses or major improvements rather than invested in a land title. The majority of Japanese tenants, as revealed by investigation, will choose land ownership. But some might elect to remain tenants under the improved tenancy practices instituted by the Japanese government. Agrarian reform in Japan should proceed simultaneously to improve the tenants' status while also extending ownership among tenants.

Limitations of Partial or Complete Agrarian Reform

The principal criterion of agrarian reform is the economic welfare of the Japanese peasantry. With improved tenancy practices and extended land ownership among tenants, economic benefits are great. The large proportion of the crop which the tenant previously has been required to deliver to the landlord would be retained for his own consumption or for sale. His increased purchasing power will benefit industry, the trades, and professions. However, the improved economic condition of the tenants should not be overestimated since there is not enough arable land in Japan for the large farm population.

If all cultivated land of Japan were equally distributed, the average holding would amount to 2.5 acres. This is nearly 2 acres short of an estimated profitable acreage. In suburban districts where truck farming is possible, the cultivation of 2 or 3 acres of land might produce an income large enough to support a farm family; in most districts, however, where farmers earn their living primarily in rice, other grains, vegetables, and mulberries, such small tracts will hardly support a family.

The lot of the tenants, as tenants, would be improved if the available acreage were large enough to enable them to double or triple their rented holdings. Acreage cannot be materially increased in Japan, unless the Japanese farm population declines sharply in numbers.

Changes in the Japanese agricultural economy through land reform will not ease the pressure on the land to any considerable degree. Given the present cropping system, land configuration, and equipment, Japan has been exploiting the available arable land fully. This is indicated by the high yields. The labor force used to maintain a maximum production is greater than required. A more economic utilization of agricultural resources in terms of land, labor, and equipment would add only to the surplus labor reserve. This would create the need for alternative occupations which the farm cannot provide. Therefore, it is important to alleviate the Japanese problem of "many people on little land" not only through agrarian reform but also through greater industrialization and expansion of commerce and other measures intended to absorb and control the surplus rural population.

Annex 1. Types of Written Agreements

Tenant's request for land

Location of the tenanted land . . . .
Number of plot of land sought . . . .
  Aza (buraku)[39] . . . .
  Village (Yamagata prefecture) . . . .
  Gun[40] . . . .
Acreage of the tenant land . . . .
  Tan[41] . . . . Se[42] . . . . Bu[43] . . . .
  Land value (yen) . . . .
  Rent paid in rice (bales—to be wrapped twice; each bale containing four "to"[44] of rice) . . . .

Having not enough land to cultivate, I earnestly request you to rent me your land as mentioned above, as well as shown in the attached map from the . . . . . .th year of Taisho Era.[45] If you grant my request, I shall never fail to pay

39. Aza—political term for buraku. A buraku is a social-economic unit consisting of about twenty households. One aza may include one or more buraku.

40. Gun—county unit.

41. One tan—0.245 of an acre or 1/10th of a cho.

42. One se—0.0245 of an acre or 1/10th of a tan.

43. One bu—0.000816 of an acre or 1/30th of a se.

44. One to—0.5119 U.S. bushels or 0.1 koku.

45. Taisho Era—from July 31, 1912, to December 25, 1926.

rent in rice of superior quality on or before the end of November of every year. If the payment is late, my guarantor will surely pay in my place, so that you may not suffer any loss or trouble through this delay. When using your land, I swear to you that I shall never trespass its boundary as specified in the map. You may at any time recover your land if you need it. I will sign this document for future reference:

Dated ..... in the .....th year of Taisho
Name and address of tenant ...... (Seal)[46]
Name and address of guarantor .... (Seal)
Address of the landowner ....
No. .... Aza .... Village .... Gun ....
His name ......Yamagata prefecture ....

Outline of tenancy contract

Location of the tenanted land ....
No. .... Aza ....
Village (Kanagawa prefecture) ....
Gun ....
Paddy field ...... tan .... se .... bu ....
Upland field ..... tan .... se .... bu ....
Paddy field ...... tan .... se .... bu ....

Total
Upland field ..... tan .... se .... bu ....

Rent paid in rice per annum
Unpolished rice ....
koku ..... to ..... sho ..... go .....

Terms of tenancy
Five years: From (month)
in the ....th year of Taisho
To (month)
in the ....th year of Taisho
Terms of tenancy may be extended upon expiration by mutual agreement.

1. Tenancy begins on April 1 of every year and ends on March 31 of the next year. Rice to be paid as rent shall be of B class, examined, and delivered to the place designated by you on or before December 25 of every year.
2. If there should be an unforeseen decrease in harvest caused by inevitable calamities, a discount rate of rent paid in rice shall be determined by mutual negotiation.
3. If tenancy is terminated at the will of the tenant before the agreement expires, the tenant must inform the landlord to that effect during the period from August to October in order to get his consent. The land shall be recovered in January of the following year.
4. If the agreement is terminated at will by the landowner, the tenant must be notified of the contemplated action three months in advance. The landlord may recover the land by paying a penalty which has been previously and properly determined by the two parties. The termination of the lease will take place only after the harvesting of crops.
5. The use of the land for purposes other than those specified or the transfer of the right of using the land to a third person is not permitted unless with permission of the landlord.
6. All taxes or public imposts on the tenant land shall be borne by the landowner and other expenses by the tenant.

Pledging you to abide by the contract as mentioned above, I will sign the document for future reference.

Dated .... in the .... year of Taisho
Name and address of the tenant (seal)
Name and address of the guarantor (seal)
Name and address of the guarantor (seal)

Name and address of the landowner

Complete tenancy contract

Revenue
Stamp (seal)

Location: No. 157, Tate aza, Shiki machi, Kitaadachi gun, Saitama prefecture.

Area: Paddy field 1 tan and 3 bu.

Rent: Paid in rice 2 bales (each containing 4 "to") and 2 "to" 1 "sho." I rent your above land under the contract as follows:

1. The above rent should be paid before December 20 every year.
2. The term of tenancy is from January 20 until January 20 next year. But, prior to the expiration of this term, if neither of

46. A Japanese name on a formal document is accompanied by the seal of the individual.

the parties proposes the dissolution of the contracts, it may continue for another year.

3. Tenant cannot transfer the lease nor sublease the land nor have the land cultivated by others.
4. Landowner can, at any time he desires, recover the tenant land regardless of the tenant's circumstances. The tenant cannot object.
5. When payment is delayed, both guarantor and tenant should pay the rent jointly.
6. Such acts as the change of footpaths between paddy fields or transformation of the land or transferring the soil of the field are prohibited.
7. It is agreed that inside of the embankment the medium rice maturing plants must be cultivated, and early rice maturing plants must be cultivated outside of the embankment. If the tenant does not carry out these terms, the rent must be paid as mentioned regardless of the tenant's difficulties.
8. Bales must be wrapped carefully as specified.
9. The rice for rent must be prepared carefully, never containing green or smashed or moist rice, etc. The rice must be dry. Even the unwrapped rice must be of the same quality.
10. Even if the rent is less than 4 "to" but more than 2 "to," the tenant must deliver it to the landlord wrapped in a bale.
11. Regardless of tenant's circumstances, if the total harvest is less than the specified rent, the tenant cannot request any reduction in rent.
12. If the tenant does not remove his farm equipment or other tools and objects immediately when the time comes for the tenant to return the land, the landowner can appropriate them or get others to remove them at the expense of the tenant. In such event the tenant cannot raise any objections, and also no objection will be made even if these objects are disposed of by the landowner.
13. When the time comes for the tenant to return the land to the landowner, the tenant cannot ask the landowner for any compensation.
14. If the tenant breaks the contract, it will lapse without notification or any other proceedings.
15. In case the payment of rent is delayed, the landowner can legally compel him to make the payment.
16. If the landowner transfers the land to another tenant, this contract will lapse automatically.

Pledging you to abide by the above contract, I will sign my name.

Dated January 21 in the thirteenth year of Showa (1938)

Tenant: No. 2720, Shiki machi, Kitaadachi gun, Saitama prefecture, *Kensaburo Mikami* (seal)

Guarantor: No. 2882, Shiki machi, Kitaadachi gun, Saitama prefecture, *Tokuiiro Nagashima* (seal)

Landowner: *Mr. Nishikawa*

9. *Trial Balance in Japan*

The special interest of this piece is that Ladejinsky undertook in it a broad assessment of the accomplishments of the occupation authorities, as he viewed these toward the end of 1948. In the light of the Allied Powers' aims for postwar Japan, Ladejinsky examined the occupation's programs and accomplishments in the fields of demilitarization, the adoption of a new constitution, the dissolution of holding companies, the enactment of antimonopoly legislation, the reconstruction of the labor movement, and educational as well as agrarian reforms. Concluding that "the military occupation as we have known it need not continue much longer," Ladejinsky envisioned a continuing need for a very different kind of occupation "made up primarily of experts, advisers and teachers." Some years later this sort of thing would be called the Point IV program of technical assistance. The concluding paragraphs of the article are presented here.

SUCH ARE THE ATTEMPTS to translate the aims of the occupation into reality. It is the tale of a moderate middle-class revolution, designed to create a stable system of capitalistic democracy. The changes that have been made have so loosened the fetters that held the Japanese people that reactionary forces would find it difficult to tighten them again. The American occupation has rendered an historical service in daring to give new direction to the economic and political arrangements of the defeated country. Such reforms as the widespread ownership of land, the recognition of the right of labor to economic security, and the measures for the control of disease and improvement of public health so effectively carried on by the Public Health and Welfare Section of SCAP have given tangible evidence to the common people of Japan that we are actively espousing a new way of life and not merely opposing the old. It is this that distinguishes the American occupation of Japan from the occupation of other defeated countries by victors in wars, past and present.

The road upon which the occupation has set the Japanese is sound and promising, and the Japanese have gone a considerable distance along it in the last three years. But a generation must pass before we can know how deeply its spirit has been accepted and how lasting its effects will be on Japan. The economic rehabilitation of Japan is essential to the strengthening of the social reforms introduced by the occupation. There must be a degree of economic well-being at least as great as was possible under the old régime. The reforms do not inhibit the fullest utilization of Japanese labor, managerial skill, and indigenous natural resources; nor are they of a kind which would prevent extension of American credits sufficient to cover imports of raw materials and the cost of new capital equipment as well as the investment of foreign capital.

But we must face the fact that, even if the economic climate is favorable, the reforms may not be wholeheartedly accepted. And here we touch on a crucial problem. In the past the keynote of government and society was authoritarianism. The spirit of Japanese culture has been communal and disciplinary. Tradition and group opinion, not individual initiative, have been the mainsprings of Japanese behavior. There is no reason to believe that the cake of custom is so hard that it cannot be broken, for the Japanese were amenable to indoctrination in other times and are more open to it now. The overwhelming defeat shook the established political and social order to its foundations and shattered many illusions, chief among them the belief in the unique mission of the Japanese people. The fact is that the Japanese are a practical and imitative people, and large groups are willing to follow a set of principles used by the country which is apparently the most successful in the world. But though the balance of forces in Japan's society after three years of the occupation favors the sloughing off of the worst excrescences of Japanese feudal life, the Japanese people have not themselves fought for these democratic reforms. They have been a gift. The Japanese can make them their own only by using them. The processes of democracy are not simple, and the Japanese cannot hope to employ them successfully without much further guidance. But the military occupation as we have known it need not continue much longer. If the Allied Powers are satisfied that the military structure of Japan has been dismantled (and there is no doubt that it has been), if the auxiliary military guarantees for the future have been fully secured, and if, as has been sufficiently demonstrated, the purely police problems are small and the overt opposition negligible, then there is no reason why the military occupation in its present form and extent cannot be terminated shortly.

What will be needed is an occupation which emphasizes training in running the newly established institutions. Now that we have taken the Japanese nation in tow, we cannot abruptly set it adrift; the main effort of education has only just begun. This is a task for an occupation made up primarily of experts, advisers, and teachers. American aims in Japan and the novel meaning imparted to them by SCAP constitute a test of the universality of certain Western precepts. The future of Japan as a useful member of the society of nations depends upon America's determination to continue aid and counsel to the Japanese.

10. *The Rent Reduction Program in Taiwan: Field Observations*

This was the first of three field studies Ladejinsky carried out for the Joint (U.S. and China) Commission on Rural Reconstruction (JCRR), typical in the logical nature of its organization and in the thoroughness of its execution. He exhibits fairness in weighing landlord complaints about the manner in which rents have been adjudicated and gives an immediate action response to the inadequacy of tenant representation on local land commissions. Concerning this, T. H. Shen of the JCRR wrote the editor as follows: "He (Ladejinsky) strongly recommended re-organization of the committees after the pattern of the Japanese land commissions where the tenant farmers accounted for at least one-half the total membership. The Government approved his proposal and put a new provision for this re-organization in the Rent Reduction Law on May 25, 1951, which stipulated that tenant membership should not be less than one-half of the total committee membership." In the opinion of Raymond Moyer, then a U.S. member of the JCRR, this proved to be a major contribution to the reform in Taiwan.

This paper took the form of a September 1949 memorandum addressed to Chiang Mon-Lin, then chairman of the Joint Commission on Rural Reconstruction. It appeared in the JCRR's General Report no. 1.

BETWEEN SEPTEMBER 9 AND 22 I TRAVELED extensively along the western coast of Taiwan in order to gather firsthand information on the application of the rent reduction program, which went into effect in April 1949. The rent reduction program stems from conditions under which the tenants worked the land and their economic and social position in the rural scheme of things. In order, therefore, to place the rent reduction program in its proper context, it might be useful, first, to touch upon tenancy conditions as they existed prior to the promulgation of the program and, second, to outline the principal provisions of the rent reduction regulations. Against this background the evaluation of the progress of the program as observed on the trip will be made in the subsequent paragraphs.

Land Tenure Condition

Agriculture is the backbone of the economy of Taiwan. Before World War II the value of the crops in addition to sugar and other processed foodstuffs made up four-fifths of the total value of the island's production. These commodities also accounted for four-fifths of the total value of exports. Nearly 60 percent of the population is directly involved in agricultural production. Despite this important role played by agriculture in Taiwan's economy, the industry does not spell anything like prosperity to the great majority of the farmers. On the contrary, the field trip just completed made it abundantly clear to me that most of Taiwan's farmers suffer from undernourishment, disease, and poverty.

Many factors have contributed to this, but

the principal one is that in Taiwan, like in many other regions of the Far East, there are too many farmers and too little land. Taiwan has a total of 823,000 hectares cultivated by 530,000 farm families—an average of 1.5 hectares per family. In the majority of cases, the actual size of a cultivated holding is smaller than the average. Fully 25 percent of all the farm families cultivate less than half a hectare and 20 percent cultivate from one-half to one. Only 26 percent cultivate from one to two hectares, and those who cultivate more than 10 account for less than 1 percent of all farm families.

As for land ownership, almost 60 percent of all the farm families of Taiwan own less than a hectare per family and the total amount of land they own is only 14 percent of all the cultivated area of Taiwan. On the other side of the scale, less than 3 percent of the farmers own 10 hectares or more. This group has 36 percent of all the land. In short, a great deal of land is concentrated in relatively few hands, and the holding cultivated by a Taiwanese farmer has little relationship to the amount of land he owns. Because of the unequal distribution of land ownership, many farmers own no land at all and must cultivate land that belongs to others. Some own so little that in order to improve their economic status they are compelled to rent additional land. This explains the three groups of farmers in Taiwan: owners, part owners and part tenants, and tenants. At the present time 32 percent of the farmers are owners, 28 percent are part owners and part tenants, while those who have no land at all represent 39 percent of all the farmers. The tenanted acreage accounts for 54 percent of the total.

Tenancy as an agrarian institution is not an evil, but it becomes one when tenancy conditions are heavily weighted in favor of landlord against tenant. Taiwan is a case in point, as my trip throughout the island and talks with the farmers and landlords so clearly prove.

A sound tenancy system is a fair, joint partnership of landlord and tenant which provides the tenant with a measure of economic well-being and security of tenure. Neither consideration held true in Taiwan prior to the rent reduction program. In discussing the tenant situation with landlords, I did not get the impression that as a group they are bent on squeezing all that they can get out of the tenants—but they have taken full advantage of the hunger and the intensity of competition for the limited amount of land and the cultivation right of that land. This resulted in land values and rental rates far above the level justified by the productivity of the soil. It resulted also in a situation where the tenants have been willing to bind themselves to the land regardless of the terms of tenure.

At the farm meetings that I attended, I raised numerous questions on the subject of tenancy conditions prior to rent reduction. With one or two exceptions, the answers were given freely and I have no reason for doubting their accuracy. The landlords present in those meetings did not question the evidence furnished by tenants, and their own answers to my queries checked with those of the tenants. The general pattern of the conditions left no doubt in my mind that the improvement in the position of tenant farmers is one of the very important factors in improving the agricultural economy of Taiwan.

There was a widely prevalent view that a tenant in Taiwan paid 50 percent of the crop as rental. This notion is not altogether correct. In most of the rural districts visited, and particularly in such important rice-growing districts as Taichung and Tainan, a tenant paid more than 50 percent. The following figures gathered at Panchao District are typical of the situation in most of the districts.

YIELDS BY LAND CLASSIFICATION AND RENTS BEFORE RENT REDUCTION

Yields	*Catties per hectare (rent)*	*Percentage*
First-grade land (10,000 catties)	6,000	60
Second-grade land (9,000 catties)	5,000	55
Third-grade land (8,000 catties)	4,500	56

Rentals as high as 60 to 70 percent of the crop are not unknown. In one district it was estimated that only between 5 to 8 percent of the tenants pay less than 40 percent of the crop.

The rental is not the only charge on the tenant. He must buy very expensive chemical fertilizer; pay for part of the irrigation fees; and provide his own seed, equipment, farmhouse, and farm buildings. Only in two out of the eleven districts visited was I told that landlords share with the tenants in the cost of chemical fertilizer. I tried to find out what the rent, fertilizer, and water fees mean in terms of a volume of rice harvested. The replies seem to indicate that they quite frequently represent as much as 70 to 75 percent of the crop. The share from which the tenant draws sustenance represents not more than 25 to 30 percent of the crop.

The rent paid by a tenant is not a share of the crop but is one of so many catties of rice per hectare of land regardless what the yield might be. I have come across cases of landlords who willingly reduce rents in years of bad crops, but this is not the usual procedure. Normally, when a fixed amount of rent is paid, neither exemption nor reduction is recognized. It is fixed and unchangeable, and such a rent is known among the farmers as "death rent."

Rent represents only a part of payment that a tenant has to make to his landlord. The other payment is the so-called "deposit" or "key money" every tenant must make to the landlord for securing the lease. All the tenants questioned concerning the deposit system unanimously confirmed its existence. The size of the deposit varied from one- to two-years value of the rental. The deposit is kept by the landlord as a security fund as long as the lease is in force. If a tenant fails to pay part or all of the rental when due, the landlord has the right to deduct the unpaid rental from the deposit. The deposit system has prevented many poor farm laborers from becoming tenants, let alone owner farmers.

The rental and other charges would be considered high even if the tenant rented a holding of 10, 20, or more hectares. Actually, and very significantly, 35 percent of the tenants rent less than one-half of a hectare of land. In Wan-Li, for instance, thirty-five out of thirty-six tenants present at the meeting cultivated less than a hectare; and at Wu-Hong the average amount of cultivated land was 0.8 hectare. Assuming that all of the land is paddy rice land and high-yielding land—both ideal conditions but not always present—the tenant will harvest from 11,000 to 12,000 catties of rice. Rental and fertilizer deduction will reduce his share to between 3,500 or 4,500 catties. Considering that a family of six requires 3,500 catties of rice, the tenant would have little or no rice left for sale in order to cover other living expenditures. In reality, of course, he cannot afford to eat all the rice he needs; he subsists on potatoes and other cheaper foods.

The consumption pattern was made quite clear to me by the farmers. When asked how many of them had enough rice to take care of their needs between crop seasons, the usual answer was that a great many had little or none. It is rather difficult to reduce the replies to statistical figures, but my impression is that in the rice-growing areas probably one-third of the tenants fall into that category. In the southernmost part of the island in the village of Shin-Yin, this group constituted approximately 70 percent of all the farmers.

Without attempting to express any moral judgment on the landlord-tenant relationships, one is warranted in saying that the tenant gets too little and the landlord gets too much. As the tenancy practice was examined and sifted, it became obvious that prior to rent reduction the tenants as a group suffered from insecurity of tenure. Under conditions of keen competition for a limited acreage, the right of the tenant to remain on the land is of utmost importance. Written contracts before the rent reduction program went into effect were few and far between. The life of the lease in a written contract ranged from one to three years, the yearly lease being the most common type. Oral contracts prevailed with no duration of tenancy specified. Whether written or oral, the leasing arrangements afforded the tenants little protection. I talked with a tenant who had cultivated the same piece of land for nearly half a century (incidentally, he assured me that he could not afford to buy that piece of land even if he worked it for another fifty years), but I have listened to many more who complained about the case with which their landlords had cancelled leases in order to get a higher rental from other tenants. Some agreements permitted the landlord to recover the land before the expiration of the contract with no compensation to the tenant for the inconvenience. Other agreements called for payment, but in actual practice few tenants ever received compensation. Whether the tenant was compensated or not, he lost the right to cultivate the land.

What tenancy in Taiwan meant in terms of economic well-being prior to rent reduction cannot be determined statistically. The few references available are out of date, and no attempt at such studies has been made in recent years. The lack of this type of information, however, offers no insuperable difficulty in evaluating tenants' economic status. Even a brief visit to a tenant's homestead throws much light on this point, and I have made many such visits in the course of my trip. What they have most are children; what they have least are things that spell material well-being. Of all the farmyards I have seen in the Far East, Southeast Asia, and in the Middle East, that of the average Taiwanese tenant is among the worst, both in appearance and in equipment. Tenants' huts, so-called barnyards, equipment, and livestock, as well as their health point to nothing but poverty.

The touchstone of a tenancy system is whether a tenant, if willing, can become an owner cultivator. Historical data and my own discussions with farmers reveal that in Taiwan the shift is most difficult. Tenants and landlords were unanimous on one point: The income of the average tenant is so small that he cannot buy—exceptions notwithstanding—the rented half hectare, hectare, or 2 hectares of land. I encountered only one exception, that of fifteen tenants who had pooled their own and borrowed resources and acquired a total of 8 hectares of land. The question whether tenants want to buy land was greeted in many places with good-humored laughter. They would then turn the tables on the interrogator and ask him how he would go about buying land with no money. Many a time a workworn tenant stated that, if he lived twice as long and worked twice as hard, he could not save enough to acquire the land he now cultivates as a tenant. The number of tenants who succeeded to the status of owner cultivator were very few. The successful ones were beneficiaries of the following conditions: a benevolent landlord; a larger than usual tract of good land; and a small family or a large family with a preponderance of hard-working adults. This was amply confirmed by the farmers during the field trip. The overwhelming majority of the tenants of Taiwan have never experienced a climb up the agricultural ladder, from tenant to owner cultivator, not to say from tenant to landlord. The farthest they climbed is from farm laborer to tenant. Having reached that situation, the chances are that they will always remain tenants. Therein lies the greatest weakness of the agricultural system of Taiwan.

The tenancy practices in Taiwan prior to rent reduction assume added significances when one considers the great number of people adversely affected by them. Two-thirds of all the farm families—tenants and part owners and part tenants—are subjected to the inequities of the system. The inequities are not only economic, they are also social and political. One does not have to be a believer in the theory that a man's economic position determines his social and political status, but this is certainly true in tradition-bound rural Taiwan. One need not be primed to the existing conditions by elaborate studies to realize that the community is divided into first-class (the owners) and second-class citizens (the tenants). It is sufficient to attend a meeting of landowners and tenants and a meeting made up exclusively of tenants to realize who rules. The attitude of the officials—of every level—toward the various types of farmers is also symptomatic of the fact that as long as the tenant continues to remain in the lowly economic position, the social barriers within the community will persist. Under these conditions it is idle to seek for any semblance of economic, political, and social stability in rural Taiwan. What the searcher is more likely to find is fertile ground for political extremism and civil dissension.

Effort to Improve Tenancy Conditions

Such were the principal tenancy practices and their consequences prior to the rent reduction program inaugurated in April 1949. The system is Chinese in origin and closely resembles land-tenure arrangements in present-day China. Neither before nor after the occupation of Taiwan by the Japanese has the land tenure system been changed in any way. It is for the first time, therefore, that an effort is being made to revise some of the worst features of the system through rent reduction and greater security of tenure.

The program rests on a set of regulations issued by the provincial government of Taiwan in order to enforce the rent reduction program of the executive Yuan. The main provision of that program is that farm rent must not exceed 37.5 percent of the total annual output of the main crops—those customarily accepted in payment for rent. If the rent is less than 37.5 percent it remains unchanged. In the event the yield is less than 20 percent of normal, the tenant is free from rent payment.

For the purpose of rent collection, all the cultivated acreage is divided into twenty-six grades, and a standard yield is determined for each grade. If grade I yields 10,000 catties, the rental is 3,750 catties; if grade V yields 6,000 catties, the rental is 2,250; and so forth. Rentals, therefore, are not to be collected on the basis of actual yield of every rented plot of land. Rents must not be collected in advance. In cases where the landlord supplies the tenant with work animals, seed, and fertilizer, he may collect an additional charge not exceeding 10 percent of the value of those items. Irrigation costs are to be borne by the landlord and tenant in the following manner: Special charges in connection with the improvements of the canals, dams, pumping facilities, and so forth are to be paid by the landlord; ordinary year-in-and-year-out fees for the use of water should be paid by the tenant as an item in the cost of production.

Deposit or key money must not exceed one-fourth of the annual rent. Anything above that must be returned by the landlord to the tenant. The tenant is to receive an appropriate interest on the deposit; the accumulated interest should become part of the rent and deducted from it annually.

The landlord's former right not to renew a lease for the purpose of rerenting it to another tenant is outlawed. The landlord's right to dispossess the tenant for the purpose of cultivating the land himself is restricted. He is permitted to take back (after the normal expiration of the lease) only part of the rented land subject to the consent of the "37.5 percent rent campaign committee" and final approval of the provincial government. The approval in turn rests on the following conditions: That he must cultivate the land himself in order to provide a living for his family and that the land is within 3 miles of his residence. In the event a landlord wishes to sell or mortgage the land, he must notify the tenant of the acreage and price involved and inquire if he wishes to buy the land. The landlord may sell the land to another person if the tenant fails to reply within ten days of notification.

Old contracts, whether written or oral, are invalid. New written agreements must take place of the old ones and must be registered with the proper authorities. In addition to being of at least three-years' duration, every contract must contain the following main points: (a) the crops grown and amount of rental; (b) the amount of the deposit money, not to exceed one-fourth of the annual rental; (c) types of water fees paid by landlord and tenants; and (d) the extra rental received by landlord if the tenant is furnished with the means of production, such as work animals, seed, fertilizer, and farm equipment.

The instrument of enforcing the program is centered in the Enforcement Committee for Farm Rent Reduction. A committee is composed of the following fifteen or nineteen members: (a) magistrate or mayor and land administration chief of the district or municipal government; (b) one representative from each of the following organizations—law court, district or municipal council, and local farm associations; (c) three to five principals of secondary schools and three persons of high standing not connected with government or farm organizations; and (d) two landlords, two owner cultivators, and two tenants.

I familiarized myself with the provisions of the rent reduction program in anticipation of the field trip. I was not unmindful of the inadequacy of some of the items, as I shall have occasion to point them out later on, yet the body of the regulations as a whole seemed reasonable and adequate to meet the purpose for which they were designed. But even the ideal law is no better than the manner of its application and enforcement. And in China a great deal of agrarian legislation has remained on the statute books since Sun Yat-sen's days without a thought of implementation. This explains my skepticism about the rent program in Taiwan and the little credence I placed in official statements that came to my attention and in the bright pictures painted to me by the official in charge of the program. Firsthand observation of the application of the program seemed the best way of gaging the status of the rent reduction program. With that in mind I visited every important agricultural region of the island.

Methods of Field Observation

In the course of the trip a total of twelve meetings were held, each meeting lasting an average of three hours and each meeting attended by a minimum of 40 and by as many as 150 farmers. The assembled represented more than a village; they usually represented a district made up of a number of villages. In all but one instance, the farm groups were composed of landlords, tenants, and owner cultivators. Since there are many more tenants than there are landlords and owner cultivators combined, it is but natural that the tenants were in the majority. In the instance referred to above, the group was made up of tenants only. The gathering of this group was not intended as a measure of discrimination in favor of the tenants as against the landlords but merely to find out if tenants would state their views with greater freedom and ease when not inhibited by the presence of their landlords. This experiment was induced by the observation that on more than one occasion many a tenant's attention was divided between the "investigator" and the landlord, not to mention a side glance at the few local officials usually in attendance. It fully confirmed the suspicion that, without the benefit of the presence of the landlords, the tenants were more loquacious and the process of eliciting meaningful information was smoother and more successful.

In addition to group meetings, I talked with individual tenants and small groups of tenants in the fields, the rice mills, in the fertilizer distribution centers, in wayside eating places, and in "general stores" where farmers the world over gather to discuss the events of the day. A special effort was made to secure the reaction of individual landlords, big and small, to the rent reduction program. I have had a number of talks with such landlords as members of the premier landowner family of Taiwan, the Lin family, whose founder succeeded in obtaining all the land "within the sight of his dwelling." It should be noted that the gentleman in question was not doing things by halves; he succeeded in accumulating 10,000 hectares of land. I also discussed the reform program with landlords who deserve that designation by courtesy rather than because of the amount of land they own. The scale of operations of an average Taiwanese farmer is so small and the demand for land so great that one need not own a hundred or even 10 hectares to become a rent collector. A couple of hectares might serve the purpose. Nor were the officials concerned with the carrying out of the program neglected. As a group they were the most talkative, eager to help, but not too informative. All in all, in the search for an answer on the status of the rent reduction program, every source of information available in the field was tapped with various degrees of success.

Observations and Conclusions

At the time of this writing the program is only five months old. Its test at this early stage is not whether all its pertinent parts are being carried out but whether most of the tenants paid their rentals for the first 1949 crops on the basis of 37.5 percent instead of on the traditional basis of 50 or 60 percent of the main crops. The answer is that the great majority of the tenants availed themselves of the opportunity to pay rent in accordance with the new formula. The conclusion I therefore reached upon the completion of the field trip is that the rent reduction program is being carried out successfully.

Many of the provisions of the rent reduction program are unknown to most of the tenants. I have not met one who has read all the regulations, but it is equally true that all of them are thoroughly familiar with the all-important 37.5 percent part of the program. I know of no exception. Of course, to know of this provision does not necessarily mean living up to it, and a tenant's statement that he paid 37.5 percent of the main crop as rental was not by itself acceptable as bona fide evidence. By questioning tenants regarding their crop yields and rents in the past and in 1949, I concluded that most of them honor the provision. The tenants were not prepared for such on-the-spot queries, and I am certain that they were not improvising answers merely to prove that they comply with the main provision of the program. My notes contain a great many answers in terms of so many catties of output per unit of land and so many catties paid out as rental. The pattern reveals payment of rent in accordance with the provisions of the law. As a matter of fact, in many instances the rentals this year were even less than might have been expected under the 37.5 percent provision. I shall have occasion to discuss this point elsewhere when the subject of the chief grievance of the landlords against the program is considered.

The benefits of the rent reduction program become obvious as soon as one begins to compare the rentals before and after the enactment of the program. The comparisons show a difference in favor of the tenant ranging between 1,000 and 2,000 catties of rice from the first crop. More often the figure closer to 2,000 and 1,500 catties seems to be a reasonable average. Tenants who raise two rice crops a year will benefit to the extent of approximately 3,000 to 4,000 catties. The significance of the gain cannot be overestimated. A basket of 50 to 60 catties of rice has a definite place in the life of a tenant who is habitually short of rice between crops. It may mean a week's supply of food for a family or 50 to 60 catties of the indispensable chemical fertilizer. When hard pressed, as is usually the case, he borrows a basketful of rice at a 15 percent monthly interest rate, thus depleting still further his supply in the coming season. Now for the first time he will have an additional twenty or forty baskets of rice, thus assuring enough for his family and for a supply of fertilizer.

Despite the advantages that a tenant derives from the application of the rent reduction provision, not all of them have availed themselves of the opportunity. Seldom did a tenant admit paying rent at the old rate, but they all know about tenants who engage in secret deals with their landlord in order to evade the intent of the 37.5 percent provision. In one village a courageous tenant stood up and related his "deal" with the land, which meant a nominal payment of 37.5 percent but the customary rent in practice. On second thought he changed his mind as he realized that he was placing himself in the position of a double loser, paying the old high rent as against the new lower rent and at the same time subjecting himself to criminal prosecution in the event the illegal agreement with the landlord came to light. And so he said that he had reported the case to the authorities. His account was being interrupted by hostile shouts from all over the hall. As nearly as I could make out, the assembled group felt that such episodes should not be paraded before outsiders; they had best be settled within the family. There are a number of other evasions that one learns about as he shifts from village to village, questioning farmers on the progress of the rent reduction program—illegal dispossession of tenants; failure to change the old contracts for new; changes in the wording of the contracts; advanced payment of rent; and irregularities in the payment of the water fees, in assigning correct yields, and in calculating rentals in accordance with the new regulations. All these, in addition to the first one mentioned, constitute the catalogue of evasions. Some of the evasions are landlord made (dispossession) while others (black market rent) are committed by landlords in connivance with the tenants.

When one considers the fact that never before in the history of Taiwan was any effort made to ease the burden of the tenants and that the tradition of landlordism and its domination of the countryside has been unimpaired for many generations past, then there is nothing surprising about deviations of the kind referred to. That in itself is no cause for alarm. The real question is how widespread the evasions are and whether they can undermine the program. The answer is that they can neither endanger the life of the program nor impede its progress to any considerable degree.

Because tenants are reticent about illegal dealings, I cannot venture a statistical estimate of the magnitude of the evasions. The investigations of the provincial government, however, throw considerable light on this subject. The August report to the JCRR shows the corrected evasions by prefecture and for Taiwan as a whole. Some of the figures appear high. The 14,000 instances of failure to renew contract or the 10,000 instances of incorrect (purposeful or otherwise) estimates of the size of the rent are cases in point. Many more cases have not been examined and corrected simply because they have not been uncovered.

The significance of this information need not be exaggerated. As against the "no contract" farmers, 285,000 tenants were making contract arrangements. Even if the cases of "no contract" and "incorrect calculation of rent payments" were twice or three times as large, they would account for a small percentage in both instances.

By far the most serious evasion is the black market rent. The report registers 1,800 such cases. Actually the number is much larger. But in that event, too, there is no reason to fear for the success of the program. Assuming that the evaders of legal rent are not 1,800 but 10,000 or 20,000 strong, they would still represent approximately 3.5 or 7 percent of all the tenant families of Taiwan. Above all, I am of the opinion that the current trend of the program points to greater compliance and away from the emasculation of the basic provision of the rent reduction program.

The rent evasions were not expected. The close relationship between landlord and tenant and particularly the fear to incur the displeasure of the former who still owns the land the tenant cultivates is the main cause of the evasions. A good many tenants look critically upon government actions even if designed to benefit them. They must be convinced in a very practical way, and over a long period of time, that the efforts of the government in their behalf are not in the nature of "Greeks bearing gifts." Hence the cautiousness, bordering on mistrust, with which some tenants approach the rent reduction program. In meeting after meeting, I raised the question if in their opinion the program will "stick." The answer was typical: "Yes, if the government so desires." But there was considerable doubt in the minds of the tenants whether the governments succeeding the present one will "so desire." And so long as tenants entertain such doubts, there will be a number of "fence sitters" who will continue to pay the traditional rent—just in case.

It is very likely, however, that as the payment of rent for the second crop draws nearer (December–January), the number of "wait and see" tenants will decline considerably. They note with more than academic interest that the great majority of their neighbors are getting ready to pay a much reduced rental for the second time. The most ignorant of them can calculate the difference between having or not having an additional 1,000 catties or more of rice, and the presumption is that the "economic man" will prevail.

Tenants may be quite right in questioning the interest of the future government of Taiwan in land reform matters, but the current very vigorous efforts by the Taiwan provincial government is there for all to see, and it cannot be underestimated. There has been no slackening of the drive between the two crop seasons (July to November) to insure the success of the program. Its net effect is the weakening of the landlord opposition and the growing conviction among the tenants that the program and its benefits are there to stay. My own impressions gathered in the course of the field trip lead me to the view that most tenants already have that conviction. This explains why the evasions are not of a magnitude that can impede the program, let alone undermine it. The meetings attended by large and small groups of farmers or talks with individual farmers seem to indicate that the great majority of them have given the program a hearty approval—and not in a perfunctory manner. One might say that they voted for the program by paying 37.5 percent of the main crops, compared with the customary rents. The rent reduction program stood the test well in its infancy: there is reason to believe that it will pass the test even better as it matures as the rent payment for the second crop takes place.

One of the questions raised at every meeting is the way in which tenants utilize the benefits derived from rent reduction. Judging by the answers, the extra catties of rice will be used to improve living conditions and to raise agricultural production. More specifically, tenants will eat more rice; they will be in a better position to repair their homes and farm buildings, add to or renew their transportation facilities, give their children a measure of schooling, buy more fertilizer, farm implements, livestock, and discharge a small debt. In discussing the benefits and their utilization, a few voices were raised on the subject of small versus "big" tenants. It was stated that the big tenants are presumably the real beneficiaries of rent reduction; that is, they will be able now to acquire land.

It is true, of course, that the bigger the tenant the greater his benefits under the program. The fact is, however, that there is no reason to believe that many tenants will be able to purchase land. The exceptions will merely prove the rule.

Prior to the enactment of the rent reduction program, a Taiwanese landlord, normally, had no land for sale. On the contrary, he was in the business of buying land and converting small owner farmers into tenants. This attitude has undergone a basic change since the introduction of the rent reduction program. I heard many a landlord say that he wishes to sell all or part of his land, the only stumbling block being lack of buyers. These were not mere statements for tenants, too, were pointing out names of landlords unsuccessfully trying to sell land. Some of the big landlords disposed of their land in anticipation of some kind of land reform. A good example of this is Lin Tsu Wei, one of the biggest landowners of Taiwan. In an interview he told me that, anticipating that some changes in the land tenure arrangements were inevitable —and at the expense of the landowners—he decided to sell his land. And so he did, and at the peak of the market, in late 1948. He received 15 ounces of gold per hectare of land. In the fall of 1949 similar land was vainly offered for sale at 9 to 10 ounces of gold. The significant point about the sales of land prior to rent reduction was that merchants and other city people of means were the buyers. Tenant participation was rather limited. Mr. Lin's case is illustrative of the situation: He told me that 90 percent of his land was bought by merchants and only 10 percent went to the tenants. And even the decline in prices and the benefits of rent reduction have not effected any change in the buying, or rather lack of buying, capacity on the part of the tenants. It is worth noting that in the six months after the inauguration of the rent reduction program the tenants acquired only 537 hectares of land. In short, the rent reduction benefits, real and significant though they are, will have little effect upon the innermost desire of every tenant in Taiwan—to own the land he cultivates.

While the tenants fully realize that ownership of land is beyond their means, at least for the time being, they all want the next best thing—security of tenure. This point was emphasized in meeting after meeting, and the tenants felt that the new contract under the rent reduction program falls short of security of tenure. The three-year life of a contract was much too short, they insisted. Tenants would like to see the tenure provision changed from three to six years. The argument for the extension of the life of a lease is unimpeachable. The willingness to invest in improving the land and raising agricultural production would be materially heightened. I could not help but agree with them, since it is a well-known fact that the longer the duration of the lease the better the chances are that the tenants will put the land to the best possible use. Dr. Tang of JCRR and the provincial officials responsible for the program are cognizant of this situation. Since the extension of the contract from three to six years would benefit both tenant and landlord, there is no reason that the duration of the contract, as part of the reduction program, should not be revised accordingly.

In 1947 the total cultivated area of Taiwan was estimated at 834,000 hectares. The area under irrigation was placed at 600,000 hectares, 72 percent of the total arable land. It follows that irrigation in Taiwan is as important as fertilizers as a means of increasing agricultural production. The rice and sugar culture of the island, the basis of its wealth, is dependent upon maintaining the existing and constructing new irrigation facilities. In the years prior to rent reduction, irrigation facilities have been well maintained, but in certain circles in Taipei there was much concern about the state of these facilities in consequence of the application of the rent program. It was felt that the landlords, with reduced incomes, will lose interest in maintaining the old and adding new irrigation facilities.

The matter of the future of irrigation came up for serious discussion at every farm meeting; and if I am to take the statements of the tenants and landlords at their face value, a deterioration in the irrigation facilities need not be anticipated. The financial burden of providing water for the fields was formerly carried by the landlords, but according to the new regulations irrigation expenses are divided into two categories: (a) "special water fees," paid by the landlords, to cover maintenance and repairs of dikes and water channels and (b) "ordinary water fees," paid by the tenants, to cover the water supply for irrigation. The provincial government apportioned the expenses in this manner on the ground that the landlord who owns the land must see to it that the overall productivity of the land is maintained, while the tenant's payment for the water is just another item in his cost of production. The tenants questioned on this point have taken no exception to this expense. They are fully aware of the reduced income of the landlords and, above all, of the importance of irrigation and are willing to pay for it. Their real objection is that, although tenants (as well as the landlords) must make payment to the water conservancy committees, which determine the size of the fee, they are not as yet represented on the committees. It is something in the nature of objecting to taxation without representation, and tenants fear that committees dominated by the landlords will pass on to tenants some of the "special" fees in one way or another. It is obvious that once the problem of representation is taken care of, tenants would bear the cost uncomplainingly, even if fees had to be revised upwards. The provincial government is aware of the fact that the irrigation expense provision of the rent reduction program calls for changes in the composition of the water conservancy committees, which until recently were made up by landlords only.

Landlords are less happy about the irrigation provision than tenants are. Neither in the presence of tenants nor privately have landlords objected to their share of the charge as far as repairs and maintenance of the existing facilities are concerned, but I was left with the impression, nevertheless, that they would much rather be spared the charge now that their income is reduced. On the other hand, it is also true that, like it or not, the landlords will have to keep the irrigation facilities in good condition in order to maintain the productive capacity and market value of their land. The tenants who expressed these views actually spoke for the landlords who understand the economics of irrigation just as well as the tenants do. The conclusion that landlords will maintain the irrigation facilities and tenants will pay for the water supply is therefore warranted.

The situation described above refers to observation gleaned from a field trip in September. Since then new water fee standards have been set up, and a perusal of the list of landlord-tenant irrigation fees indicates that the share of the tenants has been increased. They must bear, for instance, 20 percent of the cost of "restoration of damaged construction" and all of the cost of "maintenance of small-scale construction." As in the past, the cost of new construction projects will be charged to the landlords, but whether they will be willing or able to develop new irrigation facilities is open to question.

In order to facilitate the enforcement of the rent reduction program, its authors provided for the creation of special rent campaign committees, composed of landlords, tenants, owner cultivators, school principals, village heads, police officials, and so forth. Committees of this kind, bent on the execution of the program, can be a great boon not only for the immediate purpose for which they were designed but also for the general improvement of the economic, political, and social welfare of the community. The land commissions of Japan, created for the purpose of enacting the land reform there, are an excellent example of that. In Japan the work of the commission has given rise to a form of adult education and new village leadership from among the tenants. A Taiwanese village would benefit enormously from the participation of the tenants in important decisions affecting life of the community. Unfortunately, the trip through Taiwan in September of 1949 revealed that the rent campaign committees were more paper organizations.

For all practical purposes the committees were not functioning. In village after village I found that they seldom meet, that the application of the rent reduction program is dependent upon the drive of the government rather than upon the work of the committees. In short, the program designed for the benefit of the great masses of the people is being carried out without the participation of the people. Moreover, the very composition of the committees does not augur well for the training of new leadership in the villages. The number of tenants on a committee of approximately eighteen is usually not more than two, while the others are representing groups that are not necessarily in sympathy with tenant interests. It is extremely difficult under such conditions to change even slightly the customary conduct of and the thinking about village affairs. It should be added also that the members of the committees are not elected but are appointed by the village heads, at the village level. Too much responsibility—not always justifiable—is lodged, therefore, in the hands of the village heads.

The findings concerning the "work" of the rent campaign committees were called to the attention of the responsible officers of the provincial government. As an aid to the reorganization of the committees along sound lines, I prepared a detailed statement on the nature and work of the land commissions in Japan, urging that they serve as a pattern for Taiwanese committees. I left with the assurance of S. K. Shen, director of the land bureau of the provincial government, that immediate steps would be taken to give real meaning to the committees—that less emphasis would be given to government edicts and more to the participation of the farmers themselves in the enforcement of a program which affects them deeply.[1]

The attitude of tenants towards the rent reduction program was what was to have been expected and so was that of the landlords, favorable in the first instance and one of opposition in the second. The whole idea of rent reduction, greater security of tenure, and abridgement of their former rights and privileges made no sense to landlords. It constituted a break with tradition, a break that came rather suddenly and for which they were not prepared. Even the actual promulgation of the program in April 1949 was not taken seriously by the landlords (nor by the tenants for that matter). Only when Governor Chen Cheng made it clear that he intended to enforce the program in order to secure a measure of political support from the people, and only after a few recalcitrant landlords were sent to jail, did landlords bow to the inevitable.

The reluctant acceptance of the program was apparent in all my meetings with landlords, whether in the village hall or in their homes. Their views were tempered, of course, by the fact that I was a representative of JCRR, an institution supporting the program. This explains in a large measure why I never encountered a clear-cut expression of opposition to the program. On the other hand, they spoke freely in opposition to the land-grading system upon which the 37.5 percent rental is based. In the main, their criticism of the unfairness of the grading system was justified.

During the Japanese administration, all land in Taiwan was graded (classified) according to location, irrigation facilities, fertilizer application, type of crop, and so forth. Basically, grading of land was in accordance with its fertility. Altogether there were twenty-eight grades of land, and the standard yield of a given grade of land determined the rent and the rate of taxation. The war and postwar dislocation caused a decline in yields, especially in the immediate postwar years. With the increase in fertilizer supply and improvement in irrigation facilities, yields have risen in 1948 and 1949. It is the contention of the landlords that in revising the grading system the provincial government set the standard yields at levels considerably below the actual yields. The result is that in practice many a tenant pays not a 37.5 percent rental but something lower than that; I know of instances where the rental was closer to 30 percent of the crop.

The question of grading came up at every meeting, and it was quite obvious that tenants were on the defensive insofar as this very important question was concerned. I had the impression that they were a bit embarrassed by the very favorable yield standards set by the government. Talks with local and provincial officials pointed to the fact that the grades and yield standards were a "slapdash" affair, contrived without much investigation or reference to actual conditions.

Much was said on both sides on how to solve the problem. From the point of view of the landlord, nothing could be fairer than 37.5 percent of the actual yield in any given year. From the point of view of the tenants, the solution lies in the adjustment of the grades and existing standard yields that would conform more nearly to the actual yields. The tenants recognize, therefore, the unfairness of the grading system; and, to the extent that they do, it seems to me that their idea of an adjustment is much

1. Ladejinsky's note on the nature and work of the land commissions in Japan follows this paper. [Ed.]

sounder than the one propounded by the landlords.

The latter want, in effect, a flexible rent which would have to be determined every year, depending upon the yield. The procedure would lead to no end of disputes between landlord and tenant, and, above all, it would tend to thwart the tenant's incentive to improve the land and raise agricultural production. It would seem, therefore, that a fair adjustment of land grades and standard yields and periodic revisions approximately every five or ten years would meet the criticism of the landlords and particularly of the small landlords whose source of income is rent from very few acres. Before my departure from Taiwan I was given to understand that the provincial government will make an earnest effort to correct an admittedly bad situation.

Since the end of World War II, I have visited a number of regions in Asia formerly occupied by Japan; but only in Taiwan did I hear peasants say: "Oh, yes, under the Japanese we were better off," or words to that effect. I heard such statements on numerous occasions in the course of the field trip. The answer to my questions as to why peasants felt that way would be given in terms of ample availability of fertilizers, credit, and technical services rendered by the cooperatives and experiment stations. These are the things that have been largely responsible for the development of the island's agriculture in the past fifty years, and these are things the scarcity of which is keenly felt by Taiwanese farmers.

It was neither within my province nor did I have the time to delve into the problems of fertilizer supply, farm credit, cooperatives, and experiment stations. Moreover, JCRR is fully aware of the existence of these problems and much is being done to solve them. A good example of that is the marked improvement in the fertilizer supply. It is worth emphasizing, however, that even under the rent reduction program a great many Taiwanese farmers are marginal operators. They know little of "fat" years, and of reserves to meet the "lean" years—much less. Taiwan has no farm credit system of any consequence. The so-called credit system means borrowing from loan sharks in the guise of friends, relatives, and merchants at interest rates ranging from 150 to 200 percent a year. No figure of less than 150 percent was quoted to me. Under the Japanese the credit system, which enabled a farmer to secure a loan at a reasonable interest rate, was part-and-parcel of the widespread cooperative network. The latter was the center of dissemination of agricultural knowledge, of fertilizer, and farm equipment distribution on credit, and so forth.

JCRR is concerned with the reorganization of the farm cooperatives, but to give meaning to the best outline of a co-op system it must be provided with initial funds and efficient management. These are very scarce now and will probably remain so for some time to come. Yet it should be reiterated that even the important benefits derived from the rent reduction program will not suffice to insure for the post-reduction rent tenants a real measure of economic security. Somehow sound credit and cooperative systems must be created in order to avoid the burdens of indebtedness and usury. These, plus the technical improvements that are being effected by JCRR in Taiwan, are the sureties that will enable an industrious and efficient tenant to maintain his newly acquired gains.

One of the most prevalent grievances of the tenants of Taiwan is the handling of the public lands by the provincial government. The question of public land is a complicated one, but for our purposes it is sufficient to point out that the provincial government is in the business of renting and selling some of this land. Without exception, the tenants are critical of the government in both instances.

The public land in question is land rented to the tenants by the Taiwan Sugar Corporation, which is government owned. The rental is 25 percent of the crop, or 12.5 percent less than the rental on privately owned land. The seeming difference in favor of the tenant is more apparent than real. This stems from the fact that, while the tenants grow sugar cane, the rent is based on so many catties of rice per unit of land. A tenant then sells the sugar to buy rice in order to make good the rental. It happens that the price of sugar is low and the price of rice is high; thus, the net income a tenant derives from a sugar cane crop is much smaller than the income from a similar area planted to rice. Hence the complaints that 25 percent rental on public lands is, in effect, higher than 37.5 percent on private lands. The tenants insist, therefore, that it is unreasonable to be forced to grow sugar cane and to have to pay rent in rice, a crop they do not grow on this land.

The provincial government is interested in promoting land ownership among tenants. Therefore, early in 1949 it set aside 10,000 hectares for sale to the tenants. The price was fixed at 250 percent of the main crop yield of land, payable in five to eight years, depending upon the condition of the land. This means that in the first case a tenant must pay 50 percent of the crop and in the second, 33 percent. Throughout 1949 only 4,000 hectares had been sold. The purchasers and would-be purchasers take exception not to the price but to the short period of repayment. They would like to have it extended to ten to fifteen years.

It is axiomatic that a new owner can be as badly off as a tenant paying high rentals if the payments for the land fall due in a relatively short period of time. His chances of firmly establishing himself on the land through greater accumulations in the initial years of ownership are thereby reduced. The suggestions of the tenants to pay for the land over ten to fifteen years is well taken.

I believe that the tenants have a justifiable grievance on the score of leasing as well as of purchasing public land. The government is eager to collect as much rice as possible. Its motives are quite understandable: it must feed an army and an urban population (particularly the government servants) and accumulate emergency reserves. The government welcomes any additional volume of rice, small though it might be, that comes into its hands. On the other hand, there is the danger of minimizing the effects of the larger issue involved in the rent reduction program, which is to raise the economic welfare of the tenants and to gain their goodwill and support.

Because of the prevailing lower sugar cane prices, tenants on private land and owner farmers have been shifting from sugar to rice culture in increasing numbers. Tenants on public land would like to do the same, but they must grow sugar on that land and have no other occupation than farming. It seems reasonable that the government should not take advantage of this situation and should, instead, adjust rentals to something approximating the benefits of the tenants on privately owned land.

Such, in the main, are my observations and conclusions of the operations of the rent reduction program in September 1949. The outstanding impression is that the program came into being not at the whim of General Chen Cheng, governor of Taiwan, but because the need for it was rooted in the rural conditions of the island. The work and life of the tenants of Taiwan, as I observed them firsthand, testify to the need for overhauling landlord-tenant relations in order to alleviate some of the worst features under which they are carried on. It is therefore to the great credit of Governor Chen and the few men around him who had the wisdom to initiate the rent reduction program and who later displayed the tenacity of purpose so basic to the overcoming of the opposition and to the carrying out of the program.

No observer of the program in action can fail to notice numerous failings, and practically all of significance have been recorded in the foregoing pages. They do not constitute insuperable problems; even the more serious of them, such as the land-grading system, can be solved. There is one exception, however: the opposition of the landlords, especially the small ones. An improved grading system will not change their attitude.

As land reform programs go, the one in Taiwan may well be placed in the category of the mild ones. But no matter how mild, it calls for a reduction in the income of landlords. In no other way can the standard of living of Taiwanese tenants be raised dramatically and quickly. Nor has the government any other means of making a serious effort to secure their political support at this time. Considering the premises underlying the rent reduction program in Taiwan on the one hand and the conditions of the traditional patterns of landlord-tenant relations on the other, the landlords' dislike of the program is inevitable. This was true also in Southwest China, where I observed the application of the rent reduction program in late fall of 1949. It is true of the Far East and of the rest of Asia where land reform programs have either been instituted (Japan) or are about to be instituted (Korea, India), or are yet to be introduced. This opposition alone is no argument against land reform so long as the need for it is urgent and so long as the program is not confiscatory in nature. Both are true in Taiwan. The real concern then of the sponsors of the rent reduction program should be only with the correction of its shortcomings. I left Taiwan with a strong impression that an earnest effort will be made in that direction.

By the end of September the program was only five months old, and while examining it I was struck not so much by its deficiencies (these were to be expected) but rather by its accomplishments. The inescapable impression was that the program though new was being carried out, that the tenants were reaping considerable benefits, and that the evasions were not impairing the progress of the program. This is not to say that the program is fully established, although indications are that it will be. The driving force behind the program is, of course, the abiding interest and energy displayed by the provincial government. A change in attitude coincident with the rather negligible participation of tenants in promoting the program would seriously retard the entire effort. Only a careful appraisal of the rent reduction program in early 1950 (i.e., after the second rent payment) will tell how firmly the program has become rooted.

Assuming that the program had been carried out successfully, the immediate economic gains to tenants will be considerable indeed. But will the realization of the program secure for the government the political support of these tenants? At first glance the answer should be affirmative on the ground that "you don't shoot Santa Claus." Yet, my impression is that to most Taiwanese tenants the provincial government is rather an exacting and, above all, a demanding taskmaster. Government distribution of fertilizer only in exchange for rice serves in itself to minimize in the eyes of the tenants the benefits from the government rent reduction program. Official rice prices and rice requisitions swell the list of grievances of the Taiwanese farmers against the government. Very significant, too, is the unfortunate political climate that pervades the island, and, rightly or wrongly, the average Taiwanese holds the Chinese administration responsible for it. For these reasons the task of the provincial government to gain the popular support of the tenants and make political capital out of the rent reduction program is rather difficult, if not impossible, at this stage.

The rent reduction program in Taiwan, in contrast to that of Southwest China, was initiated by the provincial government. But the role of JCRR in helping to execute the program was as important as it was in Southwest China. This came about through JCRR's expert technical advice and through financial assistance. The latter was devoted entirely to such items as printing new contracts and paying salaries of registrars and supervisors over a limited period of time. Watching the results of this type of assistance, I believe that, had JCRR aid been lacking, the rent reduction program would not have gone as far as it has, either quantitatively or qualitatively. I doubt very much that so many new contracts would have been drawn up and that so many tenants would have paid rents on the new basis. And yet, the amount of money spent by JCRR from the inception of the program (May) through (September) was only $30,000. It is no exaggeration to say that approximately 300,000 farm families benefited from this expenditure. JCRR has demonstrated how much can be accomplished with a little money when it is spent for a good purpose and judiciously.

JCRR's contributions in the field of rent reduction far transcend the benefits now enjoyed by the tenants of Taiwan. It has broadened the concept of technical agricultural aid by recognizing the importance of dealing with Asian land tenure problems as a vital part of agrarian reform. Closely allied is the fact that even the best results of technical aid will be limited if landlords continue to receive the larger share of the increased output. The tiller of the soil will not have the necessary incentive to sustain him in his effort to improve the land if the burdensome conditions under which he rents his farm remain unchanged. This then is the significance of the JCRR pattern of agricultural reconstruction in Taiwan. It has concerned itself with all principal elements that tend to make for higher production and for a fairer distribution of farm income. This is a pattern of pioneering work worth studying and applying, with modifications where necessary, in poverty-stricken rural areas everywhere.

11. Land Commissions in Japan

This is the note referred to by Ladejinsky in the preceding paper. It took the form of a memorandum addressed to Raymond Moyer, one of the two U.S. members of the JCRR. In transmitting the memorandum, Ladejinsky observed: "I might add that the commissions amply justified their existence, for the land reform program in Japan could not have been carried out without their existence." This helps to explain, therefore, the great importance Ladejinsky always attributed to the role of implementation in agrarian reforms. Also noteworthy is his view of the function served by local commissions in developing a new generation of leaders in the countryside. Nearly twelve years later, after a visit to Indonesia, Ladejinsky transmitted a copy of this memorandum to Dr. Sadjarwo, minister for agrarian reform, with whom he had discussed agrarian questions. In his letter of transmittal he wrote: "In asking you to examine it, I am not suggesting that the Japanese method is necessarily applicable in Indonesian village conditions. . . . However, I do wish to reiterate once again that even in Japan, with a well-functioning administrative machine, the reform could not have been carried out without the existence of the commissions. In a real sense, they insured the success of the reform."

This memorandum, undated, must have been finished shortly after the field trip in Taiwan was completed in late September 1949.

RESPONSIBILITY FOR THE ADMINISTRATION of the land reform program and for the actual transfer of land ownership through purchase and resale was vested in the land commissions. The enacted legislation provided for the election of local and prefectural commissions and the appointment by the Cabinet of a Central Land Commission.

Elected members of local and prefectural land commissions were to be composed of representatives of three categories of voters, defined as follows: (a) tenants: persons engaging in cultivation who owned no agricultural land or who cultivated twice as much land as they owned; (b) landlords: persons who did not cultivate the land they owned or who owned twice the amount they cultivated; and (c) owner cultivators: those who cultivated the land they owned but who did not belong in either of the previous categories.

Local Land Commissions

Elected members of local land commissions in each city, town, or village would include five tenants, three landowners, and two owner cultivators who would serve for a two-year term at a nominal salary of ¥200 per year. Prefectural governors had the right to appoint three additional commissioners, provided that these appointees were requested and unanimously approved by the elected commissioners. The chairman would be chosen by and from the ranks of the elected commissioners. In the event of a deadlock, the prefectural governor would appoint a chairman from among the three additional appointed members. The composition of the local land commissions could in special cases be altered; however, the proportion of representation would always remain the same.

Persons eligible to vote or become candidates for office on the local land commissions had to be at least 20 years of age and be immediate members of households cultivating one tan (one-fourth of an hectare) or more of land in Japan proper and three tan in Hokkaido.

Voters dissatisfied with their representatives could, under the provisions of the law, demand their resignation and schedule new elections.

The procedure for recalling commissioners was as follows:

1. Petitions requesting the resignation of a particular commissioner had to be submitted in writing to the mayor or village headman by a majority of the eligible voters within the same category as the commissioners in question. The recall petitions would affect all the commissioners within any single category.
2. Recall elections would be scheduled by the mayor or village headman within two weeks after public announcement of the recall petition. Commissioners elected in recall elections would serve the remainder of their predecessor's term of office.

Neutral appointed members of the commission could be dismissed from their posts by the prefectural governor but only with the unanimous consent of the popularly elected commissioners.

The actual purchase and sale of lands would be carried out by the local land commissions. As executive organs of the land reform program, these commissions had broad powers which they could exercise with a minimum of governmental interference. These included (a) drafting the purchase plans for each village, (b) determining the suitability of lands to the purchased, (c) establishing the eligibility of purchasers, and (d) deciding cases requiring unusual or special treatment and approving cases of exemption from the purchase provisions. In addition to purchasing lands subject to compulsory sale under the terms of the law, local land commissions were empowered to purchase for the government other agricultural lands offered for sale and approve the purchase by the government of pasture lands, reclaimed lands, housing sites, buildings, and equipment in order to facilitate land ownership among tenants. All transfers of ownership and transfers and cancellations of leases and superficies or other rights had to receive the approval of the local land commission or of the prefectural governor. The commissions also mediated tenant disputes and determined the amount of farm rents. The meetings of the commissions were open to the public, and the minutes were available at all times for public inspection.

Prefectural Land Commissions

Prefectural land commissions were composed of ten tenants, six landlords, and four owner cultivators elected by and from the membership of the local commissions. The Minister of Agriculture and Forestry could appoint up to ten additional members. The governor of the prefecture would act as chairman. Prefectural land commissioners received ¥60 for every semimonthly meeting attended.

The prefectural land commissions functioned primarily as agencies for reviewing the decisions of the local land commissions. They were required to approve the purchase and sale of farm facilities, building sites, buildings, and grassland and to pass on the cases of exemptions allowed by the local commissions. They functioned also as boards of appeal to which landlords or tenants dissatisfied with the decisions of the local commissions could present their cases for review. If tenant arbitration proved unsuccessful, the prefectural commissions submitted detailed opinions necessary for settling disputes concerning the continuation of tenacy and the modification of the terms of tenancy agreements. Other duties included: conducting the exchange, division, and pooling of farmland; drafting purchase plans for unclaimed lands and fishing rights; establishing acreage standards within prefectural districts in connection with the average retention rates allowed by the Central Land Commission; and passing judgment on all other matters referred to them by the local commissions.

Central Land Commission

The Central Land Commission consisted of eight tenants and eight landlords elected by the prefectural commissioners plus two representatives of national farm organizations and five agricultural experts appointed by the Cabinet on recommendation of the Minister of Agriculture and Forestry. The Minister was chairman of the commission. Central land commissioners received traveling expenses of ¥200 plus ¥105 for each monthly meeting attended.

The Central Land Commission was responsible for formulating nationwide policies on land reform, such as (a) establishing prefectural breakdowns of national retention averages allowed by the law; (b) determining the basis for setting prices of agricultural facilities, unclaimed lands, grasslands, and buildings; (c) determining the ratio of yearly installment payments to the value of the crops; and (d) establishing rent scales within the limits provided by the law.

Pre-election Registration and Information Campaign

The first step towards the implementation of the basic land reform legislation was the registration of all eligible voters in the categories of owner, owner cultivator, and tenant in preparation for the land commission elections. Prospective voters registered their names, date of birth, permanent residence, and the area of the land owned or cultivated by them at village, town, and municipal offices throughout Japan. The lists of electors were thrown open to public inspection for a fifteen-day period after November 5, 1946, and copies were made available by village headmen at all polling places. On November 30, 1946, the nationwide registration of electors was completed.

In preparation for the elections, which were scheduled to take place on December 20, the Ministry of Agriculture and Forestry issued instructions to all prefectural government officials concerning their responsibilities in giving adequate advance publicity to the forthcoming elections. Each governor was informed on December 12, 1946, that he would be held legally responsible for the proper dissemination of information on election procedures within his prefecture. He was further directed to call meetings of all subordinate regional heads in order to ensure adequate distribution and circulation of 630,000 posters, circulars, and pamphlets prepared by the ministry for the purpose of informing farmers of their rights under the land reform legislation. This printed material included (a) 300,000 circulars explaining the functions and duties of the agricultural commissions and the procedures governing the elections; (b) 300,000 circulars explaining the basic provisions of the law; and (c) 30,000 pamphlets explaining the text of agrarian reform legislation, for the purpose of instructing public officials responsible for its administration. This information campaign was supplemented by releases to newspapers throughout rural Japan and by a series of radio programs.

Local Land Commission Elections

Rural land commission elections began on schedule and were completed December 31, 1946. The procedures followed were in general the same as those governing the election of city, town, or village councillors. Surveys of the election returns revealed that a total of 10,777 local land commissions had been formed but that only 52 percent (5,600) had actually been chosen by election. The remaining 48 percent (4,732) had been seated by proclamations issued by villages headmen, since the number of candidates equalled the number of posts to be filled, in accordance with the Agricultural Land Adjustment Law Enforcement Order, which allowed candidates to be elected unopposed. Subsequent field investigations conducted by occupation authorities revealed that from 75 to 80 percent of the land commissioners were accepted as satisfactory by their constituents, that in the majority of cases in which no elections had been held the selection of candidates had been accomplished according to democratic procedures, but that there had been a few instances in which there were evidences of voting controlled by vested community interests. As a result of this survey, the Ministry of Agriculture and Forestry took immediate steps to acquaint farmers through all available information media of their rights under the law to oust commissioners who did not represent them. Prefectural governors were given specific instructions relative to the circulation of 300,000 circulars in all rural areas explaining the procedures for requesting and scheduling recall elections in communities where the majority of voters were dissatisfied with their representatives.

Prefectural Land Commission Elections

In order to ensure the proper dissemination of information on the procedures for selecting prefectural land commissioners, the date of the prefectural land commission elections was postponed until February 20. The elections, which were completed throughout the forty-six prefectures by February 25, were conducted according to two alternative methods for establishing election districts. One method treated the prefecture as a single voting district which elected a commission of ten tenants, six owners, and four owner cultivators. The second method provided for two voting districts, each of which elected five tenants, three owners, and two owner cultivators as representatives on the prefectural commissions.

During February 1947 the Ministry of Agriculture and Forestry requested all prefectural governors to submit the names and biographies of eligible candidates for consideration as neutral members of the prefectural land commissions. All elected members of the prefectural land commissions were screened in accordance with the purge directive before they were permitted to pass on the appointed members of the commission. All appointed members in their turn were screened before assuming office.

Central Land Commission Appointment

Candidates for appointment to the Central Land Commission were selected by the Ministry of Agriculture and Forestry and screened in March. The list of the names and biographies of the proposed appointees was then forwarded to the Cabinet for approval. Final appointments of the central land commissioners was confirmed on March 26, 1947.

Adult Education Aspects of the Land Reform

The duties performed by the land commissions have given rise to a form of adult education, the significance of which would be hard to overestimate. Farm tenant members of the commission, many of whom were obviously ill at ease and insecure in their new posts in early 1947, were seasoned performers in 1948. For in the meantime they had taken a responsible part in numerous transactions and decisions. The selection of an agricultural land commission afforded all the adult farm population, tenants as well as owners, an opportunity to vote on a matter of real interest to all elements of the village. The very composition of the commission proclaimed the fact that the tenants' interests were to be protected by the tenants themselves rather than by someone acting on their behalf within the traditional pattern of rural Japan.

Soon after the agricultural land commissioners had been selected in December 1946, they began to explore their assignment. The first phase was the purchasing of the arable land of absentee owners and corporations, for these purchases did not involve the complicated matter of determining retention rates before purchase. Later followed the purchases of lands rented out by temples and shrines and then the lands of resident–owner operators and temple and shrines lands operated by priests or groups of parishioners. The maximum retention rate in most villages was about 3 cho (7.4 acres).

In early 1947 there was considerable doubt in some of the villages, especially those in southern Japan, as to whether or not the land reform would actually be carried out. There was evidence that many tenants wanted to play the game safe. They wanted to own the land, *if* it were possible; but, not being absolutely certain that it would be sold to them, they wanted to maintain their traditional, accepted relationships with the landowning families from whom they rented.

As the commissions went forward with their program of purchasing on schedule, and especially when the reselling of the land started, the village importance attached to membership on the commission increased greatly. Then, and only then, did it become plain that the tenant members of the commissions actually had authority. Every week added to the prestige of the commissions, and especially to its tenant members. Up until the resale of the land began, many of the tenant members didn't know just what their situation was. Most of them seemed to have assumed that the landlord members of the commission were cooperating in the program under duress and that, as soon as they had the opportunity to undo what was being done, their own sense of self-interest would lead them in that direction. This attitude vanished with the resale of the land actually under way.

The purchase and resale of the land were complicated affairs. In each village hundreds and hundreds of small tracts were involved, each with its own size and shape; and, in addition, many tracts with forest or other rights had to be dealt with. More than this, some tenant families could not purchase because the plots they were tenanting were not for sale; that is, the fields might be those which an owner operator could hold within his retention right, above and beyond the area he was cultivating. Such tenants often wanted and were eligible to buy land. The commissions sometimes resourcefully worked out exchange of plots between absentee owners and those that could still be rented by resident operators. In these cases two or more transactions had to take place before the tenant could buy, which meant additional detailed work by the commission. The vast amount of detail involved becomes understandable when it is noted that in one village there were approximately 11,000 pieces of land involved in purchase and resale, while in others the number was around 2,000 to 4,000.

The clerks of the commissions played a most important part in the whole program. It was they who figured out how much land was to be bought and from whom; they kept the records of the transactions, and they received and first studied the instructions that came to the commission office. In many villages they were real teachers of the commission members.

By the time it had been agreed which families could buy which tracts of land, it began to be widely believed by the tenant families and by the families from whom land had been purchased that the initial phase of the program would be completed on schedule. By then the villagers were generally aware of the authority which the commission actually had. This authority became especially convincing as the village offices began arrangements for formal entry of the sales in the land ledger in the registry office under the jurisdiction of the attorney general's office.

People other than tenant members and clerks who received valuable new training through the execution of the land reform program were the buraku (subdivision of a village) representatives, who took an active part in the work of the commissions, especially when the land in question was located in their particular part of the village. There was an average of at least three clerks per commission, and perhaps not less than a half dozen buraku representatives. Thus, about fifteen to twenty men in each village had unique firsthand leadership training within two years; and at least a third of these were from the farm tenant group, who within the old tenure system could hardly have hoped to play such a role.

With the same type of local land commission functioning in more than 10,600 villages in Japan, 150,000 or more people received this leadership training and 50,000 of them were from the farm tenant group. So quite aside from the many benefits which the new owner farmers are securing from the land they have purchased under the land reform program, the way in which the transfers were carried out has in a very short time produced a sizeable group of potential new leaders in rural Japan.

12. *Field Trip in Szechwan: Tenant Conditions and Rent Reduction Program*

The field trip of October 13 to 20, 1949, on which Ladejinsky here reports, took place shortly after his first field trip in Taiwan. It possesses a special interest in that it represents the only work Ladejinsky ever did on the mainland of China; it came at a time when the Communist forces had already taken over most of the country and were threatening the southwest, the only territory still controlled—though not for long—by Chiang Kai-shek's National Government. This lends poignancy to Ladejinsky's observation, toward the very end of this paper, in which he was impressed by "the enthusiastic response of the tenants to the reform recently instituted." It is quite clear, he states, "that a wise government could have made a good deal of political capital and perhaps achieved military stability if the farm aid made its appearance in good time. After all, China's armies are peasant armies." Unfortunately, as Dr. Moyer has pointed out, by the time the land reform in Szechwan was instituted, it was already in the nature of "a rear guard action." Discussing the nonimplementation of earlier rent reduction measures decreed by the Executive Yuan, Ladejinsky observes, dryly, "Nothing was done about this very limited of measures to ease the burden of the tenants; the record has not a blemish of accomplishment."

Ladejinsky, it is shown here, could be a doer as well as an observer in the field. On discovering a weakness in the enabling regulations, he went straightaway to the acting governor of Szechwan and persuaded him to issue a corrective order within ten days after termination of the tour.

Since he was aware that this report would be read by a number of U.S. officials, Ladejinsky closes the report by drawing some plainspoken lessons from this experience for U.S. policy.

This paper, like the Taiwan paper preceding, took the form of a memorandum addressed to Chiang Mon-Lin, chairman of the JCRR, dated November 7, 1949. It appeared in General Report no. 1 of the JCRR.

General Observations

ON OCTOBER 13 I SET OUT FROM Chungking in the company of Mr. Tang and Commissioner Sun Lien Chuan to observe tenancy conditions and the application of the rent reduction program in the third prefecture. In the course of the field trip, which lasted seven days, we stopped at numerous contract registration points, village offices, tenant homes and talked about the program with a great many tenants landlords and local officials. The observations on tenancy in the third prefecture and the rent reduction program to improve those conditions are contained in the pages that follow, but the basic conclusion may be stated at the outset: while the rent reduction program is only at the early stages of application, it is quite certain that, if the current policy of brooking no opposition from the landlords is continued, the program will be carried out in the course of the next two to three months (that is, the majority of the tenants will reduce their customary rental payments by one-fourth).

The first and third prefectures are in Szechwan province, but they are miles apart and not only because of the 300 miles which separate them. The first prefecture extends over the entire flat-as-a-table, well-irrigated, rich soil of the Chengtu plain. The third prefecture on the other hand extends over the mountainous Chungking area. The flatlands are, in most cases, mere narrow wedges between two mountains. Practically the entire cultivated land of the prefecture consists of a series of elongated or semicircled terrace strips of 1 to 2 mou in size. Infinite toil and care are required to prevent the washing away of the terrace boundaries. The fertility of the soil is poorer than that of the Chengtu plain; artificial irrigation, which is one of the outstanding features of the first prefecture and made the prefecture's rice crop failure proof, is almost absent here. The rice crop in the third prefecture depends entirely upon rainfall which occasionally falls short of requirements. The result is that the yields here are approximately 50 percent below those enjoyed by the farmers of the first prefecture.

In one important respect, however, both prefectures are very much alike: the tenantry is extremely poor. The tenants of the third prefecture pay a lower rental than do the tenants of the second, but because of lower and uncertain yields and smaller holdings they are worse off than the Chengtu tenants. The tenant's problem of securing food for his family is more acute than in the first prefecture. All other "amenities" of life are in roughly the same proportions. It is obvious, therefore, that a reduction in rent by one-fourth is a most important measure to increase the tenant's food supply and to strengthen him economically. And, as will be pointed out elsewhere in this memo, the reduction in rent is likely to bring in its wake at least the beginnings of the much needed but long overdue social and political changes.

Tenancy Conditions

The third prefecture, with a population of over 5 million and an area of 7,000 square miles, is one of the larger prefectures of Szechwan. The nearby Chungking, the wartime capital of China, places this prefecture in a special category. But more important is the fact that the mass education movement and the Peipeh special district are both in the third prefecture. The mass education movement stresses the organization of farmers into cooperatives through the use of literacy classes to explain the advantages and desirability of cooperative action. The Peipeh special district is worth noting for the remarkable work carried out by its administrative commissioner, Lu Tze-Ying and his brother, Lu Tzu-fu, in the creation of the modern city of Peipeh as well as in the carrying out of a land ownership project, which, though modest in scale, was of national significance. But neither of the mentioned developments succeeded or could have succeeded in creating conditions from which the mass of the peasantry could derive even a semblance of well-being. Peipeh remained a small island within the vast ocean of a peasantry, maintaining its existence in the traditional way. A trip from Chungking in any direction, a visit in any village, a talk with any number of farmers—all bear ample testimony to that effect. The reasons underlying this state of affairs are partly nature (climate and soil); partly man-made (rising population pressing on the land); and, finally, lack of interest on the part of the Nationalist government in the welfare of those who till the soil of China.

The third prefecture is one of the hilly regions of Szechwan. The remarkable system of terraced agriculture almost overcame this handicap; it explains why 2.3 million acres or half of the total area of the prefecture is cultivated. But terraced agriculture calls for enormous application of labor, which is difficult to compensate even under favorable soil, climatic, and ownership conditions—the more so when, as in the third prefecture, the cited conditions leave much to be desired.

The soil lacks the natural fertility of the heavy, black loam of the first prefecture, and from the point of view of water supply it is still worse off. Szechwan as a whole has good irrigation possibilities, and the first prefecture has realized them through an irrigation system created some 2,000 years ago by the famous engineers Lee Pin and Lee Erlong, father and son. No such condition prevails in the third prefecture; the farmer depends almost entirely upon rainfall, which is known to fail occasionally at the critical periods. The combination of soil and water conditions is chiefly responsible for rice yields of only 0.8 to 1 picul per mou—and less than that in years of low precipitation—as against the virtually never failing 2 to 2.2 piculs in the first prefecture.

But even in conditions of higher yields, most of the farmers of the third prefecture have to contend with a land tenure system which placed them on one of the lowest rungs of the economic ladder. Data on land ownership for the entire prefecture are not available, but information dealing with the Peipeh special district, which we visited, is revealing. About 60 percent of the population (10,000 farm families) lives on land (a rather low proportion for the prefecture as a whole). Of these farm families, approximately 70 percent are tenants, 14 percent part tenants, and only 16 percent owner cultivators. Inquiries at fairly large farm meetings indicated a somewhat lower proportion of tenants, but in no case did they represent less than 55 to 60 percent of the farm population. More than half of the land is owned by absentee landlords who reside in nearby towns or in Chungking. The land is farmed by tenants who pay a minimum of 60 percent of the principal crops (chiefly rice, and to a much smaller extent, wheat). Rentals as high as 70 percent are not exceptions.

These terms of land tenure are typical not only of the Peipeh special district but also of the other hsiens. Coming as we did from the first prefecture, we could not but notice the difference in rent. Since 80 to 90 percent of the main crop is the customary rent in the first prefecture, the 60 to 70 percent in the third prefecture did not seem exorbitant at the first glance. Yet a further probing into the farm practices of the prefecture revealed that they are, in effect, just as high and with an edge in favor of the tenants of the first prefecture. Because of scarcity of rainfall in the spring for rice planting, few of the rice fields of the third prefecture are utilized for winter crops; they are mainly water reservoirs for the planting of rice in the usually dry spring. The riddle of the huge pools of water instead of fields, first caught sight of from the plane, became readily understandable when the farmers explained the whys of agricultural practices in this region. In the first prefecture on the other hand, the never-failing irrigation system provides the land with all its water requirements; the fields are free, therefore, for the planting of winter crops, which belong to the tenants. Only a small minority of the tenants of the third prefecture have such crops to fall back on. The winter crops of the first prefecture, while of relatively small importance when compared with the rice crop, more than offset the rental difference between 80 to 90 and 60 to 70 percent of the main crop.

The landlord-tenant relationship is regularized by a written contract. Oral agreements are rare, and we did not encounter a single tenant who rents land on that basis. A number of contracts were examined and discussed with tenants and landlords at considerable detail. The characteristic feature of a written contract is not what it contains but what it omits; most of them, for example, fail to specify their duration. The tenant is thereby denied security of tenure. The contract specifies the rented acreage; the rental; the deposit or key money (about which much more later); and the extras on New Year's day (although not in every instance), which may consist of such assorted items as so many catties of beans, rice straw, chickens, and ducks. Neither landlord nor tenant was familiar with the practice common in other countries which entitles a tenant to certain compensations for improvements on the land.

A time limit does not necessarily insure for the tenant security of tenure. The landlords have been in the habit of changing contracts, with or without duration clauses, and tenants submit to it for fear of dispossession. The latter means total dispossession since the house and farm buildings occupied by the tenant belong to the landlord. The impression we carried away is that tenants have no bargaining power whatsoever; that the contract as a means of securing certain rights for the tenants was worth no more than the flimsy sheet of paper it was written on; that the tenant-landlord arrangements are primarily a matter of the landlord's discretion; and that the courts, the magistrates and local officials have, in the main, given sanction to that discretion. No moral judgment of the landlord is intended here; what is recorded is an expression of the behavior of the "economic man" in the midst of a continuous struggle for survival among the tenants. That struggle is the keenest competition for the privilege of cultivating a few mou of the landlord's land. The landlord, free from any legal and moral restraint, merely takes advantage of the opportunity.

In China as in a number of other countries with a large farm population pressing on a limited amount of land, the lot of the tenant is conditioned not only by the high rental but the fact that he cultivates a few rather than many acres. The third prefecture is a good illustration of that truism. Data on the amount of land rented by a tenant are not available, but inquiries among tenants and landlords offer fairly reliable information on this point. It appears that, whereas in the first prefecture a tenant cultivates an average holding of approximately 20 to 25 mou, in the third prefecture the tenant must be content with about 10 to 15 mou, or 2 or slightly more than 2 acres. Assuming that he cultivates the upper limit and all of the land is devoted to rice (usually not the case) and the yields are normal, the tenant can harvest a maximum of 15 piculs of rice. A minimum rental of 60 percent would leave him with 6 piculs, admittedly a small return even under the assumed favorable conditions.

The economic consequences of the mentioned conditions are quite apparent without any recourse to standard-of-living studies, with all their paraphernalia of measuring the immeasurable in China. The naked eye and the sifting of the statements of the tenants concerning the things they live by are quite sufficient. The matters of clothing and shelter did not call for much investigation. The attire spoke for itself, and it became clear to us why and how a tenant makes a cotton shirt do five or more years of service. Their so-called homes ("so-called" even from the point of view of their neighbors, owner cultivators) and their contents can boast of no worldly goods worth mentioning. We visited a few, and the living quarters, tool shed, and storage room all spoke the same story: on how little man can suffer through a relatively long life of privation.

The matter of sufficiency or insufficiency of food on observing the tenants about us was not as easily discernible. We could not tell without questioning whether tenant Chiang Mon-lin lives on rice that he wants most, plus a dash of meat and fish, or on sweet potatoes plus rice and no meat and fish. Numerous recorded answers show, and with no exception, that the principal rice producers of the third prefecture have but one thought in mind: how to secure enough rice for themselves and their families. A tenant called upon to tell just how much rice he must have to go about his work presented, what seemed to us then, a startling bit of information since confirmed by many others: he needed fifteen fair-sized bowls of rice a day, or five bowls per meal. With a family of six, tenant Shen Tsung Han requires 6 piculs of polished rice a year. The same tenant was cultivating 15 mou of land, of which 12 were under rice. He harvested 9.6 piculs, the landlord receiving 5.7 piculs as rent, and the tenant 3.9 piculs of unpolished rice, or 2 piculs of polished. His rice deficit was 4 piculs.

The tenant admitted that he could not afford all the rice that he wants and needs, and his problem is how to secure the major part of it. Here is where, in addition to the secondary crops, the ducks, the chickens, the occasional goose, the eggs and the hogs raised by some of the tenants, come into the picture. The hasty suggestions that in all probability they are consumed on the farm was met with an outburst of laughter, as if to underscore the naiveté of the visitor. They eat meat only on special occasion. Mainly they raise fowl and hogs for sale and exchange the proceeds for the much needed rice, salt, and few other basic items without a modicum of which a tenant cannot maintain his substandard of living.

It became obvious to us that few are the tenants who can make both ends meet from agricultural activities alone; the great majority must engage in every known nonagricultural activity that serves the needs of the community. At one meeting we had with us a tenant tailor, a tenant shoemaker, a tenant chairmaker, and in fact some twenty-odd dual occupation tenants. Some of the occupations were typically Chinese; when we pointed to a tenant, asking him how he manages to pay his way through tenancy, the tenant admitted, hesitatingly and under the good humored prodding from the audience, that he also is chair coolie, specializing in carrying brides on their wedding day!

All in all, on a field trip through the third prefecture one fails to observe a "healthy rural life," or a rural life as the much-talked about way of life. One rather observes an ill-fed, ill-clothed, ill-sheltered, overworked, and over-exploited peasantry, chained to farming for the lack of alternative occupations. Hardly anyone can fail to see that if unaided much longer they would sink completely in the mire of poverty. But aid is finally making its appearance in the form of a rent reduction program.

Remedial Efforts

The rent reduction idea is not a new one. It dates back to the Kuomintang Party Conference in October of 1926, when a 25 percent reduction in rent was officially made the most important item in the Kuomintang's agrarian platform. The maximum rent was assumed to be 50 percent, and the 25 percent reduction was to have created a standard rent of 37.5 percent of the main crop. The realization of this platform plank would have eased the rent burden of the tenants, but it was never given a fair trial. Between late 1926 and the end of 1927, Chekiang, Kwantung, Hunan, Hupeh, and Kiangsu provinces have actually issued rent reduction regulations; but the split between the Kuomintang and the Communists in 1927 and the fierce opposition of the landlords to the scheme served to impart to it a red and fatal tinge. This and the deliberate obstruction of the scheme by the provincial and local administrations caused the idea of rent reduction to die aborning.

The immediate postwar period witnessed a flurry of interest in rent reduction. In October of 1945 the Executive Yuan decreed a one-fourth reduction in rent for all of China, for one year only. This was by way of counter-balancing an item in the same order that relieved landlords from paying taxes for two years, one year in Japanese-occupied China and one year in unoccupied China. Taxes were not collected, but it appeared too late to do anything about rent reduction. On September 20, 1946, the Executive Yuan decreed a two-year rent reduction, one-eighth each year, and back to the customary rent beginning with the third year. The tenants of the formerly Japanese-occupied China were to become the recipients of the benefaction in 1946 and 1947, while the tenants of the nonoccupied China were to get their share in 1947 and 1948. The provincial assembly of Szechwan reviewed the order of Executive Yuan and decided to shift its application to 1948 and 1949.

It is a matter of public record that the Executive Yuan's decree was honored only in breach. Nothing was done about this very limited of measures to ease the burden of the tenants; the record has not a blemish of accomplishment.

What the Nationalist government would not grant the peasants in the fullness of time, at the peak of its prestige and power, it is granting now when time is short and the regime is at its lowest ebb. But regardless of the setting, the rent reduction plank of the Kuomintang is at long last being put to a test by General Chang Chun, former Prime Minister and currently High Commissioner of Military and Political Affairs of the Southwest. The outline of the scheme is contained in the "General Regulation on the Enforcement of Farm Rent Reduction," issued on July 31, 1949. The central point of the program is that from 1949 on, the amount of farm rent originally agreed upon by the landowner and the tenant or collected according to custom shall be reduced by one-fourth, and no increase in rent shall be allowed henceforth. The regulation to enforce the program in Szechwan was promulgated by the provincial government on August 29, 1949, while a separate regulation or registration and exchange of the old contracts for new ones was also promulgated on the same day. These three documents constitute the basic rent reduction legislation.

The Program in Action

Mindful of the past experience and of the fact that in Szechwan rent reduction was never even attempted, it was but natural to view the current effort of application and enforcement of the program with considerable misgivings. For the same reason official statements on the progress of the reform were heavily discounted. The observations and conclusions are based mainly on talks with the tenants and landlords; unannounced visits to contract registration points and village offices; examination of contracts, old and new; and close questioning of local officials who deal directly with the farmers.

The third prefecture is made up of ten hsiens and one special district, containing a total of 770,000 farm families, or 240,000 more farm families than in Taiwan. We visited only three hsiens and a dozen village offices and many more contract registration centers. The sample may not be considered overly large, but we feel it was quite representative of the activities, problems, successes, or failures of the program an investigator could have noted in most parts of the third prefecture.

In the middle of October it was much too early to answer the main question: Is the rent reduction program in the third prefecture a success or a failure? The real test of the successful application and enforcement of the program is whether the majority of the tenants have paid rent in accordance with the provisions of the regulation. If they did, the conclusion is obvious: The implementation is an accomplished fact. We found, however, that hardly any rent, on the old or new basis, has yet been paid. This finding was one of the first clues that called our attention to strength of the campaign to carry out the program.

Normally, in the third prefecture tenants pay their rent between September 15 and October 15, but the program had not really gotten under way until about the 10th or 15th of October. These time factors were a source of considerable apprehension concerning the outcome of the program. Talks with the farmers proved that the fears were groundless, since the bulk of the rents remained unpaid.

Many a tenant was questioned on this seemingly abnormal development. The answers can be summed up as follows: They heard about a rent reduction scheme even prior to August 15, when the High Commissioner of Military and Political Affairs of Southwest made public the July 31 regulation. The publication of the document on the day when the payment of rent usually begins created hesitancy among the tenants. To be sure, Szechwan landlordism is sufficiently aggressive and powerful that it could have overcome the wavering tenant, particularly since the landlords knew that the government of Szechwan would issue its own rent regulation only at the end of August.

This likely development was frustrated, however, by Commissioner Sun of the third prefecture, a leader of the mass education movement. He took matters into his own hands and on August 20, nine days before the government of Szechwan acted, issued a statement to the people outlining the main provisions of rent reduction, pointing out that rent reduction applies to the third prefecture, and that he intended to carry out the program with all the means at his disposal. Seventy thousand copies of his statement were distributed throughout the prefecture. Commissioner Sun called a conference of the magistrates of the ten hsiens, the chairmen of the hsien assemblies (landlords all), and the chiefs of the land tax bureaus. He instructed the magistrates to tell the people of their respective hsiens that: (a) rent reduction will and must be carried out in accordance with the law, (b) all old contracts must be registered and new ones issued, (c) failure of the landlord and tenant to renew the contract is subject to punishment, and (d) if any rent already paid is in excess of the rent stipulated in the new contract, that excess must be turned back to the tenant.

The energy displayed by Sun and his insistence that the program must be implemented is not synonymous with compliance on the part of the landlords. But the activities of the chief officer of the third prefecture are significant for two reasons: First, they are a far cry from the sabotage of similar programs indulged in by officers in similar positions in the past. The example set by the commissioner has percolated down the lower layers of the administration. One had the feeling that the administrative machine is geared to the successful promulgation of this program. The officialdom, which can greatly influence the course of the program, is for once in a constructive rather than destructive frame of mind. Second, the commissioner's timely steps have had much to do with keeping the tenants from paying the traditional rent during the latter part of August. Such payment could have had a fatal effect upon rent reduction in 1949. It is fair to conclude that the current attitude of the administration augurs well for the execution of the reform.

There are other guideposts that point to the possible outcome of the program. These are: the registration of the contracts, the exchange of the old contracts for the new ones, the trend of rent payment for the 1949 crops, and the efforts of some landlords to dispossess tenants. All these elements offer a good test of the attitude of landlord and tenant toward the rent reduction reform, and they measure the degree of success or failure of the reform.

Article 2 of the August 29 "Regulation Registration and Revision of Farm Contracts" directs landlord and tenant to appear jointly in the village office for the registration of the old contract. A copy of a registration blank gives the name of the landlord and tenant, rented acreage, original rents, and new rent. This is the first step of giving legal sanction to the new program, and for the first time in the history of Szechwan (and the prefecture in question) there will be gathered the raw material from which an accurate statistical picture of land ownership can be drawn. Article 3 of the same regulation stipulates that, having completed the registration, a new contract shall be drawn up and the original contract cancelled by "chopping" thereon: "This contract has been registered and thus nullified according to law." Article 6 provides that if the landlord fails to make a joint application for registration, the tenant may do so himself; he cannot, however, secure a new contract unless the landlord and tenant jointly apply for it. What the tenant can do is withhold the payment of rent until the landlord consents to a joint application.

The regulation on registration and exchange of contracts was promulgated on August 29, but the actual work had not begun till late September–early October. In one village of the Chia Yung district, registration only started on October 17, the day we visited it. Under the circumstances, the work cannot possibly be completed by the end of October, as called for by the regulation. This in itself is no cause for alarm. The program is new, and much more time should have been allowed to this phase originally. The selection and briefing of 506 registrars and 103 supervisors is in itself a time-consuming chore.

More significant is the fact that, now that the work is under way, the progress is being impeded by the reluctance of an inestimable number of landlords, chiefly absentee landlords, to come to the village office and apply for registration and exchange of contracts. The resident landlords do not present a serious problem; most of them live up to this provision of the regulations. Those who hold out cannot resist very long the pressure generated by the program activities all about them. It should be noted, too, that the absentees in the third prefecture are not nearly so numerous, rich, or influential as those of the first prefecture. It is easier, therefore, to overcome their opposition.

There is no question as to the attitude of the tenants. They throng the village offices or specially created registration points with or without their landlords, old contracts in hand, eager to make the change. We watched many a group—landlord, tenant, and registrar—in the process of examining the old contract, registering and voiding it, and issuing a new one. It takes about 30 to 45 minutes to complete the transaction. The tenant goes through this novel experience with the seriousness befitting the occasion. For altogether different reasons the landlord, too, is a picture of concentration. He is engaged in a most unorthodox act, signing over for the benefit of the tenant a few piculs of rice which normally would belong to him and he cannot but realize that the new contract is a more businesslike, less one-sided, landlord-dictated document. To the tenant the registration and new contract is the one proof that rent reduction is a reality. A copy of the contract is deposited with the village office, and this automatically reduces the arbitrariness with which a landlord treated a contract in the past. The most ignorant and tradition-bound tenant appreciates the significance of the event. The only problem is how to induce all the landlords to register and exchange contracts.

Many offices we visited were working overtime, writing notices to landlords that they must join the tenant for registration and exchange of contract. Some offices, in fact, have begun their activities in this manner. This is particularly true in the hsiens close to Chungking, where most of the absentee landlords live. Chen Tsi Liu, chief of the land division of Chungking municipality, himself a landlord and knowing something about the behavior of landlords, told us that he notified the absentees in writing on October 3, the first day of registration. Why? "The landlords are so busy in Chungking with their nonagricultural activities that the entire rent reduction program may have passed them by," said he with his tongue in his cheek. They must make their appearance between the 6th and 18th of October, but on the 13th, Chen was already planning to send out a second notice on the 18th—not called for by the regulation—in order to give them, as he put it, "no excuse whatever for not registering." Having done that, the tenants will register the contracts unilaterally, and registration will be completed on November 10. This was the only instance where a second notice was resorted to; in every other place we visited the one notice prescribed by the regulation was being issued when needed, and tenants were permitted to register when the landlords failed to respond.

The attitude of the absentee landlords can do no more than delay the registration. Few tenants are influenced by the recalcitrant landlords. It is safe to say that, with or without the benefit of the presence of the landlords, the tenants will have registered their old contracts by early November. The same cannot be said about the exchange of contracts, since it calls for a joint application. Tenants have resorted to such tactics as sending out sedan chairs to carry the landlords to the villages. At least in one place two landlords succumbed to this type of persuasion. There is a more telling way of making landlord comply; this is the strict enforcement of the regulation provision of "no contract—no rent."

The commissioner of the third prefecture and his administrative machine are actively pursuing that line. The mass education movement through its adult education classes and cooperatives has been enlisted for the same endeavor. Willy-nilly, a landlord must come down or send his agent to collect the rent. A tenant could then exact the price of rent, the contract. We are not prepared to say that all tenants thus involved would necessarily do that, although questions and answers on this point show that they intend to do that. It is well to recall at this point that in the third prefecture the absentees represent a minority of the landlords and that there is no village where an exchange of contracts had not taken place. Moreover, the energy with which the government is implementing the program will in all probability strengthen the resolve of the tenants to secure a new contract. By the same token the opposition of the landlord will be weakened. The commissioner of the third prefecture may be quite right in stating as he did that all the contracts will be exchanged by the end of November, but while we were there not all the evidence was in.

The exchange of contracts would not have been in question if Article 6 of the "Regulation on Registration and Revision of Farm Contracts" had granted the tenant the right to apply for a new contract if the landlord refused to make the application jointly. This limitation may not have serious consequences in the third prefecture, but its adverse effect upon the program in the first prefecture is beyond question. It was urgent that corrective measures be taken promptly. With that in mind, on October 29 we conferred with Acting Governor of Szechwan Men Kwang Pei, Executive Secretary of the Land Commission Yuan Su Chun, and Commissioner of Social Welfare Wang Tsung Chian. After a thorough examination of ways and means of overcoming the continued opposition of some of the absentee landlords, it was finally agreed that an order be issued by the governor, stating that, if within one week of the publication of the order a landlord fails to make a joint application for a new contract, the new contract will be given to the tenant upon his application. The tenant shall not pay the rent until such times as the landlord signs the contract. The order was issued by the governor on October 30.

A student of law would probably raise his eyebrows at the rough-and-ready manner of solving contractual relationships. But the fact of the matter is that the current rent reduction program is just one order after another, issued by the governor of Szechwan, or the High Commissioner's Office of the Military and Political Affairs of the Southwest. Time does not wait any longer in Nationalist China, and this certainly applies to the rent reduction program in Szechwan. Hence the advisability of quick action to cure the soft spots of the reform. There is reason to believe that the governor's order and the publication in the Chengtu press (October 26) by the order of the governor of the names of the leading obstructionists of the rent program will speed up the exchange of contracts and that of the program as a whole.

In the third prefecture as in the first, in Kwangsi, and in Taiwan, the tenants are not familiar with some of the most important parts of the rent reduction provisions, let alone with the details. They all know, however, the most significant point of the program, which is that customary rents must be reduced by one-fourth. Even the tenant in the hinterland of the prefecture, who admitted that he had already paid his rent on the old basis—even he knew of the one-fourth reduction.

Perhaps no other question came in for such detailed consideration as the one on rent payment. We raised it on all occasions and in all manner of places where a farmer or farmers were present. We were mindful of the fact that normally no tenant would admit to paying or intending to pay anything but the legal rent. The tenant referred to in the preceding paragraph is no exception. But compliance on the part of every tenant (or landlord), while desirable, is not absolutely necessary to insure the success of the program. During our stay in the third prefecture, which was the early stage of the program's implementation, the valid question was whether a considerable majority of the tenants intend to live up to the rent provision.

Developments in the next month or two will furnish the answer, but in the meantime we are inclined to the view that the answer will be in the affirmative. From our discussions with the tenants, we carried away the strong impression that the tenants appreciate in a very practical way the difference between paying less rent rather than more—that the economic position of the tenants, particularly the need to buy rice during the greater part of the year, will compel them to take advantage of the benefits of rent reductions. Their actions to date support the conclusion. For the time being most tenants have either paid no rent at all or have made token payments. They expect to pay the rent or the balance sometime after the renewal of the contract. If some of the contracts are not renewed and the rent is paid, the transactions are likely to be on the basis of the one-fourth reduction. The tenant will insist upon it, for there is the growing awareness that the landlord's weapon of realization is blunted.

A rather significant straw in the wind is the episode that took place in the Tayi hsien of the fourth prefecture. A considerable acreage in that hsien belongs to General Liu Wen-hui, governor of Sikang, and a very influential man in Szechwan. In the middle of October General Liu sent one of his estate managers with a detachment of troops to collect rent, and on the old basis. The tenants being familiar with the general's practice, did something they have not done before: they armed themselves for the coming event. When the manager and the troops appeared on the scene, the former was told that they intended to defend their rights under the rent reduction scheme and will not permit the collection of the customary rent. Whereupon the manager proceeded to collect the reduced rent. This is indeed an unique experience for both the general and the tenants, an experience that speaks well for the progress of the program.

One of the unfortunate by-products of the rent reduction program is the widespread attempt on the part of the medium and small landlords to dispossess the tenants from all or part of the rented land. This development became apparent immediately upon the publication of the rent reduction regulations and before the implementation of the program had begun. At virtually every meeting we encountered tenants who are faced with this problem. The number of such cases in all of the third prefecture is difficult to determine, but if the Chungking municipality with its 8,000 tenant families out of a total of 11,000 farm families is anything like a representative sample, then the situation is indeed serious. Chen Tsi Lin, chief of the Land Division, informed us that approximately 30 percent of the landlords have requested the return of land, presumably because they wished to cultivate it themselves. Dispossession in any part of Szechwan would inflict upon the tenant a major disaster, since it meant leaving the land and the buildings and with hardly any opportunity of acquiring another rented holding.

Any drafter of a rent reduction program in Szechwan, if at all familiar with landlord-tenant relationships, should have had no difficulty in foreseeing the eviction attempts and draw the regulations accordingly. Actually, the regulations are most deficient on this all-important point. Article 2 of the "General Regulation of the Enforcement of Rent Reduction" of July 31, 1939, states that "As to matters left unprovided in this regulation, the provisions of the Land Law and other relevant laws and decrees shall be binding." The Land Law has laid down a set of provisions, which, if enforced, would make the termination of farm lease difficult, if not impossible. Thus a lease can be terminated only in the following instances: (a) when a tenant dies and leaves no heirs, (b) when the tenant waives the right of cultivations, (c) when the tenant sublets the land to another person, (d) when the rent is in arrears for two years, and (e) when the landlord decides to cultivate the land himself. In the latter case the tenant must be given notice a year in advance.

The provisions are reasonable, except that even a superficial knowledge with landlord-tenant lease arrangements shows that they have never been enforced; the tenants have only the haziest notion about their existence, and failure to spell them out now leaves matters pretty much where they stood before. Nor does Article 12 of the "Regulations on the Enforcement of Rent Reduction in Szechwan Province" issued on August 29, which merely states that upon the expiration of the lease the landlord can take the land back if he decided to become an owner cultivator, render the tenant much protection. It is a well-known fact that tenant contracts in Szechwan seldom contain a time limit and that the life of a contract is largely determined by the landlord. For this reason the opening part of Article 12, "If upon the expiration of the *contracted period of lease* . . ." [italics W. L.] has little relation to the existing conditions, and deliberately or otherwise has given the appearance of sanctioning the efforts of the landlords to dispossess tenants.

We are glad to record the field observation that there is an enormous gap between wanting to dispossess a tenant and actually dispossessing him. The act of omission on the part of the provincial government was being corrected while we traveled about the third prefecture by the refusal of the tenant to move and by the announcement of the chief officer of the prefecture that no landlord will be permitted to take back the land unless with the consent of the tenant. The commissioner's statement was no idle threat; this was apparent as we listened to a number of landlords pleading before the land commissions for the right of eviction. It is worth noting that such incidents point up a break in the traditional freedom with which the landlord treated the tenant. Therein lies an important consequence of the application of the program even at this relatively early stage.

The step taken by Sun was an expedient to meet an emergency situation; it is not a strong foundation upon which the tenant's security of tenure could be built. The foundation is now being provided posthaste by the provincial government of Szechwan.

When the "Regulation on the Enforcement of Farm Rent Reduction" was first published (August 29, 1949), Chiang Mon-lin, chairman of JCRR, noted the inadequacy of Article 12; and our own observations of the progress of the program in the first prefecture amply confirmed Dr. Chiang's misgivings. The very first day in the third prefecture told the same story of threatened evictions. In order to arrest this development, JCRR had suggested to the provincial government a change in regulations that would secure for the tenant a three-year lease from the date of issuance of the new contract. On October 21, Governor L. D. Wang of Szechwan acted accordingly. The three-year lease provision must be inscribed in every contract. The new contracts already issued must be corrected.

A great deal more work and no little confusion is in store for those closely connected with the promulgation of the program. This is a small price, however, compared with the value of the enacted measure. The latter, coupled with the actions already taken by the prefectural officials and the resistance displayed by the tenants themselves, are quite sufficient to deflate the landlords' threat of eviction. When the dust of the rent reduction reform finally settles, the tenants of Szechwan will have achieved a measure of security of tenure. And next to rent reduction that is the most important thing.

One of the items that seems to exercise tenants most in the process of exchanging contracts is how much deposit, key, or security money should be inscribed in the new contract. The deposit money is an old Chinese practice which at one time meant paying the landlord a sum of money in exchange for obtaining the rights of permanent tenancy. More recently it has become a payment for a leasehold to be returned to the tenant upon the termination of the lease, although the ostensible reason for the deposit is to insure the payment of the rent. This system of rent deposit is particularly dominant in places where industry and commerce are poorly developed, communications difficult, and the number of tenants unusually large. Szechwan fits all these conditions. The amount of the deposit varies from region to region as well as within region. A landlord sets the deposit on the basis of what the traffic will bear. In the third prefecture it ranges from 10, 20, and more silver dollars per mou. In all cases where we inquired into the value of the deposit when originally made, it was almost twice the value of the crops harvested by the tenant.

It is difficult to describe the utter confusion and bewilderment among the tenants on the subject of deposit money at the present time, and certainly on no other matter have they been so eager to unburden themselves and with so much bitterness. The cause is not far to seek: Of the few tangible assets that a tenant possesses the deposit money is the most important one. In recent years that asset has been melting away; and now that the time has come to state its amount once again, many a tenant realizes what he has known all along but refused quite to admit, namely, that his deposit is all but gone.

The importance of a deposit to the tenant and particularly the role it might play in the future became quite clear in the course of a dinner conversation with the commissioner of the third prefecture on the eve of the field trip. Commissioner Sun was enlarging on the possibility of a land purchase scheme after the completion of the rent reduction program. There was no disagreement with the soundness of his suggestion, and we inquired how he was going to finance the project. "Why," said he, "we shall do it through a land bank." "Well," said we, "and what are you going to use for money?" "Simple," said the commissioner. "We shall put the deposit money to work." He proceeded to tell us that a survey of a district made some years back revealed that the total amount of deposit money represented from 15 to 20 percent of the price of the land and that was typical of the entire third prefecture. We agreed with the commissioner that such a volume of deposit money could serve as a good foundation for a land purchase program. We agreed also to inquire in some detail into this problem in the course of our field trip. We did just that, and, sad to report but true, the resources of the would-be land bank kept on dwindling as we continued on our trip; and by the end of the trip it became clear that the commissioner's information was quite dated. The money was not there in a great many cases. Inflation and, above all, the privilege of the landlord of changing contracts at will have seen to that.

Take, for instance, the case of one of the biggest of tenants, Yen Yan Tsu of the Taiping village of the Chungking municipality. In 1929 he rented 50 mou of land, equally divided between paddy (rice) and dry land. He paid the landlord 700 silver dollars as deposit money. At the then prevailing purchase price of rice land, tenant Yen could have bought 10 mou of land, while in terms of the value of the entire holding the deposit money represented from 25 to 30 percent of the value.

And now for the tale of woe as revealed by the dispute between the landlord and tenant before signing the new contract. It developed that in 1946 the landlord decided that 700 silver dollars he received in 1929 depreciated to a point that their real worth was no more than four piculs of rice, which could be purchased at between 30 to 40 silver dollars. Having reached this decision, he cancelled the old contract and made the tenant sign a new one, where he entered the tenant's deposit as four piculs of rice. Being a man of sound business principles, the landlord did not fail to take back the tenant's receipt for the 700 dollars. All reference to tenant's 700 dollars has disappeared, and with it the tenant's most important tangible assets. Now that tenant is about to get a new contract as part of the rent reduction progress, he insists that the item under "deposit" should read "700 silver dollars," and not "four piculs of rice."

This case is quite typical, although there are many variations on the same theme. The amount of land and deposit are usually smaller. The shift from the substantial deposit to almost no deposit may have come about not in one change but in the two or three changes of contract over a period of years, each change reducing the value of the original deposit. Every change in contract was the landlord's insurance against inflation.

The number of deposit cases is difficult to estimate accurately; a show of hands at well-attended farm meetings indicated that more than a third of the tenants present were deposit sick. Officials of the Chungking municipality believe that the percentage is much higher; they estimate that in the municipality 80 percent of the contracts were changed by the landlords immediately after the war. Whatever the landlords' motives, the fact is that large numbers of tenants in the third prefecture became separated from their deposits. Commissioner Sun had to admit finally that a land purchase scheme through the use of deposits could not possibly succeed. The affected tenants lack the means of renting a few mou of land if faced with the necessity, let alone of making a substantial initial payment toward the purchase of the same mou.

Whether through ignorance or design, the rent regulation contains no reference to the deposit problem. It accepts the practice, a practice that has not improved with age. The theory that deposit money serves as a security for nonpayment of rent is not in accord with the facts. We heard of no disputes on that score, and it is quite understandable; failure to pay rent means dispossession, and no tenant wishes to face that prospect. It would have been the better part of wisdom if the rent regulations provided for some restrictions upon the deposit system with an eye to its eventual elimination. The rent program of Taiwan provides for that. It is startling, too, that the drafters of the rent program failed to take notice of the wave of postwar change of contracts and its real meaning.

Tenants are most vociferous now about the contract changes and changes in the deposit value, but in those days they accepted them with hardly a protest. They were given a contract to sign and that was that. It is in the plus side of the current program that they feel safe in calling for the settlement of the old deposit score, and with some measure of success.

Commissioner Sun, on his own initiative, issued an order providing for the change of contracts on the basis of those in existence prior to 1946. The difficulty with the order is that many of the pre-1946 contracts and the original deposit receipts are not obtainable. There is no end of dispute as to what those contracts did or did not contain. More far-reaching is the order of the High Commissioner of Military and Political Affairs of Southwest, issued on October 7. It states that the dollar of a deposit made prior to 1938 is equivalent to the silver dollar now in circulation. Deposits made from 1938 on should be converted to an equivalent of rice that a deposit could have bought at the time it was made. That volume of rice in turn should be converted into silver dollars on the basis of the current price. The end product should be inscribed in the new contract as the deposit money.

From the tenants' point of view this is a sound decision, but it is much easier to state the process of conversion and reconversion than to carry it and with a considerable measure of success. The landlords and the tenants are poles apart on the amount involved. Yet, on balance, the measure will be of benefit to many tenants. It will never restore to them anything like the volume of their former deposits; most likely, the measure will lead to numerous compromises out of which a tenant's new contract will show a larger deposit than otherwise. There is the possibility of another benefit. The deposit money issue, which comes into the open under the impact of the rent reduction program, has placed the deposit practice itself in an unfavorable light. It should be easier now to revise it out of existence if an attempt in that direction were made. And that would not be a day too soon.

An important and sound aspect of the rent reduction scheme is the creation of land commissions in every hsien (county), municipality, hsiang (village), or tsen (market town). Their primary functions are to assist in the work of rent reduction and to settle tenancy disputes. The hsien or municipality land commission is essentially a government body. One finds on the commission the magistrate or mayor; a representative from the hsien assembly, court, and farmers' association; a rent supervisor; and two representatives from the Bureau of Agriculture and Forestry and two from the Bureau of Social Affairs of the hsien. Almost everybody is there but a working farmer; the representatives of the assembly and farm association are landlords.

The composition of the hsien or municipal land commission is rather unfortunate because every land commission, regardless of its level, could and should become a training center for the participation of the tenants in the affairs of the community, out of which a new farm leadership might in time arise. The failure of the hsien commission to include tenants among its members is partly corrected by the hsien or tsen land commission. The village or land commissions are supposed to consist of: the village head, who is the chairman of the commission; one representative of the village assembly; one representative from the village farm association and/or a representative of the people known for his scholarship; two landlords; two owner cultivators; and four tenants. A total of eleven or twelve members as the case may be. In practice we found that the number of representatives of a local commission is rather flexible; some have six members, others nine, and still others twelve as required by the regulation. But whatever the number, the tenants are fairly well represented. One of the commission we watched in action was made up of twelve members including four tenants; in this particular instance the commission was divided into three groups in order to expedite a large number of disputes then pending.

A feature typical to all land commissions is that they are appointed bodies, the hsien magistrate appointing the members of the hsien commission and the village head doing the same for the village commission. A great deal of power, which may not always be exercised for the welfare of the tenants in whose behalf the reform has been designed, is thus concentrated in the hands of the village heads and magistrates. This matter was touched upon repeatedly in the talks with the officials, and the suggestion was made for the introduction of an elective system whereby each group of the farm population—tenants, landlords, and owner cultivators—elect their own representatives to the land commissions. The suggestion received polite nods; in the eyes of the administration the idea of a secret ballot in the hands of the farmer is even a greater break with the traditional way of doing things in rural China than the rent reduction is or a land purchase program might be. It is understood, too, that, even if the provincial government were convinced in the soundness of the proposal, it is much too late now in the midst of the program to make any changes.

In the middle of October the local land commissions were barely a month old; a good many were only a week old, while in some instances the village head was only making the appointments to the commissions. Yet, even at this early stage it is fair to say that they are not paper organizations. The commissions already organized, which means the great majority, are actually engaged in the important work of examining and attempting to settle disputes that have arisen in consequence of rent reduction. The disputes cover the entire field of landlord-tenant relationships, but during October deposit money and efforts of landlords to evict tenants were the chief causes. Instead of taking their troubles to the court, a time-consuming and costly procedure, both landlords and tenants are displaying a strong tendency to make good use of the commissions. The decisions of the latter are not binding, but in the majority of cases the parties to the dispute accept them. One village land commission dealt with twenty-nine disputes, and the decisions in twenty-two were acceptable to landlord and tenant.

In the course of the field trip in the third prefecture, we had occasion to watch two local land commissions in action. We are quite certain that the proceedings were not put on for our benefit; we just happened to stop at two village offices where the land commissions were in session. The commission of Yang Chia Tza village was deliberating on a would-be eviction case, in the presence of landlord Hsaio-Tung Fei, pleading for the eviction and tenant Chin-I Chang, arguing against it.

Landlord and tenant were quite remarkable in the presentation of their respective arguments. The landlord was very much the "beaten" man, trapped as he was by the issuance of a new contract and the statement of Commissioner Sun that no dispossession will be permitted (the three-year provision came into the picture a few days later). He was clearly in the novel and uncomfortable position of not being able to dictate to the tenant in the manner of the very recent days; the tenant, on the other hand, was "feeling his oats," as it were and probably for the first time. He seemed quite certain of his rights, evidently entertained no fear of retaliation on the part of the landlord, and with calm deliberation was telling the landlord that he was not going to move from the land. Both parties were subjected to close questioning by the members of the commission, and after an all-morning session rendered a unanimous decision in favor of the tenant. Since the decision is not binding, the commission spent a good deal of time trying to convince the landlord of the futility of taking the case to the court. While we were there the landlord did not say "yes" and he did not say "no," but regardless of the next step this particular landlord might take, the decision of a commission cannot be disregarded without incurring at the same time some censure of the farm community. This explains to a degree why in the majority of cases the disputants accept the decisions as final and binding.

To observe a local land commission in session is to become aware of one of the brightest features of the rent reduction program. For the first time in the history of rural Szechwan tenants are beginning to participate in affairs that affect deeply the life and work of the largest group in the community. In so doing, the tenant members of the commission not only protect the interest of their group but receive a practical training in leadership, which prior to rent reduction was the sole preserve of the landlords and of local and district officials, largely drawn from the landlord class.

The work of the village land commissions goes, therefore, far beyond the specific intent of the rent reduction regulations; even as constituted at the present time, they bid well to provide in time a farm leadership which is lacking in Szechwan and without which the full benefits of the most sound programs of economic, social, and technical improvements cannot be realized.

Conclusions

Such are the principal observations on the progress of the rent reduction program in the third prefecture. Much of the same can be said of the first prefecture, where we visited in early October. If they describe fairly accurately tenancy conditions and efforts to remedy them in the other prefectures of Szechwan, then it is reasonable to conclude that rent reduction, a measure of security of tenure, and the beginnings of a change in landlord-tenant relations and attitudes are taking place. The question whether the rent reduction program has been carried out will have to wait for an answer until the end of the year when all the evidence, favorable or unfavorable, will be in. But what is available now shows that a great deal has been accomplished and, barring the Communist occupation of Szechwan within the next few weeks, a successful completion of the program may be anticipated.

It is rather surprising that so much has been achieved in so short a time. The last regulation was made public only at the very end of August. The regulations themselves have been devised in the greatest of haste (and carelessly) with hardly any time allotted for the assimilation by the farmers of its principal points and with very little time for its implementation. In short, what the Nationalist government failed to do in the twenty-three years since rent reduction became a plank in its platform, the government is attempting to push through now in less than six months. It is in the nature of a minor miracle, therefore, that the program is being carried out.

The rent reduction program in Szechwan, as in Kwangsi and Taiwan, was devised by the government and is being implemented by the government with the aid of JCRR. In order to carry out the program, the provincial government draws on a huge number of people for assistance. A breakdown by administrative categories, from the highest to the lowest, shows that more than 100,000 people are devoting to the program at least part of the working day. The field trip in the third prefecture offered a good opportunity to observe the work of some of this personnel. The impression is a favorable one. Most of them are not zealots for the cause of the betterment of economic conditions of the tenants. But among the highest strata of the administrative machine one finds conscientious workers like Commissioner Fong of the first and Commissioner Sun of the third prefectures, while Chen of the Peipeh special district is the shining example of a man who day in and day out works with and for the farmers. The bureaucrats in between take their orders and act upon them if sufficient pressure is applied from the top. That they have had in good measure in recent weeks. The moral of it is that even in Szechwan, China's most landlord-ridden province, not only a rent reduction program but a reform that goes far deeper than that can be carried out if there is the will at the top to do it.

Another observation gleaned from personal contacts in the field in Chengtu (capital of Szechwan) and in Chungking is the close cooperation between Chinese officials concerned with the program and JCRR. It is expressed in deed and on important issues. The measures taken to fill in the gaps in the rent reduction regulations are proof of that. Moreover, JCRR did not have to exert strong pressure to bring about changes deemed desirable. A suggestion made around the table and a statement of the reason prompting it were quite sufficient. A cynic might say that the Chinese officialdom sees the writing on the wall, and what it is doing in effect is fixing its fences in relation to the Communists and the tenantry. This is probably so in many cases, but whatever the motivation a reporter on the current rent reduction activities in Szechwan cannot help but record the spirit of cooperation between the provincial and prefectural administrations of Szechwan and JCRR. In a large measure the successful application of the program to date can be traced to the harmonious working of the two agencies.

In addition to expert advice bearing on the rent reduction program, JCRR has contributed $218,000 for the purpose of helping to carry it out. The latter part of the statement may sound like a stock phrase, but the fact that the money is being spent for that purpose only is illustrated by the following allocation of the financial aid:

	Silver dollars	*Percentage of total*
Supervisor salaries and travel expenses	99,732	30.8
Registrar salaries	67,120	20.8
Subsidy staffs land commission, hsien and hsiangs	65,750	20.4
Printing contract forms	40,000	12.4
Printing publicity material	21,061	6.5
Contingency fund	29,366	9.1
Total	323,029	100.0

The supervisors, the registrars, some of the staff of the land commissions, certain technicians in the hsien and hsiang offices, printed contracts, and publicity material—these are the people and things on which the money is spent. More specifically, the financial aid of JCRR in five prefectures[1] of Szechwan made it possible to engage the services of a total of 2,249 individuals, or 2 percent of the total number of people concerned with the program. The 2,249 individuals are composed of 17 technicians of the provincial land committee, 554 supervisors and 1,678 registrars—persons who are most instrumental in the daily work of making the application of the program possible. Having seen them in action and the use to which the printed contracts and publicity material are put to, this observer is firmly of the opinion that the rent reduction program could not have been applied and surely not carried out without the services of the people and printed word provided by the financial aid of JCRR. Hence the conclusion that, with all the goodwill on the part of the provincial government and with a perfect set of rent reduction regulations, the chances are that the program would have never seen the light of day but for the financial aid of JCRR and the efficient manner in which that aid is being utilized.

The financial contribution of JCRR toward the realization of the rent reduction program in Szechwan (and in Taiwan) points to a pattern of aid that we do well to remember: A great deal can be accomplished with relatively small means in the right place. Approximately 350,000 tenant families in Taiwan (67 percent of the total), 1,700,000 tenant families in the five prefectures of Szechwan (roughly 60 percent of the total), and an unknown number of tenant families in the other eleven prefectures are recipients of the aid. A dime per family is what the aid amounts to. It has been a source of constant wonderment to this observer that $30,000 in Taiwan and $218,000 in Szechwan could do so much for so many tenants in helping to fill their rice bowls and in widening their horizons. General Liu Wen-hui's defeat at the hands of the tenants is a good example of that; it is a break in the cake of the customary landlord-tenant relations that transcends the significance of a reduction in rent. It is reasonable to conclude, therefore, that a JCRR financial aid many times larger would have been amply justified by the results already achieved.

Another corollary of the rent program bears on United States foreign policy in and outside of China. The program and the aid rendered by ECA through JCRR are shaped with the view of lasting benefit to the Chinese farmers. In doing that, it has demonstrated in a practical way that the Americans can be not merely against certain ideas, which the Chinese farmer does not begin to comprehend or has any interest in, but also for ideas which are close to the heart and are understood even by the most ignorant of Chinese peasants. The rent reduction work, therefore, in its immediate and direct impact upon the farmers of Szechwan or Kwangsi, leaves a mark that may not be expunged. The seeds thus planted may not all be lost when the Nationalist government has disappeared from the mainland of China and the Communists have taken over.

An observer of the current efforts in rural Kwansi and Szechwan cannot help reflect on how much more could have been achieved by the Nationalist government, particularly in the political field, if the reforms were promulgated

1. The total number of prefectures in Szechwan is sixteen, but the aid extended to the eleven prefectures is rather small. The five so-called "designated" prefectures account for 44 percent of all the farm households of Szechwan and 40 percent of the cultivated land.

at a more propitious time. All it had to do was to recognize, earlier in the day, that in peasant China (or peasant Philippines, India, Siam, Middle East, and so forth) political power must rest upon peasant support and that the condition of the support is filling in the empty rice bowls or adding to the half-empty ones. The evidence gathered on the field trips is that the quickest and most expeditious way of alleviating China's discontent grounded in riceless bowls is to change and improve the existing land tenure arrangements. Hence the conviction that the failure to implement the Chinese agrarian legislation of the 1920s has not only tended to keep a huge number of peasants on the ragged edge of penury without any hope of relief but has, by the same token, widened the breach between the Nationalist government and the peasants. This point is brought up not in the spirit of what might have been; the lost political opportunities in rural China and their adverse effects upon the fortunes of the Nationalist cause cannot be undone now. It seems reasonable, however, to draw attention to the lesson of this experience, obvious though it may seem, if the repetition of the same errors are to be avoided elsewhere.

One of the outstanding impressions of the trip through southwest China and Taiwan is the enthusiastic response of the tenants to the reform recently instituted. Their attitude does not stem from any final solution of their problems, final in the sense that most of them were about to become owners of fair-sized holdings that would enable them to live happily ever after; there is simply not enough land in relation to the multitude of farmers to bring that condition about. But farm relief in China or in Asia in general is not a question of a whole loaf or none at all; as far as the farmers are concerned, any efforts to ease their burdens—even if it be of a limited nature—is welcome. The tenant of the Chengtu Plain of Szechwan who produces the rice crop but is compelled to buy most of the rice he consumes is grateful to his government for the aid that will add so many more basketfuls of rice to his meager larder, give him greater security of tenure, and curtail the arbitrary actions of his landlord. Any number of cases observed on the trip can be cited to illustrate the point. The tenant in the village office in Kwangsi, half crazed with joy over the receipt of a deed that gave him ownership of one acre of land that he worked as a tenant for thirty-two years, spoke without words for all others in similar positions. Nor is there any doubt about the state of mind and willingness of the tenants of Southwest China and Taiwan to support a government that has finally brought them hope of a better day through rent reduction. It is quite clear, therefore, that a wise government could have made a great deal of political capital, and perhaps achieve military and political stability, if the farm aid made its appearance in good time. After all, China's armies are peasant armies.

What is true of China is essentially true of other parts of the Orient. Every densely populated farm area of Asia counts land-hungry or just-hungry tenants by the million. Most of them would sell their souls to their governments for a piece of land in fee simple or for reasonable tenure conditions. The only question is whether those governments will heed the lesson of China's disastrous experience. It behooves us to make sure that they do. If for no other reason than enlightened self-interest, in the contest with the Communists in Asia, the United States cannot be friendly to agrarian feudalism simply because we are against Communist totalitarianism. Our attitude should be one of positive support of agrarian democracy. We should lend our influence and prestige in whatever form possible to the agrarian reforms already in being and those yet to come. We shall thereby help cut the political ground from under the feet of the Communists and aid the forces that make for a middle-of-the-road, stable rural society.

13. *Too Late to Save Asia?*

This short piece, written not long after Ladejinsky's return from Taiwan and Szechwan, is characteristic of his occasional attempts to inform and influence public opinion in constructive directions. It came at a time (shortly after the Berlin blockade, the creation of NATO, and the Communist takeover of China) when the cold war was at its chilliest and fears were widespread that the rest of Asia would shortly be lost to Communism. Amid the hysteria of that time, Ladejinsky's remained a cool voice, affirming that it was not too late—that democratic agrarian reform was the way to save Asia.

This article appeared in the *Saturday Review of Literature*, July 22, 1950, and is reprinted by permission.

THE CHINESE PEASANT SEEMED out of his mind. He kept jumping up and down, making strange gestures with his fingers and palms, working them in a semicircular fashion around his mouth. My interpreter chatted with him briefly, then explained the situation to me. It was the paper the peasant held in his hand that caused his hysteria. That paper was the deed to the single acre of land which he had worked as a tenant for the past thirty-two years. Now that precious acre of land belonged to him!

This scene, which I witnessed not long ago in the courtyard of the land office of the small village of Kwangsi, deep in the heart of China, epitomizes the problem and promise of all Asia. It sums up the present and perhaps the future of a continent where agrarian discontent is gnawing at the vitals of the social order. It is on this strife the Communists have been able to capitalize so successfully by posing as advocates of reforms designed to benefit the peasantry.

If the Chinese peasant of Kwangsi who seemed mad with joy is the symbol of the promise, the threat was stated by Kipling:

> His speech is of mortgaged bedding,
> On his kine he borrows yet,
> At his heart is his daughter's wedding,
> In his eye foreknowledge of debt.
> He eats and hath indigestion,
> He toils and he may not stop;
> His life is a long-drawn question
> Between a crop and a crop.

Kipling was not one to neglect his poetic license, but his picture is not actually overdrawn in hopelessness. For in 1950 the Asiatic peasant is no better off than he was in the late nineties when Kipling wrote. The Communists are translating Kipling into political language. They know that Asia's problem is the problem of land.

Four-fifths of the continent's vast population are peasants. Agriculture, not industry, is the pivot of its economic life. Industry has made but a small dent in the character of Asia notwithstanding oil gushers in Iran, tin mines in Malaya and Siam, jute and cotton mills in India. The ambitious postwar schemes for industrialization throughout Asia as yet remain mostly blueprints. The factory may bring material advancement to the Asians someday, but that day is in the distant future. The heart of the problem of Asia today lies in the countryside. It is on the farm where solutions must be sought and found.

In the Chengtu Plain of Szechwan, the richest granary of China where I traveled last fall, the farmers who had just harvested excellent crops seemed concerned with just one idea—how to secure enough rice for themselves and their families, how to fill the empty rice bowls and add to the half-empty ones. The discontent I sensed in Szechwan is typical of the many regions I visited in recent years in other parts of the Far East, Southeast Asia, and the Middle East. An overworked and overexploited peasantry that for centuries was inertly miserable is now alertly miserable.

The reasons for the age-old wretchedness of the Asiatic peasant can be summarized in a brief sentence: too many people, too little land. Pitifully small holdings, inadequate tools, and institutional arrangements over which the peasant has no control intensify his plight. A large part of the misery and hunger arises from the relationship between the peasants and the landowners, a relationship which the courts and local officials have given official sanction through the centuries. The peasants have been obliged to pay exorbitant rentals, for they have no bargaining power. Nor is their contract of tenancy worth the paper it is written on; it may be altered or abrogated at any time at the whim of the landlord.

The peasants of Asia have never been satisfied with this state of affairs. They have often raised horny hands against persons and governments which they believed were the causes of their distress. But in the main, until very recently the conservatism and inertia of the farmer and his ingrained, feudal subservience to the state and to his landlord kept the pot from boiling over. Now the forces that keep the peasant within well-defined bounds are breaking down under rising agrarian discontent. The peasantry is at last in motion. The Communists have exploited this fact and placed it in the center of Asiatic politics. It is only one side that has known how to exploit this overwhelming question and to place it in the center of Asiatic politics where it belongs.

The catastrophe of China is a case in point. Many are the reasons that explain the victory of the Communists and why in 1949 Nationalist China could not boast of a single Leonidas holding a single Thermopylae. But one cause seems to me beyond dispute: Nationalist China was pressed and pushed over not so much by force of arms as by the Russian Communist tactic of giving land to the poverty-stricken, landless, hopeless peasantry.

Is there any wonder that peasants living on the ragged edge of penury are easy marks for Communists who are masters in the art of exploiting agrarian discontent for their own political ends? The peasants know nothing and care less about Marxism, Leninism, and Stalinism, and they are surely not eager for collectivization. The Communists, however, promise them not collectivization as it exists in the Soviet Union but land with which the peasants may do as they please. And the peasants, in sheer despair, believe the promises, if for no other reason than that thus far in Asia the Communists have not yet had the opportunity to betray their promises as they did in Russia when they eventually nationalized and took away the land from the peasants and herded them into collective farms at the point of a bayonet.

Every revolution invokes the name of Liberty, but the concept of liberty is always shaped by the existing economic and social needs. For the Russian peasants liberty meant the ownership of the lord's land. The Communist promise in 1917, which ultimately was broken, that the land would belong to them fell on the eager ears of peasant soldiers who promptly left the front lines, rifle in hand, and went home to divide it. As Lenin put it, taunting his opponents, the peasants voted with their feet for distribution of the soil. Thus, Lenin and his party succeeded in "cornering" what Karl Marx called "the peasant chorus without which the (proletarian) battle cry will degenerate into just another swan song." The Communists would never have obtained power in Russia had they not successfully exploited the peasants longing for the landlords' acres.

Toward the end of his life Lenin despaired of a Communist victory in Western Europe. He visualized the final crucial battle as a conflict between a Communist East and a capitalistic West. In this struggle China and India were to join Russia as the forces of Communism. To win Chinese and Indian support, Stalin developed a program for those countries consisting of three stages: a struggle against foreign imperialism, an agrarian revolution under the leadership of the Communist Party, and finally a proletarian dictatorship. The key step was to be the wooing of the peasants.

It is in connection with this step that the Chinese Communists have assumed a role which has led many observers mistakenly to assume that they are "mere agrarian reformers." Actually they played this role because, according to Communist theory, the road to political power and dictatorship in China lay through the sponsorship of agrarian reform—"land for the landless." To satisfy the innermost needs of the peasantry was only a means to a large end which has little to do with the welfare of the peasants.

Stalin's three-page program of revolution, as exemplified by China, is now being exported to Indo-China, to Burma, the Philippines, and Indonesia, and even into little-known Hyderabad. The revolt last year in Hyderabad was a peasant uprising, about the largest, and for a brief moment perhaps the most effective in Asia outside of China, and it was under Communist leadership. The uprising was finally crushed and the Communists jailed, yet it is worth noting that as in China the peasants of Hyderabad were not enticed by Communist dogma but merely followed the welcomed Communist bait of "land and liberty in Asia."

There are fertile fields for the Communist strategy in other parts of India and Pakistan. I found plenty of evidence of this in a visit to a typical Pakistan village during the spring of last year. All one hundred families who resided in the village were tenants of a single absentee landlord. Yields had been declining. I asked the assembled villagers for the reason. "Our land," replied one of the tenants, "is like a pitcherful of water from which we keep pouring out while nothing is being poured in." "Why don't you pour something in?" I asked. "How can we?" countered the tenant. "The landlord's share is so high that we would benefit very little." The tenants had no incentive to improve the land. The landlord received too much and the tenant too little.

Many people wouldn't hesitate to approve of a revolutionary movement if it is the only way the common man can secure his elementary wants. But we must realize how serious a threat an agrarian revolution could be at this point of history, even if the upheaval seems justifiable from that point of view. The only way to thwart Communist designs on Asia is to preclude such revolutionary outbursts through timely reforms, peacefully, before the peasants take the law into their own hands and set the countryside ablaze. But reforms, if they are to have a lasting effect, must come not only from opposition to Communism but from an honest purpose and plan to raise the status of the peasantry.

Encouraging in the seemingly dismal situation in Asia is the fact that the American agrarian tradition of "forty acres and a mule" or an Asian variety thereof offers a better solution of the agrarian problem than the new Communist ideology. This is quite apart from the consideration that the latest events in China and the implied threat to the rest of Asia oblige us to bolster our position there with something more lasting and effective than we have done thus far. The positive and prompt decision to place the peasant "in the center of the piece," as Nehru once put it, is essential if non-Communist regimes are to survive.

According to Nehru, India and Asia as a whole are confronted with no more urgent tasks than agrarian readjustment. This presupposes: (a) the improvement of agricultural techniques in order to raise farm production; (b) an approximation to an adequate standard of life through raising the output per man, providing good credit and marketing facilities, fair taxation, health, and educational facilities; (c) proper utilization of land and water resources; and (d) a basic change in landlord-tenant relationships, which in certain parts of Asia is immediately the most important objective, both intrinsically and because of the threat of Communism. Technological improvements alone will fall short of our goals. Increased production must be accompanied by a fairer distribution of farm income. Improvements in the use of land can come about most effectively if the tiller knows that the land or a reasonable share of the product of the land will belong to him. Otherwise not all prerequisites will exist for the technical progress envisioned, for instance, by the Point IV program, and our political aims will be made more difficult to attain.

Douglas MacArthur in Japan, Nehru in India, and, more recently, General Chen Cheng of Formosa have understood the urgency of taking the wind out of the Communist sails in a peasant ocean. MacArthur knew how to do it and did it. A successful land reform under his direction has created in Japan a new, large class of private owner cultivators and has rendered rural Japan practically impervious to Communism. Nehru is trying hard to do it. India's needs for technical farm improvements are overwhelming, and Nehru is well aware of it. But he is just as conscious of the urgency for a concurrent land reform program which will give the Indian peasant an incentive to improvement and a sense of responsibility.

Nehru's motivation is economic, social, as well as political: "If we don't do it, they will." And "they" are the Communists.

Suggested farm reform can become a powerful political instrument. The native governments friendly to us would be more likely to win popular support, and popular support in Asia is "peasant support or nothing." An owner cultivator or a reasonably satisfied tenant would acquire a stake in society. He would guard that society against extremism. Private property would be strengthened where it has been weakest—at the huge base of the social pyramid. The common man of Asia would become a staunch opponent of Communist economics and politics—not necessarily to favor the interests of the United States but simply because his own interests lay in the same direction.

Secretary Acheson summed up the causes of Asia's tensions in the San Francisco address on U.S. policy toward Asia. "They (the Asian people)," he said, "have been striving for independence, better education, more widespread ownership of land, and control over their own destiny." He continued, "It is no accident that their goals and our goals are the same." But while we are busy enriching the economic, political, and social institutions the American people have already achieved, Asia is yet to create the beginnings from which a democratic society may evolve. Such conditions cannot be handed to Asia as a gift; they must grow out of Asia's wishes, opinions, and activities.

Nor is this easy. For Asia, unlike Western and perhaps also Central Europe, lacks almost entirely the traditions, institutions, habits of thought, and experiences which are essential to democracy. That indeed is the principal political and psychological reason why Communism is a much greater threat in Asia than in Europe. The natural defenses against Communism that a tradition of individual dignity, the rights of man, and of democracy sets up in different degrees in various countries are yet to be created in Asia. We are still to demonstrate in that part of the world that democracy is better and more profitable too. And it is through helping to solve the problem closest to the heart of the peasant that such a demonstration can be made. That is how the rural program here outlined, if realized at least in its major parts, would bring the Asian people closer to the goals set out by Secretary Acheson.

The need for U.S. material and technical assistance in Far Eastern programs of agricultural improvement is accepted as an integral part of our policy in Asia. The Point IV program is a good example of that. Another instance in the same direction is the economic aid being rendered now to Southeast Asia by ECA. Less widely accepted and often even unrecognized is the necessity to dissuade the recipients of our aid in Asia from allowing narrow institutional interests to stand in the way of a progressive rural policy, through which the maximum economic and political benefits of our technical and financial assistance could be realized.

Whatever we may contribute to Asia's advancement and stability—be it in the form of dollars, of technical guidance, of organizational advice, or of military assistance—our policy and all our diplomatic competence and tact should be actively and sympathetically guided by the knowledge that the foundations of the social structure stand or fall in the countryside and that the peasant and his interests and aspirations must be in "the center of the piece." We must make an effort to persuade the more conservative Asian groups that rural reform is essential to their own preservation as well as in the interest of the peasantry. Provided such basic attitudes are developed, here and in Asia, the United States could begin to supply the mechanisms of reconstruction and effectively employ them.

Secretary Acheson is addressing himself to the core of the problem when he urges that "We must increasingly in all we do and say affirm the positive goals of a free people." If Mr. Acheson's words are to carry meaning for the common man of Asia, we cannot remain indifferent to the dominance of the traditional beneficiaries of the prevailing feudal systems. From ambassadors and ministers to foreign service clerks, we must begin to feel and act in terms of the common man, and in Asia he is the peasant. We must make a special effort to seek out and encourage in every way possible the native liberal groups who might otherwise be lost to our common cause. It is in the vast numbers of the cultivators of the soil and in the small yet important elements concerned with the welfare of the people where we would be well advised to find and win our most potent allies. A strong, fairly contented peasantry and progressive native forces supporting them provide a more solid base for cooperation with the West and for the struggle against Communism than the forces at work in many parts of Asia to whom the very words "reform," "change," and "concession" partake of the devil. They suffer from myopia, a disease likely to be fatal. The landlords of Szechwan refused to support a mild reform program even though the Communist armies had already penetrated the gates. In the rest of Asia, too, landlords also may be their own gravediggers—and of their own governments.

In country after country one sees feudal classes bent on maintaining the status quo unsullied. They cannot gain popular support. They neither benefit the rural community nor have they proved at all effective against Communist penetration. They are the unwitting and unwilling allies of Communism, for they are the creators of the revolutionary situation. It is these and such forces that today hold great and decisive power in Asia. Without their consent and support no progressive organization can function. One of the basic tasks of the United States foreign representatives abroad should be to convince these groups and their governments to follow the example of Tzar Alexander II, who saw in time that his only chance of keeping the throne was to initiate reform. He warned the Russian landowners a hundred years ago that "it was better to begin the abolition of serfdom from *above* than to wait for it to begin itself from *below*." He abolished it from above, and the Romanov dynasty retained the throne for another half a century until Alexander's less wise grandson lost it.

Fortunately, we already have a technical and social agrarian program that has been developed and successfully tested by the Joint Commission on Rural Reconstruction (JCRR). This organization, created in China as part of ECA under authority provided in the China Aid Act of 1948, held its first meeting in Nanking on October 1, 1948. The work of JCRR came too late to alter the course of events in China itself. But what I saw of the JCRR's operation in China and in Formosa, where the operation is still going on, convinces me that it was one of America's most fruitful ventures. It points the way to what might be done on a larger scale and in better time elsewhere.

JCRR is not an American but a joint American-Chinese commission, with a Chinese chairman and two Chinese and two American members. We Americans furnish the money but do not call the tune. With nationalism in Asia at white heat we must be careful not to do so. The commission works through local organizations and finds local sponsors among them for its projects. These projects are not based on preconceived theoretical notions but on indications as to what the farmers themselves wish most. From such grass-root needs JCRR has evolved a practical national plan with this order of priority: (a) land reform, (b) irrigation, (c) fertilizers, (d) farm organizations, (e) farm credit, (f) plant and animal disease, (g) seed multiplication, (h) animal breeding, (i) rural health, and (j) audiovisual education.

The response of the Chinese and Formosan peasants to JCRR has been enthusiastic, although the commission has actually spent little money —only five million dollars from its inception to date. As I traveled from village to village to see JCRR in action and talked with people in all walks of life, I was impressed by the prompt results of this grass-roots diplomacy. It made more friends for the American cause and did more actual good than many an elaborate diplomatic act backed by the richest of treasure.

Most significant was the reform of the tenure system. "Why," I asked Chiang Mon-lin, chairman of JCRR, "are you engaging in this work when the landlords are certain to oppose you?" "To give meaning to the nine other branches of our work," Chiang Mon-lin replied. "We know that using science to increase production is relatively easy, but solving social problems is difficult. If we had not developed a land reform program along with our production program, the better part of the good results would have gone to the landlords and not the tenants. That would defeat our purpose."

JCRR did not create or impose agrarian reform upon China. Its Chinese chairman convinced the Nationalist government that rent reduction and land purchase programs, which had already been on the statute books for more than two decades and never applied, should be given a trial if only as enlightened self-interest.

Within the short space of a few months, the land reform produced substantial and immediate results in terms of greater security of tenure and more rice for the tenants. It is questionable if years of farm improvements would have given the tenants benefits anywhere near as great as these.

Remarkable also was the manner in which tenants asserted their new-found rights. This was impressed upon me as I watched the land commission of Yang Chia Tza village, in the province of Szechwan, as it considered an eviction case in the presence of a landlord and tenant. Calmly and with deliberation a tenant contended that under the reform he could not be evicted—something that would have been unthinkable a few years ago. Or take the case of General Lin Wen-hui, governor of Sikang, one of Asia's largest landlords, chief purveyor of opium throughout Southwest China, most recently in the service of the Communists. Reforms were not meant for him, he thought. He sent one of his estate managers with a detachment of troops to collect rents—the high, old rents. The tenants did something they had never done before—they armed themselves. When the manager and the troops appeared they announced they intended to defend their rights under the new rent reduction scheme and would not permit collection at the old rate. Whereupon the manager hastily agreed to accept payments at the reduced rates.

This was a unique experience for the general and the tenants. It betrayed a crack in the cake of custom and the American people have had much to do with it. In the province of Szechwan alone more than two million tenant families were direct beneficiaries of JCRR. Yet JCRR disbursed but $218,000 on a reform that went to the heart of China's rural and political problem.

The reform came too late to give the Nationalist government even a measure of popular support. But our last-minute efforts in Southwest China—and at present in Formosa—nevertheless carry with them a most important lesson. ECA through JCRR has demonstrated in a practical way that the United States can act not merely *against* Communism, which has little meaning to the average Asian peasant, but also *for* certain ideas understood by the most ignorant of peasants.

It is true that a final solution of problems of the Asiatic peasant—final in the sense that most of them will live happily ever after—cannot be achieved. There is simply not enough land for the great multitude of farmers. But progressive rural reform in Asia is not a question of the whole loaf; any effort to ease the peasant's burden is welcome and lays a foundation for a middle-of-the-road, stable rural society. The tenant in the village office of Kwangsi, half-crazed with joy over the receipt of a deed that gave him ownership of but one acre of land, spoke for Asia's millions.

"Land and liberty" has ever been the ideal of all peasants. It is the American ideal. We must lend all means at our disposal to bring it closer to realization. If the struggle against Communism is to succeed, it must be inspired by an ideal, a broad and bright vision of the future, that will find an instant response among the disinherited of the largest of the continents. It was Goethe's Faust who said so wisely:

> To many millions let me furnish soil
> Though not secure, yet free for active toil.
>
> * * *
>
> And such a throng I fain would see,
> Stand on free soil among a people
> free.

14. *Rural Reconstruction under the China Aid Act*

Ladejinsky was impressed during his advisory service with the JCRR in Szechwan and Taiwan in the fall of 1949 with the way this agency, created by the Aid to China Act, was functioning. He advocates here that it serve as a model for other U.S. technical assistance programs. This model was, however, not the one subsequently adopted for implementation of President Truman's Point IV program, unfortunately, in my opinion.

This article was published in *Foreign Agriculture* in August 1950.

EVER SINCE PRESIDENT TRUMAN announced his Point IV program on January 20, 1949, there has been a great deal of discussion in and out of government circles on how best to give technical aid to economically underdeveloped countries. Much of this discussion has centered on agriculture for most of the people in these areas depend on farming for a living. The effectiveness of Point IV aid will depend on several things, including its scope in a given country, the number of countries involved, and the material means and technical personnel available. Much, too, will depend on how well we apply the lessons and experience of U.S. agricultural aid in Latin America, China, and Formosa.

Technical assistance in Latin America has been going on for a decade; the work in China and Formosa began less than two years ago, but it is there that I saw first-hand a pattern of agricultural assistance that may be applicable to the Point IV program. This is the pattern worked out by the Joint Commission on Rural Reconstruction (JCRR), set up in 1948 under the China Aid Act.

The purpose of this act (Public Law 472, 80th Cong.) was to improve and stabilize China's economy. Four-fifths of the Chinese are farmers, most of them barely making a living. Therefore, no improvement in living conditions was possible, and hardly a measure of political stability could be achieved without giving some aid to the country's farm economy.

The act recognized this, and section 407 provided that up to 10 percent, or $27.5 million, of the amount appropriated for economic assistance to China could be devoted to the betterment of agricultural conditions. This fund was to be made up of U.S. currency, or Chinese currency from sales of commodities made available to China from funds under the act, or both. To carry out this work, a Joint Commission for Rural Reconstruction was created, with authority to formulate and execute such a program in China.

The JCRR was to be composed of two American and three Chinese citizens, the chairman of the commission being a Chinese member. By September 16, 1948, all members had been selected. The caliber of the membership deserves notice, for they were all people with intimate knowledge of China's rural conditions. On the Chinese side there were Chiang Monlin, former minister of education, one-time secretary general of the Chinese cabinet, and a student of rural China; James Y. C. Yen, for almost thirty years head of the mass education movement among Chinese farmers; and Shen Tsung-han, China's foremost plant-breeding expert. On the American side were Raymond T. Moyer, for fifteen years before the war engaged in agricultural work in China and chief of the Far East Division of the United States Department of Agriculture, during and immediately after the war; and John Earl Baker, engineer, former adviser to the Chinese Ministry of Railways, and for years a relief executive in China. This was the group that formally assembled for the first time in Nanking on October 1, 1948, elected Chiang Mon-lin chairman, and launched the formidable task of agricultural reconstruction in China.

The life of the commission in China was a short one. As the Communist armies moved through China in late 1948 and 1949, the commission was compelled to shift its headquarters from Nanking to Canton and finally to Formosa in August. By the end of the first year of the commission's existence, all its work on the mainland had to be suspended and carried over to Formosa, still outside the reach of the Communists. Notwithstanding the extremely difficult circumstances under which the commission worked on the mainland of China, it gained useful experience in the field of agricultural reconstruction along the following broad lines:

1. Establishing basic principles underlying a sound reconstruction program in underdeveloped areas;
2. Establishing criteria for selecting projects;
3. Organizing an agricultural reconstruction program;
4. Achieving results quickly; and
5. Accomplishing much with little money.

In working out the agricultural program in China, a number of problems had to be treated simultaneously. This is illustrated by the nature of the commission's work. From its inception through November 1949, the commission assisted a total of 216 projects: agricultural improvement, 91; irrigation, 51; rural health, 25; rural industries, 18; farmers' organizations, 16; land tenure reform, 8; and citizenship education, 7.

One of the principal tasks of the commission was to determine what parts of the overall goals were realizable, especially considering the rapidly deteriorating military and political conditions on China's mainland. Much thought, therefore, had to be given to implementing a program that would yield results quickly. At the beginning the commission decided that an integrated program should be built on a selected county basis. The idea behind it was that a successful program of this kind might, by force of example, spread to other counties and provinces. However, according to Raymond T. Moyer:

> The program as adopted at the beginning could not influence the situation significantly within the time which we foresaw would be allowed for the commission's work. The integrated program centers developed too slowly, and assistance on a piecemeal basis did not help enough people solve their most basic problems. . . . We decided that a great deal still might be done by action in a more limited program, emphasizing felt needs of vital importance to the rural people, carried out on a large scale.

Actually, the "limited program" meant handling a limited number of the most important problems on a wider scale. For what followed was a bold rural program in the provinces of Szechwan and Kwangsi on the mainland and later in Formosa.

Out of the shifts and changes, induced primarily by experience of working with the farmers, the JCRR developed the following national program, its constituent elements given here in order of priority: (a) land reform, (b) irrigation, (c) fertilizer, (d) farmers' organization, (e) rural industries, (f) plant and animal disease, (g) seed multiplication, (h) animal breeding, (i) rural health, and (j) audiovisual education.

Placing land reform first among the items was no accident. The enthusiastic reception of the commission's help in carrying out what is known as the Lungyen land reform project and the exceptionally bad tenancy conditions in the Kwantung province induced the commission to pay close attention to the problem. It was equally clear to the commission that the chief beneficiaries of the technical projects would be the landlords and not the majority of the farmers—the tenants—unless tenancy conditions were improved.

To help farmers get credit and market their produce, the commission encouraged the organization of farm cooperatives through which rural industries might develop and which generally might serve as centers of agricultural rehabilitation.

The audiovisual education efforts stemmed from the illiteracy of the farmers on the one hand and the need to provide a readily understood medium to arouse the cooperation of farmers on the other. To solve this problem, the

JCRR developed such media as graphic portfolios, posters, film strips, simply written short pamphlets, and comic books.

Generally speaking, on the technical side of the program as a whole, the commission has been supplying specialists, both Chinese and American, to formulate sound plans and establish means of carrying them out. On the financial side, the commission has been furnishing a certain amount of funds for specified purposes essential to the success of the program, supplementing them with funds supplied from provincial or local sources.

Basic Principles

Perhaps one of the most important features of the JCRR, in itself and in relation to any Point IV program, are some of the principal ideas that have stamped its character. For example, the commission ruled out costly projects with modern equipment—good for American agriculture but definitely too "high hat" for Chinese agriculture. The commission also decided not to set up new enterprises that would compete with those already established. If, for example, a local serum-manufacturing business was found to be inadequate, the commission was ready to assist in enlarging and improving it but was not willing to start a completely new, competing enterprise.

The commission was careful, too, not to approach its work with a preconceived notion as to what is good for farmers. Such ideas, it was felt, must come from the farmers themselves as a result of close contact with them. As Chiang Mon-lin put it: "Regardless of how good are our intentions and how sound a program, if the rural people do not want the program it cannot be imposed upon them."

In addition the commission avoided long-range projects but did not hesitate to spend a considerable amount of money on irrigation when it was certain of higher yields in the immediate future.

Selecting Projects

As the commission got under way, it received hundreds of applications for assistance. Between November 1948 and February 15, 1950, 1,178 applications were received from China's provinces and from Formosa. Only 297 applications were approved. In Formosa alone by May 31, 102 project applications out of a total of 429 were acted upon favorably. Clearly, the commission had to make choices, and it made them by establishing a number of rather significant criteria, which might well serve as a basis for a similar approach in agricultural assistance work in other areas.

The commission proceeded on the assumption that in China (as in any other country) when one engages in farm aid, one must win the goodwill of the farmers. To do that, those projects that the farmers want most must be promoted. No project can succeed without the active participation of the people; it must have the support of a local group. Past experience has shown that money expenditure alone does not create lasting reconstruction; what is important is that the aid should help the people to help themselves by utilizing their own resources to a maximum degree. Not only must the people take active part to insure the success of a project, but the cooperation of the government is equally vital. Without its support and without its general concern for the welfare of the people, the development of a project is practically impossible.

In addition, the educated youth of the country must be drawn into the work of agricultural reconstruction. In China as elsewhere, they represent the country's hope of the future, and to secure their support for agricultural reconstruction means training a new leadership so vital to the lasting success of a country's agricultural economy. Too, the commission has felt that a project must benefit the greatest number of farmers. And finally, the commission's criterion for giving aid and assistance to a project has been based on the idea that physical reconstruction must be accompanied with social justice, which, in effect, means fair distribution of farm income among all the parties connected with the land. This idea is closely associated with improving productive technique and with the business side of farming as well as with the attitude of the farmers toward agricultural reconstruction.

The JCRR, for example, favored irrigation as a means of increasing agricultural productivity, but it found it necessary to raise this question: "Whose land?" The commission was afraid that failure to ask this question and *act* upon it might bring about a situation where, in its words "the rich became richer and the poor poorer." Hence the criterion that while one must begin with physical improvements, *social* reconstruction must not be neglected if the true goals of rural reconstruction are to be achieved.

Organizing the Program

These ideas and criteria were being translated into action first on the mainland of China, and currently in Formosa, through the actual operation of the national program. For this purpose the program of the commission has been divided into five main divisions: agricultural, land tenure, irrigation, rural health, and audiovisual education. The division heads are either Chinese or Americans with a great deal of experience in their respective fields; their staffs screen the projects, make recommendations to the commission, and supervise field operations. When the commission operated on the mainland of China, it had regional offices headed by a representative of the commission. The representative served as a liaison between the commission and local agencies, and he was responsible for supervising the projects.

Much of the technical work within the commission is done by Chinese. It has been the commission's policy to hire American technicians only when absolutely necessary. Their role is one of guiding and training, and in the experience of the commission such efforts can be accomplished with few Americans. In Formosa the commission employs 200 people; of these only 17 (including the two commissioners) are Americans. The commission can function with this relatively small staff because it has succeeded in inducing the agencies sponsoring the projects to do most of the work.

The role of JCRR in controlling the rinderpest outbreak in Formosa in the fall of 1949 is a good example of that. On learning of the outbreak of the disease, the commission secured the complete cooperation of the Formosa Department of Agriculture in the campaign to eradicate it. The commission provided the vaccine, an experienced veterinarian, and obligated itself to pay half the indemnity (Formosa's Department of Agriculture paid the second half) for the slaughtered cattle. This, together with an intensive educational program aroused the farmers to the danger of the epidemic and secured their active cooperation. As of February 1, 89 percent of the 67,000 cattle were vaccinated, and by the middle of February the operation was virtually completed. The monetary contribution of JCRR to this program was only $6,192, but the other and real contribution, hard to evaluate in terms of dollars, was the drive the commission generated to insure the success of the fight against rinderpest. The actual work was carried on by the local department of agriculture and other provincial agencies in close cooperation with farmers.

The program of JCRR as indicated is made up of a great variety of projects. At present they are confined to Formosa; as of May 31, 1950, the commission was assisting 102 projects. To cite but a few:

1. Rehabilitation of rural domestic water supply plants, sponsored by Provincial Department of Reconstruction; commission subsidy, $189,000.
2. Multiplication and extension of rice seeds, sponsored by Formosa Provincial Department of Agriculture and Forestry; subsidy, $30,000.
3. Production of hog cholera crystal violet vaccine, sponsored by Formosa Provincial Veterinary Serum Institute; subsidy, $20,000.
4. Supervision of distribution of ECA fertilizer, sponsored by Bureau of Food, Formosa provincial government; subsidy, $15,000.
5. Health survey, sponsored by Public Works Administration and Provincial Health Administration and National Institute of Health; subsidy, $1,000.

How is a project initiated, contracted and paid for, and inspected? And how are the results evaluated? A project is generally initiated by a sponsoring agency, governmental or private. The appropriate division of the commission reviews the project from the standpoint of technical feasibility, reality of cost estimates, conformance with JCRR policies, and then makes its recommendation to the commission.

Schedules of payments are included in the recommendation if the project is to be approved. Upon approval, the commission enters into a written agreement with the cooperating agency in which responsibilities of both parties are clearly stated. In every instance the commission retains the overall supervision of the work of the project.

The agreement provides for a careful audit of money spent by the sponsoring agency. All payments, other than the initial payment, on the part of the commission depend on the progress of the work. The sponsoring agency submits a monthly progress report and the commission, in turn, investigates developments.

Results

What of the results achieved by the commission? Because of the brief period of the commission's activity on the mainland of China, it is difficult to determine the accomplishment of the projects initiated in areas now under Communist domination. The results of such projects as the organization or reorganization of farm cooperative associations or the projects centered around the mass education movement in Szechwan could be fully determined only in time and under normal political conditions—conditions not obtainable. Despite these handicaps, I believe that where the projects were initiated, the commission usually planted seeds that may not necessarily be lost even where Communists hold sway. Farmers were not slow in appreciating the value of the efforts in their behalf. The enthusiasm of farmers for the practicality with which the Commission went about developing a project was quite apparent as was the enthusiasm of many Chinese agricultural officers participating in this work. It would not be surprising, therefore, to discover at some later date the survival and continuation of some of the work and ideas introduced into rural China by the commission.

In the field of land reform, which in some areas was instigated by the commission while in others by the Nationalist government, 15 million farm people are bound to raise their standard of living above a bare subsistence level through the 25 percent reduction in rent. In Szechwan and in Formosa I saw how successful the land reform program was, totally out of proportion to the financial assistance given by the commission: In the former, for example, JCRR was able to assist the rent reduction program from which 1,700,000 tenant families benefited at a cost of only $218,000, or not much more than a dime per family.

In Kwantung, Kwangsi, and Szechwan, more than 200,000 acres were irrigated or existing facilities improved in less than a year. The increase in rice production was estimated at 164,000 tons. In the province of Hunan, where the commission allocated $1 million for irrigation, matched by a similar contribution by the local people, it was estimated that the first post-irrigation crop would increase the value of the output by $12 million.

The multiplication seed project in Formosa, already carried out, involved the commission in a cost of $30,000. The results of the work on the foundation and stock seed farms cannot as yet be determined, but, from the seed on the extension farms and that distributed among farmers, even a 5 percent increase in yield will raise production by 13,000 tons, valued at more than $1 million.

In addition to its accomplishments, both in China and Formosa, in the field of increasing agricultural production and of distributing farm income more equitably, the commission made considerable headway in improving rural health, eradicating animal disease, organizing farm cooperatives, and aiding rural handicrafts with which peasants have supplemented their income from the land.

Some of the real achievements of the commission fall into the category of the so-called intangibles, which cannot be easily measured by any merit point system. Every country of Asia has a reservoir of native talent, but often it does not know how to go about improving the welfare of the people. The JCRR tapped a measure of that talent in China. The few but well-chosen Americans on the commission's staff performed that function in addition to their described duties as they appear on an organizational chart. It is fair to say that one of the commission's important projects, albeit never listed, is the talent multiplication project.

It is difficult, too, to evaluate in concrete terms what seems to me one of the commission's greatest achievements—the philosophy, as it were, shaping the work of the commission. Frequently, Westerners operating in non-Western surroundings have been all too prone to carry over notions foreign to the new environment, thereby reducing the effectiveness of the effort. The JCRR has been free from making such mistakes. On the contrary, it has formulated a set of workable principles in the light of China's and Formosa's rural needs, which in turn have given meaningful direction to its work. Such nonmaterial developments cannot be easily measured and it is difficult to assign specific values to them. But that they are of great significance, nevertheless, was quite obvious to me as I watched the commission at work and studied its decisions at close range.

Cost of the Program

How much did the program cost and how well did the commission spend the money? It was originally agreed that the commission would finance its operation from three sources: U.S. funds and funds furnished by the Chinese government and by Chinese farmers. The contribution of the latter was to have consisted mainly of payments in kind and labor and repayment of the JCRR loans contracted by them for a given project. But from the outset, and particularly in view of the worsening military situation and financial chaos, it became difficult to secure the agreed upon share from a sponsoring agency. The commission was compelled to draw more and more upon U.S. resources. Yet, despite this and notwithstanding the considerable tangible and intangible results of the work of the commission, it spent only $4,755,500 from October 1, 1948, to May 15, 1950.

Whether considered alone or compared with other China aid programs, the sum is small indeed. The commission has been most conservative in allocating money for projects and, once allocated, careful in insuring that the money be spent as intended. The chairman of the commission, Dr. Chiang Mon-lin, has repeatedly stated in public that the U.S. taxpayer furnishes the bulk of the money and that it must be used only for sound projects yielding significant results. The largest amount of money was spent on irrigation projects scattered in various parts of China, often difficult to control closely. Yet there is reason to believe that between 90 and 95 percent of the money was properly spent. My own field observations on the JCRR's expenditure of money on the rent reduction programs led me to the conclusion that, when money is judiciously spent, a great deal can be accomplished with relatively small means.

The care with which the commission was spending other people's money was also due to the fact that the commission drew a sharp line of demarcation between a program of expenditure for capital equipment and the commission's program of technical, economic, and social rural reconstruction designed to help the farmers to help themselves. Moreover, the commission realized that "the number of tested improvements which can be introduced on a large scale in China's rural areas still is limited." Where it found them, as for example in irrigation, in control of animal disease, in multiplication and distribution of improved varieties of important field crops, or in rent reduction, the commission did not hesitate to give maximum aid. At the same time, the experience of the commission shows that, unless it can be usefully absorbed, money alone cannot solve problems. And this is a point worth remembering in all rural aid efforts in underdeveloped countries.

Such in the main are the aims, organization, work, and results of the work of the JCRR. It came too late to alter the course of events in China, but I believe that this is one of America's most fruitful ventures, for what it attempted, actually did, is still trying to do in Formosa, and, above all, for what might be done, on a larger scale and in better time elsewhere.

The Joint Commission on Rural Reconstruction is truly a *joint, cooperative* organization, where Americans and Chinese work together in a common task. This device points the way to a similar pattern, whenever possible, in other parts of the world. Such a pattern goes a long way toward obviating the accusation that we harbor ulterior motives in offering assistance, and, above all, it insures an effective working relationship.

The commission has demonstrated that in rural reconstruction a little money well used can go a long way, providing, of course, that the cooperation of the government is enlisted to a point that it is willing to take the required action to put the necessary measures into effect.

The commission has pioneered the idea that in order to meet the fundamental causes of rural discontent, a rural program must be fashioned in the light of the economic, political, and social conditions existing in a country. For this reason, the JCRR program does not limit itself to technical problems pertaining to raising agricultural production. The commission has broadened the concept of technical agricultural aid by recognizing the importance of raising the questions: Whose land is being improved, how will increased income be shared between landlord and tenant, and how do tenants fare in existing conditions?

This novel concept of combining technical aid with that of remedying social inequities strikes at the fundamental causes of peasant discontent. The peasants with an improved economic status and a brighter hope of the future acquire a stake in the countries in which they live. Therein lies the foundation for a middle-of-the-road, stable rural society. The political corollary of this type of rehabilitation cannot be overestimated, for it is fair to assume that peasants will be more prone to support a government that is contributing its share to making the economic and social benefits. And what many a native government in Asia needs most is popular support, and popular support in Asia is peasant support. My first-hand observations of JCRR lead me to the conclusion, therefore, that its program as a whole deserves serious consideration as a means of meeting the problems of poverty-stricken rural areas everywhere, particularly wherever the countryside is an easy prey to Communist propaganda and infiltration and wherever Communists are trying to harvest crops they have not planted.

15. Observations on Rural Conditions in Taiwan

Not long after Ladejinsky's return to Tokyo as agricultural attaché in October 1950, his advice was again sought by Premier Chen Cheng of Taiwan, and he revisited Taiwan in the spring of 1951. This paper, the most substantial of his China-Taiwan papers, reported on his observations. These embraced farm prices, agricultural taxation, farm credit and fertilizer distribution, as well as rent reduction and the government's land purchase program. This revisit to Taiwan, some eighteen months after the institution of the rent reduction program he observed during his 1949 visit, must have impressed on Ladejinsky that basic land redistribution or rent reduction and security of tenure measures cannot by themselves ensure the well-being of their beneficiaries and that complementary measures (on fertilizer and other inputs, taxes, prices, and so on) were essential to round out a successful agrarian reform program. One immediate result of his recommendations was a government order to the Taiwan Sugar Corporation to sell to the government, for its subsequent redistribution, two-thirds of the acreage it operated under lease to tenant farmers. Because of the paper's length, only the opening paragraphs and the summary and conclusions are presented here.

This report to the National government of China and the Joint Commission on Rural Reconstruction was dated June 1951.

THE WEEK OF APRIL 27–MAY 7, I spent on a field trip in the company of Tang Hui-Sun, chief, Land Division, JCRR, and N. M. Lin, interpreter and assistant, and member of ECA, Taiwan. The original purpose of the trip was twofold: (a) to examine the status and consequences of the land rent reduction program, initiated in early 1949, and (b) to inquire into the attitude of the tenants and landlords toward a land purchase program, currently contemplated by the government. However, in the course of the trip virtually every other important farm problem came up for consideration.

Method of Investigation

Our party traveled through the densely populated agricultural region of the western coast of Taiwan, starting from Taipeh in the north to the southernmost part of the island beyond Pingtung. We visited a total of ten counties; these included the principal rice, sugar, and sweet potato-producing areas, some two- and one-crop areas—both well-irrigated and poorly irrigated land sections of the island where a landlord stands for vast acres as well as sections where a landlord is one chiefly by courtesy, as it were. We held a total of ten meetings with farm groups composed of tenants, owner cultivators, and landlords, with the attendance ranging from 40 to as many as 300. We made numerous stops along the road to talk to individual farmers, visited farm homes, and looked into many a pigsty of contented hogs and into a few bins to see how the rice was holding out one or two months before the new crop is harvested. County, township, and village officials were not neglected; we could not escape them.

Nature of Report

The report doesn't claim to be more than it is: firsthand impressions of farm conditions and farmers' views buttressed, wherever possible, with what seemed fairly sound facts and figures. The linguistic difficulties may have resulted in some erroneous impressions and occasionally misplaced emphasis, and I am not unmindful of the propensity of the farmers to complain of bad times. On the other hand, I covered roughly the same ground in 1949 and was in a better position to compare farm conditions now and then. There was, for example, no mistaking the mood of the farmers as they gave vent to their pent-up emotions on the subjects of farm prices, cost of production, taxation, land prices, the Sugar Corporation, and so on. If in 1949 I noted a certain reticence on the part of the Taiwanese farmer, in 1951 I was struck by the eagerness to "tell all" and to insure that the visitor understood and would present their case. Opinions of disgruntled farmers are no substitute for a careful study of village conditions, but I carried away the impression that notwithstanding an obvious degree of exaggeration, the statements of the farmers reflected what ails the rural areas of Taiwan at the present time.

The Basic Problem

What is it, then, that disturbs the farmers regardless of strata? Why the note of urgency in the recitals of their problems? The answer appears to be that from the latter part of 1950 on the farmers of Taiwan have been having difficulty making both ends meet. The farm community, so the farmers insisted, is worse off now than in 1949, despite the extremely valuable aid given them by the Joint Commission on Rural Reconstruction and ECA.

Summary

The purpose of the trip was to study land reform problems, but after one day in the field and two farm meetings, it was clear that the farmers were deeply concerned with low farm prices coupled with continuously rising prices of nonfarm products.

The undeniable existence of the "scissors" problem is affecting adversely—even if in different degrees—every class of farmers. Theirs is a daily and painful experience when they buy or wish to buy a catty of salted fish, a catty of soya sauce, matches and tobacco, a few yards of piece goods, a pair of shoes, a plow, a harrow, some building materials, and a host of other necessities a farmer needs to live. This unbalance of prices is at the bottom of the current farm depression. From the farmers' point of view, the unreasonableness of the situation is accentuated by the knowledge that the world market price of rice is much above the price they receive. Rightly or wrongly, the farmers are of the opinion that the unfavorable price relationship is not an act of God but is a deliberate policy of the government to serve nonfarm ends which they do not comprehend.

The farmers were not only complaining; they were ready with remedial suggestions. What they want most is real parity. They did not demand a boost in the rice price for fear of inviting an even greater rise in other prices. In meeting after meeting they laid much emphasis on a balanced relationship between farm and nonfarm prices. The other remedy suggested by the farmers is the exportation of rice and in a manner whereby they, too, would benefit from the high world prices.

To the unbalance of prices must be added a number of other factors which add to the burden already shouldered by the farmers. They stem from all manner of governmental economic and political considerations. The principal grievances were: (a) some dissatisfaction with a particular aspect of fertilizer distribution, (b) increasingly heavy taxation, (c) absence of adequate credit facilities and dependence upon loans at usurious rates of at least 10 to 15 percent per month, (d) sale of green crops (standing crops) forced by lack of capital at far below the market price, (e) burdensome conditions under which the sugar cane farmers operate, and (f) low returns to farmers from nonagricultural activities.

On the positive side and from the point of view of the tenant, the rent reduction program has been successfully carried out and rather significant consequences came in its wake. One can point to evasions and continued complaints about the unfairness of the land-grading system, but basically the program has been translated into action. Tenants readily admit that they benefitted from the rent reduction program. In the present depressed state of Taiwan's farm economy, a large number of tenants would be in dire economic straits but for the reduction of rent. Precisely because rice is so cheap and other commodities are so expensive, the extra thousand or two catties of rice the tenants may retain assume added importance.

Under the impact of the worsening farm conditions, however, tenants are inclined to minimize the benefits from the reduction in rent. This attitude also manifests itself in relation to many other benefits derived from JCRR and ECA aid programs. The farmers are well aware of the positive results of irrigation, seed multiplication, pest control, animal husbandry, rural health, fertilizer distribution, dissemination of information on agricultural improvements, and of the better practices introduced in other fields of farm activity. They point out, however, that this assistance is countered by economic and fiscal policies, which, whether by design or not, siphon off the gains. If this condition continues, the pattern of U.S. aid to the rural areas of Taiwan may not achieve all of its purposes, nor can the present standard of living raised by JCRR and ECA be maintained. We cannot expect to harvest the kind of political capital among the recipients of our aid as might have been the case under more normal conditions.

The consequences of the rent reduction program extend far beyond the immediate benefit of so many more catties of rice to the tenant. The program is responsible to a large extent (taxation and low prices are the other factors) for the drastic reduction in the market value of land worked by tenants as distinct from land farmed by owner cultivators. Most landlords are eager to sell their land, but people with means wouldn't invest in lands. The only would-be buyers, the tenants, lack the means; very few among them will be acquiring land in the foreseeable future through their own efforts.

The land distribution program for Taiwan has been under discussion for almost a year. The results to date are promising in one respect and rather meager in another. The government decision of May 30, 1951, and subsequent steps to engage in a large-scale sale of public land to the tenants is an important step in the right direction. This should have a good effect on the contemplated land purchase program of privately held land. The planning to date with respect to privately held land is only tentative. Much spade work must be done before an effective program can be drawn up. The prospects of selling the greater part of the public land are good, as it now seems likely that government corporations controlling it will be induced to give the land back for eventual sale to the tenants. Premier Chen Cheng's sincere and active concern about land reform augurs well for an effective land transfer program for the benefit of the tenants of Taiwan.

* * *

Conclusions

General

The poor exchange value of farm products in relation to nonagricultural products casts a shadow on the rural economy of Taiwan. The purchasing power of the farmers is at a very low point. The consequences are quite obvious. The agricultural recovery of the island, in full swing in 1949 and early 1950 under the well-conceived and effective ECA and JCRR aid programs and the Chinese government's own efforts, is suffering a setback. In 1951 the tenants are inclined to minimize the significance of the rent reduction program; they tend to overlook the fact that they would be much worse off now without the extra rice gains. Farmers as a group appreciate U.S. economic aid which penetrated practically every field of farm endeavor. But the results of that aid cannot counteract the effects of a sharp decline in the real value of rice, not to mention the effects of onerous credit and taxation conditions. If the current farm depression is not alleviated, the American economic efforts on behalf of the farmers cannot achieve the levels of economic and social betterment originally anticipated.

It is easy enough to present the evidence of depressed farm conditions, but it is not easy to make valid and enforceable recommendations to remedy the situation. This is particularly true in the case of Taiwan, which in recent years has had to support more of a burden than its mainly agrarian economy could normally manage. In a large measure the unbalance of prices, heavy taxation, and lack of credit are the consequences of requirements that could not be fully met from domestic and foreign (aid) sources. For this reason, most of the problems cannot be solved in the immediate future. The general and specific conclusions stated below aim only to ease the farmers' lot.

The one outstanding lesson that emerges from the present agricultural situation is that if a farm aid program is to achieve its aims, government policies must be in tune with it. Both in philosophy and action, JCRR is the best example of a well-integrated American system of aiding another people. Measures to increase agricultural production in its broadest sense are accompanied with measures to insure better distribution of income among those who actually till the soil. And yet, it is apparent that to the extent that government price, marketing, credit, taxation, and other economic and fiscal policies are out of joint with basic needs of the farm community, the positive results of our aid may not contribute to any real improvement in living standards.

This leads to a general observation applicable to Taiwan—and beyond. In dispensing aid we must realize that economic policies of a government in their many ramifications have a direct bearing on whether the rice bowl is full or empty. The ECA China mission and JCRR are well aware of this and are acting now to help the government evolve economic and fiscal arrangements essential both to effective economic aid and to a farm program that safeguards for the farmers the fruit of their labor.

Premier Chen Cheng has an abiding interest in land reform measures as a means of raising the status of the farmers, and he and Governor K. C. Wu are concerned with other farm problems as well. They welcome criticism, sound advice, and act upon it. This appears not to be true, however, of the majority of government officials who take the farmer for granted. They believe that regardless of his economic condition, he will continue to provide food for the people, the army, and the exportable surplus to raise foreign exchange funds and will contribute the lion's share of government revenue. It is held that he should do so uncomplainingly, since most everyone on the island is poor.

This hangover from traditionalism is bound to give rise to serious economic and political repercussions. Even if for purely selfish reasons, the ranks of the government must come to appreciate the fact that in 1950 more than 90 percent of Taiwan's exports were of agricultural origin. If this is to be maintained, healthy agricultural conditions are indispensable to the welfare of Taiwan as a whole. Without an appreciation of this basic fact, it is impossible to lay down general agricultural policies which really meet the needs of the farmers and to carry out specific measures to make those policies meaningful.

Specific

With regard to *farm prices*, the financial position of the government does not permit a balance or real parity between farm and nonfarm products at the present time. However, farmers could benefit from the following measures:

1. The first crop is expected to be larger than that of 1950, and 1951 rice exports could be increased to 150,000 or 200,000 metric tons. Farmers would consume more of the super-abundant sweet potato crop and less rice if a good share of the export proceeds were returned to them in terms of imported consumer goods at prices commensurate with the rice prices paid the farmers by the government.
2. A firm understanding on that score between the government and the farmers now that the first crop is being harvested would have a salutary effect on the second rice crop planted later in the summer.
3. Purchase of surplus rice by the government immediately after the harvest when prices are lowest, based on cost of production and 75 percent parity of general price level, would help to stabilize rice prices. It also would provide the government with adequate rice for domestic purposes and increase the amount available for export purposes, and the supply and price of rice could be adjusted by resale of rice in case of shortage. This operation could be financed from the revolving fund of the Food Bureau and provincial government appropriations.
4. Credit should be extended to the rice producers before the harvest with the loan repayable in rice at a fair price after the harvest. Farmers would obtain some badly needed cash, and the glut of cheap post-harvest rice would be reduced. This is a type of short-term loan that can be relatively easily provided and advantageous to both parties.

With respect to the *cane grower*, the Sugar Corporation and the government might profitably give consideration to the following:

1. Reduce the sugar tax of 37.5 percent to the original 25 percent.
2. Limit the corporation processing charge of 50 percent of the delivered cane to 45 percent.
3. Compensate farmers for the by-products of cane refining to the extent of 50 percent of their value.
4. Increase credit to the farmers on a basis similar to expenditures made for maintenance of the plants and the corporation's own cane plantations, since farmers produce more than four-fifths of the cane used by the corporation.

With regard to *farm taxation*, it is recognized that the government now is in urgent need of all possible revenue. However, in determining the taxation to be paid by the farmers, the following points should receive the most careful consideration:

1. The farm tax burden should be eased if for no other reason than that continued exploitation of farmers will eventually decrease agricultural production and revenue.
2. The "take" from the farmers should be determined in the light of the government's contribution to the farm economy.
3. The government should carry out its long-projected thorough examination of the overall tax structure of the island.
4. Taxes levied on industry and commerce should be tightened up and the loop-holes plugged. According to official sources, they do not pay their share.
5. The land surtax and the compulsory sale of rice tax should be reduced but not eliminated altogether.
6. Compulsory purchase of victory bonds and coupon savings should be [altered] to a point more nearly in accord with the financial position of the farmers.
7. In general, a careful sifting of the twenty or more tax, fee, and contribution items farmers now pay might suggest the elimination of some and a reduction in others.

With regard to *farm credits*, it is generally agreed that credit, both short- and long-term, at low interest rates, is urgently needed; but everyone is at a loss as to how to secure the requisite funds. Private sources cannot be counted on; the moneylenders' credit is impoverishing the farmers. It follows that:

1. The government might consider furnishing an additional NT$100 to 200 million to create a credit system that would serve the needs of a greater proportion of the farm population. This proposal need not be summarily dismissed on the grounds that the government has no money. The rising financial burdens to which the lack of credit is basic jeopardize the maintenance of even the present levels of agricultural production.
2. The JCRR should give thought of setting up a credit revolving fund equivalent to U.S.$1 million to $2 million in local currency. This fund may be as good an investment as, for example, an equal amount of money extended to finance irrigation projects.
3. A percentage of the foreign exchange received from export of agricultural products might be allocated to the credit fund.

With regard to *fertilizer distribution*, the 1951 ECA agreement with the government (fertilizer for the second rice crop) provides the formula that will serve the best interests of the farmers and of the government. It may be repeated here that:

1. Any farmer eligible for allocation of fertilizers has the privilege of deciding whether he will take his fertilizer by direct barter or by loan barter.
2. Fertilizer distribution is not merely a device to collect rice from the farmers. Its timely distribution as an aid to increased rice production is the major, continuing objective of the fertilizer program.

As to the *rent reduction program*, it is an accomplished fact. The evasions do not endanger the program, but they should be corrected. Continued success lies in:

1. The immediate election of local land commissions to be charged with the main responsibility for the enforcement of the May 25, 1951, rent reduction law.
2. Sufficient government supervision and, if necessary, pressure to insure strict interpretation of the letter and spirit of the rent law by local administrative authorities.

With regard to the *public land purchase program*, the government should:

1. Launch the next phase of the land reform program with the sale to tenants of the government-owned public land.
2. Enlarge the present scope of the public land sale scheme.
3. Take possession of the public land except that portion without which the operations of the government corporations would be seriously endangered.
4. Deposit all the revenue from sale of the public land, whether in the national or provincial categories, in a fund from which the landlords would be paid for their land.

With respect to the *sale of privately held land*, it is recommended that:

1. The acreage available for purchase by tenants should be determined on the basis of the JCRR studies of landlord categories.
2. The terms for compensation should provide for an initial down payment in cash
3. The cash portion of the compensation should be prorated—the bigger the holding the smaller the per-acre cash payment, or vice versa.
4. The government should provide the cash fund through the sale of the public land, sale of government-owned enterprises to private individuals, and through whatever other means that can be devised.

A number of the conclusions or recommendations are made with considerable reluctance. Though they are proposed with the island's precariously balanced economy in mind, some improvements will involve an immediate decrease in government income and an increase in government spendings. The extent to which the government's financial position can permit this will, of course, be made the subject of very careful study. At the same time, one must bear in mind the fact that, if the basis of Taiwan's economy and principal source of revenue is to be protected against further deterioration, funds to aid agriculture should be given a priority commensurate with its importance.

16. *From a Landlord to a Land Reformer*

In June 1951, upon the completion of his second visit and field trip in Taiwan, Ladejinsky attended a dinner party given by the Chinese Association of Land Reform, apparently in his honor. These remarks, which surely he never intended for publication, were found in typescript among his personal papers. They may have been typed from a hastily drafted handwritten script or, equally likely, from shorthand notes taken by the association's secretary and forwarded to Ladejinsky as a memento of the occasion. The language suggests the latter. It is presented here partly because Ladejinsky recounts in it a bit of his personal history and partly because it presents him, after one or two unaccustomed dinner party drinks, in an untypically gay, romantic mood.

MR. HSIAO, THANK YOU FOR THE very kind words you have spoken about me. As a general custom, I am supposed to say something in reply to the remarks just made by our host, but I would rather prefer to have our guest, General Chang Chun, tell us some stories instead or to have Dr. Chiang Mon-lin say something. However, I am not a dictator, so I do not think I could command them to talk.

Just minutes ago, Dr. T. H. Shen joked on me by saying that I am an absentee landlord in the United States. Yes, I am, because I have recently bought seventeen acres of land in Virginia at the cost of $1,000 per acre, mainly for the purpose of preventing any loss to my savings due to the inflation of U.S. currency. But Dr. Shen did not know that at one time I was a real landlord. Recently, I have been called a Communist or a radical in Japan. I believe I would not be so called in Formosa. I am also known as a land reformer. But, as a matter of fact, I was at one time a staunch conservative.

I have worked hard the whole day today and felt very tired. But, having had several cups of "cheap" beer tonight, which, I hope, would not cost our host as much as (the) six silver dollars a bottle it did cost me in Chengtu in 1949, I feel refreshed again in spite of a whole day's hard work. So I am ready now to tell you something about the story of how I have changed myself from a conservative to a land reformer.

To begin with, I should say that I had two teachers who made me undergo such a change. One was Russian Communism and the other was General Douglas MacArthur.

I was born in Ukraine in the southwestern part of Russia. When I left Russia, I was only seventeen years old. Ukraine was the richest province of Russia at that time and my family owned a lot of land and several flour mills as well. But in spite of all the richness of the region, most Ukrainian peasants were very poor because most of the land was in the hands of big landlords like my father and grandfather.

You remember that in 1863 there was a liberation of serfs in Russia and many peasants were set free. But the liberation did not do much good to these liberated peasants because the land still belonged to the landlords and was not set free. The life of these peasants was still so hard that a great many of them went to Siberia to look for land in order to make a bare living from it.

In 1917 a revolution broke out in Moscow when Russia was still at war with Germany. At that time, Kerensky, the leader of the Menshevik Party, was at the head of the revolutionary army and came into power. He tried to democratize the country but did not know how to do it. While he was still wondering about what he should do for the country, another revolution was started in November of the same year under the leadership of Lenin, who was the head of the Bolshevik Party.

In starting the revolution, Lenin, without sitting down to write fine laws, immediately told the Russian soldiers at the front that, if they laid down their arms and came back to Russia, they would be allowed to divide up all the big landlords' land and have it for themselves. The soldiers immediately responded to Lenin's call and left the battle ground. The land was duly divided and given to the soldiers, and my family (property) was completely liquidated. You know Ukraine was a part of Russia. But the Ukrainian people were radically different from the Russians in custom, language, and culture. In spite of these differences yet, the Ukrainian peasants supported the Bolsheviks when they received the land. In this way Lenin got a huge army of supporters from these soldiers and Kerensky was thrown out of power immediately.

Later, Lenin was criticized for his way of doing things. Someone asked him how he knew that the soldiers wanted the land so badly. Did they vote for it? "Yes, they did," replied Lenin. "How did they vote?" asked someone. Lenin replied: "They voted by their legs. The fact that these soldiers immediately left the battle ground in response to my call was ample proof of what they really desired and was more meaningful than regular voting."

Soon after Kerensky was driven out of Moscow, he organized an army in South Russia and began to fight back northward. At one time Kerensky's army was very near to Moscow and it looked as if Lenin were to be completely defeated and Kerensky would come to power again. But Kerensky's generals made another grave mistake. When the territories lost to the Bolsheviks' army were recovered one by one by Kerensky's army, they wrote many measures, left to right, to dispose those lands divided by the Bolsheviks. But by the measures they wrote and adopted, they insisted that anything done by the Bolsheviks was no good and must be changed. As a result, they re-installed all the old laws in replacement of the new ones introduced by the Bolsheviks in all places recaptured from the enemy. The land which had already been divided up by the soldiers was returned to its original owners by force. In this way, Kerensky soon lost all the sympathy and support of the people and soldiers, both of his own and of the enemy's, and his whole cause was lost immediately.

This was the first lesson which I have learned from the Russian revolution. There was at that time only one support in Russia—peasants' support or no support at all. Lenin would not have succeeded in permanently seizing the power of whole Russia had he not called for the soldiers to leave the front line and come home to divide up the landlords' land or had Kerensky and his followers not been so foolish as to try to take back the land from the soldiers and return it to its original owners. Kerensky was completely defeated by Lenin and was never able to rise again because he did not understand the psychology and desires of the peasants.

The second lesson I have learned was from General Douglas MacArthur.

I have known General MacArthur for many years, but I did not really know him until 1946 when I went to Japan and worked with him. Before that time, I thought that General MacArthur would not be interested in such matters as land reform because generals usually do not have time and interest in such matters, especially a five-star general like MacArthur. But after I worked with him, I found that MacArthur was not only interested in land reform but also in social insurance, in local government, and other matters.

After the Allies occupied Japan, General MacArthur was the first one to see the urgent need of reform in Japan's land system. In Japan as elsewhere in Asiatic countries, most of the land was owned by landlords and the majority of peasants were leading a poor life. Soon after occupation, General MacArthur instituted land reform and carried it through very successfully. The reform has given land to the majority of the Japanese peasants. Before the reform, Japanese peasants were worn out economically and the whole peasantry was a hotbed for Communism. These peasants were indifferent to the Government and tried to seek relief by siding with the Communists. As a result of the reform, the picture was entirely changed. Most peasants who formerly supported the Communist Party turned their support to the government, as evidenced many times during elections. This was one of the great things that General MacArthur had done for the Japanese people while he was the Supreme Commander of the Allied Forces in occupied Japan. His interest in other social reforms was also widely known.

One day when I was with the General in his big office in Tokyo, I asked him how he became interested in land reform. He told me that his interest in land reform had grown out of three lessons he learned from history, his father, and his own experience in the Philippine Islands.

The first lesson that made him interested in land problems, the General told me, was the teaching of history about the ancient Roman Empire. As everyone knows, the Romans had conquered practically the whole of Europe and Africa and succeeded in establishing a great empire. But the Romans didn't know how to govern the empire. They paid no attention to agriculture or to the interests of the peasants. As well-described by Professor Sinknovich of Columbia University in his book entitled *Hay and History*, the Romans lost the empire because of their negligence in matters relating to agriculture and peasantry. Without hay there will be no agriculture and without agriculture there will be no government. Despite the changes of time, the lesson from the fundamental cause of the downfall of the Roman Empire is still applicable to present world conditions, especially in this part of the world where the economic structure is still similar to the hay economy of Europe in the Roman time.

The second lesson, the General told me, is a story he learned from his father, General Arthur MacArthur, about forty-five years ago when the latter was the military Governor of the Philippines. At that time, the Philippines was, as it is still now, essentially an agricultural country. In the country the landlords often owned land ranging from several hundred to several thousand acres, while comparatively few peasants possessed any land at all. The general economic condition was poor and the social order was bad. Trying to improve the situation, General MacArthur proposed to divide up those big landed estates and introduce the American family-size farm system into the Philippine Islands, distributing the small pieces of land as a result of land division to the landless peasants. His proposal failed because of strong opposition from the big landlords, with the result that most of the Philippine peasants are even today still toiling on somebody's farm without any hope or relief. What happened has had a profound effect later on the social development of the islands, in that the most important cause for the present agrarian unrest in the rural areas of the Philippine Islands could well be traced back to the failure of that program during General Arthur MacArthur's time.

The third lesson was the General's own experience in the Philippine Islands. During World War II the Philippine Islands were occupied by the Japanese. After the Japanese had surrendered and the Philippine government was reestablished, it was found that many rifles were still left in the hands of the guerillas. This would make a nuisance to the government. To recover those rifles, President Quezon issued an order saying that fifteen pesos would be offered as a reward for every rifle surrendered to the government, but the result of this order was that not many rifles were forthcoming. The President then asked General MacArthur how to recover those rifles. General MacArthur suggested that, in addition to every fifteen pesos offered, another fifteen acres of land should be rewarded for every rifle surrendered to the government. If a peasant turned in two rifles, his reward should be doubled. But President Quezon did not accept the General's advice. The result is that the Philippine government is now still paying a tremendous amount of money and life in fighting for those rifles without much success.

It was these two teachers, Russian Communism and General Douglas MacArthur, who have changed me from a landlord to a so-called land reformer today.

In conclusion, I would like to add that, if the majority of the people in a community could not live a contented and satisfactory life, the minority people who own large estates would not have a comfortable living, sooner or later. It is not only for the sake of others but for the enlightened self-interest alone that those minority people should share their lot with the majority in the economic life. It is true that land reform would not cure all the evils of an economically underdeveloped community. It is also true that it will not necessarily even improve or raise the living conditions of the majority of peasants. But it will at least give them some satisfaction of their basic needs, and it is the struggle for the satisfaction of those basic needs that has led Communism to grow to such an extent as to threaten world peace today.

I left Russia thirty years ago, as you left the Chinese mainland two years ago. This is the lesson that I have learned in my life and also the very lesson for you to learn. Where there is life, there is hope. When I left Tokyo, General MacArthur expressed his sincere hope that I should help the Chinese government, not only in carrying out those necessary land reforms on this island but also in instituting those reforms to be applied to the Chinese mainland. Now in Formosa I am glad to know that Premier Chen Cheng is an energetic land reformer. He has carried out the rent reduction program and will continue to carry out other necessary reforms. All these reforms will have a profound effect on this island as well as on the Chinese mainland. I believe that these reforms which have been done in Formosa in recent years will be at least partly applicable to a land reform program on the mainland when you people go back.

I thank Mr. Hsiao and you for the kind attention you have given me this evening.

17. *The Plow Outbids the Sword in Asia*

This short article, the subtitle of which is "How General MacArthur Stole Communist Thunder in Japan with Democratic Land Reforms, Our Most Potent Weapon for Peace" is a companion piece to the "Too Late to Save Asia?" article of a year earlier.

This appeared in *Country Gentleman* (now the *Farm Journal*) in June 1951. Reprinted with permission from *Country Gentleman*, © 1951 The Curtis Publishing Company.

TO AMERICANS THE WORD "FARMING" conjures up thoughts of prosperous communities, vast panoramas of wheat and corn, herds of fat cattle, bulging silos, and good family living.

To Asians, where four out of every five till the good earth, "farming" is synonymous with crushing poverty, a bad case of landlordism, heavy taxation, usury, and debt without end. For these nameless millions the equivalent of $25 can mean the difference between comparative security and total catastrophe. This poverty on the land has been intensified in many areas by political unrest and war.

The Communists, with a quick eye for the main chance, have been making the most of peasant discontent by holding out that most enticing of baits—ownership of the land. The Communists succeeded in China. They have failed utterly in Japan. There, General MacArthur stole the Communists' thunder and made the landless peasant's dream of a piece of land he could call his own come true. In Japan we have forged an economic and political weapon more potent in Asia than the strongest battalions and blandishments the Communists can put forth.

It is difficult for an American to appreciate the Japanese farmer's hunger for land. Over 6 million farm families are crowded onto less than 15 million acres of land, broken up into 110 million plots of all conceivable shapes. In the United States there are 160 acres of farmland for each farm family—in Japan only 2.3 acres. And two-thirds of Japan's farmers work *less* than 2.3 acres each.

Four-fifths of Japan's cultivated land are devoted to food crops; more than half of that land is in rice. The farmers are among the best in the world. They garden rather than farm; the land never lies fallow; some of it is double- or triple-cropped. They are among the world's heaviest users of chemical fertilizers and manure. They get 65 bushels of rough rice and 26 bushels of wheat per acre, or more than twice the yields in other parts of Asia. Truly, they grow two blades where normally one would grow.

Japan's intensive farming is a never-ending cycle of toil. Yet, even the owner operator has had little to show for his pains because his holding is too small and because before the war Japan's economic, political, and social structure pressed down on the backs of the farmer and his wife. This was particularly true of the tenants. And there were plenty of them. In pre-occupation Japan, only 30 percent of the farmers owned the land they cultivated; the other 70 percent either owned no land or were part tenants and part owners.

In the United States the tenant-to-owner ladder is much traveled but not in pre-surrender Japan where once a tenant, always a tenant was the rule. The terms of tenure were much too burdensome. The tenant paid 50 percent of the crop as rental; he provided, in addition, his own expensive chemical fertilizer, farmhouse, farm buildings, implements and seed, and numerous assessments other than the land tax. A tenant was lucky to keep 30 percent of the crop for himself. Eviction was a common practice. A typical agreement said to the landlord, "You may at any time recover your land." And there is the old Japanese saying, "A farmer is like a sesame seed—the more you squeeze, the more you get." This well described his life and work.

This plight of the majority of the Japanese farmers was not lost on General MacArthur. He remembered the failure of the Philippine government in 1945 to act upon his advice to fight farm unrest among the Filipinos through more widespread ownership of land. He understood that any real chance of cutting the political ground from under the feet of the Communists, of bringing even a semblance of democracy to Japan, depended on the improvement of the lot of those who worked the land. He knew that there was no point preaching democracy to empty stomachs.

On December 15, 1945, General MacArthur issued the historic directive, ordering the Japanese government "to take measures to insure that those who till the soil of Japan shall have a more equal opportunity to enjoy the fruits of their labor."

In compliance with this directive, the Japanese government drafted a land reform program, formally enacted into law in October 1946. The reform provided for the purchase by the government from the landlords of 5 million acres for resale to the tenants who had priority of purchase. Absentee landlords were required to sell all their land; noncultivating resident landlords could retain 2.5 acres, and owner operators could retain 7.5 acres—or 30 acres in Japan's northernmost island where land is more plentiful. In cases where subdivision would lower the productivity of the land, the owner operators could cultivate a larger acreage.

The price of land was fixed by the Japanese government at per-acre rates of 3,000 yen for rice land and 1,860 yen for dry land, when the rate was fifty yen to $1.00. The government, in addition, paid a subsidy to the landlord. Tenants who cultivated the land acquired by the government had priority of purchase and could pay for the land at once or in thirty annual installments.

There was no attempt to abolish tenancy completely; this was thought neither feasible nor desirable. Only the abuses were to be corrected. Cash rent ceilings of 25 percent of the value of the crop were established; a written contract was introduced which specified the rights and obligations of landlord and tenant.

The administration of the land transfer was in the hands of local, prefectural (state), and national land commissions. The local land commissions were the pivotal agencies to decide who was to buy which plot and at what price. They were democratically elected, each village commission consisting of five tenants, three landlords, and two owner operators.

The occupation helped to formulate policies but the legislation, administration, and enforcement were initiated and performed solely by the Japanese. The scope of work was enormous, for it meant buying and reselling 30 million individual plots in 12,000 communities.

There was serious opposition to the program to be overcome. Some landlords attempted to impede the program by illegal eviction of tenants, lawsuits against the Japanese government alleging that the reform law was unconstitutional, and through propaganda to dissuade the tenants from buying land. The courts ruled the law constitutional, and in most cases tenants refused to be moved from the land. Lingering doubts as to the intention of the government to press the program were dissipated as the first purchases and sales of land became known.

The fact that the occupation backed the program had much to do with the overcoming of the opposition. But even General MacArthur with his authority and the confidence of the people could not have altered a traditional land structure if conditions had not been ripe. They were indeed. Land reform was not an occupation whim but a well-judged response to the needs of the great majority.

Three years after the inauguration of the land purchase program, the task was accomplished. A Japanese farmer once told me that "A farmer without his land is like a man without a soul." There are now many more farmers in Japan with "souls"; more than three million, in fact—the number of households which at last have acquired a stake in their communities by purchasing 5.5 million acres of land. The acreage operated by the tenants has declined from 46 to 11 percent. The owner operator is now Japan's typical farmer. This all came about without a single loss of life, the shedding of a drop of blood, or a yen's worth of damage to property.

Less obvious, yet very important, is the occupation's effort to improve farming practices. Much attention has been paid to insect and plant diseases which cause in Japan an annual loss of not less than 10 percent of all crops. This is an enormous loss in a country that must import nearly one-fourth of her food requirements. American technical aid has also gone into experiment station work; farm taxation; crop insurance; home life improvement; and farm cooperatives, which are important agencies in Japan, handling most of the farm products sales, half of farmer purchases, and much farm credit business.

One typical farm Americanism exported to Japan is the rural youth clubs, patterned after the 4-H Clubs and Extension Young Men and Women Clubs. Only one year old, they have a membership of 500,000, and indications are that Japanese rural youth take to the idea like ducks to water. They are the boys and girls on the "green bicycles," as they call them in Japan, carrying the extension work message from village to village.

It is too early to strike a balance sheet of the American grass-roots diplomacy in rural Japan, but the outlook is encouraging. No reform can add to the very limited arable acreage of Japan. Even under a system of equal distribution of arable land, a farm family would own only 2.3 acres. This is almost 2 acres short of what is needed to give the Japanese farmer a reasonable assurance of a decent standard of living. The occupation farm policies were designed, therefore, not to seek unattainable final solutions but to alleviate burdensome economic conditions and to forestall the Communist peril that feeds on them. This much has been achieved.

The landlords, naturally, do not like the program. The great majority holds that reform is bad for their own families, but the same majority admit that widespread land ownership is definitely good for the village. The landlords realize that the reform has strong roots. However, he would be a bold man who would assert that a direct or indirect attack on the reform is out of the question after the occupation.

The acceptance of the new system among the new owner operators is general, of course. They have finally acquired the one thing the farmers the world over want most: a piece of land of their own. A new song one hears about the village shows how they feel:

The same paddy field, but now my own;
The richer soil yields golden blades of rice.
Heavy is the crop and light is my heart;
Sharp is the sickle, and blue is the sky.
Hail this happy harvest day!
Bright shines the sun on my own piece of land!

The American export of ideas and deeds is narrowing the traditional differences between social classes in the village. As one Japanese farm owner told me in early 1948, "It is high time for the landlord to step one down and the tenant to step one up so that the two may meet for the first time." They are meeting now on the farm committees, cooperative and school boards, and village offices. They rub elbows, dealing with common problems. Here lies the hope of new leadership and the idea of citizenship rights—both foreign to the pre-MacArthur village. This along with the ownership of land spread among the multitude of farmers is the foundation of a more satisfying rural life and the beginning of political and social democracy.

The occupation, by meeting the farm issue squarely, has destroyed one of the fertile sources of Communism in Japan and has given the country a good measure of political stability. In the view of one like myself who saw the Communists rise to power in Russia, failure to ease the farmers' lot would have turned the Japanese countryside into a hotbed of unrest and possibly delivered it to the Communists. The occupation's farm aid strengthened and multiplied the independent land-owning peasants, the group that makes for a middle-of-the-road, stable rural society, lending no ear to political extremists. For this reason, we—and not the Communists—have made political capital from the rural diplomacy we conducted here. Ours was the gain because General MacArthur grasped from the beginning the need to fight Communist ideology with an effective version of American farm tradition.

The Communists of Japan tried hard to exploit the agrarian difficulties of the country. They regarded the village as their very special oyster, and they cherished the ambition to use it as a base from which they would infiltrate the entire Japanese economy. Japanese Communists frankly admit now their lack of popular support in rural areas. This was eloquently proved by the September 1949 election of prefectural farm commissions when only two Communists were elected out of a total of 456 members. Today the Communists' role as a political party in the village has almost vanished.

What has taken place in Japan is food for thought in meeting the farm problems in the rest of Asia, in the Philippines, and in the Middle East. It is a commonplace that no government in Asia can hope to survive without peasant support. No armed force can keep the pot from boiling over if the needs of the farmers are not met. However, the Communists in Asia still hold the stolen trump cards of "every farmer on his own patch of land—every rice bowl full."

This slogan is in the American, not Communist, tradition. Senator Thomas Hart Benton of Andrew Jackson days, the man most responsible for shaping the land policy of the United States, had this to say: "The freeholder . . . is the natural supporter of a free government; and it should be the policy of republics to multiply their freeholders, as it is the policy of monarchies to multiply tenants. We are a republic, and we wish to continue so: then multiply the class of freeholders; sell for a reasonable price to those who are able to pay; and give, without price, to those who are not . . ."

In Japan General MacArthur didn't wait for the Communists to usurp this heritage and appear as the champions of social betterment.

What holds good for post-MacArthur rural Japan holds good, with variations, for all of Asia. That continent, too, wants a demonstration that what we call democracy is useful and profitable. To carry conviction, this idea must be wrapped in a package of economic and social well-being; it must combine technical know-how to increase the productivity of the land with reforms that insure the farmer "the fruits of his labor." Our efforts in that direction may well help fill the empty rice bowls and win the firmest allies in the clash of ideas with the Communists. Then we will have won the peace in Asia.

18. *Field Observations in the Punjab*

Sometime in August or September 1952, Ladejinsky arrived in India and remained there for about three months, during which he made field trips in the Punjab, Madras, and Kashmir and prepared reports on each for the Planning Commission. The setting for his temporary duty visit is provided by the following telegram of March 17, 1952, from Ambassador Bowles in New Delhi to the Secretary of State:

> GOI heavily committed, particularly after recent elections, to program land reform. However, initiative in hands of states and so far progress slow. Very little experience, knowledge or understanding among legislation most state assemblies of this complicated difficult problem and results likely be rather disappointing and many mistakes probably inevitable.

Discussed situation with Deshmukh recently and told him particularly our experience Japan and South Korea. He said deeply appreciate Ladejinsky now Tokyo come in India for say ninety days study what has been done, go over projected plans and make recommendations to state and central governments.

Earnestly hope Ladejinsky's assignment here this period can be arranged as it vitally important for economic development India land reform gets off on right basis. Since state assemblies start to meet in May, time element important.

Also wish Dept consider possibility assigning Ladejinsky here permanent basis to deal this problem.

Ambassador Murphy in Tokyo was reluctant to release Ladejinsky for the temporary duty requested and insisted that he could not be spared until mid-August at the earliest, hence the time interval between Bowles's request and Ladejinsky's arrival in India. Although Bowles had also asked for consideration of Ladejinsky's assignment to India on a permanent basis, a request on which the department initiated preliminary action, Ladejinsky did not come to India on a permanent assignment until some fifteen years later, when he was posted to New Delhi by the World Bank. Ambassador Murphy cabled from Tokyo on May 16, 1952: "In my opinion permanent transfer Ladejinsky would constitute gravest loss to this Embassy and unfortunate diversion of unique talent and experience . . . I recognize India's need expert advice land reform and Ladejinsky's special talent. Therefore would reluctantly concur in TDY maximum ninety days. His permanent transfer, however, would be over my strongest possible protests."

In the absence of any stated explanation for the selection of the three states chosen for Ladejinsky's field observations, I assume the Punjab was chosen because it is the heartland for wheat production in India, Madras because it had recently been the scene of violent peasant demonstrations, and Kashmir because it had recently carried out a radical land reform.

Ladejinsky was attached to the Indian Planning Commission for the purpose of these studies. The reports were made directly to the Planning Commission and discussed in each case with its members and staff. The following report is dated October 8, 1952. I do not have the dispatch number or date of its transmittal to Washington.

ON SEPTEMBER 20 THE WRITER of these notes and A. N. Seth, research officer of the Planning Commission of the government of India, set out to visit a few villages in the Punjab to observe the implementation of Punjab Tenants (Security of Tenure) Act, 1950, as amended on December 27, 1951. Time at our disposal being very short, we could not delve into the problem as deeply as we should have liked. We visited only three villages, but there we met with a goodly number of farmers as well as with a few officials concerned with land matters. What follows, therefore, is not a definitive account of the reform in the Punjab but rather a set of impressions, the soundness of some of which may be questioned. If, on the other hand, our brush with the land reform situation in this state has any value, it will be because the villages we looked into and the people we talked to were largely a matter of chance; and statisticians say that, when selection is left to chance, a small percentage may serve reasonably well to gauge the whole.

Apart from the fairly large proportion of owner operators, the tenants in the Punjab fall into two categories: first, occupancy tenants with heritable rights, a comparatively small group cultivating from 7 to 10 percent of the land; second, the so-called tenants-at-will who cultivate between 30 to 40 percent of the land but enjoy no permanent rights and can be evicted by the landlord at his pleasure. It is for the sake of this group that the act was enacted.

Subject to the qualifications made in paragraph one, then the following main conclusion may be drawn with a fair degree of certainty: The legislation falls short of its goal "to provide for the security of tenure to the tenants." This in turn is based on these findings:

1. Some of the provisions of the act and developments to which they are giving rise are endangering rather than strengthening security of tenure; the number of tenants ejected from the land is likely to be greater than the number of those who will achieve security by virtue of the same act.
2. The act is an example of legislation which failed to take into account large-scale displacement of tenants and preemption of the land by the landlords on real or fancied grounds of self-cultivation.
3. The right of land purchase with which the act endows the tenant is a provision of extremely restricted applicability, and few tenants are likely to benefit thereby.
4. The reduction of rent from one-half to one-third of the group will, at best, apply only to those with legal right to remain on the land undisturbed for a period of five years. Those with no security of tenure lack due bargaining power to enforce the rent reduction.
5. It is questionable if the patwari register reflects the true landlord-tenant arrangements. The patwari is heavily biased in favor of the landlord.
6. Even if the act were so drawn as to insure security of tenure and widespread ownership of land among tenants, it would still need to provide for an enforcement mechanism without which no legislation can be translated into meaningful action.

The original act of 1950 is made up of numerous provisions, but the significant one gave the tenant-at-will security of tenure for a period of four years. But it applied only where the landlord had land in excess of 100 standard acres or 200 ordinary acres. The acreage below this limit (100 or 200) the landlord could claim for self-cultivation and evict the tenants from that portion of land.

Upon the promulgation of the act, it became apparent that it would not serve the purpose for which it was created. In the words of a principal officer of the government of Punjab, "The number of such [large] landowners in the state being small, the act thus, in fact, touched only a limited number of tenants-at-will"—and "the relief granted by the act . . . was extremely meager." The legislation was, therefore, drastically revised and reenacted a year later under the name of the Punjab Tenants (Security of Tenure) Amendment Act, 1951.

The main points of this act are:

1. Security of tenure was extended to five years;
2. The area a landlord can keep for self-cultivation was reduced from 100 standard or 200 ordinary acres to 50 standard or 100 ordinary;
3. A tenant who has been in continuous cultivation of the land for more than twelve years has the right to purchase the land; and
4. Rents were fixed at one-third instead of the customary one-half of the gross crop.

The revised act is indeed an improvement over the original one and, if accepted at face value, would seem to give the tenants-at-will a number of significant privileges. But in practice this is not the case, judging by the application of the provisions, their consequences, and the expressed views of the tenants.

In the past the landlord could and did shift the tenant-at-will from one piece of land to another or evicted him altogether, even though the original conditions of tenancy were not in question. Interviews with the farmers revealed that, ever since the land reform became a subject of governmental concern, shiftings and evictions became a common practice to which the landlords resorted in order to prevent accrual of any claims in favor of the tenant. It follows that, from the point of view of the tenant, the single most important provision of the act relates to security of tenure, and its test lies in whether or not the tenant obtained the much sought-after security.

The privilege of remaining undisturbed on the same land for five years in the light of the former lack of security is a gain and is generally acknowledged by the farmers. On the other hand, the time period is a limited one and does not compare with Bombay and Hyderabad, where a ten-year period of fixed tenure was adopted. Moreover, what a tenant really wants is permanent occupancy with all the rights it entails, but this the act does not give him. The purchase provision will be discussed in a separate paragraph, but it should be noted here that the five-year security period does not augur well for land purchasing because the act confers upon the tenants the right to buy land only after more than twelve years of continuous occupation of the land.

From the point of view of the tenant, the real flaw in the legislation lies not in the limited security but rather in the principle of "permissible self-cultivation." This has in effect given the landlords the right to displace tenants and replace them with hired labor or with erstwhile tenants willing to remain on the land on terms laid down by the landlords. Before the amended act came into force, the landlords appear to have largely exercised the right of resuming self-cultivation of 100 standard or 200 ordinary acres, i.e. they have displaced the tenants who cultivated the land. The amended act reduces the area for resumption to 50 standard or 100 ordinary acres and provided for restitution of lands to tenants who were evicted under the original act from land in excess of 50 standard or 100 ordinary acres. For instance, if a landlord owned 150 standard acres of which he cultivated 60, under the original act he could evict tenants from 40 standard acres in order to bring 100 standard acres under personal cultivation. Under the amended act the 40 standard acres would be restored to the evicted tenants. However, we were told that if he had owned only 99 standard acres instead of 150 and had evicted tenants from the remaining 39 acres, the evicted tenants could not get protection under the amended act since the original act did not apply in cases of landlords holding less than 100 standard acres. If so, a large number of tenants may not get the benefit of the restoration clause under the amended act.

If the act had originally called for 50 standard acres of self-cultivation, the degree of displacement would have been smaller; but displacement was bound to take place for two reasons: the fear of the landlord lest further changes in the legislation conferred more rights on tenants and the fact that the act calls for a significant reduction in rent. Since there is no lack of farm labor eager for employment, the landlords can do better working the land with hired hands. It follows then that tenants on holdings of 50 standard acres or less get no security of tenure at all. Tenants on land in excess of 50 standard acres receive security of tenure for a period of five years, subject of course to the qualification noted in the last two sentences of the previous paragraph. After the five-year period they can be evicted at the option of the landlord.

In order to determine how large a group of tenants stand to benefit or lose from the amended act, one would have to know the distribution of landlords' holdings of more than 50 standard acres and the number of tenants in relation to the total tenanted area and total number of tenants. Data on these points are lacking, but it is safe to infer that, as the bulk of the area is made up of holdings of 50 standard acres or less, a large number of tenants do not receive protection under the act. The fear complex engendered by the recent legislation among the landlords has caused large-scale evictions which may not have otherwise obtained. Hence the conclusion that the act may have harmed more tenants than it has benefited.

The displacements noted above are not mere logical inferences; the references are based on our first-hand village observations and uncontested statements made by the tenants in the presence of landlords and local officials. Village Maina of the Rohtak district is a case in point. It is a small village with but 200 families and 2,100 acres, of which 1,700 are under cultivation. Owner-cultivated land was around 1,200 acres, and until not so long ago the remaining 500 acres were worked by tenants. The major portion of the tenanted area was held by three landlords.

One of the most surprising statements made by the patwari[1] was that the tenanted acreage has declined so sharply that at the time of our visit it was a mere 20 acres. The first reaction was to congratulate the villagers on the virtual disappearance of the tenancy system. It would have been a premature gesture, for in reality nothing has changed in the land ownership pat-

1. Village head.

tern of the village despite the seeming elimination of tenancy. The landlords still have the land; only the status of the tenants has undergone a change—they continue to work the land either as hired hands or under unchanged conditions, except that their names no longer appear in the patwari register[2] as bonafide tenants. The dozen or so assembled farmers were rather reluctant at first to account for the change, especially in the presence of a couple of landlords. Gradually, however, as answers followed questions, it appeared that the land had been taken back by the landlords. A landlord provided the answer in reply to a question as to what he knew and thought of the act. He stated rather bluntly that he was familiar with its provision and that was why he had taken his land back. But this was only part of the explanation that lies behind the movement of resuming land for self-cultivation.

That the process of "proletarization" of the village of Maina was greatly accentuated by the self-cultivation provision of the act is undisputed, but it had evidently gone on for some time before the enactment of the original 1950 act. The mere land reform talk preceding the legislative enactment convinced the landlords that it simply was not "safe" to rent land to tenants or, at any rate, to have the patwari register bear witness to that. Nobody present denied what is obviously a common practice in Maina village. The procedure is simple: the landlord tells the patwari to reregister the tenanted land in his name as a self-cultivator, and the request is usually complied with. Ethical restraints hardly enter into the picture. The act simplified matters considerably; landlords in the category under consideration do not have to resort to the patwari register technique or to the shifting of tenants from holding to holding or plot to plot. All he has to do to accomplish the same end is to stand on his rights under the personal cultivation provision of the act.

While the landlords know what to do and are competent to cope with the new developments generated by the act, the tenants appear helpless and bewildered. This is particularly true of those who have no land of their own to sustain them to any degree. Those who do, as exemplified by the owner tenant of the village of Bapori who enjoyed the privilege of sitting alongside with the landlords, have apparently no interest in the act. In reply to the question as to what he thought of the new legislation, he stated that he knew nothing about it and did not care to know. But to the harijan tenant who squatted on the floor just outside the room where the rest assembled, and to the nonharijan tenants who were squatting in the room just inside the door, the act is not a boon. For an undetermined number of tenants who may even be the majority, the act failed to bring security of tenure. By the same token they have moved one step down from the already low rung of the agricultural ladder on which they had a precarious hold.

The corollary of this is that the tenant is hungering for land on any terms. Under the circumstances, the main concern of those we interviewed was not a fixed period of occupancy but how to retain a foothold on the land. Conditions of tenancy assume secondary importance. The statement of one of the tenants that "the law should make certain that I have some land" is quite understandable. Since more often than not such tenants have no land, particularly in the villages we visited, they are ready to cultivate land on any terms. Failing that, they are in danger of becoming hired hands with all the disadvantages that the status entails.

The more specific and inevitable consequence of the right of displacement is that the positive provision of the act, the reduction of rent from one-half to one-third of the crop, cannot but be subjected to wholesale violation. To have force, the rent reduction must go hand-in-hand with real bargaining power, which in turn can come only with unquestioned security of tenure. To the extent that only a certain number of tenants are likely to obtain that under the act, the rest will not benefit from it. It is quite possible that some landlords will not claim all the permissible land for self-cultivation, yet it would be idle to expect that the tenants will insist on the rent reduction, realizing as they do that the landlord can always eject them on the ground of personal use of the land. It may be concluded, therefore, that while the rent reduction is indeed a welcome step in the right direction, its effect will be limited in scope.

2. Village head land register book.

In the course of our trip the question of land ownership came in for a good deal of consideration. If there is one thing on which all tenants are unanimous, it is the wish to own land. This is the highest goal to which they all aspire. The act is not unmindful of that, and so it gives the tenant the right to purchase his holding after at least twelve years of continuous occupation. But this right is illusory because for reasons given above many tenants find it difficult to prove in law continuous possession. The village of Ghdhaam-Kalanaur, where all of the land is tenanted, provides another illustration of this difficulty. There the tenants have been in continuous cultivation of the land for many generations, yet they are faced with the problem of establishing occupancy of specific holdings over a period of twelve years. At this time the village of 200 families is in the midst of fighting a total of 145 eviction suits.

Even if the twelve-year period were eliminated, there is nothing in the act, as it stands now, which is particularly helpful to a tenant desiring to purchase land. The act states that "A tenant shall be deemed to have purchased the land if he deposits the value thereof as so fixed with the Assistant Collector within such time as that officer appoints." In practice this means that the tenant is expected to buy the land at the market price (paragraph 13 of the "Introduction" of the Punjab Tenants [Security of Tenure] Act, 1950) and pay in one lump sum. Under such conditions of land acquisition, most schemes are doomed to defeat, no matter by whom initiated and where they take place. If a tenant could buy land at the prevailing market price and pay the full sum initially, there would be, of course, no need for a government-sponsored land purchase scheme. But when a government does become concerned with land ownership among tenants, two conditions are basic to such a program. First, the price fixed by the government must be reasonable—that is, it must be adjusted to the ability of the tenant to shoulder and discharge the financial obligation without any undue strain. Second, the payments must be spread over a number of years. The tenants of the Punjab are not in a position to buy land on conditions other than those just cited. They told us repeatedly that they wish to buy land if the price and terms of payment are reasonable. The act does not provide for that.

The ineffectiveness of the Punjab legislation is heightened by what seems to us the lack of an enforcement mechanism. The act imposes the responsibility for implementation upon the magistrates and revenue officials—the collectors, revenue assistants, tehsildars, naib tehsildars, circle kanungos, ending with the officers closest to the villagers—the patwaris. It is our observation that these individuals are not in a position to insure for the tenants whatever rights do accrue to them under the act.

Rohtak district, for example, is made up of about 1,000 villages with a population of about 1 million and supervised by a small number of revenue officers of all grades (other than the patwaries). Their normal, prereform duties are so complex and time-consuming that they cannot be expected to devote much effort to the enforcement of rents and occupation rights affecting large numbers of tenants and a good many landlords. Moreover, even if they could spare the time, our impression of the officials we encountered on the trip is that they are at best indifferent and at worst hostile to the subject of land reform and its implications. Many of them belong to the landowning class against whom the reform is or will increasingly be directed. It would be idle to expect them to administer a law which is not in tune with their own convictions and, often, material interests. The patwari at the village level sheds much light on this point. He could serve as a most important link in administering the act, but as against that there is the readiness with which he has been serving the cause of the landlords, as exemplified by his falsification of the patwari register. On a higher level, the very articulate and voluble magistrate of the city of X represents the same attitude. Himself a landholder, he made no secret of his opposition to the act and reform in general. It is a fair surmise that, in a landlord-tenant dispute, the tenant is at a considerable disadvantage from the very start if the case is to be adjudicated by the type of magistrate referred to above.

It is our belief that, if the act is to be properly administered, the villagers themselves must be entrusted with the lion's share of the task. By this we mean active participation of the tenants, landlords, and owner cultivators, represented on a village commission or council, with the authority to carry out the provisions of the act according to prescribed rules. The watchdogs of the act should be the people who know intimately the past and present conditions in the village and are in a position to prevent malpractices which stem from an age-long superior position of only one group of the village, albeit a minority group. Not only is this type of local enforcement desirable on practical grounds but from a moral and political standpoint as well. The community participation would ensure the changes of certain rights under the act, and something else besides—a more equitable social relationship among the divergent groups of the village. The fact is that group participation, even if based on "class" lines, is a unifying rather than divisive force. It may be further observed that the reform in land tenure relationships and mode of executions are not entirely separable. It is a cardinal principle of democratic thought that ends and means of socioeconomic action may not be considered apart from one another. In these terms a freer and more democratic rural India would not come in the wake of a land reform program conducted entirely, or for the greater part, along normal administrative lines with decisions rendered from on high. What is therefore called for is less governmental and more citizen participation in the enforcement of a program intended for those very citizens.

Remedial measures are implicit in the analysis of the act and the mentioned problems arising from its application. They may be summed up here as follows:

1. If the right of self-cultivation must be extended to the landlord, it should be so restricted as to reduce displacement of tenants to a minimum. While it may be necessary to protect the interests of small peasant proprietors who have had to resort to leasing for short periods for a variety of reasons, the absentee landlords with large holdings do not seem to us to have a case for resumption. Tenants on lands which have not been cultivated by owners for a considerable period, say five years, may therefore get a right of permanent and heritable occupancy. At the same time, in order to exempt the small holders from the effects of this provision and as a measure of rehabilitation of others, this provision might permit self-cultivation of a subsistence holding by landlords of all categories—be they absentee or resident.
2. Rent reduction is meaningless without a degree of permanency of tenure. To deny the tenant the latter is to deny him the benefit of reduced rents.
3. Land reform legislation, whether in the Punjab or elsewhere, must provide a legal framework for the active participation of the people in the implementation of the program. Land reform cannot be carried out from the cities of Delhi, Madras, Hyderabad, Bombay, Nagpur, and so forth. Decisions concerning the correct status of individual landholdings, ownerships, and occupancy rights, both past and pending, should be made on the spot by the villagers themselves who have thorough knowledge of the existing conditions. Illiteracy is not ignorance; it need not preclude assumption of responsibility and exercise of authority—development most conducive for the kind of leadership presently absent in rural India. The very nature of a land reform program designed to benefit the large group of underprivileged demands that properly constituted groups of farmers, equally representative of all village interests, must be entrusted with the responsibility for doing the job. Only on this basis can the letter and the spirit of the legislation be fulfilled. On economic, political, and social grounds there appears to be no substitute for this approach.
4. If a land purchase program sponsored by the government for the benefit of the tenants is to be successful, it must be based on a price which is financially manageable from their point of view. This automatically precludes buying land at the prevailing market price which in the context of keen demand for land is a competitive price with little or no relation to the capitalized productive value of the land. High market prices since the war have increased manifold; and, even though agricultural prices have followed suit, there is certainly no evidence that the tenants' capacity to assume obligations under high land values has improved any. Assuming that a measure enabling tenants to buy

land must be designed so that they can discharge the responsibility of ownership with relative ease, concurrently improving the land and maintaining a decent standard of living, then it follows that the purchase provision of the Punjab Act would benefit only the landowner and help maintain existing land values and rents.

An alternative is not easy to suggest for the simple reason that fixing land prices is normally a difficult and thankless job particularly in India where formation of land prices is characterized by a set of rather complex factors. "Scientific" approaches to fixing prices and fair rents have been tried in India in the past by all kinds of rural investigation commissions, but in the end they resorted to something less than one might call precise. This explains why the Floud Commission, after a painstaking inquiry into the determination of a fair rent, concluded by quoting approvingly the following: "The Commission could only define fair rent by the rather indefinite description—such a share as shall leave enough to the cultivator of the soil to enable him to carry on the cultivation, to live in reasonable comfort, and to participate to a reasonable extent in the progress of improving prosperity of his native land."

What is true of this key to fair rent seems to us equally true with respect to fixing land prices, namely, the welfare of those for whose sake a reform is initiated. A dose of arbitrariness or compulsion on the part of the state becomes essential in redressing economic and social inequities which can no longer be adjusted by the contending groups themselves. The Indian land reform legislation to date abounds in examples of this type of compulsion. It may very well be extended to price of land and payment for it. If this approach is adopted, and we think it should be, the point of departure in fixing land prices should be drastically reduced and enforceable rentals in conformance with the Floud Commission philosophy. Having done that, the price of the land should be based on the cash equivalent of a number of annual rentals and should be made payable in small annual payments.

In short, the terms of sale should place the land securely in the hands of the tenants. If the sales plan required an appreciable initial payment or a short period of amortization, the new owners would have only a tenuous hold on the land. The new owner who feared foreclosure would have an instability of tenure similar to that of tenants. The owner who paid too high a proportion of his income to meet payments on his newly acquired land would be in the same position as a tenant who paid too high a proportion of his farm income as rent. Also, we must remember that *fixed* annual payments have helped nullify efforts of farmers to achieve permanent ownership. They have thereby increased tenancy rather than diminished it. A land ownership plan should provide flexible payments to allow for variations in crop yields and in prices received for agricultural products and allow deferment of payments to enable the new owner to build up a reserve and strengthen him in his new enterprise.

These are some of the principles that constitute a realistic government land purchase program to insure land ownership among tenants who have no means of their own to attain that status. If this outline is basically valid, then the Punjab Act and its land-purchasing provision must have been written for a very "rara avis" of a tenant. But for the run-of-the-mill tenant who populates the villages of the Punjab, a new act will have to be drawn up if the state has any serious intentions of assisting him into ownership of the land he cultivates.

19. *Land Reform Observations in Madras*

These observations in Tanjore, Chingleput, and Malabar districts, all in the one Madras state, help one to understand the tremendous diversity of conditions which complicate the efforts of those whose duty it is to deal with India's problems. Especially in Tanjore, where the Communists had gained a strong position in the countryside, the conditions made an indelible impression on Ladejinsky.

This report, dated October 31, 1952, was transmitted to the U.S. Department of State in Dispatch 1150 dated November 6, 1952.

Purpose and Method of Inquiry

FOLLOWING THE TRIP TO THE PUNJAB, this observer, accompanied by M. S. Menon, official of the Ministry of Agriculture and Food, spent the week of September 23 to 30 in the state of Madras. The main purpose of the trip was to study the Tanjore Tenants and Pannaiyal (Protection) Ordinance of August 23, 1952, and the reaction of the several interest holders in the land to the measure. We had little time at our disposal, but we managed to visit not only three villages in Tanjore but also two villages in Malabar and three in Chingleput district. Our visits were never prearranged; the tenants, the pannaiyals,[1] and casual landless laborers were met in the villages or gathered at random from the fields as we went along. The answers were elicited by easy, informal talks from which all air of authority was carefully kept out, although we were in most cases accompanied by revenue officials. It should be noted, however, that on virtually all occasions the farmers made no secret of their views. When in the village of Inrungattukottai it looked as if everybody was satisfied with his lot, even there, as we were about to take our leave, a farmer upset the apple cart by spilling some of the real conditions existing there. Considering that the landlords are never reticent to state their case, the results of our inquiries in terms of what we were told and noted must be considered satisfactory. On the other hand, we are conscious of the fact that neither the time spent was long enough nor the range of the investigation sufficiently wide and detailed to lend infallibility to our interpretations and conclusions.

1. "Permanent" farm servants and often sharecroppers.

Summary and Findings

This report concerns itself with tenants and agricultural workers only—the terms under which they cultivate the landlords' land, their consequences, and attempts to remedy them:

1. In Tanjore, Malabar, and Chingleput agricultural efficiency, the output of the land, and the rural standard of living are all at a very low ebb.
2. This is only partly explained by unfavorable climatic conditions as in Chingleput, where the pressure of population on the land and the indigenous methods of cultivation are common to all areas. A good deal of the explanation lies in the rights and privileges of the different interest holders in the land.
3. Absentee landlordship is rampant: in Chingleput due to the very smallness of the holdings, in Tanjore and Malabar due to the concentration of a large acreage in a few hands. In both cases the cultivators and the land have suffered.

4. Exceedingly high rents ranging from 50 to 80 percent of the crops, cost of cultivation imposed upon the tenants, and lack of security of tenure are the main features of the landlord-tenant relationship. The tenants have no incentive (and no means) to improve the land, while many landlords who do have the means are content with their returns and conditions of the land as they are.
5. The region had long been in urgent need of far-reaching revisions of tenancy conditions, but little had been done in that direction. For this reason, tenants and agricultural workers, the latter largely harijans,[2] are increasingly and actively discontented with an inflexible status quo of poverty and social degradation. Tradition and authority will not prevent them for long from taking the law into their hands, as they did in Tanjore, to redress grievances not otherwise assuaged.
6. The Tanjore ordinance was enacted to improve relations among the various groups on the land, but this may not materialize. The basic rent and security of tenure provisions are inadequate, while large groups of tenants and agricultural laborers are excluded from its provisions altogether. The ordinance is an involuntary and meager concession made many years too late.
7. Chingleput has no tenancy legislation whatever, and in Malabar successive legislation over the years has not in actual practice benefited the cultivating tenant and landless farmhand, and the land is badly neglected. Evasion of the letter and spirit of the provisions is common practice. The suspension of the 1951 act, which attempted to give tenants certain rights, has left them with no legal redress against the prevailing tenancy conditions.
8. In all areas visited we found no trace of village activities by the Congress Party on behalf of the underprivileged. Tenants and agricultural workers do not identify the Congress Party and the government with efforts to assist them. In their mind the Communists play that role in Tanjore, while in Malabar and Chingleput the ground is well-prepared for a similar development unless state-initiated effective and timely economic measures will forestall it.
9. The basic land reform issue in Tanjore, Malabar, and Chingleput is that the cultivator should become the owner of the land he tills. But the more immediate issue is one of improving the terms of tenancy through reasonable rentals, fixity of tenure, and so forth. Little progress has been made; although there is no lack of sound recommendations to correct that. What is lacking is the recognition that the landless are entitled to a stake in the land and that failure to act is fraught with serious consequences. What is lacking above all is a resolute leadership willing to initiate and enforce long-overdue changes in the land tenure system.

Uniqueness of Madras

A mere glance at the map, supplemented with a few basic facts and figures, shows that Madras is an important state from many points of view. It forms the greater part of South India, stretching as it does from the borders of Hyderabad and Orissa to the tip of the peninsula. The state comprises nearly one-sixth (57 million) of the total population of the country and is noted for its local cultural patterns, including four distinct languages, and a high literacy rate. With nearly 41 million acres under cultivation (including current fallows) out of a total area of 82 million acres and relatively little industrialization, Madras is essentially an agricultural region.[3] As regards the great variation of land tenure arrangements and difficult problems surrounding them, Madras can hold its own with any other state of the union.

Madras in many ways offers an interesting study in contrasts. Primarily an agricultural state with some very rich soil but with uncertain rainfall, Madras is normally a food importer, and the standard of living of a large

2. "Children of God," in Gandhi's term, or the "untouchables."

3. Only 12.3 percent of the population are directly dependent upon nonagricultural production for a livelihood. The all-India average is 10.5 percent (1951 Census).

group of the actual cultivators is abysmally low. The state is noted for its educational progress and as a breeding ground for some of India's keenest leadership, yet poverty is rampant as if these are unrelated phenomena. There are regions in Madras where the level of education of women is one of the highest in India, and yet in those very regions the growth of population is most marked. Rural society is very backward, caste and religious prejudices still rule supreme, and yet this center of orthodoxy can also be the center of rebellion; it is in Madras that Communism has gained its most noted victories. Madras was the first state to carry out zamindari abolition, and the last to attempt—or rather to be forced—to deal with the urgent problems of private ownership of land. Until the very recent enactment of the Tanjore ordinance, only one district in the state, Malabar, had the benefit of tenancy legislation; but that, too, is dormant now for reasons pointed out elsewhere. In sum, nothing can be taken for granted in Madras, for reality and appearance there are often worlds apart.

Tanjore District

The rural setting

The real significance of the Tanjore ordinance can be judged only against the background of the conditions which induced its enactment. The preamble of the ordinance itself provides a partial answer: "Whereas in the district of Tanjore the relations between landowners and their agents on the one hand and tenants and farm laborers on the other have become strained, resulting in displacement of tenants and the dismissal of farm laborers and in agrarian crimes and disturbances: And whereas the situation threatens to cause deterioration in agricultural production . . . the Governor of Madras hereby promulgates the . . . Ordinance." But in addition to this helpful elucidation on the pre-ordinance situation in Tanjore, we must note a number of pertinent facts underlying the "crimes and disturbances," facts without which no rounded-out appraisal of the measure can be made.

For one recently arrived from the Punjab it was particularly striking to see the green sweep of the paddy fields disappearing in the horizon. They closely resembled the Japanese rice fields with hardly any fallow, all the land being laid out in checkerboard fashion, one crop growing directly against the next; only the ridges separating the fields were too big, not carefully tended, and almost bare of catch crops. Instead of the sandy plains of the Punjab with the uncertain rainfall, the ground slopped away in almost constant greenery, the fields that had been recently plowed presenting a moist and rich brown loam. For the water, too, was there in the wells and reservoirs, chutes, and spillways, although at the time of our visit not in the usual abundance.[4] The outstanding impression was one of rich soil and rich country where farmers are relatively well off, certainly better off than in the Punjab. The more surprising therefore were the realities described below.

Although nearly 40 percent of the area of Madras is in crops, cultivated and cultivable lands are so scarce in relation to the competing demands for their ownership, ownership and cultivation, or merely cultivation that there are large unsatisfied demands for land. This land hunger spells also acute competition among landowning noncultivators, landowning cultivators, and landless laborers of the farm community. Land ownership is of course ill-distributed: small holders who constitute 97 percent of the total number of holdings have 59 percent, and big holders who constitute 3 percent have as much as 41 percent of all wet lands with assured water supply. This situation is fully reflected in Tanjore, where the average size of a holding is only 2.6 acres and where 2.4 percent of the holders with 18 acres and more own 36.6 percent of the land. In some areas of Tanjore, as for instance in the Shiyali Talua, the concentration of land ownership is especially pronounced. One landlord owns about 10,000 acres of land, and not a few own thousands of acres each. Official statistics on acreage under tenancy are not available, but sample investigations show that more than half the land in Tanjore is worked by tenants and part owner–part tenants. More startling is the large

4. Tanjore is one of the best irrigated districts in Madras. It produces surplus rice for shipment to deficit areas of the state.

number of agricultural laborers on these and owner-operated farms. This is supported, even if indirectly, by official data on the occupational breakdown of the farm population of Tanjore. Tenants with no land of their own constitute 22 percent of the agricultural population and agricultural laborers 35 percent. To this must be added a large but undetermined number of part owner–part tenants who are included in the category of "cultivator of land wholly or mainly owned." It may be concluded, therefore, that from two-thirds to three-fourths of the farm population in Tanjore either leases or works the land of the noncultivating owners.

Under such conditions coupled with a landlordism concerned only with the maximum possible gain, terms of tenancy and farm wages cannot but be extremely onerous. Any random inquiry among the tenants of Tanjore reveals extreme poverty. Prior to the ordinance, rentals amounted to as high as 75 to 80 percent of the gross produce, not counting the cost of cultivation for which the tenant bears most of the responsibility. It might be noted that the cost of a tenant's own labor is not taken into account. When asked why, then, he leases land, the answer is that he has no other alternative. What he doesn't say is that the only alternative, that of a pannaiyal or casual laborer on the same or some other holding, is even worse than the lowly status of a tenant. In the end, many do make the break, even when rents are lower. The tenant in the village of Poosivakkam of Chingleput district cultivated less than six acres, but he told the Brahman landowner in our presence that he is giving up the land because all his work is for the benefit of the landowner. Discontent and unrest on the one hand and impoverishment of the land on the other are the inevitable consequences.

As noted earlier, the farm laborer in Tanjore is more plentiful than the tenant. We met both pannaiyal type and the casual daily laborer. It is obvious that the problems of the tenants and laborer with respect to the land are closely interrelated. Many an evicted tenant told us that he was displaced by a pannaiyal, who in turn is displaced by a casual laborer. This trend has been in evidence for some time; it is actively promoted by the landowners, whether big or small, and chiefly for the sake of reducing the number of would-be claimants to occupancy rights. The customary wage of a pannaiyal is cited in the ordinance; the wage of the casual laborer is not very much more and is seasonal. Socially, too, they and the pannaiyals are in the same position: they are mostly harijans, the most depressed caste, and as such are at the very bottom of the economic and social pyramid of the village. The shifts in the agricultural ladder are instances of the gradually weakening position of the marginal interests in the land, which nevertheless have the important function of raising the crops. And therein lies yet another cause leading to the ordinance, for the total lack of any stake in the land is conducive to rural unrest and political agitation; and the two have direct bearing on each other resulting in intensification of both.

In the postwar period the apparently inflexible order of things was beginning to give way under the rising tide of agrarian tensions. They assumed the form of strikes, riots, setting crops on fire, a burned-down landlord house, or an occasional murder of a landlord or his agent. The collector of the Tanjore district, Mr. Palaniappan, related some of his experiences with pannaiyals who refused to till the fields on the old terms, which he admitted were low, and refused to admit outside labor to take their jobs. The Mayuram Award (October 28, 1948), Mr. Palaniappan told us, accorded the tenants a 5 percent increase in the share of the crop and higher wages to farm laborers, but the landlords responded through evictions. It might be added that subsequent proposals for agrarian reform were rejected by the government, and it resorted instead to the maintenance of law and order through police action. Little wonder that in the context of these developments the very floodgates were opened to successful Communist agitation based on the popular slogan of "land to the tillers." We spent all too little time in Tanjore to study the role of the Communists in exploiting agrarian grievances for their own political ends, but we did have the privilege of an excellent account of their activities by a politically astute police officer. He felt that, since only the Communists had become the active promoters of agrarian reform, farm leadership was theirs for the asking. They didn't create the grievances; in the absence of any effort by the Madras government to correct the maladjustments breaking into the open, the

Communists articulated the grievances to the obvious satisfaction of large groups of non-political farmers. This explains the successes of the Communists and losses of the Congress Party in Madras during the general elections held in November 1951 through February 1952.

Tanjore on eve of the ordinance

Our talks with all manner of people directly or indirectly concerned with village conditions left us with the impression that by the spring of 1952 Tanjore was anything but just another rural district carrying on in the traditional way. The class lines became more clearly etched, and the smoldering discontent tinged with a pronounced red coloration was there for all to see. The enjoyment if not the possession of some of the most fertile land of Madras state became the subject of the bitterest controversy and struggle. The mirasdars (landlords) form a most powerful landed aristocracy in the state, conspicuous as much for their accomplishments in the arts and letters as for their all-or-nothing economic and political power stance—facts which have sharpened the discontent and animosity of all the rural have-nots. The mirasdars had tried to meet the danger by evicting the tenants, dismissing their customary farm servants (pannaiyals), and replacing both with casual laborers. Determined opposition was natural. It was with this economic threat and political danger that the government of Madras finally decided to deal when it enacted the Tanjore ordinance.

Tanjore ordinance

The main provisions of the ordinance are as follows:

1. After deducting the harvesting charges, the crop is to be shared on the basis of 60 percent to the landlord and 40 percent to the tenant. All cultivation expenses are borne by the tenants.
2. A tenant evicted after December 1, 1951, is entitled to reinstatement. A tenant can be evicted if he fails to pay rent within one month of the stipulated date or when, in the opinion of the Revenue Court, the tenant's management of the land is not in the interest of good husbandry.
3. A cultivating tenant is given security of tenure for a period of five years; if the landlord decides not to renew the lease, he must notify the tenant to that effect a year before its expiration.
4. Pannaiyals are given the option to choose between the following: (a) compensation on a daily basis, when employed, of 2 marakals of paddy per day per man, 1 for women and three-fourths for those not adults (a marakal is 2 Madras measures or 6.4 pounds), or (b) the old system of 1 marakal of paddy per day for male, three-fourths for women, plus one-seventh of the crop at harvest time. When a landlord dismisses a pannaiyal, he can be reinstated if the act is deemed unjust by the conciliation officer, provided the pannaiyal made an appeal within a week after the dismissal.
5. The ordinance applies only to landholdings above 6.5 acres of wet or 20 acres of dry land.

Tanjore ordinance and official attitude

Preparatory to our field trip we had preliminary talks with B. Natarajan, economic adviser to the government of Madras; S. K. Chettur, revenue secretary; and K. Ramuni Menon, chief secretary to the government. In the main they all agreed that the ordinance, which received a great deal of publicity in and outside of Madras, was a step in the right direction. It granted the tenants and pannaiyals important concessions at the expense of the landlords and, in doing that, the officials felt that the ordinance improved relations among the contending groups in an important food-producing area which has been a hotbed of successful Communist activities for the past few years. One of the officials stated that the tenants received more than they had anticipated; and it was the view of all of them that landlords, tenants, and pannaiyals seem to be satisfied with the measure; and that if the effects of the Tanjore ordinance live up to expectations, the same type of reform will be extended to other parts of the state. The subsequent paragraphs are devoted to these queries: How significant is the ordinance and will it achieve its goal?

Tenants and the ordinance

In quest for an answer, we visited a number of villages near Tanjore town and the town of Mannargudi, which had been a trouble spot some time back, and questioned farmers regarding the prevalent practices in the area before and since the enactment of the ordinance and what they thought of it. In one case supported by a number of others, the tenant said that he had been the continuous cultivator of 2 acres for which he has been paying 45 kalams[5] and retaining 15 for himself, or a rental of 75 percent of the gross crop. As nearly as we could determine, he was responsible for the greater part of the cost of cultivation. He knew of the ordinance, but he wasn't satisfied with the 60–40 division of the crop. The old 25 percent share didn't furnish him enough food, and he had to look for outside employment to make good the food deficit. Now he might have enough food, but it is not likely to leave him any surplus to invest in land improvement. Asked what he proposed doing if the landowner insisted on more than 60 percent, he replied that the duty of those who passed the ordinance was to see that its terms are honored by all parties.

Another tenant out of 135 kalams of paddy, which was the yield of 5 acres of rented land, paid the mirasdar 120 kalams, or 89 percent. Out of his 20 percent share he had to furnish the cost of cultivation. He did better with respect to straw: he retained 21 bundles per acre and turned over 6 to the landowner. He lived in a house site furnished by the landowner free of charge, but it was the business of the tenant to take care of the field channels and, in this instance, to apply costly ammonium sulphate at the rate of 100 pounds per acre. Clearly, the cultivation cash expenses besides the occasional wage to a farmhand were not negligible. He, too, stated that the ordinance was not fair to the tenants; asked what he would consider a reasonable rental, he replied 60 percent to the tenant and 40 to the landowner. Cultivation expenses and the post ordinance practice among the landlords not to extend the customary seed loan were cited as the reasons. Since he cannot provide his own seed and must look for a loan elsewhere at a high interest rate, his cost would go up. Actually, more than seed was involved; it turned out to be seed plus an amount of rice he (or they) lack between seasons. We asked the tenant what he would do if the landlord should insist on more than 60 percent. He replied without hesitation, in a natural and matter of fact way, that he would turn for help to the Tamil Nad Kisan Sabha of Tanjore, or the Red Flag Association organized by the Communists, with a reputed membership of 200,000. The association, he was certain, would take steps to redress his grievance.

The significance of the remark cannot be gainsaid, and not only because in the eyes of the nonpolitical tenant the Communist Party is his sole spokesman. Government officials in and out of Madras and private individuals versed in rural matters all noted that Congress leadership in the village was not there when it was needed most and when it could have assumed that role with no effort. Of course, leadership implied the advocacy of measures to forestall the development of an uncontrollable crisis. But whether for this or other reasons, no one was there to fill in the vacuum but the Communists. When in 1948–50 after an outbreak of sporadic violence the Communists (and many a villager) were put in jail, nothing further was done. Once again the Congress Party failed to move in to take over leadership of the masses. The situation was allowed to drift until the Communists, freed from jail on the eve of the elections, again easily stepped into the place left vacant for them. At the time of our visit nothing much has changed in this respect; we didn't hear of any Congress group working with the farmers in the villages we visited and rightly or wrongly, large groups of farmers appeared to have faith in the promises of agrarian reform held out by the Communists. This explains the gap that exists between the government and the majority of the farmers. Whether the ordinance is going to bridge it, time alone will show; but our fears, based on what we saw and felt and on a cold analysis of the several provisions of the ordinance, are that it may not.

The reaction of other tenants was pretty much in accord with the cases cited, although

5. A kalam is equal to 80 pounds.

the Communists as the sole protectors of the tenants in face of the landlords and the ordinance did not come to the fore as in the cited case. It is important to note, however, that to date the tenants look upon the ordinance in a much more critical way than we were led to believe by officials in Madras and in Tanjore. Some of the officials propounded the view that the ordinance was giving them more than what at one time the tenants had asked for, obviously referring to the 5 percent increase under the Mayuram Award, which, incidentally, never materialized. The share is indeed greater, but what may have looked substantial in 1948 doesn't look equally substantial in 1952. Also, it is pertinent to mention here that some of the dissatisfaction stems from the clear-cut manner with which the ordinance specifies that cultivation costs must be borne by the tenant, the sole exception being harvesting cost shared by the landlord. Furthermore, it is not conclusive, as we think back of our field experience, that before the ordinance *all* tenants were in practice responsible for *all* the cultivation costs. In the light of the categorical provision regarding expenses of cultivation contained in the ordinance, it is to be feared that the special arrangements which previously were in practice between landlord and tenant, which were helpful to the tenant, will be discontinued.

Farm labor and the ordinance

The Tanjore ordinance is the first enactment in India which recognized the need of easing the lot of the agricultural laborers. The casual workers are not covered by it, but it may be assumed that better conditions for the pannaiyals may reflect favorably upon their status also. The casual laborer works from four to six months at a daily wage of nearly 2 rupees in recent months; even the higher compensation could not possibly provide them with a living for twelve months of the year. Their main concern at the moment of our visit was how to keep out the considerable influx of farmhands from the adjoining district of Ramnad, where working conditions were particularly bad. The pannaiyal works during the greater part of the year, about nine months, and, aside from the monthly wage of about 20 rupees a month, lives in the landlord-owned hut and is given a third of an acre of land for personal use and an occasional bonus if he is in good standing. Yet, contrary to the prevailing impression, some pannaiyals told us that they would prefer the casual worker status with its greater freedom to move about. The fear of unemployment is, however, a weightier consideration and he stays on if he can. In recent years, though not all of them could be evicted, evictions have taken their toll. A conciliation officer told us that since the enactment of the ordinance he has been flooded with applications for restoration. The former pannaiyals will judge the ordinance primarily on the basis of whether or not they will be restored in their rights. Our attention was called to the power of the landowner to upgrade or downgrade a pannaiyal at will. We actually listened to accounts of pannaiyals upgraded to a tenant status, but such cases are rare; the common practice is to downgrade the tenant to a pannaiyal and the pannaiyal to a casual laborer. Whether the ordinance will put an end to this unsettling factor in the community it is too early to say.

Reactions on land ownership

Tenant, pannaiyal, or farmhand—they all wish to own land. But it doesn't take much probing to find that they totally lack savings or credit standing to enable them to acquire any land, unless on the easiest possible terms, with initial cash payments virtually ruled out. Land is being bought and sold but by landowners only, and good paddy land is commanding a price of Rs1,000 to Rs1,500 an acre. For the time being the ordinance had no depressing effect on land values, although the more recent landlords are inclined to sell rather than buy land. The Communists are beginning to show greater interest in land ownership in contrast to the former emphasis upon better tenancy and farm labor terms. In the villages we visited, the Communist slogan is "five acres for me and mine." The slogan is addressed now not so much to the tenants as to farm labor. This tactic is evidently based on the theory that should the grievances of the tenants be redressed through the ordinance or new legislative enactments, there will still be the farmhands who will remain their staunch allies because even the best-intentioned of governments must find it extremely difficult to solve what is almost an insoluble problem, the surplus farm labor problem.

Landlords and the ordinance

We did not chance to meet as many mirasdars (landlords) to ascertain their points of view as we would have liked. But we did meet a few, and the one in Mannargudi Taluk of Tanjore district, with whom we had a long talk, spoke for the many. He expressed the prevailing sentiment of the mirasdars when he stated that the ordinance is undermining their position; that it denies them their livelihood by making land ownership uneconomic; that it is unjustifiably biased in favor of the tenants and agricultural workers, is inopportune, and unnecessary. The reinstatement of tenants was a particularly bitter pill: it would be tantamount "to throwing in a man's lot again with the divorced wife." The evicted were "bad" tenants, unruly, and allied with the enemies of law and order. Above all, the landlord must have absolute control of the management of his land if agricultural progress is to be attained.

The landlord could not say, and especially in the presence of the revenue officials, that he would disregard the provisions of the ordinance; but he made no secret that he intended to "take it out" on the tenant in some other way. A case in point is that he and the other mirasdars are no longer willing to make seed and manure loans to tenants because: (a) the ordinance states that the tenant is to furnish it; (b) with a reduction in the share, landlords can't afford to make advances; and (c) the mood of the tenant and the continued political agitation make such loans risky. "But if you fail to lend him the seed," we asked him in the presence of one of his tenants, "how is your tenant going to plant the crop and how are you going to collect the rent?" At first the answer was a sly smirk, but gradually the real cause was divulged. Since the tenant's chances of securing any seed or the requisite volume are poor, land productivity is bound to suffer. For the landlord this is a blessing in disguise, for if that should come to pass, he is in a position to evict the tenant under clause 10 of the ordinance. He will have then preserved his rights as he conceives them to be and through the very same ordinance he is inveighing against.

Evaluation of the ordinance

Judging by the reaction to the ordinance on the part of the tenant, agricultural worker, and landlord, it is clear that each group has some misgivings about it. From our point of view, the landlord should have none, because as Chief Minister C. Rajagopalachari has correctly pointed out, if the landlords did not make any concessions they would be facing the danger of losing everything. He therefore exhorted them to accept the ordinance in order to save themselves from a worse fate. We agree with the revenue officers that the landlords' attitude will become clear three months from now when the new harvest will be in. But it seems to us that, even if the landlords accept the ordinance, the tenants will not be content with it; its provisions are inadequate and will not prove to be lasting because they do not satisfy the basic requirements of the tenant and farm worker.

Rent reduction from between 75 and 80 percent to 60 percent is indeed a welcome measure, but it is our strong impression that the tenants will not rest content with a 40 percent share of the crop. The rumblings for revision are already in the air: if it is not 60 percent for the tenant, then it is one-third for the landlord, one-third to the tenant, and one-third to whoever bears cultivation expenses. The Communists have already seized on this and nailed it to their banner for the next round of the fight. One of their leaders stated that: "Their (the mirasdars') calculations are simple. They expect that the peasant, given only 40 percent of the share and saddled with cultivation expenses, will not be able to pay the landlord his share and will default. Then once more they will proceed with eviction as freely as before." The statement ends with these portentous words: "So a new stage in Tanjore Kisan Movement opens up. And no doubt, strengthened by his victory in the battle against eviction, the Tanjore Kisan will acquit himself as creditably as he has done in the last decade."[6]

6. "Rajaji and Tanjore Tenants' Ordinance," by Mohan Kumaramangalam in *The People*, September 7, 1952.

It is not important that the prediction of the Communists about defaulting and eviction will not necessarily come true. The fact is that with 75 to 80 percent rentals the tenants somehow paid their rentals. What is important, and it creates a serious issue even without any assistance from the Communists, is that all expenses (other than harvesting which is shared equally by landlord and tenant) are borne by the tenants. The impression that the ordinance "saddled" the tenant with such expenses is not correct. As stated by Chief Minister Rajagopalachári, "The Ordinance only put down in words the customary practice in regard to expenses incurred for seeds, manure and cultivation. The charge laid on him now under the Ordinance has not been laid on him for the first time. There is no variation from the customary practice in this respect."[7] We stated elsewhere that the issue of who paid what expenses is not clear-cut; some tenants may not have shouldered all of the burden. Furthermore, in practically all cases the landlords extended them cultivation credit. The ordinance legalized what is supposedly a custom with no exceptions, and, indirectly, it created a mood among landlords of not advancing credit any longer.

Assuming that this is the custom, a land reform intended for the benefit of the tenants need not necessarily protect what is customary, especially when the custom is bad. The custom plus the high, preordinance rentals left most tenants with barely anything; the custom with reduced rentals will leave him a little something if credit is not denied him. One of the most responsible revenue officials and a supporter of the ordinance told us that with the ordinance in force a tenant can count on a net share of the crop of about 10 percent. It would be idle for the government to expect any increase in production when the tenant's share is much smaller than the landlord's and the responsibility for cultivation also rests on him. It gives the Communists a rare chance to keep up the agitation for a further reduction in rent or sharing of cultivation expenses or both.

The ordinance attempts to deal with the evil of eviction at the whim of the landlord and to give the tenants security of tenure. To that end, it lays down that all tenants who were cultivating the land on December 1, 1951, shall have tenancy rights for five years and that those who have since been evicted must be restored in their rights. They are seemingly good measures, but upon an examination their effectiveness appears rather limited. To be reinstated, a tenant must file an application within thirty days of the enactment of the ordinance. This right is likely to go by default in many cases since many of the tenants are illiterate and find it difficult to get from the village officers the survey numbers of the land they need in order to file the application with the conciliation officer. Second, no restoration can take place unless the applicant makes good any arrear of rent and reimburses the landlord or the tenant in possession of the land for any expenses incurred on the land since the applicant's eviction. It is questionable if an evicted tenant is financially capable to cover such expenses before he had raised a crop. The timing, if nothing else, is not helpful. Third, the ordinance doesn't apply to tenants evicted prior to December 1, 1951. This is a significant omission, for most of the evictions had taken place in 1949–50, the years of active disturbance in Tanjore.

The five-year security period entitles the tenant to limited tenancy rights. Every landlord can take back his land after five years. And he can take it back within the five years if the land is not cultivated well, perhaps because of the difficulty of securing seed and manure or other unavoidable causes. This is a severe limitation on security of tenure; the threat of eventual eviction is a weapon of coercion in the hands of unfriendly landlords, a category into which most of the landlords might fall. The rent reduction provision can be gravely compromised because it cannot be enforced without real bargaining power, and the tenant doesn't get it under the ordinance.

The pannaiyals dismissed prior to March 1, 1952, have no redress under the ordinance. Those dismissed between March 1 and the enactment of the ordinance may apply for reinstatement, while those dismissed after the promulgation of the ordinance are given one week within which to apply for reinstatement. The time limitation is much too short to do the pannaiyal much good. Moreover, the legal procedure of reinstatement as outlined in clauses

7. *Indian Express*, September 9, 1952.

2, 3, and 4 of section 12 is too complicated and involved to assure speedy settlement of the case, and the practical validity of the reinstatement provision may be questioned. Better results could be achieved if a problem of this kind were resolved by an elective or appointive committee made up of the representative groups of villagers.

With respect to pannaiyal wages, the ordinance gives him the choice of the old or the new schedule. If he accepts the new it means more wages, but he will get paid only for the days he works and he is not entitled to any bonus, which, under the old system, meant one-seventh of the crops of 6.6 acres (only) if the pannaiyal harvested the crop. This was an incentive measure to insure the pannaiyal's interest in raising a better crop. It is of some significance that the landlords accepted this provision with practically no protest. The pannaiyal, on the other hand, doesn't quite know how it will all work out; it will take at least a season or two to determine the effect of the measure. The ordinance makes no attempt to consider the casual labor problem.

Clause b of section 3 provides that wet holdings of 6.6 acres or less or dry land of 20 acres or less are not subject to the provisions of the ordinance. The average holding in the Madras state is 2.6 acres of wet land in wet districts and 6.5 acres of dry land in dry districts. The number of titleholders made up of 10 acres of wet land and 15 to 20 acres of dry land are about 6.5 million, or roughly three-quarters of the existing number of proprietors.[8] Roughly calculated, this means that approximately half of the proprietors escape the jurisdiction of the ordinance. In view of the extreme mildness of the ordinance, it is difficult to understand why it has allowed the small landlords who constitute the bulk to continue the tenure practices it is in the process of denying the bigger landlords. The reputation of the small landlord as a rack renter is notoriously bad. Their exemption, therefore, doesn't seem reasonable or justifiable.

The ordinance doesn't compare well with the main recommendations of the Land Revenue Reforms Committee of Madras of 1950, known as the Subramanian Committee. Those recommendations, which were never applied, were by no means bold or comprehensive; but at least they concerned themselves, even though from our point of view largely in a negative way, with such important issues as peasant proprietorship, acquisition of land beyond a certain limit, ceilings on personal cultivation, and restrictions on sales of land to noncultivators. In addition, the report recommended the sharing of rentals on the basis of 40 to 45 percent (depending upon the type of land) to the landlord and 55 to 60 percent to the tenant. The ordinance doesn't go even that far. For these and other reasons, the ordinance betrays a fear to stand still, a fear induced by the events of the past few years, at the same time betraying an even greater fear to move ahead. Hence its failure to deal drastically with the causes which made for the disturbed conditions in Tanjore.

Nevertheless, the ordinance served its purpose. It was a reminder to the landlords that the status quo cannot be maintained and that more substantial concessions from the landlords are perhaps in the offing. While the landlords oppose the very conservative provisions of the ordinance, actually it is the shape of things to come that troubles them more than what they are called upon to grant now. The desire to sell out their holdings, which is quite prevalent among landlords, is an expression of that fear. The ordinance whetted the unassuaged appetite of the tenants and farm workers for terms more nearly in accord with their desires and needs, and this must be counted among the welcome developments.

Chingleput District

General and tenure conditions

Chingleput and Tanjore districts differ greatly in many respects. Tanjore is a breadbasket of Madras by virtue of the available water supply and predominance of paddy culture. Chingleput, on the other hand, is a deficit food area, being a chronic sufferer from lack of water. This has left its mark upon the types of culture, tenure conditions, and living standards.

Our observations relate to Sriperumbudur

8. Subramanian Committee Report, 1951.

Taluk, a section of the district which has been drought ridden for a number of years. There are no river or spring channels in this taluk except the Kambakal, irrigating some of the villages in the western part. In the few places with an adequate water supply, the wet land yields two crops. Successive failure of rains had dried up most of the "eries" (storage tanks) when we visited the area. Varadarajapuram is a small village of some 300 cultivated acres, the bulk of it owned by small noncultivating owners.

The tenants are all poor tenants-at-will, enjoying little security. The lands are taken either on waram or guthagai basis. Under waram, generally the landlord and the customary rent is half of the gross produce. In practice, the landlord's share is larger. The tenant is entitled to the straw in most cases. Generally, the terms of tenure depend upon the fertility of the land and the availability of water. Where water has to be lifted or is scarce, the tenant sometimes gets two-thirds and the landlord one-third of the crop. Under guthagai, the tenant makes a fixed grain payment, which varies from village to village. Under both systems all cultivation expenses are borne by the tenants. Guthagai is the prevalent system in this village. The landlords are paid 2 jodus or 4 measures per unit of first-class wet land and 2.5 measures per unit of dry lands. Half of the rental is paid for the second crop on the same land. Where permanent farm laborers are employed, they are paid Rs20[9] per month plus one meal a day, one bag of paddy for each crop, and the remnant left on the threshing floor after the grain is removed; casual laborers are paid one rupee for males and 12 annas for female labor and one meal at harvesttime. Generally, casual laborers work only about three months in the year. Of the 150 families in the village, 100 are harijan families segregated from caste families and eking out a miserable existence.

The next village was Irungattukottai, by far the biggest, with a cultivated area of 1,200 acres. Here, too, most of the land belonged to a large number of noncultivating farmers. Generally, lands are rented out on Guthagai and the share of the landlord is about the same as in the previous village. Two tenants, when questioned, said that they have been cultivating the same land for four and eight years respectively, although evictions are common. They complained about this rent arrangement since it compelled them to pay the landlord a fixed amount of grain regardless of the yield. In an area of recurring monsoon failures, this rent system is indeed a burden hard to shoulder. The permanent farm servants get only Rs12 per month with one meal a day. The harijans accounted for nearly a third of the village population.

The third village, Pusivakkam, had about 500 acres under cultivation. There were 200 families in the village, half of whom were harijan families. Out of 500 acres, 100 acres were cultivated by the landowners themselves and the rest by tenants. Where the tenant has to bail water, he takes two-thirds of the produce. Generally, waram tenure is on half-and-half basis. But on being closely questioned, it developed that the landlord deducts 12.5 percent of the produce in the first instance and only the balance is divided half and half. The landlord's share, therefore, is closer to 60 than 50 percent. For this reason the tenants would consider half-and-half division just if strictly adhered to and if the landlord also contributed half the cost of manure.

Conclusion

Agricultural efficiency is at a low ebb in the areas visited. Small landowners owning less than 5 acres predominate. Water supply is uncertain, and with pumping sets to irrigate about 10 acres costing more than Rs1800, the land is ill-supplied with the prime requisite of cultivation.

Between 30 and 50 percent of the landholders are absentee landlords. In some cases this percentage is said to go beyond 70 percent. They are small holders and they seem to be forced out of the village mostly because of the inability of agriculture to support them. This leads us to the conclusion that absenteeism may result from the possession of too little as well as too much land.

Perhaps this accounts for the fact that even though tenure conditions are bad, there is less evidence of that purposeful harsh exploitation

9. One rupee equals U.S. 21 cents.

of one class by the other so prevalent in Tanjore and Malabar. Both appear to be the victims of a situation which affords few employment opportunities and which has grown from bad to worse by the niggardliness of weather over long periods. The poor condition of the tenant is exceeded only by that of the large class of harijans, who are mostly landless laborers.

There is much to share in Tanjore, there is little to share here. This emphasizes the fact that under such circumstances the land cannot support both landlord and tenant, and steps should be taken to combine ownership with cultivation, if only in order that the people who work the land get a living out of it.

In the Chingleput district the landlord-tenant strife, which came into the open in Tanjore, is not in evidence, but the ingredients for a similar development are there. We found no tenant satisfied with his share. The growing awareness of the need of getting a larger share is evident from the demand that the landlords should bear half of the expenses of manuring. He will, of course, not get it unless the landlords are compelled to make the concession. As yet the demand for even this one measure of relief is not very strident, but the tenant who had to relinquish part of the acreage because it yielded him nothing is a symbol of a situation which might deteriorate fast. The demands will gather momentum so long as the tenants are responsible for the stated shares and all costs of cultivation, while the contribution of the landlords—small owners though they are—is that of titleholders. But serious as the tenant problem is, an even worse one is the large segregated class of harijans subjected to unrelieved economic and social hardship. Both problems can be ignored only at the peril of the established government.

Malabar

General and tenure conditions

Our trip to Malabar helped to underscore once more the diverse and common features one finds within the confines of Madras state. The one thing common to Malabar, Tanjore, and Chingleput is that the welfare of the agricultural community depends to a considerable degree upon the terms under which the cultivating tenants and agricultural workers till the landowners' land. But in many other respects, including the very special type of tenure arrangements, Malabar is as different from Tanjore and Chingleput as the last two differ from each other. Malabar is a part of the state where the monsoon almost never fails and, although a deficit food area, famine is unknown. The soil is naturally fertile, and no one who travels through Palghat Gap can fail to be struck by the rich paddy fields, luxuriant gardens, and the pleasant view of the seldom-to-be-seen in India scattered homesteads in the midst of farmers' gardens and farms. Paddy culture predominates, but the striking feature of Malabar is the large acreage under cash crops, such as coconuts, pepper, jack, areca nuts, and a great variety of other fruit crops, which is a source of higher income than paddy or other cereals. Add to this a somewhat better-than-average sanitation, health conditions, and native intelligence and educational advancement, and the rating of Malabar as one of the more progressive places in India is understandable. And yet, it takes little searching to find that the actual cultivator of the soil is not better off economically than is his opposite number in Tanjore or in Chingleput.

In the course of our visit we were being told by officials and landlords that part of the explanation of the realities which belie the appearance of prosperity lies in the pressure of population in the land. This is undeniably so. In all of south India, Malabar district with a density of 640 people per square mile is second only to that of Tanjore. The average amount of cultivated land per person is three-fifths of an acre. This is in itself a welfare-limiting factor. Another factor called to our attention is the low productivity of the land. There is a saying in Malabar that "the yield is on the plough." But whether meant in a literal or figurative sense, the plough doesn't yield much notwithstanding good soil and an adequate water supply. The commonly suggested explanations are the cultivator's indolence; being certain of a crop, he doesn't exert himself; he doesn't plow deeply, doesn't weed, neglects to repair water courses, manures little or not at all, and, generally speaking, readily accepts the idea that what was good for his forefathers is good enough for him.

These and other handy explanations have some validity, but only when it is added that higher yields call for all manner of improvements, which in turn call for capital investment; and capital, regardless how small the amount, the cultivator of Malabar hasn't got. The landowners do have it, but as an agricultural official noted, the capital they spend "will be limited to the strict requirements to ensure their usual income from the land."[10] He further stated that "There is no incentive to spend more, because, the surplus yield is not to be his exclusively but has to be shared with others. Unless the man who spends money gets the full benefit of the expenditure, he has no incentive to do so."

The correctness of the quoted observation was all too apparent when we inquired into tenancy conditions prevailing in the Palghat section of Malabar. With little time at our disposal we stopped at only two villages, discussing matters with farmers, landowners, and revenue officers. We saw, however, enough to convince us that reforms cannot long be withheld both in the interests of land improvement and higher productivity as well as that of social peace and stability. An accepted view in certain circles of Malabar is that its tenure system represents a combination of land, capital, and labor in which cooperation could be productive of good results. But in the course of our trip we did not see any evidence of the consummation of such a happy development.

The land of Malabar is split into minute holdings, but its ownership is reported to be in the hands of a small number of persons, a smaller number in fact than in any other district of the state. Malabar, and especially South Malabar where we traveled, is commonly referred to as the land monopoly region; and the great majority of the farm population cultivates somebody else's land. The results of individual interviews with a number of tenants working in the field are representative of Malabar. Just prior to our visit, the crop had been harvested and the rent collected. Under the Malabar Tenancy Act of 1929 the tenant is entitled to the customary one-third of the net produce, but the practice reveals wholesale violation of the custom and the act. The tenants couldn't tell us the exact percentage of the crop paid to the landowner, but they all knew how it was shared. One tenant told that he paid 15 paras (a local measure) out of a total yield of 20, or 75 percent; another paid 16 paras out of approximately the same yield, or 80 percent; and the same was true of a third tenant, a woman renting a piece of land, which, incidentally, appeared to us to be no bigger than a fourth of an acre. They were all responsible for all the cost of cultivation. We were informed by the landlord of these holdings, who happened to come by in a shiny Ford as we were interrogating the tenants, that the expenses amounted to little more than their labor since the land received very little manuring.

The tenants didn't contradict his statement, and the landlord didn't contradict theirs, when in his presence, we asked them to state once again the terms of tenure. Their net couldn't have been more than 10 percent of the crops. The tenants were very hazy about the act, and it would not have helped them much even if they were familiar with its provisions. The landlord would not grant him one-third unless the tenant appealed to the Civil Court. This involves expenditure and delay plus the risk of displeasing the landlord and eviction. Whatever the reason, it was apparent that the tenants had little to show for their pains. In Malabar as in other districts their only alternative was the uncertainty of a casual laborer, and so they continued as tenants.

The landlord was rather uneasy about the airing his stewardship of the land was getting in the presence of a foreigner; but, having collected himself, he attacked the tenants on the ground that they did not improve the land and, failing that, theirs must be a small share. "But how could they spend money raising higher yields," we asked him, "the rental being so high?" In reply he shrugged his broad shoulders—he happened to be at least six feet two inches tall and nearly half that in the waistline—and remarked that this had been the customary practice and that the tenants till now

10. "Report of the Special Officer for the Investigation of Land Tenures on the Recommendations of the Malabar Tenancy Committee," May 1947, p. 31.

had not expressed any dissatisfaction. He himself was not investing money in land improvements. He was a good landlord, he continued, and in proof thereof he pointed to a tenant saying that on the occasion when his rent was in arrears he didn't evict him. It developed then that the tenants had no security of tenure and that the landlord could change them at will, even though the act conferred upon them qualified fixity of tenure through the right of demanding a renewal of their leases.

Malabar kanamdars

Upon further questioning, it appeared that the landlord was not the real owner of the land. He was a kanamdar, or a holder of land from the original owner, the janmi, in return for certain compensations about which more will be said in a subsequent paragraph. Suffice it to say here that this intermediary is the prevalent type of landholder which distinguishes the Malabar system of tenure from that of Tanjore and Chingleput. He, rather than the legal owner, is the rack renter, exploiting both tenant and original owner. The kanamdar in question told us that although he pays the janmi only Rs120 a year for an unspecified acreage in the area we visited, the rentals he receives from the same land range from Rs2,000 and up.

Not all kanamdars of Malabar are cut from the same cloth. We had met two who were less greedy and more foresighted. One of them built a spacious and modern-looking farmhouse for the tenant, drained leveled and reclaimed land, dug channels from small reservoirs to insure double cropping, and constructed a modern well in his garden. According to the landlord, he shared with the tenant the crop on a fifty-fifty basis. We were not in a position to find out all the financial ramifications between landlord and tenant, but the former was indeed concerned with land improvement. Another landlord of the same category expressed the view that it was high time for a rent reduction in Malabar. Moreover, he believed that the entire land tenure system of Malabar will be subjected to a very drastic revision. Acting upon that conviction, he has already met that eventuality in two ways: his tenants continue to work on his land as of old, but in the land register book he appears as a self-cultivator; the second step was to divide the land among his three sons. He thereby insured himself against would-be occupancy rights and/or ceilings on individual holdings. He merely behaved like an "economic" man. In the context of Malabar tenure conditions, he and his colleague may be rated as "good" landlords. But they are the exceptions. The "80 percenter" with no interest in the land and in the tillers is the rule. The long history of the land tenure legislation in Malabar is revealing in this connection.

Tenure legislation and effects

As mentioned earlier, the tenure system in Malabar is more complicated than in the rest of the state because of the existence of three distinct rightholders in the land: The janmi, the kanamdar, and the verampattamdar. Janmi is the landlord, and janmam denotes proprietorship. Under the janmi is the kanamdar who holds land by advancing a loan to the janmi for a specified period, usually extending to twelve years. The kanamdar pays rent to the janmi after deducting the land revenue and the interest on the money advanced. When the lease is renewed, the kanamdar must pay the janmi a renewal fee, mainly because the rent he pays is generally small, and is in some cases a nominal sum. The fee in itself is considered a portion of the profits, sometimes a large portion, which accrue to the kanamdar. The right of eviction which the janmi enjoyed in the past was the weapon of compelling a kanamdar to accept what was demanded of him. The kanamdar seldom cultivates the land himself; he leases it to the verampattamdar or cultivating tenant on terms already described.

This system with its numerous variations has been a source of antagonisms among the three parties leading to outbreaks first recorded in 1836 and culminating in South Malabar in open rebellions in the 1920s. The principal cause was frequent evictions of kanamdars by janmis and of cultivating tenants by kanamdars. This in turn must be attributed to rise in population and competition for the land and the consequent disintegration of such basic, old-established tenure terms such as a tenant's share of one-third of the net output, reasonable renewal fees, reasonable fixity of leaseholds, and compensation for land improvement. The principal victims, however, were and are to this day the cultivating tenant and the agricultural laborer.

The effort to cure uncertainty of tenure with its concomitants of sky-high rentals and reckless evictions dates back to 1887 when the right to claim compensation for improvements was conceded to the kanamdars and the cultivating farmers by the Tenants Improvements Compensation Act. A somewhat similar piece of legislation was promulgated in 1900, but with equally ineffective results, the reason being that what the "inferior" interests in the land wanted was not so much compensation on quitting the cultivated land, which, incidentally, was hardly ever granted in full, but a right to continue to cultivate the same land on payment of a reasonable rental. Little had changed by 1915, although in that year the revenue collector of Malabar, Sir Charles Innes, held that there was a case for legislation on the following grounds:

> 1. The prevalence of rack-renting and arbitrary and capricious evictions.
> 2. Inadequacy of compensation paid for the improvements.
> 3. Insecurity of tenure.
> 4. Levy of exorbitant renewal fees.
> 5. The social tyranny and miscellaneous exacting of the Janmis.
> 6. The ownership of land being a monopoly of a small landed aristocracy who do not cultivate and who rarely contribute to the improvements of the lands.

However, this view was not shared by others who maintained that there were no political or economic reasons for undertaking legislation. In the years following they became all too compelling, and the Malabar Tenancy Act of 1929 was finally enacted.

The principal beneficiary of the act was the kanamdar. It effectively curtailed the janmi's right of eviction and fixed a maximum fee for lease renewal, which was much below that paid in the past. At the same time, the act in no way weakened the superior position of the kanamdar over the cultivating tenant. "Qualified" fixity of tenure to restrict evictions by the kanamdar and "fair" rents for the tenants mentioned in the act proved to be paper provisions. The kanamdars, unlike their own tenants, were financially and politically strong to enforce their new rights, and in the years following they rather than the janmis became the symbol of typical landlordism in Malabar. And yet, totally unconcerned as they are about the circumstances in which their own tenants live and work, the successful struggle of the kanamdars against the janmis was a step in the right direction. It meant that the center of gravity of land interests has finally shifted from the higher to the lower group of landholders. Therein lay the promise that eventually it might shift from the kanamdars to the cultivating tenants and agricultural workers.

To date, that promise remains unfulfilled. The enactment of the 1951 Malabar Tenancy Act, currently not in operation, might have improved the condition of the cultivating tenant, although the kanamdars would have benefited most because the act abolished the system of renewals and renewal fee. For the kanamdar this meant low or nominal rental and fixity of tenure subject to resumption of the land by the janmi in case of bona fide self-cultivation. The rent was fixed at one-half of the net produce, and the kanamdar is denied the right to demand of the tenant any advance of rent or security for its due payment. The tenant was given qualified fixity of tenure—that is, he could be evicted only if the kanamdar could prove that he wished the land back for self-cultivation.

The act was subject to a spirited attack by all parties concerned, the janmis on the ground that the abolition of the renewal fee and renewals in general violates the fundamental right to the holding and enjoyment of property guaranteed by articles 19 and 31 of the constitution. The kanamdars attacked the rent reduction provision for obvious reasons, while the tenants opposed the act chiefly because from their point of view the security of tenure provisions were not sufficiently binding on the landlords to prevent illegal evictions. The agricultural workers were not involved in the controversy; as in the past, the act didn't concern itself with their status. In view of the three-corner criticism of the act, it has been suspended altogether and a new one is under consideration. Whether the 1951 act would have resulted in a new chapter in the landlord-tenant relations is anybody's guess; the degree of enforcement is the great unknown, and lack of enforcement in the past nullified the few well-intentioned measures. But for the time being and for all practical purposes, the tenants and agricultural workers carry on as of old under highly disadvantageous terms.

A New Approach

Such, in the main, are landlord-tenant relations in Malabar, Tanjore, and in Chingleput districts, the attempts to improve them, and the attitude of the principal parties toward them. The measures do not go far enough to meet the problems in the parts of the state considered in this report. It is reasonable to infer that the rest of the state is not free from the same type of problems. It may be inferred also that the overwhelming majority of the tenants and agricultural workers of Madras are no longer satisfied with the existing state of affairs. Until very recently their conservatism, inertia, and ingrained feudal subservience to the landlord kept the pot from boiling over. Now the forces that held them within well-defined bounds are breaking down under rising agrarian discontent. Some of them are at last in motion and others are bound to follow.

It may be anticipated that before long the Tanjore ordinance will be revised, and the Malabar Tenancy Act of 1951 will have to be revitalized and put in operation. But before the Madras government does that, it might consider a number of propositions so far overlooked. To begin with, there must be the realization that Tanjore district is not a special case. Only the Communists spotlighted it, and that made Tanjore special. Malabar district is equally bad, and its kanamdar system is an element of further exploitation of the tenant. Conditions in Chingleput district are perhaps not as critical, but they have other attributes which could stand doctoring. In sum, it is Madras state rather than this or that district which is in need of overhauling of the conditions relating to man and the land he cultivates. Second is the recognition that the time of hasty patchwork typified by the ordinance has long past. The content of the legislation must be such that would place the peasant "in the center of the piece," as Nehru so aptly put it. This presupposes the resurrection of the old Congress slogan of "Land to the tillers." While this doesn't mean that all land must be turned to the tenants forthwith, it does mean that the various measures must be designed with a view of achieving that goal. Third and most important is that whatever the scope of the reform, its letter and spirit must be actively pursued and enforced. This is seemingly a truism hardly worth reiterating, but the inescapable fact is that the progress of land reform in certain parts of the country is gravely impeded by the enormous gap between the aim of the legislation and performance.

It is not necessary to spell out the details of a sound agrarian reform based on the above principles. The critical review of the ordinance implies, at least in part, what it should have been or what the forthcoming legislation should be. More important is the fact that the rich literature on agrarian reform in India doesn't suffer for a lack of sound recommendations. It abounds in them and the principal ones may be easily culled: occupancy rights; rent reduction; rent not to exceed one-third of the crop or less, with cost of cultivation shared by landlord and tenant; grounds for eviction carefully circumscribed; full compensation for land improvement; ceilings on existing holdings; land purchasing at fair prices and reasonable terms of repayment; drastic limitation of absentee ownership; prohibition of land acquisition by noncultivators; and curtailment of resumption of land for self-cultivation. More recently, the Planning Commission has set forth comprehensive recommendations in its latest report which, if adopted by the state governments, would go a long way toward the rehabilitation of the tillers of the land and of the rural economy.

To repeat, the reform roadblocks are not due to either ignorance of the existing state of affairs in the countryside or absence of expert advice on how to correct it. They lie rather in the hesitancy of the authorities to face the issue that quick action is necessary, a situation so well illustrated in Tanjore. What the peasants want most is land ownership, but they would settle for much less, provided, however, other reform measures such as rent reduction, occupancy rights, and farm wages, for example, are not half measures merely to meet Communist penetration into the countryside but carry lasting meaning in terms of hopes and aspirations of the cultivating farmers and agricultural workers. The landlords of Madras, judging by their attitude in Tanjore and Malabar, are by and large dedicated to the proposition of maintaining their rights through thick and thin, and further opposition to reform must be anticipated. The remedy for that is determined and enlightened government leadership. Much can be done with it and nothing without it. There is no substitute for that kind of leadership if the psychological climate of rural Madras is to undergo a radical improvement and if in the minds of the peasants the government is to become identified with their welfare.

To deal with agrarian reform in Madras (as elsewhere) is to be confronted with the unpleasant fact of change, the one unalterable constant in human affairs. With but one exception, zamindari abolition, rural Madras has not changed from time immemorial because the people who "count" believe that things are very well as they are and that there is no need to be overly zealous to change things around. This attitude is understandable, but it cannot be supported any longer in the face of the undisputed fact that the peasants are actively unwilling to accept the "normal" place to which they are relegated by custom and tradition. For this reason it would be worse than folly if in the year 1952 the privileges of a minority group were to continue to stand in the way of economically and politically necessary changes. And it is well to keep in mind that even substantial changes initiated by an intelligent government leadership will be less disrupting and infinitely more stabilizing than if carried out from below, by the farmers themselves, by taking the law into their own hands.

Finally, when the desirable legislation is enacted and enforced, the millenium will not have arrived in rural Madras. There simply is not enough land to satisfy the cultivating owner proprietors, tenants, and agricultural workers whose numbers are rising year by year. The most egalitarian distribution of land or the fairest of tenure terms cannot negate the ill effects of the multitude of competitors for a cultivated area which is almost stationary and an acreage of relatively low productivity. Only by industrialization and the consequent siphoning off of some agricultural labor, coupled with a sharp rise in agricultural productivity, can a real dent be made in the vicious circle of more people—less land and growing poverty. Such developments, especially the first one, still remain in the future. This, however, doesn't in any way argue against the validity of reform measures within the existing village condition. The issue is not of *solving* the rural problem, but of palliatives capable of wiping out the worst features of a cake of custom which condemns the farmers to a below-subsistence level of existence. The immediate problem is not one of a whole loaf or nothing; a good chunk of the same can go a long way if given in time. When that is brought about, the very size of the loaf will be increased to the greater satisfaction of all concerned.

20. Land Reform Observations in Kashmir

In the third of his 1952 India papers, Ladejinsky examines the rare phenomenon of a radical and successfully implemented land reform in India—that in Kashmir under Sheikh Abdullah. It has a special interest if only on that account.

The report, itself undated, was based on a field trip made in October 1952. It was transmitted to the U.S. Department of State as Dispatch 1276, dated November 25, 1952.

IT WAS IN KASHMIR THAT this observer witnessed a far-reaching land reform program in India. Its character will be described and analyzed in subsequent paragraphs, but it may be stated here that it differs from all other land reforms of India in three respects: content, enforcement, and government's role in the program. Whereas virtually all land reforms in India lay stress on the elimination of the zamindari tenure system with compensation or on rent reduction and security of tenure, the Kashmir reform calls also for the distribution of land among the tenants without compensation to the erstwhile proprietors; whereas land reform enforcement in most states of India is not too effective, in Kashmir enforcement is unmistakably vigorous; finally, whereas in other states one seldom encounters among the farmers a close identification of the government with their needs, in Kashmir the preponderant majority of them knows that the government and more specifically Sheikh Abdullah is the source of the reform and of the benefits they are deriving from it.

The land reform legislation of Kashmir may be divided into three phases. The first one occurred in April 1948 when the government abolished the jagirs, muafis, and mukarraries. The first two stand for privileges of collecting revenue in kind or cash from lands given to certain individuals by a ruler or the state in compensation for services, real or fancied, rendered by them. Some of these privileges were temporary while others were held in perpetuity. Whether short- or long-lived, all the benefits from the assigned lands went to the holder of the jagir (the jagirdar) or of the muafi (the muafidar) and none to the state. Not only did they pay no revenue (land tax) on these freeholds, but none was paid for the lands they owned. Along with abolition of these two types of rights went those of mukarraridar, a class who received from the state treasury fixed cash grants every six months.

The tangible and intangible consequences of this enactment are quite apparent. As to the first, the jagirdars (396) and mukkararidars (2,347) have disappeared from the rural scene. This observer has met some in Srinagar, the summer capital of Kashmir, in a state of deep perturbance over their lost privileges, partly expressed in revenues estimated at more than Rs1 million, and a few thousand acres of land owned by them outright. By the same token the state revenue increased, although not by this amount. From the point of view of the government and of the cultivating farmers, the intangible gains are greater because in practice a jagirdar ran his land domain as it were a state within a state. He was an authentic feudal lord, with rights over the tillers far exceeding the revenue collecting exercised by the former zamindari in India, and almost not accountable to the state for his actions. In the words of one of the principal land reform promoters in Kashmir, Revenue Minister M. M. A. Beg, this order of things "had reduced thousands of peasants to tongueless 'hewers of wood and drawers of water'." A government whose leaders have repeatedly stated that "building a Socialist order is our objective" could not but begin with the elimination of the very special kind of intermediary endowed with all manner of rights and few obligations either to the cultivator or the state. In doing that, the government hoped to give currency to the idea that the cultivator and the land must be linked closer to each other.

That the elimination of the jagirdars and similar rightholders in land was only the first step in a new land tenure policy became evident in October 1948 when the government initiated the second land reform phase by amending the State Tenancy Act of 1924. Like most states in India, the farm holdings of Kashmir are small (frequently only 1 to 2.5 acres), fragmented to a point perhaps unequalled in India, and many farmers work the land as tenants. The total cultivated area of Jammu and Kashmir (excluding Pakistan-held territory) is estimated at about 2.2 million acres, of which the tenanted acreage is roughly one-third of the total. This land is in the hands of 120,000 noncultivating owners, but among them about 500 own as much as 300 acres each, and less than 2,000 own as much as 75 acres. These landlords are substantial only in relation to the predominant small cultivator. The number of landless tenants is estimated at 300,000. To this must be added an estimated 250,000 part owners–part tenants who also cultivated part of the tenanted land. Only a small part of the tenants enjoyed occupancy rights (protected tenants); the great majority were tenants-at-will—that is, they could be evicted at the landlord's pleasure. Rentals compared well with other states of India; they ranged from 50 to 60 percent of the gross crop, the tenants bearing most of the cost of cultivation.

The principal provisions of the Amended Tenancy Act of 1924 to strengthen the position of the tenants are as follows:

1. Rentals for wet land, including those growing wheat, corn, sugarcane, and linseed, are fixed at one-fourth of the gross crop. Rentals for dry land are fixed at one-third of the crop. These rent reductions apply to tenanted land on holdings exceeding 12.5 acres. Rents on holdings of 12.5 and less are fixed at half the crop. Under no condition can the fixed rentals be enhanced.
2. Recognizing that no rent reduction can be enforced unless the tenant has real bargaining power, which is security of tenure, the act provides that in the province of Kashmir and the frontier districts of Ladakh and Gilgit a tenant shall have permanent occupancy rights on 2⅛ acres of wet land and 4⅛ acres of dry land; in the Jammu province the respective figures are 2⅛ and 8⅛ acres. This is roughly the maximum acreage cultivated by a tenant. A landlord can still appeal to the court for the eviction of a tenant under certain conditions, but, as will be pointed out elsewhere, unwarranted ejectments have come to an end, and the tenants of Kashmir have finally achieved security of tenure.
3. The act provides also for reinstatement of tenants evicted after April 1947 and prohibits the enforcement of a court eviction order once the tenant has acquired protected status.

Just as the elimination of the jagirdars was a prelude to the restriction of the landlords' rights in relation to the tenant, in the same manner the latter was another step leading to land distribution among the tenants. This was not a new idea conceived by the government since its accession to power in November 1947. It goes back to 1944 when Sheikh Abdullah, then the president of the All Jammu and Kashmir National Conference, in presenting the program of the "New Kashmir," made this statement: "The peasant sons of the valleys and the mountains have scratched only nine inches of top soil and eked out a bare existence. Now the time has come when they must dig into the bowels of the earth and yoke the techniques of modern science to task of getting for themselves a bigger and better morsel of daily bread."

To bring this about, Sheikh Abdullah felt that the following must be accomplished: "Abolition of landlordism—because such a step would be the pivot of all progress. . . . Land to the tiller—because after the abolition of the landlord class it will be possible for the first time to satisfy the land hunger of the landless peasant, and insure efficient working of the land."

This is the background which led the government headed by Sheikh Abdullah to promulgate the Abolition of the Big Landed Estates Act of July 13, 1950. The outstanding provisions of the act are:

1. No landlord can hold more than 182 kanals (22.5 acres) of cultivated land. The ownership of the excess is to be given to the tenants. No cultivator can, with the land so obtained and that which he may own, possess more than 20 acres. The act does not apply to orchards (which are important in the agricultural economy of Kashmir), pasture, and forest lands. The landlord was given the right to select any 22.5 acres which suited him best. To prevent evasion of this provision, the act decreed that the transfers of land effected by the landlords after April 13, 1947 subject to the provisions of this act are invalid.
2. With the excess land go all rights in trees, wells, tanks, ponds, and water channels free from all encumbrances; and the interest of the new owner in such land is not liable to attachment or sale in execution of any court decree. Having taken over the land, the cultivator pays to the state the land revenue and other customary dues.
3. On the all-important question of compensation, the act provided that the national assembly of the state should resolve it at a future date. At the outset, the government agreed to pay the following annuities for the expropriated land: three-fourths of the land revenue during the first year; two-thirds during the second year; and one-half during

the third and subsequent years. It was further laid down, however, that under no condition can the annuities exceed Rs3,000, nor will any annuity be payable for land which was at one time in the category of village common land. The government has ultimately decided that the landlords are not to be compensated at all, and it is this rather than any other action of the government that became the target of landlord opposition and makes the Kashmir program unique in India.

These are the essential provisions of the land reform legislation in Kashmir. To this observer they are basically sound, except for the failure to pay any price for the land and the failure to make a distinction between the significance of 22.5 acres of wet land and 22.5 acres of dry land, fertile or nonproductive. In Jammu, for instance, where dry land predominates, this amount of land an owner can retain is worth infinitely less than a similar amount of wet land an owner can retain in the Kashmir valley. But vastly more important than a foreign observer's strictures against these particular provisions of the reform is whether, from the point of view of the tenant, the legislation as a whole is being enforced and what benefits he derives from it. With this in mind, we visited a number of villages and land revenue offices in various parts of the Kashmir valley where we discussed land reform matters with a great many farmers; we met also with two of the leading landlords of Kashmir, one Hindu and one Muslim, who stated their side of the story in no uncertain terms. Prime Minister Sheikh Abdullah, Revenue Minister M. M. A. Beg, Chief Land Reform Officer R. C. Raina, and a number of local officials were readily available for a statement of their position.

While in Srinagar the government of Kashmir put its "best foot forward" to present its case, but at no time did it attempt to convince the visitor of the excellence of its program by resorting to anything resembling a conducted tour. Nothing was prearranged; we traveled north or south, east or west as the spirit moved us. Sometimes the jeep would halt at a village, at other times at a threshing common in the open field, at the sight of a few farmers by the roadside, or at a district office. The very competent Mr. Raina accompanying this observer was most agreeable to this rough-and-ready method of "feeling the pulse" of the farmers. And we don't believe that he permitted his bias in favor of the reform to creep into the English rendered from the Kashmiri for our benefit.

Our findings on the very first trip our typified much of what we found on subsequent trips. We stopped at the revenue office of the township of Chrorisharif where we chanced on a group of sixty to seventy farmers. On the wide expanse of the lawn, ranged in a circle, all manner of questions were raised and answers given. They spoke eagerly, interrupting each other lest the visitor failed to get a point. This was the most loquacious group of farmers this observer has yet met in India. This feature was characteristic of all other people whom the land question touched in one way or another. Most of the farmers in question were there at the revenue office to realize a lifetime ambition: their ownership titles to the land obtained under the reform were being registered that day, and only at a cost of a rupee or two for stamps to legalize the deed. Thus our first encounter with the "Land to the tillers" idea translated into action.

"Would any of the assembled care to explain what this reform is and how it came about?" There was no end of willing exponents. They couldn't cite this or that section or subsection of the act, but they knew a good deal about rent reduction, occupancy rights, and ownership of land, judging by individual recitals of rents paid, permanency of tenure enjoyed, and land owned before and after the reform. The often-heard statement that illiterate farmers—and 96 percent of the farmers of Kashmir are illiterate—cannot presumably assimilate rapid changes of status, doesn't hold true in Kashmir. They are illiterate but not ignorant; they know full well the meaning of a larger or smaller share of the crop, of eviction as against noneviction, a piece of land they can call their own as against rented land. In the mind of would-be recipients of land reform benefits, the changes in their status are blurred only when reform enforcement is nonexistent or is badly lagging. In Kashmir reform legislation is enforced and the drive of enforcement permeates all layers of government from the top down.

When the farmers were asked how the reforms came about, the answers were unanimous: "Sheikh Sahib gave them to us." Throughout the Kashmir valley the refrain to this question was the same. Many a farmer answered the query in a tone of incredulity; to think that the visitor doesn't know that Sheikh Abdullah is the source of this munificence! "But how do you know that Sheikh Abdullah is responsible for the reforms?" The answer is that they heard him speak about the reform, explain its essence, and assure them it is meant for them. And to the question why Sheikh Sahib gave them the reform, the answer is that "he is of our own stock and he took pity on us." Visits to various parts of the valley made it abundantly clear that Sheikh Abdullah carried the reform message on all conceivable occasions. In the peasant-proprietory village of Droru, the farmers told us that they, too, knew that it was Sheikh Abdullah's reform and that some of them traveled to a neighboring district to listen to "Sher-i-Kashmir" (the lion of Kashmir) talk about the land and the farmers. A farmer interjected saying that "even our children know about him, and we all pray for his health." In this village the farmers' concern was centered on lack of livestock, better seed, and fertilizer to raise agricultural production. Their faith in Sheikh Abdullah being so great, we asked them why they did not appeal to him for assistance. One of the farmers answered without hesitation: "Sheikh Sahib is busy now fighting our war; when that is over we shall ask him for help."

Sheikh Abdullah is a well-known political leader of Kashmir of many years' standing. Among his own people he is fast becoming a legend, and many a farmer who had never heard or seen the Prime Minister will nevertheless insist that he beheld him in the flesh. What is important, however, is that he did raise the land issue to a place of highest importance, and he did set the tone for the rest of the government to follow. We met with local officials who are themselves landowners, but most of them would not dare to water down the enforcement of the provisions. The climate is against such proclivities. The tenants know that they can find redress in the higher governmental levels and so do the officials. The overall impression one gets is that in the minds of the farmers the reform is closely identified with Sheikh Abdullah and his government. In the circumstances, enforcement is inevitable; the farmers themselves are becoming the enforcers although formally they are not delegated with such duties.

In the village of Kanispara we witnessed a change in the attitude of the farmers which is easily traceable to the conviction that the reform was there to stay and that the government is on their side. One of the tenants stated that his landlord refused to accept one-fourth of the crop due him in accordance with the provision of the act. In response to the question how he resolved the problem he said: "I gave him nothing; I shall pay him the legal rent when he changes his mind." Whether all tenants behaved similarly is difficult to judge, but the whole pattern of the questions and the manner in which the answers were given points to this conclusion: the rental provisions are being carried out.

As with the rentals, so with the occupancy rights. In Kanispara and in the other villages these rights are in being, and eviction of a protected tenant—a status most of them enjoy now—is extremely difficult. Both parties to the land, tenants and landlord, are aware of it. To recover the land, the owner must prove to the satisfaction of the revenue court a good deal which would be hard to prove. He must satisfy the court that the land in question is being misused, or that the tenant failed to pay the rent without due cause, or that he doesn't cultivate the land in the customary ways of the locality, or that he wants the land back for self-cultivating. The latter point must be proved beyond a shadow of a doubt.

To date, there have been virtually no eviction cases on non-self-cultivation or self-cultivation grounds. Tenants are not in the mood of moving off the land even when the owner establishes a good case of self-cultivation. In their minds a prereform noncultivating owner should not become a cultivating one now. This stems from a growing belief among the tenants that as cultivators they are the real owners of the land, legal rights to the contrary notwithstanding. This is not surprising. Leading members of the government are on record that the land belongs to those who cultivate it. Revenue Minister Beg stated that "The land of God shall belong to the men of God." In a personal interview with this observer, he repeated what he broadcasted far and wide throughout Kashmir: "So long as the land doesn't pass into the possession of the self-tilling peasants nothing can better their lives and give light to their so far dark and dismal existence. The peasant must feel that the land on which he toils belongs to him, and that he is not just sweating for somebody else. Of course, then alone agriculture will begin to yield the maximum harvest." A good deal of this preachment has reached the village, has stirred up the tenants, and is creating a state of mind which assumes that the land he cultivates belongs to him. As one tenant put it: "Before the reform the landlord could do anything; now, with Sheikh Sahib, we own the land." There is nothing academic in this attitude; its practical significance becomes apparent no sooner the question is raised of compensation versus noncompensation for the land acquired by the tenants under the Abolition of the Big Landed Estates Act.

Even a casual reading of the legislation reveals that the gains of the tenants are unequally distributed, depending on the type of land cultivated, and, above all, upon the size of an owner's holding from whom a tenant rents land. In the village of Pampur we asked a number of tenants what they gained under the reform. All of them acquired occupancy rights. Some of them had their rents reduced from one-half to one-third; others from one-half to one-fourth of the gross crop; while those who rented land from owners with holdings of 12.5 acres or less continued to pay half of one crop. Unless a tenant's rental was more than half of the crop, his sole gain was the right of remaining on the land undisturbed. The more fortunate ones, those renting from holders of more than 22.5 acres, actually acquired land, the amount varying from one acre to as high as six acres; and a few were given state-owned land. Clearly the range of the gains is very considerable—from occupancy rights only to actual ownership at no cost to the tiller.

The inequality of gains could become a source of dissatisfaction and jealousies among the tenants, and their reaction on this point was elicited. They stated that such fears were groundless because of the strong conviction that sooner or later *all* of the land will belong to them. They looked upon the present retention limit as a temporary development and were of the opinion that it will be drastically reduced if not wiped out altogether. The question then was raised about the moral justice of the step they looked forward to. By way of a reply a tenant stated that they were entitled to the land because "our ancestors reclaimed it and we till it." The tenant spoke for the many in his village, and subsequent experience proved that he voiced the view of tenants in other parts of the Kashmir valley as well. Revenue Minister Beg confirmed this. He stated that Kashmir as a welfare state has its primary obligation to the tillers, and, under certain circumstances, abridgment of property is an accepted principle exercised even in capitalistic countries. As far as he was concerned, expropriation of land of noncultivators falls in the same category as, say, the redistribution of wealth in England through taxation, the infringements on private enterprise through price controls, or eminent domain concepts, and so on. The attitude of Sheikh Abdullah is essentially the same, even if stated differently. Hence the strong impression that in Kashmir, where the government rules without the benefit of an effective opposition, the current land ownership provisions of the act are not the last word on the subject; the retention limit will in all probability be revised downward in order to satisfy the economic and political ideologies of the "New Kashmir" on the one hand, and the hunger of the peasants for the land on the other.

If all the 120,000 noncultivating owners were denied the right of land ownership, the tenants would stand to gain an estimated total of 660,000 acres. But under the provisions of the existing act the immediate question is threefold: how many of them will be affected, how much land will be turned over to the tenants, and how many tenants will benefit from the land transfer? The answer depends upon whose set of statistics one accepts—those furnished by the government or those supplied by the spokesmen of the landlords. Not being in a position to venture an opinion about the soundness of either, both versions may be mentioned.

The government and landlords agree as to the total acreage of the noncultivators; but, while the former claims that 9,000 landlords or 7.5 percent of the total are in the 22.5 acres class, the latter set a figure of only 5,000 or 4 percent of the total. If the first variant is correct, the landlords can retain over 200,000 acres in addition to a fairly large acreage under orchards, or a total of approximately 250,000 acres. This would leave 400,000 acres to be distributed, according to official sources, among a few hundred thousand landless tenants and part owners–part tenants. The second variant presents a different picture. It assumes that the 5,000 own 250,000 acres, of which they can retain 110,000, leaving only 140,000 acres to be shared by a much smaller number of tenants than that anticipated by the government. We are not in a position to check the accuracy of either version, but the discussion with the landlords left a strong impression that they were bent on minimizing the scope of the reform and its benefits to the tenants. They go so far as to claim that not 140,000 but 80,000 acres is all the tenants can get. The idea behind this approach as expressed by the landlords is simple: Since relatively few tenants will benefit from the Abolition Act, why bother with the scheme at all? The partisanship of the landlords in presenting their case is understandable, but the data upon which it rests are questionable. The official data, too, are in all probability subject to a considerable margin of error and for the very opposite reasons which motivate the landlords. On balance, however, many more landless and near-landless will be favorably affected by the current land distribution program than the landlords care to admit.

The Kashmir landlords do not wish to part with any of their land even for a price. No wonder, therefore, that they are dismayed with the decision of the government to deprive them of the land with no compensation whatever. This feature, much more than the ceiling on landlord holdings, places the Kashmir program in a very special class. The landowners advanced a number of reasons for compensation. These included constitutional rights, the fundamental right of a person to own property, moral and social obligations of the state towards the citizens regardless of class, allusions to Sheikh Abdullah's "New Kashmir" program with its stated aim that no one should be allowed to starve, the historic development of landlordism, and the legality of land acquisition. They even invoked the shades of John Stuart Mill and Harold Laski to bolster up their case; they quoted approvingly from the latter's *A Grammar of Politics* to the effect that "Men will sooner, as Machiavelli said, forgive the death of their relatives than the confiscation of property. Nothing is more likely to poison the spirit of body politic than the sudden disappointment of financial expectations." Last, but not least, they argue that the reforms partake of religious discrimination because the landlords are mostly Hindus while the administrators of Kashmir are Muslims. But all the arguments availed them nothing.

The government never intended to pay for the land. The appointment of the Land Compensation Committee to examine the issue and make recommendations was the way of giving a semblance of legality to nonpayment. Sheikh Abdullah foretold the turn of events before the recommendations were in when he spoke as follows: "Substantial portions of the landed property came to be owned from such land as was the property of the State before, and in every case the acquisition of the land was free from any encumbrance or payment of any consideration. It is in the light of this historical backgrounds that the Honorable Members of this House shall have to consider whether there is any justification for payment of any compensation to such landowners for lands from which they are expropriated under the Big Landed Estates Abolition Act."

The commission under the chairmanship of Minister of Revenue Beg handed in its report on March 22, 1952. The report noted that the land originally belonged to the tillers; that the intermediary interests were created by the maharaja's rule; that present-day proprietors are the descendents of the rent collectors; recepients of government gifts, and leaseholders who succeeded in getting themselves recognized as landowners even though the land was reclaimed and improved by the tenants. The commission held that, as "sleeping partners," the landlords had no right to own land and must be contented with their retentions under the act. As to those who purchased land in recent years, "most of them have made large profits and in any case all these are mere speculators in land in the ultimate analysis." In the commission's view the compensation clause of the Indian constitution did not apply to Kashmir[1] and the state has a right to acquire private property on any conditions it deems proper if the act is in the best public interest. Based upon these and other arguments, the commission arrived at the following conclusions and recommendations:

> The tillers, to whom the excess land from which the big proprietors are expropriated is transferred in ownership right, are an indigent, impoverished and much exploited class. No question of recovering the price of the lands from them does arise. As a State with limited resources we are too poor to pay compensation from out of the State Revenues. The financial liability will be of a very serious nature and payment itself will prove incalculably mischievous. Apart from these considerations, there is no moral, economic or social basis for compensation. We therefore recommend, both on principle and policy, the payment of compensation to the expropriated proprietors is not desirable. The payment of compensation would perpetuate the present inequitable distribution of wealth.

Four days later on March 26 the Constituent Assembly of Jammu and Kashmir decreed that no compensation shall be paid for the land distributed among the tenants under the Big Landed Estates Abolition Act.

The commission stretched the "principle" issue of noncompensation too far, and the arguments of the landlords on this point have not been met. Nor is the lack of means a valid reason for noncompensation. Yet anyone who has seen at close range the poverty of the Kashmiri tenant and is familiar with the overall depressed economic conditions of the state can readily appreciate the dilemma which faced the reformers of Kashmir: a token payment at best or wiping the slate clean by shifting all of the burden of reform on the big owners. They did the latter and for a number of economic and political reasons, each one affecting the other. Sheikh Abdullah told this observer that his government initiated this type of reform partly because it was of utmost importance to divert the minds of the majority of the people from communal (religious) strife which came near to delivering Kashmir to its enemies. "Let them," he remarked, "concentrate their thoughts and energies on problems which really matter, the economic problems." Another factor is the apparent dedication to fashion a "New Kashmir." The theoretical underpinnings of this last factor are revolutionary slogans. In Kashmir, only a day removed from feudalism and yet still steeped in it, Revenue Minister Beg prefaces his series of speeches on the "whys" of reform with these famous lines from Shelley's "Song to the Men of England":

> Men of England, wherefore plough
> For the lords who lay ye low?
> Wherefore weave with toil and care
> The rich robes your tyrants wear?
>
> Rise like lions after slumber
> In unconquerable number.
> Shake your chains to earth like dew,
> You are many, they are few! [2]

But it is not only slogans; the actions, too, are resolute and uncompromising with an eye to achievement of the goals. One wonders only why the government of Kashmir devoted time and effort investigating the issue of compensation versus noncompensation when in the opinion of the government it had no other way out and when, ideologically speaking, the government has always been in tune with noncompensation.

The Big Landed Estates Abolition Act is far from completed; as of July 1952, a total of 132,000 acres were distributed among 129,000 cultivators, while 48,000 acres temporarily vested in the state are earmarked for distribution. The enforcement is proceeding apace, and it may be said that it will not be relaxed until the last acre subject to the act finds its way into the hands of the tenants.

The process is of necessity a slow one. Notices must be served on the big proprietors to select the land they can retain before the excess is distributed. Changes in the revenue books must be made accordingly. The transfer of the land to the tillers entails numerous alterations in the village records. The distribution of the land calls for reassignments of revenues assessments, which in turn are based on the establishment of an accurate record-of-rights. The

1. Kashmir's accession to India in 1947 was with reference to only three subjects—defense, foreign affairs, and communications.

[2. The second stanza is not from "Song to the Men of England"; its source has not been identified.]

culmination of this painstaking and time-consuming process is the registration of title of ownership of the new owner cultivator. In the village register it reads as follows: "Field measuring ——— canals of land in holding Number ———, of which the right of ownership of so-and-so as against payment of land volume and cesses amounting to ——— per annum, in accordance with the Act of October 17, 1950." It is surprising, therefore, how much has been accomplished when one examines the village records, as we have done repeatedly. Some of the reasons underlying the progress have been dealt with. Part of the explanation lies in the chief land reform officer's able planning and direction of the work in the field. Important, too, is the fact the government for its part augmented the old land revenue staff with a special contingent to step up the implementation of the program. It aimed to select a group of people who know what the policies are about and are in sympathy with them. On both scores, the test paper entitled "Departmental Examination of Deputy Commissioners and Tehsildars for 1951," prepared by Revenue Minister Beg, is revealing. The paper consists of four questions, and the first and fourth are reproduced below:

Subject D: Agrarian Reforms

I. "The history of freedom movements from the Middle Ages to modern times had only one lesson to teach that freedom from all forms of economic exploitation is the only true guarantee of political democracy, and without it political freedom is a mere shibboleth. Even the winged ideals of 'Liberty, Equality and Fraternity' of the French Revolution bought at the price of the people's blood, degenerated into the autocracy of Napoleon, because privilege and exploitation persisted in their old strongholds. Freedom and Privilege are the two sides of a pair of scales: as Privilege gets lighter, Freedom gets heavier." (Quotation from *Introduction to New Kashmir* by S. M. Abdullah.)

In the light of the basic idea underlying the above passage, state what concrete steps did the present regime of S. M. Abdullah's government take toward ending the economic exploitation by the privileged class of the feudal system in the state.

IV. A. Under the provisions of Abolition of Big Landed Estates Act, 2007, a certain proprietor after selecting his 182 kanals leaves for transfer to the tillers 100 kanals which is under a mortgage. How will the rights and liabilities of the mortgages be determined;

if

1. the mortgagee is in cultivating possession of the area, and
2. the tiller of the area is a person other than the mortgagee.

B. "A" is an ex-proprietor of 100 kanals, the ownership of which has been extinguished under section 4 of the Abolition of Big Landed Estates Act. To whom and to what extent will you transfer the area in the following cases:

1. Where the tiller of this area has already 100 kanals under his own proprietorship,
2. Where the tiller of the area is a non-state subject,
3. Where the area in question was owned by an evacuee as defined in the Jammu & Kashmir State Evacuees (Administration of Property Act) 2006, and
4. Where the area is untenanted.

To this observer this is just another illustration of how concerned the government has been with the land program and how a combination of all kinds of measures engendered by a determined leadership serves to insure the completion of the program.

Land reform is the most publicized of all rural improvement measures, but the attempts to relieve farm indebtedness, as we observed them in action in the Kashmir valley, are also worthy of notice. Along with the rest of India, the farmers of Kashmir are "born in debt, live in debt, and die in debt." About four-fifths of the indebtedness is incurred for nonproductive purposes at rates ranging from 12 to 50 or more percent a year. Normally the debt continues increasing automatically because it could not create any means of its own repayment. The moneylending practices and their consequences have been treated exhaustively by students of Indian rural finance, and it should suffice to say that M. L. Darling's *The Punjab Peasant in Prosperity and Debt* surely applies to Kashmir.

The Kashmir government decided to act drastically with the burden of accumulated indebtedness, and for that purpose it enacted the Jammu and Kashmir Distressed Debtors' Relief Act of 1949.

The act consists of thirty-seven articles, but the following are of real importance:

1. Special conciliation Boards set up throughout Kashmir to engage in settlement of debts not exceeding Rs5,000.
2. Within four months of the establishment of a Board, debtor and creditor are expected to submit to it an application for debt settlement. When the creditor fails to do that his debt claim is automatically extinguished.
3. The Board is given the right to "presume" that the debtor has paid 150 percent of the principal where a debt has remained outstanding for six years or more.
4. All claims against a debtor are to be dismissed where he has paid back to his creditor in cash or kind or both an amount equal to 150 percent of the principal. Any payment above that may be credited by the Board as a sum the creditor owes the debtor.
5. Any settlement of a debt arrived at after the establishment of the Conciliation Boards are considered void unless approved by a Board.

The rules laid down by the act certainly favor the debtor, and a few hours spent watching the Srinagar Board in action made it clear that the members interpret not only the letter but the spirit of the act as well. One notices at first glance that the debtors feel at home there, while the creditors are ill at ease, not only because they know that the "outstanding" as they calculated it will be cut sharply but also because many sordid details of creditor-debtor relations come into the open. The board procedure is of the simplest kind. The request for settlement is accompanied by a board fee of Rs2, and seldom does debtor or creditor appear with a lawyer to argue a case. The members of the board examine the appeals and often render a decision on the spot. A farmer walks in heavily indebted and walks out free of all debt or the debt drastically reduced. In one instance the creditor sued the debtor for Rs500, but having examined the case and with great dispatch, the board decided that the creditor owes the debtor Rs20. Unpalatable though the decisions are, the creditors usually accept them. The conciliation boards not only scale down debts but they also fix easy terms of repayment, the average period being twelve years. Between July 1950 and the end of 1951, nine conciliation boards examined 41,000 and settled 35,000 cases. The total debt of these cases involving an estimated Rs10 million was scaled down to Rs2.3 million or 77 percent.

The Debt Conciliation Board scheme did not originate with the Kashmir government. The Punjab, Bengal, the central provinces, and Madras have applied this method to the same end in the years past. The only difference between the one under discussion and the others is that in Kashmir the scaling down of debts has been more drastic. It should be noted, however, that in Kashmir as in the other states, relief from indebtedness, important though it is, is not a lasting solution. Even though the farmer is relieved from debt, his needs for credit, whether they be productive or nonproductive, do not slacken, and the boards tend to dry up the normal credit channels. They are not substitutes for a sound, government-financed credit system. And this is what Kashmir needs (as do all other states of India) if indebtedness is to be maintained at reasonable levels and interest rates.

Assuming that the reforms are effectively carried out, what will be their effects in Kashmir and in the rest of India? The confiscatory nature of the Kashmir land distribution program will be felt, if it is not felt already, throughout India. The dilemma which confronted Kashmir leaders is confronting many states in India not so much with respect to zamindari abolition,[3] which has been facilitated by a recent amendment of the constitution,[4] as

3. Zamindari abolition refers to the elimination of intermediary interests who hold land on payment of land revenue to the state and the person with the right of cultivation.

4. Article 31 of the constitution has been amended to provide that the validity of any compensation for an intermediary right, which may be decided by any state government, will not be questioned in any court. This amendment doesn't, however, cover the cases where a tenant holds land from owners who have the right of cultivation, a right associated in most cases with that of transfer, sale, mortgaging, subletting, and so forth.

with respect to giving ownership of land to tenants of persons who own it. The slow progress of this direction is due to the fact that neither the tiller nor the government is in a position to buy off the landlord interests.[5] And the constitution of India prohibits the acquisition of such property rights without reasonable compensation. The result to date is extreme cautiousness in undertaking far-reaching measures involving large-scale compensation. For this reason in certain circles—and not Communist circles—the Kashmir approach is envied, even if not followed. It is worth noting that upon the recent publication of the Planning Commission's report ("Five-Year-Plan: Draft Outline") voices have been raised about the need of reexamining article 31 of the constitution (which provides for compensation of property) with a view to speeding up the land reform movement. The latest straw in the wind is the request upon the chief ministers of the states, made by the All-India Congress Committee, to report as soon as possible on legal impediments to land reform in their territories arising from the constitutional provision regarding compensation. This point was raised at the last meeting of the Congress Working Committee by the Minister of Revenue of Bombay who called the attention of the Executive Committee of the Congress Party to the difficulties experienced in his state. It was then decided that the question should be examined by the AICC for the purpose of suggesting ways and means of amending article 31 of the constitution.

Whatever the future repercussions of the Kashmir reform upon India, in Kashmir proper they are significant. The economic gains have already been outlined. They are uneven now, but most tenants can point to some benefit. The retention limit is bound to be reduced, and more tenants will enjoy the more profitable status of owner cultivator. In sum, in the light of prereform conditions, they are unquestionably better off economically. The social gain is also apparent. To listen to the farmers of Kashmir talk is to realize that they are shedding the traditional subservience to landlords and governments and that they are aroused to greater efforts for further betterment. But even in Kashmir where the land reform has reached out deeper than in any other part of India, the result is an economic palliative rather than the solution of the new owners' problem. The beneficiaries are mainly farmers with only 1 to 2 acres, and therein lies the crux of the difficulty.

Sheikh Abdullah knows that tiny holdings limit the effectiveness of the soundest of reforms, but he doesn't underestimate the beneficial changes attained and correctly surmises that the Kashmir reforms provide a suitable foundation for land improvement and higher agricultural production. This observer carried away the impression that his government is ready to tap all available indigenous resources to that end. Sheikh Abdullah stated that Kashmir could learn much from the experience of other countries, and he wishes to take advantage of opportunities in that field as they present themselves. The farmers, too, understand that the tenure changes signalize the beginning of a land improvement program. On numerous occasions they voiced the opinion that as owners or protected tenants they will improve the land, but the need of continued government assistance is mentioned in the same breath. To date the government has demonstrated its pledges to assist the farmers in a manner the landlords never did. Six new irrigation canals have been opened and two more are under construction. Of the six canals, two are already functioning, and they have brought under cultivation 10,000 additional acres of land. Next year it is expected that 13,000 more acres will be added. But much more will have to be done, and the pace of making the small farms large through expanded agricultural production will depend upon the financial means and technical assistance, domestic and foreign, to further such old and new undertakings as reclamation, irrigation, drainage, animal husbandry centers, experiment stations and extension service, agricultural credit, and a host of other measures. Whether the necessary means will be available is another matter, but since the government and the farmers are as one on the tasks ahead, the environment for further progress has been created.

While economic results of the reforms call for qualifications, the political need none. When

5. The very, very few who could pay are reluctant to do so because of the widespread feeling that eventually they will get the land for nothing.

the Kashmir leaders talk or write about agrarian reforms, the political motivations and their political consequences are hardly mentioned. They are there, nevertheless, chiefly because the Muslim religion of the farmers doesn't loom any longer as a serious issue in the final disposition of the Kashmir problem. Large segments of the farm population have been successfully induced to shift their attention from religious to economic matters. This in itself is a significant political development from the point of view of India-Pakistan difficulties over Kashmir. Even more important is that the most striking by-product of the reforms is the widespread conviction among the farmers that Sheikh Abdullah and his government are responsible for the improvements in their economic and social status. With this goes political support. The other side of the picture is that the reforms have alienated Hindu landed aristocracy. For this reason, the agitation of the Praja Parishad (People's Party), as opposed to Sheikh Abdullah's National Conference, has taken on a communal or religious turn, essentially Hindu. But the fact is that the number of affected and disaffected landlords is small, whereas the number of those who benefited is large. Moreover, in the final analysis the Hindu landlords cannot side with Pakistan for religious reasons, if no other. Some individuals in Srinagar tried to convince us that these conclusions are unwarranted and that the Kashmir government is so unpopular that even Hindus, let alone Muslims, are pro-Pakistan. It is our strong belief, however, that they have been expressing strong political prejudices which will not stand up under test. When the real test comes of whether or not Kashmir should opt permanently for India, the majority of Muslim farmers of Kashmir along with the Hindu population including the Hindu landlords will in all probability vote for Sheikh Abdullah and his preference. Therein lies the political consequence of the land reforms in Kashmir.

21. *Comments on Land Reform in India*

Almost immediately after concluding his late 1952 visit to India, Ladejinsky participated in an in-house Conference on Land Tenure conducted by the Ford Foundation in New York in early December of that year. The main object of the conference, apparently, was to help the foundation define program objectives for its work in Asia on land reform. The comments presented here, taken verbatim from the conference transcript, represent Ladejinsky's longest single intervention at the conference. He obviously spoke extemporaneously, the session was quite informal, and the communication between speaker and audience—with frequent question interruptions—was close. Ladejinsky here draws basic conclusions from his Punjab, Madras, and Kashmir observations in fundamental terms; stresses the futility of any external official attempts to influence land reform policies in India; and suggests to the foundation that it can contribute most usefully in this field by supporting relevant work at Indian universities, where respected scholars can do much to improve the climate for necessary reforms. An interesting revelation is his account of how he came to be invited to make his 1952 visit to India and the reception he encountered. This was not the only time in Ladejinsky's career that he arrived to find something less than a warm welcome. His participation in this conference is said to have influenced considerably the shaping of the Ford Foundation's programs in Asia.

MR. LADEJINSKY: To begin with, I will deal with India, as that country is the freshest in my mind. Or how do you feel about it? Shall I talk about Japan or Formosa?

VOICE: I think really what we're interested in are the issues that are involved rather than the particular countries. I see no great difficulty in your moving back and forth from one country to another.

MR. LADEJINSKY: . . . India is forty-eight different countries. Madras alone is six.

Well, as some of you probably know, I spent the last three months in India looking at the question that we call land tenure or land reform. We will spend some time finding out what they are doing and perhaps come up with some conclusions as to what they are doing or what they might do.

As to the matter of looking at it, I pretty much followed in the footpath of my predecessor sitting in the chair here, and I toured a good deal of the country, north and south, east and west. Since the country is an enormous one, I couldn't visit all of the principal states that one should perhaps look into. But I have seen enough to venture an opinion as to the goings-on there.

When one reads the Indian press—and by this I mean the English language press—and follows it day after day for the past two years and certainly during the period I was there, one gets the impression that a good deal is being done in India. Seldom is the day that one does not come across an item about land reform distribution in this state, land reform legislation in another state, and all along the line.

One does get the impression on the basis of these reports that a good deal is going on in the country at the present time. In reality, when you begin to look behind the scene—and by this I mean when you go down to the villages and talk to the people, talk to the officialdom, examine the legislation—then you begin to realize that something is going on, but there is not very, very much. One couldn't say, for instance, that in India at the present time a land reform program is being carried out. One could say that some initial steps have been taken along certain lines.

We might divide the whole proposition into three basic points. This is the one of legislation enforcement and the climate surrounding the reforms. Each in itself contains a number of items, and I shall try to touch upon them.

As to the legislation itself, it may be roughly divided into two parts. One is the attempt to abolish the so-called zamindari, which is the group of landowners who, by virtue of certain provisions made by the English as far back as 1793 were given the right to collect rent for the British and by virtue of that service were later on given what amounts to ownership of that land. A part of the legislation is dealing with the abolition of the zamindari system, taking them out of the picture and making the cultivating farmer pay directly to the state.

At the present time, or before reform in certain states, it is the zamindar who collected the rents. The zamindar paid part of the revenue of those lands to the government. Now, as a result of some of the changes that will be of help to the cultivating farmers, they pay rent directly to the government and come into direct relationship to the government.

As far as that goes, something has been done. In the . . .[1] province Ken visited, there is a law to the effect that the zamindari as such are abolished, that the rents as of a certain date—which are in effect the land tax—will be paid by the tenant to the government. The significant fact about the zamindari abolition is this—that it is really not as profound a reform as one was led to believe. I myself was under the impression at one time that the abolition of the zamindari in India would be doing a great deal.

You haven't done a great deal for the simple reason that the status of the cultivating farmer really does not change in any way. That is, what he paid to the zamindar he continues to pay to the government, except that the government, by virtue of that change, will get a much larger revenue than it was getting in the past because in the past the zamindar would retain a good share of that revenue.

I will illustrate my point. When this so-called permanent settlement of the British took place in 1793, the rule was that the zamindar would collect one-eleventh of the revenue as his for his service and ten-elevenths of the revenue would be given to the government. In time it came about that the relationship has changed. The zamindar retains ten-elevenths and the government receives one-eleventh.

Now, with this change, the government will

1. Evidently missed by the transcriber. [Ed.]

be getting the full revenue. The question arises: What is the tenant getting out of it? And the answer is that basically he does not get much out of it because he is paying the same revenue to the government. Perhaps the only positive aspect of it is that he comes in direct relationship with the government. Perhaps it may lead to some other changes. He does not have an overseer in the form of a zamindar.

But when you talk to the farmers about it—as I did recently in one of the provinces in Bihar, for instance—they will tell you that as far as they are concerned, abolition of the zamindar did not make much difference for the simple reason that the government can be much harder a taskmaster than the zamindar can be. The idea that every zamindar is a rapacious landlord is not really correct. Some of them do have a good deal of noblesse oblige, and it does come to the fore now and then. I know many instances of when the tenant is remiss in the paying of his revenue, over the years he can make some deal with the landlord. The zamindar would permit him to pay on an installment basis. All in all, he is closer to the zamindar than he is to the government. The fear on the part of the cultivating farmer is that he may not get the same sort of treatment from an impersonal government which sends around a revenue collector to collect the rent. If you don't pay, you simply have to pay.

So there is no enthusiasm, I would say, on the part of the cultivating tenants.

VOICE: Is the situation in Bihar very much like it is in U.P. [Uttar Pradesh] with reference to . . .

MR. LADEJINSKY: Very much, very much indeed.

And so I was rather struck at the beginning to find that the much talked-about zamindari abolition is not all it might seem to be. Moreover, even [with] the abolition of the zamindari system in effect, a cultivating owner is not in effect the owner of the land. Ownership of the land is vested in the hands of the government. And so the government, as is the case in Uttar Pradesh, told them, "Yes, you can become owners providing you can pay ten times the revenue"—is it Ken?

In actual practice no tenant is willing to pay ten times the revenue to the government for the ownership of the land. In part, as I shall point out later, there is the feeling on the part of the cultivators that sooner or later they will have the land anyway. What is more, they intend to get it—although they don't quite say it—but the feeling is prevalent that when they do get it they will get it for nothing. And this is something that you can find no matter where you go in India at the present time. Although there is nothing in the legislation or climate surrounding the land reform program, anything that indicates that they will get the land for nothing. However, this is the feeling. It has a very important inference and the cultivating farmers, as a body, are not buying the land from the government at ten times the revenue—and I might tell you that ten times the revenue really is a small price, much smaller, I imagine, than the price in Iran or, for that matter, in Formosa or Korea or in any other country I know of.

VOICE: Would that word "revenue" be interchangeable with "tax"?

MR. LADEJINSKY: With land tax.

So this is as far as the zamindari part of it goes, and I am by now convinced, though I do not want to be dogmatical about it, nevertheless I have the strong feeling that it is really half a measure as far as the cultivating farmers are concerned. Sooner or later the government will have to go the whole hog and somehow effect a deal whereby they will become the virtual owners of the land and all that goes with it.

But the real problem in India is not the zamindari, the ownership of the land by the zamindari. It is the other and much larger group of tenants and agricultural laborers who work the land of the people who own the land as distinguished from the zamindari who received the land by virtue of that special deal with the British. There you find an entirely different situation. Before I go into that, I think it should be made crystal clear that you have a very special situation in India. Perhaps a third of the farm population has no stake in the land whatsoever, and I am talking now about agricultural laborers—so-called permanent laborers and casuals. These constitute, they say—nobody really knows what the figures are—incidentally, as far as statistics are concerned, I think India is about as badly off as Iran is or any other country in Asia—but on the basis of what I estimated in going from village to village, I find that at least a third belong in that category. They have, as they say, no stake in land of any kind.

As to the cultivating tenants themselves, one must make very clear that the very term "tenant" is altogether different than the term as we use it in the United States or in any other Western country. I found, for instance, that the worst exploiters in India were not the landlords, the people who actually owned the land, but secondary groups who are called tenants but who do not cultivate the land. They sublet the land to somebody else, and these do the cultivating.

Some of my experiences in Madras might be enlightening on this point.

VOICE: There were once subzamindars, or something like that?

MR. LADEJINSKY: What you actually have is this. Take the case of Bihar: In this province I found in a number of villages at least six different rights to the land between the one who claimed ownership and the one who actually cultivated the land.

VOICE: Or the top zamindar and the intermediates.

SECOND VOICE: You have this stratification whether you have zamindari or noncultivating tenants.

MR. LADEJINSKY: As the top zamindari began to get more and more than this one-eleventh and gradually shifted over to getting ten-elevenths, they decided really that they didn't need to bother with the land. They farmed out the land to, say, B; B then decided to get his cut and would not bother with the land, and farmed it out to C, then from C to D, and so on all along the line. In Bengal you can count as many as seventeen different rights to the land.

This points, incidentally, to the tremendous complexity of the problem and to the fact that you deal not only with a particular individual but, when you talk about a land reform which would eventually give the land to the man who actually cultivates it, you somehow have to take care of all the various spheres which are superimposed upon the man who cultivates that land.

But to go back for an illustration to Madras, for Madras is not unique in this respect. In a state such as Madras—and when I say Madras I must refer only to two big districts which I visited primarily because the Communist influence has become very pronounced there and very important—in those districts I found situations which are, for instance, as bad if not worse than, say, in the Chengtu province in the fall of 1939.

In these districts the tenant may pay a rent of from 75 to 80 percent of the crop. At the same time he bears virtually all the costs of cultivation, and the net result is, of course, next to nothing. If he gets the straw he is very lucky.

Then the question is, of course, what's the use? Why go on in this particular situation? The answer is simply that he has no alternative. The only alternative is to become a casual laborer. A casual agricultural laborer works from three to six months a year at a wage of twenty rupees, which is four dollars a month. So, having no other alternative, he keeps on cultivating the land.

But to go back about this peculiar relationship that you have in India. The men doing this rack-renting are not really the owners of the land. They are men, so-called comindari, who rent their land from the people who actually own the land. As one of them told me, "Yes, I rent this piece of land for 120 rupees a year. I get from it roughly 2,000 rupees a year." He is the one who is really getting the lion's share of the profit.

This situation is found with variations all over Madras and many another province.

VOICE: Does he perform any function?

MR. LADEJINSKY: As to this—and I am glad you raised the question—I am of the opinion, supported by a good deal of evidence (the landlords themselves provided evidence) that they perform no function other than collecting the rent. When in one particular instance I went to talk with the farmers in the fields and the landlord happened to come around in a shiny new Ford, it happened that the land belonged to this particular landlord. By the time he came there we roughly knew what the terms of servitude were. As he walked out we called over the tenants and they repeated in his presence—and he agreed that those were the terms—the landlord became rather uneasy. He didn't like foreigners asking such silly questions, but nevertheless he became on the defensive and said, "Well, but they don't improve the land!" The inevitable answer was, "How could they improve the land under those conditions? What do you do about improving the land?"

"No," he said, "I'm not supposed to improve the land. It is their job, for they bear all the costs of cultivation. It is assumed that on their profits the tenants do the improving of the land."

And so we had a situation where the landlord in general—of course there are exceptions—does not perform a function, either managerial or otherwise. They perform the managerial function only in this sense, that they are most careful to see to it that the same tenant should not remain on the same land year after year. He keeps on shifting them all over the place. No tenant knows which land he will cultivate next year. Again I say that this may not be true in every case. It is an outgrowth, really, of all the land reform agitation in the last six years. The idea of that is that if you stay on the land over a certain number of years, then you can claim occupancy rights. This means that you limit the powers of the landlord to evict the tenant. In order to prevent that, the landlord shifts the tenant from one piece of land to another. The result is that, in one village in the Punjab province way up north, I came across a situation where the farmers went for seventy-five years working in one area, but no farmer could prove in court that he had been on the same land for seven years. If he could, he would then get occupancy rights. But in no instance could he prove that he has those rights, because, in the first place, the records are nonexistent.

When the talk about land reform began, the landlord saw to it that no record would exist that A cultivates this particular piece of land. So you have this situation that is, as they say, a managerial function. But it is largely addressed to the idea that you keep the farmers shifting from one piece of land to another. When it comes to improvement of land and such matters, he doesn't function there at all.

VOICE: Before you get too far away, this owner who came out with the bright new Ford —is he the real owner?

MR. LADEJINSKY: No, he is not the real owner. He is the man who is renting the land . . .

VOICE: Is he A?

MR. LADEJINSKY: No, it happens that he is B.

VOICE: So there were only two . . .

MR. LADEJINSKY: In that case there were only the cultivator, B, and A. The elimination of the other intermediaries is due to custom and tradition, which goes many, many years back. For many years B has been renting land from A. Now, as a result of legislation, A cannot evict B. He can evict him only after a certain number of years and under very special conditions.

For all practical purposes, B is the one who runs the show.

Now this sort of a situation—there is no need for me to go from state to state. Rents are not as high in other provinces as they are in Tanjore and Malabar. They are particularly high there because you do have this tremendous concentration of population. It is the best land, and where you have these conditions rents are normally of that type.

Now, what does the Indian government do about this?

VOICE: This is also rice land, isn't it?

MR. LADEJINSKY: In this instance. In Tanjore it is rice land, in Malabar it is the area for garden land. Basically they are good, wet lands. . . .

Now the question is as to what the government is doing about some of these things. You know that the Indian government has committed itself to what they call a land reform program. The Congress Party, which is the governing party of India, has had much to say on that subject—in the past before they assumed power and now when they have been in power for the last five years.

I might also point out here that there has been recognition—and I think we have to agree with it—that you cannot enact an overall legislation for India as a whole. The differences are so terrific, not only between state and state, but even within a state, that the talk about a national law wouldn't be practical. The best you can do is lay down certain general principles, and we'll talk about that later.

And so there has been recognition that the states themselves will have to enter the picture and do something about these landlord-tenant relationships. In the process, during the past five years, something has been done. To begin with, I mentioned the zamindari abolitions. In the U.P. I described what the situation is. In Madras the situation in this respect is much better, partly because the zamindari group is a very, very small one. Moreover, whereas in the U.P. the state government has committed itself to the payment of compensation to the zamindari group, in Madras the government has started to take over the land with virtually no payment for those zamindari rights. But I must repeat that it is a small problem in Madras and that is why they were able to do it.

So on one hand you have the zamindari legislation that was enacted. Its consequences I described. The tenants are not particularly happy about it unless and until the government goes a step further and actually gives them the land and they become the real owners.

But the more significant aspects, again, is what the government is doing about improving the conditions of cultivators as I described them a few minutes ago. For that purpose they have enacted a number of laws in Punjab, in Madras, in Bihar at the present time, and the central provinces. And incidentally, for this reason I went down to the Punjab to find out—this was my first experience as to what the government did in that particular province. There again, when you read the announcements and even when you read the legislation, you get the impression that the government did a worthwhile thing, namely, it has reduced rents by one-third, it has also given occupancy rights to the so-called tenants-at-will who could have been evicted by the landlord at will and that is why they are called tenants-at-will; and so the government decided to give them certain rights in the land, both by making them occupancy tenants and also by reducing the rent.

When I went down to the villages to find out what the situation really is, I found out that the legislation is nothing but legislation on paper. It is a mere blank as far as the village community is concerned. I will give you the reasons. It is not applicable; and, if it were applicable, then in effect the tenants would perhaps be evicted in greater numbers as a result of that particular legislation than would be given the right of remaining on the land.

The reason is a simple one. The law states—I am not giving you the details, but just one item—that, in yielding those rights, the landlord has the right to resume, say, 50 acres of land. By that is meant that he has the right to take back into personal cultivation 50 acres of land, of a certain type of land. The tenants on the landlord's land above his 50 acres get those occupancy rights.

What happened as a result of the legislation is this: that in giving the landlord the right to resume possession of 50 acres of this certain type of land and 200 acres of another type of land, thousands upon thousands of tenants have been dismissed from the land as tenants. That is, all the tenants on the land that he took back are not tenants any more. They have become agricultural laborers. Do I make that clear? It is true that the tenants on the land above 50 acres have those rights, but the point is that the drafters of the law knew, for instance, that if you set the ceiling at 50 acres there are relatively few landlords in the Punjab who have 50 acres or more. In granting them that right of resumption, the landlords immediately decided to evict the tenants because he is afraid that if they remain on the land the tenants will claim occupancy rights at a future date and some other legislation will come into the picture. So the net effect of the legislation was that more tenants were evicted from the land and have had to become agricultural workers than tenants were given security rights. So as far as this very important aspect of the legislation is concerned, nothing was really accomplished. In fact, more harm than good was done.

As to the reduction in rent, it happens to be a fiction for the simple reason that no rentals, no reduction in rent can be enforced, unless you give security of tenure, unless you give the tenants real bargaining power. To the extent that they have no security of tenure, they cannot have their rents reduced. Many and many a tenant has told us—incidentally, in the presence of the landlords—that they don't bother about the rent reduction because the issue now is to be able to remain on the land. That is the important issue because they hate like the dickens to become agricultural laborers for employment as casuals. And so the whole legislation, by virtue of these particular provisions which really have no relevance to the conditions as they exist in the Punjab, has resulted in the tenants having neither a reduction in rent nor did they get security rights. And so we had to conclude that this type of legislation is mere paper legislation. It certainly effects no changes other than the changes as I described them and certainly does not benefit the community as a whole.

Another very important aspect of this Punjab situation which is reflected throughout India is that, even if the legislation were perfectly sound, if the tenants were given rights and the reduction in rent had real meaning, there is really no way of enforcing it for another reason. There is no machinery in India, there is no state, which would enforce any legislation because the enforcers of the legislation again are the so-called revenue officers, the tax collectors. The British have left an excellent tax collection system in India, and it is these people who are entrusted with the job of implementing the land reform legislation, whether it is in the north in the Punjab or in Madras in the south.

Who are the people who are enforcing the legislation? The people who are enforcing the legislation are mainly people [who] themselves have an interest in land, a proprietary interest in land in one way or another. I spoke with a local judge, a magistrate as they call them there, who, incidentally, is the one to pass on all problems arising out of this legislation. Well, it happens that he made no bones about it. He is a man who owns a good deal of land. He believes that the whole reform business is essentially a bunch of nonsense. It is against tradition, it is against custom, it goes against the financial structure of the community. He made it quite clear that he is not the man—his implication was very obvious—to go ahead and to enforce that particular legislation.

But you can go a step further down the ladder and you will see that the rent collector himself comes of a relatively well-to-do family. He too is involved in the land, and I think it is too much to expect that these people, the enforcers, would be willing in a sense to legislate themselves out of existence. They really would not behave like an "economic man." The result is that you have no machinery to go about enforcing the legislation about tenure at the present time in India.

Taking another aspect of it and comparing with the Punjab, in Madras—I am sure many of you read it—you had a situation where the Communists have really penetrated into the villages. In the past elections some of the districts went almost completely to the Communists. They sent the greatest number of candidates to the Madras state legislature. The reasons—their appeal is a very simple one. When you pay 80 percent rent and you bear all the costs of cultivation, and with the landlord performing no special duties other than the ones I described, with the Congress government—and this is a story in itself—really not operating in the field at all, there you have the perfect vacuum; and anyone can walk in and take it over if he has the energy and the willingness to do it. The Communists are very willing. They walked in and they took it over.

To illustrate my point, I talked to a number of villagers in the presence of the revenue officers. I asked one of the peasants, "What would you do if the landlord were to refuse to accept this particular rental as ordained by the state legislature? To whom would you go for redress?" The reply was very simple. "I would go to the Red Flag Association." Everybody knows what the Red Flag Association is in the Madras and Tanjore districts, for it has a membership of about 200,000. This is the group that fought for the tenant's rights, represented him, are the spokesmen of the community. And so you have in Madras the Communists as an important part of the rural districts. And it was only the impact of that development—particularly in the elections of the summer of 1952—that the Madras legislation was acted on with the intent of improving the lot of the farm community.

And so they enacted an ordinance called the "Tanjore Ordinance." Basically it means that the rents are reduced from 80 percent to 60 percent. The tenant still continues to pay 60 percent of the [crop as] rent, and it is made perfectly clear that the tenant is to continue to pay all the costs of cultivation. That is the sole legislation in the state of Madras, which has 60 million people, 80 percent of whom are farmers. The legislation applies to only one district—to the one which was worst affected by the penetration of the Communists. The state as a whole is not affected by it. This is the sum total of the effort on the part of the Madras government to improve the lot of the peasantry.

It is, I think, quite obvious that it isn't going to stay [that way] because the tenants are very dissatisfied with the provisions, even though they are an improvement. But the improvement is not enough, really, to redress the balance. The landlords, incidentally, are all [up] in arms against it because they felt and still feel that any legislation infringes on their rights, and they are dead set against it.

So you have a situation that, if it continues, is really a cold-war situation. You do not have the open rebellion that you had in Madras two or three years ago. The government settled that issue by jailing a good many Communists. I was also distressed to see on the trip to Madras —I don't think I told you about this—a number of prisoners at certain stations, handcuffed, and they looked to me like the kind of peasants I had seen all over the state; and later on I inquired and it turned out these are not really Communists but dissident peasants who were dissatisfied with the conditions as they are and had something to say about it. In short, police measures in Madras are still the most significant measures as far as the government is concerned to allay discontent and distress in the community.

Also, as to enforcement you have roughly the same situation. It is the same in Madras as you have in the Punjab. I could go on repeating this story state by state, but the picture is the same. The legislation is totally inadequate. Granted that the situation is a very difficult one, that they would have had to step on the toes of many an important landowner, the legislation is inadequate and the enforcement provisions really aren't there at the present time.

You find an entirely different situation when you go down to Kashmir, where I visited. There the picture changes radically. It is a long story, but I'll give you the gist of it. You have Sheikh Abdullah, the leader of the Kashmir government, a very determined leader who set himself to carry out a program which would give the land to the cultivating tenants. He is doing it, but he is able to do it, basically, through one measure only; namely, he is in effect confiscating the land. The law provides that the landlord can retain only a certain amount of land, which is 22 acres. Anything above that goes to the tenant who cultivates that land, and there is no compensation for the land confiscated, plus the fact that all tenants on the land below 22 acres get occupancy rights, which means that they can inherit the land, they cannot be evicted from the land, they can even sell their right to the land. Naturally the tenants are overjoyed. The program is actually being carried out because the climate in Kashmir in this respect is altogether different from the one that I encountered in the other provinces. Also, there are the benefits that the tenants get—of course they vary from tenant to tenant. If he is on the land of a big landlord, he becomes the owner of the land. And this is definitely a great gain because he makes no payment for that land. Others get a certain reduction in rent. Others can get only the occupancy right in the land. I cannot go into any detail, but actually there are three types of gain made by the various tenants.

I did raise the question with them, "Isn't there any jealousy among you people? Here your next-door neighbor succeeded in getting all the land for nothing and you get only a small reduction in rent." They said, "This really doesn't bother us because we know that sooner or later we are going to own all the land of Kashmir." And this is something which prevails throughout the valley of Kashmir.

How is it, then, that all the commitments on the part of Nehru, all the commitments on the part of the Congress Party, result in so little being done? And what I am going to say is certainly off the record, although I have told this to the people in the planning commission. . .

[Off the record.]

VOICE: Wolf, how long would it take to. . .?

MR. LADEJINSKY: Very briefly.

Here is the situation. I may have sounded pessimistic, perhaps unduly pessimistic, but the point is this. I believe that there is still time to do something, but, number one, no government should be doing it. That is, no foreign government. No ECA, no Point IV. Such things are completely out of the picture. If we do, then I would say God help us. When we consider the sensitiveness on the part of the Indian, the kind of sensitiveness that I haven't encountered in any other country—in any Asian country at any rate—then you must be doubly sure and clear that in no way, form, or shape should U.S. agencies of one kind or another, no matter how well intentioned, participate in anything that might lead to a change in the situation.

It comes down to this: That a private agency would have to do it. Now as to the question of what it might do. The difficulties I have mentioned throughout my talk really come down to this, that the climate isn't there for the reform at the present time. This for the reasons I have indicated. However, I think a much better climate could be created. I am not saying that this alone will do it, but unquestionably much more can be done. The only ones who could do it, of course, are the Indians themselves.

Now the question arises as to who are those Indians we are talking about? Remember this, that no Indian official, regardless of how high a position, is really particularly concerned with this issue. Some are. Some of the members of the Planning Commission are deeply concerned about it, but the mere fact that they are officials of the government prevents them from making very clear their interests in these problems outside the small government circle in which they operate. What is needed is awareness on the part of the public at large that here are these issues and something must be done about it.

I can't think of a better group than some of the outstanding Indian universities, the professors and teachers in them, who could do that particular job. On my trip to Bombay I talked about these matters with some of the people who, incidentally, right now are very much on the sidelines because the government doesn't consult them. [Interjections as to who was seen where.] But people who would like to do something about it are not in a position to do it in the sense that the government is not really concerned with their efforts. The business that we had in the United States, especially during the New Deal days, when the government would call in Professor A and B and C and all along the line for competent advice—that sort of thing is pretty much absent in India. However, they would like to do it. And it is this group that I think should be and could be induced, with little difficulty and, incidentally, with not too much money, to constitute themselves as a group that would study, write, publish, lecture, deal with all these issues. And they have great standing. It is this group that could create the kind of climate which is lacking and the kind of climate which is needed in India at the present time.

And so, in the roughest, my suggestion would be for the foundation—and I might spell it out, for the Ford Foundation—to do something along that line. There is nothing particularly vague about it. It is a very concrete issue. We know the universities, who the people are. A group of this kind could be created. The necessary work could be done—and I might say that in India it isn't so much a question of finding out what the conditions are. They know what the conditions are. Nor is it a question of finding out what constitutes a sound reform. They know very well. India has an enormously rich literature on the subject. If they wanted to, they could easily establish a sound program. When the Planning Commission wanted to, they wrote a section on land policies which offers most of what is needed for a good program. But that is not the issue. But the group I propose could generate this kind of interest, follow up enforcement, know what is really going on, bring it to the attention of the government as no leader in the government could possibly do. I would like to see the creation of such a group within the universities who would dedicate a good deal of their time to this one basic proposition: how to keep the issue alive.

To know at the present time, let me tell you something which to me was unbelievable and Ken knows about it. You take the central government of India, a government of 400 million people. How many people in the central government, do you think, know or deal with land reform matters? I would like to take a vote around this table.

Two persons, no more than that. It sounds unbelievable but it is true.

VOICE: Are these two members of parliament?

MR. LADEJINSKY: No, I am talking about the commission. You take the Ministry of Agriculture of India. There is one man. You know him, Ken. On the Planning Commission there is another person who does know about it . . . another one, and . . . But actually, I couldn't find six people in the central government of

India who are in one way or another concerned with this problem on a day-to-day basis. No more than that.

VOICE: Would you answer this question? Were your presence and your activities there resented? You mentioned a moment ago that this is something Point IV can't do, a foreign state can't do, and so forth. Yet you were there and you did a tremendous thing in developing this report.

MR. LADEJINSKY: I am afraid you are making assumptions which are really not warranted. The point is this, and I think Ken will support me on this. You hit a very, very touchy subject as far as I personally am concerned. Number one, when I came to India, my opinion was that the Indian government wanted me there. In effect, the Indian government didn't want me to come to India. I found out how these things happen. Mr. Chester Bowles met [a certain high Indian official] at a cocktail party. He said, "You know, Mr. ———, we understand you have a land reform problem. We know so-and-so. He might be of some help to you." He puts Mr. ——— on the spot and Mr. ——— says, "Well, that may be a good idea." And so I am invited by the Indian government.

Then I come and, of course, I realize the deal. As far as I was concerned, I wasn't particularly welcome. I was subjected—so it seemed to me—for the first two weeks essentially to a cross-examination as to my intellectual standing, as to how much I know about a chapter of English history, whether I can turn a phrase, whether I have intellectual ideas—matters of that kind. And after I had apparently passed the examination, their attitude somewhat shifted. But one had to be forever careful not to recommend anything. I recommended nothing. No recommendations of any kind except in terms of discussing experiences here, there, and other places, when certain things fall as they will and some of the members of the commission picked them up here and there. Later on these might be incorporated in a report. Only in that sense, and in no other sense, were there recommendations. . . .

22. *A Comment on the Report of the Ford Foundation Conference on Land Tenure*

Ladejinsky's comments on the report of the Ford Foundation's 1952 Conference on Land Tenure take the form of a letter to Carl Spaeth, director, Division of Overseas Activities of the foundation, who had requested them. His response from Tokyo was dated March 23, 1953. It contributes significantly to our understanding of his views on external assistance.

I HAVE YOUR NOTE OF MARCH 9, and I hasten to comply with your request for a comment on the report of the New York Conference on Land Tenure. I am not certain that what follows will ease your labors, but I assure you that I am not trying to add to them.

I find that the report is a very good summary of the proceedings, and this applies particularly to part II and to some sections of part III. Part I, "Do We Call It Land Reform?" is a bit discursive, but it finally does get to the point, and the conclusion that we call it "land reform" is well taken. I am less certain about part III and, as you will note in a subsequent paragraph, I should like to see some modifications in part IV.

It appears to me that part III is too drawn-out and academic and that some of the issues raised may lead to misinterpretations of the character of land reform movements, which should be considered as they are rather than as we wish them to be. The ten principles of land reform policy set forth in the report are essentially correct, logical deductions, primarily as a Western agricultural economist conceives them. They tend to minimize the economic, political, and social realities which determine the manner in which the hunger for land is being satisfied now and will be, increasingly so, in the years immediately ahead. It would seem, therefore, that the land policy as set forth in part III is not necessarily germane to known land reforms beginning with the one achieved by French peasants under the French Revolution and ending with the Japanese reform of more recent date. In short, the principles constitute something akin to a perfectionist scheme, upon the fulfillment of which no land reform waited in the past or will wait in the future.

The report should make clear that in present-day Asia the landless peasants will in all probability acquire ownership of the land even though their native lands may lack "trained professional personnel, purposive educational systems, skilled laborers," and so forth. Similarly, and whether we like it or not, it is well for the foundation to be aware of this crucial point, it is not true that "every land reform must deal with the problems of acquiring the land, and if from private owners, of compensation," or that the "purchase" terms for peasants involve not only the essential rates of peasant savings and investments (and the consequences for the total rates of savings) but also the peasants' relation to the financial responsibilities of the entire economy. The subject of this report is not land reform in the United States but in the backward areas where such a concept hardly applies. In Kashmir where "land to the landless" made notable progress, no one troubled about compensation or rates of peasant savings; essentially the same happened in Japan; India is likely to follow suit if the tenants are to acquire land; and the chances are that the peasants of Iran will acquire the crown land without paying for it, despite the elaborate system of payments we helped to set up and despite our own position that "The prompt and faithful payment of the purchase debts is considered . . . to be absolutely essential to the success of the program."

Some of the principles in part III are eminently sound in the sense that they are relevant to reform efforts in Asia. Items 1, 2, and 3 exemplify that; but point I is the truly significant one, for most everything else hinges upon it. The writer of the report is indeed right when he states that "Wherever there is a recognition of a real need for land reform there may be an opportunity for sympathetic and competent counsel and assistance in making orderly adjustments, with a minimum social cost." This is where outside assistance might come in. But "The General Issues" section as a whole could stand drastic condensation, partly for the reasons stated and chiefly because in its present form it tends to obscure the real issue, which is whether to aid reforms already in being or actively under consideration.

Since the main purpose of the report is to help the officers of the foundation determine a course of action, the content of the entire report could stand considerable trimming for the sake of greater succinctness; in addition, it should be so modified as to bring out in sharp relief some of the features which typify the agrarian reform movement in Asia. With respect to the latter, the first item is that widespread ownership of land among tenants is a desirable condition for economic, political, and social reasons; such often-mentioned arguments that it might or would result in a decline of agricultural production are essentially not valid, and they could be very well omitted. Second, that despite the opposition of the landlords or of governments dominated by them, the transfer of ownership is inevitable; what is in question is the pace—quicker in a country where the government is bent on implementing a reform, slower where a government has neither the strength nor the desire to activate the issue. Third, that the reform movement is a revolutionary one even though landlords' heads do not roll and noblemen's nests are not set afire. The acquisition of the land is not and will not take place in the spirit of the due process of law as understood and practiced in the Western world. More specifically, the present trend is in the direction of depriving the landlords of their land for relatively little or no compensation at all. Fourth, while the officers of the foundation should be fully cognizant of the confiscatory character of the movement, they should be equally aware of the fact that in Asia the infringement upon the concept of private property at the top strengthens the very same concept at the base of the social pyramid where it is weakest. The economic and political corollaries which flow from this change are too obvious to require further elucidation. If I appear to stress too much the fact that the movement is revolutionary and not reformist, I have in mind possible criticisms leveled against the foundation; I recall the time when General MacArthur was very closely questioned by an important American public figure whether the Japanese land reform was in consonance with American notions about private property and free enterprise.

Assuming that after an examination of the pros and cons the foundation decides in favor of aiding land reform movements, I wish to touch upon the character into which it might evolve. Part IV of the report outlines various courses of action, and they are quite in line with our exchange of views in New York. However, the four guiding principles seem to tend in favor of a kind of assistance which is not of crucial importance to Asian countries. The principles imply that we know a great deal and would-be recipient countries very little. My own experience tells me that once a government is determined to carry out changes in land tenure arrangements, our aid cannot but be limited. The reforms have been slow in coming, not because of the fears on the part of governments that they don't know how to formulate policies, administer, execute, and evaluate them but primarily because they don't want reforms in the first place. When Sheikh Abdullah of Kashmir decided it was time for peasants to get the land, he did a pretty good job with a limited staff working with a peasantry which is 95 percent illiterate. We know too little about land arrangements in Asia to presume that we can determine policies and so forth. The analogy of an American land tenure expert service with that of the financial experts of the World Bank does not hold any water. Every country in Asia has an articulate group with knowledge of the country's agrarian problems and ways of solving them. The difficulty is that often they choose to ignore them. To be of real aid we could do no better than try to persuade this group that it is perhaps later than they think. This is where under propitious circumstances we could, with subtlety and experience, render useful service.

The above doesn't decry the fact that a private agency should give financial assistance for such purposes as the creation of a favorable reform climate, for helping gather basic land tenure data, or to stimulate relevant research in the country in question. What I question is the nature of the role we assign ourselves, based on the mistaken theory that we have the answers (see i, ii, iii, and iv under (G) p. 22) and that the mainspring of our expertness will come through research. You will note that every item under (G) is research and yet more research. To repeat, the biggest single factor that handicaps land reform is the absence of the climate which spells the difference between action and inaction. But I doubt that the answers lie in the overemphasis on research and in the unwarranted assumption of our own omniscience. A good "salesman" or two might be used to greater advantage.

At one point the report states, correctly, that we have only a modest role to perform in the land reform field. I would add that that role derives not so much from our expertness, which is indeed meagre, but rather from the financial aid—even if limited—we might be in a position to extend under favorable circumstances. It behooves us therefore to act accordingly. From my point of view this means that the foundation should eschew "nerve centers," big programs, special land reform divisions, or anything that might imply that on the subject of land reform we are big time, eager to teach the people of Asia how to reform their land. The mistakes committed in this regard by our technical assistance programs (including FAO and other UN organizations) are ample warning to a private agency not to repeat them. The fact is that the land reform issue in Asia is at once so complex and delicate that the outsiders must approach it with caution, making haste slowly. For these reasons the recommendations of the report should lay stress on what can, rather than what should, be done, and a good deal of exploratory work in the field appears to be the prerequisite. Until much more knowledge and experience are acquired, the report should recommend an operation on a piecemeal basis, each request for *financial* assistance judged on its own merits. Our fund of practical expertness is all too small to offer that as our main stock-in-trade. I feel rather strongly that land reform institutes and such, underwritten by the foundation, are a matter for the future—if at all.

I don't believe that assistance to the land reform program of a particular country is within the purview of the report, but I cannot refrain from mentioning India. Here is an instance where the government is deeply involved in it, but progress is slow. One of the reasons is the failure of the university people to put their shoulders to the wheel and thereby help create the climate so often mentioned in this letter and without which the process cannot be speeded up. India also represents an instance where the initiative for assistance comes from the central government; we are not drumming up trade, as it were—a very important condition for the soundness of our aid. I therefore repeat the plea I made in New York that financial assistance be given to a few selected universities in India in order to articulate the agrarian issues for the benefit of willing and unwilling state governments there. And in India, no American, however knowledgeable, could do that kind of job.

Another reason for giving that type of financial assistance to India is to safeguard the sum total of the foundation's investments (and TCA's) in the community development projects. It is no secret that thoughtful Indians and Americans in that country are deeply concerned over the fact that these projects are built on the old land tenure arrangements, which is not a solid base. This explains the disparaging remarks of many a tenant about the community project. On more than one occasion I listened to such remarks as: "What is there in your project for me?" Pretty little, if the lion's share of the increased output remains in the same hands. If that condition should remain unchanged, then our efforts will fall far short of their goals. That is what I mean by safeguarding our investment, and that is why I am taking the liberty of bringing up the subject once again.

I am afraid I have been too discursive, but in justification I can say that I read the report carefully and tried to treat it with the seriousness it deserves.

23. *The Status of the Land Reform Program in India*

Ladejinsky made another visit to India from his Tokyo base to assess the progress of land reform in the summer of 1954. His report, he notes with regret, was necessarily based, for want of time, on discussions and reading; "of field observations there were relatively few." This perhaps explains the sense of restrained optimism one notes in the piece. Although Ladejinsky sees all too clearly the inadequacies of both legislation and enforcement, he is obviously pleased by the government's attempts to prod the state governments into more effective action and finds "the mere raising of the tenancy issue into the open . . . a positive development. The trend in India is toward reform; once begun, it cannot be stopped." He insists that security of tenure and reasonable rents rather than land ceilings and redistribution are "what the struggle should be about," that substantial progress will require time and patience as well as effort, that measures of land reform in the Indian setting can by themselves achieve but limited results, and that the Indian universities can and should "render a signal service in the field of reform developments." There is space here for only the conclusions, which Ladejinsky presents at the outset of his report.

This report, dated August 4, 1954, was transmitted to the U.S. Department of State with Dispatch 236, dated August 18, 1954.

Introduction

THIS REPORT ATTEMPTS TO MAKE a brief assessment of the current status of the land problem in India. The reporting officer's stay here has been all too short (three weeks) to permit him a more thorough analysis of developments since early 1953, when he left India after a prolonged study of the problem. The subsequent paragraphs are based primarily on talks with Clifford C. Taylor, agricultural counselor, American Emabssy, New Delhi; officials of the Planning Commission and of the Ministry of Food and Agriculture; revenue officers of the states of Uttar Pradesh and PEPSU; and on the perusal of relevant material, published and unpublished. Of field observations there were relatively few. Considering the fact that the observer attaches great importance to firsthand information gathered in the field, the discussions must be judged in the context of this limitation.

Conclusions

The principal conclusions which emerge from the reexamination of the reform program are as follows:

1. Despite the sharp criticism levelled against the land reform program in India on the ground of meager accomplishments, the program is making progress.
2. If progress is to be judged in terms of the accomplishment of the sum total of the goals enunciated by the Congress Party or the Planning Commission's five-year plan, the achievement is limited; if, on the other hand, progress is to be judged from the point of view of the existing balance of political power and prevailing economic conditions, the program is moving forward.
3. The impatience with the slow pace of India's economic development stems from the unwarrantedly exaggerated notions as to what can be accomplished in that country within a short period of time. This fallacy has special bearing on the land reform programs, for in no field has the issue between the "haves" and "have-nots" been so sharply drawn, and, for this reason, the half-hearted concessions so grudingly granted.
4. One segment of the program, although not the most important one, the elimination of the zamindari land rights, has been completed in some states and is under way in others. This program, as all other reform programs, suffers from acts of commission and omission but, above all, from explicit and implicit promises some of which cannot but fall short of realization. However, on balance, the zamindari abolition is an important step in the right direction.
5. The other, by far the most important part of the program, relating to the measures designed to improve the well-being of the tenants, is under way, but very haltingly. If the volume of legislation be a measuring rod of achievement, India would be in the forefront of such movements. The fact is, however, that the legislation is inadequate in important respects, while the enforcement is riddled with a "thousand gaps."
6. The government of India is becoming increasingly aware of these gaps and is prodding the state governments into more concerted action. This is not an easy task since the progress of the program calls for avoidance of political compromises leading to questionable economic results and the creation of a "climate" for the effective enforcement of the legislation. The states which are the legislators and enforcers would have to rise above the political bias of their legislative bodies in order to eschew the first and attain the second. This stage will not be reached in a day.
7. The sharp contrast between appearances and realities need not obscure the fact that there is solid merit even in this admittedly very "spotty" program. The mere raising of the tenancy issue into the open, let alone attempting to clothe it in a legal framework, however limited in character and emasculated in application, is a positive development. The trend in India is towards reform; once begun it cannot be stopped. And a start has been made.
8. The philosophic bases of the reform and the enormous body of legislation are replete with double-talk about land purchase

schemes and ceilings on land ownership, goals which are hardly attainable and which at the same time tend to divert attention from the points that truly matter. Stripped to their essentials, the fundamental issue is not so much land ownership as security of tenure. The cultivator does want a piece of land that he can call his own, but since he must pay for it even under the reform, for the vast majority this is unattainable. The sound alternative, therefore, is effective government action that would leave him on the land undisturbed, paying a reasonable rental and making the most of the residue. Currently this is what the struggle should be about, and it is in this connection that the reform legislation most needs propping up.

9. The land reform program in India has a long road ahead and a hard row to hoe. The class of people whose rights and authority the reforms seek to restrain is a class which has deep roots in society. It is the recruiting base for the intelligentsia and its members play an important part in the universities, government offices, press, and every important sector of public life. Persuasion and pressure may reconcile them to the seemingly inevitable sharing of property rights in the context of India's mass poverty and underdevelopment. But this will not be achieved in a year or two; the impatient ones would be well advised to steel themselves for a decade or so of patient waiting. The results should prove to be rewarding.

10. When that time comes—that is, when legislation is soundly drawn up and the implementation well carried out—the reforms by themselves will not have solved the riddle of better living for the farm community. The poverty of the peasant is a formidable enough problem; far more formidable is the poverty of the 40 to 50 million landless farmers (agricultural laborers and their families) who have no stake in land whatever and with whom the reforms are concerned only indirectly. There simply is not enough land to satisfy all claimants whose numbers are rising year by year. For this reason, without accelerating the rate of economic development in agriculture and industry, measures of land reform will achieve but limited results. Industrial development is particularly important as new jobs could be created to siphon off some agricultural labor and thus make a dent in the vicious circle of more people, less land, and growing poverty. This, however, does not in any way argue against the validity of reform measures; the issue in India is not one of *solving* the rural problem, but of palliatives capable of wiping out the worst features which condemn the farmers to a below-subsistence level of existence.

11. One of the many regrettable aspects of the lagging reform movement is the failure of the individuals and groups most capable in this field to be articulate concerning reform problems. With but few exceptions the Indian universities, for example, are very much on the sidelines, at best in the roles of impartial and inactive observers. They need not become propagandists for the land reform cause; students of the problem would be more to the point, for India suffers from a lack of information on the elemental yet basic features of postreform agricultural economy. This applies to prereform agriculture as well; despite all the talk about fixing ceilings on agricultural holdings as part of future reform measures, no data are available on the size of landholdings and land ownership. The first census is only now in preparation. Above all, the effects of the reform legislation already enacted can be only surmised through personal observations because only a few of the twenty-eight states of the union of India are taking the trouble now of finding out what they are. To be effective, this work of many years' duration must be done in India and by Indian agricultural economists, sociologists, and political scientists. It is a tailor-made job for the universities. In doing that, they could render a signal service in the field of reform developments, which, incidentally, might also serve as a basis for shaping future reform policies.

24. *Advancing Human Welfare*

I have already attributed a great significance, in my introduction to this book, to this letter to Kenneth Iverson of the Ford Foundation, as a broad expression of Ladejinsky's views. The letter, undated, would have been written sometime in November or December 1954. In transmitting to Ladejinsky the two copies of it he had requested, Iverson's secretary noted: "For us, it was a case of having another run made because your paper had been so much in demand by staff members here." No further comment is required, except to urge the reader not to pass this statement by.

THE CONTENTS OF YOUR LETTER of October 13, 1954, have been very much on my mind and a subject of deep concern to me. It is obvious that you did some soul-searching before you committed your thoughts to paper, and I assure you that my delay in commenting on it was not for lack of soul-searching on my part as well. But I must admit that you set before me a formidable task. The subject we are talking about—even if not expressly stated—is the sum total of what constitutes human welfare in the broadest sense of the term and, above all, how to advance its cause. The task of how to achieve this is the more difficult because "The critical problems which obstruct advancement in human welfare and progress toward democratic goals are today social rather than physical in character. The problems and opportunities of our time arise out of man's relation to man rather than his relations to physical world."

I shall have occasion to return to these meaningful words which I culled from the "Report of the Trustees of the Ford Foundation," September 27, 1950. I am citing them here because they are germane to what I shall have to say in subsequent pages. I am not certain, though, whether their content will meet your needs. I am not setting forth a program for the foundation in the sense suggested in the first part of the next to the last paragraph of your letter. Instead, I propose to put down such thoughts as I have on the setting and the magnitude of the problem which we face in underdeveloped countries; on the real danger that, if our aid to these countries is lacking or wasteful, the Communists will do the "job in their own way, and to the irreparable loss of the West and of the United States;" on the nature of the assistance practiced to date with a suggestion for a more inclusive program extending to the realm of measures of social significance; on the nature of the more specific problems relating to the matter (and of possible interest to the foundation) such as land tenure and credit; and, finally, a discussion of the proper kind of personnel and proper approach in our relations with underdeveloped countries.

Much of what I have to say will sound to you like a repetition of well-known truisms, but it may help me to explore more fully the issues under consideration. Naturally, I shall be the happier if some of the presentation also makes sense to you.

The Setting and the Magnitude of the Problem

The history of the more advanced areas of the world has shown that good government, electric power, transportation, and the social services are the four primary elements for economic and social progress. A high standard of living and so much else that distinguishes the advanced from the backward countries are their corollaries. The underdeveloped countries we concern ourselves with comprise two-thirds of the world population and are characterized by mass poverty, which is chronic and not the result of some temporary misfortune, and by obsolete methods of production and social organization, which means that the poverty is not entirely due to poor natural resources and hence could presumably be lessened by methods already proved in other countries. Moreover, an underdeveloped country often has an underdeveloped government. It lacks an effective civil service system, an effective budgetary control, a sound fiscal policy, and, naturally, it lacks an efficient governmental structure. In short, most underdeveloped countries lack almost entirely the traditions, institutions, habits of thought, and experiences which are essential to modern democratic society. That indeed is the principal political and psychological reason why Communism is a much greater threat in Asia than in Europe. The natural defenses against Communism that a tradition of individual dignity, the rights of man, and of democracy set up in different degrees in various countries are yet to be created in Asia. This goes a long way to explain why the Russian Communists succeeded in seizing power in Russia and why the Chinese Communists met with the same results in one of the most underdeveloped countries in the world.

Poverty, hunger, disease, and the lack of opportunity for self-development that these imply have been the lot of the overwhelming majority of the people in underdeveloped countries. What is new about this poverty is that it has become a source of discontent, and an overworked and overexploited common man who for centuries was inertly miserable is now alertly miserable. The peasants of Asia have never been satisfied with this state of affairs; but in the main, until very recently, the conservatism and inertia of the farmer and his ingrained feudal subservience to the state and to his landlord kept the pot from boiling over. Now the forces that keep the peasant within well-defined bounds are breaking down under the impact of what has been aptly termed "the revolution of rising expectations."

The "shrinking" of the world in recent years has made these people conscious of the existence of a way of life different from their own. They have been brought into contact with more advanced countries where scientific and technological progress has raised the standard of living of all classes of the people to unprecedented levels. They are awakening to the realization that there is hope for change and improvement. They are being told that they can produce more with new instruments of production and that the fruits of their labor should be more equally shared. There is greater awareness that it is possible for them to be rid of disease, epidemics, and illiteracy and that they have a right to some education provided by their government. These new ideas are only beginning to take shape, but they are powerful and call for translation into reality.

Part of the setting and the magnitude of the problem is the social order prevailing in the underdeveloped countries. Government authority is controlled by the very forces likely to be adversely affected by progressive economic and social development. In country after country these groups, largely landlord groups, are determined to maintain the status quo. They hardly benefit the rural community, and they are not proving at all effective against Communist penetration. They are, in effect, the creators of a revolutionary situation. To them, the very words "reform," "change," and "concession" partake of the devil; and yet without their consent, no progressive organization can function. It would be a mistake, however, to single them out as the sole obstacle to progress on the ground that they aim to hold on to their privileges regardless of the impending changes. The situation is more complex, partly because even among these conservatives there are individuals with progressive proclivities, willing to learn and to be persuaded, and, above all, because the traditional climate—nurtured in an age-old culture in the broadest sense of the word—dulls the interest in social and economic reform of the people as a whole.

While the ruling groups are unwitting and unwilling allies of Communism, there are also those consciously accepting the deceptively simple solutions of the Communist line. They come not from the poorest sections of the people but from middle class intelligentsia. In this respect the Communist leadership in the underdeveloped countries of Asia is cut of the same cloth as the early leadership of the Communist Party of Russia or China. From my own observations in Asia, members of the intellectual proletariat (college graduates, for example, all dressed up with no place to go) swell the ranks of the Communist chorus out of sheer frustration induced by lack of suitable occupations, lack of sense of participation, and no promise of leadership thought to be their due. The dynamism of the oversimplified Communist "solution" is alone sufficient to recruit them for its cause.

Fortunately, there are other groups and individuals eager to use their talents towards modernization of the setting we are talking about. They are found in the academic halls; in the liberal professions; as well as among the few spokesmen for the farmers, the progressive traders, and industrialists. They are the ones to be sought out and encouraged as part of the process of assisting the underdeveloped countries to achieve new goals. Given an opportunity, their contributions are assured.

Western or Communist Assistance to Underdeveloped Areas

The mere desire for change is undermining the old order. The passivity and fatalism of the masses of Asia (or in other roughly similarly situated areas) are beginning to yield to the desire for higher standards and the determination to acquire them. Industrial technology has already found its way into underdeveloped countries, and fundamental changes in the political and social structures of the underdeveloped areas are unavoidable.

What is of vital importance to the West is not the inevitable stresses and strains accompanying the economic development efforts but the general direction of the change from feudal or semifeudal to something different. It is pertinent in this connection to inquire as to the source of basic ideas in human relations which must shape the technological changes. Will the process of change proceed along Western or Communist lines? The Communists have already exploited this type of situation and, as in the case of China, turned it into a central political issue and thereby seized political power in that country.

From the point of view of the West and the preservation of its most cherished values, it is of the utmost importance whether, in their current efforts to modernize, the underdeveloped countries will lean towards the West, adapting its technology and political ideas to suit their special needs or, instead, accept the Communist promises and eventually the Communist system. It is therefore no exaggeration to say that the existence of the Western, democratic world will depend upon the choice, free or accidental, of the underdeveloped countries between following in the Communist path or proceeding with Western aid.

Raymond T. Moyer, in his reexamination of Point IV in Asia ("A Re-look At Point Four in Asia"), states very succinctly the problems confronting the underdeveloped countries. "Their fundamental problem," he writes, "is to eliminate the conditions that make Communism possible, and that would be a potential source of danger whether Communism existed or not. The essence of their problem, in other words, is to bring about the progress on which depend their continuing freedom, the achievement of their rightful place among the nations, and the welfare of their people."

The West has a vital role to play in achieving such goals because the future course of the underdeveloped countries, in cooperation with the West, is a necessary part of the conditions for Western survival. By the same token, how soon and by what means Asia deals with its problems successfully will determine what Asia's position will be in the struggle between dictatorship and democracy. Equally true is that the economic, political, and social structure with which Asia will eventually emerge will be that structure which comes nearest to giving a satisfactory answer to problems obtaining in the underdeveloped societies.

What Kind of a Program—Technical Assistance or Technical Assistance Plus?

The point has been made that an underdeveloped country often has an underdeveloped government. This is another way of saying that it is not generally true that a country is backward in some single phase of its life. Low literacy, poor health, low agricultural productivity, and lack of industrial development all go hand in hand.

A program cannot concentrate on some particular element of development without running the risk that it may be rendered useless by failure to carry on parallel developments in other fields. Developments along some line may actually create serious problems in other fields. There must be some balance in the elements comprising a program intended to aid an underdeveloped country. This does not presuppose that an aid program must be all-inclusive, but it presumes that if a serious effort is to be made to revise the standard of living of a backward area, the program must not be limited to what is commonly known as technical or economic assistance.

Not all realities are economic, and in the underdeveloped countries many of the motivations behind the drive for improvement are social, political, and psychological. There is no gainsaying the importance of a rise in industrial or agricultural output, but it does not by itself meet the needs and it is not at all certain that achievements in technical assistance will automatically or necessarily insure a greater sharing of economic welfare, greater sharing of political power, free public schools, the emergence of representative governments, and other developments which denote progress in democracy. Technical assistance may or may not foster such results; the industrial development in the West has in most cases created the climate for a healthy democratic spirit, but the sudden transplanting of industrialism in a radically different environment does not at all argue for a similar outcome. Its matter-of-factness cannot by itself fire men's vision and stir men's minds and hearts. The real test of technical assistance is what happens to men's minds. The nonmaterialistic appeal for the change in social institutions has always been the more powerful weapon. Even well-conceived and well-carried out programs of technical assistance in the economic field may not succeed without successful modification of educational, public administration, and other institutions bound up with economic development. For this reason, narrow concentration on material output, with its overemphasis on "know-how," typical of Point IV and most other U.S. government programs, runs the risk of not providing the people of the underdeveloped countries with the lasting values which in the long run determine the success of the program of the underdeveloped countries. The conclusion is warranted that "The needs of underdeveloped countries for better techniques are as great in the social as in the mechanical field."

A program for underdeveloped countries must, of course, begin with economic development. The physical resources they do have (and a good many of them are richly endowed) must be placed at the service of the people with an eye towards a higher standard of living. Part of a successful economic process is the creation of institutions for the utilization of the energy, skills, and character of the people. But if, as is correctly contended, even successful economic development cannot by itself create a democratic society, then a program which also addresses itself to the sum total of human welfare is imperative. What the ingredients of such an overall program should be, in addition to a higher standard of living, was outlined in the "Report of the Trustees of the Ford Foundation" (September 27, 1950) with reference to the achievement of "democratic strength, stability, and vitality." The report states that "To work toward these objectives means attack upon many subsidiary problems, all interrelated, all urgent":

> The need for governments, national and international to be more truly responsive to the people, to be more efficient and at the same time to be grounded more firmly in the active participation of its citizens: The need to achieve a relatively stable and more healthy economic system with greater opportunity for personal initiative, advancement, and individual satisfactions; the need to develop more able and public-spirited leaders in all fields of responsibility and endeavor; the need to improve our educational system for the better development of such leaders and for the preparation of men and women everywhere for the increasing tasks of citizenship and for the conduct of more purposeful and better-rounded lives.

The authors of the report may not have specific reference to Asia or other undeveloped regions, but it is significant to note that Eugene Staley in his "The Nature of Underdeveloped Countries," written in 1954, comes to the same conclusions in his choice of objectives of a successful program. These objectives, strikingly similar to those of the foundation report, even if differently worded, are:

1. An adequate living,
2. A sense of security,
3. A sense of freedom and participation,
4. A sense of belonging,
5. A sense of purpose, and
6. Creative opportunities.

Every Asian will buy these program objectives. Every Asian, however illiterate, consciously or unconsciously aspires to a measure of this welfare which consists of better living conditions, better health, better social status or greater equality of status, better government, greater participation in local or national affairs, and of a host of other values which spell out the ideas of human welfare and of "the dignity of man." Such objectives, outlined by the foundation in 1950, may have appeared at the time a bit utopian, achievable at best only in the dim and distant future. But in view of the shape of things in the underdeveloped countries in 1954, they have acquired a sense of reality and urgency. The survival of the underdeveloped countries along Western lines dictates the application of these objectives, for progress in democratic techniques and ideals is a basic prerequisite for economic assistance.

A Neglected Objective— Land Tenure Arrangements

It has been my observation in India that the foundation is attempting to carry out a program based on objectives described here. My familiarity with the work of the foundation in Burma, Pakistan, and the Middle East is meager and secondhand, but I am under the impression that, taken as a whole, the foundation aims to shape a many-sided plan for sound economic and social development. Yet, one important aspect has not been given the consideration it deserves. I have reference to the crux of the agrarian problem, namely, the land tenure arrangements and how the foundation can assist the peasant of the underdeveloped countries in securing the most democratic of assets: freedom to farm.

Four-fifths of the people who populate the underdeveloped areas are peasants. Agriculture, not industry, is the pivot of their lives in all its principal manifestations. Industry has made but a small dent in the character of Asia, notwithstanding the industrialization of Japan, the oil gushers of the Middle East, the tin mines of Malaya and Siam, and the jute and cotton mills of India. The factory may bring material advancement to the Asians some day, but that day is in the future. The heart of the problem of Asia today lies in the countryside. It is on the farm where solutions must be sought and found.

The threat inherent in the rural problem both to Asia and the rest of the world was well stated by Kipling when he penned the following lines describing the life of an Indian farmer:

> His speech is of mortgaged bedding,
> On his kine he borrows yet,
> At his heart is his daughter's wedding,
> In his eye foreknowledge of debt,
> He eats and hath indigestion,
> He toils and he may not stop;
> His life is a long-drawn question
> Between a crop and a crop.

Kipling was not one to neglect his poetic license, but his picture is not actually overdrawn in hopelessness. For in 1954 the Asian peasant is hardly any better off than in the late nineties when Kipling wrote.

Land ownership in most underdeveloped countries is the source of economic wealth, political power, and social prestige. Conversely, the millions upon millions of peasants who own no land enjoy none of those attributes. The relationship between the man who owns the land and the tenant (or land laborer) who works the land is too well known to require detailed description. Suffice it to say that in most cases rack-renting is the order of the day, the right to till the land is minimal or nonexistent, usury is prevalent, the incentive to increase agricultural production is lacking, and political discontent exploited by the Communists is inevitable and is actually taking place.

Some of this was epitomized in a personal observation in a Pakistani village in the Punjab. Most of the families who resided in the village were tenants of a single absentee landlord. Upon examination it turned out that yields had been declining. I asked the assembled villagers for the reason. "Our land," replied one of the tenants, "is like a pitcherful of water from which we keep pouring out while nothing is being poured in." "Why don't you pour something in," I asked. "How can we?" countered the tenant. "The landlord's share is so high that we would benefit very little." The tenant had no incentive to improve the land. The landlord received too much and the tenant too little.

The changes in land tenure arrangements will enhance the political power of the peasants and very possibly endow them with rights and responsibilities resembling those of the rural people in a democratic society. On the other hand, failure to come to grips with the problem is likely to advance the cause of Communism in the countryside. The significance of the latter development cannot be overestimated.

The Communists are exploiting this issue and they place it in the center of Asian politics where it belongs. The peasants living on the ragged edge of penury are easy marks for the Communists. The peasants know nothing and care less about "Marxism, Leninism, and Stalinism," and they are surely not eager for collectivization. The Communists, however, promise them not collectivization as it exists in the Soviet Union but land with which the peasants can do as they please. And the peasants, in sheer despair, believe the promises, not knowing that they will eventually be betrayed, their land nationalized, and they themselves herded into collective farms at the point of a bayonet.

The potency of the Communist battle cry of "Land and liberty" has not been lost on all the leaders of Asia. India's needs for technical farm improvements are overwhelming and Nehru, for example, is well aware of it. But he is just as conscious of the urgency for a concurrent land reform program which will give the Indian peasant an incentive to improvement and a sense of social responsibility. Nehru's motivation is economic, social, and political: "If we don't do it," I heard him say, "they will." And "they" are the Communists. He stressed, therefore, that the land tenure problem must be placed "in the center of the piece."

If this reform problem is dealt with and with reasonable success, it would become a powerful political instrument. The native governments friendly to the West would be more likely to win popular support, and popular support in Asia is peasant support or nothing. An owner cultivator or a reasonably satisfied tenant would acquire a stake in society. He would guard that society against extremism. Private property would be strengthened where it is weakest—at the huge base of the social pyramid. The chances are good that in that event the common man of Asia would become a staunch opponent of Communist economics and politics—not necessarily to favor the interests of the West but simply because his own interests would lie in the same direction. So much for the reasons why the issue is important and why a foundation concerned with human welfare should include it within the scope of its activities.

What Part Can and Should the Foundation Play?

The foundation is already engaged in rural improvement work through community development projects. Sounder land tenure arrangements, including sounder farm taxation and farm credit systems, would help to meet urgent needs without which a rise in agricultural production is but a half measure. How should the foundation treat such issues? The subsequent paragraphs relate to land tenure (with special reference to India) because, while credit and taxation are indeed important in the rejuvenation of the agricultural economy of Asia, the focal point of attack must begin with the land tenure systems.

Land reform, unlike technical agricultural improvements, is a highly controversial issue. The landlords are bound to resist it and are doing so in countries where such reforms are in the making. The class of people whose rights and authority the reforms seek to restrain is a class which has deep roots in society. It is the recruiting base for the intelligentsia and its members play an important role in the universities, government offices, press, and every important sector of public life. Foreign "interference" may add fuel to the already burning issue by antagonizing not only the conservatives but also the pro-reform supersensitive nationalists.

The above may be true (although not as a rule) when a foreign government espouses controversial reforms, and this would be doubly so if the implementation of such reforms were a condition of aid. The stigma of interference least applies to a private organization, and this is well-illustrated by the relationship between the government of India and the foundation. Moreover, the kind of contribution the foundation might be willing to make would preclude altogether the charge of interference. Significant in this connection is the fact that India has already a reform in being; regardless of the difficulties encountered since its initiation, it cannot be stopped and eventually will be carried out in its main outlines. What is in question is the slow pace of the movement and the one fundamental fault of the legislation. The foundation could provide a significant measure of assistance in both regards.

The role of the foundation in a land reform program cannot but be a modest one measured in financial terms. On the other hand, it can play an important part as a stimulator of the reform idea, of new ideas which have validity in national and local terms, and as a means for the creation of a friendly climate among important groups of people currently aloof from the reform endeavor. The foundation could not and should not be concerned with the payment for the landlords' land earmarked for distribution among tenants, nor need the foundation engage in the financing of the actual implementation of the program. The measures relating to these two points are solely the responsibility of the country in question, assuming, of course, that the country is willing to carry out a reform. Nor should the foundation take upon itself the writing of a land reform law. Such a move is politically dangerous and, besides, almost as a rule the Western expert does not know enough of Asian conditions to undertake the chore. Suggesting ideas, based on close field studies, is something else.

My experience in Asia leads me to believe that the reforms have been slow in coming, not because governments do not know how to formulate policies, administer, execute and evaluate them but primarily because they do not want reforms in the first place. When Sheikh Abdullah of Kashmir decided that it was time for the peasants to get the land, he did a good job with a limited staff, working with a peasantry which is 95 percent illiterate. When the governor of Formosa decided to put through the land reform program on the island—something the Nationalist government refused to do on the mainland of China—he found the means to do it. In India the issue is not one of reform or no reform; the issue is one of the slow pace and how to accelerate it, and of certain vital errors in the land reform provisions. The reform being a highly controversial subject, the problem is one of helping India create a more favorable reform climate, extending to groups whose support could make the difference between determined action on the part of the state governments and halfhearted effort, or none at all.

More specifically, one of the many regrettable aspects of the lagging reform movement in India is the failure to support the movement by individuals and groups most capable of performing that service. The Indian universities, for example, are very much on the sidelines, at best in the role of disinterested or impartial observers. They could be won over to the side of reform and at the same time become students of the problem. India still suffers from lack of information about the elemental yet basic features of postreform agricultural economy. This applies to prereform agriculture as well. Despite all the discussion about fixing ceilings on agricultural holdings as a part of future reform measures, no data are available on the size of landholdings and land ownership. The first census is only now in preparation. The effects of the reform legislation already enacted can be only surmised through personal observations because only a few of the twenty-eight states in the union of India are taking the trouble now of finding out what the effects are. To be effective, this work of many years' duration must be done in India and by Indian agricultural economists, sociologists, and political scientists. It is a job tailor-made for the universities.

The role of the foundation in such an undertaking is equally tailor-made. Its aid would avoid any stigma of imposing a reform upon India and yet stimulate interest in it where it is lacking and thus contribute to the implementation of the reform program as a whole. The financial assistance need not be overly large, depending upon the number of universities, the number of people willing to undertake such work, and the number of projects or studies involved. Significant in this scheme is the person representing the foundation and charged with the responsibility of promoting this task. What his qualifications should be are reviewed in the section on "Personnel and the Correct Approach." Suffice it to say here that he must have close familiarity with the problems about which he is concerned in a given country, that he must appreciate that his approach to the problem should be framed in terms of needs of the country under consideration, and that to establish these needs clearly he must understand the value of field investigations and consultation with those who "live with the problems" day in and day out. He must be able to distinguish between need and opportunity because, while important human needs are easily ascertainable, real opportunities to meet a need by the application of knowledge are not always clearly defined.

These are some of the principles one would expect a disseminator and stimulator of ideas to apply. The one essential not mentioned, but all-important, is that he and his ideas be accepted by those in a position to make policy decisions. This is not necessarily a test of his success or failure to advance the cause of the reforms, for conditions beyond his control may block his efforts. The fact remains, however, that whether or not a suggestion is accepted at a given moment, the consultant has rendered a service by providing something approximating a forum for the idea, thereby assisting, even if indirectly, in the ultimate solution of the problem.

In no two countries are the land reform problems alike, but the ideas as to how the foundation could stimulate the progress of a reform program—with variations—in most countries are closely related. India is far more advanced than Pakistan in this regard. In Pakistan the readily ascertainable basic facts are yet to be gathered, described, and analyzed; and the willingness of the government to launch a reform program is yet to be secured. But the approach of the foundation need not be basically different. In both instances the financial aid is but a reflection of the foundation's successful planting of ideas among the policymakers. My own experience in Nepal points to the same conclusion even though the idea of a land reform does not loom large in the minds of the country's leaders.

In summary, aid of the type discussed here and the type of person to help utilize it are fundamental prerequisites for the safeguarding of the sum total of the foundation's investments in effort, time, and money devoted for the promulgation of other reforms, particularly those related to community development projects. It is no secret that thoughtful Indians and Americans are disturbed by the fact that these projects and many of the other agricultural improvements are often built on the old land tenure arrangements. This is admittedly a weak base. It explains the disparaging remarks of some tenants about the community projects. On occasion I have listened to such remarks as: "What is there in this project for me?" The answer is obvious: rather little, if the lion's share of the increased output remains in the same hands. If that condition should remain unchanged, the efforts to give new life to India's agricultural economy would fall far short of the goals. That is what is meant by safeguarding the foundation investments, and that is why the role of the foundation, in enhancing the pace and where possibly improving the content of land reform programs, is of vital importance.

Current Status of Land Reform in India and Consultants' Immediate Tasks

Aside from the general task in this field that the foundation might undertake, there are some specific issues highly relevant to the future of the reform which call for a consultant's immediate attention.

The land reform in India is more than six years old; most of the legislation has been enacted, and certain of its important provisions have been or are being implemented. And yet, the Planning Commission, which gave the form and substance to the reform, has never quite decided what the principal aim of the reform should be or how far the reform should reach out in its effects upon the landlord groups.

The philosophic bases of the reform and the enormous body of legislation are replete with double-talk about land purchase schemes and ceilings on land ownership, goals which are hardly attainable and which at the same time tend to divert attention from the issues which truly matter. Stripped to their essentials, *the fundamental issue in India is not so much land ownership as security of tenure.* The cultivator does want a piece of land that he can call his own, but since he must pay for it even under the reform, for the vast majority this is unattainable. The sound alternative, therefore, is effective government action that would leave him on the land undisturbed, paying a reasonable rental and making the most of the residue. Currently, this is what the effort should be about, and it is in this connection that reform legislation needs drastic revision and propping up.

Since the land purchase provisions have hardly any practical meanings, the whole concept of the Congress Party slogan "Land to the tillers," is but a mere expression of intent of the reform legislation. Nor is the concept of rent reduction, one of the main pillars of the reform, especially strong. Rent reduction is meaningless without security of tenure, and the fact is that the security of tenure legislation is the Achilles heel of the entire tenancy reform. Such is the case because part and parcel of the tenure legislation is the right of the landlord to resume a fixed area of land for so-called self-cultivation by evicting the tenants. This provision has been made with the intention of striking a balance between the claims of the landlord and the tenant. Its net effect is the weakening rather than the strengthening of the security of tenure idea. In the Punjab, for example, the number of evicted tenants is probably as great as the number of those remaining on the land. The net result is a new type of tenant in India—the "evicted tenant." Resumption or permissible self-cultivation is a basic flaw of the land reform in India. The new self-cultivators work the land with newly hired hands or with the erstwhile tenants turned farm hands.

Another immediate issue, although not as important as that of resumption, is the proposal of ceilings on holdings as a means of distributing land among tenants. This idea stems from considerations of social justice, but there are other considerations which cannot be dismissed, such as the economic criterion for the fixation of ceilings, the economic efficiency of reduced holdings, and the matter of compensating the landlords. Perhaps the most important consideration is whether the fixation of ceilings would release an acreage large enough to satisfy a fairly large number of farmers hungering for land. The evidence is that little land would be made available by the imposition of ceilings.

Very germane to the entire land reform movement is the question of rural credit. Hardly any beneficiaries of the new land tenure arrangements could continue so for long without the aid of a rural credit system which is virtually nonexistent at the present time. The will to create one is as yet absent; and, until that comes about, the Indian farmer, whether old or new owner or tenant secured in the right to cultivate the land, is bound to have a hard row to hoe to maintain, let alone improve his status.

The Planning Commission is not unaware of these fundamental shortcomings of the land reform movement. The Congress Party, which staked its political life on "Land to the tillers" promises, is also beginning to recognize their urgency. The discussions of this observer with the members of the commission and a number of members of the Congress Party bear this out. But it will take much soul-searching and thorough appreciation of the issues at stake to make the decisions which cannot be delayed much longer. Advice from the outside has its place and will be accepted if based on knowledge and proffered in the spirit which induces Asians and Westerners to work together for a common goal.

Personnel and the Correct Approach

The assumption has been made that what happens in Asia in coming decades is as decisive for the future shape of the world as anything that happens in Western Europe. If this is correct, then the individuals concerned with the various aspects of development in Asia must shoulder great responsibilities.

All our past affinities have been with Europe by origin, culture, trade, and history. Americans emerging from classic isolationism only in the past few decades have come to place their relations with Europe in a new perspective. This has not been easy, but it has at least been made easier by the fact that we have been tied to Europe in one way or another by our entire past. Not so with Asia, and it explains in good measure our bewilderment over recent events there. In a real sense we are facing people we have not met before, and this necessitates a very special type of person to carry out certain tasks.

That he must have technical competence is assumed as a matter of course and needs no further affirmation. In the light of the goals our aid should pursue, he must have more than technical competence. The field-worker must be able to understand the position the people of a given country are in, to grasp something of their attitudes, their feelings, their state of mind, their view of events and of the world; he must have the ability to see local problems in their logical framework, and possess the willingness to work within this framework towards stated objectives. Working as he often must in a country formerly dominated by a Western power, he would do well to start with a look at its legacy, although, obviously, that power is not its sole creator. He must see what it is by way of economics, politics, and attitudes that these people bring with them onto the world stage at the present time. He must appreciate the fact that the economic legacy is often poverty, as illustrated by low per capita income, short life expectancy, low food intake, and a high degree of illiteracy. The psychological legacy is fear, suspicion, and hostility. It stems from the imposed military, economic, legal, and racial superiority of the ruling power and all that resulted in the imposition of enforced inferiority. The political legacy is authoritarianism, even when it is embellished with modern democratic forms. All of this and more are the content within which the worker must carry on his task, and to a large degree the success or failure of his labor will depend upon the extent to which he understands and accepts these fundamentals.

It has been my experience in Asia that much can be achieved by "indirection," as it were. Even if it were possible for a technically competent person from the West to write sound legislation relating to land tenure problems, farm credit, or farm taxation, there is no assurance that it will be accepted or, if accepted, enforced. It is easy to say: "I recommend this" or "I recommend that," but when dealing with proud, sensitive, and suspicious individuals, shot through with nationalistic proclivities, this is not a fruitful approach. If the disseminator of ideas in Asia is to be successful, he must first be *accepted* by the people who ultimately are likely to act upon them. And the acceptance in turn presupposes not only technical knowledge, but also a keen understanding of the propositions discussed earlier and the ability to articulate them. Having succeeded in that, the ideas relating to the specific subject, expressed orally or in writing and with a humility born of true knowledge and understanding, will find their mark. The ideas then are adopted by the recipient country as if they were of its own origin, and their ultimate application is insured. This is what is meant by assistance by "indirection." In a microcosm, this method serves to illustrate the thesis enunciated by the Ford Foundation, which states in part, that "The problems and opportunities of our time arise out of man's relation to man—rather than his relations to the physical world."

In "A Re-look at Point Four in Asia," Raymond T. Moyer argued that, although officials in underdeveloped countries are often not concerned with public welfare, this does not exclude successful attempts to achieve certain objectives if "the right kind of approach" is applied. He points out that "Governments, like people, are seldom wholly good or bad. Nearly every government has some responsible officials with far better intentions than would be suspected from the actions of their governments as a whole." It is the responsibility of the person in the field to seek them out and encourage, educate, and persuade them, and always with tact and humility.

The possession of the latter attitude is the most invaluable of assets. This is particularly true in Asia because "Asians," as Dr. Moyer correctly pointed out, "have had enough of rough-shod treading on their feelings by boasting Westerners, who go blithely ignoring the fact that in most cases they must learn before they can teach. Of such, Asians want no more. But Asians will welcome gladly those who come from Western lands with a reasonable sense of humility, prepared to work as associates in a task in which both believe."

Last, and surely not least, the proper person with the proper approach must understand the significance which Asians attach to personal relations. Admittedly, in all human relations there is no substitute for the personal element. This is even more so in Asia where most activities are regulated on the basis of the man-to-man relationship.

These are the principal requisites a person working in underdeveloped areas should possess when he embarks on the task of advancing their development.

Such, in the main, are my observations and the way I approach the problem I believe you have in mind. The picture drawn in the first part of the paper is of necessity a generalized one. I understand that each country presents a separate problem, requiring its own case analysis, study, and understanding. Consequently, a general type of program may have to be tailored to the peculiar requirements of separate areas. But the nature of the principal issues and the way of dealing with them have application in most underdeveloped countries. This is particularly true of the personnel most suitable for the task.

III. THE VIETNAM YEARS, 1955-1961

LADEJINSKY ARRIVED IN SAIGON in January 1955 and served there until the middle of 1961. From February 1956, when he terminated his service with the U.S. aid mission, he was in the direct employ of the Vietnamese government as personal adviser to President Diem. The general setting in which he was to work and how he came to serve in this seemingly unlikely place merit some description.

Ngo Dinh Diem had himself returned to Vietnam, initially as premier under Emperor Bao Dai, only six months before Ladejinsky's own arrival there. This was only months after the defeat of the French at Dien Bien Phu and shortly after the Geneva Conference. The country was war torn; French troops were still present in considerable numbers; the influence of the Viet Minh in the countryside was still very strong; and the authority of the new, weak, French-sponsored government in Saigon was flouted by the powerful Cao Dai, Hoa Hao, and Binh Xuyen sects. "The south had become a political jungle of warlords, sects, bandits, partisan troops, and secret societies."[1] This condition had caused the United States to post Colonel Lansdale to Saigon in June 1954 as head of a military advisory mission. Lansdale had gained fame as an antiguerrilla fighter in the Philippines not long before, where he helped Magsaysay subdue the Huk insurgents. There he had learned to appreciate the importance of winning over the peasantry, because it was they who determined whether the environment in which guerrillas had to operate would be friendly or hostile.

According to Delia and Ferdinand Kuhn, the American journalists who interviewed Colonel Lansdale shortly after his arrival in Saigon, he told them that what he needed most were a few Americans who could go into the villages and talk to the peasants in terms that would help win their confidence and support. The Kuhns, who had come to know and admire Ladejinsky during his Tokyo years, told Lansdale, "It sounds to us as though Wolf Ladejinsky is just the man you need." By August the U.S. embassy in Saigon was requesting the Department of State to send Ladejinsky out for at least a short period to assist in an urgently needed land reform effort. He was not made available at that time. But a few months later, when a major incident erupted over Secretary of Agriculture Benson's allegation that Ladejinsky was a "national security risk," the U.S. aid agency (then, the Foreign Operations Administration) promptly offered Ladejinsky a job as land reform adviser in Vietnam and he just as promptly accepted. Colonel Lansdale made sure to introduce him personally, on arrival, to President Diem and strongly advised the president to keep his door open to this man at all times.

1. Frances Fitzgerald, *Fire in the Lake* (Boston: Atlantic Monthly Press, 1972), p. 93.

Twenty-four of Ladejinsky's writings during these Vietnam years are known.[2] Fourteen of these are concerned with Vietnam and are concentrated heavily in 1955–56. Three of the five 1955 papers are devoted, typically, to field observations in central and southern Vietnam, in which Ladejinsky sought to learn at first hand the nature and dimensions of the problems involved in implementing successfully the rent reduction and security of tenure measures (ordinances 2 and 7) which had been promulgated in January and February of 1955. These masterly analyses of conditions in the Vietnamese countryside are bound to be of great value to future historians of the period. Other 1955 papers describe a meeting with a group of landlords and another with President Diem. In 1956 Ladejinsky's emphasis turned to the important Cai San resettlement project and to shaping the U.S. aid program along the lines he thought necessary. But after 1956 the only Vietnam papers available are retrospective in nature. I presume that Ladejinsky became heavily involved, operationally, in the implementation of ordinance 57, promulgated—surely on his advice—in October 1956, which introduced a large-scale land redistribution program in the heavily tenanted southern part of the country. This program, fully described in his 1961 paper, "Agrarian Reform in the Republic of Vietnam," bears many of the characteristic Ladejinsky hallmarks—not least, the painstaking determination of the acreage subject to redistribution, the determination of land purchase prices which would at once be fair to the owners and feasible for the tenant buyers, and the creation of provincial and local committees to supervise and administer the program. Because of delays in negotiating problems of payment for the French-owned lands which comprised a substantial part of the total to be redistributed, most of the land transfers were not completed until the end of 1960. Although such an operational preoccupation would suffice to explain the drying up of Ladejinsky's Vietnam-related papers during 1957–59, it seems a fair presumption that after some time in 1959 he became progressively more disillusioned with the evolution of events in Vietnam and with the prospects for his further useful service there.

Remaining papers after 1956 show his interests turning elsewhere. He comments on the agrarian revolution in Japan and in Asia to a Catholic conference in 1957; he examines in the same year Communist China's new push toward collectivization; and in 1959 he reviews independent Vietnam's first five years and does a final assessment of the agrarian revolution in Japan. But the 1960–61 papers are more forward looking and reflect Ladejinsky's preparations to leave Vietnam. One letter, discussing a prospective work relationship with the Ford Foundation, contains a revealing self-description and self-assessment. Others show him, on leave for short intervals from his Saigon job, exploring in Nepal a post the Ford Foundation has suggested to him there and conveying to the prime minister his misgivings about Nepal's new five-year plan. During the same period he is seen reporting on an interesting visit to Indonesia, responding to an enquiry from Japan, and reviewing Communist China's new thrust from agricultural collectives to communes. There is, finally, his considered assessment of agrarian reform in Vietnam, prepared for a conference at Michigan State University.

Of the twenty-four papers plus one fragment produced during the Vietnam years, fourteen are presented here. Of these, eight are concerned with Vietnam, two each with Nepal and Japan, and one with Indonesia. Also presented are his candidly engaging self-description and the important fragment on Asia.

2. See the Chronological Bibliography.

25. *Field Trip Observations in Central Vietnam*

Ladejinsky could not have been in Vietnam more than a few weeks before he set off, characteristically, on an exploratory trip into the countryside on March 7, 1955. Although warned by officials of the Ministry of Agriculture against such a trip, unless accompanied by a large party, because the countryside was "insecure," he preferred to go out with an interpreter only. He found, among other things, that the government's new rent reduction program lacked appeal to the farmers because the Viet Minh had instituted even lower rent ceilings and the tenants were not paying any rents at all to their landlords; some landlords welcomed the reform because for them it meant at least some rent rather than none at all; the tenants had enjoyed more status and power in their villages under the Viet Minh; and, because of small-sized farms and poor soil conditions, land reform in central Vietnam "must give way to 'agrarian reform,' a term which stands for the improvement of all economic and social institutions connected with farm life." Moreover, Ladejinsky compared the political atmosphere then prevailing with that under the Viet Minh, and the behavior of the national army he compared with that of the Viet Minh soldiers—in both cases a comparison highly favorable to the Viet Minh. Upon his return from this field trip, Ladejinsky concludes, he "had the privilege of discussing his impressions with President Diem. The principal points at issue, including the role the president himself might play in dealing with them, have been conveyed to him." This last could scarcely have found a sympathetic reading at the U.S. Embassy. But this would not serve to explain why Ladejinsky's report, dated April 2, should not have been transmitted to Washington until May 23 (dispatch 410).

Ladejinsky noted that "The writer of this report appreciates the danger of generalizations based on a small sample. And the sample was small. Future events may prove his fears groundless and suggestions uncalled for; but, be that as it may, the observations reflect existing conditions."

Introduction

ON MARCH 7 THE WRITER of these notes started on his first field trip in Vietnam, a trip lasting but four days. It was intended, in part, as a "feeler" to test the reports in Saigon on the difficulties of travel in the countryside. The officials of the Ministry of Agriculture, for example, took a dim view of the undertaking when first approached on the subject. If their advice had been followed, it would have meant travel in the company of the minister of agriculture, local bigwigs (both civilian and military), and all the fanfare attached to such a "safari." Security, presumably, was the excuse for their concern. Fortunately, wiser counsel prevailed and the assistance of the ministry was dispensed with. Instead we were accompanied by the English-speaking Do Trong Chu of the refugee commissioner's office and later joined by Tran Duc Nhuan of the USOM [U.S. Operations Mission] regional office in Hue, who proved himself to be a young man with superior knowledge of central Vietnam.

The purpose of the trip was to see something of the lay of the land of Vietnam, to note the character of the village, and to get firsthand impressions of the attitude of the farmers toward the land-rent control program about to commence. However, in the course of the trip, other problems pressed themselves upon the visitors. They are part and parcel of the economic and political conditions of the countryside.

The party traveled north and east of Hue, then south of Tourane and from Tourane further south to the districts of Faifoo and Tam Ky. We visited four villages, of which two were under the Viet Minh for a number of years; three district offices; and paid calls on two chiefs of provinces. In the light of the experience gained on this trip, it appears that fears expressed by Saigon officialdom were grossly exaggerated. But it must be admitted that the time has not yet come when a curious foreigner, no matter how well-intentioned, can drive into a village in search of information and gain the immediate cooperation of the farmers. This was particularly true of the former Viet Minh villages, where people look at you with suspicion and say very, very little. Under these circumstances the assistance of the district officer is invaluable. His aid smooths the rough road toward any valid impression of past and present conditions in a given village. Talks with district officials (as with the chiefs of provinces) preceding the trips into the villages were in themselves invaluable. Officials gave the impression of close familiarity with local conditions, some of them having lived under the Viet Minh for years. Without their aid, the value of this trip would have been limited indeed.

The subsequent paragraphs do not claim to be more than they are—firsthand impressions of rural conditions in a particular locality, hastily gathered, and based on facts and figures which leave very much to be desired. The villages have no written records to speak of, especially those villages formerly under the Viet Minh. Land and tax records and district registers were hidden or taken away by the Viet Minh when they left the district. The district offices do have certain overall figures, but they must be treated with extreme caution. The main ingredient of village data is the old-timers' ability to recall fair approximations of population and acreage, livestock numbers, current prices, and so forth. The density of population on the land is in itself an important clue to basic conditions in a village. The important data denoting the economic and social structure of the community are a more difficult matter. Probing for such information causes a great deal of headscratching, and answers finally elicited are no more than approximations at best.

Another factor worth noting is that the area under consideration is not necessarily typical of other parts of Vietnam. It does not call for profound knowledge of the country to realize that both regional differences and differences within regions are very great in Vietnam. The existence of public or communal land and its periodic subdivision among all the farmers are characteristic of certain parts of central Vietnam, but this practice is not universal even there, much less in the rest of the country. Equally important is the fact that while landlordism—in the commonly accepted meaning of the term—is widespread in southern Vietnam, it is not, as will be pointed out elsewhere, an important feature in central Vietnam.

A jeep ride from Hue to Tourane in any direction, over a long or short distance, serves as a good briefing about the war-torn country. Barbed wire, perhaps more than any other feature, dominates the scene. Protective barbed wire obstacles of all forms and shapes, encircling watchtowers, shot-up bunkers, and troop installations dot the main line of communications as far as the eye can see. Some of the watchtowers and bunkers seem to be deserted now, but they served their purpose in the past. Their past utility was brought home to us by a district officer as we watched the plain from the top of a bunker. "Before Geneva," he remarked, "this outpost helped us to secure the countryside in the daytime, while the Viet Minh controlled it in the nighttime."

The farmers, whether pro- or anti-Viet Minh, were caught in the middle, for they had two sets of forces to contend with. "Whenever I think of the road leading into our village," a farmer remarked, "I think of the two masters—the Viet Minh by night and the French by day. During the night we had to dig it up and during the day we had to fill it in with dirt and debris. Digging out for the Viet Minh and filling in for the French." But not everything could be fixed by day which was destroyed under the cover of darkness.

When one travels through the country and

listens to accounts of the recent past, the illusion of "security by day" becomes quite obvious. One could not help but face blown-up bridges; torn-up roads, particularly in the areas controlled or contested by the Viet Minh; and irrigation canals and dikes in disrepair. Everywhere we heard of the sharp decline in acreage and yields, and concern about the shape of things to come. As against the effects of recent events, we encountered much evidence of roads and bridges being rebuilt or new community houses replacing the ones burned down by the Viet Minh. We saw the teeming market towns and hosts of farmers streaming down from their hamlets, with lively waddling steps, carrying over their shoulders baskets of foodstuffs to the market. These signs of quickened activity were heartening to behold; but, on balance and from the point of view of the region's farm economy and the struggle against the political legacy of the Viet Minh, the aftermath of the years of strife is very much in evidence and the effort to cope with it is only beginning to take shape.

The Village and the People

What is the village like and what are some of its problems?

A fundamental one, the small holding, was apparent even before we landed in Hue. From the plane we observed a land configuration strikingly similar to that of Japan, Korea, or Formosa—small fields fitted together like the pieces of a jigsaw puzzle. What was not familiar were the pockmarks with which some of the fields were crowded; upon closer examination they proved to be graves. Talks with farmers and local officials readily revealed that this section of central Vietnam, and indeed all of central Vietnam, is characterized by that well-known feature that plagues so many other countries of Asia and the Middle East—too many people on too little land. Holdings of one hectare, a half hectare, or even less, broken up in a number of pieces and scattered in all directions, are the prevailing pattern.

Thua Thien province, for instance, has an estimated 40,000 farm families, subsisting before the war on an estimated 33,000 hectares of cultivated land. If it is true that the prewar area has shrunk to but 15,000 hectares due to war dislocations, the average size of a holding is small even by the crowded Asian standards. Conditions created by the war aside, the normal pattern of a single hectare or less is amply supported by evidence gathered in the villages. La Chu village, for example, is made up of 350 farm families who divide among themselves a total of 200 hectares, of which 150 hectares are cultivated. In the village of Su Lo Dong, 400 families subsist on a total of 220 hectares, more than half of which is rice land.

While the cultivable acreage in this section of the country has never been large, it was reduced materially during the war years. The deterioration of irrigation facilities and of dikes and dams are cited as the primary cause. In the district of Phu Loc, 40 percent of the land (1,800 hectares) is out of cultivation because salt water has seeped in through broken dams. The decline in the buffalo population is another important factor leading to the same result. By the same token, it is a significant index to the worsened economic condition of the farm community.

For many a farmer, a buffalo is the most important source of wealth, its value being estimated from $100 to $200, depending upon the quality of the animal. For a tenant farmer, his buffalo is the principal tangible asset. His farm equipment doesn't cut much of a figure. The wood plow with the iron tip, the wooden harrow, the wooden roller, the brush-cutting knife, the hoe, the scythe and the threshing basket—all these call for an investment of no more than $25. And the information points to the fact that the buffalo population is roughly one-half to one-third of what it was before the civil war began in 1946. A decline in cultivated acreage was inevitable and, along with it, a decline in the "wealth" of the community.

The land of any given village is not all privately owned. Much of it is community-owned land. As in other instances involving quantitative delineations, the proportions of community and private land are not easily determined. If the guesstimates are taken at their face value, the percentage of community land can be as low as 20 and as high as 80. Modern developments, mainly those tending to disrupt the old village pattern, account for the differences. Community land is redivided every three years and, theoretically, the standard of distribution depends upon the number and ages of the male members of a family. Every male between 18 and 60 is entitled to one unit of land, and males above 60 to one-half of a unit. In reality, there have been many instances when the village notables (a term applied to the appointed leaders) charged with the task operated on the theory that "To him who has, it shall be given." As a result, some farmers have hardly any land at all and rent private or community land from those who control it. The one restriction on community land is that it cannot be sold.

The economic well-being of the village revolves not only around the land it possesses as a unit but also around the relationship of the cultivators to the land. But information on this crucial point is hard to get. Neither district officers nor villagers will shed much light on the subject. When a district officer was pressed for information, he finally resorted to his "Bible," a dog-eared statistical compilation published in 1931, with figures for the year 1929. The villages we visited had no recorded data, modern or ancient. Nevertheless, a farmer would shift his weight from foot to foot, think hard, and produce a figure. It was on this basis that we were left with the impression that approximately one-third of the farmers are tenants working somebody else's land.

Big landlords are a rarity in central Vietnam. The biggest we encountered was a chief of a province who owned 40 hectares rented out to eighty tenants. The run-of-the-mill landlord is often one by courtesy. Two to 3 hectares will put him in that category, while men with 10 hectares are indeed substantial owners. Rentals, in normal times are approximately 50 percent of the main crop if the landlord furnished seed and fertilizer, or 25 to 30 percent of the crop if these items were furnished by the tenants.

The area here considered is not a typical landlord-tenant area; yet, in view of the fairly large number of tenants, one of the questions raised was whether or not a tenant is ever in a position to ascend the higher rungs of the agricultural ladder. The answer was invariably in the negative. The very question seemed to be out of order. And indeed it would be surprising, considering prevailing land prices on the one hand and income on the other, for a tenant to acquire a piece of land on his own. Farmers assured us that there were virtually no land transactions of any kind. Clearly, the uncertainties of war and the Viet Minh position with respect to the land question, about which more later, were in themselves powerful factors discouraging transfer of land. It is all the more surprising, therefore, that land prices are very high, judging by such quotations as 100,000 to 140,000 piastres per hectare or roughly $600 to $800.[1] When the farmers are asked "Why so high?" the reply, as one put it, was that "a box of matches now costs 2 piastres, whereas before the war the price was only one centime." But whatever the cause of high land values, it is quite clear that in the decade since the end of the Second World War no tenant could possibly accumulate enough to buy even a portion of a hectare. He could not do that not only because he works a small holding but also because the yield of his main crop is only one ton per hectare, or approximately one-third the average of Japanese rice yields. Moreover, because they work such small holdings and the yields are so low, tenants as well as small farm owners produce only enough rice to last them from four to six months. They buy whatever additional rice they can through the sale of other products on the local market, but a great many of them cannot get their fill of the foodstuff they want most.

Taxation is a burden about which farmers the world over complain. It was a novel experience, therefore, to find that the farmers we talked to felt taxation was no problem. The reason is simple: While the farmers pay indirect (sales) taxes or stamp charges when they file or register an occasional document, in 1955 they will pay no direct taxes. Even the land tax, which is only 150 to 200 piastres per hectare, will not be collected this year. The entire prewar taxation system fell into disuse in the past decade or so, and in Viet Minh–occupied areas the new taxation practices disappeared with the departure of the Viet Minh. However, the Viet Minh did succeed in collecting taxes in 1954; hence the decision of the national government not to impose any taxes in 1955 on the ground that the farmers should not be taxed twice.

1. Piastres converted into dollars at the unofficial rate of exchange.

The same practice will apply in non-Viet Minh–occupied villages. The restoration of the taxation system will come with the return of normalcy, but for the time being this is one item which doesn't seem to trouble the farmers.

Farm indebtedness, too, is evidently no great problem at the moment. We were not prepared for this situation because traditionally the farmers carried a heavy burden of indebtedness. The whole population, a noted scholar once remarked, is caught in a "tightly drawn network of loans and debts." This statement referred to Indo-China in general and central Vietnam was no exception. And one of the classic reasons for this prewar state of affairs was: "The people of Annam (central Vietnam) would try to borrow up to the extreme limit of their credit." How is it then that the question of indebtedness is no great issue now? The answer lies not in the prosperity of the village but in the virtual drying up of sources of credit. In saying this, no attempt is made to equate availability of credit with heavy and burdensome indebtedness and thus argue against the creation of sound credit facilities; the fact is that it is not indebtedness that is the evil but excessive indebtedness for unproductive purposes.

Whether the farmers employ credit wisely or whether the existence of credit creates burdensome debt were not the points discussed with the farmers. What we were trying to ascertain was whether they needed credit and, if so, how they could secure it. Opinion is unanimous on the critical need for: (a) short-term credit during the growing, harvesting, and marketing periods; (b) intermediate-term credit for investment in livestock; and (c) long-term credit for a variety of land improvement activities undertaken individually or collectively. For most farmers, however, especially the small marginal cultivators, securing a loan is well-nigh impossible under present conditions. An occasional loan is still obtainable in the form of rice, secured three to four months before the harvest at the rate of 130 kilograms for every 100 kilograms borrowed.

In prewar days the moneylender was the principal source of credit. But, if the farmers are to be believed, he has practically disappeared in the turbulence of civil war. Also, the few well-to-do have left the villages for the cities for security reasons, and the net result is a vacuum which no one has yet begun to fill. Occasionally a farmer does buy a buffalo or manages to obtain a loan in an emergency; in such cases the farmer must rely on family connections to tide himself over. The poverty of the average farmer is an additional handicap in securing a loan for the simple reason that "no one but a fool or philanthropist will lend to a pauper." In the villages we visited there are probably some fools, but evidently not monied ones, and surely no philanthropists.

A striking feature of the trip was the market centers. They convey the impression of a bustling money economy; but as far as the majority of the farmers are concerned, the impression is more apparent than real. Little money changes hands except when a buffalo is sold or at the cock fight on the fairgrounds of the city of Tam Ky where a relatively small but obviously dedicated circle of spectators flashed wads of bills as tokens of their confidence in the fighting spirit of this cock or the other. But as we watched the average farmer selling fruit, vegetables, eggs, chickens, or an occasional piglet and exchanging the proceeds for other necessities, it was obvious he would return to his village with little cash, if any. A visit to a farmstead completes the picture. Exceptions notwithstanding, in most cases the familiar hut made of local grass and brush is bare of worldly goods. A rough-hewn bedstead is the most likely piece of furniture, while a bench, a few earthen pots, a change of clothing and a brood of half-naked children complete the picture. "Porgy and Bess" could provide the theme song for many of the farmers of central Vietnam—"I've got plenty of nothin' and nothin's plenty for me." It doesn't call for an elaborate study of the standard of living to say with fair certainty that the Vietnamese standard is certainly way below that of the Japanese farmer and below that of the Korean, Formosan, Burmese, and Filipino farmer. They come closest to the level of certain groups of Indian farmers.

Free Vietnam is ablaze with slogans, and one of the most commonly encountered is that "Literate people make a powerful nation." This emphasis, if backed by deeds, is well-taken, for the educational facilities of rural central Vietnam are conspicuous by their absence. In this respect, the area does not stand alone; other

Asian and Middle Eastern farmers share the same problem with their Vietnamese counterparts. Yet in the light of the drive for elementary education in many Asian countries, it seems that the region we traversed is remarkable for its lack of educational facilities.

If we are to believe the notables of La Chu village, there is only one school to every forty villages of the district! It is not so bad in other districts where the proportion of schools to villages was variously estimated at one to five, one to ten, and one to fifteen. Nobody voiced any objection to school or schooling, and one of the elders gravely observed that "the fight against illiteracy is as important as the fight against hunger." Yet, even in the relatively well-off village mentioned above (nobody there, the notables assured us, is either too rich or too poor), there did not seem to be any interest in spending any money on teaching the three R's to the young. The school we saw in the city of Tam Ky was a very simple affair: a blackboard, a few benches, and a few tables, all open to the weather, since the thatched roof was supported only by four corner posts. The arrangement did not appear to be an expensive one even for the poor villages of central Vietnam. The real reason for lack of schools, one suspects, lies in the inertia of the farmers themselves and of their leaders.

It is worth emphasizing in this connection that farmers and officials alike have words of praise for the efforts of the Viet Minh. According to their accounts, the Viet Minh was the most active in creating school facilities and finding local talent to spread literacy. This is reflected also in official statements. A Vietnamese document "Concerning the Economic, Political and Cultural Problems of Tam Ky and Request for Aid" has this to say: "Tam Ky is a large district with a population of 300,000. Before 1945 there were four primary schools (one school for boys and for girls at Tam Ky, one for boys at An-Tan, and one for boys at Chien-Dan). Under the Viet Minh regime there were three secondary public schools, two private secondary schools and 100 private and public primary schools. The people need more and more education, and their standard of education must become higher and higher. If we cannot satisfy the people in this matter, it will result in a political defeat for the National Government because people are comparing the educational activities of the Viet Minh with that of the National Government." Our informants bemoaned the fact that, with the departure of the Viet Minh, educational activities slackened. It appears, nevertheless, that neither the local administration nor private citizens has seen any need to fill the breach.

Such, in the main, were our hasty impressions of this particular rural section of central Vietnam. Land holdings are small and not too fertile, and the people cultivating them eke out a poor living. They are faced with all the problems associated with such conditions. If these chronic rather than temporary conditions were bad before the war, they are unquestionably worse now. This is true also of the areas which were never physically occupied by the Viet Minh. The Viet Minh affected the countryside as a whole, although their influence was felt to a greater degree in the areas under their occupation. In the latter, the small number of well-off farmers have gone through a leveling process which placed them alongside the mass of the farmers. Mr. Chao, deputy chief of the district of Tam Ky, formerly a Viet Minh stronghold, was probably right when he observed: "In our villages there are no more rich and poor people; there are only equally poor people."

The economics of equalization downward is quite obvious and needs no comment. Less obvious but nonetheless significant are the effects of the "Vietminhization" process on the multitude of farmers and their attitude toward the national government and its activities. The land-rent control program initiated by the national government is, at least in part, a case in point. What the Viet Minh has done in this field has important bearing upon the program considered in the subsequent paragraphs.

The Land Reform Program

In January and February 1955, President Diem issued a number of ordinances inaugurating the land-rent control program in Vietnam, commonly referred to as the "land reform program." Its principal provisions are: (1) Rentals from 15 to 25 percent of the major crop; owners of the poorer land are to be compensated at the former rate while owners of good land are to be compensated at the latter rate. (2) The tenants must pay additional charges for the use of the landlord's work animal (not to exceed 12 percent of the crop) and the actual cost—plus 12 percent annual interest rate—of seed and fertilizer, if such are furnished by the owner. (3) The rental terms will be specified in a written contract, covering a period of three to five years. (4) The implementation of the program is to be carried out by elected committees of tenants and landowners in each of Vietnam's 350 cantons, 102 districts, and 34 provinces. (5) Resettlement of farmers on private lands abandoned by the landlords or on government-owned land. Priority rights are given to refugees, former tenants, heirs of war heroes, war veterans, and nonfarmers who wish to take up that occupation. Considering the exorbitant prewar rentals, the implementation of the program should benefit the tenants and provide them with a considerable measure of relief from arbitrary actions on the part of the landlords as well as a degree of security of tenure.

What are the realities in the light of this reasonable assumption? How important is the program to the farmers? How significant is the program politically? Is the program being implemented? These were the questions addressed to landlords, tenants, and local officials.

It is too early to give definitive answers. The fears and hopes engendered by the recent past have so proliferated in the minds of the tenants that they are hardly in a position now to see the reform in clear-cut terms. Yet a preliminary assessment may be made subject to correction implicit in the limitation just noted. Weighing the pros and cons, it appears to this observer that the tenants do not look upon the program as one of overriding importance. The economic advantages seem to them less impressive than one might have expected at first glance. The political aspects do not concern them, although the officials believe that politically the national government stands to gain from the application of the land reform ordinances. The position of the landlords is one of favoring the reform on both grounds; if the ordinances are enforced, they will naturally benefit from them. They collected next to no rent under the Viet Minh and the chance of actually receiving rent, albeit a reduced one, is indeed welcome.

The program was inaugurated in February and by early March the farmers had only a hazy idea what it was about. Few farmers could state the most essential provisions of the program. Some had vague notions—and no more. If they knew more about it, they probably would not overlook a weakness of the program—the relative ease with which the landlord could recover the land after the expiration of the contract. On at least two occasions farmers assumed that the program dealt with land ownership and land distribution, which, of course, it does not. The printed posters with the brief outlines of the rent control scheme had yet not reached the villages. District officials, on the other hand, were familiar with the law's provisions; but, being few in number, they had no time to spread the word or to explain and impress upon the farmers the meaning of the reform. To correct this, the chiefs of provinces are beginning to select groups of young people for future work on the local level. The time element and shortage of personnel, rather than deliberate lack of interest on the part of the officials, were partial causes for the lack of activity and the impression that these measures are not of very great significance. Time may correct these shortcomings which, incidentally, have also shown themselves in countries with much better administrative organizations, where there had been a more thorough preparation for tasks of a somewhat similar nature.

More disturbing is the lukewarm attitude toward the program on the part of those on whose behalf it was initiated and who do know that rent reduction is its essential part. When a farmer is asked to state the needs which truly matter to him, rent control is virtually ignored as if the rent problem did not exist. The number one item the farmers (and the officials) invariably mention is better irrigation facilities, followed by requests for aid in reconstructing dikes and dams and in purchasing water pumps. Fertilizer and credit are next on the priority list. It has usually been necessary for the inquiring observer himself to raise the land rent question in order to solicit answers. Hardly ever did the farmers react to the reform as if it were a "felt need."

Why this appearance of unconcern, a position sharply in contrast with that of other farmers in Asia when the land question is under consideration? The answer to this question appears to be twofold. The first reason—one of relatively small importance—is the type of landlordism prevailing in that section of the country. The second explanation, and by far the more important one, lies in the consequences of the land reform and political activities of the Viet Minh.

As to first—and at the risk of repetition—it is well to stress that in this region landlordism of a kind that gives rise to great political upheavals is absent. Excessive concentration of land ownership is unknown, and absentee landlordism is not a common feature. Landlords' holdings are too small to place them in a category of exploiters living in ease at the expense of the tenants. Even in appearance they can hardly be distinguished from other farmers working in the field or taking part in village affairs. To this should be added that much of the land is community owned—not privately owned. A landlord can rent out the land, but he does not own it and his instinct of unbridled proprietorship of land is somewhat watered down. Under the circumstances, the average landlord in this part of central Vietnam is far from the classic Asian landlord symbol wielding unlimited economic and political power in his community.

The tenant's attitude is but a reflection of the above conditions. Community-owned land provides him with a holding, even if it is only one-tenth of a hectare. Since the total cultivated acreage of a village is very small, the additional land he rents is also small. As a renter, he has grievances against the landlord; but they are tempered by the fact that he deals with a man who may be only one step removed from his condition. The tenant will surely be helped by reduction in rent; such gains, however, depend not only upon the newly fixed rentals but also upon the size of the holding he rents. Since that is notoriously small, from an economic point of view the gain cannot be very great. It is not unnatural, therefore, that to some tenants the importance of measures resulting in increased production on his small holding seems more meaningful than rent reduction. For the same reason, they would point to southern Vietnam where land is more plentiful as the place where a rent reduction program should have much wider scope.

The above is only one minor aspect of the problem. The land-rent reduction need not have great urgency in order to be received very warmly by the tenants of this part of central Vietnam. Every additional measure of rice is a gain; every deficit rice producer is cognizant of that—and they are all deficit producers. The ordinances would have ushered in a highly significant event but for the fact that the Viet Minh had taken the edge off this greater expectation by the implementation of "reforms" of their own throughout most of this area. This is probably the most telling reason why the idea and content of the national government reform measures are not new and, above all, lack the impact they would have had in central Vietnam, even though land reform there is not a do-or-die issue. A tenant's list of priorities would have been quite different if the present reduction scheme had immediately supplanted the prewar arrangement whereby the rent charges amounted to about half of the crop.

Landlords, tenants, and officials are well aware of the Viet Minh's agrarian policies. While they cannot describe the changes and twists of the Viet Minh's agrarian policies covering nearly a decade, they have had a great deal of experience with Viet Minh rent reduction schemes, outright confiscation of land belonging to "traitors" and to absentee landlords and the like. Thus, the Viet Minh reform became a household word in occupied and nonoccupied areas. Different groups were differently affected by their measures, but the attitude was not formed solely by economic considerations. This is an important point in any attempt to evaluate the position of the tenants and their halfhearted acceptance of President Diem's measures. It explains also their reluctance to talk about their experience under the Viet Minh.

The landlords were willing to share their experience. The big landlord of 40 hectares lived under Viet Minh for nine years and now is chief of one of the provinces. As a resident landlord, he was never dispossessed of his land. Soon enough the land became a burden he wished to shed, but the Viet Minh would not permit it. They preferred him (and other resident landlords) to continue renting out land to tenants, but, of course, on the Viet Minh's own terms. These terms, according to his account, meant a rental of 15 percent of the main crop, out of which he had to pay a tax in kind amounting to 90 percent of the rental. For all practical purposes, he received almost no rent at all and was hardly better off than the absentee landlord whose land was confiscated and distributed among the tenants. His continued titular ownership was a source of unrelieved hardship to the landlord, but it served the Viet Minh well: he was maintained as a handy political target and a source of the economic wherewithall to help keep the Viet Minh going.

The tenant's testimony is less explicit and more contradictory. He admits to the burdensome Viet Minh terms under which he worked the land, but he is not often critical of his former masters. When the information is pieced together, it would appear that tenants of resident landlords paid rents ranging from 15 to 20 percent of the crop, while tenants who received plots of confiscated land paid no rent at all to the Viet Minh. In both cases, however, the Viet Minh imposed taxes in kind. The size of the tax burden was difficult to determine, for no two accounts agreed. Some tenants stated that it was as high as 50 percent of their output, while others said it was no more than 10 percent, depending upon the family's size. There were occasional extra "contributions" when the Viet Minh was especially hard pressed for grain. Taken altogether, the inescapable impression is that the Viet Minh exacted a heavy price for the privilege of owning a piece of land or renting land from virtually dispossessed resident landlords. It is the more surprising, therefore, to note the following comments by the anti-Viet Minh chief of a district: "The poor farmer, although paying 50 percent of the crop as a land tax, is much better off than the landlord. Thus the poor farmer has benefited from the Viet Minh regime. Poor farmers support the Viet Minh regime because its army is made up of poor farmers. Even if the farmers are no better off economically, they are much better off politically. They are the power in the village." A variation on the same theme is contained in another comment picked at random from the observer's notebook: "Under the Viet Minh regime the poor farmers are in no better economic condition, but they are happy because they are more important in the village, while the landlords are becoming poorer and have lost their former prestige and importance."

These and similar statements emanate from the local officials as they attempt to describe the effects of the Viet Minh upon the farmers. Farmers who have been through the Viet Minh school tend to keep to themselves their innermost thoughts on Viet Minh practices. The suggestion that the fear of possible reprisal from Viet Minh agents dictates this position is only a partial explanation, according to the local officials. There are deeper causes.

The assumption that a fuller or full stomach is the sole mainspring of a poor farmer's attitude towards a political entity is of doubtful validity. The Communists, who glory in the principle of economic determinism, have enough appreciation of realities not to overlook the fact that not all realities are economic; that noneconomic, psychological, and political factors may in the final analysis motivate man's behavior. It has been correctly said that "Where political psychological motivations come into conflict with desires for better economic well-being, the priority of the political is generally rather clear." More specifically, ideas of equality, respect, and status in the community weigh heavily in the scales of the awakened nationalism which has been exploited by the Viet Minh over the years. Illiteracy is no barrier to an emotional acceptance of these ideas. The fact that these concepts are only tactics to be discarded by the Communists at a later date either does not occur to the farmer or is beyond his comprehension. But the assiduous hawking of such palatable notions has created a "Viet Minh legacy" now shared, according to estimates of local officials, by approximately 50 percent of the farmers of central Vietnam. This may, in large measure, explain the seeming paradox of farmers who view with skepticism the economic benefits offered them by President Diem's reform shortly after they have escaped the economic burdens imposed upon them by the Viet Minh.

Even if these observations are perfectly valid, it would still be premature to conclude, at this early stage, that the implementation of the program is doomed to failure. It is possible that the anti-Viet Minh feeling among the landlords, owner cultivators, and a segment of the tenants may yet play an important role in shifting the balance toward acceptance of President Diem's reform by the majority of tenants. It would be a mistake, however, to assume this would take place without strong prodding. The program is getting off to a slow start, while the Viet Minh agents and Viet Minh ideas continue to exert considerable influence in the countryside. Speeding up the program calls for an enormous amount of effort on the part of the national and local governments. This work must be essentially of a political nature, if for no other reason than the standoffish position of so many of the tenants is essentially political, too.

Promulgation of the land reform ordinances by the national government was in itself a political action. Local officials are agreed that this action will counter to some degree the claims of the Viet Minh that they alone are the spokesmen of the poor farmers. But the Viet Minh hold the advantages of having been the first in the field. To overcome this handicap, the national government must move beyond the mere promulgation of the reform measures. There is no evidence that the government had wholeheartedly thrown itself into a continuing campaign of demonstrating its abiding interest in this program. Apathy is the dominant note, and the best explanation that can be offered is that that attitude is an expression of the political vacuum so characteristic of the countryside. The current political difficulties and the concentration of all top executive effort on immediate, urgent problems is understandable. On the other hand, this limited land reform program is also a political matter not unrelated to "pure" political issues.

These observations are not aimed at an assessment of the rent reduction scheme as such, yet it may not be out of place to note the program's limitations. One may venture to say also that the same criticism would hold true even if the Viet Minh had not staked their claim to being the original land reformers in Indo-China and even if the scope of the national government's measures went so far as to make land ownership and land distribution its primary aim.

The statement is based on the very strong impression that in the section of central Vietnam under review one can see an excellent example of the crude oversimplification which presents "land reform" as a cure-all. The Communists have given currency to this particular idea, and the idea has been uncritically accepted in non-Communist circles chiefly because it offers a simple solution easily propagandized. The small size of holdings and relatively poor soil conditions are in themselves warnings against the all-embracing proposition that all benefits flow from ownership or reasonable rental terms. Such measures, no matter how liberally drawn, can neither increase the size of the holding nor materially raise the productivity of the soil. Yet these are fundamental. In order to overcome these drawbacks, "land reform" in central Vietnam (as in other regions and other countries) must give way to "agrarian reform," a term which stands for the improvement of all economic and social institutions connected with farm life.

Rural Japan is case in point. The Japanese land distribution program has been quite successful, not merely because the tenants became owners of the land but chiefly because such ownership has been accompanied by a strong farm cooperative movement, a sound credit system, and a network of highly developed agricultural policies (particularly price policies) designed to protect the agricultural economy of the country. The net result of the incentive offered by ownership plus the means of maintaining the productivity of the land at high levels have made the land reform meaningful economically, politically, and socially.

Under the prevailing farm conditions in central Vietnam, a farmer cannot expand the size of his holding through increased agricultural production. This is not an argument against land reform of a limited or wider scope; it merely stresses the fact that to this important precondition must be added the other means for bringing about more efficient use of the land and of human resources. The marginal farmers' pleas for irrigation facilities and credit are an extension of the idea that independent, self-reliant producers and citizens are not created simply by shifting the pattern of ownership. Unless the much improved land-rent situation or even outright ownership go hand in hand with all the other necessary improvements, the farmers may not be much better off than they were in prereform days. In the Vietnam of today, this idea may be easily dismissed on the ground of more urgent priorities, but only at the risk of a doleful shape of things to come in the country's economy and politics. It would be wiser for the national government to consider this problem sometime soon before it is too late.

Political Note

Land reform and farm conditions are political issues. The Russian and Chinese Communists have succeeded in projecting the land problem into the very center of Asian politics. The Communists have been able to capitalize on the landlord-tenant strife with startling success by posing as advocates of reforms for the benefit of the peasantry. The Viet Minh followed in the footsteps of their masters, successfully preaching the same gospel. The accounts of officials and landlords may have exaggerated the Viet Minh's hold on the people, perhaps, in part, to underscore the magnitude of the task facing the new administrators. However, even when this factor is taken into consideration, the persistence of ideas implanted by the Viet Minh cannot be denied.

It has been suggested that noneconomic ideas were largely responsible for the Viet Minh's strength in the community. Their method of "selling" their wares was one of a tremendous and unceasing political activity which kept the countryside seething with excitement; even anti-Viet Minh officials and landlords expressed grudging admiration for the Communists' organizational ability and their political acumen in exploiting to the fullest advantage every exploitable issue. As one official remarked: "They made greatest political capital of the land issue, but it was not only the land. Everything the Viet Minh touched was political: the creation of educational facilities, repair of roads, tax collections, army behavior, or capital punishment meted out to a 'criminal' farmer—all of these and much else were carefully explained on the ground of political necessity." The Viet Minh were, in effect, engaged in a continuous process of trying to convince the common people that their interests were identical with those of the Viet Minh and vice versa. Significant also is the evidence that the Communists were able to mobilize local talent to carry out their multifarious activities.

If the above observations gathered in the field are correct, there is cause for concern on the part of the national government regarding the current state of political activity in the countryside, or rather the lack of it. A brief visit in central Vietnam convinces one of the existence of a political vacuum as wide as the countryside we traversed. Worried officials recognize the fact that no new content is being poured into the vacuum created by the physical disappearance of the Viet Minh. The province and district offices and city streets are decorated with all kinds of slogans and homilies on the virtues of hard work, patriotism, good citizenship, thrift, honesty, literacy, and the like. These are all desirable attributes, but in the light of the pressing realities the inescapable impression is that this sloganeering does not move much beyond the space the slogans fill.

The more one listens to the tales of woe of an intelligent and well-intentioned chief, the more one comes to believe that his office is but an island within a peasant ocean, the latter hardly affected by the former. There is no evidence that the persistence of Viet Minh influence has induced the national and local governments to attack the crucial political tasks with all the urgency and seriousness they deserve. The often mentioned handicap, shortage of trained personnel, cannot be taken seriously as an excuse. This is an unsubstantiated charge—and it did not take the Viet Minh to disprove it. Trained personnel are not born. The Vietnamese do not lack the main ingredient, native intelligence; what is needed is leadership to harness ability for the building of the new state. It will not be harnessed and put to work with zest and dedication unless the top leadership takes the people into its confidence in words and in a spirit which neither overlooks the harsh facts of life and the people's deeper aspirations nor fails to inspire the people to a willingess to come to grips with their problems regardless of the cost. Mere slogans and propaganda directed against the Viet Minh are not enough to save the day in the coming battle for Vietnam.

The chore will not be an easy one in central Vietnam. The French are responsible for the security of that part of the country. Their presence is conspicuous throughout the area. Without referring to security problems, it suffices to say that the presence of the French army makes it much more difficult to convince the people that Vietnam is truly independent. Within sight of continuous traffic of French troops of many colors and origins, a Vietnamese nationalist has a difficult time convincing the people that the Viet Minh has bartered away the country's independence to a foreign creed espoused by Russia and China. Hence the difficulty of the task and the need of a supreme effort to put the best face on an apparent contradiction.

This political problem will persist as long as the French are in central Vietnam, but there is yet another immediate problem which is very much on the mind of virtually every local official who is eager to establish closer relations with the farmers. This is the unfriendly attitude of the farmers toward the national army. According to available information on the spot, the political arm of the Viet Minh army saw to it that the relations between the Viet Minh soldiers and the farmers were close and friendly. In this effort the Viet Minh met with success. The propaganda slogan that the army fights the peasants' battles might not have sufficed, but the behavior of the soldier evidently carried conviction. Although they taxed the farmers and lived off the land, the officials would have us believe that many a farmer forgave the Viet Minh soldier as "a father an erring son." This is apparently no mere figure of speech, for the Viet Minh soldier entering a farmhouse would address the farmer as "father"; his wife, "mother"; and the children as "brother" and "sister." It would appear that the Viet Minh admonition to their soldiers that "you are the fish and the people are the water" was not without effect.

On balance, in the eyes of the people we talked to, the national army suffers by comparison. One of the district chiefs spoke bluntly when he stated that "the soldiers of the national army maltreat the people and they naturally do not like it. They do like the Viet Minh army, which is looked upon as a people's army protecting the people." To substantiate the charge, he pulled out a dossier of cases of soldier misbehavior, some of which were trivial and others most serious.

The deep concern with the position of the national army in the community is understandable. They equate the national government with its army. With the local administration uneasily feeling its way into the village, the army could become a powerful link between the government and the people. How can this be achieved, we asked a very troubled chief of a province? Without hesitation he replied as follows: "If you see President Diem upon your return to Saigon, deliver this message in my name: What the national army needs today is not so much military training as political training and of the latter it gets none. Military training is relatively easy, but political training is difficult. It will be successful only when the people accept the army as their own. The army will have fulfilled its mission only when the farmer invites a soldier and treats him as another farmer's son away from home." These are sentiments born of a tragic experience and of an appreciation of the urgent political tasks and tests which lie ahead.

There are other means of making political capital. The land reform of the national government is one of those "other means" as well as the first attempt to translate one of the aspirations of the farmers into political language. For reasons discussed elsewhere, this land reform phase may fall short of the anticipated results. Hence the pressing need to raise the status of the farmers by attempting to satisfy some of the human wants about which they feel so keenly.

It is not the purpose of this "political note" to outline a plan of action, although if the picture of a village in central Vietnam is close to reality, then the question of what to do is fairly clear. What needs to be emphasized here is that the conscious and yet unsatisfied wants constitute, what one writer aptly termed, "the revolution of rising expectations." In a politically conscious Vietnam engaged in a bitter struggle with the Viet Minh, it would be doubly dangerous to delay indefinitely the partial realization—at the very least—of some of the expectations. The price of inaction might eventually be Communist action.

This observer is not without awareness of the serious political conditions now prevailing in Vietnam: of the "state within the state" phenomenon exemplified by the so-called religious sects; of the national government's exercising only limited power in the countryside; of provincial governments governing poorly; and of local, grass-roots administration honeycombed with Viet Minh holdovers and steeped in Viet Minh influence. Nor is the country in a good economic state, depending as it does upon American economic aid to a major degree. The normal reaction to this state of affairs is reluctance to seize and direct the "revolution of rising expectations." The position is a mistaken one because the political vacuum cannot be filled unless the questions of a more adequate standard of living, a greater sense of security, a greater sense of freedom of participation, a deeper sense of belonging, and similar basic human wants are at least actively pursued. The availability or lack of money is indeed important, but at this particular juncture the inspired expression of understanding these issues is of the highest importance. Not everything must be done for the people of Vietnam; there is much that they themselves could do by themselves if their sense of responsibility were aroused and they were given opportunities to exercise it. But to achieve that end, the national government must come closer to the people; it must put all its cards before the people—the good, the bad, and the very worst—and seek their support and advice. It must, of course, convincingly identify itself with the things that truly matter to the people. If that is done, the pro-Viet Minh sentiments might well become a memory; and the people might even come to understand and accept the fact that not all their deeply felt needs can be realized in short order. The faith and hope that they are moving in the desired direction and are taking part in a task in which they believe—these matter most, even if the process is a slow and difficult one and the immediate rewards are meager.

Upon return from the field trip, the writer of these observations had the privilege of discussing his impressions with President Diem. The principal points at issue, including the role the president himself might play in dealing with them, have been conveyed to him.

26. *Field Trip in Southern Vietnam*

From central Vietnam, Ladejinsky proceeded to make this undated field trip in southern Vietnam. The April 29, 1955, date of the dispatch transmitting it to Washington suggests that the field trip itself was completed in March or early April at the latest. The report finds a dismal state of affairs in the southern Vietnam countryside, with the national government held in disrespect, the local authorities lamentably weak in relation to the sects and the Viet Minh, the farmers confidently expecting the Viet Minh to return victoriously in 1956, and in general a state of "quiescent anarchy" in the "political and administrative vacuum in the countryside."

Although the U.S. Department of State was able to supply the covering dispatches for this paper and the earlier one on central Vietnam, it was unable to provide the papers themselves, which had somehow been lost. These important historical notes can be presented here only because they were found among Ladejinsky's personal papers.

This paper was transmitted to Washington with Dispatch 366 on April 29, 1955.

Introduction and Summary

IN THE REPORT on central Vietnam, the writer of these notes underscored the fact that that part of the country was not typical of Vietnam as a whole and that the findings, if valid, may not necessarily apply to other parts of the country, notably southern Vietnam. This point was driven home by a friendly critic who suggested that the reluctance of the tenants of central Vietnam to accept President Diem's reform may have been born of purely local conditions and that the attitude of the tenants in southern Vietnam may be quite different. In fact, on more than one occasion the central Vietnamese voiced the opinion that the land reform had much greater relevance to southern Vietnam, where holdings are large, where large-scale proprietorship prevails, and where the triangular landlord-tenant–farm laborer struggle for rights and positions has been acute. And so we went to southern Vietnam in search of answers to the following questions: Are the tenants interested in President Diem's reform program? Is the program being implemented? Will the position of the national government be strengthened by the promulgation of this reform?

It might be stated at the outset that the findings are even less encouraging than those of central Vietnam. In southern Vietnam, too, large groups of tenants show little interest in the reform, and its implementation has hardly begun. The principal causes are as follows: the agrarian activities of the Viet Minh before Geneva and the strong influence they continue to exercise; the inapplicability of the reform in the large areas controlled by the religious sects; the 15 percent rental decreed by the former President Nguyen Van Tam in 1953; the interest in land ownership rather than in rent reduction; and, finally, the most important cause—the political and administrative vacuum in the countryside which stands in the way of the enforcement of most measures sponsored by the national government. This is of supreme importance and no discussion of land reform in southern Vietnam is meaningful unless placed against the background of a woeful lack of a reasonably authoritative administrative machine and a political stagnation that leaves the countryside wide open to the enemies of the national government.

General Observations

Southern Vietnam and central Vietnam are two regions apart. Both are agricultural, but the former is more typical of the country's agriculture. The first thing that strikes the traveler only a few miles out of Saigon is the absence of the mosaic of tiny plots so characteristic of central Vietnam. The plots are large, the rice plains stretch into the far horizon, and it takes little initiation to realize that this is the region of surplus rice production. As we crossed from the province of My Tho into that of Ben Tre, and then into Can Tho and Vinh Long, the lay of the land and the cropping pattern are about the same, except for the presence here and there of large coconut groves, notably in the province of Ben Tre. It is this section of the country where we first encountered individual or corporation holdings measured in thousands of hectares, some few tenants cultivating as many as 5 hectares and more, and numerous farm laborers working for tenants and landlords alike. The first two groups of farmers we chanced to talk to were farmhands working in the field. They knew nothing, or pretended to know nothing, about a land reform program of any kind.

For the first time, too, the sects began to assume significance more real than one normally derives from reading or hearing about them in Saigon. When we reached the city of Can Tho, we received our first object lesson on the "state within a state" phenomenon. When the chief of the province discussed his problems, he made no secret of the fact that, as a representative of the national government, he is hemmed in on all sides by the Hoa Hao sect and that the greatest part of this richest of provinces belongs to the Hoa Hao to do with as they please. The boundary between his city and the Hoa Hao territory is just across the river, but for all practical purposes it might have been a thousand miles away: He could not penetrate it even if he tried. This point is mentioned here, for the Hoa Hao sect as well as the Cao Dai sect limit the application of the land reform severely. Putting it more directly, in Cochin-China, which encompasses most of southern Vietnam, the sects account for approximately one-third of the area and the land reform ordinances of the national government are completely disregarded there. Of course, no other form of control normally exercised by a national government applies there either.

Our first talks with officials and farmers substantiated the roadside impressions and the statements heard in central Vietnam that southern Vietnam offers ample scope for land reform measures. In some respects, however, the two parts of the country are alike. The first is the decline in acreage under cultivation, in consequence of the civil war. We found no case where the cultivated area now and ten years ago was about the same. In Can Tho province the rice area shrunk from 200,000 to 130,000 hectares, in My Tho province from 115,000 to 85,000 hectares, and approximately the same holds true for other provinces. This does not mean, however, that officials and farmers concede that their provinces have surplus land for the settlement of refugees. This is one of the few points on which the two are in agreement. The argument is that the temporarily abandoned land will be repossessed once a period of normalcy has finally set in.

The second similarity is the lack of data; on the provincial, district, and village levels the story is almost the same. Here, as in central Vietnam, approximations are the rule and they may or may not be fairly accurate. It is clear, however, that one does not deal here with one-acre tenants or less, or with "two by two" landlords. My Tho province is not the most typical of provinces where the biggest holdings have been carved out in the past 60 to 70 years or so, but the pattern of concentration of land ownership with its large tenant and land laborer classes is indeed there. In My Tho province a big owner is measured in hundreds of hectares, but in Can Tho province one corporation alone has title to 10,000 hectares, twelve owners dividing among themselves 25,000 hectares, and many more owning from 500 hectares and up. Such concentration of ownership falls into the category of "big time"; it can stand comparison with other countries noted for their land ownership disparities, and it certainly dwarfs the size of the landlord holdings in central Vietnam.

Most landlords here, whether big or small, rent the land to tenants; in cases where they manage the land themselves, they resort to hired labor. Pre-civil war rentals varied from a low one-third of the crop if the land was poor to half of the crop on good-quality land; farm equipment, including the buffalo, was the responsibility of the tenant. His net share of the crop was considerably below the indicated figures. We could not judge the position of the tenants in those days, but today they appeared to be much better off than their opposite numbers in central Vietnam. According to their testimony, they not only have a sufficiency of rice for their own consumption, but a disposable surplus as well. They do not give the impression of grinding poverty, although here and there one sees a sight testifying to a seamy side of the life and work of the cultivator. We watched a few tenants "plow" fairly large rice fields with nothing more elaborate than a hoe. Theirs was clearly a long and arduous task undertaken for lack of the basic motive power of the countryside—the buffalo. These scenes, a throwback to the most primitive agricultural techniques, were not a common sight; but they were, nevertheless, most incongruous in the midst of the richest agricultural region of the country.

Even on the basis of these sketchy observations, it is obvious that southern Vietnam presents a fertile ground for the application of a land reform program. Large landholdings in relatively few hands and a huge tenant class and normal tenure terms which do not favor the tenant presuppose that. While no one seems to be in a position to state how many of the broad acres of southern Vietnam are rented by tenants, an estimate that they work three-fourths of the land is probably not far fetched. But even if this figure is exaggerated, it is true that the proportion of tenanted land to the total is considerably larger than in central Vietnam. The acreage per tenant is also larger, and the normal conditions of tenure are apparently no better and no worse than in central Vietnam. Working a larger and more productive acreage, the tenants of southern Vietnam are in a stronger economic position than those in central Vietnam, but this in no way made "land reform" less appealing to them. It is natural that the state of relative well-being only whetted their appetite for further improvement. In a couple of sessions with fairly large groups of farmers, approximately half of them spoke out in favor of rent reduction as the number one item on their list of basic needs; credit, irrigation, and drainage facilities as well as lack of buffaloes did not loom as important. The reverse held true in central Vietnam. How, then, do the tenants react in the face of the rent reduction scheme inaugurated by the national government?

Negative Reaction to the Reform

My Tho province, only 40 miles from Saigon, was our first stop and the first to provide the answer. The deputy chief of the province made no attempt to sugarcoat existing conditions. As of March 20, only a few land reform committee elections were held in the 116 villages comprising the province. Since the implementation of the reform is predicated on these committees, the election failure was certainly a setback and the officials did not disguise their disappointment. When the question was raised about the very spotty picture, the reply was to the effect that, whereas the landlords are willing to take part in the elections, the tenants are most reluctant. How reluctant became apparent when we looked into this problem in the Binh Phan village, where an election had actually taken place. It is a village of many small landowners and a large number of tenants with holdings averaging 2 hectares. The village council felt that it has done its duty and by way of proof displayed the name lists of landowners and tenants who participated in the elections. A quick counting of heads revealed that the first totaled sixty-six and the second forty-three, or 50 percent more landowners than tenants. When this peculiar reversal of the normal—fewer landowners and more tenants—was called to the council's attention, it developed that of the more than 300 tenants in the village only forty-three have seen fit to take part in the elections; most of the landlords, on the other hand, entered their names and were at the polls on election day.

With relatively minor variations, the story appeared to be the same in the provinces of Ben Tre, Can Tho, and Vinh Long: either no elections at all or elections poorly attended by the tenants and well represented by the landlords. In late March no provincial office had detailed information on the election of the committees, the first basic step towards the enforcement of the reform program. The chief of Vinh Long province went so far as to say that, to date, he had no reaction at all from the peasants of the forty-eight villages which made up his province. His contact with the villages was limited in the extreme, and not because of lack of interest on his part. The provincial office depends for its information on the district office, which in turn must secure it from the village councils. This evidently is easier said than done, for so far the information is only a trickle furnished by villages which are located close to the district office; the goings-on in the outlying villages are anybody's guess because, as one district officer put it, contact with such villages is often avoided "for the fear of the tenants." There is, however, always an official's reluctant estimate; when these are put together, it might be said that, in the provinces here considered, about one-third of the tenants have shown an interest in the program. Whether it is a third or one-half or even more than that, the fact remains that there are many tenants who are as little concerned with the land reform program as are the tenants of central Vietnam.

The Landlord and the Reform

At first glance the readiness of the landlords to accept the reform may seem surprising because its implementation would cut down their share of the crop from one-half to 25 or even 15 percent of the crop. The explanation of this seeming paradox is not far to seek. During the long civil war period, the landlords of southern Vietnam collected little or no rent from their tenants. Certainly the big landlords, who disappeared from the villages for security reasons, collected no rent at all. The smaller ones, having remained in the villages, received only such token rents as were permitted by the Viet Minh. Eight months have passed since Geneva, but the situation has not undergone any material change for the resident or for the absentee. Some landlords have had the courage to visit and inquire into their holdings. The receptions vary: In villages close to the seat of authority they are likely to collect some rent, but in villages not within effective administrative control they receive only a cold shoulder. Moreover, some tenants are known to have refused to acknowledge the landlord's right to the land. The reason for this fundamental departure from the traditional landlord-tenant relationship will be dealt with in another paragraph, but that it has come about has not been lost on the landlords. Taking a practical view of the situation and attempting to salvage what they can, they are understandably eager to settle for a rental of 15 to 25 percent of the crop. They are motivated by a still more significant consideration: In accepting the rental terms of the reform, their title to the land is given full recognition. The importance of this stems from the fact that many a tenant has come to look upon his rented land as if it were his own. In short, while in the pre-Viet Minh days the landlords would have viewed President Diem's reform as a diabolic plot against them, in the spring of 1955 they are ready to make the best of an admittedly difficult situation. What troubles them is the thought that their willingness to adhere to President Diem's reform is no guarantee that the issue is settled. One landlord spoke for many when, in reply to the question whether he is ready to accept the reform, stated: "Yes, I shall be happy to sign the contract but my tenants won't."

The Tenant and the Reform

Not so with the tenants. The very reasons which induce a landlord to favor the reform cause many tenants to look upon it with disfavor. In the years of the Viet Minh domination—and they ruled the countryside in southern Vietnam as effectively as in central Vietnam—the tenants' rent was, as distinct from the tax levies, about 15 percent when the landlord was around to collect it. Many, as noted above, did not bother collecting any. Important too, is the fact that the Viet Minh have distributed some of the land among the tenants. For them to have to pay rental on such land is particularly distasteful. The physical departure of the Viet Minh has not changed the tenants' attitude toward the landlord; the tenant is still very much under the Viet Minh influence, as will be pointed out at some length in another connection.

The negative position of a good many tenants toward President Diem's reform stems from yet another cause. In June 1953, President Diem's predecessor, President Tam, decreed a rent reduction program which called for a rental of 15 percent of the crop regardless of the productive capacity of a given plot of land. It is interesting to note that, while most tenants disclaimed much knowledge of the principal provisions of President Diem's reform, not a few knew about President Tam's 15 percent rental. This is the more surprising since President Tam never attempted to enforce his program. The fact remains, however, that some of the tenants we met used it as an argument against President Diem's "exorbitant" rentals of 15 to 25 percent of the crop.

This range has been a source of much concern to local officials; and at least one of them, Nguyen Van Hoi, chief of the province of Ben Tre, decided to clarify the issue in his own way. On February 2, 1955, he called together a group of tenants and landlords and worked out with them the following compromise: If one hectare of land produces less than 40 units (a unit equal to 40 liters), no rent is to be paid; if less than 80 units, the rental is to be 15 percent of the crop; if the yield is above 80 units, the tenant must pay 25 percent of the crop.

Mr. Hoi displayed commendable initiative, and the scheme he devised is weighted in favor of the tenants. In normal conditions—in conditions which would enable him to exercise real administrative authority in his province—the refinement of the rental provision would have been a sound measure towards the enforcement of the entire program. In reality, Mr. Hoi realizes that his action will not meet the situation—not until he can effectively extend his authority beyond the city of Ben Tre and the few villages close to the city. At the time of our visit he readily admitted that neither his provincial office nor his district offices were in a position to enforce measures emanating from him or from the national government. Therein lies one of the crucial problems that impedes the progress of the land reform program in southern Vietnam.

The program suffers not only from the negative stand taken by the tenants, such as refusal to participate in the elections; occasionally this negativism assumes the form of a "positive" demonstration against the reform. Thus in a few villages of Ben Tre province the tenants appeared at the polls as organized groups—and then refused to vote in protest against the elections and the land reform in general.

It is still too early to judge what this passive or active disinterestedness on a part of an unknown number of tenants will do to the program. There are simply no reliable data—or hardly any data—to help one form a reasonably correct opinion of the progress to date. Willy-nilly one must resort to such unorthodox indices as the "long" faces of the officialdom and the mien of the farmers which betrays none of the enthusiasm associated with a long-awaited goal about to be achieved. If these criteria are permissible under the circumstances, the land reform in southern Vietnam has a long and hard row to hoe.

The Farmers versus the Government

One of the encouraging features of the trip through southern Vietnam was that the chiefs of the provinces were wide awake to the conditions prevailing in the areas they are supposed to administer. They could have painted a silver lining in their sad accounts of their stewardships, but they eschewed that. They gave the impression of people anxious to unburden themselves and let the chips fall where they may. What they were saying did not reflect their success as administrators; but as the chief of Vinh Long province put it, "Since you are not after appearance, I shall call your attention to the realities, painful though they are."

What are those realities?

The principal one, as the administrators see it, is the "ugly" mood and behavior of the farmers who live and work outside the precincts of the religious sects and are theoretically subject to the control of the national government. This behavior antedates President Diem's reform; the latter brought it only into sharper focus. Their unwillingness to pay taxes, for instance, is another facet of the same problem. This has an immediate and crippling effect on budgetary requirements all along the line. When two opposing forces are ranged against each other, the charges and countercharges may be wide of the mark. This may be true of the charge of the chief of My Tho province that he will collect from the farmers only 2 percent of the estimated 1954 tax bill. It may turn out ten or fifteen times that, but the point he made remains valid. Farmers deny the charge, but their statements carried no conviction and they were unable or unwilling to produce receipts in support of their contention. Another item is cited that bespeaks the same attitude: this is their alleged refusal to furnish paid labor for public work. Since we observed farmers repairing public roads, the above charge is subject to serious qualifications. To the local administration, however, the reluctance is one more expression of the farmers' tendency to avoid direct contact with the government wherever and whenever possible.

All of this points to cleavage between the government and the farmers. The existence of this condition is one of the most vivid impressions of the trip. Officials contributed to this by their outspoken reflections on the state of affairs in the provinces, while farmers contributed more by their inactivity rather than by what they said or did. This apartness of the farmer is barely discernible near the seat of government; but, once outside of the immediate reach of authority, it is by all accounts undeniable. One high official used the term "anarchy" to describe the realities in the countryside of southern Vietnam. "Quiescent anarchy" might be more accurate. On the surface everything is quiet: no flag waving, no open exhortations to disobedience, and no reported shootings of any kind to force the farmers' will upon the government. Nor does the government use force to make the farmers do its bidding. Each one is left to his or its own devices. The net result is what another much-concerned official called "immobilisme," that is to say—no forward motion!

Causes and Remedies

The inevitable questions that come to one's mind are: How did this state of affairs come about, and what can be done to remedy the situation?

The first question has been partly answered, but it can stand repeating. Prior to Geneva the farmers in southern Vietnam as in central Vietnam had to contend with two forces—the French and the Viet Minh. They lived in fear of both—and this applied even to those who accepted the Viet Minh leadership. Now, so the story goes, they have no fear at all of the national government, which has supplanted the French and the Viet Minh. This is an oversimplified version of what has actually taken place. It is true that the French are out of the picture in South Vietnam, but it is not true that the influence and the fear of the Viet Minh have come to an end. If the information gathered is at all correct, the Viet Minh leadership in the village councils is greater than in central Vietnam and shows no signs of diminishing. Orders are still being issued and obeyed, and no anti-Viet Minh farmer is willing to stick his neck out in opposition to the Communists. Much of the Viet Minh activity is geared to 1956. A report on some negotiations for the sale of land, presented to us in the office of the chief of Can Tho province, illustrates the point. The land prices there are relatively low and a few tenants had agreed to buy the land they have been renting from the landowner. At the last moment the deal fell through; the local Viet Minh "advised" the would-be buyers to save their money, for come 1956 the land would be theirs for the asking. As with the purchase of land, so with President Diem's land reform, with paying taxes, and with much else that constitutes the normal pattern of relations between a government and its people. No use signing a contract or taking on an obligation, however legitimate, or closely associating with the government when in 1956 entirely new arrangements will prevail. The "1956 wait-and-seeism" line of propaganda is undoubtedly serving to draw a formidable barrier between the government and the farmers. The local authorities know this; they recognize that if the course is not reversed the political fortunes of free Vietnam may be gravely affected.

The obvious question we raised with the local authorities was: "Do you combat this trend, and, if so, how?"

Without exception they stated frankly that they do not, although ideas on how to combat the "invisible enemy" were not lacking. Nguyen Van Ke, chief of My Tho province, in an early morning breakfast session delivered himself of the following observations: "The influence of the Viet Minh is very great and we are not really opposing them in the village. You need no army to spread your ideas. A few agents in a village is enough, and of these the Viet Minh has enough. Their main theme is that they will be back in 1956 despite the French and the Americans. This colors the attitude of the farmers to a point that they follow the directions of the Viet Minh, and the village councils are largely pro-Viet Minh just as they were before Geneva. Our real problem is to convince the people that the Viet Minh will not return to power in 1956. The first step in that direction is that the national government must stop being too lenient, like a kindly father spoiling his children. The government must give the local authorities real power to deal with the Viet Minh agents. Failing that, even anti-Viet Minh farmers don't believe that the government can extend them security. The task of tracking down the Communists becomes impossible."

These ideas have been reiterated by all our official contacts. They all discarded the notion of "authority" without a show of strength and the means to back it up. Nguyen Van Dinh, chief of Vinh Long province, cited his own experience in support of this point. He decided to put into practice the force idea in the villages of his province. With this in mind, he stationed six soldiers in every village, thereby trying to give "backbone" to the anti-Viet Minh farmers in general. The soldiers are still there, but they are the laughing stock of the villages; Mr. Dinh could furnish only two rifles for each unit, and the rifles are all of World War I vintage. The soldiers, he noted ruefully, are afraid to discharge them for the fear that they might blow up along with the rifles. When asked why he could not provide them with more formidable armament, he replied that his appeals to the national government through proper channels have never been acknowledged. And this, incidentally, is a major complaint of every chief of province. They are all ready to exhibit numerous communications to the central government but have little to show in response. Their link with the villages is indeed tenuous, but it is evidently not too strong with the top either.

To insure for the farmers security from the Viet Minh mistreatment is indeed a prime prerequisite in any successful attempt to undermine the Communist domination of the countryside. However, naked force even in the form of soldiers with *modern* rifles is only a small part of the answer, and this the local authorities appreciate very well. They turn their attention to other means. The first mentioned is the possible role of propaganda, but they dismiss it in the same breath. They felt that the people were fed up with propaganda and promissory notes on the future. Besides, "Anything we can promise," said Colonel Nguyen Khanh, chief of Can Tho province and military commander of the western sector, "can be outpromised by the Viet Minh more convincingly, and they are under no obligation to perform whereas we must make good our promises." They are all agreed that one of the important keys to the question of authority that would carry weight with the people is to convince them that the government is stable and is here to stay for a long time to come. To be effective, mere reiteration that the government is stable will not do. It is this observer's impression that the local authorities themselves are not convinced that the Diem government is here to stay. The people know (and Colonel Khanh knows only too well) that, although he is in command of 9,000 nationalist troops, he is not the master of his province. The province is largely in the hands of the Hoa Hao sect, and he is only a mere reminder of the existence of the national government. This fact alone would impose a severe strain upon the local authorities if they were faced with the task of selling the people the idea that they represent a stable and lasting government.

The sect problem is indeed one of the great and unfortunate realities referred to earlier. More than the Viet Minh, at the moment, these "religiously enlightened bandits" are the open challenge to everything which stands for good government, and then literally and figuratively lie athwart any possible re-creation of a government with authority in the cities and in the countryside. This explains why Colonel Khanh would rather talk about the Hoa Hao and Cao Dai sects than the Viet Minh. So far as he was concerned, the elimination of the sects as a military power is the only possible stepping stone to the successful solution of the problems, including the Viet Minh, discussed earlier. His account of the growth of the sects, the major role the French played in the past decade in raising them to the present position of prominence, the treatment of the farmers under the control of the sects—all this and more is part of another story. What should be underscored here is that the land reform and its relation to authoritative, stable government is tied up not only with the Viet Minh but with the sects as well. Without in any way underestimating the already baleful effects of the Communist cells in the villages, their tasks are made easier because of the presence of the sects and the paralyzing results they have on the position of the national government in much of South Vietnam. The local authorities cannot actively concern themselves with the sects. They are powerless to treat with the villages and would not dream of tangling up with the organized force of the sects. All they could do was to call the observer's attention to one more important item the national government must deal with in order to insure for the countryside conditions under which a land reform and other measures would have a better chance of implementation.

When we consider the recent and current political developments and the further weakening of the national government, the suggestions so far on how to regain the countryside have a touch of unreality about them. And yet the local authorities realize that sometime soon a firmer link between the farmers and the government must be forged. And that link must revolve around the land question. Independently of each other, three out of the four chiefs of provinces expressed the belief that, since rent reduction holds out little promise for a good many tenants, a land ownership program might be in order. This is one area where the national government could outdo—let alone outpromise—the Viet Minh; the fact is that the Viet Minh engaged in relatively little land distribution, and the bulk of the land is still owned by the landlords.

Months back the government had expressed its intention to do just that after the implementation of the rent reduction scheme. Whether it is willing or capable to initiate that kind of a program is a matter not quite germane to these observations. What needs to be pointed out here is that the Vietnamese officials in the field, civilian and military, understand that the political fortunes of Vietnam are inextricably tied up with the countryside. Mr. Hoi of Ben Tre province has his hands full with the rent reduction ordinances. And yet he spoke most unbureaucratically when he said: "The ordinances are not meeting with success, but they deal with land and so they are my only contact with the people. They may not like the provisions of the program, but land is on their mind and it gives me a chance to talk to them. I cannot talk propaganda to them." For a firm contact with the people and to lessen the vulnerability of the Communists, he strongly advocated a land ownership and land distribution scheme.

Colonel Khanh of Can Tho province is young, intelligent, every inch the military man, with heavy military and civil responsibilities. Before, during, and after dinner he was given vent to his pent-up emotions about all manner of things: about his government which tied him down with strict instructions not to engage in any armed dispute with the Hoa Hao sect; about his government which he believes to be quite inactive and has demonstrated little interest in the area he has been entrusted with; about his consuming hatred of the Hoa Hao, for there he was with 9,000 troops yet so boxed in by the sect that he could cross a bridge only at the pleasure of Hoa Hao troops guarding it; about the Americans who decreed, so he stated, that his force be cut by 54 percent as a means of reducing and yet strengthening the national army—a procedure he could not fathom at all; and finally about the Communists who infiltrated the rural districts and much else besides. In all of this he perceived grave danger to the survival of Vietnam. But his enemies are the Hoa Hao and the Viet Minh, the latter an enemy with a "principle," the former an enemy always looking for the highest bidder. He felt that in a pinch and for a price they would not disdain to rejoin the Viet Minh, despite their alleged religious scruples.

Colonel Khanh knows that "land for the landless" had been one of the most potent slogans of the Communists in wrestling political power from a government already in power. He watched the process in Indochina, and he had the advantage of a visit in the Philippines where he observed the struggle against the Huks. His solution, therefore, and before it is too late, is not unlike that of Mr. Hoi: namely, a program based on the slogan of "land for the landless." This, as he put it, "is the stone with which to kill two birds"—the Hoa Hao and the Communists. As he saw it, the Hoa Hao sect is the authentic exploiter and rack-renter of Vietnam, and the subjugated peasantry would rise to the occasion and support the liberators. To achieve his end quickly he expressed himself against a belated, carefully worked out program, with all sides satisfied, all the "t's" crossed, and the "i's" dotted. He wanted *action* without delay, even at the price of temporary chaos. In his own words, he wanted "an economic and social revolution" in order to arouse the government, the army, and the people from lethargy and stagnation and into purposeful activity so that the political and administrative vacuums might be filled with political ingredients commensurate with the problems facing free Vietnam.

Conclusion

Such, in the main, are our impressions of the trip through four provinces of South Vietnam in search for information on President Diem's reform program. Normally, this great farm area, with a high degree of concentration of land and a high tenancy, should be fertile ground for the scheme. But times are not normal, the implementation of the reform is barely off the ground, and its ultimate realization is highly questionable. The countryside, partly enslaved by the religious sects which are laws unto themselves and partly dominated by the invisible yet ever-present Communist Viet Minh, is tense and waiting. And in between the sects and the Communists are the local authorities representing the national government. They are not authorities in the real sense of the word, for their power to govern is so limited as to be almost nonexistent; and the sects, the Communists, and the farmers behave accordingly.

The great impediment to the land reform outside of the sect areas is unquestionably the Viet Minh farm policies before Geneva. For this reason alone the program of the national government was bound to suffer from the loss of novelty and impact. The other handicap lies in the Viet Minh's continued hammering that in 1956 all of Vietnam will be theirs. Little wonder that even the anti-Viet Minh farmers must keep their second thoughts a deep, dark secret—unless the local authorities are in a position to provide them with security and with the conviction that President Diem's government is here to stay. This, by their own admission, they are not capable of doing. Were they in a position to demonstrate convincingly both propositions, President Diem's program would be acceptable even by the pro-Viet Minh farmers, let alone the anti-Viet Minh or the "neutralist" fence-sitters. The fact that in the years of civil strife the tenants paid little rent would not have weighed so heavily in their calculation (Viet Minh also levied taxes) when set against the reasonable terms of the reform in question. The comparison has become unfavorable only when considered within the context of the other all-important conditions and circumstances which have little to do with President Diem's program as such. The obvious lesson is that no program, however technically perfect, can be implemented in conditions of political stagnation and administrative vacuum.

Mr. Hoi and Colonel Khanh are quite right in advocating a land ownership and land redistribution program. The national government has recognized its inevitability and sooner or later, whether with alacrity or not, the matter will have to come up for active consideration. When it does, the politicians willing, the technicians can devise a reasonable program. There is the danger, however, that if the political basket of worms of present-day Vietnam is not cleaned up, either through Colonel Khanh's "economic and social revolution" directed from the top or through other actions which go beyond the self-deluding temporary "deals," the national government may find itself once again all dressed up in very elaborate finery but with no place to go. The above is not intended as an argument against action that might help the government to wean away the farmer from the Viet Minh influence or undermine the position of the sects in their fiefdoms. It merely intends to stress that at this particular time the overriding problem is one of achieving the essential precondition for the successful implementation of any worthwhile economic, social, or political reform. That beginning of all wisdom is a strong, authoritative, and acceptable government at all levels of administration—from the seat of government in Saigon to the village councils in the countryside. This is, of course, a truism; but an observer recently back from South Vietnam cannot emphasize it too strongly or repeat it too often. One may well take the trouble of shouting about it from the housetops. At stake are not so much the land reform program or the American economic aid but the very survival of free Vietnam. This observer does not share the view of those writers who contend that the countryside is a case of "all hope abandoned." It is not that yet, but one must record the fact that nothing is taking place there that might reverse the disturbing trend of events.

The precondition cited above cannot be created overnight. But a beginning must be made. In the "Field Trip Observations in Central Vietnam," the writer of these notes attempted to state what the national government might do to lay the foundation for a closer relationship between itself and the people. His experience in South Vietnam demonstrates the urgency of that need even more dramatically. The realities are that the sects overtly and the Communists covertly dominate that part of the country. In a country such as Vietnam, it is peasant support or no support at all. The status quo, which to date means little or no peasant support, plays into the hands of the enemy and may eventually lead to a political debacle. The Vietnamese and non-Vietnamese friends of the country must find ways of enhancing the prestige of the national government and helping it find a common ground with the people.

This assistance may well begin by a clear demonstration that President Diem is the undisputed "boss" of Saigon, a development which, if realized, would have incalculable value reaching out beyond Saigon. He must be encouraged to take hold of bold measures even if termed by some "revolutionary," and one of those may in time spell out an agrarian reform which would not be to the liking of the landlords at all but acceptable to a grateful multitude of farmers. In suggesting this, one must not lose sight of the distinct possibility that if the national government doesn't do it, the "others" will do it and set the countryside ablaze. We know who the "others" are. Once a start has been made on the road towards a creative state, the sects can be mastered; and the fairly widely accepted inevitability theory of a Communist victory in 1956 may turn out to be a dragon with drawn teeth. If, on the other hand, nothing far-reaching is ventured, not even against the gangsters of Saigon, on the theory that one must not "rock the boat," then the "quiescent anarchy" in the countryside is indeed explainable; the failure to accept President Diem's reform understandable; and a revolution from the bottom rather than a controlled one from the top is perhaps inevitable.

27. *A Visit with President Ngo Dinh Diem*

This is a memorandum of conversation which Ladejinsky prepared for U.S. Ambassador Reinhardt after a meeting with President Diem on June 1, 1955. The discussion ranged widely from land reform to refugee resettlement to local administration and other matters. It is clear that this was merely the latest of a number of similar discussions he had had with the president, with whom he had obviously established an easy and direct relation and gained the all-important "acceptance." Historians may find this insight into Diem's thinking during that period of more than slight interest. The paper also illustrates the useful intermediary role technical personnel can sometimes play in political situations.

Ladejinsky's memorandum of conversation with President Diem was addressed to Ambassador Reinhardt on June 7 and transmitted to Washington as Dispatch 453 on June 15, 1955.

AT THE INVITATION of President Ngo Dinh Diem, I called on him on June 1 to report on the latest agrarian developments. In the course of the visit, which lasted from 5:00 to 6:30 P.M., a number of other topics were touched upon such as settlement of refugees on land, local administration, the struggle with religious sects, and the U.S. position in regard to France and Vietnam. Recalling previous visits, the wide range of topics, mostly raised by the president himself, did not surprise me; there was one item, however, which was altogether novel to me: I have in mind the President's statement that Japan should pay reparations to Vietnam for the harm Japan inflicted upon the country during the occupation.

By way of comment, it may be pointed out at the outset that President Diem does not intend to effect any drastic changes in the agrarian reform field and in local administration; he may take a harder look at the critical refugee resettlement problem; but, in the main, the president's attention is riveted to the immediate day-to-day internal political developments, the struggle with the Hoa Hao sect being the center of the piece. While such issues continue to dominate the scene, he is inclined to look upon most other problems as essentially peripheral. Suggestions that might result in a closer government-people relationship, in the dissipation of fears, and in the creation of a sense of belonging and oneness with the national government are not actively received on the grounds of other pressing business. One gets the impression that the ominous fact that, in most instances, the people and the administrators are still poles apart does not loom large as a truly serious matter. The urgency of igniting the spark that would inspire the people to greater deeds on behalf of an independent and free Vietnam is not apparent.

The president is confident of settling the Hoa Hao issue on his terms through negotiations, if the outcome is the disestablishment of the Hoa Hao army organization, or by force of arms if necessary. The defeat of the Binh Xuyen and the neutralization of the Cao Dai has given him confidence. He is not overly concerned about the French, although he continues to deplore their tactics. He spoke of the French intransigeance more in sorrow than in anger. As to the U.S. position in regard to France and Vietnam, the President appeared to be both understanding and grateful.

I found the president in a relaxed and self-confident mood. He did not minimize the crucial problems facing free Vietnam; but, while three months ago he was unmistakably a much worried leader about to enter a period of trial of strength with his numerous opponents, now he could speak, as he did, with quiet self-assurance born of recent successful tests. He did not, for example, underestimate the seriousness of the involvement with the Hoa Hao; but he reminded the listener with quiet satisfaction that three months ago he was faced with the "united front" opposition of the Binh Xuyen, Cao Dai, and the Hoa Hao, whereas now he was concerned with the latter only. In one other familiar respect he did not change: He still prefers to talk rather than listen. Nevertheless, this time he seemed to have relaxed this presidential prerogative in favor of a balanced give-and-take.

The nature of my report was threefold: The sad state of the agrarian reform; the poor progress of settling refugees on land, and the problems presented by local administration.

As to the agrarian reform, I told the president that the appointment of a minister for agrarian reform was certainly a step in the right direction but that, to date, the reform is not off the ground and that there are no indications that the situation will undergo a favorable change in the immediate future. I pointed out why the tenants show no interest in his program, the potent influence of the Communists in this connection, the preference for land ownership to rent reduction, and the political and administrative vacuum in the countryside which prevents the enforcement of most measures sponsored by the national government. I took the liberty of suggesting to the president that, if his government is to make any political capital of the agrarian reform, then the time has come to reexamine the entire problem in the light of the current state of affairs. More specifically, I suggested (a) that, as long as the program is in being, the chiefs of the provinces in the nonsect areas be held responsible for the implementation of the program and (b) a national conference of the interested parties with an eye to determining whether the existing program can be implemented or whether it should be abandoned altogether and a land ownership and land distribution program devised in its stead.

The president did not dispute the impasse reached by his program, but he does not intend to give the land reform question a national hearing. He has confidence in his new minister of agrarian reform and is evidently willing to let him try his hand at it for a while longer. The president's reluctance to review the issue with all the care it deserves may be traced also to, what appears to me, a lack on his part of a truly abiding concern with this matter. I am not prepared to say whether the president's attitude stems from the fact that he is not a "land reformer," or that the more pressing day-to-day issues bordering on the very survival of his government have primacy in his thinking and his effort to the exclusion of much else, including agrarian reform. Whatever the real reason, this much can be said with certainty: On the occasion of this talk, as during all previous talks, the land reform problems did not appear to loom large in his scheme of things.

As to the settlement of refugees on land, the president was informed that as of the end of May few refugees had been settled on land; by this is meant settlement in the sense that a refugee willing to farm has more than shelter and a garden patch, namely, well-demarcated holdings of their own, with farm tools and animal power to put the land to use. I pointed out to him that most of the farming by the refugees in this crop season, commencing with the rains, will have to be delayed until the crop season of 1956. The refugee problem being close to the heart of the President, he was visibly disturbed by the account and prospects. It is my impression that he had not been fully aware of these developments. I explained to him that I have not been too close to the problem and that I do not know all of the reasons for the delay; I did point out that perhaps the basic one, a *firm* evaluation of the total acreage available for settlement, its location, and quality has not been made and that this is the first order of business if the program is to be successfully implemented. With that in view, I suggested to the president that he immediately appoint a small commission with instructions and powers to determine the land availability, based on the commission's firsthand observations in the field. I suggested a time limit of one month and assured him that the mission (FOA [Foreign Operations Administration]) will be happy to render its technical assistance to this urgently needed survey. The president responded favorably. Whether he will act accordingly or, having discussed the matter with appropriate agencies, will act in favor of another method of moving refugees on land remains to be seen. Whatever his future course of action, he is fully cognizant that the permanent solution of the refugee problem is yet to take shape and that further delays may postpone the solution indefinitely.

As to local administration, I stated that, with few exceptions, it is ineffective and that no application of any national legislation is possible unless the administrators themselves become conscious of the fact that a free and independent Vietnam demands of them a zeal and zest of performance over and beyond the customary. I expressed the view that the real difficulty with the administrators is not their lack of formal public administration training but rather the lassitude, disinterestedness, and seeming failure to sense or comprehend the critical transitional period Vietnam is passing through. I made the point that, just as the national army is in need of political training, the administrators are surely in the same need. The president countered by saying that this problem has been on his mind, that he ordered the delegate for South Vietnam to prepare a secret report on local administration, and that the findings justified his worst fears. However, he did not believe he can deal with the problem outside of the overall issues relating to the country's pacification and stabilization. Serious though the administration question is, for the time being he intended to leave it in the hands of the three delegates of Vietnam.

The visitor did not argue the presidential preference, nor did he tell him of his most recent experience in one of the most important provinces of South Vietnam, where a new chief of province appointed by the delegate is unmistakenly anti-Diem, antireform, and pretty much anti-everything that spells deviating from the current state of inaction.

I did tell the president that the standoffish attitude of the farmers toward the government is not unrelated to the local administration; above all, I suggested that it is closely related to the weak link between the national government and the farmers. I elaborated on an earlier statement under somewhat similar circumstances and suggested that the president himself might devote some time to help create among the farmers a sense of freedom of participation, a sense of belonging with the government in the business of creating a new state, basing this approach on the government's convincing appreciation of the people's fundamental needs. I attempted to impress upon him that he more than any other Vietnamese is in a position to articulate these ideas, which in the long run should prove to be the effective weapon against the Viet Minh and for the stability of Vietnam. But now, as in the past, the president pleaded extreme preoccupation with urgent matters.

As usual, the sect problem came in for considerable comment. The president knew of my visit to Tay Ninh only a few days ago and of my meeting with Pham-Cong Tac, the "pope" of the Cao Dai sect. I told the president that for once I was not a bearer of bad news, since the pope stated (a) that although he opposes President Diem he intends to eschew politics; (b) that he favors the elimination of the army as a military arm of the Cao Dai; and (c) that from now on he, the pope, was going to devote himself to the propagation of the Universal Church, which in his opinion is synonymous with Cao Dai.

The president took this good humoredly, remarking on the pope's past criminal foibles and total unreliability. He agreed, however, that the pope's new line—even if only a "line"—was out of the question three months ago when he was the guiding spirit of the "united front" opposition to the president. He agreed also that the pope's relative inactivity and his loss of army control was indeed a dividend of the successful fight against the Binh Xuyen. The dividend is bigger than that, according to the President: He is of the opinion that the Cao Dai army is decidedly on his side and that by integrating 8,000 of the best Cao Dai troops, the remainder, which he estimates at no more than 3,000 to 5,000, is no threat to him. He could have them whenever he wanted them, and he had no immediate plans for incorporating the nonintegrated Cao Dai forces into the national army.

With the Binh Xuyen almost eliminated and the Cao Dai army leaders pledging their support to the national government, the president spoke with confidence of the struggle with the Hoa Hao sect. They were a dangerous enemy, he thought, but he was sure of victory if the Hoa Hao should accept open battle. They were dangerous as guerilla fighters and as terrorizers of the countryside. Another aspect of his struggle against this sect that gave him pause was the fact that he was compelled to divest some sections of the country of its troops. He mentioned specifically the PMS (Pays Montagnards du Sud)—the ex-Bao Dai stronghold; it appeared that most of the imperial guard troops are in the south facing the Hoa Hao. In answer to the question if the fight against the sect calls for a military decision, the president replied in the affirmative; but he stated also that he was ready to negotiate if it resulted in a peaceful elimination of the military power of the sect and if the elimination is not to add measurably to the personal fortunes of the Hoa Hao generals. In this connection, he reminded the visitor of a remark he made three months ago, namely, that the political stability of Vietnam cannot be achieved unless the military power of the sects is eliminated. "Only then," he continued with a touch of irritation, "will I be in a position to pay greater attention to the numerous other problems involved in the creation of a politically and economically strong national state." His remarks were interspersed with praise of the national army, and he spoke disparagingly of the French who questioned the loyalty of the army to the national government. The listener was all ears, for he well-remembered the president's unflattering comments on his army not long before it became obvious that the government and the sects were rapidly approaching a showdown. On that occasion the visitor suggested that with a strong national army he could deal with the sects singly or collectively, depending upon the shape of things at a given moment. His comment was: "Ah, the national army, an army of married soldiers . . ."

The president was not nearly as bitter about the French as was his custom in the past, but he was not friendly either. He does not give much credence to the official pronouncements that the French do not interfere in the internal affairs of Vietnam. To illustrate his point, he mentioned the fact that very recently the French furnished the pope with a million piastres. He thought the sum was far in excess of the officially acknowledged figure, and he spoke with derision of the French explanation that the money was intended for the pope's social service activities. To President Diem, social service in this instance is but a "euphemism for a new armed force the pope wishes to create, now that the Cao Dai army has slipped out of his control." He was firmly of the opinion that the French have a hand in the armed rebellion of the Hoa Hao, and at this point he delivered himself of this apparent oversimplification of French views and attitudes: "The French are property-minded people; with Vietnam politically stable, the value of their property is bound to go up; with Vietnam unstable, the value of their property suffers accordingly. What purpose then do the French serve in helping to keep the country in turmoil?" I did not comment on this statement, nor did the President offer an answer to his own query. He merely concluded on the note that the French are poor losers, that their disruptive tactics will prove to no avail, and that the era of amicable Franco-Vietnam relations is still in the future.

The president spoke warmly of the American attitude toward Vietnam, but he also remarked that the United States is so heavily involved in Europe and is so anxious to maintain friendly relations with France that it cannot see the French position in Vietnam as the Vietnamese see it. He added laughingly that he may be expecting too much.

It is worth noting that sometime back he spoke with asperity of Americans in Saigon who, so he said, didn't understand him and whom he, too, failed to understand. He complained without being specific that Americans gave him contradictory advice and that they were occasionally subjecting him to "shock treatment," offering suggestions at the last moment when he no longer had time for careful consideration of them. I carried away the strong impression that whatever may have given rise to those views, he hardly shares them now. When I thanked the president for his consideration, especially since I had seldom anything cheerful to contribute, he replied that Americans have a way of stating their views straightforwardly, which he appreciates even though he may not agree with them. In general, the United States was the part when the president was at his most mellow and friendly.

I was halfway downstairs, on my way out, when the President called me back. He expressed his regrets for detaining me and asked to bring up a point which had been on his mind for some little time. He began by asking me what I knew of reparations exacted from Japan by countries under its occupation. I told him that I was not well informed on the subject and added that Japan had settled the reparations problem with Burma and has been negotiating for some time with the Philippines and Indonesia. He proceeded then to tell me that Japan ought to pay reparations to Vietnam for the damages caused the country during the occupation. He didn't mention any figures nor whether his government has given this problem serious thought. He inquired into Japan's capacity to pay reparations and I replied that Japan's economic position is a precarious one, although the outward appearances seem to belie it; that Japan makes both ends meet only with the financial aid of the United States; and that, generally speaking, reparations as a method of settling international disputes should be subject to most careful scrutiny. He did not pursue the subject further, and I was happy to be relieved of the discussion of this unexpected and potentially troublesome subject.

28. *South Vietnam Revisited*

Uncertain about the reliability of the small sampling on which his two earlier papers on central and southern Vietnam had been based, Ladejinsky made three additional sorties into the South Vietnam countryside in May and June of 1955. These provide the basis for this paper, the third of his Vietnam field surveys. In transmitting it to Washington, the U.S. embassy stated: "Together with the previous reports in the series, this account by an alert and penetrating observer of what is going on in the Vietnamese countryside and in the minds of those who dwell there is notable as virtually the first such insight which has become available in a score of years."

After summarizing Ladejinsky's analysis of the agrarian situation and his recommendations for a bold emergency credit program, the embassy called attention to a number of other matters on which Ladejinsky throws a passing light, including "a remarkable interview" with the Cao Dai pope. Among these, however, it failed to mention the accepted presence of former Viet Minh among the government's officialdom in the field, the prophetic emphasis on the critical importance of the peasantry in the government's political future, the fact that Diem's government was a weak reed for the United States to lean on, and his emphasis that U.S. aid could at best be marginal in determining the basic political issue.

But all this only begins to suggest the extraordinary richness of this paper, which I believe must be ranked among Ladejinsky's most important and best.

The paper, dated July 16, was transmitted to Washington with Dispatch 61, August 29, 1955—somewhat belatedly for a paper which recommended emergency action on an urgently needed agricultural credit in time to affect the 1955–56 crop.

Introduction

THIS IS THE THIRD in a series of reports on rural Vietnam. The first two, "Field Trip Observations in Central Vietnam" and "Field Trip in South Vietnam," written in late March and late April, had this conclusion in common: the land reform program had hardly commenced in either area, for essentially the same reasons. This observer could not be absolutely certain how accurately these reports reflected local conditions or whether the latter, in turn, reflected developments in a much wider area of the country. Fact gathering in Vietnam is difficult. Nothing is arranged in neat columns from which one may deduce reasonably accurate information. The attitude of the farmer is often even more difficult to fathom; it is not easy to determine whether he speaks his innermost thoughts or the piece implanted in his mind by years of Viet Minh indoctrination. One is therefore never absolutely certain that the picture itself and the conclusions drawn from it are what they appear to be. The only corrective to this unsatisfactory situation is further probing, in the hope that it does not merely compound original errors. It is with these uneasy thoughts that I set out for South Vietnam twice again after my first two reports to observe conditions in the provinces of Tay Ninh, Soc Trang, Bac Lieu, and Can Tho.

The time of the visits, late May and middle and late June, is worth noting because the month of May is likely to be remembered as a crucial one in the history of modern Vietnam. It was the month of the defeat of the Binh Xuyen gangsters; the neutralization of the Cao Dai sect; the emergence of the national army as a fighting instrument; and, by the same token, the emergence of President Ngo Dinh Diem as the undisputed leader of Vietnam. These are all elements pointing in the direction of unification and stabilization of the country.

On the face of it, one might have expected to find that the national government had extended its authority and that the land reform program was being more readily accepted. In reality, recent observations in South Vietnam do not support all of these seemingly logical assumptions. The salutary political effects of the victory over the Binh Xuyen was apparent in Tay Ninh, the stronghold of the Cao Dai; but it has failed as yet to have a similar effect upon the attitude of the farmers toward the reform and the national government. Nor has local administration been visibly affected by the significant events of the past two months. The prominently displayed motto in local offices reading "Make it brief, time is money," is strictly for a laugh and helps only to underscore the lack of zeal and zest on the part of most officials encountered on this trip.

With few exceptions, chiefs of provinces and district officers are disturbingly unconcerned; even the exceptional official tends to engage in merely verbalizing the need for the enforcement of the law rather than concentrating on what little he could do to gain the confidence of the farmers and persuade them to accept the national government as *their* government. The emphasis is on the word "little," for, in justice to this better type of local administrator, he cannot do much until and unless the national government does its part in helping create the climate of "acceptance." The much vaster job of the national government lies beyond the successful trial of arms—it is to demonstrate its appreciation and understanding of the fundamental aspirations of the farmers. Of that there is only the merest beginnings; the people close to the grass roots must shift for themselves as best they know how, and the best is none too good. Unless the situation is radically improved, it will continue to benefit the antigovernment forces.

The Setting

Tay Ninh, Soc Trang, and Bac Lieu are all in southern Vietnam. The first lies about 60 miles northwest of Saigon; the latter two neighboring provinces stretch southward to the very tip of the peninsula, the minimum distance from Saigon being 65 miles and the maximum 200 miles. As settled communities, the three provinces are relatively new; they are all products of French colonial enterprise which created a surplus rice culture and, here and there, rubber plantations, out of swamp and jungle. The process of settlement has long since been completed, but in the remote sections of Bac Lieu, notably in the district of Ca Mau, it is not hard to visualize that only yesterday, as it were, man put his stamp upon nature. The farmer seems to be lost in the vast expanse laced by a few roads and many canals. This is the authentic new land carved out and divided up by the very few who had the opportunity to claim the land and the means to cut the canals needed to put the land to use. In parts of Tay Ninh the rubber plantations, whether in excellent condition or in a state of decay, bear witness to a determined French effort which the small peasant enterprise does not convey. In Soc Trang and Bac Lieu it is not the carefully laid-out rubber plantations that strike the traveler but the huge and unobstructed stretches of rice fields—and nothing but rice fields—flat as a table, reaching out to the horizon and beyond.

All of South Vietnam is a huge rice bowl. Regardless of the output of a given province, the unit of cultivation is small, ranging from 2 to 10 hectares. The standard equipment is the iron-pointed plow, the hoe spade, the coupe-coupe or brush cutting knife, the wooden rake, the weed cutter, the small sickle, and threshing basket; the motive power is the all-important buffalo. Transplanting the rice seedlings is the accepted methods of putting in the crop. Many farmers are familiar with chemical fertilizers but few can afford them. Improved seed is little used, and there is hardly any weed control. The land yields only one crop a year; artificial irrigation facilities are the exception rather than the rule, and the rainy season is the principal source of water supply. The rice yields are low, only 1.2 metric tons per hectare, or approximately half of that of the southern U.S. rice-producing areas, and less than half of the Japanese rice yields. The inescapable impression is that improved farm practices do not exist. The French have indeed helped to open the land of Cochin China (South Vietnam), but it does not appear that the ideas or techniques behind the scientifically managed plantations have been transmitted into the peasant sector. Rice experiment stations exist, but it is fairly clear that very little has been done to place the results of their work at the disposal of the farmers. All of the responsibility for this failure does not necessarily rest with the French, but as direct administrators of Cochin China, theirs was the main responsibility. In any case, the net result is in the total separation of the plantation from the peasant economy and, above all, a low level of farm practices in South Vietnam.

The tractor and the combine are familiar, and most farmers without buffaloes would like to see the tractors do the jobs for them. To find the tractors, one must look for the few French-owned, thoroughly mechanized large estates, measured in thousands of acres and operated with the help of hired labor. The long row of tractors engaged in plowing up 20,000 acres out of a total holding of 65,000 acres owned by the Domaine Agricole de l'Ouest stands out in sharp contrast with the neighboring small holders; they do the work with the buffalo or, lacking the animal, turn the soil by hoe spade. There are many such farmers in Soc Trang and Bac Lieu and in other provinces as well. Rich and relatively abundant land and poor farmers are a common phenomenon there.

By and large, the farmers of these provinces are slightly better off than their opposite numbers in central Vietnam. They cultivate more land and their food supply is more abundant and varied, fish being an important ingredient available without too much effort or expense. Nevertheless, there are pockets of grinding poverty. A trip along the canals of Ca Mau (a district of Bac Lieu) can be a pleasant experience—if officials don't dog your steps, if you are immune to the myriad flies of all sizes and perseverance, if you forget about the security problem in the guise of a dozen armed soldiers in the hold of the boat—if you are oblivious to all of this and your attention is consumed by the soothing motion of the boat, the water and the sky, and the lush vegetation lining the banks of the canal. But these make-believe thoughts snap the minute you leave the boat and make for a peasant hut. Once you are in it, you are reminded of the worst of conditions of central Vietnam. Since the measuring rod is the eye, peasant statements, and one's memory, it is difficult to judge the degree of poverty of one set of farmers as against another. And it is not really necessary; poverty in both areas is a threat to the political stability of Vietnam. What should be stressed is that in Ca Mau (as in most of Bac Lieu) the threat is greater than in many other provinces, for Ca Mau was and still is the most Communist-ridden region of free Vietnam.

When we talk of farmers in South Vietnam, we refer to tenants. In all of South Vietnam the majority of the farmers do not own the land they cultivate. The French claim that "No country in the world could have done so much for the native population as France has done in Indochina." This statement alludes above all to the development of rice culture in Cochin China. Viewed in this light, a good measure of the claim is not without validity, although in the process of opening the land the French did not fail to stake out for themselves hundreds of thousands of choice acres in Cochin China and a total of about 1,300,000 acres in all of Indochina. Rich Vietnamese followed suit, many of the cultivators who followed the "stakerouters" found themselves with no land of their own, and became tenants of other men's land. The depression of the 1930s helped to swell the tenant ranks. They often continued working the same land, but on somebody else's terms.

The size of the tenant group and the acreage they cultivate is important to this account for economic and political reasons, but reasonably correct data are no more available for Tay Ninh, Soc Trang, and Bac Lieu than they are for other provinces. However, certain approximations can be deduced indirectly for the last-mentioned provinces. According to data collected nearly thirty years ago, before the depression had taken its toll, two-thirds of the land of Bac Lieu was in the hands of owners of more than 50 hectares each; if the middle-sized proprietor (from 10 to 50 hectares) is added to the above group, the two controlled 90 percent of all of the land of the province. On the opposite side of the scale are the owners of 10 hectares and less controlling 10 percent of the land. In some districts of Bac Lieu, such as Long Thuy, the big owners alone possessed 85 percent of the cultivated rice land. Since few proprietors cultivate more than 10 hectares of rice land—and it is virtually all rice land—it is reasonable to conclude that perhaps 90 percent of the land is worked by tenants. This explains why provincial and district officers, landlords and tenants simply assume that almost all the land of Bac Lieu is cultivated by tenants. In the district of Gia Rai a tenant was asked to comment on the farmer categories he was familiar with; said he, "We are all mostly tenants and some of us are farmhands." He did not exaggerate by leaving out the exceptions.

Similar clues to the situation in Soc Trang are not available, but on-the-spot sampling shows that tenancy is in the widely prevailing system although not as overwhelming in scope as in Bac Lieu. Like Bac Lieu, rice is virtually the only crop; there are 2,000 to 3,000 hectares of nonrice fields compared to 200,000 hectares under rice. Populationwise, Soc Trang has a strong non-Vietnamese element; of the estimated 250,000 people, 60,000 are Cambodians and 30,000 are Chinese.

Of the three provinces, Tay Ninh appears to be the least tenancy ridden. More than half of the province is under forest; part of the remainder is under rubber plantations (14,000 hectares); and, of the prewar 60,000 hectares under rice, 40,000 to 45,000 hectares are presently under cultivation, providing for a population estimated from 150,000 to as many as 250,000. From the point of view of arable land, Tay Ninh is not in a class with Soc Trang, and surely not with Bac Lieu. It fits better with central Vietnam.

But Tay Ninh is a very special case. It helps to throw light upon the relationship of the Cao Dai sect to the national government and upon the future of the Cao Dai. More than half of the province is controlled by the sect, with its own religious dogma, economics, and politics. The other part of the province is under the control of the national government. Taken together, Tay Ninh is an instance of two coexisting states within one geographic entity. The cold war of recent months between the government and the Cao Dai has subsided; the final settlement is yet to take place, but the odds are in favor of the national government.

So much for the general setting of the three provinces where we searched for answers on the state of the land reform program; the role of the Viet Minh; and, as in Tay Ninh, on the effects of the successful fight of the national government against the sects.

Nationalist Tay Ninh and Land Reform

Tay Ninh province is closest to Saigon, so that is where we first headed. The ease with which we reached and traveled through Tay Ninh was in itself a sign of recent political changes. Only a while back it was almost forbidden territory; leaving the confines of Saigon was not considered safe, and the reaction of the Cao Dai to an inquisitive stranger was open to question. But with the defeat of the Binh Xuyen came the neutralization of the Cao Dai sect, and relative freedom of movement followed.

Since Tay Ninh is largely Cao Dai and the remainder nationalist, the land reform there and the Cao Dai land policies were of practical interest to us. There was another immediate reason for the trip to Tay Ninh. In early May in the village of An Hoa of the district of Trang Bang, there took place a much publicized ceremony of tenants signing contracts. This is not a common occurrence in Vietnam, and for that reason alone a visit to the village was in order.

The trip through the three provinces was not a prearranged one. The reaction of the officials to an unannounced visit varied from friendly in Bac Lieu to barely disguised hostility in Tay Ninh. On one such occasion we could not help but recall an old Russian saying that "an unexpected guest is worse than a Tartar." The explanation was obvious; they did not welcome the questions which were posed and they did not relish answering them. The questions were simple. They bore upon the knowledge of the chief of the province of conditions in his territory and his compliance with requests made upon him by the Ministry for Agrarian Reform regarding the implementation of the land reform ordinances. From the district chiefs we sought more specific information about the villages under their control. We did not profit from the answers in either case.

Do Tuong Thanh, chief of the Trang Bang district, has only seven villages, three of them Cao Dai. In the villages under his supervision the farmers worked, so he informed us, a total of 2,000 hectares of rice fields. He did not know how many tenants there were or how much land they cultivated, but he assumed that there were relatively few tenants. Having made that assumption, or perhaps for other reasons he failed to reveal, he evidently decided that the entire matter was not worth bothering with. There was not a piece of paper—posters of government reform regulations—one might have expected to find in an office which is supposed to have close contact with the farmers. The very subject of land reform seemed foreign to him, and talk about it was clearly an annoyance he did not trouble to hide. Since in Mr. Thanh's view most farmers in his villages were proprietors, the taxation question was raised. More specifically, were farmers paying taxes? He started by saying that since the May events it has become easier for him to deal with the village authorities who are responsible for tax collections. However, he was not in a position or perhaps unwilling to illustrate this shift in attitude. He could have done that by a reference to the tax returns, but this he failed to do and we, naturally, did not press him. The more likely presumption upon which my capable assistant and interpreter, Nguyen Ngoc Linh, agreed with me was that the district chief was no better informed about the tax problem than he appeared to be about the land reform program.

One thing he knew and it was beyond dispute: We could not enter a Cao Dai village without permission of the Cao Dai authorities. Some of the villages, as mentioned earlier, were within his district; but such is the division of authority, enforced by the presence of Cao Dai troops, that those villages might have been in another part of the country rather than within a radius of a few miles. Neither the district chief nor the chief of the province pretended to have any knowledge or interest in the arrangements prevailing in a Cao Dai village.

The chief of the much larger Go Dau Ha district, sixteen villages with an estimated population of 40,000, did not deny the existence of tenants and boldly stated that "we follow closely the government and all our tenants have signed contracts." The crude overstatement of the situation gave rise to a more detailed discussion, the results of which can best be summed up in the words of Mr. Linh, noted for his knowledge of American jargon: "He is giving you a snow job, and he is the biggest damn liar I have ever met." Mr. Linh sized him correctly. He was not familiar with the provisions of the ordinances; there were no land committees in the district; he couldn't exhibit a single signed contract; and he declined to accompany us to any village of his choice on the ground of alleged lack of security. Nor was he forthright in his comments on the Cao Dai villages, of which there were seven in his district.

The chief of the province proved to be a match to his district officers. It is in his office that we first noticed the motto of making it brief because time was money. His own inordinately long siesta, way past the normally generous one, belied that slogan. No great show of activity was being displayed in the other offices as we looked into them while pacing the corridor in anticipation of his arrival. And we wondered about the real value he attached to time and brevity and the efficient, businesslike administration they imply. When he finally appeared on the scene, he proceeded to make a number of observations on how busy he was consolidating village administration through appointments to the village councils, providing greater security for the villages, promoting the land reform program, and how careful one must be furnishing credit to farmers because "it is easy to lend them money but hard to get it back."

Even with the excellent assistance of Mr. Linh, it was not clear to this observer what this official was really thinking or doing about the points he raised. A number of specific questions followed: Is there any person in his office charged with overseeing land reform developments? What progress has been made in the five months since the ordinances had been promulgated? What, precisely, is the position of the tenants regarding the reform? What are the farmers in his province concerned with most? What were his own views on the land issue? If farmers cannot be trusted with money, what other means of credit could be provided for those in need? Are the farmers aware of the events in Saigon, and what is their attitude toward the national government—before and after? To these questions we were unable to elicit answers worth recording. His interests, whatever they were, lay in other fields. It was a relief to part company with him, as with his deputies in the districts, and get into the open, away from this type of negatively inactive administrators.

In the field, among a group of farmers, the errors of commission and omission of the local administration became quite apparent. These were farmers of the village of An Hoa, referred to above; a good many of them were preparing the fields after the first few steady rains. We stopped near one of the fields and, true to custom, farmer after farmer laid down his tools, gathered around us, and puffed away on American cigarettes; they had their say, some guardedly and others freely.

Mr. Thanh's insistence that there were hardly any tenants in his district was not true; the farmers we chanced upon were either tenants or farmhands. It is true that the landlords are not big proprietors; they are just above the category of the landlords of central Vietnam. In this particular village they could boast of only four landlords with 20 hectares each. Most of the others owned less than 10 hectares each. The secretary of the village council present with us could not tell the number of tenant families but ventured a rough guess of about 225. The tenants in the field disclaimed any knowledge of the land reform program; they paid 25 percent rental to their respective landlords and intend to do the same in the future without the benefit of a contract under the land reform. Private deals with the landlords were evidently not uncommon. The farmhands would like to rent land, but they have neither farm tools nor buffaloes nor hope of ever acquiring them. The tenant sets his sights much higher. Renting land under the ordinances leaves him in roughly the same position as before. What the tenants of An Hoa village wanted is ownership of the land. The question they raise was whether the government could not help them buy the land through some credit arrangement. Barring that, they must continue to cultivate the land as of old. Talks with farmers in two other villages revealed a similar pattern. The implications of all of this will be touched upon in another connection, but at this point it is sufficient to note that, in the government-controlled area of Tay Ninh, rent reduction is not an issue. Hardly a tenant will take a local official to task for not keeping him informed about the program, and no official will call a tenant to task for not pursuing his rights under the ordinances.

At first glance the above is in contradiction with the earlier reference to the signing of contracts by tenants of An Hoa village. But it is not. The publicized account of the event bears little resemblance to the existing state of affairs. It would appear that only 20 out of the 225 tenants signed contracts, presumably because those 20 tenants could not get along with their landlords and a written obligation on the part of the landlord was a measure of security.

We did not quite accept the story of the twenty, for we could not find farmers who had signed contracts, while the secretary ignored the suggestion of sampling them. Of course, some few tenants and landlords may have invoked the provision of the land reform legislations, but it would not have changed the picture of what is essentially, to date, only a reform on paper.

Reluctance of tenants to accept the reform is not a novel story; what is different about it in Tay Ninh is that there attitude cannot be attributed to the influence of Viet Minh agents. According to all accounts, the Communists are few in number; perhaps 5 percent of the villages are affected by them. This is a very low proportion compared with other provinces we visited in central or southern Vietnam. The explanation is that the province is dominated by the Cao Dai sect; and the latter, unlike the representatives of the national government, is in active touch with the people. Moreover, for some years the Cao Dai have not seen eye to eye with the Viet Minh, and the Viet Minh agents can flex their muscles only at the risk of swift punishment from the military arm of the Cao Dai sect. The same conditions do not as yet apply in most of the government-controlled provinces.

Cao Dai–Controlled Tay Ninh and Land Reform

What of the arrangements in the Cao Dai–controlled villages?

The local administration of the national government could speak only in the most general terms about the Cao Dai renting out absentee landlords' land to tenants who in turn paid rent and taxes to the Cao Dai. The natural thing to do was to go into such a village to gather the information, but this was easier said than done. While a curious visitor would in all likelihood not suffer bodily harm if he entered a village and addressed himself to the farmers, the fact is that the "curtain" is there against all the non–Cao Dai. Only the Cao Dai authorities could raise it, and this they did without much ado after we stated the purpose of the contemplated visit.

Before we started on the trip and during its early stage, the very helpful Major Luc (and later on the Cao Dai pope himself) had quite a bit to say on the agrarian polices of the sect. In brief, it was a picture of a system under which the man who cultivates the land owns that land. This did not prove to be the case as we observed conditions in new villages near Tay Ninh city established by the Cao Dai, in areas of the so-called "shifting cultivation" or "fire farming" at the foot of the extraordinary mountain with a truncated conical mass which dominates the plain of Tay Ninh, and in old settled Cao Dai communities on the very border of Cambodia. The nearest to the officially expounded policy, and only in a very limited sense, are the few new settlements. There, the faithful receive from the sect pieces of land large enough for a house and a garden patch. The houses are built at the expense of the individuals; they seem to be in fairly good shape . . . The communities give the appearance of planned enterprises; the principal feature is a house site of equal size strung out along both sides of a straight and broad street ending in the rice fields. More often than not the people in the communities neither own nor tenant those fields. As farmers who settled there only three or four years ago, very few of them own land and the only land available to them, aside from the garden plots, is the land they rent upon terms imposed by the owners of the land. They don't mind paying the rent; the problem is that very little land, and particularly good rice land, is available for rent.

The case of one farmer, one of the oldest inhabitants of one of the communities (five years), is fairly typical. He owns one hectare of poor land which he recovered from the forest on which he plants manioc, peanuts, and sweet potatoes. He has a family of eight and ekes out a poor living. He and his like consider themselves "landless" and are eagerly accepting employment at 20 piastres a day, or less than 10 cents a day at the open exchange rate. What he wants is to rent a hectare of rice land, but finding one is not easy, although the land is generally sandy and yields only 0.7 tons of rice per hectare. There are a few landlords in the neighborhood with 30 to 50 hectares each who live in Saigon or in the city of Tay Ninh. They rent the land to middlemen who in turn sublet it at 35 percent of the crop—plus 300 piastres "key money" per hectare. Cao Dai though they are, the farmers complain against the failure of the government, and presumably the national government, to come to their aid. More than one farmer stated with bitterness that "government only benefits the rich and the middle-class farmers." Perhaps out of deference to the major who was accompanying us, no criticism was directed against the Cao Dai leadership.

The new villages, which are few in number, strike the observer as resettlement projects, shelter and garden patches being the principal ingredients. The lack of arable land imposes a severe limitation upon their economic stability, yet they are the authentic Cao Dai villages if for no other reason than that the sect created them. Quite different are the numerous old settled villages controlled by the Cao Dai. There is nothing Cao Dai about them, except for certain levies imposed by the sect.

Official statements to the contrary notwithstanding, the same land arrangements prevail here as elsewhere, with landlords, tenants, farmhands, and small proprietors. There is the same lack of knowledge about the reform program but with this difference: Since landlord and tenants are Cao Dai, Viet Minh influence is virtually nonexistent. Relative political stability has been existing since 1947, so the landlord-tenant relations are more normal, harking back to pre-civil war days. The village councils are made up of the substantial landowners who evidently have had no difficulty in keeping the rest of the farmers in line. Considering the quality of the soil, the rents, which average a third of the crop, are high. In addition, the faithful are assessed 10 piastres a month as special tax, occasionally contribute as much as 10 percent of the rice harvest, and help to cultivate Cao Dai church land where such is available as a special contribution to the church. These "extras" were not treated by the tenants as burdensome; they were rather looked upon as contributions to sustain the Cao Dai. Except for these features, there is nothing especially characteristic of the Cao Dai about these villages. Rent reduction is a minor issue; finding land to cultivate is the big issue. So is credit. "We don't borrow money," said one tenant, because "we tenants are poor and nobody lends to poor people." To some the lack of a buffalo is probably the most important problem. Without the buffalo, the farmer is near the very bottom of the agricultural ladder. The conclusion is warranted that these Cao Dai villagers want the very same things non-Cao Dai farmers are after: land, credit, irrigation facilities, and fertilizers. Outside of their religious proclivities, they constitute just another group of Vietnamese farmers struggling to make both ends meet.

The Cao Dai officials have striven to create the impression that their villages were something special and different. The local officials of the national government have helped to give widespread currency to this false premise by not visiting the Cao Dai villages on the ground of exaggerated security fears. For reasons already cited, the official Cao Dai line on land has no basis in fact. Perhaps the weightiest evidence against the official view is that, despite the presence of Cao Dai troops or guards here and there, in the minds of the farmers "government" is not identified with the Cao Dai but with something not quite comprehensible residing in Saigon. Many farmers failed to identify President Diem; but, when they complained about the lack of land, credit, buffalos, and so forth, they did not address themselves to the Cao Dai as a source of possible fulfillment; instead, they talked of the government in Saigon as a possible source of largesse. This fact is not without political significance.

The universe of the Cao Dai farmers extends beyond the Cao Dai realm; the isolationism imposed by the latter upon the farmers thwarts their hopes of achieving certain minimum requirements and is bound to give way at the first opportunity. That opportunity will come some day soon when the national government will be in a position to remind the sect of the injunction to "Render therefore to Caesar the things that are Caesar's, and unto God the things that are God's." The government will then extend its authority into the Cao Dai countryside. The farmers there who haven't been affected by the Viet Minh blandishments will welcome the move, particularly if the government makes it clear that their aspirations will not go unnoticed.

The Cao Dai and American Aid

When farmers search for betterment and do not find it, inveighing against government is natural. The surprising part was that the United States was not spared from criticism either. We came across one "isolated" Cao Dai farmer who was critical of American economic aid, or rather the alleged lack of aid. A well-spoken farmer stated that for all the talk of such assistance, he had seen no evidence of it. We could not very well convince him that in some way, sooner or later, he may benefit from the agricultural projects and other assistance now being rendered by the United States operations mission. It was all too nebulous for one looking for immediate relief. But the first to mention the subject critically was General Nguyen Thanh Phuong, supreme commander of the Cao Dai forces and major general of the Vietnamese national army, whose headquarters we visited.

The general, currently a supporter of the national government, remarked that he has approached the government for funds in order to create two model villages. His purpose is twofold; to demonstrate that improvements can be achieved and that the job can be carried out without outside aid. By "outside aid" he meant American aid, because almost in the same breath he voiced the opinion that the Vietnamese peasants have never received anything from the United States and that they are beginning to think that "Americans are supplanting the former French colonialists." The role of the Americans is misunderstood, he thought, and added that the United States should try to dispel this impression which adds fuel to the Viet Minh propaganda. We argued the point at some length; it was clear in the end that he used the peasants as a peg on which to hang his own prejudices. It is quite natural for a farmer in the presence of an American representative of USOM to raise the question about American aid in relation to his needs and, what seems to him, the inadequacy of that aid. General Phuong, on the other hand, left the impression that he was speaking a political piece all of his own for motives not too clear.

The Cao Dai Pope— His Economics and Religion

In Tay Ninh the Cao Dai continue to enjoy temporal as well as religious power over the greater part of the province. At first glance, it appears as if nothing has changed since May, after the defeat of the Binh Xuyen sect by the national government. This impression is deceptive. The neutralization of most of the military forces of the Cao Dai is the beginning of the change and more is bound to follow. In the end, only the exercise of the religious influence of Cao Dai will be preserved and probably expanded; Cao Dai economics and politics and military power to sustain them will not survive the accelerated unification of the country under a national government. This is the impression we carried away after a lengthy talk with a former secretary of the customs administration, now His Holiness, Pope Pham Cong Tac, the supreme leader of Cao Dai and a staunch opponent of President Ngo Dinh Diem. One suspects that it was his opponent's resounding victory over the Binh Xuyen that has set the pope to thinking about the unrewarding risks a religious sect must shoulder when it gets actively involved in the tangled politics of Vietnam.

The meeting with His Holiness was a by-product of the arrangement to visit Cao Dai villages. While we were at the military headquarters of the sect to secure the necessary clearance, Major Luc inquired if we wouldn't care to meet the pope himself. We readily agreed. By way of preparation for the session, we visited the Cathedral of the Holy See, designed and built by the pope, over which he presides. The writer's thought was that the cathedral itself might give us some clue about the pope as a person, as turned out to be the case. There are numerous accounts of this unfortunate architectural marriage between a pagoda, a southern baroque church, and something else that defies description. The words employed by visiting commentators range from "fascinating horror" to plain "horror." There is probably no religious edifice in all of Asia which is so altogether out of keeping with the culture of a given people and its physical surroundings. However, when the next morning we sat listening to the pope's melange of ideas on economics, religion, and politics, the architectural aesthetics of his vatican did not seem altogether surprising.

It is a sign of the times and of the road this religious sect has traveled in the not quite thirty years of its existence that the immediate entourage of the pope is not what one would expect of a spiritual leader. We were escorted to his comfortable, upper-middle-class European style house by the military, and the people who greeted us before we were ushered into the pope's presence were the deputy commanding general of the Cao Dai armed forces and his chief of staff. There was not a cardinal, archbishop, bishop, or a priest about. It is quite possible, of course, that his military double in brass.

His Holiness is an elderly man, of slight build, with grey hair carefully combed back, a clean-shaven animated face, dark eyes of the high-powered headlight kind, an ebullient air, and quickness of movement. He was dressed in a long, white silk robe and Chinese slippers; the most prominent part of his attire was a big ring which proclaimed his position as religious leader. Whatever dignity he attaches to his calling, he kept the formal part well hidden. He was all-benign friendliness, his face was wreathed in smiles, and he made every effort to put his visitors at ease. It was hard to believe that this man, the acknowledged leader of some two million faithful, has been in the thick of the country's opium trade and every political shenanigan practiced in Indochina these past ten years.

At our suggestion, the pope addressed himself to three topics: land policies, religion, and politics. He spoke with alacrity, never at a loss for words, playing with them as a good pianist plays the piano. His contradictions did not trouble him; perhaps he was not aware of them. In the lengths he went attempting to prove that Cao Dai is the only universal church, he displayed a prodigious imagination and an aptitude for making assumptions which stagger one for their audacity. His views on politics were both forthright and of the tongue-in-cheek kind. All in all, like the physical appearance of his Holy See, they added up to a "fascinating horror."

As to property, which the pope naturally thought of in terms of land only, he took a leaf out of Blackstone without, we may suppose, having read him. Said the pope: "From a legal point of view only the state is the legal owner of land. If the state did not exist, landlords would not exist." The pope draws the logical conclusion that there are no absolute rights in property and urges the distribution of land "among those who cultivate it." He also introduces an equalitarian principle: "Divide it equally in accordance with the farmers' varying needs." His Holiness believes that his ideas of private property in land are particularly applicable to Vietnam because "the landlords never bought the land; it was all concession land and the sole payments were in the form of land taxation. The government should take the land back and repay the tax in annual installments." But what about those owners who purchased the land? His answer is that "the same system of annual payments should apply."

What about the land reform of the national government in the light of the above disquisitions? He has a ready answer: "It alleviates conditions but doesn't go far enough. What the peasant needs is land ownership. The government should do what we have done—divide the land among those who can work it. The Cao Dai religion forbids the purchase or sale of land among peasants. The land of our farmers is theirs for life unless they move into another part of the country. If a farmer has more land than he can cultivate, we take it away from him. We have carried out our reform not by word but by deed—go and see our villages and find out for yourself." We had gone and we had seen, but we spared him the embarrassment of telling him that he does not practice what he preaches.

The pope is against "capitalism," for the latter, he stressed, "is against justice and humanity." He spoke feelingly of the "Confucian golden mean" as the ideal economic and social arrangement. Above all, he had words of praise for the economic system of the Incas in Peru. What that system was he did not tell us, and we did not press. Since he had utterly dismissed capitalism, and the Inca civilization and Confucianism were but memories, we asked His Holiness what economic system he would recommend for present day Vietnam. He had an answer: "Let the government furnish the land, seed, and implements in return for half of the crop. This would also take care of the taxes normally paid by the farmers." Like so many other Vietnamese, he has a penchant for modern mechanical equipment, but with this difference: he wants it entirely owned and controlled by the government. But would not this enable the government to exercise control over the farmers and perhaps exploit them? Such would be the case, he admitted, "but ours would be a wise government, not a Communist government; and besides this type of equipment efficiently managed would raise the power and the prestige of the government." So much for his economics.

The religious beliefs of the pope, important though they are, are outside our scope, but his theological views, and particularly the feats of imagination which permeate them, throw light on the character of the man who until recently wielded so much political power. Westerners make much sport of the Cao Dai pantheon of

Jesus, Buddha, Lao-tse, and of the other luminaries such as the great Chinese poet, Li Tai Po, Joan of Arc, William Shakespeare, Victor Hugo, Sun Yat Sen, and, rumors have it, Charlie Chaplin. But if we are to believe His Holiness, the heart of the matter, the spiritual heart, is the thing. What his religion stands for is "worship of God and worship of humanity. All other religious leaders are God's creations to do just what we are aspiring to do—except that we do more. Ours is a universal church; we unite all religious beliefs and practices, and thus we return to God's inspired unity of God and man. The spirit of Cao Dai has existed from time immemorial unbeknownst to us; all religions have flourished in Vietnam, and that is why the third revelation of God has taken place in Vietnam and its incarnation is Cao Dai."

We asked His Holiness about the two other revelations. In reply, he laid stress not so much on the nature of those revelations as upon their geographic distribution; a factor pregnant with meaning but allegedly overlooked by all students of religion. Said he: "Moses in Palestine was God's first revelation, followed by the lesser signs of His mercy in the form of Brahma, Siva, and Vishna in India and Laotse in China. The second revelation was Jesus Christ, again Palestine, but extending to Europe as well. And then Confucius in China, Buddha in India, and Mohammed in Arabia." But there was one missing link, said the pope, unaccountably untouched by God's grace, and that was "Indochina, the center of the universe." The seeming omission was understandable because the two major and the number of minor revelations "were only preparatory for the third major revelation which took place in Indochina in the year 1926. It is here that God revealed himself under the name of Cao Dai, uniting all creeds to serve the world and give humanity a chance to regain its conscience. That is why ours is the universal church."

The Pope's Politics and President Diem's Fight Against the Sects

We are too ignorant in matters theological, nor is it within our province to comment on the pope's presentation of the religious meaning of Cao Dai. We shifted with regret to politics, a field in which His Holiness is adept.

He spoke with the same ease on politics. There is much that he left unsaid, and some of his presentation was less than candid, but the sum total of the pope's views have much relevance on the point we set out to test: the effects of the successful fight of the government against the Binh Xuyen.

Throughout the rapid-fire discourse, he made no bones about his pet peeves, the principal one being President Ngo Dinh Diem. His bill of particulars against the president was not novel: (a) he does not represent the people, (b) he cannot work in a coalition, (c) he is dictatorial, (d) he refuses to take any advice other than from members of his own family, and (e) he lacks the flexibility so indispensable to a head of a government. "He is not doing anything right," was his conclusion. When we reminded His Holiness about the coexistence of President Diem's government and the Cao Dai, he allowed that "*We* support it because it is the lesser of two evils, and we must avoid chaos and anarchy."

Bao Dai did not emerge any too well either. To the pope he is a man of lost opportunities, devoid of political sense, soft and self-indulgent. He admits that the partition of Vietnam was Bao Dai's political undoing, but he felt that Bao Dai could have retrieved much if he had done, what he, the pope, urged him to do. He advised him to strive for a constitutional monarchy with a popular base. He wrote to Bao Dai that "When people are in a revolutionary frame of mind out of despair, it is the duty of government to create a good democratic form of government." This particular letter, the pope remarked, was never acknowledged. "After eighty years of French domination," His Holiness continued, "Bao Dai had a good chance to lead the people, but now the South is against him, and so is the North, and the partition of the country finished him." But evidently not quite, in the pope's mind, for at one point he expressed the belief that Bao Dai would return, from his own government, and fight President Diem. What will Bao Dai fight with? The pope groped for an answer and finally observed that: "Not all national troops are pro-President Diem, and other revolutionaries (the term he used) like Hoa Hao and the remaining Binh

Xuyen would flock to his support." He did not mention the Cao Dai for reasons to be explained presently. Generally speaking, the possibility of Bao Dai's return was one of the least convincing parts of the pope's pronouncements; it may well be that he brought this subject up only because the wish is the father of the thought.

The pope was particularly vigorous about recent events in Saigon, and apparently on the defensive. He would have us believe that he created the National United Front (in March) made up of the Cao Dai, Hoa Hao, and Binh Xuyen to strengthen the government, "for everyone must play a part in creating good government." He was not unaware of certain problems inherent in his move: "I recognize there are bad elements among the Binh Xuyen, Cao Dai, and Hoa Hao; but we can best be rid of them within the united front organization," and, he added, "The best way to disarm the sects is only when they are represented in the national government."

He deplored the fighting in Saigon. Bao Dai acted illegally when he sold the rights of policing Saigon to the Binh Xuyen, and President Diem should have demanded the right to supervise them. But, he said, "The President had no business trying to disarm them by force and then create his own police force." Now that this particular fight is over and important military Cao Dai leaders have proclaimed their allegiance to the national government, "Is it correct to assume," we asked His Holiness, "that the Cao Dai will throw in its fortunes with the national government, incorporate its forces with those of the latter, and fight a common battle against the remnants of the Binh Xuyen and the Hoa Hao?" He did not relish the question, but what he did say complainingly was as follows: "The government has integrated only 8,000 out of our total armed force of 25,000. This puts me in a quandry about the remaining 17,000. Had the government integrated the entire force, my position would have been clearer." "Would not then the real integration of the Cao Dai forces cause the elimination of the political power of the Cao Dai?" Without hesitation he replied in the affirmative. "Is it in your interest," we further queried His Holiness, "to give up all the temporal power after the years of effort to build it up?" His reply was the high point of the interview, although it calls for qualification. He stated that, "Yes, it is in our interest to eschew politics. It is time that we go back to religion and to the original aim of Cao Dai. By giving up the army we shall extend our international position as a religious force. We do not wish to tie the destiny of Cao Dai to the destinies of Vietnam. Cao Dai is the universal church, and we shall go back to fulfilling our original mission of a creed which unites all creeds."

This was indeed a remarkable statement from the lips of a man steeped in politics, the man who only two months back was the prime mover of a scheme to unseat President Diem. His expressed desire, therefore, to go back to religion is something more than meets the eye. That "something more" is the price that he must pay for a political failure. Without impugning his devotion to the fortunes of the universal church, it must be stated that should he, this time, begin to practice what he preaches, it would be a direct consequence of the defeat inflicted upon the Binh Xuyen by the national government. This in turn has played a part in the neutralization of the Cao Dai armed forces. If this state of affairs continues, the political arm of the pope becomes virtually nonexistent. Viewed in this light, the pope's current aspirations to devote himself to the fulfillment of the original mission of his church are understandable. For the same reason, it will not be surprising if the national government strips Cao Dai of its temporal power. When that comes to pass, the "state within a state" practice will be eliminated, and along with the elimination of the Hoa Hao sect free Vietnam may at long last achieve political unity. This is the principal impression we carried away from Tay Ninh as we were winding our way down south to Soc Trang and Bac Lieu.

Soc Trang and the Land Reform

The province of Soc Trang, unlike Tay Ninh, is not half nationalist and half sect. It is all nationalist controlled. So is Bac Lieu—at least on the surface. Therefore, one is in a position to inquire into agrarian developments of a province as a whole. We are dealing here with areas of big holdings, a vast number of tenants, and relatively small groups of owner cultivators.

On the face of it, these conditions are much more conducive to land reform than in Tay Ninh or provinces similar to Tay Ninh. How then did the reform fare during nearly six months since its inauguration?

The answer is that it did not fare well at all. The reasons are not dissimilar to most of those already described in previous reports. But there is a new disturbing element in the picture: whereas in other provinces the tenants were the more reluctant to sign contracts under the reform provisions, here many landlords feel similarly. They do not appear to be eager to insure for themselves a 15 to 25 percent rental merely because they have been getting little or no rent in the past ten years. On the contrary, some of them insist on retroactive rent payment as a condition of leasing the land; others wish to take back that portion of the land which the Viet Minh redistributed among the tenants, while most of them feel that the time has come to secure a degree of control over their land similar to that of pre-civil war days.

Essentially, their attitude is an expression of their belief in the stabilization of their provinces and the unquestioned support they are getting from the local administration. Their confidence in the security of the province is a welcome sign of changing political conditions, but from the point of view of a countryside rent by social conflicts, the position they are taking is deplorable. Turning the clock back may be good for the landlords, but it does not take much political astuteness to note that what is good for the landlords is not necessarily good for free Vietnam. The strains and stresses this attitude is creating might prove too much for the welfare of Vietnam.

Once in the capitals of Soc Trang and Bac Lieu, it is apparent that the officialdom is not engaged in the promotion of the land program. It is Tay Ninh all over, even if with variations. In Soc Trang officials were candid enough to say that they lack the administrative machine to cope with the problem and that the district officers do not venture outside of their offices because "their influence does not warrant the trouble of a trip." What was more disturbing is the actively negative attitude of the new civilian chief of the province. He told us with utter frankness that as a landlord he is opposed to the reform and doesn't intend to help enforce what he does not believe in. "We have been robbed by the Viet Minh over the years," he stated, "and we resent similar treatment from the national government." He was going to appoint new village councils and insure their protection with detachments of auxiliary provincial guards. He spoke like the economic man pure and simple, undisturbed by all the other issues which gave rise to the reform. He was speaking for many others, the difference being that "others" are not in the crucial administrative position he is in.

One need not be a chief of a province and a landlord to view the program with disdain or to be actively unconcerned with it. The chief of Bac Lieu province, a very self-confident major in the national army, ready to take on any subject with equal glibness, told us with evident relish that everything was under control. Naturally, he did not control the situation at all. He called a rally of landlords and tenants, urging upon them the acceptance of the program, and then promptly concluded that they have accepted it. A trip in any direction from his capital would reveal, as it did to us, total noncompliance and much more besides. The seething discontent in an area notorious for its Viet Minh infiltration stood out in sharp contrast to the verbiage of the happy-go-lucky, back-slapping major.

When pressed for an estimate of program fulfillment, the Soc Trang officials ventured a figure of 5 percent. It is a low figure and may be a correct one, although in our trips out into the country finding a signed contract was like finding a needle in a hay stack. Perhaps the encouraging part was that in the first village we visited we met with a group of farmers who had a very good idea of what the reform provisions were about. In this case, familiarity bred contempt for the reform, the main reason being that the range of 15 to 25 percent rental meant 15 percent to the tenants and at least 25 percent to the landlord—with no common ground for agreement whatsoever. There was not a contract in this village. The landlords, for their part, blame the impasse on the farmers' union (Nghiep Doan Nong Dan) to which a good many of the tenants of the village belong.

We visited the union in its offices in Soc Trang. It was obviously performing a union function judging by the prominently displayed signs: "Have you paid your dues?" But there were more pertinent indications of its activity. Many farmers were there to seek advice; we listened in on the proceedings, and it was clear that the union took a position for a rental of 15 percent or less, depending upon the quality of the land. With or without the union advice, the tenants were concerned with matters beyond rent control. They pleaded for ownership of land. The spokesman of the village group made the point that "The village needs peace and that the landlord-tenant conflicts will never cease under the rent reduction program." They pleaded for credit and for lower-priced fertilizer. They blamed the government for not doing any of these things. Said one, "We can support the government only if it supports us."

This was one of the most articulate groups of farmers we met. It was not an ex-Viet Minh village with a Viet Minh overtone but one with deep-rooted problems about which the farmers were actively concerned. The official tendency is to brand such outpourings as Communist inspired; we have encountered this attitude on a number of occasions, but officials know too little of their own villages, where they seldom appear, to permit them to judge the motives behind the demand of the farmers. Unquestionably, the tenants of the village of Tai Sun argue for the lowest rent possible, but the failure to comply with the reform cannot be placed solely at the door of the tenants. More security as a means of scaring the tenants into accepting the unpalatable provisions of the rent law is not the answer. The issues go beyond rent reduction and beyond any attempts to ensure law and order in the countryside by relying upon larger armed forces, local or national.

Soc Trang is a province with large French landholdings. What of the application of the reform on such properties?

One of the principal French holdings is the Domaine Gressier of nearly 70,000 acres of rice land, scattered in a number of large chunks. The one we visited in Thanh Tri district has an area of 10,000 acres. We approached the administration headquarters, or the manor house, through one of the worst rural slums we have seen in Vietnam. The manor house is like the slums, both outside and inside. The stagnant canals which crisscross the estate, the rickety bridge on the main approach to the house, the filthy yard with farm equipment strewn in helter-skelter fashion, the huge boilers as the only remains of the rice mills fired by the Viet Minh, the three enclosures of skillfully arranged barbed wire surrounding the house, and the three-story house pock-marked with bullets—all these bear witness to a formerly besieged fortress and an enterprise in the last stages of decline. But to the French resident manager, who has watched over this estate for ten years, things are looking up. The Viet Minh have gone and so have the two army units which were posted on the estate. Only a small company of twenty armed guards still keep an eye on the immediately surrounding countryside.

Contrary to what might have been expected, a semblance of economic activity never ceased on the estate during the civil war days. Some of the land was cultivated and rents were collected by posting guards at the canal exits; since the only means of transporting rice from the field was the canals, rent payment could not be easily evaded. There were more important reasons why the French owners held on to their estates. What they failed to collect in rent and whatever damages they sustained through the depredations of the Viet Minh, the French government made good through a special war damages fund. The provision that such funds did not have to be reinvested in Vietnam was another reason explaining why the owners held on so tenaciously in the midst of unfriendly surroundings. Monsieur Archier admitted that, while the exportable surplus was way down, "we weathered the civil war in good shape." He was sympathetic with the Vietnamese owners who had no claims against the French government for economic assistance. Unlike some other French rice plantations, the entire acreage is rented out to 700 tenant families, rents, so he told us, ranging from 15 to 20 percent, depending upon the quality of the soil and the distance from the seat of the plantation. In addition to collecting rent, the administration runs two rice mills and maintains a few tractors to plow the fields for those who have no buffaloes.

Monsieur Archier seldom leaves the estates; his only connection with the outside world is teletype communication with the Gressier interests in Saigon. He was indeed pleased to see us and the chief of the district who accompanied us. He discussed the eventful years calmly and matter-of-factly; but, when we touched upon the land reform program, he grew livid with indignation. "The government," he said, "is more revolutionary than the Viet Minh and it is too bad that the land reform program is not accompanied with a propaganda campaign to remind the farmers that not everything is permissible. Doesn't the government know that the farmers are only waiting for 1956 in the hope that the Viet Minh would come back?" We countered by saying that since the rents the estate charges compare well with the rent reduction provisions, why not accept them. He would have nothing to do with the reform, not because of the official rent rates, which, upon examination, proved to be lower than those charged by the estate; it was the mere idea of bargaining with the tenants which was so repugnant to him. This meant a break with a seventy-year-old custom of setting the rules of the game and expecting the other fellow to accept them without question. Now he was baffled and distressed, not by active opposition of his tenants which, incidentally, is not yet apparent, but by the inescapable feeling that they harbor new thoughts, that they listen to the Viet Minh propaganda and consort with the tenant union agents, people with whom they should have no truck.

Monsieur Archier spoke warmly of Vietnam; his wife is Vietnamese, his children are raised in Vietnam, and he expected to spend the rest of his years in Vietnam, but evidently only on his terms. Since these are not likely to obtain any longer, he expressed the hope that the Gressier would sell the land, cheaply if necessary, and wind up their business in rural Vietnam. The question which had been raised but left unanswered was, who was going to buy the land and pay for it, even at the reasonable price of $50 a hectare? Who indeed?

On this note we left the Gressier rent-collecting institution and drove into the Gressier slums for a talk with the tenants. It was not easy at first; they could not see why we should be interested in their relations to the Gressier and clearly suspected our motives, until Mr. Linh convinced them that our motives were above suspicion. It should be noted here that the chief of the district was not helpful at all. He was parked behind our jeep in his Land Rover, sitting there in his gleaming white "sharkskin," bored with the whole business but with no intention of getting out of the Rover. This former employee of the Ministry of Public Health, turned district chief since June 1954, had never visited this or any other village; the farmers did not know him and he did not know them. The two never met, not even while we were there.

As we visited hut after hut, the Gressier-tenant relationship became obvious. Every tenant had a little green, paperbound notebook, with "D.G." (Domaine Gressier) printed on the cover. Inside, it merely stated that farmer so-and-so rented so much land at so many units of rice per hectare. These figures revealed that the actual rent greatly exceeded that cited by the manager. In some cases they were nearly twice as high. However, the farmers admitted that they were lower than in the pre-civil war days. The rent notation was the only visible evidence of terms under which the tenant worked the land. This is quite understandable if the tenants are to be believed that numerous other conditions that normally govern such a relationship are decided solely by the management. A tenant, it appeared, is on a particular piece of land one year; he may be given other land the next year or put off the land altogether if he fails to comply, whether by design or not, with the unwritten conditions—old or new—prescribed by the management. This may explain why this company village contains not a few farmhands who were formerly tenants. Above all, the existing rentals, if nothing else, explain why the management is so adamantly against the rent reduction program. The tenants had heard something about it from a union agent, but not one of them could cite the rent provisions. This is not surprising, for the prime sources of such information and education are the actively disinterested chief of the province and his local arm, the chief of the district, the bored Buddha in sharkskin, intent on his dignity to the very end. Not all officials are cut of the same cloth; there are exceptions. The chief of the Ca Mau district of Bac Lieu, an ex-Viet Minh adherent, claimed no progress for the land reform and had a good idea where the difficulties lie and where the first remedies should be applied.

Bac Lieu and Land Reform

In an earlier section of this report, an attempt was made to sketch the place of Bac Lieu in South Vietnam. It dealt mainly with the historical and so-called permanent factors which conditioned the development of the province. There are also disturbing elements of recent origin, all of considerable importance. The first, the visible one, is the inundation of tens of thousands of acres of land by sea water in the wake of the washed-away dams which once kept the sea in its place. Like so many other dislocations, this is one of the effects of the civil war. A trip from the city of Bac Lieu down to Gia Rai and on to Ca Mau bears ample witness of stricken land. As one farmer put it, "A wet bottom is good for our rice land, but a salty bottom is good only for brush." Thousands of farmers have been compelled to look for land elsewhere. The draining and rebuilding of the land is a top-priority task, the realization of which cannot be postponed indefinitely.

Point number two is that the buffalo is very much on the mind of the tenants—and almost all farmers here are tenants. It is not easy to convey to the reader the single-mindedness of a tenant who has no buffalo unless he has watched—let alone tried—the "plowing" of a rice field with a hoe spade. Without the buffalo, we pointed out elsewhere, the tenant considers himself landless though he has land to cultivate. When, in our presence, an official once suggested to a group of farmers that tractors might take the place of buffaloes, the reply was: "Give us buffaloes." If it is true that in Ca Mau the pre-civil war buffalo population of 30,000 has declined to about 3,000, then the problem is indeed formidable. This estimate is probably greatly exaggerated, but the theme "we want buffaloes" has recurred much too often not to be a problem of immediate importance and urgency.

Point number three is that, to the farmers we talked to, rent reduction as such is not a great problem. We have met with this situation in other provinces; but in these districts, or rather sections of districts, not many landlords have felt safe enough to return and take stock of their holdings. Where they have returned, the situation is quite different. On one occasion, we asked a farmer whose land he cultivated. His reply, "I don't know," was not surprising. When he abandoned his land to the salt waters, he moved to another piece of abandoned land without troubling to find out whose land it was. In reply to a question, another farmer had this to say: "A week ago a man came by and told me that I cultivate his land. He did not speak to me about rent but told me to work the land and take good care of it. I don't know the man's name and I don't know where he lives." No one present doubted the veracity of his statement. These farmers are not concerned about rent—not yet. Those who know the owners of their land had no opinion on the subject, having heard very little about the reform program. Every one complained about the lack of something; if it wasn't the lack of a buffalo, it was the lack of farm equipment, seed, credit, or drainage facilities. One farmer summed it all up: "Our condition is difficult." We asked him if he was better off ten years ago; he said he was, and that then his landlord supplied him with all the necessary means to work the land. "Would you like to see the landlord back?" "No, let the government give me three hectares and a pair of buffalo, and I will pay for them in installments." Clearly, he did not want the landlord back; his ideal was a version of the American "forty acres and a mule."

The outstanding impression was the poverty of the farmers, although most of them paid little and some paid no rent at all. There is no evidence that the activities of the Viet Minh among the farmers resulted in any substantial benefits despite the redistribution of some confiscated land of absentee landlords and the admonition of the Viet Minh—"work all the land you want." A couple of farmers were bold enough to state in the presence of officials that the Viet Minh were good to them; no rent, and tax on output was only 5 percent. This does not accord with other evidence. Moreover, it is obvious that after nearly ten years of Communist rule over the countryside the farmers are indeed in bad shape.

They look to the government for assistance; the question is whether the national government will take the opportunity and help fill in the gaps. If it does, it must go beyond rent reduction. All other needs the farmers mentioned are by far of greater importance. The farmers believe that the government has not yet made known its concern for their welfare. So far, the government is identified here with military pacification activities since the takeover in February. A farmer expressed this idea, in this manner: "National government is active only when we see national government troops come through." Nor can one place the burden of responsibility upon the local officials. The offices of the chiefs of province and the district offices are not equipped to carry out government policies in the villages even if they wanted to. They haven't the people, to begin with. Bac Lieu with its enormous rice area of 300,000 hectares has one agricultural officer, and there is not another in any of the four districts into which the province is divided. This fact has not been lost on the district chiefs of Gia Rai and Ca Mau, particularly the latter. While they are not agricultural specialists, they know only too well the effects of the Viet Minh upon the farmers; they know that the eradication of that influence is not a military task.

Ca Mau and the Viet Minh

Throughout the civil war Ca Mau was one of the principal centers of Viet Minh activity in South Vietnam. It was also the staging center of Viet Minh armed forces preparatory to their removal to North Vietnam in compliance with the Geneva agreement. According to local officials, they left behind very potent ideas and agents among the farmers. How deep this influence runs or how numerous the agents are, no official would venture to guess; but the district chiefs of Gia Rai and Ca Mau agree that in a free election held today the majority of the farmers would vote for the Viet Minh. Wherein lie the causes for this support and how can they be eliminated? An attempt to answer these queries had been made in the previous reports. However, since Ca Mau and other sections of Bac Lieu are special cases of Communist infestations, the problem can fruitfully be touched upon again, particularly as the district chief of Ca Mau, a former laborer in the Communist vineyard, was willing to air his views on the subject.

At one point, early in our discussion about Viet Minh tax practices, he casually inquired of one of his assistants: "Can you recall how we taxed the farmers in our Viet Minh days?" It might be appropriate to note here that ex-Viet Minh in the service of the national government is a common practice. Nobody raises an eyebrow; nor did we on this occasion but simply proceeded with the business at hand. It was clear that the ex-Viet Minh official and two of his assistants, also ex-Viet Minh, had a good knowledge of the Viet Minh tax practices. Without referring to the Communist tax bible, "The Agricultural Tax," published by the Communists on July 15, 1951, with that singular title page inscription, "The Democratic Republic of Viet Nam, Seventh Year, Independence-Freedom-Happiness," they launched into jotting down on a sheet of paper the ABC's of the Viet Minh tax practices. They coincided fairly closely with the items contained in the Viet Minh tax book.

Tenants, owner cultivators, and landlords all were subject to taxation under the Viet Minh, the scale ranging stiff-stiffer-stiffest. Exemptions were almost nonexistent, since the Communists levied a 6 percent tax even on those so-called producers whose output was only from 61 to 75 kilograms per capita, far below their own consumption requirements. To be sure, hardly any farmer in Vietnam produces that little, but the Viet Minh position that even the smallest of the small producers is subject to tax is significant. What this district chief told us confirms the official Communist dictum that "The agricultural tax concerns everyone who has crops, be he landowner or tenant farmer . . . It is obvious that the agricultural tax must be levied on the vast majority of farmers who will pay a rational part of their yield, a part which will not affect their normal standard of living and at the same time will concentrate a relatively large amount of food." What is that "rational" part which, in the words of an old saying, would keep the sheep alive and the wolves happy?

Leaving the landlords out of the picture, these are the tax burdens a tenant and an owner cultivator have to bear under the Viet Minh. If, for example, an owner cultivator works 5 hectares yielding 500 units of rice or 10,000 kilograms and has a family of five, he must pay a tax on output of 2,000 kilograms per capita. This puts him in the 31 percent tax bracket, which in turn is supplemented by a basic 15 percent tax, so that he pays a total of 46 percent of the output. If a farmer (tenant) cultivates 5 hectares of somebody else's land yielding 10,000 kilograms, he is taxed on the production of 7,500 kilograms, the rest being charged presumably to rent. On this basis and with the same number of mouths to feed, he pays 27 percent plus the 15 percent basic tax, or a total of 42 percent. The majority of tenants of Vietnam produce from a 1,000 to 1,500 kilograms per capita. The Viet Minh enforced the tax law with their customary vigor. The tenants were paying a tax ranging from 36 percent (21 plus 15) to 41 percent (26 plus 15). Assuming that they paid no rent at all, the burden was perhaps only slightly less than that in the pre-Communist days when the landlord's rent was the main charge.

The claim of two tenants that the Viet Minh were "good to them" because of the alleged small, 5 percent tax seemed questionable to us, even if allowance were made for a margin of error in favor of the tenants. The ready official answer that they must have been pro-Viet Minh did not explain why their allegiance lies in that direction. More particularly, it did not explain why the local officials were so certain that the majority of the tenants would vote for the Viet Minh despite the economic burden imposed upon them, especially in view of the fact that the organized military force of the Communists has gone north.

We addressed these questions to the ex-Viet Minh official. The principal answer he gave is roughly the same we heard in central Vietnam and in other provinces of South Vietnam: the farmers did not look upon the rice tax as economic exploitation. If our informants' statements were correct, the farmers themselves practically administered the rice tax because of their supposed conviction that theirs was but a voluntary contribution to the People's War for Liberation. To promote that particular goal, the Communist leadership succeeded in convincing large sections of the people that they, the people, truly matter and that the economic imposts and other hardships were self-imposed by the multitude rather than by a small group with interests all of their own. This altogether new approach of not merely telling the people what to do but of taking them into the Viet Minh confidence and consulting with them impressed and flattered the farmers. It gave them a sense of participation and belonging never experienced before, and the support of the Viet Minh was a natural development. The chief of the district, certainly above the average cut in intelligence, stated that until 1953 he, too, a non-Communist opponent of the French, accepted the gospel of the Viet Minh. As if in self-justification, he cited the life of extreme austerity of the Viet Minh soldier combined with a discipline that was much admired by him as well as by the common people. "The Viet Minh soldiers," he said, "lived with the people; they lived frugally, and people pitied them and loved them. The people remember that the national troops fought under the French leadership and did not behave well. It will take time to dispel these memories."

Terror as a method of enforcing the "will of the people" was among the noneconomic factors he cited. But in his opinion leadership and organization as means of identifying the Viet Minh with the people were by far the more potent weapons. He called this to our attention when we raised the question why the Communist influence continues to persist at a time when to all outward appearances the Viet Minh have gone. In his view the Viet Minh organization is intact in Ca Mau and in other parts of Bac Lieu. Only the methods of operation change some. "They are not overtly active," he continued; "they behave in a manner calculated to minimize our activities. I don't know the leaders—the roots; I know only some of the agents and the followers—the leaves. To combat them we must have the people to eradicate the roots. Leaves grow with every season even if we pluck them between seasons."

Our final question was the natural and most difficult one: "How would you combat the Viet Minh influence?" The district chief began by pointing to the rent reduction as a good first step if all other needs are dealt with simultaneously. Although the tenants display no enthusiasm about the reform, he believed that with the return of the landlords, land reform may assume considerable significance in the eyes of these very tenants. However, this was merely an introduction to his main theme that the Viet Minh are operating in the villages without any opposition, that it was high time for the national government to combat them on their own ground, employing a method as simple and subtle as that employed by the Viet Minh. The method which he expounded may be summarized as follows: In appearance a Viet Minh agent looks like any other villager of comparable age. He lives and by-and-large behaves like most farmers. He is undistinguishable from his surroundings. He is always concerned with public opinion. Because he is in underground status, he sells his views and actions retail rather than wholesale. Said the district chief: "When he walks the village street and sees an old farmer bending under the burden of a heavy rake, he will be right there saying 'Father, may I help you?' When he learns that a widow is short of help in the field, you will find him deep in the muck of the rice paddy working his way into public opinion. No job is too hard, too big, or too small for the agent."

To the listener this sounded like an idealized view of the Communist agent in action; there may have crept into his account echoes of his own former association with the Viet Minh and the frustrations his current job entails. Allowing for all these, the undisputed fact is that in Ca Mau and Gia Rai the Viet Minh agents have the run of the countryside for their efforts. Making friends and influencing people is surely not a Communist monopoly. Their hold can be broken and eradicated. More troops and more security have only limited value; they fall far short of solving or eradicating the tensions which are so easily exploited by the Communists. The nationalist government must challenge their monopoly by giving the kind of aid the farmers hope to get as well as by doing the kind of work the district chief was talking about—work for which there is no substitute if the government and the farmers are to come closer together.

Minister Thoi Meets the Farmers

The observations made thus far are a product of brief visits without prearrangement or fanfare and of discussions with individuals or small groups. The subsequent paragraphs, however, which may well serve as an indication of what is on the mind of the tenants and landlords of South Vietnam, derive from an entirely different and, to us, unusual set of circumstances. The reference is to a trip undertaken by Mr. Thoi, the minister of agrarian reform, in order to study at firsthand agrarian reform developments in the provinces of Can Tho, Soc Trang, and Bac Lieu.

This observer was part of the minister's entourage. It was a big one: five carloads of officials who had not been out in the field since the inauguration of the program six months ago. Everything was prearranged with pomp and circumstance—the priority crossing of the ferries, the flag-bedecked cities (notably Bac Lieu), the brass bands, the honor guards, luncheons and dinners (printed menus on two occasions), and hundreds of landlords and tenants attending the meetings. We had some grave misgivings about the value of the enterprise—until the first tenant had had his say. Our fears were understandable, for as we looked at the visitors and local dignitaries ranged around the speakers' tables and the first row of the audience, we saw mostly impeccable white "sharkskin" contrasting sharply with the nondescript dress of the mass of the people occupying the rest of the hall. The thought that the proceedings might degenerate into just another variety of a "Potemkin village" was very strong. But our fears were groundless. Times have changed. Landlords *and* tenants spoke their minds with utter frankness and ease. The presence of the notables did not inhibit the tenants in the least. Stiffness and occasional double-talk was largely a contribution of officialdom. In retrospect, the trip was one of our most rewarding experiences to date, even if deeply troubling.

From the very outset, following the usual exchange of amenities between the minister and the chiefs of province, one sensed that the atmosphere was charged with tension, suspicion, and a seemingly irreconcilable conflict between the tenants and the landlords. All the hopes, aspirations, and fears were written large in the faces of the contending parties; the landlords argued their position with considerable restraint, while many of the tenants spoke with passion and anger, leaving no doubt in anyone's mind that for them land rents and security of tenure are matters of life and death.

From the experiences of this trip, a number of facts have emerged. The principal one, that the land reform program is not off the ground yet, was not new. It was not necessary to take a vote as to how many tenants and landlords present at the meetings have signed contracts under the provisions of the ordinances. Much to the embarrassment of the major, who had two weeks earlier claimed that everything was under control, a landlord of Bac Lieu stood up and stated that in his estimate more than 80 percent of landlords and tenants have signed no contracts. He underestimated the degree of noncompliance, for we found no landlord or tenant present who was willing to admit the existence of such contracts. It was apparent, too, that both parties were poorly acquainted with the provisions of the program. This was obvious from such questions as: Has the proprietor the right not to renew a contract upon its expiration? Can he take the land back for nonpayment of rent? Can the landlord recover the land on the ground of self-cultivation? Is a verbal landlord-tenant agreement valid? Has the tenant the right of cultivation in case his landlord sells the land to somebody else? Have nonabsentee owners the choice between contract A and B under the ordinances? The answers to these questions are in the text of the provisions; the responsibility for the lack of familiarity is largely that of the central and local authorities who have made virtually no effort to publicize and explain the ordinances. Failure to answer these questions in the field are evidence of the total absence of an administrative machine to implement the program.

A host of other questions was raised, for which the text of the law is no help but which furnished ample food for thought to the very capable deputy director for agrarian reform, Nguyen Manh Tu. Some of these are: How can the ineffectiveness of the joint committees at the various levels be remedied? How can the conflicts between the tenants and owners be resolved in case of nonconciliation through mediation? Considering the current state of the committee, is it advisable to have the agrarian disputes settled through the juridical courts? What is the meaning of the term "principal annual crop"? Can the owner charge rent for both crops of rice if there are two successive crops of rice a year on the same land? What is the minimum acreage a tenant may work by himself? How can the rent provision be changed to minimize the landlord-tenant conflicts? Can a single rate be effectively applied, and what of the consequences of such a move at this date? In cases where the Viet Minh distributed land among the tenants, can such land be taken back by the landlords if they refuse to sign contracts or pay rent? Can the landlord recover such land if the tenants wish to sign contracts and pay rent? While the dispute is on, who enjoys the right of cultivation?

These unresolved questions and those mentioned earlier indeed handicap the implementation of the reform. It would be a mistake, however, to assume that they are the principal cause of the failure to accept the program by landlords and tenants alike. Perhaps the principal reason is the rent provision calling for a payment of 15 to 25 percent of the crop. This is the major factor that divides the tenants and landlords into two camps. The former wish to pay no more than 15 percent and the latter will not lease the land for less than 25 percent. What is meat for the tenants is obviously poison to the landlords, and vice versa. References to the past, when rentals were closer to 40 or more percent of the crop, make no impression on the tenants; the immediate past of little or no rent is what the tenants remember best. And besides, "How can we pay a higher rent," says a spokesman for the tenants, "when most of us have no credit, no animals, or seed. Our children are hungry because they have no rice and our buffaloes have died. Landlords always say tenants don't cooperate; how can we when the landlords don't understand our plight? Let the government understand that our three needs are land to cultivate, credit, and buffaloes. If we don't have them and the Viet Minh invite us to their parties, we go to get a bite; but if we had enough to eat we'd rather stay home." The fundamental split on the rental provision is also a sign of the changing times. The 15 to 25 percent rent provision is closely tied up with the productivity of a given piece of land and who determines the final estimate. "In the old days," a tenant remarked, "what the landlord said the land produced was law and I paid accordingly. Now I know what it produces and I don't accept his estimate any longer." He voiced a break with a practice which formerly demonstrated to the tenant that the landlord had undisputed control over the land. His challenge of this custom is bound to widen the gulf between landlord and tenant. It all adds up to one big conflict which, if not averted, may be at white heat before much longer.

The position the tenants took did not surprise us; it merely underscored what we have observed in other provinces of Vietnam. What was strikingly new about these meetings was the much stiffened position of the landlords. Whereas only a short while back we met landlords willing to abide by the provisions of the land reform program on the theory that a bird in hand is what counts, while the tenants were the reluctant ones, the landlords' more recent stand in these provinces reveals that they are in no hurry either. Their insistence on 25 percent (or more) is not the sole factor. Not a few among them insist on retroactive payment of rent as a condition for signing a contract; in one case the landlord demanded eight years back payments of rent. Not a few are trying to get land back from the tenants for a variety of reasons. In the process there have been instances where local officials exercised their authority to the point of jailing tenants for failure to clear off the land. These revelations were the most dramatic incidents of the sessions. They were pin-pointed by tenant after tenant reciting personal experiences. On a few occasions tenants addressed their landlords present among the assembled, accusing them of attempting to deprive them of the land or of demanding retroactive payment of rent. Such confrontations resulted in charges and countercharges mostly to the embarrassment of the landlords. It would appear that the landlords are trying to reassert their pre-civil war position with regard to the control of their land. This seems to be a recent development, particularly in areas where security has been re-established and help from local authorities can be safely counted upon.

Such moves on the part of landlords lead to the extreme polarization of the village community and, by the same token, to serious political consequences. This is not simply the writer's speculation. A tenant suggested the possible outcome. Without raising his voice and addressing himself to the landlords, he made this observation: "In my village three tenants were recently deprived of their rented land because they refused to pay the landlords a rental of 35 percent of the crop. Ninety percent of the farmers in the village are tenants. If the landlords should continue to dispossess them in order to get higher rent and prosecute them for disobedience, the conflict will end in bloodshed." This was a threat which could not have gone unnoticed. Other speakers followed; and although no one else mentioned bloodshed as a means of settling landlord-tenant conflicts, enough had been said to give pause to the most unreconstructed landlord bent on maintaining the previous status quo. On the other hand, as one watched the landlords defend their interests, without hardly a suggestion of enlightened self-interest, one was ready to doubt whether the lessons of the past ten years have left any serious imprint upon those whom the events should have affected most.

One other development was typical of all the proceedings, namely, the question of land ownership versus tenancy. In Can Tho, for example, this came in for more discussion than rent reduction; the same, if to a lesser degree, was true of Soc Trang and Bac Lieu. The idea of land ownership has the keenest appeal to the tenants. Once it is mentioned, they don't let go of it. Under their prodding the visitors and the local authorities responded by saying that land redistribution is one of the ultimate goals of the national government. One chief of a province was carried away by this theme to greater lengths than he had meant to go. He tried to recover himself when he concluded on this note: "Don't you go home beating your breasts and telling the other tenants that the land is theirs."

It was a fair warning, for there will be many a slip between wanting the land and getting it. The landlords, particularly the bigger ones, are interested in selling out. Their bold front, as if little or nothing has changed, is more form than substance. There is not a landlord in Vietnam who would not concede that land ownership of the traditional sort has fallen on evil days. There are probably not many big owners who do not believe that sooner or later they will have to part with much, if not all, of their land. Above all, they know that the days of undisputed control of the land are on the way out—hence the desire to sell. The stumbling block, of course, is the method of payment. The price itself is no great deterrent; the going price of good land is relatively cheap, $50 to $60 a hectare at the open rate of exchange. Tenants also assume that the land will have to be paid for; but just as the rental provision in the rent reduction program divides the two groups, so does the method of payment for purchase of land, although the issue for the moment is academic. The landlords know that they cannot get full payment in cash to effect a sale. The consensus among them was 50 percent cash and the remainder in five yearly installments. A number of tenants spoke of ten annual payments with no cash down; a few expressed willingness to pay one-third of the annual crop, and at least one tenant said a 20 percent initial payment might be reasonable. At this point somebody raised the embarrassing $64 question—where will the tenants get the cash? It was clear as day that they had none. The landlords did not contest this fact.

It was then up to the minister to speak for the government. This he did in generalities, partly because the government has no land purchase policy of its own and partly because he knew that the government has not a nickel in the till for such purposes. Nevertheless, he and his assistants understood that the tenants were vastly more interested in land ownership than in rent reduction. As on similar occasions when something had to be said to meet the plea for assistance and to allay the sharp criticism of the government for inaction, the minister fell back on a 225 million piastres credit fund (furnished by the United States) as the would-be miracle performer. In view of the overall credit needs, this is not a large fund; moreover, it is an emergency fund intended primarily to assist refugees and other farmers ready to take up abandoned land rather than to assist farmers already on land. Being hard pressed, the minister disregarded this distinction and freely played with the 225 million piastres as if it were a magic wand, spreading cheer in all directions. He *had* to say something to smooth the very much ruffled spirits of the farmers he was addressing. There is a strong possibility, however, that the chickens will come home to roost a while later, with feeble prospects that his promissory notes will be made good.

Summary and Conclusion

The principal impressions of the recent trips through South Vietnam are as follows:

First, the rent reduction program is six months old, but in terms of effective implementation it is hardly out of the paper stage.

Second, while the ordinances contain flaws, they are not the major impediments. The outstanding and overriding handicap is the total absence of any administrative organization to implement the program. In Japan, in conditions of a well functioning government and a host of capable technicians, the land reform program was carried out with the assistance of nearly 400,000 paid and volunteer workers. In all of Vietnam there are not 400 people directly and actively engaged in this task.

Third, the tenants are more interested in ownership of the land than in rent reduction. Rent reduction and security of tenure have already begun to embroil the tenants and the landlords in a bitter fight. It is doubtful whether it can be settled to any one's satisfaction. The discontent and unrest adversely affecting the political stability of the countryside will cease only when the tenants become owners of the land they cultivate.

Fourth, neither rent reduction nor land ownership can meet the immediate and urgent need of the farmers for credit, irrigation and drainage facilities, buffaloes, and fertilizer. In the minds of a good many farmers these items loom larger than the land question.

Fifth, the land in South Vietnam is plentiful and rich and the condition of the land and of the farmers can be improved. Failure to improve it means greater poverty, feeding greater political and social tensions to a point of endangering the very stability of Vietnam. This is a possibility, for in the final analysis the future of the country will be shaped not so much by the events in Saigon as by the events in the thousands of villages where the people live and in the rice paddies where they work.

Sixth, the greatest impediment to harnessing the elements which make for progress is the character of the national and local administrations. Chiefly for historical reasons, their efforts in the field of agricultural improvement have been inconsequential.

Seventh, the effectiveness of American technical assistance depends in large measure upon the government doing its part. For this reason, our aid cannot be more than marginal to the effort put forth by the government and the people in order to achieve their aims.

Eighth, relating our future activities to the specific conditions in Vietnam, it is well to appreciate the fact that for some time to come we may have to work in conditions where the effort is perfunctory, inefficient, or fainthearted and where the will of those who must help themselves is but a slender reed on which to pin our will and our hopes.

Despite the strong reservations, a fresh attempt should be made to consider a program which, by a process of peaceful change, would help better the conditions of the farmers. No attempt is made here to set forth the details of such a program but rather to raise a few points which may serve as guidelines for the national government and USOM, both for immediate and future courses of action. The program is to serve a twofold purpose: one with a possible immediate impact upon the village community, the other of a long-range character and equally concerned with the welfare of the farmers.

The well-known internal political conditions amply justify a program of this type. Its content is not an issue. The farmers have provided the answer when, time after time in central or South Vietnam, they have stated that their basic requirements are credit, buffaloes, fertilizer, and farm tools.

Needs are relative, but the credit need is a desperate one. In face of this, the traditional credit sources have dried out and nothing has taken their place. The American fund of 225 million piastres can reach only farmers or refugees taking up abandoned land. They are but the great minority of the farm population. The majority of the farmers are just as anxious to become recipients of such assistance. If this were to be done with telling results in 1955–56, another American credit fund of $20 to $30 million will be necessary. Having provided the money, we shall have to be content with a makeshift distributive mechanism and with the realization that little of this will be repaid. These disturbing assumptions stem from two conditions: the speed with which the credit fund must be put into operation and the absence of a tried and tested credit organization. Viewed as an emergency effort with an eye mainly on making political capital, the abnormal conditions should not deter us from furnishing the money quickly in the hope that it will reach the farmers without the usual delays. Any return above that, such as the foundation of a lasting farm credit system and repayment of most or the major part of the loans, will be most welcome indeed. It would be a mistake, however, to pin high hopes on such an outcome.

Providing farmers with buffaloes belongs in the same quick-impact category. This operation must be a selective one with an emphasis on those parts of the country where the requirements and Communist penetration are greatest. The districts of Ca Mau and Gia Rai of Bac Lieu province are a case in point. There the farmers suffered great losses of animals, and the distribution of approximately 8,000 to 10,000 buffaloes by the national government would go a long way towards a closer and more friendly relationship between the people and the government. Whether the farmers should pay for them in full or should receive them at a much reduced price or as an outright gift is open to debate; in this instance a middle course seems to be the sounder approach. Regardless of the terms, if we decide to supply animals to the farmers, the execution of this project must be carried out at the expense of red tape at every administrative level. This remark is prompted by the slow motion buying of buffaloes from Thailand, in which we have been engaged in recent months. If the process is to be repeated because every "t" must be crossed and every "i" must be dotted, the animals will not get there in time to carry out effectively this limited program encompassing only Ca Mau and Gia Rai.

Fertilizers and farm tools could be put to the same use. That we have, for example, unexpended funds for fertilizer, or fertilizer in warehouses which has not been distributed, doesn't argue against the great need for fertilizer. It only means that either the terms of distribution or the process of moving it from the warehouse to the end user call for revision and improvements. This is true of credit and animal distribution and of virtually every other program we are trying to promote in Vietnam. These are tasks for the national government to resolve, and it has been resolving them ever so slowly. Nevertheless, this being an emergency we cannot but assume that we shall succeed this time in prodding the government into action. Any other assumption would preclude the success of the program even before we start.

As regards the long-range program, credit and fertilizers are a part of it in addition to irrigation and drainage, extension, cooperatives, and land reform. But we cannot do everything at once. Precisely because free Vietnam is here to stay, we ought to be selective and concentrate only on those items about which the farmers feel keenest. In the American farm tradition an extension service is all-important, but in Vietnam it is peripheral. A pilot project here and there is all we need to do for some time to come. The same is true with cooperatives. On the other hand, credit, fertilizers, irrigation and drainage, and land reform are the things the farmers want most; and these are the heart of the long-range program.

With respect to land reform, it may be assumed that rent reduction will give way to land redistribution. For reasons already stated, even if the rent reduction program were implemented at some future date, the political gains of the government and the economic gains of the tenants would be minimal. Since ownership of the land by the tenants of Vietnam, particularly in South Vietnam, is inevitable (the change of status being only a matter of time), we should urge upon the government the initiation of this measure. The price of inaction may well be tenant action without due process of law.

USOM's position toward a land ownership program implies twofold assistance: technical advice in drafting the program and financial assistance to help create the administrative arm of the program. Naturally, the government expects much heavier financial support. In this connection it should be made clear that the real land distribution problem is neither the willingness of the owners to sell the land nor the price; the problem lies in the terms of payment. The government has stated that it intends to do justice to the landlords and tenants alike. This means that the purchase of the land will call for initial heavy outlays, a down payment of at least 25 percent of the agreed price of the land. The tenants cannot make any such payments, and it is up to the government to find the money if this is the scheme it prefers. Unofficial suggestions that we should underwrite this type of financing have been discouraged. Official requests should meet with the same response. If we don't, many another Asian country will ask for similar aid in order to bail out their landlords. Finally, it must be stressed that in this field pregnant with political and social significance we cannot substitute our will for their will; we cannot carry out the program for them. The government and the people must have the resolve to do it. The ingredients of our effort must be limited to financial aid on the one hand and endless prodding to stimulate initiative, self-reliance, and responsibility on the other.

As to irrigation and drainage, it should be noted that only a judicious control of water can raise production on the double-cropped fields of central Vietnam or introduce a measure of double cropping into South Vietnam. Preceded by a careful study of irrigation possibilities, we should be prepared to concentrate ample means and a large number of technicians to further such a program. Other useful but nonessential projects may go by the board, but this need not trouble us so long as the problems that truly matter are dealt with resolutely. Fertilizer requirements are not only top priority for the moment. The consumption of fertilizer could and should be expanded, and it need not be a giveaway scheme. The distribution mechanism and terms of distribution are the problems. A program akin to that in Formosa would go a long way to satisfy this need. The rich experience in Formosa and the experience of the American specialists who have been instrumental in developing this program would be invaluable in Vietnam. We are already involved in farm credit as part of our aid to the refugee resettlement, but the expansion of credit facilities to cover some of the requirements of the nonrefugee farmers is certainly on a par with the other items mentioned here. What is involved is the creation of a national farm credit system, and that is a long, drawn-out affair.

To date, the efforts of the government to improve the condition of the farmers are hardly worth recording. It is not surprising, therefore, that our own strivings have not been too successful. We are not without a small share of responsibility for the meager results of our aid. Part of the difficulty may be traced to the idea that our main function is to give advice to officials who, unfortunately, either do not care for the advice or are indifferent or are incapable of translating it into action. There is yet another reason. With small resources in men and money, we have engaged in an emergency program of "bits and pieces"; in the eagerness to accomplish a great deal, we have spread ourselves so thin sponsoring a multitude of projects in virtually every major field of agriculture that the effort has left only a few marks upon the face of the land of Vietnam.

If the above comment is valid, it would seem that we might do well to shift from trying to do a great variety of things with few people and little money to a much-reduced number of projects and only two or three major fields upon which the very existence of the country's agriculture depends. And we should undertake the task with a good many more people and a good deal more money than we have at our disposal at the present time. This approach also calls for a change in the emphasis on the advisory role played by our agricultural specialists. We have been here long enough to appreciate how important it is to advise or to prod the officials in Saigon as a prelude to any action. And yet, fundamentally, the place of the technician is not Saigon but the field. The process of transfer of knowledge and techniques are certainly not easy in a country such as Vietnam. Unless the specialist can demonstrate to the farmer, the experiment station man, or the would-be extension-service man how this or that is done and the advantages in new ways of doing things, the old ways will remain intact. In Vietnam, where the few native-trained agricultural specialists are practically all concentrated in Saigon or in the few other larger provincial capitals, there is no substitute for the day-to-day field work of the American specialist. Once the advantages of the new ways are convincingly demonstrated, the farmers will make them their own; and, as they succeed in doing so, they will gradually forget both the old ways and foreign origin of the new ways which have supplanted them. This type of service in conjunction with concentration on a few well-chosen objectives might well spell the difference between success and failure.

The degree of success will depend upon the internal stability of the country and the competence with which the national government will administer the country. But whatever one's misgivings about the applicability of these criteria, it is clear that if Vietnamese farm economy is to be moved off dead center in the desired direction, it is time that we re-examine searchingly our objectives, our methods, and our resources. The few suggestions outlined here may be found wanting; they are intended primarily as points of departure for further discussion, and they have been motivated by the belief that our current program does not hold promise for greater accomplishments in the future.

The simple, unadorned truth is that in Vietnam the peasant is the center of the piece. If we are to help him to become a useful, self-reliant citizen, making the most of his land, and giving his allegiance to the national government, then we must make an effort equal to this goal, particularly since we are faced with the excruciatingly difficult problem of inducing the Vietnamese to fulfill their major part of the undertaking. We cannot therefore achieve the objective with a program and human and material resources which are inadequate to the task. We cannot afford to forget that much of our economic and noneconomic aid to Vietnam stands or falls in the degree to which the great majority of the people share or don't share in it. Nor can we afford to forget that the welfare of the peasant is the welfare of Vietnam. A wise but unnamed Chinese philosopher summed it up in these words: "The well-being of a people is like a tree; agriculture is its root, manufacture and commerce are its branches and its life; if the root is injured, the leaves fall, the branches break away, and the tree dies."

29. *U.S. Aid for Land Reform in Vietnam*

By early July 1956 President Diem had decided to carry out a land redistribution program in the southern part of the country, where large landholdings and tenancy were most pronounced. In anticipation of this, Ladejinsky here writes to his old friend, Raymond Moyer, to ask for U.S. assistance in financing the pending program. Moyer had been Ladejinsky's chief in the Washington years in the Far East division of the Foreign Agricultural Relations Office; subsequently, he was a U.S. member of the Joint Commission on Rural Reconstruction in China and responsible for invoking Ladejinsky's assistance there. In July 1956 he was regional director for the Far East in the U.S. aid agency (by now, the International Co-operation Administration). He was therefore on two counts the "right" person for Ladejinsky to turn to. A third reason would be Moyer's sympathetic understanding of the values and views Ladejinsky espoused.

Ladejinsky's letter is dated July 10, 1956.

THANK YOU FOR THE NOTE. I was glad to hear from you, but, as I mentioned on another occasion, you needn't bother acknowledging the occasional items I send along; I know you are busy.

The purpose of this letter is to apprise you of an impending development with which you are, I believe, already familiar in a general way. Only the other day it appeared that on July 20, the second anniversary of the Geneva Agreement, the government of Vietnam would state publicly its intention to carry out a land redistribution program. For reasons I describe in a subsequent paragraph, the announcement is being delayed until October 26, the first anniversary of the founding of the Republic of Vietnam.

The background of this development is a bit complicated, but the essence of the matter is this: the rent reduction program is no great shakes, and the president has come to appreciate the significance of a "Land to the tillers" program. He is thoroughly convinced of the urgency of a measure of this kind. He has discussed it with the two top American officials here, and he has also stated the case for the reform for the benefit of a very important personage who has just passed through Saigon.

According to the current thinking, the government will buy the land from the landlords at a fixed price, give them an initial cash payment equivalent to 10 to 15 percent of the price of the land, and make the remainder payable in bonds. The tenants in turn will pay for the land in a number of annual installments. The immediate problem is to find $30 million to make good the initial cash payment to the landlords. The government hopes that the United States would provide the sum as a grant or loan, preferably the former.

The reform is aimed at the southern part of Vietnam where holdings are large and the tenants are legion. Central Vietnam is a different case altogether and will not be dealt with for the time being.

Having had something to do with the matter in my past and present capacities, my own comment may not be out of order. First of all, I share my chief's view that the land must be given to the landless without delay and for well-known reasons. Procrastination will not do any longer—not if we are interested in the survival of Vietnam on the side of the free world. This may sound alarmist, but in the southern part of Vietnam, which is the agricultural heart of the country, the landless who cultivate 75 percent of the land are not in a happy frame of mind. The landlords feel no better, and the upshot is that government tax collections are at a vanishing point. On the political side, the repercussions may be worse and the government is cognizant of that.

Second, while $30 million is a lot of money, it is a modest sum when viewed in the context of the purpose to which it will be put and the anticipated effects. I am told that it costs hundreds of millions of dollars to maintain one U.S. division in the field. As against this, $30 million will enable the government of Vietnam to secure the support of the overwhelming majority of the people; in time of need they may be expected to make up many a loyal armed division.

Two other considerations enter into the picture: the rapid pace of economic development in North Vietnam, assisted by the Russians, Chinese, East Germans, Czechs, and Poles, and the wave of neutralism which is engulfing all of the mainland of Southeast Asia. In these circumstances, South Vietnam is the only uncompromisingly anti-Communist, anti-neutralist state in Southeast Asia. It is an outpost we cannot afford not to strengthen in every way. The scheme under consideration is perhaps the most important step in that direction. It appears to me, therefore, that the $30 million is not an excessive contribution towards the realization of a *must* program about to be undertaken by the government, in the face of known landlord opposition.

Third, my own concern is not so much the money as the virtual lack of a land ownership census, lack of personnel capable of dealing with the many other technical problems connected with the preparation of the program, and the general absence of an administrative machine to translate the program into action. These are the real, tough problems and they are very much on my mind. This explains why, as we talk about money, a coherent program is yet to be drawn up. It explains also why most of the Vietnamese agrarian reform attempts in the past five years have remained largely on paper. It explains the postponement of the announcement. We hope to make good some of the deficiencies in the interim.

Skepticism about the new attempt is understandable, but it would be a mistake not to resolve the money question now instead of waiting till the program is all complete. The granting of the money would in itself be a great stimulus in approaching more realistically the problems I mentioned above. Moreover, it is clear that none of the funds would be spent unless there is something to spend them on.

We must be prepared for rough going, but there is another side of the picture which need not be overlooked. Knowing something of the president's determination and drive, I believe that this time it will be different. It will be a case of waging a war that must be won. In the tradition-ridden Vietnamese society this is an Herculean task, but the president views these difficulties as so many challenges.

He needs assistance to overcome some of the landlord opposition, *but not assistance to bail the landlords out*. The sum involved is much too small for that. Moreover, the president is motivated primarily by the welfare of the landless. "The reform," as the president pointedly reminded a number of his ministers—not a few of them big land proprietors—"is not an accommodation for landlords; it is a revolution." The anticipated aid from the United States is to speed up and smooth the process of an orderly reform from the *top*. The stakes in this undertaking are greater than the risks; the former are nothing less than meeting the Communist challenge at a time when all opposition to Communism is softening, of broadening the base of political power, and of creating some sort of economic order where little is to be found now.

There is nothing theoretical about these propositions. They are the realities of present-day Vietnam—and I am not unmindful of the political achievements bordering on the miraculous. Hence the plea to be *bold*, to use President Eisenhower's special aid fund if possible, and, by the same token, to give added meaning to U.S. foreign aid. The president's request falls into that category.

I know that in the established practice of official American economic aid it is safer to spend money on no end of orthodox projects, as if numbers truly matter and as if "know-how" must concern itself with physical phenomena only. There is enough evidence that compellingly calls for at least a dash of a new approach. I have in mind certain aspects of the so-called great debate relating to U.S. foreign aid practices in general.

I am under the impression that the president

will attempt to see the program through even if he should fail to receive financial assistance from the United States. I share his attitude; and, if he needed my encouragement to that end, I would cheerfully give it to him. For it is time that, with or without foreign assistance, Vietnam must come to grips in earnest with this all-important problem. If I judge the Vietnamese temper correctly, an American refusal might have certain undesirable effects. I don't wish to imply, however, that U.S.–Vietnamese relations will be affected to any serious degree by our failure to assist the country to resolve this crucial issue.

Vietnam is not going to trade on both sides of the street in the manner of other recipients of U.S. aid. Because of that, in any consideration of the problem discussed here it is well to keep in mind that, on the mainland of Southeast Asia, Vietnam is just about the only country which acts like a true ally. Vietnam's anti-Communism and anti-neutralism are not subject to bargaining. Vietnam doesn't expect special rewards for its fierce opposition to Communism. This attitude derives from inner compulsions, and this explains its strength. And precisely for these reasons we can do no less than treat Vietnam's "felt needs" with boldness and imagination. Whether Vietnam can convincingly explain how the very last of the $30 million worth of piasters will be spent or how the government intends to repay the loan, if it is a loan, on due dates is important but not crucial. Despite my great uneasiness about conditions mentioned under "Third," the important thing is to see that this reform battle is won. When that day comes, there will be satisfaction and glory for all concerned—the people of Vietnam and the people of the United States.

You know my pro-reform bias well enough to know that I am writing in this manner not because I have the privilege of working for President Diem. I would have done the same if I had remained with ICA [International Cooperation Administration]. They and we need Cai Sans in agriculture and in other fields of economic development. They and we must support reforms which go to the heart of the matter. They and we must eschew the sin of scatteration, which is nothing less than the pernicious idea that we must concentrate on every, or almost every, phase of a country's requirements. It should be replaced with the highest degree of *selectivity*, backed up by *massive* application of aid. Boldness and imagination are the inevitable ingredients of this approach. The agrarian reform very much on the mind of the president is an example of what I have in mind.

I believe that Ambassador Reinhardt and Mr. Barrows favor this approach, and I am happy to note that they decidedly favor the land reform idea sponsored by the president. The fact that they have their reservations about lack of technical preparedness merely points to their appreciation of local conditions.

Finally, I repeat that I am well aware of the difficulties. To the best of my abilities, I am conveying their magnitude to the president in the hope that something will be done about them. With the best good will and determination, they will have to be battled all the way, from the initiation to the completion of the reform. But there is no point throwing up one's hands in despair. If we did that there would have been no Cai San. Happily, we took the bull by the horns, paid no heed to the skeptics, and helped turn wilderness into human habitation.

We must try to repeat the process even though in this instance class antagonisms are at razor's edge and all other problems are infinitely more complicated than those faced at Cai San. As in Cai San, but on a much larger scale, much improvisation will have to be resorted to, hoping at the same time that the mistakes will not be irreparable. Time is not always on our side. Vietnam cannot wait until all "t's" are crossed and all "i's" are dotted. The time element being what it is, the day will probably never come when Vietnam will be able to present a "bankable" reform program on which money can be lent. And so, it comes down to this: Something must be taken on faith; but, above all, we must dare and carry on with the thought that our daring will be rewarded.

I like to think that you will not hold against me this discourse. I am not writing in any spirit of criticism. I have no axe to grind other than the lasting bond between the two countries. All the rest is peripheral.

30. *Making the Pending Land Redistribution Program More Practicable*

I have given the above title to a memorandum Ladejinsky addressed to President Diem on October 9, 1956, under the subject heading "Agrarian Reform." The draft ordinance then pending was issued later that month as ordinance 57. Ladejinsky requested a number of modifications in the draft, before promulgation, to make it more practicable. Among these, he urged a landlord retention ceiling lower than the 100 hectares proposed and a ten-year tenant repayment period rather than the six years proposed. Ladejinsky's 1961 review article reveals that the president did not accept these suggestions.

YOU HAVE BEFORE YOU the final draft of the "Ordonnance No. 57 du ——— Portant Réglementation de la Réforme Agraire." Your approval of the ordinance is indeed a step of major national importance. It is with this in mind that I take the liberty of a few comments on the land redistribution program about to be inaugurated. In doing this, I am concerned only with the soundness of the program and its successful implementation.

My comments relate not so much to the ordinance itself as to the measures that should be taken along with it or even prior to its enactment. But first some observations about the ordinance.

The ordinance is a well-drawn document, outlining the basic provisions of a land transfer program in general terms. It suffers, however, from one weakness already noted on another occasion but worth mentioning once again now that the ordinance is about to become the law of the land. I have reference to article 14, which states that the tenant should pay for the land in six annual installments.

In a memorandum of August 17, 1956, I stated that from the point of view of a more successful implementation of the program, a ten-year period of payment is the more valid approach. I need not repeat all the reasons for the position I have taken. I do wish to note, however, that, since the reform is intended primarily for the benefit of the landless, it would be dangerous to overestimate the financial position of the tenant who takes on new obligations as a purchaser of land. I do not question the expertness of the committee in deciding upon six years; nor do I pretend to have intimate knowledge of the current economic position of the tenants; but, unless they are indeed well off, I doubt if the six-year period can be enforced. For administrative, economic, and political reasons I believe it would be wiser to extend the repayment period to ten years. There seems to me no compelling reason to risk the political support of a great many farmers on an issue of this kind and to endanger the ultimate goal of the program. As I shall point out in subsequent paragraphs, every precaution must be taken to ensure the success of this program—the first land redistribution program in Southeast Asia. The longer period suggested here is one of the vital precautions.

Article 4 (paragraph "b") provides that a landlord has a right to retain 100 hectares of land. The figure seems high. The permissible retention determines the scope of the program or the total acreage available for redistribution. Because of the lack of precise statistical data, it is not clear just how much land will be available for sale to tenants. This much is fairly obvious: If landlords may retain 100 hectares each, approximately 70 percent of the tenants will fall outside the scope of the program. If this estimate is roughly correct, then the question may be raised if the retention figure should not be lowered. The reasoning behind this suggestion is as follows. The principal purpose of the agrarian reform is to create as many owner farmers as possible. Since it is evident that the tenants can buy land only with the aid of the agrarian reform, the lowering of the retention becomes a necessity.

The members of the Agrarian Reform Committee have undoubtedly had sound reasons for setting the permissible retention at 100 hectares. Nevertheless, it is reasonable to assume that the demand for land by the tenants who fail to qualify under the 100 hectare retention will compel a downward revision of the retention. If the Agrarian Reform Committee recognizes this as a possibility, would it not be preferable to make the change now by reducing the retention limit to 75 or 50 hectares?

Another related consideration bears on the same problem. In a democratic country, no agrarian reform can or should eliminate tenancy altogether. At the same time, if it is designed to achieve economic and political stability in the countryside, it must apply to the great majority of the tenants. When this is done, free Vietnam will not only fulfill the real purpose of the reform, but it may also help set the pattern for the rest of Southeast Asia—and beyond.

A revision of the two items would affect favorably the application of the ordinance, but it must be stressed that the ordinance by itself, no matter how sound, cannot implement the program. The execution of a program as far-reaching as that embodied in the ordinance and the revitalizing of the existing rent-reduction program involve a major task of administration. The success or failure of these programs will depend upon the creation of a well-functioning administrative machine in Saigon, in the provincial capitals, and in the villages. The ordinance recognizes the need for an administrative body, as evidenced by Titre III, Des Organes de la Réforme Agraire, Articles 25, 26, 27, and 28; but, for the time being, a land reform administration barely exists in Saigon and is virtually nonexistent at all other levels. It is yet to be created, and it must be created if the ordinance is to have any meaning.

The first step in this direction is the appointment of a secretary of state for agrarian reform who can devote *all* his time to the organization of his department and to the administrative duties with which it is charged. At first glance, this may appear superfluous because the Department for Agrarian Reform has been in existence since May 1955. However, this is more apparent than real. For many months now, the secretary of state for agrarian reform has been devoting most of his time to the Cai San project; under the circumstances, the department cannot be as effective an organization as it should be. The execution of a land transfer is a most demanding and time-consuming task; and so long as the secretary's attention is concentrated elsewhere, the principal tasks suffer accordingly. The Department for Agrarian Reform has never been fully staffed and organized to meet the needs of effective implementation. This is reflected in the rather unsatisfactory results of the rent reduction program inaugurated in early 1955. With these lessons in mind, the execution of the land redistribution and of the rent reduction programs calls for an administrator thoroughly in accord with the purposes of agrarian reform, concentrating only on rural reform activities. It will be his task to create what will be in effect a new organization.

In view of the urgency of the land redistribution program and of strengthening the implementation of the rent reduction program, the full-time administrator for agrarian reform should be given powers wide enough and budgetary support large enough to create a department capable of carrying out the provisions of the ordinance. In this connection, the suggestions of the MSUG on the reorganization of the Department for Agrarian Reform are of great value, and proper use should be made of them. The authority of the Department for Agrarian Reform is laid down in general terms in the ordinance, and it is the responsibility of the department to develop the specific tasks. It is not the purpose of this memorandum to deal with them. Suffice it to say that the Department for Agrarian Reform must initiate overall policies and translate them into operating functions in order to achieve a coordinated program of land transfer and rent reduction within a certain time period.

The reorganization of the department cannot be achieved overnight, and the enactment of the ordinance cannot be delayed until that has been realized. On the other hand, once the individual responsible for the program has been selected, his immediate and most profitable order of business is to deal with the problems mentioned below, whether the ordinance has been enacted or not. Some are rather simple and others are complicated; but, taken together, they constitute the important groundwork of the agrarian reform.

The implementation of a land distribution program will be a new experience for Vietnam. For this reason, it is well to utilize the knowledge of those who implemented similar programs. I have in mind Japan and Formosa. In both countries and particularly in the latter, there are valuable administrative lessons to be learned. Both countries, under different circumstances, using different financing schemes, but with quite similar administrative organizations, successfully achieved their goals. Vietnam need not copy their methods but only adopt what is useful and applicable. In both countries the village land commissions were mainly responsible for the implementation of the program; in the final analysis in Vietnam, too, the success of the program will depend upon the effort in the villages. There is not much to be learned in Japan about the financing of land purchases, but a great deal of useful information can be gotten in Formosa on this very subject. I have particular reference to the bond issue techniques.

In the light of these considerations, a small group of Vietnamese specialists should visit both countries for a careful assessment of these problems. This would result in saving time and effort, and, most importantly, preventing mistakes.

Finally, the Department for Agrarian Reform should enlist the services of a few of the most experienced technicians who were largely responsible for the success of the Formosan program. I worked with them, and I am of the opinion that they would contribute greatly to the success of the Vietnamese program. I cannot speak for USOM, but I believe that USOM might agree to furnish the means for the study-trips to both countries as well as the necessary financial support to secure the services of the Formosan specialists.

In Vietnam, unsettled conditions for many years have prevented the gathering of adequate statistics concerning rural land ownership. Ideally, a land transfer program such as is now before you would be preceded by a complete agricultural census. It is impossible to accomplish this in Vietnam in less than a year, however much it is needed. Therefore, I suggest an alternative shortcut to securing the information which will be needed.

If the retention limit is set at 100 hectares, approximately 2,600 landlords will be affected. All these landlords should be required to register those plots they wish to retain. At the same time, landlords owning more than approximately half the retention limit should also register, listing their total holdings. In all, only some 6,000 landlords would be affected, the majority of them owning less than 100 hectares.

The tenants who would like to purchase their holdings and who think their landlord owns more than 100 hectares should register their purchase claims. Where there is a conflict about the total amount of land a landlord claims and a tenant says the landlord owns, the cadastral service would make a survey. Later it would survey all holdings affected by the transfer.

By some device such as this, the government of Vietnam can obtain relatively quickly a fresh and fairly good idea about the size of the landlords' holdings.

A measure that has the highest priority is an estimate of the financial cost of conducting the affairs of the department and of the cost of administering the land transfer and the rent reduction programs. The two are not necessarily alike. The reform agencies in the provincial capitals, district and village offices will call for a far larger expenditure than the maintenance of the department staff in Saigon. While the land transfer will be directed from Saigon, the implementation work will be done in the field, outside of Saigon. Any attempt at budgetary estimates will, of course, necessitate a consideration of the total administrative scope of the program. This in itself is an indispensable chore, even if the government had all the financial resources at its disposal. Since this is not the case, the administrator of the reform and the government must have a fairly clear idea of the necessary outlays before the program is in operation. This may call for changes in budgetary allocations within the government or requests for aid from USOM in the event the government budget cannot meet the necessary outlays. All of this is a significant part of the land reform plan which must be given timely thought if the tasks ahead are to be treated with the seriousness they deserve.

Article 21 of the ordinance provides for an initial cash payment to the landlords of at least 10 percent of the land purchase price. It is not the business of the ordinance to state in detail how this money is to be used, although the ordinance implies its use as a means of orienting the large landowners toward industrial activities (article 1). This is a complicated question tied up with the future economic development of the country. It cannot be resolved in a hurry. It would be prudent, therefore, to begin the examination of this problem now, before the application of the ordinance. It would be unfair to the landlords to block their cash payments indefinitely because of the failure to devise a practical plan for the utilization of that money. If, on the other hand, the study of this problem reveals that the cash can have no particular bearing on industrialization, then, naturally, it would be useful to know over what period the payments should be spread. The related question is whether the reform is to be carried out in one year, two years, or more.

There are three reasons why these urgent questions demand answers: the problem of the overall financing of the land transfer program, the need to minimize the opposition of the landlords to the reform, and the possibility of securing from the United States a land reform loan or gift in the amount of some $20 million. The first two are self-evident and need no further elucidation. The third calls for a bit of elaboration. Foreign aid is now subject to the closest scrutiny in the United States. This is particularly true of the type of agrarian reform aid, which has never before been extended. The $20 million belong in that category. Under the circumstances, even a convincing statement why the money is needed will not suffice; of utmost importance is a convincing explanation of how the money will be used.

Article 26 of the ordinance provides, among other things, for a National Council for Agrarian Reform. It is not too soon to put some content into this framework, rather than wait until "five minutes to twelve." The chairman and membership of the council should be selected now; and, having done that, there is the obvious need of determining in some detail the council's responsibilities and of creating the means of carrying them out. The ordinance refers to other commissions which are to take care of a great variety of tasks directly related to the reform. The ordinance is not specific about the character of the commissions. However, since they are intended as the implementors of the reform, it is well to determine at the earliest their precise nature, their composition, whether they are to function as part of the council or the Department for Agrarian Reform, and whether their functions will extend throughout the provinces. In short, as the commissions will be a part of the administrative machine, now is the time to decide how these parts fit into the whole.

As already mentioned, the duties assigned to the commissions in article 27 of the ordinance are related to the very essence of the reform. Their effectiveness would be the greater if they were clearly assigned to an administrative unit called "village commissions." Granting the importance of instructions received from Saigon or the provincial capital, the fact is that all the important decisions for the transfer of land or rights to land will be made at the village level. This work involves detailed decisions in each individual case with respect to the seller, the buyer, the tenure status of the land to be purchased, the price, the date of transfer, and so on. An outsider couldn't possibly do it, at least one of the reasons being that many village records have been destroyed by the Viet Minh. Only the local people are in a position to know the conditions of each individual and his relation to the land. It is equally true that, even with the records in good shape, only the community is in a position to decide for itself all the vital questions bearing directly on the reform. Here lies the success of the reform and of the rent reduction program. To make certain of that, it is advisable to create effective commissions in all the villages affected by the reform. They will be made up of tenants, landlords, and owner cultivators. Mindful of the purpose of the reform and of the crucial role the commissions must play, the proportional representation of each group is a subject for serious consideration. Above all, in this regard, the commissions must not be so nearly weighed with landlords as to render ineffective the suitable implementation of the reforms.

These are some of the principal points which demand immediate attention if the contemplated rural reforms are to become a reality. It is commonplace to say that even the best land reform law cannot ensure the success of the reform. What is needed is an administrator with vision, sympathy, energy, and ambition, backed by all the resources of the government, and capable of creating an administrative organization which will indeed implement the reform law. While this ideal person may be hard to find, the best available man should be appointed without delay in order, first, to lay the groundwork discussed in this memorandum and then to proceed with the execution of the program.

All this is said not in a spirit of criticism either of individuals or of what has been done until now. I am motivated by one consideration only, and one only: the success of the land reform program which free Vietnam is about to launch.

The reform must not fail; there is too much at stake—nationally and internationally: nationally, a strong link between the government and the people, made stronger by economic stability inherent in the concept of private ownership of land and reasonable tenure terms; internationally, another lesson that an agrarian reform can be achieved without enslaving the peasantry.

On the mainland of Southeast Asia, free Vietnam is the *first* country to dare a land redistribution program for the benefit of the landless. Both the Communists and non-Communists will watch the outcome of this effort. It would be idle to expect that its successful accomplishment would cause the Communists to deviate from the line they have taken; there is reason to hope, however, that for the rest of Southeast Asia this would carry the conviction that lasting social improvements cannot be found at the end of a Communist gun barrel; it would demonstrate that, without recourse to violence and bloodshed, in free Vietnam those who till the land own the land.

The reform must not fail; no effort must be spared to ensure its success.

31. *Toward a More Effective U.S. Aid Program in Vietnam*

I have chosen the above title for this letter of November 15, 1956, to an old friend in the U.S. State Department directly concerned with the formulation of Vietnam policy at the department. While Ladejinsky wrote the letter explicitly for his friend's "background information" only, he obviously hoped that it would contribute to the strengthening of U.S. aid in Vietnam along the lines he advocated. Since I have already referred in the introduction to some of the views Ladejinsky expresses here, no further comment is needed.

THANK YOU VERY MUCH for your letter of the 17th. I have often thought of you in connection with the goings on here, but I have been most reluctant to put pen to paper. This is true even now, for issues involve people, even if not labeled individually, and the chore becomes rather unpleasant. However, I shall say something on a subject which is not new to you but which the celebration of October 26 made doubly clear: a political milestone is behind us, whereas the economic one is yet to be passed.

But first about the brighter side of the picture. Politically, Vietnam has done very well indeed. You are familiar with the general and the particular and I need not enlarge on them, except to say that the country is unified, that the government governs by day and by night, and that the people believe that the government is here to stay. My on-the-spot observations lead me to these conclusions. The president is inclined to give all the credit to the Lord and only of late has there been a greater measure of awareness—even if not acknowledged—that the Lord helps those who help themselves. So much so, indeed, that the confidence engendered by the political achievements has had this result: outside advice and friendly advice are not as readily accepted as in the past. The handling of the Chinese problem is a good case in point.

Diplomatically, particularly relations with Southeast Asia, Vietnam is doing well. Cambodia is the exception. There is widespread recognition that free Vietnam is nobody's puppet, that Diem is an Asian leader to be reckoned with, and that he earned the respect of other countries through a moral force which even his enemies do not deny him. The government of Vietnam for its part has finally recognized that isolationism with respect to Southeast Asian countries is not a good policy, that an active policy as a member of the Southeast Asian community is much to be preferred, despite the neutralism of the greater part of that community. The relations with India are indeed good, and U Nu's visit is the first significant stop in establishing friendly diplomatic relations with Burma. I have reason to be particularly pleased with this development. This by-product of my visit to Burma in July turned out to be much more important than the original purpose of the trip.

All of this is in contrast with the economic state of affairs. Improvements are likely to take place in the months ahead; but, as of now and despite an aura of prosperity in Saigon, economically the country is as weak as a cat.

The past six months or more have been spent in much argument on the subject of the country's future economic development. Nobody here minimizes the role of American aid in saving the country from economic chaos, but they believe that the continuation of this type of aid (army, refugee assistance, and MSU are excepted) would compel them to perpetual dependence on American aid. The overwhelming agricultural character of the country is accepted as the basis of its economic future; what is assumed with equal conviction is that it must be diluted with a strong dash of industrialization that fits the country's needs. In a balanced agricultural and industrial economy lies Vietnam's economic independence. Merely "keeping the country going" is not enough. It does not give rise to a higher standard of living, create new sources of employment, expand purchasing power nationally and internationally, develop managerial and industrial skills, and so forth. Our aid must begin to lead in this direction. This is the principal part of the argument.

Aside from the fact that the new political structure can have meaning only if the government demonstrates by deed that one of its principal concerns is the economic welfare of the people, there are the economic schemes of the Communists in North Vietnam which are taken here most seriously and are a source of grave concern to the government. There is no attempt to underestimate the ability of the Communists to create an economic showcase; nor is there any disposition to minimize the political results that would attend it, on both sides of the border. These motivations are not new, but they are worth repeating because they have not been fully recognized on our side. This is particularly true of the political aspects of the aid. Hence, the Vietnamese restlessness and dissatisfaction which clearly manifested itself in recent months.

The controversy through spring and summer has had one good result: we recognize now that economic development, as distinct from economic scatteration, is the course to follow. It is not a day too soon—for both Vietnamese and Americans—to try to make this a reality.

On the Vietnamese side, decisions are hard to come by. This is partly because they are not sure of themselves, partly because of the great lack of experienced personnel, partly because the administrative structure is such that authority doesn't necessarily go with responsibility, and partly because nationalistic proclivities often take primacy over day-to-day realities. There is also another important reason: The Vietnamese I am talking about have no great confidence in the ability of our technicians to help develop and execute a well-rounded program.

None of this is said publicly and absolutely

nothing for foreign consumption. To illustrate my point: A group of Indian, Thai, and Burmese journalists were in Saigon, by invitation, in connection with the October 26 affair. While here they were not only sharply critical of American aid in their respective countries on all kinds of grounds but made an effort to solicit some supporting views from the government of Vietnam. In this they utterly failed. The government has taken the position that, whatever the disagreements, they must not be aired in public.

Within the family, however, it is an open secret that, by and large, the Vietnamese tend to be critical of a good many of our technicians, claiming, among other things, that the United States is not giving them the best it can offer and that they work in a world of their own, often not related to the needs of Vietnam. Some go so far as to question our interest in the economic development of Vietnam. All of this has resulted in a marked reluctance to accept our advice.

As in all such cases, the charges are greatly exaggerated. Much of it is on the familiar theme of the rich uncle, the "colonialist" manner in which the Americans live, the extreme "isolationism" of the Americans—that is to say, the pronounced tendency not to mix with the Vietnamese. Sad to report, but in this respect in the eyes of the worldly Vietnamese, the French cut a better figure. The charge goes deeper than that. The mission is a big and growing organization, professionally speaking. They want outstanding men with proven knowledge and prestige in a field of activity, and so forth. The few Vietnamese with good training and the very, very few Vietnamese economists look down their noses upon our crop. This explains why the same Vietnamese are anxious to find highly qualified specialists who would work directly for the Vietnamese government.

ICA/Washington is considerably at fault in respect to these developments. It is inordinately preoccupied, so it appears from a distance, with slots to fill, with the mechanics of a mission's work, and with keeping the books in order. One cannot gainsay the importance of these chores; yet there are other important problems, which, until recently, failed to receive their due. Whether the mission has the right kind of a program (in 1955 it had 330 projects and not one bit of a program); or whether the economics of the aid fit the political facts of life; or whether, from our point of view, Vietnam has some special role to play—all these and more are not apparently of great moment to the policymakers. Only very lately and under the impact of widespread criticism at home and abroad leveled against our aid has ICA begun to take notice that maybe there is something amiss with this aid.

A good opportunity to set our aid on a new course was last spring. At that time, a well-staffed special mission, with prestige and authority, charged with the responsibility of examining and of prescribing for the future economic development of the country, would have been what the doctor ordered. I regret that the idea died aborning. However, this is water over the dam.

In the course of the past week or so the two parties have been discussing a document known as the "Proposed 1957 Economic Development Program," involving an American contribution of $85 million and a Vietnamese input of $20 million. I don't believe there will be much disagreement on the basic items, but there are a number of stiff hurdles to overcome before the implementation begins.

The heart of the document lies in the category called "new priority projects," made up of a number of new industrial and agricultural undertakings. So far, the document merely says: "textile mill—$2.5 million," or "sugar refinery—$3 million." All this must be developed and *refined*, which in turn calls for much talent and understanding on both sides. Problem number two is that the government must make a number of decisions as part of a development program.

Will such decisions be made and followed by action? With October 26 behind us, there is reason to believe that on the Vietnamese side most of the effort will be concentrated on economic matters. The political preoccupation excuses, justifiable until recently, are likely to be kept in abeyance; but it is not going to be easy. The decisions relate to the role of the government in economic development; the place of French and "pure" foreign capital and how to attract it; tax and budgetary reform; foreign and domestic trade controls; the possible devaluation of the currency; the form of economic austerity; and, of course, the very important administrative questions relating to implementation of the reforms. Willy-nilly and on a piecemeal basis, conditions—if nothing else—may compel the government to deal with them. But the going will be rough because for reasons already cited the economic part of the job is vastly more complicated than the political one. However, I am not without hope that with a good deal of money available for a new program, procrastination will have to give way and answers will have to be given to the troublesome problem of the country's economic development.

On our side, in addition to the money that we do have, we shall need something more to do better than we have done in the past. We shall need a few outstandingly capable people to cope with a real economic development program. We shall have to become more politically minded in order to appreciate the role of Vietnam in this part of the world and why we must help bring about *visible*, not to say spectacular, achievements, and quickly. Part of our political awareness must come from the realization that the waves of neutralism wash the very borders of Vietnam. The threat of Communism from the north and the weakening of the opposition to Communism among the neutralists west of Vietnam place Vietnam in the position of being almost the only St. George fighting the dragon on the mainland of Southeast Asia. No special rewards are asked for it, but it is strongly felt that a meaningful and rapid economic buildup of Vietnam—up to and including the sacrifice of the monumental red tape that slows down every action—is in the interest of the United States as well as in the interest of Vietnam.

Fourteen months spent negotiating a contract with a private firm to make a survey of this or that is not the way to advance our cause here. It would seem that occasionally we treat an occupied country with greater consideration than a proven ally. I remember that, when in the early days of Japanese occupation roads or airfields had to be built, the Corps of Engineers moved in and did the job. For all practical purposes, Vietnam is in a state of war, the outcome of which depends on how well "they" and "we" perform before the faucet is turned on again by our "friends" up north. Such being the case, a statement by a visiting VIP from ICA that he saw no difference between Vietnam and any other aid-receiving country on the mainland of Southeast Asia is very disconcerting. The fact is that some countries are "more equal," or rather more deserving, than others. But if he expressed the prevailing attitude in Washington correctly, I can see the why of the "business as usual" kind of operation, the pride in our doing things for "them," and the clinching argument that Vietnam is getting more money than any other country—as if this in itself were an achievement. I don't believe this is cricket; moreover, it beclouds the real issue: in the technical economic assistance field it is time that we show a great deal for the money we spend. It ill becomes us to point the finger of accusation solely against the Vietnamese. We are senior partners in the enterprise; until just now the character of our technical economic aid rested mainly with us.

And so I hope that, now that we are about to turn a new page in what might prove to be a worthwhile effort, we shall leave behind the parochial mentality that governs so much of our economic work. We are here in part because we have the so-called "know-how." Then let us display it with vision and daring. It does little good under any circumstances to keep on saying with an air of superiority that "they" are so ignorant that they cannot prepare an economic plan of their own; this is particularly regrettable because there are few American specialists in Saigon capable of tackling a chore of this type and with the kind of an economic political thinking that such a plan must entail. We have never made such an effort, even on a piecemeal basis. Right now and for a year or two to come, we and they need a truly first-rate economist of acknowledged standing to help make some of the decisions mentioned earlier. I would dispense with a good many to make room for one such man. He is indispensable, for there is an urgent need to put an end to the dangerously oversimplified kind of layman's economic thinking which prevails here. I am certain that the idea would be frowned upon in certain circles, but it is worth a most earnest try.

A touch of humility, recognition of our shortcomings, ability not only to teach but also to learn from the recipients, stepping one down to get acquainted with the Vietnamese, less preoccupation with housekeeping matters and more concentration on "what we are here for," fewer but better people—all these and more should advance our cause measurably. I am not overly optimistic that the Vietnamese will overnight acquire the "proper" attitudes and I am a bit skeptical that we can, after years of habit forming, effect a similar and speedy renewal of strength. Nevertheless, new attitudes must come to the fore. I have no detailed prescription for this sort of thing; all I can say is that events cannot be denied their due, for their logic cannot be violated. In addition, one must assume, of course, that men of good will on both sides will help the process along.

On October 27 and in the midst of continuing festivities, I had the privilege of a session with my chief. After a review of the political significance of the events just behind us, I expressed the hope that a year from now Vietnam would celebrate an economic milestone. He echoed this wish. The goal is clear, and we are all agreed upon it. We must make this good and with the very best means at our disposal.

I am afraid I have written a bit disturbingly, but I assure you there is nothing personal about it. I have held to these views and made no secret of them while with USOM. Now, as then, the only axe I have to grind is that of our common cause here.

Just a word about myself. I am well. Of frustrations there is no end, but the stakes are formidable and the purely personal ups and downs are of no great moment. I am on good terms with both sides. And, of course, I like, admire, and respect my chief. He is very human and as a nonsinner he understands the ways of the sinners.

In late March I am scheduled to go to Santiago to attend an International Catholic Rural Life Congress for Latin America. Essentially, the congress is built around the theme of "man and land" in Asia, and I am supposedly the man to tell something of the story. I am happy to oblige Monsignor Ligutti, who is the prime mover of the rural reform wing of the church. Also, this will be my first trip to Latin America; under the circumstances I don't protest too much the spending of a mint of money to give me a free trip around the world. I don't believe I shall visit the States on my way back. I am fairly busy here and I am not eager for a long vacation. However, time will show.

P.S. The content of this letter is only for your background information.

32. *Agrarian Revolution in Asia*

Early in 1957 Ladejinsky attended by invitation the Fourth International Catholic Congress on Rural Life, held in Santiago on April 1–6. It was his first visit to Latin America. Ladejinsky's chief contribution to the conference was a statement on the "Agrarian Reform in Japan," not presented here because there is another, and better, on that topic. But he also made a very short, pithy statement to the conference on the agrarian revolution in Asia which merits inclusion, if only because Ladejinsky concentrates, within the compass of four paragraphs, a number of his favorite phrases borrowed from some of his other papers to hammer home his "or else" thesis.

IN AGRARIAN ASIA, land reform is no more an academic issue; and the experience of Japan and Taiwan on the one hand and that of Communist China on the other may well serve to identify the best interests of other Asian and non-Asian, non-Communist countries. The first two have shown that the tiller can get the land he cultivates with no recourse to bloodshed and chaos. Above all, they have demonstrated that a peasant can achieve his goal as a free and independent producer. He has escaped the serfdom which Communist Russia and China have imposed upon their peasants at the point of a bayonet.

It must be underscored, however, that the achievement rests on the will and resolution of a government in power to meet the land hunger of the landless, on the appreciation of the political consequences of a land tenure system where poverty and social degradation are its hallmark, on the ability to force upon the landlord class the concessions which make reform possible.

The last point is crucial. In the part of the world I come from, landlords, big or small, are not social reformers. To them, the very words "reform," "change," and "concession" partake of the devil. In their blind insistence on the status quo, in Russia and in China they were the creators of a revolutionary situation and the unwitting and unwilling allies of Communism. In the end, they dug their own graves and those of their own governments. For these reasons and in the context of the revolutionary ferment sweeping the world these many years, the concessions imposed by a government upon the landlords are essentially an affirmation of a positive goal of a free people. Failing these affirmations, the alternatives are the agrarian revolutions of Communist Russia and Communist China.

There is ample proof that sooner or later the dispossessed will take the law into their own hands, to the utter destruction of the governments and classes who failed to grant them peacefully what they otherwise try to acquire through violence. The rise of Communist Russia and China has demonstrated that in the predominantly agrarian countries a government must have peasant support; failing that it truly has no support at all. It has demonstrated that the foundations of the social structure stand or fall in the countryside and that the peasant and his interests and aspirations must be placed "in the center of the piece." Japan and Taiwan are examples of why and how the peasant was placed there and how he and the country benefited from the act. Communist China is a lesson and a warning of the shape of things to come when the economic and social aspirations of the peasant are sacrificed for the sake of an outdated status quo.

33. Agrarian Revolution in Japan

By about the middle of 1959, as already noted, with the bulk of his work on the land transfer program in Vietnam behind him and his work in that country substantially accomplished, Ladejinsky entered a period of reflection and backward assessment. This article, a decade or more after the fact, is his final review and evaluation of the Japanese land reform. One can sense his quiet pride as he compares it with the Soviet and the Chinese agrarian revolutions.

This article is reprinted by permisssion from *Foreign Affairs*, October 1959.

SINCE THE END OF THE Second World War, "agrarian reform," or "land to the landless," has been a trenchant slogan in Asia. Variously understood and interpreted, it has come to epitomize much of the problem and promise of Asia. The Communists have made their successful bid for power by claiming that they—and only they—were the reformers. But postwar free Asia did not neglect the land issue. Not all of the Asian countries placed the peasant, as Nehru put it, "in the center of the piece"; but a movement to improve the peasant's lot is under way in many Asian countries.

That movement had its inception neither in Communist Russia nor in Communist China. The general pattern was set and the stimulus was given by arch-conservative Japan before the Chinese Communists promoted their brand of agrarianism. In late 1946 Japan promulgated and within three years implemented a program which indeed gave the land to the landless. A decade later it stands as a great landmark in the history of Japan. The unprecedented renaissance of present-day rural Japan demonstrates that only free people and widely distributed private ownership of land can make the best use of the productive forces of the village. By the same token, it demonstrates that the land problem can be dealt with resolutely without the Communist gospel and free of the tragic upheavals unleashed by the Soviet and Chinese agrarian revolutions.

II

Japanese society before the reform was not nearly so monolithic as painted by the legend-makers of Japan. The village with its overexploited, insecure, rack-rented tenantry was the chink in its armor. The landlords had the money, the leisure, the culture, and the power which they did not share with others. The Japanese farmers, on the other hand, were allegedly the bearers of the nation's traditional verities, serving the interests of a feudal and industrial Japan with equal self-denial. The realities were much grimmer than the sentimentalization of the farmer deep in the muck of the rice fields. The majority of the tenants and part tenants, comprising 70 percent of the farm families and cultivating more than half of the land, had little stake in that society. They met the exactions of their many masters with the produce from fragments of overworked land not large enough or rich enough to support their families.

This state of affairs did not go unnoticed in Japan. Before the Second World War there was a body of public opinion, including the military, which recognized that the foundation of Japan, rural Japan, needed shoring up for the sake of the farmers and the nation as a whole. And much of this was recognized also by the American occupation. There was another motive: the danger of agrarian unrest as a source of Communist power and the belief that postwar rural Japan—as Japan in general—presented a fertile field for Communist penetration. The idea of taking the political wind out of the Communist sails was as tempting as it was imperative.

It was out of recognition of these factors that General MacArthur, in his now-famous land reform directive of December 15, 1945, ordered the Japanese government "to take measures to insure that those who till the soil of Japan shall have a more equal opportunity to enjoy the fruits of their labor." The legal provisions into which this directive was embodied in October 1946 called for (a) the compulsory sale at fixed government prices of all the land of the absentee landlords, (b) the compulsory sale of all the land of the resident landlords save for the permissible retention of 2.5 acres—the average size of a Japanese farm, (c) cash rentals and security tenure for those remaining on the land as tenants. These were the principal measures,[1] conceived with an eye for the widespread ownership of land among the tenants.

The MacArthur directive went beyond the redistribution of land. It was not an end in itself but rather a means "to remove economic obstacles to the revival and strengthening of democratic tendencies, establish respect for the dignity of man, and destroy the economic bondage which has enslaved the Japanese farmer to centuries of feudal oppression." The very sweep of the pronouncement against the background of a seemingly changeless agrarian society caused many to doubt its ultimate effectiveness. Nevertheless, when in 1947 the farmers of 11,000 villages of Japan went to the polls to elect representatives to administer the land reform program, the first step was taken toward a fundamental change in rural Japan.

Similar revolutions in Russia and in China have caused heads to roll and darkened the sky with the smoke of burning noblemen's' nests. The traveler passing through Japan's countryside in the years 1947–1949 would never have known that a basic economic and social change was being attempted. It is in those years that the government succeeded in buying and reselling 5,800,000 acres of land. Whereas before the reform 54 percent of the cultivated land was

1. William M. Gilmartin and W. I. Ladejinsky, "The Promise of Agrarian Reform in Japan," *Foreign Affairs* (January 1948).

owner operated, after the reform 92 percent of the land was owned by the farmers of Japan. To a Japanese farmer "a farmer without land is like a man without a soul"; three years after the beginning of the implementation approximately 3 million farmers acquired "souls." Individual ownership of land rather than tenancy became the hallmark of rural Japan.

A decade has gone by since Japan became a typical country of individual small holders. Judging by the solid core of landowning farmers, the reform has been a success. But one may rightly ask: What lies behind the bare statistical bones of so many landlord acres sold to so many farmers? Has widespread, individual ownership of land been conducive to a rise in agricultural production and peasant welfare? Have the new owners held on to their windfall, or have they bartered it away as so many had predicted? Has it loosened the domination of the old village oligarchy and has it brought to the fore a leadership more in tune with the new conditions? Has the role of the village in local and national politics undergone any change, or has the "boss" system of delivering the farm vote en bloc and at will remained unchanged? Has the undisguised favoritism of industry versus agriculture given place to a more equitable treatment of the country's basic economy? In short, has the shift in ownership stopped short of breaking the notoriously hard cake of custom of rural Japan, or has it come close to giving meaning to the lofty aspirations of the MacArthur directive? The answers to these crucial questions shed much light on what the land transfer was really about and on the character of the agrarian revolution it set in motion.

III

An interested observer cannot but be struck with the fever for improvement that has swept rural Japan with the implementation of the reform. It would appear that all of the land was in need of face lifting. The unavoidable impression is that the farmers have declared war on the severe limitations imposed by the one- to two-acre holdings, trying to eliminate widespread fear among them that "if we stand still we shall be swept away." This aspect of the agrarian revolution is well expressed in post-reform village posters: "May our agricultural production multiply the size of our holdings."

Economic and psychological factors have combined to bring this about. It is difficult to measure the pride of ownership and the heightened sense of social position as a source of greater initiative to improve and accumulate, but their effectiveness cannot be gainsaid. Even the new folksongs reflect it:

The same paddy fields, but now my own;
Heavy is the crop and light is my heart;
Bright shines the sun on my own piece of land!

The immediate result of the transfer of ownership was the sharp increase in the accumulation of rural capital. Just as the postwar inflation relieved the farmers of the burden of heavy indebtedness, in the same manner they were relieved of land purchase obligations to a considerable degree by paying for them in inflated currency. This, followed by the elimination of the former high rentals (which formerly constituted nearly half of his farming costs) together with good crops and good farm prices, enabled the farmers to invest in the improvement of their land and modernization of their equipment. In 1957 such investments were roughly four times those before the war.

The zeal to modernize and to improve, favored by good weather conditions, has resulted in higher production. By 1958 it was 20 percent above prewar. Considering that rice output per unit of land was very high even before the war, the national increase of 10 percent attained in recent years from almost the same acreage is particularly significant. The desire to "expand" the size of the holdings has led, for the first time, to a fuller appreciation of animal husbandry as a source of income.

Encouraged by recent land improvement experience and determined to reclaim approximately 1,500,000 acres of marginal undeveloped land, the Japanese government in 1957 launched a very ambitious five-year plan. If successful, farm income will increase by 25 percent. Wheat and rice production will be augmented by 12 and 8 percent, respectively; output of dairy products more than doubled; roughly the same for sugar; and raw silk production, with an eye to increased silk exports, will be augmented by 50 percent. Whatever the actual attainments, the chain reaction commencing with the land redistribution is still on, and its ultimate aim is higher production and a higher standard of living.

But the outstanding feature of the village after the reform is the upsurge of farm mechanization and the emphasis on labor productivity as well as land productivity. For the first time in the modern history of Japanese agriculture, mechanization has extended into practices which an inexhaustible labor supply could do as well. A tractor in a Japanese rice field was unheard of until after the Second World War. But immediately after the reform, in 1949, there were 10,000 small, garden-type tractors in use; a decade later there were nearly 150,000 of them. While the problem of using machines for transplanting young rice shoots and for harvesting has not been resolved yet, the Japanese label recent attainments as "epoch making" and "revolutionary." A comparison of the number of machines purchased by farmers between 1947 and 1957 is dramatically illustrated by a few figures. During this decade electric motors increased almost fivefold; gasoline motors, sevenfold; tractors, fifteenfold; power-sprayers nearly twentyfold; power-threshing machines, fivefold; and rice-hulling machines nearly fourfold.

It might seem paradoxical that the villages with surplus labor should be concerned with labor-saving devices. Part of the explanation lies in the availability of the means to buy the equipment and in the evidence that the release of the farmer from manual labor drudgery in no way inhibits higher yields. Given these two factors, the troublesome problem of a rising surplus of farm labor is mitigated somewhat in the face of improved economic conditions, particularly against the background of the changes in the peasant's outlook. The reform has stimulated the discontent with the narrow boundaries of the old way of working and living. Farmers also wish to share in the progress and comforts caused by the tremendous expansion of Japan's industry in the past decade. For this reason, the rapid mechanization of agriculture is a phenomenon far greater than its technical aspects imply.

IV

Friend and foe of the land reform are agreed that the farmers are better off now than ever before. Although the per capita farm income is lower than for the rest of gainfully employed Japanese, in 1956 average farm disposable income was 50 percent larger than before the war. Between 1950 and 1957 farm and nonfarm income per household jumped from an equivalent of $587 to $1,005. Farmers have been spending twice as much as they did before the war on productive and consumer goods, but the balance has been in their favor.

The Japanese village became an important market for Japan's industrial output, and not only of the new and expanding farm equipment industry. The village elders with memories of strain and stress may shrug their shoulders at the "indecent extravagance," but many a farmer doesn't deny himself a motorcycle, tiled kitchen and running water, a refrigerator and washing machine, sometimes a television set, and, not least, putting aside some money for a son's or daughter's college education. It is part of the new mental outlook, which may well represent a more fundamental change in Japanese society than the more readily discernible surface changes in conventions of everyday life.

One of the significant landmarks of the post-reform Japan is the recognition by the government that the long-established practice of discriminating against agriculture in favor of industry is no longer tenable. The farmers can no longer be taken for granted with impunity; the politicians know that the farmers can be bearers of political gifts. This awareness of their ability to influence government actions was pithily expressed by a farmer: "We support those who support us."

The post-reform policies of the government to assist the farmers are motivated by still other considerations. The national interest has something to do with it, and so has the changing temper of public opinion as a whole. This temper has been mainly emotional and "touchy," idealistic, and with a strong tinge of welfare state notions. Cautious, practical, and calculating Japanese régimes have had to take notice of it. The idea that only industry and commerce are entitled to official nursing has gone out of fashion. Hence the spectacle of borrowing money from the World Bank to be used, at least in part, for agricultural development; huge government investments for land reclamation, which in recent years averaged more than three times those of prewar years; the creation of the Agricultural, Forestry, and Fisheries Finance Corporation for long-term improvement loans; a revision of the tax structure more in consonance with farm interests; and the official rice price policy which guarantees the farmer a profit for his crop.

This is, indeed, a departure from all previous official attitudes toward the farmers; from the Tokugawa age, when, in the felicitous phrase of Sir George Sansom, "statesmen thought highly of agriculture, but not of the agriculturist"; from the agricultural settlement during the early Meiji era, which while abolishing some features of feudalism gave rise to wholesale peasant expropriation; and from the less blatant but none the less "beast of burden" treatment of more recent times.

V

The change in the attitude of both government and farmers and the successful attempts to put the land in the service of greater expectations are telling commentaries on the dire predictions of a decade ago. The anti-reform spokesman argued that land ownership would be a poor exchange for the economic security offered by the landlord. Worse, they foresaw a serious economic and cultural deterioration throughout rural Japan if landlords were removed and tenants became small owners, all allegedly equally poor and with no interest in the general welfare of the community. In Japan, a decade without landlord domination has not borne them out either in the economic or in the cultural field.

The most vivid impression of a trip in any part of rural Japan is the quest for education. The first thing that strikes the traveler is the extensiveness of the school facilities, the enlargement of the old schools, and the building of new ones that would do honor to any metropolitan area of Japan. Village appropriations for education range from 35 to 40 percent of the total budget. In this respect "feudal" Kagoshima prefecture does not differ from the highly literate and sophisticated Nagano prefecture: the same passionate devotion to education, the same demonstrated effort for improvement, and the same evidence of mental horizons which are not limited by the village vistas of Kagoshima.

The yearnings assume a variety of manifestations. A half dozen miles from Kagoshima city one comes across an anthropologist's treat: a village of Samurai descendants. There are none outside of the prefecture, and this is one of the very few in Kagoshima. The general appearance of the settlement with its straight lanes upon which the two-sworded men once practiced archery from fast-moving horses, the age-tinted stone walls which give the impression of surrounding the village and yet dividing it into neat compounds, the design of the gates and the layout of the living quarters so as to give time to prepare for an unwelcome intruder—all these take one back a century or centuries. Even the accompanying Japanese friend cannot help but remark on this surviving vestige of a feudal society. But in reality it is only a museum piece. Far down the end of the same lane in a large, new school auditorium, young Japanese carry on as if they were ages removed from the Samurai ancestors and what they represented. More than a thousand of them watch goggle-eyed as the Sugar Plum Fairy and the supporting cast of dancers, aged 10 to 12, perform *The Nutcracker* in classical ballet costume. Not every village has its "corps de ballet," but the one reared in the so-called citadel of feudalism provides its own commentary on the "changeless" Japan on the one hand and the people's hopes on the other.

Mass primary education is in the Meiji tradition, but the urge to expand it has never been greater than in post-reform Japan. No one is interested in "economizing" on it, and the question of "affording" does not enter into the building of a multi-million yen school. The rise of the standard of living, the consciousness of the new position of the farmer, and the generous governmental subsidies to education have all had much to do with the education rush. But there is another explanation which reveals the darker side of the Japanese village, whether before or after reform. This relates to the second and third sons, who do not in effect inherit the land, despite the inheritance provisions of the new constitution, and must shift for themselves, looking for opportunities in urban Japan.

A village mayor made this problem clear to the writer:

> My prefecture has neither raw materials nor finished products to export. All we can export is human beings. In the past we exported them to China, Korea, the United States, and above all, to the industrial and commercial centers of Japan. Myself, I was an exportee to Manchuria. The time has passed when we can continue this process with the new generations. We can only try to export them to other parts of Japan and there they are bound to meet severe competition. The very least we can do for our surplus men is to provide them with the best education obtainable. Once educated, they will not remain in the village; they will have to go to the cities and let the cities take care of them. We will have discharged our obligations towards them, and we will have done our duty to the cities by giving them the best export goods the village can furnish. If they remain with us, uneducated and with no future, there will be no end to our misery.

This is a national and a rural problem and the farmers try to meet it in their own way. However, the schools are not only for the second and third sons. As in everything else, the landlords have been the intellectual leaders of the village, but the cultural tradition to which the landlords lay special claims shows no signs of drying up in the post-reform village.

VI

Everything about the reform pointed to an alteration in the social structure of the Japanese village. Begun with the Meiji restoration upon the elimination from the village of the feudal lord and the warrior-retainer, it was virtually completed with the reform when the village became mostly owner farmer. One need not be an economic determinist to note that with the transfer of the land to the tenants at nominal prices the landlords lost their affluence and with it a good deal of their influence. Now, new and old owner cultivators are found in positions of responsibility as members of agricultural commissions, farm cooperative boards, village offices, and school boards. The former landlords have much to say about the absence of the customary homage the tenants used to bear them, feeling somewhat bewildered by the emergence of a new leadership and the need to deal with it on equal footing.

It would be a mistake, however, to assume that the new leadership has completely displaced the old one or that it will soon do so. Tradition still counts for much, and resident landlords with two or three acres leased to the remaining tenants, and the relatively small number of ex-landlords with timber holdings, continue to exercise influence. What is taking place is the sharing of power between the two. This is a new and welcome development. Japan's agriculture after the reform still is a marginal economy with acre and half-acre farmers. It needs all the available skills and social peace and stability to deal with the ever-present problem of six million farmers on 15 to 16 million acres of land.

In his classic work, "Japan's Emergence as a Modern State,"[2] Herbert Norman speaks of "The Janus head of the Japanese peasant"—he can be alternately conservative or radical. There is ample evidence to justify this view; and yet, ever since modern Japan put the ballot to use, the peasant vote has been invariably conservative. The vote brokers, the gentry, the local political bosses, the tweedledum-tweedledee type of political parties, and the conservative side of Janus' head—all these contrived to that end. Nevertheless, the old pattern has undergone a change, part of it being the rise of the powerful socialist party. Elections in the past decade demonstrate that the peasants still vote the conservative tickets, but they also vote for Socialists and independents. Local politicians are still in business, but they are in a quandary; the vote can be, and often is, divided in too many ways to suit their customary want.

A Japanese writer refers to the reform as the event that "injected the breath of democracy among the farmers." If this is the case, one of its effects is that political labels have lost a good deal of their pull and that the farmers vote their individual preference. They pick and choose,

2. International Secretariat, Institute of Pacific Relations (New York, 1940).

and the economic interest is the motivation. Aside from the candidate's own proven concern with farm conditions, the record of his party's agricultural policies is equally decisive. Whatever action a government in power may choose to take with respect to fertilizer prices, the price of rice, scope of land improvement work, and short- or long-term credit funds will vitally affect the distribution of the vote. Clearly, "we support those who support us" has a modern democratic ring.

There is only one group for which the Japanese farmers do not vote—the Communists. They are among the authentic losers of the agrarian revolution. Whether the tenantry of Japan would have flocked to the Communists' colors if the country's fundamental farm issue had not been met cannot be answered with certainty. The radicalism of the peasant might have asserted itself; but, having blindly opposed the reform because the Soviets pronounced the "MacArthur reform" as nothing more than a plot "to sell the farmers down the river," Communist political strength in an ocean of small proprietors came to an inevitable end.

VII

In the final analysis, the basis of all the changes is the stake in the land shared by virtually all the farmers. Whether these changes are sustained depends upon the retention of the land by the owners. One of the most widespread predictions by the critics of the reform was that most of the peasants who received the land so easily would just as easily dispose of it, setting in motion the old familiar process of accumulation of land in relatively few hands and the pauperization of many others. Official land prices were decontrolled in 1950, and they have increased fifty to a hundredfold since then; but, so far, the predictions fall far short of fulfillment.

Land is being bought and sold under the supervision of local agricultural commissions, but the rate of such sales at any given year has been less than one percent of the total, or one-third to one-fourth of the prewar sales. There has been a considerable change in the content of the transactions. No land can be bought for lease to others, which would create new tenancies. The sales are made chiefly by part-time, often postwar farmers whose major source of income does not come from agriculture and who became possessors of a minute holding following the reform. The elimination of such holdings, which are undersized even in Japanese terms, is a healthy development, ending as it does in somewhat larger, more efficient farms. In general, whatever the reasons leading to the purchase and sale and whoever the participants in the transactions may be, there have been hardly any cases of land accumulation reaching the permissible ceiling of 7.5 acres, let alone exceeding it.

It is not surprising that they hold on to the land; everything in the long agrarian history of Japan underlines the quest for it. The monuments erected to enshrine the agrarian reform that made it possible and the multivolume histories published in every prefecture of Japan to preserve the memory of the quiet and yet stirring days when "land to the landless" was being put into effect are the more dramatic expressions of the peasant's deep-rooted attachment to the land. This is an important answer to the most crucial of questions: Will the reform last?

Important, too, is the fact that landlords do not claim the return of the land, although their demands upon the government for "reasonable" compensation persist. They cannot regain the land; the political climate favors the preservation of the reform. While many a reform measure enacted under the occupation has since been revised or legislated out of existence, the agrarian reform is a conspicuous exception. When former Prime Minister Yoshida, no land reformer, points with pride to the fact that the enabling legislation was passed during his first postwar cabinet, he speaks for all politically conservative governments of Japan who appreciate the political capital they enjoy in the countryside. The credit claimed by conservatives is painfully annoying to the Socialists, one of whose leaders, a former minister of agriculture, Mr. Hiroo Wada, is the real architect of the revolution. From the point of view of how lasting it will be, this is a happy augury. So long as diametrically opposed political creeds vie with each other for the honor of having fathered the land transfer program, it is politically safe.

VIII

This record of the single most important change in postwar Japan would not be complete without some consideration of the role of the United States. The Japanese and foreign literature are replete with references to the "occupation land reform." The less learned farmers are more discerning. Fearful lest a visitor's remark carried the implication that Mr. Yoshida had fathered the reform, a farmer took it upon himself to set the record straight as he saw it: "No," said he, "not Mr. Yoshida, but Mr. Hiroo Wada and General MacArthur." To the present writer, a participant in and close observer of the reform, this seemingly strange partnership of a left-wing socialist and an American five-star general, and in the order cited, comes closest to defining the role of the occupation in helping bring about Japan's agrarian revolution.

Contrary to a widespread view, the land reform program was not American made, packaged, and delivered in Japan. The idea was as indigenous as the conditions which impelled it, as the numerous prewar (albeit unsuccessful) efforts bear testimony. And yet, the occupation played a crucial role in overcoming the reluctance of the Japanese government, in speeding up the process, in the active concern with certain phases of the legislation, and in the steadfast support of all reform-minded Japanese.

The entire character of the occupation was a singular demonstration of a conqueror binding the wounds of the conquered. What was true of the whole was particularly true of the attempt to help set rural Japan free. The occupation would not have incurred the displeasure of American or Japanese public opinion if it had elected to stand aloof from this issue. Instead, it had chosen to champion the proposition that those who work the land should own it, harking back, even if unconsciously, to Senator Thomas Hart Benton of Andrew Jackson's day, who had this to say:

> The freeholder . . . is a natural supporter of a free government; and it should be the policy of republics to multiply their freeholders, as it is the policy of monarchies to multiply tenants. We are a republic, and we wish to continue so; then multiply the class of freeholders; sell for a reasonable price, and give to those who are able to pay; and give, without price, to those who are not . . .

With variations, and in a different setting, this is roughly what has taken place in Japan, and this is where the American occupation has left its mark on the face of Japan. The mere daring to come to grips with the fundamentals of rural Japan carried conviction to the farmers of the American concern for their welfare. In its effects upon Japan and against the background of the general malaise which still finds large parts of the agrarian world restless and discontented, the United States acted with farsighted statesmanship. The Nestors of Japan have taken note of it.

IX

A decade after the farmers became land proprietors, the agrarian revolution in Japan stands out in sharp contrast to those of Communist Russia and China—in one, productive forces reaching out for new horizons; in the other, never-ending struggle to subdue the peasant into an acceptance of an arrangement which denies him his lifelong ambition. In Russia as in China, the old régimes were buried under the peasants' longing for the landlords' acres. Both Communist régimes have ridden to power by "cornering," as Marx put it, "the peasant chorus without which the proletarian battle cry will degenerate into just another swan song." But having cornered them, and firmly in the saddle, the Communists took away the land, herded the peasants into collectives and communes, and wrote "finis" to the short-lived but exhilarating experience of freeholding.

The forced collectivization begun in 1929–30 was Soviet Russia's second revolution and ushered in a social upheaval which nearly disemboweled the Soviet Union in the process. The opposition was drowned in blood or starved to death, but even the "victor" was nearly overcome by its fearfulness. Collectivization and communization of agriculture in China had no such immediate disruptive results, but it is undeniably true that this is also China's excruciatingly difficult and uncompleted second revolution, upon the outcome of which the future of the régime depends.

In the light of this, Japan's experience shows

that, given the will, a conservative government can become an authentic "agrarian reformer," and with no recourse to blood and thunder. A peasant can attain his goal as a free and independent producer. When a government in power has the resolution to meet the land hunger of the landless, it has the political support of the peasantry. If in seizing power the Communists have performed any service at all, it is in proving that in the predominantly agrarian countries a government must have peasant support; failing that, it has no support at all.

It is the irony and bedevilment of Communism in the countryside that it threw away that support and substituted in its place a reign which only brute force can maintain. The Japanese agrarian revolution, on the other hand, has not perverted the idea of land to the landless and has pointed the way for agrarian, non-Communist Asia as to how to satisfy the aspirations of the peasants without destroying in the process much else a nation lives by. That Korea, Taiwan, free Vietnam, Burma, India—to a degree—and more recently Pakistan have already acted upon the lessons taught by the two types of revolution is a welcome promise of the shape of things to come. But if in the process of Communist imperialism some of them are overwhelmed, the driving force will no longer stem from the once-welcome and enticing Communist bait of "Land and Liberty in Asia."

X

Such, in the main, are the consequences of Japan's land program. Even allowing for the relatively short period since the transformation, the village "after" the reform is not the same as "before." But whether from now on the attainments will be further advanced and deepened lies not with the reform. Directly and indirectly it will have added some new acreage, but this will not change the holding pattern to any significant degree. The rising farm population is bound to reduce the holdings still further and increase the number of farms as the sons and grandsons take over.

The somber fact is that, while in the United States between 1920 and 1957 the number of farm units was reduced by 22 percent and the working farm population by nearly 40 percent, in Japan during the same period the comparable categories have both increased by 10 percent, despite the country's striking industrial upsurge. The current prosperity of Japanese agriculture cannot hide these harsh realities. Japan's non-agricultural economy has served as a safety valve by providing the farmers with approximately 40 percent of their income. If there is, therefore, any lasting answer for Japan's 37 million agrarians, it must be sought outside of agriculture in expanding industrialization and its capacity to absorb rural unemployed and underemployed.

To note this is not to minimize the great ameliorative effects of the reform on the farmers of Japan and on the country as a whole. The countryside is at peace. The farmers own the land; they are unencumbered by debts, and they are a political and social force to be reckoned with. Many a bent back has been straightened, and the dignity of individual life and work has grown side by side with greater productive resourcefulness. In this lies a partial escape from the severe handicaps of small-scale agriculture pressed upon by a huge farm population.

The Japanese village is in a state of "uplift" which has enormously stimulated the desires for betterment. By the same token, the gap between anticipations and fulfillment will probably grow. The task of post-reform rural Japan is not complete fulfillment but rather to prevent the gap from getting out of hand. But whatever the future holds in store for Japan's farmers, to date the widespread influence of the reform has been sufficiently profound to make it an epoch-making event in the agrarian history of Japan and an important guidepost for postwar agrarian institutional changes in other countries where they are long overdue.

34. *Self-description/Appraisal*

By early 1960 Ladejinsky was in communication with the Ford Foundation about leaving his post in Vietnam and going to work for the foundation. More specifically, a job in Nepal had been mentioned and apparently agreed upon in principle. A letter of June 15, 1960, from Ladejinsky to George Gant of the foundation is part of this correspondence. In this excerpt from that letter Ladejinsky provides a rare bit of self-description and appraisal. I was delighted to come across it.

I AM NOT A TRAINED agricultural economist in the sense in which our agricultural colleges turn them out. At Columbia I was interested mostly in economics and history. Nevertheless, my first published work appeared in four issues of the *Political Science Quarterly* (Columbia) dealing with the Soviet system of collective and state farming. They were the first studies of this kind in English, still used by students concerned with the origins of the Soviet economy. In the many years with the Department of Agriculture, from the lowliest rank until I replaced Raymond Moyer as chief of the Far Eastern division, I worked as an agricultural economist, mainly concentrating on economic and social agrarian developments in Asia, with a bias in favor of tenure problems.

I came to this chiefly as a result of a lesson I learned from my own experience before I left Russia in early 1921, namely, that the Communists would never have attained political power if they had not dealt with the land question resolutely, by turning the land over to the peasants. But for the failure of the anti-Communist parties to settle this issue expeditiously, it is quite probable that I would be still running my father's rather extensive flour-milling and timber interests. This is a digression to underscore how I came into the profession through the back door, so to speak, rather than through technical training.

Much of my agricultural work in Japan, during my frequent visits to Korea, slight brush with the Chinese mainland, concentrated work in Formosa for short periods, work in India and in Vietnam was of a special character. It extended beyond agrarian reform matters. The latter has always been a primary concern, but beyond that was the attempt to influence governments, if I may say so, along reform lines *and* along lines which the specialists thought—and I agreed with them—made for a "rounded out" agricultural economy. I worked with technicians, "translated" them as it were into a more comprehensible idiom, and then tried, both orally and in writing, to "influence" the proper people into accepting certain propositions. In short, in making the trip to Nepal I should like to ascertain, among other things, the kind of an agricultural economist the foundation needs there. If it is a technician pure and simple, I am not your man. On the other hand, I should imagine that USOM/Nepal has enough of the other kind.

It has been my experience in this part of the world that little can be done in the economic assistance field unless there is a clear awareness of the political climate of a given country and unless the "influencers" have the confidence of the people who make the decisions. You will not accuse me of (being) lacking in humility if I say that, if I have a "specialité," I could lay claim to this one. Experience and natural predisposition to the idea that one prospers in amity rather than discord have taught me the soundness of this approach.

Part of the same is the idea that, while the grass roots must be reached through useful endeavors, there is no substitute for a close relationship with the elite. Small though this group is in country after country, its representatives mold public opinion and policies are shaped by them or through them. I do not know the situation in Nepal, but I suspect it applies there as well; I intend to play my role accordingly. I touch upon this not as a "selling" point but merely to outline, however briefly, how I view some of my chores once I get to Nepal.

35. Exploration of Job in Nepal

It was late October 1960 before Ladejinsky could manage to take leave from his post in Vietnam and pay a visit to Nepal, where he made a broad assessment of the development situation and prospects in political and administrative as well as economic terms. He reported his findings in a letter to George Gant of the Ford Foundation, written on November 21 shortly after his return to Saigon. Although Nepal is a small country, the compass of Ladejinsky's interests and observations here is exceedingly broad. Another aspect of special interest is the insight into Ladejinsky's modus operandi. In the papers preceding, this has been more or less implicit; here it becomes quite explicit.

AFTER MUCH CABLING, arranging, and counter-arranging, I finally made Delhi in the early hours of October 23. I spent four days there mostly listening to people with knowledge of Nepal. Doug knew the right people and I had no problem covering a fair amount of ground in a relatively short time. But no amount of information, no matter how accurate and well presented, can prepare one for the realities of Nepal. I found the weather superb, the scenery beyond compare, and the people charming. Yet, this time, unlike eight years ago, I did not "fall in love" with the country. A beastly cold contracted in the chilly rooms of the Royal Hotel may have had something to do with it, but that was not serious. The reason is that eight years ago the trip was mostly a lark; this time I dug deeply earning my "keep," and the great attraction which is Nepal proved to be no substitute for the realities one must face if one is to work there fruitfully. I stress this because most Americans and Westerners I encountered there "simply love" the place; I believe that it has much to do with good climate, good housing, and a few other comforts which make Nepal so pleasant to people in a position to enjoy them. I found the problems of Nepal more disturbing and the comforts of living less enticing.

I should begin by saying that the timing of my visit was not of the best. The three principal members of the Planning Board were out of the country. The deputy prime minister, who is the man directly responsible for the second five-year plan, was away with the king in Europe. The secretary of the Planning Board was attending the Colombo conference in Tokyo, while the chief economist and alleged author of the plan was there, too. This was a disappointment, for I found myself critical of the plan and yet unable to voice my views to those best able to defend or explain it. I am still smarting under this particular difficulty because, willy-nilly, when pressed and pressed hard for an expression of an opinion, I appeared critical of a plan built, as I shall point out presently, on the weakest of foundations. I have not failed to temper my criticism with praise; indeed, the authors deserve it for merely daring to deal with developmental problems. I am afraid that the version of my position conveyed to the gentlemen then away from Kathmandu will reflect upon me rather unfavorably.

I have lived in this part of the world long enough to realize how easy it is to sound "negative" about many aspects of underdeveloped countries, particularly about a country such as Nepal. One does not have to delve deeply into the administrative and economic structure of Nepal to find a lack of almost everything. We all know the reasons why India was in a position to carry on the morning after independence and why Nepal cannot carry on with anything like similar dispatch. This knowledge should not blind us to the fact that in Asia, Nepal is *the* classic example of a medieval country which literally overnight assumed many of the forms of a modern state with relatively little to sustain it.

I spent many hours in Singha Durbar, the former palace of the ruling prime ministers, the Ranas, and presently the seat of the entire government. I talked with a great many people of all ranks and position. I found them interesting and sympathetic and many not without sophistication. But it is clear that instead of a plan aping the Russians, the Chinese, the Indians, and so forth, Nepal needs not a Planning Board for a second five-year plan (Nepal never really had a first five-year plan) but a Planning Board which will attempt to create the *rudiments* of an administrative machine. The absence of this is the most striking feature of a visit to any of the agencies housed in the Singha Durbar.

I dwell upon this because of its relevance to the Planning Board and the foundation's intention to strengthen it. My base of operations was mainly at the desk of the deputy secretary of the board, an engineer by training and, I should add, a man to my liking. His intelligence is in itself an asset, and I am not worried that he is not conversant with the dismal science of economics. What disconcerted me is the emphasis he and the others placed upon a plan which is not a plan at all. By the "others" I mean the prime minister; the deputy prime minister, with whom I caught up in Calcutta; and, by all accounts, the two gentlemen away in Tokyo.

I have been asked to read the second five-year plan and comment in writing. The latter I refused to do. I did try to find out something about the foundation or "the facts and figures" upon which this ambitious expression of goals rests. In the process I came to deal with a good many branches of government in the Singha Durbar. It is as a result of this exercise that I came to believe that not only do they need something much more modest by way of a plan but that the ABC's of administration and training must come first of all. This, rather than the high-sounding goals, should be the concern of the American members of the planning board. I say this because one of the characteristic features of the second five-year plan is not only the absence of even a rough statistical base but also an absence of any consideration of how to move the planning from the paper stage into one of execution.

I am not discovering any Americas when I say that even if the Planning Board produced a reasonable sound plan that has some relationship to Nepal's own material base, revenues, and current foreign aid, there is still the problem of implementation. As I looked about me in the Singha Durbar I couldn't help but wonder as to who was going to do that. Two days spent at Pokhra—an important district seat, 45 minutes by plane or 9 days by foot—served only to reinforce my fears in this regard. With it went the feeling that the service to Nepal and to the Planning Board can be rendered best on a step by step basis, building from the ground up and eschewing fantasies.

I did not leave behind a written statement of any kind, but I conveyed some of my views to a number of people. They met with apparently little opposition in the Singha Durbar; in fact, some members of the Planning Board (appointed representatives from various agencies) were not in disagreement with me at all, but, said they, "Now try to air your views before the deputy prime minister, the prime minister, the gentlemen away in Tokyo, and make *them* see the nature of your misgivings!" This I tried to do. I had two long talks with the prime minister and a short one with the deputy prime minister in Calcutta.

I got along with the prime minister very well indeed, but I am not sure I convinced him that the plan should be toned down; that while anticipation must always be greater than realization, past a certain point the plan can become a serious political handicap for his government; that the second five-year plan is, in effect, a series of ill-conceived promissory notes and that a failure to pay up on due date is bound to spell trouble for the regime; that, without disparaging the importance of a plan, many a country has done rather well in a planless sort of way; and so on. I did not criticize the idea of planning as such; I questioned only some aspects of this particular second five-year plan. As to the need of a Planning Board, I left no doubt in anybody's mind that Nepal needs one for many reasons, not the least reason being that with so many aid-givers Nepal must have an agency to make some order of the variety of aid projects, their scope, projects being at cross-purposes with each other, and so on. In fact, I think that the whole problem of aid from whichever source should be centered in and marshalled through the Planning Board.

In reply to some of my views, the prime minister pointed out that, politically speaking, his government is committed to the second five-year plan. There may be shifts within certain priorities but, in the main, the plan is to remain as is. Clearly, this plan has become a fetish and against this I argued strongly, the main point being that Nepal must mobilize and develop her resources without the necessity of squeezing everything into the plan.

My talk with the deputy prime minister was along the same line and his reply did not differ much from that of Mr. Koirala. Coming as it did from a successful businessman, he left me puzzled. Of course, the prime minister and his deputy were in agreement about the plan's basic shortcomings, such as insufficiency of material resources, the weakness of the plan's assumptions, and the lack of an administrative machine to implement them. Nevertheless, and surely for the time being, they stand by it. I did not expect immediate results and I am not discouraged. Moreover, according to Doug's account of his and Dr. Hill's session with the prime minister, perhaps my presence was not without its mark.

I liked the prime minister very much. He is intelligent, liberal in the best sense of the word, a sensitive man, indeed concerned with the welfare of his people, and would like to achieve it by avoiding dictatorial means. He spoke freely about his difficulties, "the social tensions" with which Nepal is ridden at the present time. Their origin stems from the enacted legislation to correct the abuses of the country's land ownership pattern. It is far from radical legislation, but the prime minister cannot implement it. He places the blame upon the officialdom and the Singha Durbar, most of whom derive income from land. Besides, the owners in the countryside feel similarly. They have staged bloody riots against enforcement, while the would-be beneficiaries have staged similar riots under the leadership of the extreme left. In the meantime, local elections in the past couple of months have recorded a steep rise in the Communist vote, steep enough to give the prime minister sleepless nights. In the prime minister's eyes, the second five-year plan is the weapon with which to counteract the extreme right and the extreme left.

It is understandable why the prime minister is wedded to the plan and its implied promises of a higher standard of living, sizable expansion of industrial employment, and a change in the character of the country's economy. By the same token, it is understandable why in his opinion every economic move in Nepal is also a political move. Less understandable is his notion that sound economic development automatically yields political rewards. I took the occasion to dispute the oversimplification of the idea that a fuller stomach *always* goes hand in hand with political stability. The prime minister is intelligent enough to appreciate the fact that psychological elements, given rise by the manner in which a country is administered, may prove to be vastly more significant in insuring the success or failure of a regime in power than the economic elements. For the moment, however, he prefers to stick to a line which points, even if only in theory, to economic growth of the country.

There is another reason why the plan has become such a significant symbol for present-day Nepal. The prime minister is an important leader of the country but not *the* important leader. He cannot carry out his land legislation, not only because of the bureaucrats and local landowners. He believes he could do the job in three months through mobile teams of his own choosing. Unfortunately, the king himself, the real leader of the country, is not a wholehearted supporter of social reform measures. The same is true of the very powerful deputy prime minister, possibly the closest man to the king. There are many important unofficial hangers-on who feel similarly. It is little wonder, therefore, that being thwarted in one direction the prime minister is the more anxious to demonstrate to the country that his regime is capable of making economic progress along other lines.

While in Kathmandu I looked into the aid programs of the United States, India, the United Nations, and that of the Ford Foundation. The Americans are not too happy with their program. They realize that our aid is nine years old and the question as to what it has contributed is not easily answerable. The American aid program made up, in the words of American specialists, "of bits and pieces" is difficult to "see" or point to. Evidently much has been said on that score by the Nepalese. The projects are mainly agricultural and of "in the long run" nature, since the transmission of agricultural science and techniques is indeed a long-range affair. I carried away the impression that under the new American aid director the program will be subjected to a serious reappraisal and with an eye to more specific and more immediately observable results. The feeling is that an organization made up of eighty Americans should put in a better performance. Actually, they don't belong miles away from the Singha Durbar but right in it, beefing up what native service there is.

The Indian aid program appears to me as the most vigorous program in Nepal. For political reasons too involved to go into, the Nepalese tend to be extremely critical of the Indian economic aid. I have encountered an instance when a Nepalese official went so far as to deny that a rather important irrigation project created by India is the handiwork of the Indians. Nevertheless, my impression is that the Indians have performed very well in road building and irrigation works; and, as an American agricultural specialist told me, "the Indians may yet demonstrate that their village block system is superior to what we are trying to do."

The U.N. program is mainly of a technical assistance type. It is well conceived and applied and obviously to the satisfaction of both parties. I was deeply impressed by the chief of the group, Brigadier Divers. The group is in the Singha Durbar, where most foreign technicians belong.

I can speak as favorably of the Ford Foundation cottage industries project. I know little of the details, but even a cursory examination of what is being created there, after a relatively short period of training, leaves one with the impression of purposefulness. The multiplying effect of crafts learned is undoubtedly there. An official of Singha Durbar, mistaking me for a big Ford Foundation "wheel," pleaded with me for a continued assistance to the project. This was in response to rumors that the foundation's aid was about to phase out. Regardless of his plea, I felt strongly about the value of the project and I discussed the matter with Drs. Hill and Ensminger in this light.

If my letter betrays more concern about Nepal than enthusiasm, it is because I cannot accept the facile way of looking at the country. It is not my intention to romanticize Nepal—this is being done to death and I am not in the business of promoting tourism. I believe that a person prepared to work there in earnest should know something of the score and be ready for much frustration. This is not the way it should be, but this is the way it is bound to be for a long to come and for good and sufficient reasons.

Despite my reservations about the place, I am willing to take on and give the chore a try along the lines of my letter to you of June 15. I know that they need fewer chiefs and more Indians, but I prefer to wrestle with the so-called "broad picture." In this connection, I may even try to talk you into hiring a well-rounded agriculture economist. But this can wait till I get to New York.

36. *Nepal's Five-year Plan*

One week after his letter to George Gant about his Nepal visit, Ladejinsky, on November 28, 1960, wrote a letter to Prime Minister Koirala, with whom he had met twice during his stay in Nepal. The subject matter of the two letters is essentially the same; they both center around Nepal's second five-year plan. The graceful diplomacy in which this second letter is couched contrasts interestingly with the blunt expression which marks the first.

I WRITE THIS TO THANK you most warmly for the privilege and pleasure of meeting and talking with you and for your kind hospitality. My stay in your country was much too short, but you have, in a real sense, extended it by implanting in my mind problems and ideas I would have otherwise missed. I am grateful to you for this and your confidence in me that went with it.

I also wish to take this opportunity to amplify some of my observations about Nepal, observations I touched upon during our second session. I do this with great diffidence, for my knowledge of your country is limited and the impressions I gathered are, for the most part, fragmentary. With these reservations in mind, I should like to raise a few points.

I feel strongly that anyone prepared to work in Nepal in earnest must have few illusions about what can be done or what one must strive to do within the context of the available material and human resources, domestic and foreign. I say this with reference to the second five-year plan.

To begin with, I am not against a "plan" or "planning" in general. I do not believe that these terms popularized by the Soviets have any special Socialist or Communist life-giving mystique about them. The fact is that Soviet planning procedures in practice do not—or at least did not up to now—differ very much from those that were used during the war by American and British governments, for example. In the language of both camps, planning simply means mobilization of resources. Naturally, it is a good thing, and it is in fact a *must* for Nepal to mobilize its resources.

The second point is that the utilization of the resources must be highly selective, which is to say, concentrated. I don't believe that it is within the capacity even of nations richly endowed with material and human resources to do everything at once and according to plan. The second five-year plan of Nepal gives that impression—except that it leaves out a most important consideration: the means of implementation. The Soviets have mobilized and concentrated massive resources on a limited number of key sectors—productive investment, heavy engineering, technological education, military-scientific research, and so forth. The United States has done similarly during the war, albeit not always in the same fields. It is the process of concentration rather than any detailed overall planning which contributes to rapid growth of the selected fields. I emphasize this because in a country such as Nepal, and for reasons I need not repeat here, the plan must be devoid of any *scatteration* of resources, the leitmotif being concentration on those fields which lend themselves to the most return. The exceptions are being taken for granted.

Nepal is a classic case of an underdeveloped country in Asia. Additionally, Nepal is not a dictatorship. I mention the two in the same breath because the insufficiency of capital resources cannot be exploited or augmented in the manner practiced by dictatorships, which can do some things more easily—and with impunity—than democracies. One such thing is to hold down real wages in the interest of capital accumulation or military strength or imperial grandeur. Dictators have been doing this sort of thing from time immemorial.

I bring this up not only because Nepal is a democracy but also because it is probably true that the people wouldn't permit a further deterioration of their already low standard of living. If this assumption is correct and even recognizing the possibilities of additional tax revenue, I carried away the impression of a large gap between the capital resources required by the second five-year plan and what is likely to be available. Not unrelated to this is the fact that the ambitious second five-year plan takes little account of the problem of implementation. The carrying out of a plan of this magnitude presupposes a fair measure of administrative and technical competence of a fairly high order. Regardless of this type of assistance contributed by foreigners, the problem of administrative and technical training deserves the very highest priority; by the same token, the drafting of the plan should be reviewed in the light of this factor.

If I understood you correctly, Mr. Prime Minister, you made the point that in Nepal almost every economic measure is also a political measure and that by this token the second five-year plan is of great political significance. As a political economist I can readily understand this. And yet, and at the risk of appearing lacking in daring, I should like to repeat that, perhaps out of ignorance of Nepalese political conditions, the second five-year plan in its present form may prove to be something less than a political boon. Anticipations must always be greater than realizations; but past a certain point, when the first far exceeds the latter, the plan can become a political handicap.

A political dictatorship isn't required to deal with a free electorate possessing the power to reward or punish. Mao Tse Tung's failure, for instance, to demonstrate to the peasants and to the Chinese people in general that the communes are both an escalator to heaven and a source of inexhaustible food supplies had not endangered his rule, not for the time being. It would be quite different in a country where the voters exercise the privilege of return and recall. In your country, Mr. Prime Minister, the second five-year plan is akin to a series of political promissory notes payable on due dates. Failure to meet the promises is likely to spell trouble for the regime in power. Hence, my concern about goals which may fall far short of fulfillment and my suggestion that the question of implementation capabilities be given most serious consideration.

To the extent that "planning" is another term for concentration of resources, it appears to me that it would be advisable to welcome any individual or group of individuals, Nepalese or non-Nepalese, capable of such an effort. These may not fit into a formal plan, but this would not be of great moment for at least two reasons: (a) the insufficiency of governmental expertise to implement the plan and (b) the fact that many a country has most successfully mobilized its resources in a "planless" sort of way. Japan, Israel, and many other countries can serve as good examples. It is well to keep this in mind and act upon it by seeking out and encouraging all those who can make their contribution, in their own way, to economic growth of the country.

I left Nepal with the impression that the second five-year plan is the very symbol of the country's future economic development and political stability. No one can dispute the importance to Nepal of a rapid rate of economic growth. This is self-evident and we must all strive to that end. Nevertheless, I wish to sound a word of caution in this regard. You will perhaps recall, Mr. Prime Minister, that I cited to you the case of South Vietnam where considerable economic progress in recent years did not necessarily spell political stability. I can cite many more examples. One may retort that those are special cases, but I am inclined to the thought that it is a mistake to assume that, in most underdeveloped countries, economics (and economic growth) always provide the prime motive of political action. I prefer the proposition that, as one put it, "The dynamic of social revolution in Asia and Africa is not a mass demand for tractors or for bread but the will and social conscience of a small intelligentsia . . . Mutual aid, therefore, will not necessarily provide an antidote to Communist revolutions."

In short, while it is undeniably true that in some countries the economic issue may be decisive, we tend to overrate the importance of economic factors as a political instrument. In countries which hover on the brink of a Western or Eastern alignment, the decisive influences have usually not been rates of economic growth or aid but political actions within or outside those countries. Recent events in the Congo are potentially far more explosive for the future alignment of central Africa than hundreds of millions of dollars of aid or several percentage points on the rate of growth. I mention none of this to decry the importance of economic aid and particularly economic aid for a country with such relatively limited resources as Nepal. Nor does this in any way minimize the need for a planning board in Nepal and its efforts to stimulate the economic growth of one country. I am merely suggesting the limitations of economic factors—whether planned or not—as a cure-all.

While not a cure-all, the economic factors are admittedly important and indispensable in many circumstances. In the case of Nepal, as of certain other underdeveloped countries, the two obvious stimulants to economic growth are: (a) the building-up of an administrative and technical organization to help give reality to the country's economic resources; Nepal cannot rely indefinitely upon imported talent; and (b) availability of capital and capital formation.

As to the latter, it is axiomatic that Nepal will require initially a fairly large capital outlay. Whatever the actual sum, it is safe to assume that Nepal's capital investment requirements cannot be based exclusively on internal accumulation and savings. The reason is clear: Nepalese people live on a subsistence level and have only a limited possibility to divert part of their income from consumption to investment.

I am not sufficiently versed in the possibilities of using Indian private capital in certain sectors of Nepal's future economic development. But generally speaking, private capital imports and investments from highly developed countries shun underdeveloped countries—exceptions notwithstanding. Capital imports must therefore come in the form of government aid from those able to dispense it.

Nepal has been receiving such aid for some time and will undoubtedly in the future. However, the current attempt to attain an economic breakthrough via the second five-year plan, whether modified or not, imposes upon the planners the obligations to examine more critically the volume of aid, its composition, and the ways and means of increasing the absorptive capacity of the country. I heard not a little talk in Nepal about its inability to absorb more aid, followed by the inevitable question: "Why more aid?" The absorptive capacity issue is of crucial importance, for it can largely determine the volume of aid from XYZ countries.

All of the aid, regardless of source, will have to be examined in relation to the plan's goals. Five aid-givers, more or less, each one going his own merry way, may not necessarily lead to a significant economic growth. If the aid-giving countries and their contributions are not part-and-parcel of the country's mobilization of resources, the consequences are too obvious to call for comment.

These, Mr. Prime Minister, are some of the points to which I wish to call your attention as I reflect upon my recent visit to Nepal. I have not raised them merely to be critical. Your planners deserve much praise for daring to deal with the economic development of the country. If I sound critical, it is only because it seemed to me, and possibly wrongly, that the economic growth and the political stability of Nepal could be better served by relating the second five-year plan more closely to the country's realities.

One last word about the results of the American elections. As you gathered from our conversations, they went my way. A New Deal in America was inevitable, and we are likely to have it now. The issues for all of us are momentous enough and I would rather entrust them to a young man with courage, daring, and ability than to a man who insists, at least publicly, that all is well with the United States. The tasks and challenges Mr. Kennedy is prepared to face do not leave much room for complacency. Mr. Kennedy has promised a hard four years, but perhaps they would be even harder if he were not here.

Once again, Mr. Prime Minister, accept my thanks for your kindness and consideration. I hope to return to Nepal before many months pass. In the meantime, I intend to revisit Washington in the hope of meeting some of the people of the new administration. I shall not forget Nepal when given the opportunity to air my observations about this part of the world.

37. *Land Reform in Indonesia*

I have given this title to a letter written by Ladejinsky to A. T. Mosher, executive director of the Council on Economic and Cultural Affairs (now the Agricultural Development Council) of New York on January 24, 1961. The council, it is clear, had asked Ladejinsky to visit Indonesia and offer what advice he could to the minister for agrarian reform. Ladejinsky spent some ten days there. Although he did not this time go out into the field, he learned enough to warrant telling the minister some unhappy truths about Indonesia's recently enacted land "reform." He also sent the minister a note, on returning to Saigon, attaching a copy of his memorandum on the land commissions which had played so essential a role in the implementation of the Japanese land reform.

I AM JUST BACK from Jakarta and a I am rather rushed trying to clean up here before my departure for London and then the States. Ralph Allee, who participated in all my discussions with the Indonesians, will give you an idea of what transpired there, and I shall talk with you at length when I reach New York. All you get now is a brief summary of the promised "fill in."

The Indonesian trip, including travel time, lasted only eleven days; but, looking back, it was not without its value. I learned something and chances are that the Indonesians would admit to the same. I had no idea what they were after; and, while my ignorance is still abysmal, I do appreciate now why they are so desperately calling for help and why only a miracle can help them to make some sense out of their voluminous, disjointed, contradictory, and altogether too politically, conservatively inspired agrarian reform legislation.

One doesn't have to spend much time in the Ministry for Agrarian Reform to be reminded of the stolen horse and the locked stable. I mean to say that the damage, the enacted legislation, is an accomplished fact. What is so damaging about it is that the agrarian legislation is hardly rooted in the realities of the existing conditions and that the high-sounding purposes of the reform have no great chance of being accomplished. If I rendered any service at all, it is mainly because I couldn't help but speak out upon the shape of things to come, given the conditions outlined for us by our Indonesian hosts. In short, I tried to sober them up and they know that they are in trouble. I deeply regret the lack of time to visit the countryside. I strongly suspect that a look-see there would have served to underscore the inadequacy of the reform effort here.

The agrarian reform is intended for Java above all, where allegedly 60 percent of the farmers have virtually no land at all. The stated aim of the reform is to provide them with land. It was my unpleasant duty to tell the minister that this was not in the cards, not only because land is scarce in Java and people are many but more specifically because the legislation is so drawn up as to prevent the release of a sizable acreage for distribution.

The principal joker in the legislation is the amount of land a landowner can retain on the one hand and the minimum guaranteed to the landless on the other. The latter is a fiction because in Java there simply is no arable land, actual or potential, to insure for the landless a minimum of 2 hectares per family. The maximum can range from 5 to 7.5 to 10 hectares, depending upon the density of population. In the 64 districts of Java (out of a total of over 100), a maximum retention of 5 hectares will not create any significant excess for redistribution. Roughly the same holds true for the other parts of Java. Moreover, if the contemplated order of priorities establishing claims on the excess becomes a reality, little indeed will be available for the landless.

It was not difficult to question the soundness of the core of the intended reform. The explanation lies in the extremely conservative approach to the problem of the land hungry, an approach dictated by powerful political groups. They prefer no retention limits on Javanese holdings and as of the moment they seem to have carried the day despite the specified ceilings. The radical language of the preamble is something quite apart from the business of giving land to those who cultivate somebody else's land. I had quite a bit to say on the subject. The officials, including the minister, saw the point; but I doubt if this problem will be given another hearing. For the same reason I question much of the raison d'etre of the reform.

No thought whatsoever has been given so far to the question of land prices and of financing the program. This constituted the second main topic of discussion. I called attention to some possible ways and means of dealing with this aspect of the reform so that whatever amount of land is redistributed it would not become a burden upon the new owner. It is clear that those who drafted the program have elected, for whatever reason, to overlook this obviously touchy and difficult problem. I know of no legislation of this kind with so significant an omission.

The third area under discussion was the enforcement of the reform. Not nearly enough thought has been given to this matter. The experience of other countries came into the picture and always with the emphasis that even the best legislation could not possibly be implemented unless the government has the will to see it through and unless certain administrative devices are created to go on with the job. It is in partial answer to this problem that I outlined the role of the land commissions in Japan, and later on in Taiwan, in a memorandum of which a copy is attached.

I could say much more about this brand of Indonesian agrarian reformism, but the above should suffice as an example of how not to try to give land to the landless. The minister is an intelligent and fairly well-informed man. I did not have to use a sledge hammer to convey to him the idea that, since he and I are agreed that the landless need a New Deal, drastic revisions of the recently enacted legislation are in order. The minister is essentially a politically minded man—and at one point way over to the left. I did not have to labor the point that, despite the stated and implied promises of land to the tiller, the enabling document is shot through with conservative safeguards in order to prevent any significant redistribution of land. The minister understood that I was not moralizing the issue but merely pointing out the political risks due to the likely failure to make political capital among the many who can furnish it. And the more so since the estimated total number of landowners in Java with more than 5 hectares is only 24,000, if I recall the figure correctly.

As it stands now, and mindful of how little I know of Javanese village conditions, I don't believe that the intended reforms will provide the Javanese land workers and sharecroppers with much land. I am almost inclined to the view that it is essentially an *anti*-land redistribution program, although I am certain that it was not planned that way originally. In the long run the phraseology will deceive no one, and so I restated the obvious for the minister's benefit, namely, that if the promissory notes are not made good the chickens will surely come home to roost.

Ralph will support me when I say that our talks proceeded in a most friendly atmosphere. No offense was taken by either party at any time. The minister was eager to listen and he did not try to gloss over the difficult, if not impossible, task Sukarno wished on him. I believe he would change the character of the legislation if it were up to him, but is not up to him. He was eager to have me tell the government—meaning Sukarno—what I have been telling him, but I did not have the time to wait some extra days for an appointment. He needs assistance, I sympathize with his position; and I regretted very much that I had so little time at my disposal. There is an old Jewish saying to the effect that one cannot attend two fairs with one ass—at the same time. In my case it is three, at least right now.

Despite the brief time spent in Jakarta, I dare say that the trip proved to be useful. I judge this by the fact that I left the minister in a much more troubled state of mind. Therein may lie some beginnings of wisdom; but, and I repeat, I doubt that the glaring shortcomings of the program will be corrected. The seemingly unbelievable way in which leadership and government carry on in Indonesia argues against thoughtful and carefully considered propositions, be they radical or conservative. I carried away the impression that as of now *form* counts most in Indonesia; *substance* is often troublesome, and the tendency is to shy away from it especially when it is immediately troublesome.

But having said this, I do not deny that the minister can be influenced along certain lines. I have evidence to support my contention; I will go so far as to say that he will try to act upon some of the suggestions made in the course of our discussions. Whether it is really worthwhile investing time—assuming that I had some to spare—is another matter. My skepticism stems from the knowledge that, in such a highly personalized government as the Indonesian, the number one, Mr. Sukarno, must be influenced if real progress is to be made. Regretfully, I am not sure that I could reach him, and, even if reached, I am not sure that I could find the key to a man whose attitude toward affairs of state is a great puzzle and wonderment to me and countless others. Under the circumstances, I consider my brief venture into the agrarian reform business of Indonesia as pretty much completed.

I realize how sketchy all of the above is, and I console myself with the thought that we shall discuss the matter in some detail in New York in mid-February.

38. *Agrarian Reform in the Republic of Vietnam*

Late in 1959 Ladejinsky presented an earlier version of this paper to a Conference on Social Development and Welfare in Vietnam held in New York City. Whether that earlier version was substantially modified or merely updated for publication in 1961 is not clear. Despite the intimations of agrarian reform achievement briefly indicated in his December 1959 article, Ladejinsky's fuller account here of the substantial progress made by Diem's government in the late 1950s in improving conditions of land tenure, rent reduction, land reclamation and resettlement, and, finally, land redistribution will probably occasion real surprise. Given President Diem's inherent conservatism, it seems clear that a large share of the credit for persuading him to promulgate the land redistribution program in 1956 and to carry it through must go to Ladejinsky.

"The real opposition to agrarian reform," Ladejinsky points out, "was provided by Radio Hanoi and the Communist agents in the countryside." They could not afford, he makes it plain, to let Diem's reforms succeed and later expand, as Ladejinsky thought would happen in the absence of Communist disruption. This suggests it was the fear that Diem might in this way win the solid support of the peasantry that caused the Communists to intensify their efforts in the early 1960s to bring his government down. Was the degree of agrarian reform that was achieved by 1960–61 perhaps an important factor in helping South Vietnam to hold out as long as it did?

IN ONE WAY OR ANOTHER, agrarian reform is probably as old as farming itself. Under different names, in different countries, and variously interpreted, it has meant easing the burden of the man working the land, particularly of the man working somebody else's land for a pittance. In speaking of Asia alone, an interested observer can count the "pittance" farmers by the million. But times have changed: an overworked and overexploited peasantry that for centuries was inertly miserable has now awakened and demands *basic* alterations in its condition—in many instances successfully.

The peasant awakening is part and parcel of the post-war revolutionary ferment in which Asia has been engulfed. It is also part of the realization that in predominantly agrarian Asia the new order of things, economic or political or both, depends primarily on the solution of the land question, namely, land for the landless. In a different setting, the Russian Communists exploited this fact forty-two years ago. For the Russian peasants, liberty meant the ownership of the lord's land. The Communist promise in 1917 (which was ultimately broken) that the land would belong to the people who worked it proved to be an issue of the greatest political importance. The Communists would never have obtained power in Russia had they not successfully exploited the peasant's longing for the landlord's acres.

The triumph of the Communists in China is another case in point. Many are the reasons which explain their seizure of power, but one cause is beyond dispute: Nationalist China was defeated not so much by force of arms as by the Communist tactic of promising land to the poverty-stricken, landless, hopeless peasantry.

It would be a mistake, however, to assume that because the Communists placed the land question in the center of Asian politics, they enjoy a monopoly on agrarian reform. Aside from the fact that neither the Russian nor the Chinese Communists kept their promise to the peasants, efforts to give the Asian peasants a greater stake in the land antedate the Communists. The issue has been dealt with by the Asian countries since the war. Japan, for example, did not wait upon the Chinese Communists to point the way to reform. The movement to deal with the problem has been completed in a number of countries and is under way in others. Free Vietnam is among them.

Agrarian reform there is not merely a slogan designed to serve other ends. Large groups of tenants are becoming owners of the land they cultivate. Farmer tenants cultivate the land under vastly improved terms of tenure. While the movement to create stable individual private proprietorship of land is not yet complete, the direction and achievements to date are unmistakable. The Japanese say that a peasant without land is like a man without a soul. This holds true for the landless wherever they may be. And in Vietnam, as in other parts of Asia, there are now vastly greater numbers of peasants in possession of their own "souls." How this came about, its significance, and its consequences is the subject of this chapter.

I

No two Asian countries are alike, but they all have certain features in common and the land tenure system is one of them. Whether in the Far East or Southeast Asia, the age-old wretchedness of the peasant has the same roots. More often than not there are too many people pressing on too little land; sometimes, as was the case till recently in the southern part of free Vietnam, peasant poverty prevails in conditions of a relatively small farm population and an abundance of land. For the most part, inadequate tools, primitive methods of cultivation, and institutional arrangements over which the peasant has no control are common features of the tenant's life and work. Without moralizing about the tenant-landlord relations as they have developed, it is undeniable that much of the peasant's misery results from this relationship to which customs, courts, and governments have given official sanction over the years.

Vietnam was no exception, and this is particularly true of what was formerly known as Cochin China. This is a huge rice plain of some 64,000 square kilometers criss-crossed by canals, lying in the southwestern part of present-day Vietnam. Before the extension of French rule into this entire area in 1874, there were less than 400,000 hectares under cultivation; seventy years later the cultivated area had risen to more than two million hectares. The region had become the authentic breadbasket of the country with an exportable surplus of rice of around 1 to 1.2 million tons. It is here that Vietnam's present land redistribution program is concentrated, because it was here more than in any other part of the country that Asian landlordism with some of its worst features had taken root.

By 1955 Vietnam had a very roughly estimated total of about a million tenants, of whom about 600,000 were in southern Vietnam and 400,000 in central Vietnam. The tenancy practices in the two parts of the country differ sharply. Individuals owning many thousands of hectares are not rare in southern Vietnam—and widespread tenancy is its hallmark. In central Vietnam peasant proprietorship accounts for approximately three-fourths of the cultivated land, and landholdings are small. Of the estimated 650,000 landowners, no more than about 50 own more than 50 hectares each, and barely a dozen own more than 100 hectares. The great mass of landlords own 5 to 10 hectares each; the system closely resembles the petty landlordism once prevalent in Japan, Korea, Formosa, and a number of Southeast Asian countries; and this explains why the current land redistribution program does not apply to central Vietnam.

The south presents an entirley different picture. In a total cultivated rice area of 2.3 million hectares, the concentration of land ownership was one of the highest in the Far East or Southeast Asia. Approximately 2.5 percent of the owners, with more than 50 hectares each, possessed roughly one-half of the cultivated land. Conversely, more than 70 percent owned less than 5 hectares each, which gave them an estimated total of only 12.5 percent of the cultivated land. In the province of Bac Lieu 9 percent of the landowners had 70 percent of the land, while 72 percent of the farmers had no land at all, or at most a hectare or two. In the province of Can Tho 4 percent of the landlords possessed more than 50 percent of the land. Roughly the same pattern was repeated in province after province of what had been Cochin China.

Two principal developments explain this concentration. The French colonial administration sold huge tracts of virgin land at nominal prices or gave them away to selected individuals —French and Vietnamese. The few truly big rice holdings were in the hands of French companies of which the single largest, the "Domaine Agricole de l'Ouest," accumulated a total of 20,000 hectares. The other important reason for the land concentration was the loss of land by small proprietors who could not meet their burden of indebtedness except through foreclosure and eviction.

II

With so much land in so few hands and the loss of land by small proprietors, tenancy was inevitable. Whereas in the older, settled part of Vietnam individual proprietorship was predominant on the eve of the Second World War, in the south two out of three families had no land at all; they worked somebody else's land either as tenants or agricultural laborers. In the sections of the greatest land concentration, at least 80 percent was cultivated by peasants who owned virtually no land whatever. A tenant answering a question asked by this writer about the farmer categories with which he was familiar put the situation very simply: "We are all mostly tenants and some of us are farm hands." He did not exaggerate by leaving out the exceptions.

This pattern, even though there was an abundance of land and a moderate population, resulted in tenure conditions normally found in Asian countries where population pressure is extremely acute. Rentals were as heavy as any to be found in Asia—50 percent of the crop. The tenant had to provide his own hut, tools, and livestock and hire supplementary labor at the height of the season. Often short of rice for food or seed, he borrowed from the landlord, repaying double the amount when he harvested his crop. By the time the tenant had discharged all his obligations, his share of the crop was roughly a third of the total.

For the typical five-hectare tenant, conditions were worse. A case study cited in the work of a well-known French agricultural economist sheds much light on this problem.[1] He writes as follows:

> In the fifth month, at the beginning of the heavy field-work period, he has obtained from his landlord a loan of 35 gia[2] of rice and five piasters. His crop yields 300 gia. From this quantity he must deduct 30 gia

1. Pierre Gourou, *L'utilisation du sol en Indochine française*, p. 408.

2. A gia equals 44 pounds.

for the extra rice consumed during the harvest time, 70 gia for the repayment of the rice loan of 35 gia (100 percent interest rate), 12.5 gia for the repayment of the five-piaster loan, 150 gia as land rent, or a total of 262.5 gia. There remains 37.5 gia of his crop, which is little more than 10 percent of what he has harvested. It is not sufficient to live on and, like the coolie, the small tenant must look for a job, but still he has the advantage of being able to borrow.

Even the middle tenant with five to ten hectares of rented land was not much better off. The study concluded: "On the whole, the living standard of those we call middle tenants is not very high and is only slightly different from the living standards of the lower categories."

Those who had some rice for sale were not always free to dispose of it at will; some tenancy agreements stipulated that the landlord had the right of purchase, and usually at his price. The role of the landlord in relation to his land and tenant was essentially that of a rent collector, and often this role was performed by an agent who made part of his profit out of the tenant. On top of this was the basic handicap of insecurity of tenure with no bargaining between landlord and tenant worth mentioning.

The entire landlord-tenant relationship was shot through with exactions, and this in turn bred the heavy burden of indebtedness and usury. Almost everybody in the village was in debt to the landlord or to the moneylender. Interest rates of 5, 10, and more percent a month were common. Landlords sometimes made more profit from lending money than renting land. In general, lenders were less interested in the return of a loan than in keeping the tenant in a constant state of indebtedness. The net result of this type of tenure system based upon rack renting, lack of security of tenure, and widespread usury was a large class of landless, impoverished, and discontented peasants.

III

Thus there is little wonder that the Vietnamese Communists found fertile ground in the villages for their own political ends. When, immediately after the Second World War, they began, among other things to attack the old village order, the response was immediate. They gained control of the countryside and the support of the peasants not only by appealing to a deeply rooted desire to get rid of the French but by addressing themselves to the land question as well.

The Communists lived off the land, and the tax burdens they imposed were often no less burdensome than the rental formerly collected by the landlords. Nevertheless, they posed successfully as fighters for the national cause and defenders of the peasants against the rich gentry. They convinced the peasants that sacrifices in the form of the heavy grain tax were patriotic expressions of the "people's war for liberation." The Communists had no land distribution program, but they convinced the landless that the landlord's acres would belong eventually to them and encouraged their occupation of abandoned land. The payment of rent virtually ceased. The wealthy landlords disappeared from the countryside, fearing for their lives; and much of the pattern created by Communist agitation in Russia and China repeated itself in Vietnam. Here, too, the appeal went to the heart of the matter—land of one's own and an end to the traditional landlord-tenant relationship.

The support the Communists gained in the countryside was not lost on Bao Dai's government. The emphasis the Communists placed upon the land question and the favorable response it evoked from the peasants were at least partially responsible for some government recognition of the existence of a land problem and the need to do something about it.

In his New Year's message of February 2, 1951, Bao Dai stated that the land would not be taken back from the peasants who had occupied the landlords' property during the "troubles" and were still cultivating it. This was to be done "without harming the former large proprietors who have a claim to just compensation." Land concessions were to be limited by law, as were the terms of credit traditional between proprietor and tenant in order to protect the peasant against perpetual indebtedness. A National Committee for Agrarian Reform was organized in mid-1952; in early 1953 President Nguyen Van Tam announced that henceforth rents were in no case to exceed 15 percent of the crop, and this ruling was followed up on June 4, 1953, by a number of legal provisions.

Briefly, the ordinances called for cancellation of certain land concessions which had remained uncultivated or unleased and for the redistribution of such land among squatters and other specially deserving groups. Rent was reduced to a level not exceeding 15 percent of the crop, additional rent to be agreed upon between landlord and tenant for buildings, tools and draft animals. Land leases were to be written for a minimum of five years, and there were what appeared to be limitations on the size of holdings. However, the provisions of this ordinance (no. 21) were so watered down as to make it meaningless. In effect, the landlords did not have to sell or otherwise dispose of the land they held in excess of these limits; the sole limitation was upon the acquisition of additional land either by purchase or lease.

These provisions were poorly conceived and drawn up. They went either too far, as in the 15 percent rent rate, or not far enough, as in the so-called limitation on the size of holdings. When President Nguyen Van Tam went about the countryside promising this low rental, it was more an act of desperation than an enforceable provision. It was neither equitable nor realistic. Even the Communists fixed a maximum rental of 25 percent, and no peasant with the memory of a traditional rental of 50 percent could view the 15 percent rent seriously. It was an ineffective propaganda device.

The real difficulty, however, did not reside in these hastily drafted half measures which were never applied. The most carefully prepared provisions would have met with the same fate had the government been bent on enforcing them. The overriding fact was that the government did not hold sway over the countryside. Its power was nominal even in the so-called controlled areas. Only after the Geneva agreement, which divided the country in two, and the evacuation of the visible Communist presence from South Vietnam were the conditions created whereby the countryside could be reached through agrarian reform measures. And this is precisely what the Ngo Dinh Diem government has been doing since its assumption of office.

The government's efforts to ease the lot of tenants fall into two closely related parts. The first deals with the reduction of rents and security of tenure, the second with the distribution of land among the landless.

When the Ngo Dinh Diem government assumed power, the countryside was in a sad condition. Farmland was abandoned and overgrown with weeds; irrigation and drainage facilities had fallen into disrepair; canals, the indispensable waterways of the south, needed redredging; the greater part of the work animals had been killed off, and all of this was reflected in a sharp decline in production and the disappearance of the all-important surplus of rice for export. And the deterioration was not only physical. After years of Communist penetration into the villages by means of attractive nationalistic slogans, the peasants emerged into the new era a very bewildered lot. Their main worry was that the old landlord-tenant arrangements might prevail again. This fear was a real one, for in many sections of the country the occupants had not paid rent for a decade or more and, in the process, had come to feel that occupancy of the land was tantamount to ownership. Now, having acquired land without due process of law, they were not certain that they would be able to hold onto it.

The first official answer to those fears came shortly after Ngo Dinh Diem's assumption of power. It was unquestionably the darkest period in the seven-year history of the Republic of Vietnam. Much had to be done in a hurry, and the material and psychological conditions in rural Vietnam did not permit procrastination. With that in mind, the government promulgated two basic measures embodied in ordinances 2 (January 8, 1955) and 7 (February 5, 1955). They dealt with items directly affecting the welfare of the tenants—rent reduction, security of tenure, and putting abandoned land back into cultivation.

The main provisions of these ordinances were as follows: rentals ranged from 15 to 25 percent of the principal crop; a loan of seed or fertilizer was repayable at cost price plus an interest rate not exceeding 12 percent per year; all contracts had to be in writing; the life of a lease was a minimum of five years, and the traditional right of a landlord to cancel a lease agreement was circumscribed; village, district, and provincial committees were to be created to settle landlord-tenant disputes; finally, penalties were provided in case of failure to comply with the provisions of the ordinances.

In the years of war and civil war an estimated 1.3 million hectares of cultivated land had been abandoned. It was the purpose of the government to put this land back into cultivation as soon as possible, through the application of ordinance 7. The significant provisions were these: having determined the amount of the abandoned or uncultivated acreage, each owner must declare his intentions with respect to his land. If he refused to cultivate the land himself, he had to lease the land to tenants of his own choosing. In that event, a three-year lease was to be executed under which the tenant paid no rent during the first year, half of the prescribed rent of 15 or 25 percent during the second year, and three-fourths of the rent during the third year. The landlord, in turn, was exempted from the land tax, and so was the tenant. On the expiration of the special three-year contract, the normal contract for five years became compulsory. In the case of absentee landlords, the village council had the right to allocate land to people willing to cultivate it. The rent, after deduction of taxes, was to be held by the provincial treasury for future payment to the owner. Model contracts were drawn up, printed, and distributed by hundreds of thousands as an aid to implementation.

These early measures were undertaken, not as a means of solving all of the land problems of Vietnam but in order to alleviate or eliminate some of the worst anomalies of landlord-tenant relations in Vietnam. Some of the provisions could have been better conceived, particularly the 15 to 25 percent rental range, since it was bound to give rise to disputes. But it must be remembered that the effort was being made at a time when the new government was struggling for its life, that the odds against its survival were most formidable, and that administrative machinery barely existed in the capital, the provincial centers, and the villages. The implementation of the program in these circumstances proved to be a difficult task.

It is not surprising that the application of these measures got off to a very slow start. For nearly a year after their promulgation, the attention of the government was centered on controlling the Binh Xuyen political thugs, the Cao Dai sect, and the Hoa Hao and solving the pressing refugee problem. After years of Communist indoctrination in the villages, the position of the government was far from strong. The lack of administrative machinery to implement the newly proclaimed government land policies was in itself a tremendous handicap. And so it was that during the greater part of 1955 few tenants called the local officials to task for not keeping them informed about the program, and fewer officials called tenants to task for not pursuing their rights under the ordinances.

There were other reasons why the application at first proceeded at a snail's pace. For different reasons both tenants and landlords were reluctant to sign contracts binding them to the conditions spelled out in the ordinances.

Many a tenant had not paid rent in years, and thus even the admittedly low rent of 15 to 25 percent appeared to be an imposition. Others, whose occupation of land had been sanctioned by the Communists, believed that their ownership had already been confirmed and that signing a contract now would invalidate their claim to ownership. There were also the fence sitters, those who believed the Communist propaganda that by July 1956 the country would be reunited on Communist terms, with all the former landlords' land falling into their hands. In short, the heritage of Communist village activities hindered the favorable reception of the rent reduction program.

The attitude of the landlords was no less negative. They felt abused by having rents radically reduced after years of great losses under the Communists. A chief of a province, himself a landlord, spoke to this observer for most other landlords when he said, "We have been robbed by the Viet Minh over the years, and we resent similar treatment from the national government." Some went so far as to insist on retroactive rent payment as a condition for leasing their land; others wished to take back the land occupied by the tenants during the Viet Minh days, while most of them dreamt of regaining the kind of land control they had had in the pre-civil war days. It appeared as if tenant and landlord reluctance and lack of administrative effort and perseverance were conspiring to thwart the nascent changes in the old land tenure system of Vietnam.

IV

In reality, the difficulties proved to be temporary. As the Ngo Dinh Diem government grew in strength and political stability, so did the acceptance of its measures. With the financial assistance of the U.S. aid program, the administrative part of the program also improved markedly. Although few in number, the 200 land reform agents carried the gospel with some measure of success. In time the farmers came to know the basic provisions, and the early indifference—if not hostility—of the provincial and district authorities gave way to a recognition that the ordinances must be enforced. The landlords were told that the government's struggle for security and political stability would not result in a restoration of the old land tenure arrangements. Helpful, too, particularly in the south, was the work of the tenants' union as spokesman for the tenants. The joint committees, made up of five elected tenants and an equal number of landlords under the chairmanship of the chief of the province or his appointee, gradually, if very slowly, came to function as conciliators of landlord-tenant disputes. They have never reached the achievements of the land commission in Japan, for example; but, in dealing with the ten thousand disputes recorded so far, they have helped to lessen the tensions generated by what seemed to be irreconcilable positions.

As a result of these developments, the implementation of the program quickened in mid-1956. By the end of June 1959 there were 800,000 contracts under ordinances 2 and 7, or 80 percent of all contracts theoretically possible. It is probably not true that no tenant pays more than the specified rent; this is particularly true in central Vietnam, where the competition for small landholdings is intense. Nevertheless, there is reason to believe that the majority of the tenants pay the legal rate, while hardly any pay the high traditional rental. The benefits to the tenants are obvious. Equally important is security of tenure—duration of lease, freedom from eviction at owners' will, and so on, guaranteed by written contract. The complete dependence of tenant upon landlord has undergone a change. Even in mid-1955, when the program was under way but hardly off the ground, this observer heard a tenant say before a landlord-tenant assembly, "In the old days, what the landlord said the land produced was law, and I paid accordingly. Now I know what it produces and I won't accept his estimate any longer." His was not an isolated case; he voiced the break with the old tradition wherein the landlord had undisputed control over the land.

Perhaps the real significance of these measures is that they represent the first breach in the traditional view of landlordism as the basis of wealth, political power, and social prestige. Its significance extends beyond the immediate aim of improving the lot of the tenants; newly emerging countries such as Vietnam must cut loose from their feudal moorings as a prerequisite to their economic independence and political stability. Another consequence of great importance remains to be noted. The measures noted here and the land rehabilitation program touched upon did in fact foreshadow the land redistribution program.

The government program to bring the abandoned land back into cultivation was successful. Ordinance 7, relating directly to this problem, stimulated many owners into action; those who were dilatory or incapable of reclaiming the land saw their properties taken over by the government for refugee settlement. In view of administrative handicaps and the need to do so much with so little, this was not an easy chore. Yet to date most of the abandoned land is back in production. The reclaiming of much of the remaining land presents serious technical problems, but there is ample evidence that the priority the government gives to the rehabilitation of the country's agricultural economy will eventually put all of the cultivable land into production.

The settlement of the refugees played a great role in helping make ordinance 7 a reality, and the Cai San resettlement project is its most dramatic expression.

Cai San is a great rectangle of some 110,000 hectares of land. Bounded by canals to the east and west, the Bassac River (the southern-most branch of the Mekong) defines the northern boundary, while to the south lies the Gulf of Siam. Less than six years ago the greater part of this land was abandoned, a weed-growing wilderness, a hideaway for bandits. It is now a thriving settlement. The refugee settlers of Cai San, with the simplest of tools, dug seventeen canals totaling 125 miles in length and built their huts atop the embankments. The American aid tractors were there with ex-taxi drivers to run them.

The results speak well for the remarkable performance of the settlers, for the Vietnamese administrators who rose to the challenge, and for the American aid which boldly and intelligently provided the financial means for the venture. The promise of only five years ago is now a reality. The settlers received their 3-hectare holdings virtually as a gift. They are harvesting good crops, paying their own way, and are refugees no longer. Ordinance 7 succeeded better than was anticipated; not only is the land being rehabilitated, but the refugees have become productive farmers and full members of Vietnam's body politic.

But it is not only Cai San and the other Cai Sans which are being built on the southern plains of Vietnam. On the initiative of the president, new settlements are being carved out of the highland wilderness north of Saigon, where three years ago virtually none but nomadic tribesmen lived. While this effort is not a part of the agrarian reform program, it is, nevertheless, an extension of the agrarian policies discussed here.

Into this almost unknown heavily forested mountain area of the High Plateau, until recently little more than Bao Dai's hunting preserve, nearly 60,000 people have moved from the overcrowded coast of central Vietnam. They have exchanged their rented half-hectare plots for three-hectare holdings given them by the government. With the government's technical and financial assistance, they have cleared much of their land, built their homes, put in their crops, and are on the way to becoming economically independent. The project has proved to be a pioneering venture of more than purely economic significance. Equally significant are the political and military consequences of populating empty spaces with settlers notoriously anti-Communist, for these sparsely populated areas of the highlands have long provided the Communists with a safe route for infiltration southward. The violent reaction of the North Vietnamese Communists to the opening of these lands provides its own commentary.

V

Rent reduction, better leases, and land rehabilitation were all-important ends in themselves. Nevertheless, to one who can remember the mood of the peasantry in 1955, it was clear even then that the government land policies were not going to end with ordinances 2 and 7. The Vietnamese landless, like the landless the world over, were after one thing—land of their own. This was particularly true of the south, where the vast holdings of the landlords loomed large in the eyes of the tenants. One telling picture that comes to mind was the spokesman of a village group saying to the minister of agrarian reform that "the village needs peace, and the landlord-tenant conflicts will never cease until the tenants own the land." In province after province and village after village farmers expanded on this theme, often to the discomfiture of local and central officials not prepared to discuss the issue.

Strange as it may seem at first glance, the tenants found an ally—a reluctant ally to be sure—in the big landholders. It was true that the landlords opposed the low rentals, but it was equally true that their experiences in the past decade had had a sobering effect. They were interested in selling their land. Their bold pretense that little or nothing had changed had more form than substance. There were few big landlords in Vietnam during this period who would not concede that traditional landowning had fallen on evil days. They knew that sooner or later they would have to lose much if not all of their land, and their problem was how to make the best of a difficult situation.

This explains why, even with the Communists gone and security being established, one encountered serious debates among landlords about the best ways of disposing of their land. It was not the selling price but the method of payment that was the main stumbling block. Tenants likewise assumed that the land would have to be paid for, but just as the rent reduction program divided the two groups, so did the method of payment for land purchase, although the issue for the moment appeared academic. The consensus among the landlords was that they should get 50 percent in cash and the remainder in five annual installments. Some tenants spoke of ten annual payments with no cash down, while others spoke of 10 to 20 percent in cash and the remainder on an installment basis. The difficulty was that the tenants had no cash whatever, a fact which nobody contested.

The land distributions were not confined only to landlord-tenant circles. The idea of land for the landless in Vietnam did not originate with the Communists. In June of 1953, President Nguyen Van Tam, with the support of the chief of state Bao Dai, proclaimed a land distribution program of his own. This, in addition to his 15 percent rent reduction, he described as a "masterpiece of our agrarian reform." Aside from the impossibility of enforcing it for reasons already stated, it was less than a "masterpiece," as a summary of its main contents reveals. Under the enabling ordinance 21, a landlord could retain a maximum of 45 hectares in central Vietnam and 100 hectares in South Vietnam if he owned land in both sections of the country. The joker in the piece was that a landlord had the right to a 25 percent increase for the fourth and each succeeding child; since landlords with wives and concubines had large families, the retention limit could be watered down beyond recognition. Equally important was the lack of any purchase provision to make land transfer possible. It was left to the tenant to buy land at the prevailing market price and, in theory, with the assistance of government credit. Basically, the credit was not available, the tenants had no cash, and at the prevailing market price they could not have bought it even had they had money.

VI

It remained for Ngo Dinh Diem's government to deal with the land distribution question in earnest. Diem repeatedly expressed the view that widespread individual ownership of land is *the* condition of economic and political stability in the countryside. He found a staunch ally in his brother, Bishop Thuc, a man with intimate knowledge of the work and life of the peasantry and an uncompromising supporter of peasant welfare. On the political side, it was clear to the president that the Communist threat in the villages was real and that their political capital could be undermined by narrowing the gap between landlords and the landless if not bridging it altogether. A more equitable redistribution of land was the answer. From the point of view of economic welfare, the same answer applied, even if qualified by such additional needs as better farm techniques, development of experiment stations and dissemination of the results of their work, and the indispensable farm credit. If the notoriously low Vietnamese rice yields were to be augmented, the incentive of owner cultivation of land was important in making the best use of the factors just mentioned.

These propositions, understood and accepted by the then Prime Minister Diem, were not spoken of openly in 1955. This was partly because of numerous other pressing tasks and partly because ordinances 2 and 7 had not yet gone far enough to prepare the ground for the reform phase he had in mind. But the subject was under active consideration in official circles throughout the summer of 1956, and on October 22 ordinance 57 was promulgated and land redistribution became the official policy of the government of free Vietnam.

VII

The preamble of the ordinance states its objectives in terms of equitable distribution of land among the landless, development of agricultural production, and the orientation of the big landlords towards industrial activities. The ordinance contains a number of implementing provisions.

No landlord can own more than 100 hectares, but he may keep an additional 15 hectares of inherited land for the continuation of ancestor worship and the expenses it entails; the remainder must be sold to groups specified in the ordinance in the order of priority. A landlord may himself cultivate only 30 hectares of the 100 he is permitted to retain; the other 70 hectares he must lease or sell. The land affected by the ordinance is riceland only. The excess holdings will be sold first of all to tenants and agricultural workers who have cultivated the land for two years; next in line are war veterans, refugees and the unemployed; but, in effect, the land to be divided was meant for the tenants already on the land. Tenants acquiring land under the reform cannot lease or mortgage it within ten years of the date of acquisition.

The government was to buy land from the landlords and, in turn, resell it to the tenants. Since the tenants have no cash, article 14 provides that they can pay for the land in six annual installments. The purchaser receives a certificate of ownership prior to his payment and a clear title of ownership after he completes his payments. The price to the tenant is determined by the amount the government pays to the owner for the land. The price is fixed by regional committees and approval by the National Council for Agrarian Reform. The landlords are compensated in two ways: 10 percent of the value of the purchased land is paid in cash, the remainder in nontransferable government bonds bearing a 3 percent interest rate and amortized in twelve years. However, the bonds can be used as legal tender for paying off debts contracted with the Agricultural Credit Agency as well as for land and inheritance taxes. More important, and this is one of the basic aims of ordinance 57, the bonds can be used for subscription to securities of any enterprises created by the state in the framework of a program of national economic development. More recent amendments have not changed the tenor of this provision.

The idea of using land bonds of this type to help the landlords take part in a country's industrialization has been practiced very successfully in Taiwan. There the government controlled a number of fully operating undertakings, and the exchange of the lands for government-owned shares was a simple affair. In Vietnam the government has offered what it has already in operation—industries about to be built and those contemplated. However, industrialization there is only beginning and the changeover has not been immediately rewarding.

With a view to administering the program, the ordinances provided for the establishment of a Council for Agrarian Reform with powers broad enough to deal with all outstanding issues involved in implementation. There were also, in addition, provincial and local committees to determine the acreage available for distribution, the fixing of land prices, the issuance of titles to new owners, and related subjects. Agrarian tribunals and stiff penalties for evading the provisions of the reform law are part of the administrative arrangement.

VIII

The ordinance laid down the general principles of the program, and a comment on some of them is in order.

The retention limit determines the scope of the reform, or the amount of land available for distribution and the number of tenants benefiting from it. With landholdings in central Vietnam seldom exceeding 10 to 15 hectares, the reform, clearly and correctly, was meant for South Vietnam alone. Applied there, the retention means that approximately 30 percent of the tenants will fall within the scope of the program. The government did not feel that it should begin with a retention limit which would eliminate Vietnam's rural middle class overnight. There was also the reasonable assumption that, since landlordism has lost a good deal of its attraction, landlords would be inclined to dispose of a good deal of the remainder of their land.

The government was on sound ground in assuming the responsibility as buyer and seller of the land. In so doing it took upon itself the all-important task of fixing land prices, a task it fulfilled with a decided bias in favor of the tenants. There is little doubt that, if land transactions had been left to bargaining between the landlords and tenants, little if any land would have changed hands—there is historical precedent to show that ordinance 57 would have remained a paper ordinance.

Payment for the land on the installment basis was a realistic approach to the tenants' lack of ready means to pay for it. The issue was whether the six-year period was not too short. To this observer, a ten-year period was a more sober reflection of the tenant's financial position. It should be noted, however, that the debate of six versus ten years lacked one crucial element: neither side knew what the price of the land was going to be. In the final analysis, the price of the land determined the feasibility of one period of payment as against the other. As events proved, the price finally set does not exceed three to four times the annual production of the land acquired by the tenant. In a very few instances it will be slightly more, but in many more cases a tenant will have paid for his land for the price of two annual crops. In any event, even in the most extreme case, payment could not exceed one-third of the output. In determining the tenant's capacity to pay for the land, consideration was also given to the price he received for his rice. It was assumed that the tenants could discharge their obligations in six annual installments. The immediate problem was to create an administrative machine to go on with the task once the acreage for distribution had been determined and the land prices fixed. Because of a lack of trained staff, lack of sufficient funds despite USOM's sizable contribution to administrative expenses, and the ever-present necessity to make little do much, an agrarian reform organization comparable to that of Japan or Taiwan could not be created. The program was further delayed by the protracted absence of the capable new minister of agrarian reform, recuperating from an assassin's bullet. The inevitable obstacles and consequent delays came to an end in late 1957, and the program got off the ground in 1958.

IX

The very first task was to determine the acreage subject to distribution. This was a difficult task because modern Vietnam never had an agricultural census and many of the village records disappeared during the years of trouble. The landlords were required to declare the land in their possession, and their statements were then checked against whatever information the village registers contained or village officials recalled. The final estimate thus obtained, and very likely on the low side, revealed that in South Vietnam a total of 2,033 landlords possessed more than 100 hectares each, yielding a surplus of 425,000 hectares. In addition, there were 430 landlords of French citizenship, of whom 280 were "pure" French. The land of these two groups subject to the reform program is approximately 245,000 hectares, or a grand total of 685,000, roughly a third of all the tenanted land in South Vietnam.

The setting of land prices was virtually completed by the end of 1957. This proved to be the most crucial task, for much of the success or failure of the transfer program depended upon the price and ability of the tenants to make good the payments. Productivity of the land was a basic factor, but important also was the consideration that the reform was designed for the benefit of the tenants. In all, the prices fixed for the nineteen provinces of South Vietnam ranged from 12,000 piasters or \$170 per hectare in the province of Binh Ducng to 4,000 piasters or \$57 per hectare in the province of An Xuyen. Within the provinces themselves, the respective prices could be as low as 500 and 100 piasters, and less.

The price of the French-owned land represented a different problem. This land was also subject to ordinance 57, but the owners had another alternative. According to an agreement concluded between the government of France and the government of Vietnam (October 9, 1958) the former allocated 1,490,000,000 francs (U.S.\$2,900,000) for the purchase of all the French rice lands. This land, when finally acquired, was to have been turned over to the Vietnamese government as a gift.

The French owners had the choice of accepting the provisions of ordinance 57 or the offer of their government. They liked neither the one nor the other. The Vietnamese price was much more palatable to them, but the method of payment was not. Under the French government terms they were to receive payment in francs, deposited in France, but at a very low price. With the funds firmly fixed by the French for this purchase, the average price could not exceed \$11 per hectare as against an average price of approximately \$60 to \$65 (at the rate of 70 piasters to the dollar) received by a Vietnamese landlord in cash and in bonds.

In late 1959 the French owners accepted the terms of their government and the transfer of the land to the government of Vietnam commenced in early 1960. As of the end of September 1960, approximately 200,000 hectares out of a total of 245,000 had been formally transferred to the Vietnamese government. The remainder was to follow suit by the end of the year.

The final disposition of this land is still being deliberated, but it is quite clear that, except for a relatively small part of this total acreage that was formerly cultivated by a French company as a unit with the most modern equipment and hired labor, the land will be sold to the tenants cultivating it now. In instances where land has remained uncultivated, the government has already begun to create agricultural development centers, settled by landless farmers. As in the case of the Vietnamese landlord-owned land, the turning over of this land to the tenants will entail considerable administrative effort; but past experience should stand the government in good stead once it embarks on the redistribution of this land.

To date, the Ministry for Agrarian Reform has concentrated its rather limited means upon the Vietnamese land. How much has been achieved?

The process of an orderly and carefully worked-out scheme of distributing the land among the tenants has proved to be a time-consuming affair. The land earmarked for this purpose had to be surveyed. Much time was devoted to making certain that the titles were clear of encumbrances. The drawing up of a title itself is not difficult, nor is the procedure of determining the amount a landlord is to receive for his land or making out the checks for the cash portion of the payment and the bonds. But all these operations taken together, performed by a staff not nearly commensurate with the scope of the undertaking, explain the slow pace of carrying the program when compared with those, for example, of Japan and Formosa.

Nevertheless, by the end of December 1960, the program had been largely completed. All of the Vietnamese land subject to transfer (425,000 hectares) had been surveyed and, in effect, set aside for distribution. Approximately 300,000 hectares were already under cultivation by 120,000 tenants and half of them had received titles to the land, while the landlords had received payment for about half of the land the government was acquiring from them. A considerable amount of work remains to be done; but, with the experience and training attained since the beginning of the implementation of agrarian reform and with the land fully surveyed, there is a strong possibility that the land will be fully in the hands of the tenants by the end of 1961.

X

It was expected that the land transfer, like the rent reduction, would meet with considerable opposition from landlords despite their seeming willingness to sell the land. This "seeming willingness" resulted from their assumption that they would be paid in a manner more in consonance with their self-interest. This is not the case at all. And yet, looking back at the behavior of the landlords in the past two years, one is struck by the mildness, not to say total lack of opposition.

Little has been left of the militancy of late 1955 when the landlords spoke as if Ngo Dinh Diem's government were there to restore their former privileged position. This delusion was short-lived. When the landlords submitted to the rent reduction program, in effect they acknowledged that their position had undergone a radical change. President Diem's well-known statement that the land transfer was a necessary social revolution for the benefit of the tenants served to disarm the landlords still further. The price they received for the land and the method of payment was far less than they expected, but it was still not confiscation and sizable holdings still remained in their hands. Finally, the total number of Vietnamese landlords subject to the reform is only just over 2,000. While small numbers of privileged individuals often exercise great political and economic influence, this is not quite the case in Vietnam. After years of civil war these privileged people had become economically weak, and politically they cut no great figure in the councils of Ngo Dinh Diem's government. The combination of these circumstances goes a long way to explain the landlords' acquiescence in a reform which is not in their interest.

The real opposition to agrarian reform was provided by Radio Hanoi and the Communist agents in the countryside. The latter have exerted no end of pressure, including physical threats and violence, to dissuade the tenants from buying land. The failure of the Communists' own so-called agrarian reform in North Vietnam, culminating in an open peasant rebellion against it, is well-known to the Vietnamese tenants. This knowledge and the obvious advantages the land transfer offered them were a stumbling block to the Communist anti-reform propaganda. The tenants were willing to buy land under the terms of ordinance 57. There is no question, therefore, as to where the tenants stand on the land transfer question.

In this respect they are as one with all the landless of Asia. The inborn desire for land ownership, coupled with economic security and a heightened social status, serve to make the land distribution program a significant event.

The complete abolition of tenancy is neither feasible nor desirable if low rentals and real security of tenure can prevail. Some tenants may prefer to become owners, but not the majority. With the completion of the current program, two-thirds of the tenants will remain tenants even though under vastly improved circumstances. Those who have acquired land will undoubtedly stimulate the desire for ownership among some of the tenants on the land retained by the landlords. This observer believes that the latter will dispose of some of their land voluntarily; the economics of a maximum rental of 25 percent plus the absence of the other numerous and profitable services formerly exacted from the tenants leave much to be desired from the landlords' point of view. This will be particularly true of the absentee landlords with connections in the city in commerce or public service. If this assumption is correct, the government may very well consider legislation to speed up the process of disposal of absentee-owned land. Moreover, in building the new state, President Diem has repeatedly laid down the proposition that small-scale property ownership in general and small-scale land ownership in particular constitute the very basis of an orderly, stable society.

XI

The agrarian reform measures are part and parcel of the creation of a better agricultural economy. No agrarian reform, however successful and complete, can by itself provide the answer to a sound agricultural economy, be it in Vietnam or Japan, and less so in Vietnam where for a variety of historical reasons the productivity level of peasant agriculture has remained low. The destruction caused by a decade of civil war has aggravated the problem. That Vietnamese agriculture has demonstrated real signs of recovery more recently is a tribute to the peasants and to the president's wise policy of giving the highest priority to the rehabilitation and expansion of agricultural production. It is important to note in this connection that, unlike other leaders of primarily agrarian and underdeveloped countries, the president has not been swayed by the glittering promises of industrialization as the sole panacea of Vietnam's problems. In placing his emphasis on agriculture, the country's real source of wealth, he has indeed advocated a policy of "first things first."

For the village to develop its full potentialities, agrarian reform is not alone sufficient. Also necessary are an extensive farm credit system; widespread utilization of chemical fertilizers; restoration of the livestock population; the creation of agricultural experiment stations and an extension service to disseminate the results of their work; a cooperative system to strengthen the farmer's selling and buying position; and, surely not the least, a price policy in consonance with the legitimate interest of the country's biggest industry.

The government recognizes these needs and has acted decisively upon some of them such as farm credit and the utilization of fertilizer and has made a beginning in other directions. It will take time, money, and hard work to translate all these into a productive and prosperous agricultural economy. Given peace, there is no reason to doubt the future outcome if past performance is any indication of the Vietnamese ability to meet their problems.

To note this, to note the slow but sure development of the land distribution program and to raise the question of the future in no way minimizes the significance of the agrarian reform or of the other efforts of Ngo Dinh Diem's government in the countryside. It is the more noteworthy when one recalls that the government inherited little more than rural chaos. It is easy to find fault with this or that provision, the less-than-perfect administrative arrangements and the enforcement of the reforms, or to find fault with the cautious approach to the program of land redistribution. Within the context of a new nation struggling for survival, inadequacies are inevitable and not surprising. It is vastly more important that in such conditions some of the principal reform provisions have been implemented and others are under way.

Security of tenure and rent reduction have broken the traditional, exploitative character of Vietnam's landlordism. Here and there, mainly in central Vietnam where land is scarce and density of farm population very high, landlord-tenant relations are guided not only by the provisions of ordinances 2 and 7. But even there, and certainly in the south, the new elements predominate. The tenants know the difference between 25 and 50 percent rent and know as well that they cannot be moved from the land at the landlord's pleasure.

The land redistribution program is a logical development for those areas where conditions are most suitable as they are not in central Vietnam where landlord holdings are minute and tenancy relatively small. But where redistribution of land is feasible, as in southern Vietnam, measures to give land to the landless are being carried out and in the face of the unprecedented terrorism unleashed by the Communists beginning in early 1960. As already noted, more land will have to be distributed to satisfy the majority of the tenants. In the opinion of this observer, the success of the current phase will insure the expansion of the program after 1961 if the insecurity created by the Communists in the countryside is overcome. But whatever future reform measures have in store for the farmers of Vietnam, it is fair to say that seven years after the country's independence they are measurably better off. And not only materially. Between the new and old conditions of farming lie not only so many more bushels of rice but also the invaluable realization of proprietorship for some and a sense of security on the land for all. With it goes the feeling of independence and a spirit of equality in the community. These shifts in attitude do not blossom overnight, but even in the present formative stage they tend to make for rural progress.

Very significant for Vietnam and the rest of Southeast Asia are the methods used in effecting these changes. In his efforts to deal with rural problems, President Diem and his government have not resorted to force, setting class against class, or using any of the methods used by the Communists to impose their brand of agrarianism. This is in striking contrast to the somber realities across the 17th parallel, where for nearly three years now the Communists have been rectifying the "mistakes" of subduing the peasantry through murder and bloodshed. The Viet Minh's opposition to South Vietnam's reforms is violent precisely because they *are* successful.

The chagrin of the North Vietnamese is the greater because the non-Communist Asians cannot help but make comparisons between the enfranchisement of the peasantry in South Vietnam and their enslavement in Communist Vietnam. It would be idle to expect that the achievements in South Vietnam will cause the Communists to deviate from the line they have taken. For the rest of free Southeast Asia, however, the reforms and their methods of implementation carry the conviction that lasting social improvements cannot be found at the end of a Communist gun barrel. In the company of some other non-Communist Asian nations, free Vietnam is demonstrating anew that where there is a will to redress the injustices of an agrarian system, a way can be found, and without paying the price of the tragic upheaval of Communist agrarianism.

39. Corporate Farm Management for Japan?

The last of Ladejinsky's papers for the Vietnam years is this letter he wrote from Saigon on June 7, 1961, in response to Vice-Minister Takekazu Ogura of the Ministry of Agriculture and Forestry in Japan. This official solicitation of Ladejinsky's views, so many years after he had left Japan, testifies to the high regard in which he was still held, as his own reply testifies to his continuing interest and devotion to the causes for which he worked in Japan. The letter is substantively interesting for Ladejinsky's observations on corporate farm management and his attack on the concept that, because size is advantageous in industry, the same is true for agriculture.

I REGRET VERY MUCH that your letter of March 20 came to my attention only ten days ago, upon my return from a long visit in Washington. This is the reason why I failed to write to you sooner.

I need hardly say that I am most touched and grateful to you for writing me and sending along the report on the currently contemplated new land policies. I also appreciate the reasons you cite for the needed changes and your belief that such changes wouldn't undermine the gains attained by the farmers under the reform.

Naturally, I read the ministry's paper with the greatest interest, and parts of it I reread time and again. Having been away from Japan so long, I was unaware of the agitation in favor of the corporation type of farm management and one of its would-be consequences, namely, the drastic revision of the present land tenure system. All this came to me as a surprise and, believe me, my dear Mr. Ogura, that I am indeed sorry that I am not in Tokyo now to discuss with you and your colleagues the intended revision of virtually all main provisions of the agrarian reform. I have reference to the following statement on page 43: "Points to be amended to the land tenure system: the provisions regulating land transfer control, land retention control, strengthening of cultivating rights, tenant rent control, etc., should be amended and operated leaning upon the principles prescribed above." This means, "To subordinate the land ownership to the development of farm management."

Ever since I came to know something about Japanese agriculture, I have been aware of the problem presented by the farmers with a few "tan" of land and of the desirability to have them out of agriculture and into industry. I have been equally aware of the difficulty of solving the problem. The fact that even Japan's phenomenal industrial development cannot absorb them emphasizes even more strongly how complex the issue is. Moreover, if I read correctly the "Fundamental Problems and Counter Measures for Agriculture," it would appear that even farmers with one "cho" and a little more are not suitable candidates capable of playing a role in the new agricultural economy contemplated by the writers of this report. In that case, you are in effect called upon to find a way of eliminating more than half of all the farm households of Japan.

I have a strong feeling that too much emphasis is placed upon the priority of farm management as against the firmly established *reality* of land ownership. I do not believe that this can be legislated out of existence, as you yourself indicate in your summary. The government would have to resort to very drastic, compulsory measures, which a capitalistically democratic country such as Japan would not be in a position to execute without endangering the political stability of rural Japan—and of Japan as a whole.

My real doubts about the intended new reform lie in another direction. I am under the impression that the promoters of the new policy think that the corporate system of farm management would relieve the government of farm subsidies, price supports, and so forth and so on. This assumption is very questionable for Japan; and, as you know, it doesn't apply even in countries where large-scale farm management is highly developed. The mistake which underlies this philosophy is in the thinking that the laws of economic development which characterize industry apply equally well to agriculture. This is not the case, and chances are that corporate agriculture (as well as noncorporate) will depend upon the central government for a variety of assistance and in a total amount no smaller than in the recent past.

In saying all of the above, I fully realize, Mr. Ogura, that I am too far away from the scene to be in a position to offer something constructive. On the other hand, the reading of the "Fundamental Problems and Counter Measures for Agriculture" causes me to believe that, while remedies are in order, the proposed medicine may well destroy some of the positive gains of the past decade without at the same time creating in their place a much sounder agricultural economy. I am heartened by the fact that your own concluding remarks reveal that you fully appreciate the problems inherent in the proposals. You are indeed correct in pointing out how hard it is "to take such compulsory measures"; that in present day Japan conditions do not exist that would "deny (farmers) their cultivation rights by compulsion"; and you are quite right in saying that the "farmers after land reform are quite different from those before land reform." I also agree with you that the undersized farm management system calls for changes and improvements which "could not be expected to be carried on successfully if left alone to the natural process." Finally, you are certainly right when you say that the difficulties will be greater "in attaining agricultural structure improvement . . . than in the case of land reform."

In view of all of the above, I have no doubt that you and your colleagues will not approach the problem in the spirit that everything must be reformed at once. As an old reformer, I would be most reluctant to do that for reasons you stated so well.

From the reading of your paper I can see that *nonfarm* pressure to amend the most important safeguards of individual ownership is probably considerable, but I am sure that this pressure is of less importance than the views of the farmers themselves directly affected by the would-be changes.

A reduction in the number of so-called farmers who gain most of their living outside farming is a sound idea if it can be brought about with a minimum of pain. It calls for careful deliberation and no economic punishment visited upon any farm or semi-farm group no longer considered useful. If a way is found to eliminate the real undersized ones, prosperous Japan should pay the bill.

A "flexible" and a "free" agricultural economy is a tempting idea, but such goals are hardly attainable in Japan—as elsewhere. More flexibility, yes, but the attempt to take all the rigidities out of Japanese agriculture is a waste of time. I suspect that the idea of a "free" agriculture is city-industry-commerce inspired and nurtured by the recent striking developments in Japanese industry and trade. There is the faulty thought behind it, namely, that what one can do in industry can be done, even if roughly, in agriculture as well. Strange as it may seem, it reminds me of Lenin's argument that the trinity of the combine, the tractor, and the truck would virtually eliminate the basic differences between agriculture and industry. Even the Soviet Draconian measures over a period of thirty-three years have not succeeded (in doing this).

Finally, as I look at your report once again, one significant feature stands out: the supporters of "improved farm management," including Dr. Tobata, whom I respect so highly, seem to brush aside the remarkably prosperous conditions attained by the Japanese farm community in recent years. One gets the false impression that the farmers are in distressed conditions. More than that, one not familiar with current rural conditions in Japan might also get the impression that there is something artificial about this prosperity, so much of it, supposedly, at the expense of the national treasury. If so, how is one to explain the continued and ever-rising agricultural production, the sustained and rising individual investments in land improvements, and the dramatic rise in capital farm equipment? In short, how do the promoters of "improved farm management" explain the altogether unprecedented modernization of Japanese agriculture in less than a decade? Surely the result is one of strength rather than weakness.

There is much else that I could say on the subject, mainly by way of clarification of my own ideas as a result of the reading of your report. I hope that you will not hold against me the raising of some questions and the voicing of some doubts about the argument in favor of amending the gains of the agrarian reform much too drastically. There is the danger that they might be amended out of existence. It would be a sad day for Japan if the present economic, political, and social attainment in the countryside were seriously impaired, not to say sacrificed, in the name of a new scheme of questionable validity.

Again, I was indeed pleased to hear from you and you will, I am sure, forgive me the liberty with which I chose to comment on your paper. I wish to add that you may use my letter as you think best.

IV. THE FORD FOUNDATION YEARS, 1961–1964

LADEJINSKY'S WORK WITH the Ford Foundation lasted less than three years after his departure from Saigon, from the latter part of 1961 to the first part of 1964—somewhat longer if his work with the foundation is considered to date from his exploratory visit to Nepal late in 1960 while on temporary leave from his Vietnam post. This brief association nevertheless sufficed to produce a number of interesting papers, twelve in all.

Four of these concern Nepal and are rather unusual among Ladejinsky's papers. They deal with a very small, remote, and extremely backward country, almost medieval in character but newly disturbed by the winds of change. The ruling elite and administration are so few and the country itself so small that the perceptive foreign observer can easily acquire an acquaintance with and a comprehension of their problems. This same close concentration brings the specialist in any given area inevitably face to face with the whole complex of problems with which it is obviously intertwined. Ladejinsky therefore concerned himself not only with problems of agrarian reform in the countryside but also with problems of general administration and institution-building, with the justification for a Ford Foundation program in Nepal, and with positive and negative aspects of Nepal's development prospects. He reported directly and most persuasively to King Mahendra early in 1962 on the need for agrarian reform and was much encouraged later that year by the king's apparent determination to proceed with the reforms recommended. Obviously discouraged by the following turn of events, he then soberly warned the king in March 1963 of the serious consequences likely to ensue if the reform effort were permitted to abort.

Ladejinsky was not exclusively concerned with Nepal during his Ford Foundation connection, although that had been the original expectation. Operating first from a base in New Delhi—the Ford Foundation representative there was responsible also for overseeing the foundation's work in Nepal—and later from the foundation's regional base in Malaya, as a regional specialist he also went to the Philippines, twice to Indonesia, and again, after a nine-year lapse, to India. All these trips found expression in letters or more formal reports (see the Chronological Bibliography). Also noteworthy, toward the end of this period, are a statement Ladejinsky made to a Conference on World Tensions early in 1964, an outstanding article on agrarian reform in Asia published in *Foreign Affairs*, and his classic note on land reform prepared for presentation to a conference on Productivity and Innovation in Agriculture in the Underdeveloped Countries held at the Massachusetts Institute of Technology in the summer of 1964.

Of the twelve papers, nine are presented here, two of them in truncated form.

40. *Tenurial Conditions in Nepal*

A Report to King Mahendra Bir Shah Deva

Ladejinsky arrived in Nepal in January 1962 and straightaway undertook a number of field trips in the Kathmandu Valley and eastern Terai to study tenurial conditions at firsthand. On February 15 he reported on his findings and recommendations in this most persuasive letter addressed directly to the king.

IN JUNE OF LAST YEAR, my colleague Morton Grossman and I had the privilege of an audience with Your Majesty. During that audience and subsequently in a letter to Minister Shaha, I took the liberty of expressing my views about the land tenure system of Nepal. I wish to recall here that, while I did not then offer my specific measures for altering the tenure system, I did venture to outline a general approach to this problem, which so gravely affects Nepal. In these remarks, written after many trips to the countryside in the company of Basudeo Pekurel, secretary of the Agrarian Reform Commission, I shall deal more specifically with the difficult conditions of your farmers and the ways in which you might improve them. Additionally, I shall call your attention to the question of land revenue and to the urgent need for your government to increase this source of income. First, however, permit me to state once more the essence of my remarks of last June; my recent visit to your country and my growing knowledge of its real problems have served not merely to validate but to strengthen my earlier beliefs. My main points were these:

1. Any counsel, from whatever quarter, to delay dealing with the institutional agrarian problems of Nepal is bad counsel. Just because the cultivator has carried his burdens for so long does not mean that he can stand them indefinitely. His patience might snap at a moment most unexpected.
2. In any and all schemes related to your country's development, the cultivator must be placed "in the center of the piece." There are very few countries in the world where 95 percent of the people in one way or another make a living from the land. Nepal is one of these countries.
3. The Nepalese peasant cultivator is illiterate, but he is not ignorant. He knows that he has been denied his elementary rights and needs. Later in this note I shall enlarge on this point, using specific examples from my recent experience in Nepal. Here, it is sufficient to state that any antigovernment forces—Communist or non-Communist—can readily make political capital by transforming peasant attitudes from inertness into alertness. This has been done in country after country, and Nepal need not necessarily be an exception to the rule.
4. A government which intends to meet the fundamental needs of the peasantry should initiate necessary reforms from the top rather than wait for antigovernment elements to initiate them from the bottom. This is not an academic proposition; experience in other countries gives added weight to this point.
5. In a country such as Nepal, the most important requirement for dealing with the land tenure problem is *will* and *determination* on the part of the government to proceed with the job. There can be no substitute for that. I do not minimize the inevitable difficulties of implementing a reform which threatens landlord interests; but without the government's will and determination, everything else will be writ in water.

6. The improvement of the lot of cultivators is not merely a means of strengthening the political stability of your country. Above all, it is a means of stimulating agricultural production and thereby helping to overcome the overall stagnation of the country's economy. Even a cursory knowledge of agricultural productivity in Nepal leads one to conclude that production can be increased only if the cultivators are given the incentive to produce more. This means leaving them a greater share of what they produce, to be used as they see fit. This situation does not exist now.
7. Nepal is about to enter a new stage of economic development. The plan, now in the making, is an expression of that. However, for the plan to enjoy even modest success, your country's farming must be rejuvenated; new stimulants for greater agricultural production must be found. The fact is that nearly 80 percent of Nepal's revenue comes from agriculture. The plan calls for increased revenue, since no foreseeable amount of foreign aid can possibly meet all the anticipated requirements. The only source available to the government for this increase in revenue is agriculture—but on the condition that the currently exploited and poverty-ridden cultivators can obtain conditions under which they can produce more and enjoy somewhat better living standards. This is not possible now, and, as long as this condition persists, I seriously question whether the contemplated economic development of the country can succeed.

Your Majesty, my stay in your country which began this January has greatly reinforced these beliefs of mine. During this period, I have made a number of trips into the countryside, in the Kathmandu Valley and in eastern Terai. I have talked with numerous officials, landlords, and cultivating farmers in the search for a closer, firsthand knowledge of existing farm conditions. My sample is not as exhaustive as it might be; but it is sufficient, nevertheless, to convince me that your country's agriculture is in a state of retardation and that your peasants are in a state of latent discontent—problems which can be neglected only at the risk of graver developments.

There is a common bond between the villages almost next door to Kathmandu and the villages of far-removed Jhapa district in eastern Terai—that bond is the sheer, unadorned poverty of the cultivators. They suffer from all the disabilities one finds in the poorest rural sections throughout Asia: marginal, small-scale, fragmented farming and keen competition for the landlords' acres; absentee landlordism and intermediaries who collect rents without performing any useful managerial functions; total absence of security of tenure, accompanied by exorbitant rentals and heavy indebtedness; lack of minimum food requirements on the part of the vast majority of the cultivators; and cultivators completely subject to the will of the landlords.

In most Asian countries, measures have already been initiated to give the cultivating peasant land of his own or a reduction in rent and a guarantee that he can remain on the land unmolested. The trend has been to provide the landless cultivator not only with obligations but with rights as well. I regret to say, Your Majesty, that in the entire course of my field investigation in Nepal I did not find one instance of such efforts to aid the peasant. On the contrary, what I found in case after case was extreme poverty emanating from conditions which, in effect, prevent the peasants from enjoying any significant share of the fruits of their labor.

In the villages around Kathmandu, where the land is scarce and competition for it intense, rentals are two-thirds of the gross crop. This does not include other imposts, including personal services and loans at usurious rates levied upon the tenants by landlords. The peasants I met felt strongly about their difficult conditions even if their expression was not always articulate. I found that some peasants do have a few ropanis of land of their own, but this is seldom enough to provide them with any more than the customary meager measure of living. Most farmers I encountered had altogether too little land and were compelled to rent somebody else's land to meet their barest needs. Others had no land of their own whatsoever, and the peasant who spoke of his dream to die with a little land of his own voiced the thought of all the landless.

Despite the enacted legislation of the Land

Act of 1957 and other measures talked about in Kathmandu, most farmers were virtually ignorant of these measures. It would appear that there has been no attempt either to familiarize the cultivators with the provisions of the Land Act or to enforce any of its provisions. For this reason, reference to Your Majesty's government's efforts drew no response from the peasants of the area near Kathmandu.

Eastern Terai is indeed far removed from Kathmandu. The agricultural economy there is quite dissimilar from farming in the Kathmandu Valley. It is probably the richest agricultural section of your country, producing surpluses of rice, jute, and cane. The holdings are much larger and they are counted not in ropanis but in bighas, one bigha containing 16 ropanis. Holdings of 20, 50, 100, and even 1,000 bighas and more are no exception; the shorthand hoe gives way to the bullock-drawn plow, and the pressure on the land is not nearly as intense as in the Kathmandu Valley. And yet Mr. Sharma and I found conditions in Terai bordering on slavery.

In an area with much land and relatively sparse population, one might have expected to find a preponderance of owner cultivators. The reality is the very opposite of the expectation. In more than a dozen instances, we found that the overwhelming majority of the cultivators own no land at all. We happened to come across two villages where one man, in each case, owned all the land; the farmers living there were poor tenants with absolutely no rights in the land. With no available data for eastern Terai as a whole, it is impossible to determine with accuracy the ratio of land ownership to tenancy. However, on the basis of our observations, it is reasonable to say that three-fourths of the farm population owns no land.

Tenancy as an institution is not necessarily bad. There are countries in Asia and in Europe where renting somebody else's land is a sound economic institution, profitable to both tenant and landlord. This, unfortunately, is not the case in eastern Terai, just as it is not the case in the Kathmandu Valley. In eastern Terai with its big holdings and relatively smaller pressure on the land, the situation is surely as bad as it is in the Kathmandu Valley—if not worse. In fact, Your Majesty, it is no exaggeration to say that we found there one of the very worst land tenure systems in existence in any part of Asia.

Consider the following. We were fortunate in making our trip at a time when the tenants bring the crop to an open barn assigned by the landlord. The crop is threshed there, gathered into heaps, and divided between landlords and tenants. We visited many such places, and from watching and talking with the people gathered in the barns we were able to get a reasonably good idea of what the conditions of tenancy are like. When a tenant or landlord is asked about the rate of rent, the invariable reply is "50 percent of the crop." This "50 percent of the crop" is very important because it is the basic provision of the Land Act of 1957, an act intended to improve the condition of the landless farmer. What, in reality, is this 50 percent of the crop?

In eastern Terai (as in the Kathmandu Valley), all the landlord contributes is the land. Everything else for raising the crop is furnished by the tenant; it is even the tenant's duty to deliver the landlord's share to a place of the latter's choice. If one adds up the cost of the bullocks, implements, and seed—not to mention the farmer's labor—the supposed fifty-fifty sharing of the crop is more fiction than reality, heavily weighted in favor of the landlord who is merely a collector of rent. But there is more to it than that.

In eastern Terai, for example, the tenant must borrow seed from the landlord whether he needs it or not. For this he pays 100 percent interest. When he borrows a pair of bullocks, he must pay in rice; and the same is true when he borrows money from the landlord. What all this means was demonstrated to us by a tenant in one of the barns who had threshed 80 maunds of paddy. The division was not 40 maunds to the landlord and 40 maunds to the tenant. Instead, it went like this. Having borrowed 10 maunds for seed, the tenant paid 20 maunds to the landlord because of the 100 percent interest rate. The remaining 60 maunds were reduced by 5 maunds for borrowed bullocks. Since the tenant had also borrowed Rs80, at a 25 percent interest rate, Rs100 were deducted as 13 maunds of rice, leaving a total of 42 maunds to be divided equally between the landlord and tenant, or roughly 21 maunds to the tenant out of a crop of 80 maunds. With variations, this example could be repeated by citing probably as many tenants as were gathered at any barn of eastern Terai.

The ecoonomic consequence of this was made clear by all tenants and, on several occasions, by frank landlords. Said the landlords, "The tenants are often getting no more than the rice straw." This is probably true in some instances, and even in the majority of cases the tenants seldom have enough grain left for sufficient food from crop to crop. Their bowls of rice are never full, and borrowing grain for food is nearly universal. The same applies to nonfood needs. The farmers we met marching to Assam, some eight days distance, in search of work were just another result of the so-called fifty-fifty division of the crop.

Under the circumstances, perpetual indebtedness and total dependence upon the landlord and his whims are a natural consequence. A bargaining position for the tenant in relation to the landlord is out of question. Additional exactions of personal services to the landlord are also inevitable and much in evidence. And finally, failure to comply with conditions over which the tenant is no master means eviction from the land. The tenant is forever mindful of the dreaded expression, "ta-ta," which stands for the landlord's notice of eviction from the rented land.

All of this creates agricultural stagnation, or, to put it another way, low productivity. I asked tenants and landlords whether they use any farm practices other than the traditional ones. Judging from their responses, I have the impression that practices such as better seed, better irrigation facilities, utilization of fertilizers, and the like are not practiced at all. But aside from this, it would also appear that the very idea of using such practices is not part of the mental makeup of either landlords or tenants.

In the final analysis, this results in underproduction, backwardness of the entire agricultural economy, extreme poverty of the cultivators, and great losses to the state as a whole. One does not have to be a reform partisan to recognize that present land tenure conditions are at the bottom of this situation. The landlords who, as I shall note later, pay only nominal land revenue to the state have the means to invest in the land. But they don't. As pure and simple rent collectors they are satisfied with their large share of the crop and see no need to invest capital to raise the productivity of the land. The cultivators are too poor to invest anything, even if they wanted to. But even the very few who might make some investment would be foolhardy to do so simply because the increased production would go primarily to the landlord.

Your Majesty, I wish to call your attention to one more deplorable consequence of the existing tenure system and the submerged condition of the cultivator. In the Kathmandu Valley as in eastern Terai, land is being bought and sold. I found that virtually all the sellers are peasants with small holdings while the buyers are landowners with considerable holdings. As the poverty and indebtedness of the cultivator increase, the forced sale of land is inevitable. And this is the practice.

In meeting after meeting with the peasants, the same story was told. Many of the farmers I talked with started out as freeholders of at least some of the land they cultivated; in recent years many of them have been compelled to part with some or all of their land to meet debts or satisfy immediate requirements. This is a trend which will have to be arrested—if not reversed—for the result is an exploitative agrarian system based on a mass of farm laborers with no stake in the land and no stake in society. The economics of this development are self-evident for the state as well as for the individual. But not only the economics—the political dangers to the state are equally self-evident.

This note is not intended to present a set of detailed measures for rectifying the situation I have described; it is the work of the Royal Commission on Agrarian Reform to do just that. In fact, the remarks that follow are based on the premise that this time the commission will face the land tenure problem with all the urgency it deserves, that its recommendations and enforcement power will go far enough to ease the condition of the cultivator in fact and not merely on paper.

More specifically, the commission should move to remedy the two worst evils of the Nepalese land tenure system: insecure tenancy rights and unregulated rents. The Land Reform Act of 1957 was a step in the right direction, but this time I hope that 50 percent rental will not be its basic provision. As I have already noted, the 50 percent rental rate is unreal, since tenants in fact pay more than half of the crop. On the basis of my observations, the new rental rate should not exceed one-third of the gross produce of the main crop. As my observations indicate, even such a rental would, in effect, exceed the limit of "one-third of the main crop."

I am in no position to predict the commission's future recommendations on security of tenure, but in this regard I feel that the provisions of the Land Reform Act of 1957 are quite sound and are worth repeating. Tenancy rights were to be given to any tenant who cultivated a holding for one year; eviction of a tenant was prohibited as long as he regularly paid the stipulated rent; tenancy rights were to be inheritable and salable (according to an amendment of 1959) without the consent of the landlord; and, finally, there was to be compilation of a record of tenants protected under the law so that tenants would have documentary title to the land they cultivate. The last is a most important provision. Without it the landlords' reluctance to provide their tenants with titles and, by the same token, with receipts for the payment of rents will never be overcome.

Your Majesty, my proposal does not suggest redistribution of land among the landless. I am cognizant of the importance of giving land to the landless; in the long run it makes for the soundest possible economic, political, and social arrangement in agriculture. And in the long run this will have to be done. However, for reasons peculiar to Nepal's medieval land tenure relationships, it is not timely to deal with that problem now. I urge, therefore, the application and enforcement of the minimum remedial measures outlined by the Land Reform Act of 1957 with the amended rental rate I have just mentioned. The real problem is one of enforcement. Provisions have not been enforced in the past, and the problem of land tenure in Nepal is precisely where it was more than ten years ago when the subject first came up for consideration.

Enforcement of agrarian reform legislation, which requires concessions by the landowning groups, encounters two obstacles—technical and political or ideological. In Nepal with the administration weak not only in the center but in the districts and villages as well, the technical difficulties are striking. To overcome them, one of the best methods available is to carry out a cadastral survey of the land which will pinpoint the real ownership of the various holdings. This, as nothing else, would serve to separate, so to speak, the sheep from the goats, revealing the true relationship to the land among the groups in any given village.

The importance of such work has been understood for a number of years, and some districts have already been surveyed with gratifying results. It is most unfortunate, however, that the government has left the cadastral survey starving for both men and money. Whereas an estimated six to seven hundred people are needed for the job, only a hundred or so are actually employed and at a salary much too low (Rs2 a day) either to attract new recruits or to keep the already trained workers on the job. There are already three training schools (Kathmandu, Birgunj, and Biratnagar) for these workers; and the government should now move to strengthen these schools and the entire service, giving it the broadened functions and authority suggested by E. Himsworth, advisor, United Nations mission to Nepal, in his important "*A Report on the Fiscal System of Nepal*."

Even the understaffed and underpaid cadastral survey has already made a significant contribution. The "new" land and new sources of revenue to the state which it discovered have introduced some measure of order into a seemingly chaotic picture. But for the job to be done well and expeditiously, the government must infuse new life into the service. The workers must be treated as a truly important group, mobilized to perform heroically in time of *emergency*. Failure to do this can be justly interpreted as a deliberate attempt to prevent the application of effective land reform measures as well as to obstruct the much needed effort to increase land revenue.

Bolstering the cadastral survey is the principal technical measure needed to enforce the application of land tenure reforms. But important though it is, it alone will not be sufficient. In an ideal arrangement Nepal would have a well-organized and active central and local land administration. But this condition does not exist, and it would be a disservice to the reform efforts to wait until such an arrangement has been made. The two are never quite aligned in any case, even in those countries which do have a well-organized administrative machinery. Even in countries with highly organized administrative structures as, for example, Japan and Taiwan, the execution of land reforms had to be entrusted to the peasants themselves through village committees where majority vote vested power in the landless farmers. This proved to be the prime mover of successful execution of the land reforms.

There is no reason why Nepal should not utilize this well-tried method. The ready charge that the peasants are illiterate is true. But peasants are not ignorant. They know village conditions and are all too aware of who owns or doesn't own what land. Even well-intentioned enforcement officials can be misled about existing village problems; resident farmers cannot—unless they are in the pay of the landlords. But if the peasant has your government's support, it is in his interest not to barter away his gains. It stands to reason, therefore, that the Nepalese landless, as other Asian landless have already done, will rise to the occasion and force proper application of the new security of tenure provisions.

I shall not, at this point, enumerate the details of creating a Land Tenure Administration as an autonomous body within the Ministry of Agriculture. This is the overall administrative mechanism which must be created if the tenants are in fact to attain security of tenure. I propose to discuss these matters personally and in detail with the Royal Commission on Agrarian Reforms.

As I write this, I am fully aware of the technical, administrative problems of enforcing the legislation. I know as well as anyone that it is particularly difficult to create such instrumentalities in Nepal. However, I firmly believe that in Nepal the difficulties are more political and ideological than technical. In Nepal as in some other countries, land reforms have been delayed not because of ignorance of the mechanisms but because vested interests, wedded to the status quo, have no interest whatsoever in programs of this sort. The tendency is to wait until it is too late; and it is against this state of affairs, Your Majesty, that I must speak out.

After four visits to Nepal I have come to believe that the technical problem is being used to conceal the problem that really matters—the unwillingness of the landlords to make any concessions to the just and immediate needs of the tenants. The landlords are unaware that the status quo, based on the poverty of the great majority of cultivators, can no longer endure. I cannot help but feel that this entails a risk not only for the landlords but, more important, for the welfare of Nepal itself. The winds of new ideas are bound to affect your landless tenantry, and what are basic demands today are likely to become minimal demands tomorrow.

Not all the soil of Nepal is equally fertile, but in at least one respect its fertility is at a peak. Because of the situation of the cultivators, there is now most fertile ground for the seed of any individual or group that intends to make political capital out of your farmers' struggle for bare existence. The fact that this has not happened yet does not mean it could not happen—and sooner rather than later.

Your Majesty, the landlords of Nepal are only human and it is understandable that they wish to retain the privileges they have enjoyed for so many years. But the time has come to admit that the landlords' interests are contrary to the welfare of the great mass of your people and for the well-being of your country as a whole. It seems obvious to me that the time has come to end paper reforms. Indeed, the time has come for meaningful reforms meant to be effectively enforced. When you consider the intensity of the peasants' longing for land, the measures now proposed to improve his condition are modest. This program is the minimum that should be offered to the peasant and accepted by the landlords, for the chances are that further delays will see the peasant clamoring for land redistribution rather than for security of tenure. The landlords of Nepal would do well to ponder this realistically.

Even if the Royal Commission on Agrarian Reform makes the right recommendations and prepares the necessary draft legislation, it still remains for Your Majesty's government to leave no doubt in anybody's mind—either landlord's or tenant's—that it intends to enforce the legislation regardless of what opposition comes from what quarter. This demonstration of *will* and *determination* by your government is the primary and indispensable precondition of any effort to improve the lot of the landless. Failing that, the technical difficulties will, as in the past, become insurmountable. On the other hand, with your government's position crystal clear and well publicized, the technical problems will be less important and enforcement much easier. With these thoughts in mind, I express my sincere wish and hope that under Your Majesty's guidance the long overdue measures to help the landless will become a reality. I hope this comes out not merely as an aid to the landless but as a means to a wider purpose—to stimulate new productive forces in the village, which is the beginning of all wisdom in any development scheme for Nepal.

Your Majesty, during our trips into the countryside we also focused our attention on the question of land revenue. This was natural for at least two reasons: (a) the financial relationship of the landowner to the tenant on the one hand and that of the landowner to the state on the other and (b) the urgent need of the government to increase its revenue. Both (a) and (b) are interrelated, and I would like to take this opportunity to share with you my principal observations on this problem.

Land revenue in Nepal in 1960–61 accounted for over Rs28 million, or 24 percent of your country's total revenue. This represented the single most important source of revenue. In absolute terms, the amount of land revenue has not changed in the past decade, although revenue from other sources has increased. Without elaborating the details of how land revenue assessments are fixed, it is sufficient to say that the system has developed over the years in a confused and haphazard way. The system totally lacks uniformity and this, coupled with an archaic method of collecting land taxes, has created a maze in which only the most careful student can find his way.

Whatever the method of assessment and collection, one overwhelming impression after our talks with the landowners is that the land tax is so small that it constitutes only a small portion of the value of the crops produced. The land tax originally was meant to be equal to 20 to 25 percent of the crop, but in recent years it has ranged from less than 1 to 6 percent. All the land assessment records we examined told the same story. This is one of the lowest land rates known, typical of an underdeveloped country. Let me inquire into some of the causes for this situation.

In theory, assessments are fixed in kind, but they are actually paid in cash, based on agricultural product prices of decades ago. This means that assessments have failed to keep pace with the rising value of agricultural products, especially after the First and Second World Wars. This is true also for assessments which have all along been fixed in cash; they, too, were made on the basis of prevailing prices of agricultural products at the time the settlement was made, and have never been changed.

In short, Nepal is a striking example of the weakness of a land tax tied to assessments based on productivity, which take no account of the changes in the market price of a commodity. Moreover, while cash tax payments have remained unchanged, rents and other exactions by landlords have continued to increase. To cite two cases. In Gamcha village in the Kathmandu Valley, the owner of *Pota Birta* land pays Rs0.50 per ropani. The same land yields 6 maunds of rice, or a rental of 4 maunds. The market value of the rental is Rs70 and the land tax to the government is thus under one percent of the value of the rental and much less in relation to the value of the total crop. In the village of Katanje, near the city of Bhaktapur, the highest land tax for first class *Raikar* land is Rs6.52 per ropani. In this instance the tax paid to the government is just 6 percent of the total value of the crop per ropani. Eastern Terai presents numerous similar examples; in many cases the land tax accounts for no more than 2 percent of the value of the crop. The inevitable answer to the question of who loses and who gains is obvious: the landowner or the taxpayer is the primary gainer while the government is the principal loser.

Your Majesty, this loss has great relevance to the present financial needs of Nepal. The country is entering a stage which calls not only for foreign aid but, above all, for the mobilization and maximization of its own revenue resources. The national treasury is seriously depleted and the mounting capital requirements for the plan under consideration must be secured now. An increase in revenue from domestic sources is imperative. I have already mentioned that Nepal cannot expect to meet its budgetary deficit *and* its financial needs for economic development from foreign sources alone, however generous they may be. The need to raise revenue from internal sources is surely as true of Nepal as of other more highly developed countries in Asia, in Europe, and in North America. Furthermore, the volume of foreign aid will depend on how successfully Nepal taps its own revenue potentialities. Under these circumstances, the land tax, so grossly underassessed, is perhaps the only source of revenue which can be tapped easily and quickly to produce a substantial increase in income.

Until now, your government's policy has been to avoid an increase in the land tax assessment. Instead, it has relied on assessing new land that comes into cultivation or assessing cultivated but untaxed land discovered by the cadastral survey. This technique, although useful, is not a sufficient substitute for a general increase in the tax on land already cultivated, an increase that would *double* the current land tax revenue. The Royal Commission on Taxation and Birta Abolition spoke out in 1961 against the increase in land tax rates principally on the ground that "for a long time the people have been accustomed to paying taxes at the present level." This is hardly sound reasoning for a country aspiring to change. Even more important, there are compelling reasons why the state should reassess the land tax to conform with the needs of the country and with the capacity of the landowners to pay.

Over the years those who pay land tax have continuously profited by paying the government a nominal tax while at the same time extracting maximum rents from their tenants. They thereby exploited both tenants and government. To redress this situation by raising the land tax and reducing rents is not synonymous with dispossessing owners from the land. Rather, it is nothing more than the correction of a development which has for too long been unchecked. The land is the wealth of Nepal and must, therefore, yield a maximum return to your government. This land, be it *Birta* or *ex-Birta, Raiker* or *Guthi*, belongs primarily to the state. Does it not follow, then, that the government has the right and obligation to tax the land to secure greater revenue for the pressing needs of the country? I think it does, and particularly now when the tax rates have become nominal not only in absolute terms but also in comparison to what they were supposed to represent when they were originally fixed.

Your Majesty, even if the cultivated land of Nepal were not virtually the property of the state, the right of the state to tax it remains valid, as valid as its right to tax any source of wealth. In Nepal, the main source of wealth is the land. The government has a right and a duty to use some of that wealth in the form of an increased land tax, both for its short- and long-term requirements. And, it seems to me, this is especially true because revenue for both of these requirements is in woefully short supply.

I am aware, Your Majesty, that the effective implementation of these proposals is bound to become a political issue. Well publicized, effectively enforced land tenure reform and a significant increase in the land tax will rouse the opposition of strong vested interests. Furthermore, I am equally aware that it is easier for me to formulate these problems than it is for your government to resolve them. But as I ponder the complex problems of your country, I can see no alternative course of action. Nepal must choose between progress and stagnation, and I know that your government will choose the way that leads to progress and the welfare of the great mass of your poverty-submerged cultivators. This is a problem which permits no delay. To hesitate is merely to invite greater difficulty at a later time; and I hope, in view of this, that one can look forward to your government's earnest and unswerving efforts to implement a meaningful reform of the land tenure system and a sound reorganization of land taxes.

41. *India after the China Border Clash*

Ladejinsky had stopped twice briefly in India in October 1962 on his way to and from Nepal. In a letter dated November 14, written to a Ford Foundation official in New York from Kuala Lumpur, he made some observations about the changes in political climate which had resulted from the border clashes with China. These and related observations are worth recording here.

. . . I WAS IN INDIA to and from Nepal, but that was before *the* day. I expect to revisit India (and Nepal) before long and I can hardly wait. My impression from this far a distance is that the border hostilities have done India an incalculable amount of good—and not because Menon is no longer in power. Looking back and only partly benefiting from hindsight, it is clear that unless an all-national, unifying cause entered the picture, India was in danger of being torn apart by the internally divisive elements. Little of this has come through the world press, but the fact is that in the past year or so Hindus have rioted against Muslims and vice-versa; Sikhs have fasted against Hindus and vice-versa; non-Brahmins have threatened violence to Brahmins; Bengalis have fought Assamese; North Indian speakers have burnt English newspapers, and south India non-Hindi speakers have acted similarly toward Hindi papers; an influential party in the south would like to see the south secede from the rest of India, while Maharashtra state has a border dispute with Mysore state as if the two are foreign entities. In addition to the normal clashes of religion against religion and caste against caste, there are the relatively new developments such as the rather nasty penchant for money grubbing overlayed with a bad smell of corruption. It is fair to sum up that the normally Indian disruptive manifestations have been assuming ominous proportions. In short, the ever-present elements which tend to divide India have made themselves much more felt than those tending to unite the country. Hence the thought that the relative—though temporary—internal peace attained by India via Chinese violence is indeed a blessing.

I am glad that Abe Rosenthal is reporting from India now,[1] and I hope that in time he will rise to his customary excellence, probing into what ails India. It has been my feeling all along that for some years now reporters, diplomats, and official and nonofficial aid givers have been preoccupied with India's various economic plans to the exclusion of much else and vital that lies below the surface and is not called "planning." The Indians (I don't mean the people) have gotten away with it, and, in helping them to do this, we haven't done the country a good turn. I do not wish to raise my blood pressure to a bursting point and I shall end on this even though it calls for a lot of elaboration and detailization. I am sure you get the point.

1. For the *New York Times* [Ed.].

42. *Visit to the Philippines*

Late in December 1962 Ladejinsky visited the Philippines. In this letter of January 17, 1963, to Walter Rudlin, the Ford Foundation representative in Kuala Lumpur, Ladejinsky reports on his visit there. The report is concerned, at the outset, with his observations at the International Rice Research Institute and the Los Baños Agricultural College (sections omitted here). Then, typically, he moves once again into the field, interviewing farm tenants in the important rice-growing Pampanga province, engaging in discussions with key officials about basic problems, and going swiftly to "the root of the matter"—what the country needs to do and what kind of help the Ford Foundation might try to extend.

UPON MY RETURN FROM MANILA, Bangkok, and Saigon, I conveyed to you something of my impressions of the trip. In the following paragraphs my sole attention is centered on the Philippines. I do this partly for the sake of a record of a visit but, more importantly, because of the foundation's growing preoccupation with the country. The need to enlarge our knowledge about problems directly affecting our work in the Philippines is too obvious to merit comment; but there is also a need to keep in mind problems which at first glance may appear to affect our work only indirectly. In the Philippines the latter proved to be my main concern.

The Philippines

I arrived there on the eve of the Christmas–New Year season, a very unpropitious time. I did not realize that this season of the year in the Philippines is in one respect very much like the New Year season in Old China when most of everything comes to a standstill. Seeing people normally in Manila was difficult, if not impossible, and waiting around became a time-consuming business. Nevertheless, I have talked to a sufficiently representative group of people and seen enough to make the trip worthwhile.

I am not a stranger to the Philippines; my first visit goes back to 1948. I have repeatedly visited the country since then, even if briefly, and I have some reading knowledge of the country. This time, however, I looked at it professionally. In the process I visited, first of all, the International Rice Research Institute, Los Baños Agricultural College, and the University of the Philippines and talked with the leaders of these institutions and some of the teaching staff. I followed this up by a trip to Pampanga in search of some firsthand information on rural conditions in one of the most important rice areas of the country. The information gathered has served as a main point of my discussions with a number of leading Filipinos in public life.

* * *

Pampanga Visit

The experience at Los Baños and the inquiry into PACD[1] had left me with one outstanding impression, namely, that the concentration on community development partakes of a move, even if an unconscious one, "to buy our way around reform," and "not to rock the boat," an inevitable development if basic institutional and technical changes long overdue in rural Philippines were to take place. Whether the time has come *not* to buy ourselves around

1. [Presidential Assistance for Community Development.]

reform *and* to rock the boat were the inescapable questions of the above exercise. In search of an answer to these questions, there was only one thing to do—a firsthand look at the countryside and talks with farmers. And there I went in the company of Joe Domingo, a knowledgeable collaborator of Milton Barnett's of CECA at the University of the Philippines.

The Pampanga province is one of the most important rice centers of the Philippines, growing one rice crop a year as is customary in the Philippines on good soil. When asked why only one crop, the answer is simple: lack of water and absence of irrigation facilities. "Yes," a farmer told us, "four years ago the U.S. aid mission put through some canal ditches, but in the past two years we have been getting out of the canals more sand than water." The result is that a farmer works four or five months a year; with virtually no industries in the area, he is idle the rest of the year. In the course of the day we talked to about eight small groups of farmers, and not one of them was an owner cultivator. The land of Pampanga being 90 to 95 percent tenanted, the explanation is obvious. Worthy of note, too, is that Pampanga was the center of the Huk (Communist) fight against the established government, and it is here that the late President Magsaysay first earned his spurs and fame. So much for the background.

The peasants talked readily about their conditions, and from the very first answers to our question it became apparent that a chart on rural aid agencies in Manila or Los Baños doesn't necessarily reflect their activity in the field. This appeared to be the case among the peasants of Communist-affected Pampanga province, or certainly the part of the province we picked at random for our on-the-spot look and see. They, as pointed out elsewhere, knew nothing of PACD and just as little about other agencies. In fact, they were very much on their own with all the woes that come from small fragmented fields, lack of water, underemployment of land and labor, land tenure problems, farm credit problems, slow dissemination of known technology, "lack of civic conscience and national discipline," as well as many another disability. Since they were all tenants, we touched upon the ABC's of tenancy legislation. Most of them were totally ignorant of such; one had heard about a 70-30 division of the crop (70 percent to the tenant and 30 to the landlord), but neither he nor any of the others practiced it. What prevails is the traditional 50-50 or 55-45, with all cash or in-kind expenditures equally shared. There was nothing really new in the rediscovery of what is well known in the Philippines: the enabling tenancy legislation, of which there is a good deal, has never left the statute books. There are exceptions, but these merely prove the rule. One must confess, too, that we didn't spend too much time dealing with the obvious, for which we were well prepared upon leaving Manila.

Of greater interest to us was a question we raised with every farmer we encountered: "What do you sell and what do you buy?" Before the fairly obvious answer is given, it should be pointed out that biggest tenant we interviewed rented five hectares, while some of the others rented from five down and as little as one hectare. The tenant with five hectares, speaking fairly good English, spelled out the arithmetic of his economic position without hesitation.

He had just gathered his harvest, and we were leaning against the results of his labor—a large stack of sheaves of rice, containing some 200 cavans[2] of rice, not too bad for the area and better than the national average of 27 cavans per hectare. His selling and buying depended upon the distribution of the crop, and here is the way it went. No 70-30 division here; instead, the traditional one-half goes to the landlord, another 15 cavans to the landlord for the tenant's share of the cost of fertilizer. We must note here that, since the tenant has neither cash nor security with which to buy his fertilizer from the government-owned National Rice and Corn Administration, the landlord buys it at 8.50 pesos a bag and charges the tenant 12 pesos for the same bag. Another deduction not determined yet goes to a one-hectare tenant who has been helping out during the rush season. In all, our tenant figured that he would wind up the season with 60 to 65 cavans, which equals approximately his basic food and seed requirements. He had little to sell and little to buy. He does sell a few cavans

2. One cavan equals 95 pounds of rough rice.

of rice and an occasional product of the homestead, for he must buy salt, rock-bottom items of clothing for himself and the family, and such. What this rough-and-ready accounting points to is that for all practical purposes he is not a participant in the economic market, and surely the same applies to his fellow tenants with mostly less than 5 hectares. Again, as in the case of the prevailing tenancy arrangements, except to witness its presence, we made no new discoveries. A Philippine study on "The Structure and Present Conditions of Agriculture" has this to say on this point:

"Many farm families live on the fringes of the market economy, add little to the national product, and lack income and purchasing power for living customers in a developing economy. The question is, can economic development afford to bypass agriculture, the sector that employs two out of every three laborers?" This is *the* question, and it relates to the foundation's talked-about assistance in the economic development field in the Philippines; I shall return to it in subsequent paragraphs.

Talk with Mr. Perez

I mentioned earlier that the condition of nonparticipation in the economic market was not really a new discovery. What was news to me was the *magnitude* of that condition, and this came out in a discussion of my experience in Pampanga with the Acting Minister of Finance, Roderigo Perez. He was the right man to talk to. Young, energetic, capable, of very modest means, and "with the heart in the right place," according to all who know him. He is of the Magsaysay breed, deeply concerned with rural problems, a man who spent two years on a *walking* trip through the islands in search of information and ideas on how to "uplift" the rural folks.

When his phone didn't ring, and it rang frequently and chiefly on the subject of "my thieves in the Customs Bureau and how to restrain them," he delivered himself of a number of observations. First of all, he tried to assure me that the countryside I visited was one of "relative affluence." "Why," said he, "I know of numerous sections of the country where conditions are much worse and where the total expendable income per family per year is between 100 to 125 pesos (U.S.$25 to $30)." The second point he made was that, while the small owner cultivators are better off than the tenants at the bottom of the heap, they are just as forgotten a lot as the tenants. Finally, and by his own very rough estimate, he was telling me that perhaps "half of the rural population" is hardly on the fringes of the national economy. He added that the multitude of peasants he had in mind adds next to nothing to the gross national product in the sense that they and the country as a whole stand to benefit from their efforts. The scope of the problem was indeed news to me, and I couldn't forbear from telling Mr. Perez that, as I listened to him, the fundamentally richly endowed Philippines reminded me of the more depressed areas of India.

Talks with Other Leaders and Suggestions

There is a vast literature in the Philippine on the poverty of its farm population, on the reasons for its technological backwardness, on the demoralizing effects of its tenancy, and on what is to be done about it. The "must be dones" propounded by scholars, technicians, and politicians over the years make a long and impressive list. And yet, as Dean Umali sums it up, "Most of our rice farmers are still chained to primitive farming methods." One might add that half of the cultivated land and 40 percent of peasants are chained also to tenurial conditions which spell out even less incentive than the prevailing technical methods alone engender. The questions raised by rural poverty in the Philippines are still to be answered, and in considering them one thing is evident: Since almost everything has been said already on "what" to do, the questions "who" should take the initiative in unchaining the farmers from their present state and "how" this is to be done loom larger than ever.

I discussed these matters with a number of people in Manila, notably Vice-President Pelaez; Sixto Roxas, director-general of the Economic Implementation Agency; Dr. Virata, vice-president of the University of the Philippines; and Dean Umali; and I touched on a number of points. They were not new to this select group of intellectuals but disturbing just the same. And I meant to disturb them as a possible prelude to (a) raising the agrarian issue into a *national* issue and (b) to recognizing let alone accepting the proposition that the current preoccupation with economic development has little validity unless there is an overall scheme in which agricultural problems in their broadest sense are treated with at least as much concern as industry and commerce.

Among the points I stressed was, to begin with, the *hopeless* outlook of peasants we talked to. The absence of a glimmer of hope that conditions might be better was too obvious to leave unremarked. If Mr. Perez is right that the outlook of the peasants in other parts of the country is even more dismal, so much the worse for the country as a whole. The remark of a peasant that in Manila "they talk, talk, and do nothing" needs not to be taken idly, and no array of well-meaning agencies can convince him to the contrary. I stressed that such a poverty-induced state of mind doesn't go hand-in-hand with "self-help" as means to rural well-being. Corollary to this was the point that the central feature of a poverty-ridden community is the absence of any tendency to improvement. To the extent that the admitted stagnation in output and income exists, the inevitable result is its perpetuation rather than deviation from it, as amply demonstrated in the Philippines over a long period of years.

In the stately homes of the rich of Manila, one hears comments to the effect that the peasants are lazy; that they are not concerned about their privations, practically preferring it that way; that they are docile anyway and no danger to the status quo; and that, above all and first of all, only stepped up industrialization would eventually redress the balance between city and village, thus raising the standard of living of the peasantry.

Taking this as my cue, I tried to impress upon my listeners the idea that even in the seemingly politically quiescent barrios the "wantlessness" which characterizes the very poor can no longer be accepted as a normal condition and that the expectation of rising material standards is with them. The notion that people prefer to live in poverty is "spurious anthropology," practiced only at the risk of storing up dangerously flammable social material which might easily be ignited by the "other side." The mute peasant becomes a raging "beast," and the Jacquerie is not far to seek. I did not have to search the world for pertinent examples; I merely pointed to the home product, the Philippine-bred Huks, and the long and bloody struggle to eliminate them from the countryside.

Whatever community development, farm credit, or extension may or may not have done, the evidence is that to date little has been added that might serve the peasantry as a point of departure for economic incentive and self-enrichment. Every innovation leading to improved practices, even the tried-and-tested high-yielding seed varieties, calls for risk and investment on the part of the peasant, and the barrios are not geared to that. With immediate food, clothing, and shelter needs being dominant, there can be hardly any investment on or outside the land. What is true of Mr. Perez's description of the small owner cultivator applies to the tenant in even a more extreme degree. A tenant in prevailing Philippine conditions would act as an irrational economic man if he tried to apply to the land practices that make for higher production, knowing full well that a lion's share of this would go to the landlord, moneylender, or merchant.

I also pointed to the illusion of equating industrialization, albeit of a spectacular kind, with the well-being of rural Philippines. Desirable for any number of reasons, it doesn't follow that it is also the touchstone for a kind of automatic or inevitable resurgence of a poverty-stricken rural economy—either in the Philippines or elsewhere. Thus, in Brazil the industrial development drive of the Kubitschek era brought no relief to the pauperized countryside. It was not exactly planned that way; but, having judged economic development as if it were a choice between bread and steel and failing to modernize agriculture, Brazil failed, by the same token, to provide a market to absorb the growing industrial production and overall economic disaster was inevitable. One could mention in this context rich-poor Venezuela and closer by, the disturbing consequences of India's plans with their much greater stress on industrialization and much lesser stress on agricultural development.

The common idea voiced in many circles is that the country is "on the go." Without begrudging the rich their riches and welcoming a rising middle class and employment for a greater number of unemployed or underemployed, the "on the go" in the Philippines is essentially a narrow process as far as the great mass of the people is concerned. This holds true not only for people on the land but also for the majority of the people who live in Manila, the industrial and commercial hub of the country. The eye can see it, and statistics support it. This may explain why in Manila the new industrial plants, offices, banks, and ultra-ultra-modern insurance buildings do not dissipate the impression that, despite the valuable economic services they perform, they appear somewhat like "sterile economic monuments" in a sea of poverty. After all, the big story in the *Manila Times* of December 19, for me, was not the story on page 1 but that on page 5 with the heading: "Bleak Yule for Most of 27 Million Filipinos." The "most" ranges from 60 to 85 percent of the population, with the poorest of these in the rural sections of the country, according to the details of the story. Hence one of the main points in my discussions in Manila was based on a statement made by J. K. Galbraith, which aptly sums up the situation described above: "Until people have a part in economic progress, there will be no economic progress." Until the millions of rural Philippines become something much more than subsistence producers and modern consumers, all economic plans with primary emphasis on "steel" rather than "bread" can, in larger sense, retard the country's economic progress and add relatively little to rural betterment.

We come now to the questions raised earlier and answers partly hinted at: "Who" is to become the initiator of new approach and "how" is this to be done? With so skimpy a baggage on things Philippine, one is reluctant to venture into an area where so many a good man has met with less than success. I was encouraged, however, by my discussants who urged me to voice the issue once again and hoped that I might see the president for this purpose. It was under these circumstances that I attempted a few proposals in Manila, which I now repeat. And it should be noted here that at that time I did not—as I do not now—think it was either desirable or necessary to go beyond a few broad suggestions, made mainly with a view to arousing a public debate on the subject. My purpose, I wished it to be understood, was not to outline a plan but merely to call attention to the first steps involved, if the role of agriculture in the national economy is to be recognized by deed rather than by just another pious expression about doing something for the farmer.

Present and past administrations of the Philippines have not been oblivious to the proposition that the conditions of the rural people call for drastic remedies. On the institutional side (tenure conditions) there exists a great deal of enacted legislation. On the technical side, much knowledge, time, and money to improve farm practices and render other assistance have been invested by Philippine governments, U.S. aid missions, U.N. missions, all manner of private organizations, and of course the Agricultural College at Los Baños, to mention only the principal one. In both fields the attempts must be judged a failure. No administration since Magsaysay's has displayed more than a perfunctory interest in enforcing existing tenancy legislation. In all other fields it is partly a case, as in community development, of "putting the cart before the horse" and partly of fragmented and uncoordinated bits and pieces. But it is much more than that. Above all and for whatever reason, it has been the practice of governments in the Philippines to declare undying love for the poor peasantry without acknowledging the fact that the rural condition is a *national* problem, an emergency akin to that of a country at war, and acting accordingly.

Any effort to apply the well-known "must be dones" should begin, therefore, with the administration's public stand that, for obvious economic reasons, as well as for reasons of political stability and social justice, the deplorable state of the agricultural economy and its farmers will be treated as a national emergency. Even if the bureaucrats on the highest level favored such a move, there is much evidence that the immediate responsibility and power to act do not lie with them. Furthermore, the Philippine legislative and executive branches notwithstanding, the country is governed mostly by men rather than law, and the president of the republic is *the* man. There is much that only he can do; and, with the will that comes from an abiding interest, he, above all, can face the agricultural issue squarely in and outside Congress. Only he can dare open wide the lid of the Pandora box of all that has gone wrong, face his opponents on issues of tenancy and land redistribution (and make much political capital out of it), determine the rightful place of agriculture in any future scheme of the country's economic development, and move the country from its destructive past to a more promising future. Failing that, it is my strong conviction that the current wasteful effort will go on until the institutions which bind the country to the past will destroy those very institutions. For these reasons, the active involvement of the president is essential in any attempt to deal with the problem earnestly and lastingly.

If the president does take the lead, a number of developments can be envisaged, although not in the given order of precedence. First, an individual of the caliber of Dean Umali, and preferably the dean himself, should be given the task, based on a presidential directive, to knock many a head together and streamline the wasteful proliferation of useless agencies. Second, a careful examination should be made of all the types of rural assistance now in being and the human and material resources at their disposal; it is high time to draw a line of demarcation between what is truly important and can be made effective and the host of less important, more marginal needs. Resources, available and borrowed, would be concentrated accordingly. Third, once this groundwork is laid, the question of *what kind* of economic development program the country needs should be considered anew; the purpose of this exercise should be to make agricultural development part and parcel of an overall plan, which is not the case now. Fourth, a national economic program resting on two pillars—industry and agriculture—doesn't imply necessarily a fifty-fifty division of investment funds. Neither should it imply in the Philippines much for one and very little for the other. This problem is highlighted by the sectoral distribution of total fixed investment requirements for the five-year plan, 1963–67. According to this scheme, agriculture will account for only just under 7 percent of the total. The 7 percent allocation to agriculture is very inadequate, especially if it is to turn a new leaf. For this reason, a review of the question of investment requirements for agriculture in a unified plan is a very important step suggested here. Fifth, since not all things are of equal importance whether in industry or agriculture and since resources are scarce and some things are more equal than others, priorities should be established that indeed matter. It would be a useful and bracing departure from the past to see for once, for example, if *massive* irrigation development is not more urgently needed than community development; to see, for example, if a loan from the World Bank couldn't serve better as a means to develop the rich agricultural potential of Mindanao rather than as a means to develop part of Manila. Numerous other examples may be cited, but for our purpose the above should suffice. In brief, for the first time, the planners of the economy would deal with the city *and* the barrio, industry *and* agriculture, weighing national choices and, hopefully, making decisions on their merit.

I am only too aware of all the difficulties—political, technical, and purely organizational—that lie in the wake of any of the steps suggested here. On the other hand, they are not insurmountable. Lack of talent is not the issue. Barring Japan and Taiwan, I know of no country in Asia which can boast of all the talent it needs. It is true of the Philippines, but I have seen enough to believe that this condition is *much* less true of that country. What *is* in short supply there is a strong will, a profound and interested appreciation that the country cannot go on mostly poor, and a propelling action from the top. The great weight attached to these particular elements may sound a bit unorthodox as prime prerequisites for economic development; but, if one is to deal with Philippine realities successfully, then the seeming intangibles just enumerated must be assessed accordingly.

Relevance of Discussion, Foundation Contribution, and its Role

Such, in the main, was the nature of my observations, discussions, and general conclusions. Errors of commission and omission aside, the fact remains that the entire subject is relevant to the foundation's work in the Philippines, its work in being and work contemplated. If, for instance, the ultimate goals of the IRRI [International Rice Research Institute] are to be attained in the Philippines, this will come to pass mainly in a rural setting which has parted company with stagnation and embraced growth. It is difficult for me to see that, in conditions as they are now in the countryside, the great majority of Filipino farmers are either terribly interested or materially in a position to put to use the results of the IRRI experiments. Mere dissemination of the findings for ultimate acceptance in a manner in which they reach, say, an American farmer, will probably not do for the average Filipino farmer. The acceptance is more likely to come through an effective multipronged attack against his disabilities. In Pampanga, for example, it may mean, most of all, institutional changes in working somebody else's land and irrigation facilities as a prelude to the fullest utilization of better seed. If and when such developments obtain, the farmer will be ready to cope with improved practices, including high-yielding varieties of the IRRI.

As with the IRRI, the hoped-for re-awakening of Philippine agriculture is very germane to the foundation's attempt to raise the standards of the Los Baños Agricultural College. Dean Umali understands very well that his college must not be "an island within" but one serving the interests of the farmers. More than that, he is eager to have the college play the role of the bellwether or fountainhead of policymaking ideas. With such goals, I need not belabor the significance of the foundation's contribution to impart to the college the kind of "excellence" it still lacks. In this connection I cannot refrain from expressing the hope that the Cornellians will be in tune with the idea that the process of a better college-in-the-making may very well coincide with a period of ferment and change in the countryside—and not only technical changes but social as well. Dean Umali, who recognizes the interdependence of both and welcomes an institutional change, will need all the sympathetic understanding and active support of those called upon to help create a better college.

A final word. Thinking back on the visit to the Philippines, I find the foundation's involvement there is both timely and useful. It is already true of the IRRI and might be equally true of the Los Baños Agricultural College. If the coming American staff should prove to be a match to Dean Umali, the foundation will build similarly in this instance as well. Although the IRRI and the college are not yet affecting significantly the life and work of the peasantry, I have no doubt that their potential will come to the fore in more propitious circumstances. Besides, it is too early to expect such results from an institution such as the IRRI, which has only begun to flex its muscles.

The Philippines are at a stage where additional foundation resources could be put to great advantage. The existence of a five-year plan, 1963–67, is indeed a welcome development. The contribution to it of such an outstandingly able and public-spirited man as Sixto Roxas and the contribution of the World Bank staff amply attest to that. But, as pointed out elsewhere, it is subject to question because it fails to consider seriously the role of agriculture in the country's economic development. Hence my belief that, if in the near future the foundation is called upon to give assistance to the strengthening of the plan, it could be of crucial significance. It could be, if one of the forms of that assistance dealt with what seems to me a very basic problem: the creation of a *balanced* plan, with admittedly national agricultural needs providing the balance.

In saying this, I do not mean to imply that only the foundation can perform that kind of a service. What I do mean is that, since the people who fashioned the plan may be, although not necessarily, committed to it as it stands, the foundation is in a better or freer position to give expression to the issue repeatedly mentioned in these notes. It may be considered by some that, in "making a case," I have allowed myself too much poetic license. The fact remains, however, that implementation of the kind of plan discussed here may very well result in the following: economic development in which all people participate and all share in its benefits; and political stability and social justice, which, one hopes, are its inevitable corollaries. A picture of Utopia, perhaps, but surely worth striving for, for the Filipino people—and for the Ford Foundation.

43. *Agrarian Reform in Nepal*

In September 1962 Ladejinsky had been encouraged by the king's expressed interest and apparent intention to move forward with agrarian reform. But on returning to Nepal for another visit in February 1963 he found the program lagging. In this letter of March 16, addressed again directly to King Mahendra, Ladejinsky attempted to correct the situation. Emphasizing the basic role of agriculture in development in a country such as Nepal, and the need to introduce social justice and to override firmly the inevitable opposition of vested interests, he did not hesitate to warn the monarch of the explosive political consequences which might follow if meaningful action were not taken. "Experience teaches us that the only wise and practical course lies in timely prevention of the likely outbursts from below." This sort of advice, baldly expressed to a ruler, would almost necessarily close the door thenceforth to the adviser, whether or not the advice was accepted.

IN THE PAST MONTH I had the pleasure of visiting your country twice; and, as on a previous occasion, I take the liberty of sharing with you my impressions of these visits. If I impose upon your time and attention, it is only because I am aware of your deep interest in the matter I shall presently touch upon.

The subject is a familiar one to you, namely the problem of agrarian reform. In this connection, it is my duty to report to you that on the two recent occasions I carried away the impression of lack of progress in this field since my visit to Nepal in late September of last year. Moreover, I am of the opinion that "lack of progress" in this instance means not only standing still but possibly undoing some of the work already attained.

To be specific, in September of last year, as I had the honor of reporting to you in person, I was convinced that your country was on the road to a meaningful reform. The nature of the reform program was carefully considered; and, above all, the principal provisions of the reform had been worked out and seemingly agreed upon by the representatives of your government concerned with this issue. It was my privilege to outline the provision to you during the audience you kindly granted me. I left gratefully encouraged by Your Majesty's attention and your desire to see the provisions of the program given legal form and the enabling legislation enacted at an early date.

The realities of the situation viewed during my most recent visits are not nearly as encouraging as they appeared in September. The entire matter seems dormant. The provisions have not been given legal form and, consequently, have not been enacted. What is significant in this regard is the spirit of complacency I encountered in Kathmandu. The spirit of buoyancy, of the desire to move ahead with the measures of unquestioned importance for the welfare of Nepal, so evident in September, is hard to find now.

What is especially disconcerting to me is the apparent movement to amend and water down the principal provision of the agreed-upon legislation. I have reference, Your Majesty, to the clause specifying that the landlord is to be entitled to one-third of the crop and the tenant to two-thirds. From my information recently gathered, it appears that an attempt is being made to reverse that decision, to propose now that the crop be shared equally between landlord and tenant. For all practical purposes this would mean giving legal sanction to the existing situation, without disturbing the *customary* landlord-tenant relationship. It would also mean that the original purpose of the reform to improve the tenant's condition by first reducing his rent and thereby stimulating a much-needed rise in agriculture by providing the tenant with an incentive to produce more would not be attained.

Two reasons have been advanced for keeping the present tenant's rent proposition. The first one is that the increase in land revenue now in force, followed by a reduction of the landlord's share of rent, would constitute an undue burden on the landlord. The second reason is that the imposition of a ceiling on landholdings in addition to a reduction of the landlord's rent would be akin to beating the landlord with both ends of the stick.

Your Majesty, neither argument is valid if the reform is indeed to be a measure intended for the benefit of the tenants as well as to raise the country's agricultural productivity. As to the first reason, it is well to keep in mind that the land revenue the landlords have been paying over the years, in cash, has failed to keep pace with the rising value of agricultural products, especially after the First and Second World Wars. Whereas the land tax originally was meant to be 20 to 25 percent of the crop, in recent years it has in fact been reduced to not more than 6 percent and as low as 1 percent. I submit, therefore, that the increase in land revenue or land tax is no great burden in itself but merely a long-overdue and *mild* correction of a situation from which only the landlords have benefited over a long period of years. For the same reason, a reduction in landlords' rent at this time should not be used as a sound argument to prove that this step would impose a particularly heavy burden upon them.

The second reason advanced is, to repeat, that a ceiling on landholdings plus rent reduction is more than the landlord can bear. The fact is that the ceiling provision in the contemplated reform will, at best, achieve only a minimum of what it is normally supposed to achieve. A ceiling is a measure invoked to determine the maximum amount of land a landlord can hold, releasing the excess for purchase and redistribution among the tenants. In the vast majority of cases the latter will not happen in Nepal for two reasons: (a) Most of the landlords have already divided up the excess land among members of their families, close and distant; and (b) the proposed reform program contains no provision making the ceiling stipulation retroactive as of a date prior to the division of the land among members of the family. Under the circumstances, the ceiling provisions are hardly a threat or burden the landlords have to shoulder.

Your Majesty, all these questions which have arisen in recent months have had an adverse effect upon the proposed reform, best expressed in the absence of a drive to proceed with the task. The high hope of six months ago that Nepal was finally on the road to a meaningful reform has not so far been fulfilled. This negative development is a disservice to the problems, desires, and urgent needs of the great majority of your loyal and devoted cultivators. For the same reason it is contrary to the summary of the three-year plan. The summary, referring to the section on the improvement in the land system, had this to say:

> The major portion of the crop does not go to the farmers under the present land system and so there is no encouragement to these peasants who might be interested in the experimenting with new methods to bring about an increase in production. Hence a reform of the land system has become necessary not only to increase agricultural output but also to remove the economic drawbacks inherent (in this type of rural society).

The significance and soundness of this statement cannot be overstated, and all considerations involving the land system problem must recognize the correctness of this approach and proceed from there.

Now, as in the past, what your government decides to do or not to do about the agrarian problem of Nepal is the business of your government only. But now, as on previous occasions and as a friend and well-wisher of everything that spells the welfare of your country and your people, I take the liberty of restating a few points, which it was my privilege to convey to you on previous occasions, orally and in writing.

I am aware of the fact that a land tenure reform in your country, as in most countries, is a combination of a number of considerations: political, social, and economic. Try as one may, there is no avoiding them. The political and social justice factors are so clearly part of it in Nepal that they need little reiteration. The fact is that, as I mentioned on earlier occasions, the seeming quiescence in the areas where the tenants are economically depressed is no sign that this state of affairs will last indefinitely. Experience teaches us that the only wise and practical course lies in timely prevention of the likely social outbursts from below. The same experience teaches us that there is also a more positive aspect to the problem, namely, the greater and active loyalty and unquestioned support of your government on the part of the overwhelming majority of your people. Therein lies the basis for true political stability.

One can, therefore, readily understand why the government of Iran, for example, has recently seen fit to institute a major and drastic "land to the landless" program instead of waiting for the problem to get out of hand. Significant, too, is the result of the plebiscite in which this program was put to test; virtually all peasants voted for it and the opposition of the landlords was swept aside. Without passing any judgment on whether or not complete disestablishment of tenancy in Iran was the only course available, it appears that at least one consideration motivating the shah of Iran is that in trying to improve the living conditions of the peasantry he is also making political capital by securing the support of the majority of the people. I do not cite this as an example for Nepal to follow in every respect, but it is not without its relevance to any country determined to improve the lot of its people by rendering social and economic justice.

Aside from these issues, there is, of course, the proposition that (a) an increase in agricultural productivity in Nepal is *the* basis of the country's economic survival and well-being and (b) that under the existing tenure conditions no increase in agricultural production can take place. This is the more significant since in one way or another agriculture is providing 90 percent of Nepal's income. The drafters of the Nepal three-year plan have well understood this, as evidenced by the following references of the plan to the long-term objectives:

1. to increase production in order to meet the needs of an expanding population as well as providing resources for future plans
2. to achieve economic stability, thereby creating an atmosphere conducive to economic development for the successful implementation of the plan
3. to provide additional employment opportunities for a population which is increasing at an estimated rate of 150,000 a year and
4. to ensure social justice in the distribution of the increased production.

Imbedded in every paragraph just cited is the notion that in a country such as Nepal the revitalization of agriculture through increased productivity is a *must.* What is explicitly stated and implied here is that income from agriculture must rise to meet the mounting financial needs of the current three-year plan and those to follow. With this in mind, we must remember that the problem of how to raise agricultural productivity cannot be solved unless the cultivator is given an incentive to produce; this is where rent reduction comes in, and this is why the maintenance of the existing rentals goes contrary to the needs of a better agricultural economy, the principal source of the country's wealth.

Your Majesty, I need not repeat now all the reasons I have already cited why the cultivators of Nepal are not in a position now to invest one extra rupee in order to increase their yields. Their actual share in the crop is much too little to provide them with the incentive to invest and produce more. Only a significant reduction in rent, which would give the farmer the extra Rs100 to 200 can make the difference between investment and no investment, larger production or customary stagnation. For this reason alone a *reduction in rent* and its *enforcement* are the *only* conditions that would provide the cultivator with modest funds and incentive to improve the land and produce more. To ignore this, and for reasons of questionable validity as pointed out earlier in this letter, is to attempt the perpetuation of a situation which has not helped the country's economic development in the past and which is bound to do greater harm in the future. Hence my conclusions, Your Majesty, that any and all moves to leave existing rental arrangements undisturbed are best to be avoided.

I cannot underscore strongly enough, even at the danger of repeating myself, that stagnation in agriculture in Nepal means overall economic stagnation. Your government's decision to raise land revenue was a most salutory development, particularly in the light of the evidence that it is being collected. Nevertheless, this measure by itself cannot meet the growing financial or developmental needs of the country. Financial requirements cannot be met mainly from foreign aid sources, however generous these may be. Your financial needs will rise as the country's economic development grows. *Nepal's income can only grow as agricultural production expands.* The only solution to this dilemma: to make certain that everything that tends to increase the productivity of your agriculture is brought to bear upon the problem.

The agrarian reform proposals of last September are a modest beginning in that direction. In fact, they are a bare minimum when compared to agrarian reform measures instituted with various degrees of success in Japan, Formosa, South Vietnam, India, more recently in Iran, and currently proposed in the Philippines. Whereas a number of the countries mentioned have had to face the question of land redistribution, land purchases, and ways and means of paying landlords for the land, the reform proposals for Nepal do not yet deal with this problem. Your financial requirements for the limited reform are very modest, centered as they are mainly on the credit needs of three districts only where the reform provisions are to be put into operation. In short, I am thereby calling your attention to the fact that the reform proposals of last September are indeed modest in nature. Any effort to water them down would spell no reform at all, a possibility that must be avoided at all costs.

Your Majesty, after my numerous visits to your country and my experience in other countries of Asia, I have come to believe that *will* and *determination* on the part of your government is the most important precondition for getting on with the job of increasing the agricultural productivity of Nepal, of easing the heavy burden now shouldered by your peasants, and of boosting national development under your leadership. A few undeniable technical problems and relatively small financial requirements are being used to conceal the problem that really matters: (a) the unwillingness of the landlords to make any concessions to the just and immediate needs of the tenants, (b) their seeming contentment with the low productivity of the country's agriculture, and (c) their apparent unconcern with the economic growth and general well-being of Nepal as a whole. Such landlord attitudes are common, but they are not insurmountable. Your leadership can overcome them, realizing as you do that the status quo on the land cannot endure very much longer. In my privileged association with Your Majesty, I have carried away the impression of your awareness that the winds of new ideas are bound to affect your landless people and your country as a whole; and what appear like significant demands today are likely to become minimal demands tomorrow. The inescapable conclusion is that the agrarian reform considerations of September last should be given a new lease on life through legal form and speedy enactment.

Your Majesty, with no personal axe to grind other than the political and economic stability of your country, you will bear with me if I end this letter with the concluding lines of my letter to you of February 15, 1962. They were:

> Nepal must choose between progress and stagnation, and I know that your government will choose the way that leads to progress and the welfare of the great mass of your poverty-submerged cultivators. This is a problem which permits no delay. To hesitate is merely to invite greater difficulty at a later time; and I hope, in view of this, that one can look forward to your government's earnest and unswerving efforts to implement a meaningful reform of the land tenure system.

44. *Tenurial Conditions and the Package Program in India*

After another visit to Indonesia late in March 1963 to participate in a symposium on land reform, Ladejinsky was asked by the Ford Foundation to study the relation between existing tenurial conditions and Intensive Agricultural District (IADP or "package") Program the foundation was sponsoring in a number of districts in India. Ladejinsky accordingly returned to India, where he toured intensively selected districts in Madras, Andhra Pradesh, Bihar, Punjab, and Uttar Pradesh and prepared this report—his first on India since the four 1952 and 1954 papers.

Ladejinsky's findings reportedly created a considerable stir. The Planning Commission promptly requested the several state governments concerned to comment on Ladejinsky's findings with respect to their own jurisdictions. Their lengthy replies were published by the Planning Commission in 1965, together with Ladejinsky's paper, under the same title. A more detailed report on the impact of this paper will be found in "Agrarian Reform in India" (see the section entitled "The Latest Phase"), the first paper in Part V following. The original report is dated 1963 with no month given.

The summary section with which Ladejinsky opens the report is presented here, together with—as a sample—his recommendations for the Tanjore (Madras) district.

Summary

WE VISITED THE FOLLOWING IADP districts to study the tenurial conditions and their effect on the package program: Aligarh, Ludhiana, Shahabad, Tanjore, and West Godavary. We devoted approximately one week to each of the mentioned districts, visiting numerous villages, attending farmers' meetings, and discussing relevant problems with tenants, landlords, representatives of Panchayats and cooperatives, and the officials concerned with the implementation of the land reform and the package programs. In addition, before and after a visit to the districts, we sought the views of the officials at state headquarters concerned with the two programs. In the accompanying notes, we have set out our assessment of the problem and also our recommendations. These recommendations include suggestions for necessary modifications in existing legislation as well as administrative steps to be taken to implement the legislation.

In four out of the five districts visited by us, the tenurial situation is not satisfactory. A sizable area is cultivated by tenants in all districts except Aligarh. The problem is most acute in districts Tanjore and West Godavary, where 50 percent or more of farmers cultivate wholly or partially leased lands, mostly on oral leases. In Tanjore, West Godavary, and Shahabad, the land records do not contain any information about tenants. Ejection of tenants has taken place in the past, and the landlords still continue to change tenants from plot to plot to defeat the tenancy laws. The few tenants who were allowed to continue over a fairly long period also feel insecure. Thus, a large number of cultivators hold no title to the leased lands, pay extortionate rents, and are never certain of their status. They are left with little to subsist on and much less to invest. The reason for this state of affairs is twofold: One is the faulty legislation itself; and the second is the negative attitude of the government officials at state, district, block, or village levels.

With the exception of Aligarh, no serious attempt was made to enforce the enacted land reform legislation. If land tenure conditions were a part of the criteria for selecting a package district, West Godavary and Tanjore wouldn't qualify at all.

Such tenurial conditions are obviously not conducive to agricultural production. This situation is made much more difficult because the VLWs (village-level workers) hesitate to prepare production plans for the "pure" tenant farmer and, in respect of leased lands, for the owner tenant farmer unless the landlord agrees to it. As part of this situation, a cooperative society will not provide him with a short-term loan unless he produces a lease deed from the landlord or his signature on the loan application. Even where a tenant succeeds in all this, the maximum loan that he can get is very small. There is ample evidence, therefore, that the vast majority of tenant cultivators are poor investors and the package program aimed at reaching every farm family is bound to suffer accordingly.

A factor in the tenurial situation which needs to be singled out is that, in all the four districts where leasing is permitted, a large number of small landowners have leased small acreage to make up viable units of cultivation. In enacting land legislation, this factor must receive careful consideration.

In Madras and Andhra Pradesh, the present land reform law is of a temporary, stopgap nature; and comprehensive legislation has yet to be enacted. In Bihar the law in force is still the Tenancy Act of 1885 with some modifications which are wholly inadequate. Legislation in the Punjab is extremely defective and needs complete overhauling. Only in Uttar Pradesh has a well-thought-out comprehensive legislation been enacted and effectively implemented. There, millions of tenants and subtenants were made owners and hundreds of thousands who had been evicted were restored in their rights.

The provisions for security of tenure are crucial to the whole scheme of land legislation, and it has been the experience all over India that the right to resume land for self-cultivation by the landlord has tended to defeat the provisions for security of tenure. Resumption is all the more damaging in the case of small landowners who have made up viable units of cultivation by leasing an additional acre or two of land. It would be in the best interest of agriculture, therefore, to restrict resumption severely: no resumption should be permitted to absentee owners and others whose principal occupation is not cultivation; and it should be so regulated that the holding of the tenant cultivator is not reduced below the size of, say, a family holding.

As mentioned earlier, most tenancies are oral leases, and the tenants are generally not in a position to establish their title to the cultivation of a particular piece of land. This has tended to defeat not only the provisions of tenancy laws but also the attempts at preparing production plans for them and extending them credit facilities as envisioned by the package program. It is important, therefore, that immediate steps be taken to prepare a record of tenancies in accordance with well-known procedures. As an aid to officials in the preparation of this record, a committee should be set up in each village or a group of villages made up of two tenant farmers, one part-owner–part-tenant farmer, and one landlord. In making this recommendation we have in mind the experience of Shahabad district, where the preparation of the record by the official agency proved to be a fiasco. What the officials don't know about "who is who" in the village—who rents whose land for how much—the committee would surely know and thereby ensure the correctness of the record. The other important consideration is to give the tenants a sense of participation in the implementation of the land reform program. As soon as this basic record is prepared, each tenant farmer should be provided with a certificate indicating the plots he cultivates. In view of its importance to the package program, the preparation of such a record should receive high priority and its cost could legitimately form part of the package budget for the year.

We noticed that rents are payable generally as a share of the produce. Crop share rents are extremely difficult to regulate and should be abolished and replaced by cash rents. To facilitate this, either the rents should be fixed as a multiple of land revenue, as in Maharashtra and Gujarat; or else the state government should divide lands in each area into a few broad categories and determine the "average" produce of each such category of land and the average price at which it may be converted into cash. The average produce and the average prices so determined should be announced in each area for general information.

We are aware that, due to the proverbial weakness of the tenants' socioeconomic position, the landlords would in the long run be able to defeat even the soundest provisions for tenancy regulation. We are of the view, therefore, that steps should be promoted for the early transfer of ownership to all tenants in respect of their nonresumable lands. To facilitate implementation of this program, certain steps would be necessary: (a) The rate of compensation should be equitable both to owners and tenants. We consider that an amount equal to 15 to 20 times the net income of the owner with payments spread over a period of about ten years should be quite reasonable to both. (b) The state should actively sponsor this program either by assuming responsibility for the payment of the compensation as in Uttar Pradesh or else by giving financial assistance to tenants directly or through land mortgage banks as in Maharashtra. (c) The procedures for the transfer of ownership and determination of compensation should be simple, easily enforceable, and widely publicized.

The stand taken by the cooperatives in advancing credit deprives a large number of tenants of the opportunities offered by the package program; the totally negative attitude toward the tenancy problem on the part of all officials from state to village levels only compounds the problem. The latter is probably an "inborn" prejudice and admittedly difficult to deal with. But if the cooperatives continue to treat the tenants without regard for social justice or the practical considerations of the package program, the government should assist the tenants through taccavi loans, bypassing the cooperatives if necessary.

* * *

Recommendations (Tanjore)

Our first recommendation centers on the preparation of a basic record of tenancies. This follows from the fact that presently tenants cannot assert security of tenure rights because they lack documentary evidence that they rent this or that plot of land. Such records available in other states are presumptive evidence, and anybody disputing it has to produce evidence to the contrary. Some years back the government of Madras had passed orders that the names of tenants and the terms and conditions under which they rented land should be entered in the remarks column of the village record called "Adangal." Our enquiries indicated that such entries were not being made. Without such a record any and all provisions relating to security of tenure cannot be enforced.

In view of its importance, we recommend that a basic record of tenancies should be prepared in accordance with the well-known procedures adopted in the periodical recording operations which are held in several other states. The task of preparing such a record would call for the temporary augmentation of the staff of each firka by the appointment of an additional revenue inspector and of each tehsil by a naib tehsildar. Local officials in Tanjore estimate that, with this additional staff, a record of rights could be prepared for all the villages of the district in less than one year.

As an aid to the officials, we recommend that a committee should be set up made up of two tenant farmers, one part-owner–part-tenant, and one landlord. This recommendation is based on two considerations. What the officials don't know about "who is who" in the village—who rents whose land and how much and who owns what land—the committee would surely know and thereby insure the correctness of the record. The other important consideration is to give the tenants a sense of participation in the implementation of the land reform program.

As soon as the record is prepared for any village, every tenant should be provided with a certificate indicating the plots he cultivates.

We would like to emphasize that the preparation of this record is of crucial importance not only to the implementation of the security of tenure program but to the execution of the package program as well. Once the tenants have been recorded, it will be possible for the VLWs to prepare the farm plans for them and for the cooperatives to advance loans to them freely without much risk. Production loans being available to the tenants in adequate measure and their dependence on the landlords and moneylenders reduced accordingly, it will become possible for them to assert the rights conferred upon them under the legislation. Finally, in view of its importance to the package program, the preparation of such a record should get high priority and its cost could legitimately form part of the "package" budget for the area.

With respect to the problem of resumption, we recommend that resumption should be terminated forthwith and permanent rights conferred on tenants in respect of lands which they now hold by promulgating appropriate legislation. We reiterate that, although the right of resumption is restricted to half of the area subject to a limit of 5 acres, so long as this right continues to be operative and non-resumable lands remain undetermined, the tenant can never feel secure in respect of any portion of his holding. Furthermore, any new legislative consideration of the problem created by the existing act must recognize the following: The right of resumption has been available to the landlords for over eight years. It should be safe, therefore, to presume now that, if a landlord has not exercised his right until now, he was either not in a position to resume the land or he was not anxious to do so. Whatever the reason, the new legislation along the line indicated should put an end to the insecurity currently imposed upon the tenants of Tanjore.

Once the record of rights has been established and the right of resumption is withdrawn, we recommend that the government should enact suitable legislation for the transfer of ownership to the tenants in respect of non-resumable lands. The ceiling provisions in Tanjore have provided hardly any excess land for the establishment of farm ownership among tenants. With no other government-supported land purchase program in existence, the non-resumed land could serve as a beginning of a land purchase program. We are of the opinion that an announcement by the government that ownership will be transferred to the tenants as soon as the records of rights have been prepared would remove uncertainty among tenants and create a more conducive atmosphere for the execution of the program.

With respect to rent, we recommend that a new provision be legislated reducing rentals to one-third of the gross produce. In making this recommendation we recognize that the existing 40 percent provision is not enforced. However, we are of the opinion that any reduction provided by law, even if not fully enforced, is apt to have a leverage effect on bringing down the rents in practice. This is borne out by the experience in Tanjore; when the share of the landlord was reduced by law from 60 to 40 percent, the actual rent also came down from the earlier 75-80 percent range to the 60-65 percent range.

As a further aid to the tenant and to the cause of higher productivity, we recommend that the system of crop-share rent be abolished and a fixed quantity of produce or its value in money be established. This is not a novel measure and is well applied on temple lands in Tanjore. The merit of the extension of this system is plain; it would become possible for the tenant to invest more money and labor, the incentive being the retention for personal use whatever additional crop he produced.

The existing provision of 40 percent rent based on "normal" gross produce made it inoperative. To enjoy its benefits a tenant has to go to the Rent Court in order to determine what that "normal" is and the share thereof payable to the landlord. In Tanjore hundreds of thousands of tenants would have to troop to the courts and, incidentally, overwhelm them in the process of trying to determine what "normal" is. Besides, the determination of the normal gross produce of each tenant's holding by the Rent Court would be a time-consuming process beyond practical application. We recommend that the cultivated lands in a local area be divided into a few broad categories according to their productivity and the average or normal produce of each category of land determined and notified. If this is done, it will become possible for a tenant to avail of the provisions for fair rent by offering a fixed percentage of the average or normal gross produce so determined without resort to the Rent Court in every case. It will also facilitate the determination of fair rents by the rent courts.

We noted repeatedly our awareness of the gaps between legislative action and its execution. But there comes a time and place when the atmosphere of a rural area is surcharged with uncertainty and apprehensiveness, both among tenants and owners. Tanjore is one of those places, and the time is for *sound* legislation and *determination* on the part of all the relevant parts of government to enforce it. That such a move would divorce the landlord of some of their privileges and arouse their opposition is indeed true; equally true is that Tanjore and its tenurial problems are not unique and yet similar measures elsewhere have been carried out successfully despite opposition. Failing to legislate and enforce in the face of existing nonoperative acts is little short of the state's abdication of its sovereign rights to interfere in the economic and social process when it is found wanting. The fact that the implementation of the advocated measures will not resolve all of the tenants' problems is no argument against trying to undo the worst of the tenant's disabilities, thus improving his condition as a cultivator and as a member of the village community. His contribution to the package program would be heightened accordingly.

45. *Land Reform in Indonesia*

Early in December 1963 Ladejinsky returned to Indonesia in response to a special invitation from Dr. Sadjarwo, minister for agriculture and agrarian affairs. It was his third visit there since January 1961. This time the minister asked him to "undertake a number of field trips for a firsthand examination of the land reform program" which had been enacted in 1960. Ladejinsky remained there for nearly three months, visiting a number of villages in west, central, and east Java, in Bali, and in south Sumatra. During this period he met with the minister a number of times to report on and discuss his interim findings. In this letter to the minister dated February 27, 1964, and written just before Ladejinsky's return to Kuala Lumpur, he confirmed and amplified the points already made in the earlier discussions. With respect both to its ambitiously stated objectives of land redistribution and its more modest provisions for security of tenure and rent ceilings, the land reform, then supposedly near completion, was an almost total failure. The letter concludes constructively with a suggested course of action.

AT THE CONCLUSION OF THE TASK you wished me to undertake, I am happy to begin this letter by stating how much I appreciated all you and your staff have done for me and how much I enjoyed our association. I assure you, Mr. Minister, that I look upon my work in your country as a privilege, and I shall think back of our many fruitful discussions as memorable occasions.

I write this letter in a purely private capacity. The views expressed in it are entirely my own. For this reason, the responsibility for my acts of commission or omission in the course of my observations of your rural conditions is indeed mine only. I wish to note, however, that in carrying out your assignment I have endeavored to look at the issues in a spirit of objectivity, understanding, and sympathy for what you have been trying to do these many years.

Upon my arrival in Jakarta in early December, you were good enough to suggest that I undertake a number of field trips for a firsthand examination of the current state of the land reform program. With this in mind, I spent some time in the countryside of Java, Bali, and south Sumatra. In the course of our conversations, I tried to convey to you my impressions and findings as well as to indicate the urgent need to enhance agricultural productivity. Now that I am about to take leave of your country, I take the liberty of this private letter to you to amplify the points I raised during our talks. Here and there I refer to Bali and south Sumatra, but for reasons well known to you, the emphasis in this letter is almost entirely on Java.

In line with your assignment, I visited a number of villages in west, central, and east Java, in Bali, and in south Sumatra. I talked with a good many farmers in the villages and in the field, and I had the benefit of the observations of central and local officials involved in the reform program. I should add that the cooperation of the officials left little to be desired; but, for this cooperation, the usefulness of the trip would not have been nearly as significant. Occasionally, the number of officials accompanying us seemed excessive, but it is not a cause for censure. I take this occasion, therefore, to express to you and to the officials of Agraria, both in Jakarta and in the field, my gratitude for the very efficient arrangements connected with our work.

Despite the many contacts with villages and villagers, my sample was, of necessity, small in relation to the more than 20,000 villages Java and Bali comprise. The temptation to see one more village and yet another village in order to learn something more about this or that aspect of the program was ever present, but physical endurance and time limitations could not be disregarded. But whatever the reason, I realize that on a look-and-see field trip of this kind some significant features of the program must have escaped my attention. I note this to stress that what I have to say in the subsequent paragraphs is not a definitive picture of the land reform program three years after its commencement. Nevertheless, the observations in the course of the field trips form a pattern which provides me with a fairly good picture of the present state of land reform in Indonesia.

In setting forth my finding, Mr. Minister, I need hardly tell you that I am motivated only by the desire to be of service to you in your deep concern about the welfare of the landless. From my happy association with you over the past three years, I know that you expect from me a frank expression of views, be they in agreement or disagreement with your own. Assuming that my comments are valid, it is my hope that they may be of some use to you in the tasks that lie ahead.

In brief, the purpose of the agrarian legislation enacted by your government is twofold: (a) to provide a maximum number of landless with land of their own and (b) to create more reasonable tenurial conditions for all those who must continue to work somebody else's land. It is the sum total of these two basic propositions that constitutes the core of your land reform program, and it is from the point of view of these all-important intentions that I inquired into the program in the field. Your idea to proceed with my assignment now was indeed a timely one because official after official has told us that in parts of Java the reform had been completed or was about to be completed or that the reform as a whole will surely be completed in 1964. Under the circumstances, the questions that call for answers are these: What is the nature of the "completed" reform or what is it that will be completed in the current year? Who are the beneficiaries? How large is their number? How much land will they have received as a result of the enforcement of the program? What might be the effects of land redistribution of agricultural productivity? Finally, since the vast majority of the cultivators remain sharecroppers or tenants of owners' land, are the security of tenure provisions being enforced and are the burdens of the tenants being lightened accordingly? A number of other questions might be raised, but the above will suffice for the purpose in mind.

The impression carried away from the trips is that, to date, the results of the reform fall far short of anticipations implied by the enacted legislation. The gaps are indeed wide between aim and fulfillment. Moreover, I believe that if the present trend is not reversed, even the end of 1964 will witness little change in the results already discernible. Thinking back and reflecting upon my field experience, I am of the opinion that what I referred to earlier as a "pattern" may very well evolve into the definitive picture which an unbiased and thorough observer, with much more time at his disposal, is likely to find a year from now.

Permit me, Mr. Minister, to touch first on the land redistribution phase of the reform.

Your legislation lays great stress, and rightly so, on the idea of shifting a large number of landless to an ownership base. In the long run this makes for the soundest possible economic, social, and political arrangement in agriculture. Article 7 of the first Agrarian Law promulgated on September 24, 1960, spells out the philosophy behind the idea, while the entire law of December 29, 1960, with its maximum retention limits, provides all the important provisions under which excess land can be made available for redistribution and tenurial conditions improved. The significance of this legislation is obvious; and, for reasons I shall point out elsewhere, the arguments in certain quarters that the maximum amount of land an owner can retain is too high need not be taken seriously. What is important at this juncture is whether the legislation in its present form is being carried out.

Once in the field, it becomes clear that excess land[1] is being distributed among the landless. My notebook is replete with references to such cases in all of the three provinces of Java as well as in Bali. It is quite apparent that the recipients are pleased with their plots, regardless of how small, while the nonrecipients express the hope that their turn may come at a later date. In the case of Java and on the assumption that each new owner receive one-half of a hectare, as of the end of 1963 the total number of families who benefited from the reform was approximately 128,000; if all the land earmarked for redistribution is indeed so disposed of by the end of 1964, the number of recipients will be approximately 248,000. At first glance these figures are not small; but, if the very rough estimate of four to five million sharecroppers in Java is close to reality, then not more than 6 percent of these families will have received some land upon the completion of the program.

Two sources tell this story: information provided at the provincial capitals and samples of the actual distribution of land in the villages. According to the first source, the total amount of land—wet and dry—available for distribution, estimated as of the end of December 1963, is as follows: West Java is the redistribution leader with 28,000 hectares, or just a little more than one percent of the total cultivated land; central Java with 8,000 hectares already distributed, or a fraction of one percent of the total, considers its job completed; east Java with 6,000 hectares distributed and a possible 22,000 yet to go in the current year would represent, when completed, another fraction of one percent of the total cultivated land. Bali, with the more readily available princely land, might actually redistribute 4 percent of its total cultivated land. Such is the overall picture as seen from the capitals of the provinces.

It doesn't take much searching in the villages to comprehend the real meaning of these global figures. Fairly large sections of the country have no land to redistribute at all; and, where such land is available, it is available in small amounts, divided among the few in plots ranging from 0.5 hectares to 0.2 hectares, as is the case in Bali. It has been reported to us that in Solo Residency of central Java, for instance, not one hectare of peasant excess land was available; and villages with a total of 5 hectares of excess land were no rarity. Madium Residency of east Java, with 1,151 villages and nearly 400,000 hectares of cultivated wet and dry land, has only an estimated 1,000 hectares for redistribution spread over 200 to 225 villages. So far, only part of the land has been divided; and it is worth noting that, but for the 520 hectares of one hadji, the excess land would have been that many hectares smaller. In Pekalongan Residency, comprising 544 villages, only 75 villages are involved in the reform with a total excess land of less than 500 hectares distributed among more than 900 families. Clearly, these villages are among the better-endowed ones, for such entries in my notebook as 23 hectares divided among 126 families, or 27 hectares among 104 families, and so on, are quite common. It is not surprising that with so many villages not involved in this process, I encountered many farmers who had not yet heard about the program. The understandable and very desirable provision of the reform legislation to provide the new owners

1. This relates to land owned and rented out by resident peasant proprietors with more than the permissible limit, absentee owners, hadjis, and land of the nobility. It doesn't take account of former estate lands already occupied by squatters.

with 2 hectares of land doesn't yet accord with the reality of the situation.

Small though the volume of excess land is, tens of thousands of families are receiving land. Who are these recipients? The question is not easy to answer. It is tempting to say that the farmer who cultivated a particular plot of land as a tenant received that plot of land. In many instances, this is surely the case but probably not as a rule. At least one farmer introduced a discordant note when he remarked at a public meeting that it was the relatives of the members of the village land commissions who were getting the land. One is not prepared to say how true that statement was, but it is evidently true that, with or without the support of the land commissions, some excess land tends to gravitate into the hands of relatives of the contributor of excess land. A case in Bali serves to highlight this. In our presence and in the presence of a group of officials, a landowner stated that his 4 hectares of excess land went to his brothers-in-law and nephews. An examination of the certificate of ownership of the "new" owners proved that—much to the surprise of the local officials.

Such cases are in themselves no violations of the law—so long as such distributions find their way into the hands of relatives who cultivated such lands as tenants prior to the enactment of the legislation. One of the provisions states, for example, that "farmers who have been related to the owner for not more than two generations" have a prior claim to the redistributed land. More to the point is that the eight or twelve (not clear to me just how many) priorities and categories are too numerous and complicated to administer, even for a well-intentioned village land commission. This alone can very well be a source of evasion, although not necessarily deliberate evasion. A village land commission is presented with too many choices, made more difficult by the fact that the acreage to be redistributed in most cases is very small. And it is the latter factor more than any other factor or combination of factors that lies at the root of redistribution program difficulties. The same is true, as I shall presently point out, of the security of tenure part of the agrarian reform program.

Even a brief familiarity with the work in the field leads one to the conclusion that, regardless of how small the distributed acreage, this part of the program is *the* land reform program. Whether one talks with the governor of a province and his assistants or a bupati and his co-workers, the theme is always the same—redistribution of excess land. An impression is thus created that the second part of the program, namely, the security of tenure part, is of little importance. At best, one hears occasionally a brief comment to the effect that written contracts, rent reduction, and the variety of provisions that form this legislative enactment are difficult to enforce. Developments in the countryside reflect this attitude; but, for reasons I hope to make clear, the fault is not altogether that of the officials. It is a well-known fact that, in conditions of overwhelming pressure on a limited acreage of rented land, enforcement of security of tenure provisions is extremely difficult.

The above statement, Mr. Minister, may give you the impression that the enforcement of this part of the program is largely a failure. Actually, we have encountered farmers with written agreements; and, in one village, referred to by officials as a "pilot" village, we were told that 75 percent of the tenants work under written agreements. Assuming that this is so and that the agreements have been entered in good faith on both sides, fully reflecting the provisions of the legislation, the fact remains that the pilot village is the exception rather than the rule. In the vast majority of cases tenants have no written agreements; moreover, my inquiry into the subject leads me to the conclusion that the written agreements do not necessarily reflect the spirit of the legislation. When a tenant or landlord is asked about the rate of rent, the invariable reply is "50 percent of the crop." This "50 percent of the crop" is very important because it is the basic provision of the law. But what, in reality, is this 50 percent of the crop? I was not convinced that the 50 percent of the crop is what the tenant receives after all the expenses are shared equally between owner and tenant. The picture in this regard is very mixed. The answers to questions on sharing are "yes" and "no," with emphasis on sharing the cost of fertilizer where such is used. All else appears to be as of old. If one adds to this the total number of written agreements in all of Java and Bali is only 20,000 out of a possible few millions of potential agreements, it is quite obvious that the provisions of the security of tenure legislation are being honored mostly in breach only.

Such, in the main, Mr. Minister, are the observable results of the agrarian reform program as of the end of 1963. And not only "observable," for the official figures on land distribution and written agreements lend added support to my conclusions. What of the consequences of these developments?

There are a number of them and one worthy of the most serious consideration. I have in mind the fact that the land reform as it stands now is causing a decline in the number of tenants or sharecroppers and a corresponding increase in the number of agricultural workers or coolies. The reason is not far to seek. Even though the reform has affected the owner class very little, this same class knows (a) that the reform legislation exists and (b) that it can be so amended as to reduce the permissible retention to a lower figure. The owners know also that a tenant cultivating a given piece of land over a number of years might claim occupancy right to that land and that some day a legislative enactment might sanction that. This explains the tendency on the part of the owners in Java—as elsewhere in somewhat similar circumstances—to proclaim themselves cultivators, or rather managers of the land once worked by tenants. The latter become agricultural workers, descending to the lowest rung of the agricultural ladder. As such, they have no possible claim to the land as tenants. I repeat, Java and Bali are not unique in this respect; postwar Asia is full of instances where legislation enacted to elevate the position of the tenants economically and socially has had the very opposite effect on the fortunes of many tenants.

The sample gathered in the course of the trip through Java illustrating the above point is not large. Nevertheless, small though it is, it is relevant to the issue. An owner of five hectares in Linbongan village of central Java has no tenants any longer. Instead, he prefers to work his land with 25 agricultural laborers, a number of whom are his former tenants. Other examples could be cited, all telling the same story. While I am not prepared to say how widespread this movement is, it is in being, and, if unchecked, is likely to grow with the passage of time. It is quite clear that more agricultural workers and less tenants are not in the interest of the tenants, of the village community, and of any effort to raise agricultural production.

The conceivable remedy, Mr. Minister, is additional legislation giving the tenants "occupancy" rights, thereby preventing the owner from removing him from the land (unless for special and valid cause), or from shifting him from one plot of land to another at will. If such a measure is enacted and enforced, the bargaining power of the owner would decline, that of the tenant would rise, and the basis for the implementation of all the legislation's security of tenure provisions would be greatly strengthened.

Experience of other countries demonstrates that owners oppose the "occupancy" right measure as much as they oppose the excess land provision. But despite this prospect, and precisely because of the threat of the wholesale replacement of tenants by coolies, the early enactment of the legislation discussed here and in the course of one of our talks is of the highest importance. The fact that this measure may not help the tenants already eliminated argues very strongly for the speediest protection of the remaining tenants.

The inevitable consequence of what has been described in the preceding paragraphs and of the small acreage actually distributed and about to be distributed in 1964 is that the reform's anticipated changes in the social composition of the villages constitute another serious problem. This is far-reaching and significant, for where such changes have taken place, particularly in Japan and Taiwan, they have had immense psychological and economic significance. Writing to you about this, Mr. Minister, I need not go into any detailed explanations why this is so. Suffice it to say that an owner of a piece of land of his own or a tenant secure on his rented land is the best surety of an attitude which can, if properly utilized, determine the community's primary pattern of wants, consumption, production, investment, farm management, technological innovations, credit, marketing, and, above all, the best use of available resources. In the final analysis, what we mean by a good land reform is not so much the redistribution of land or security of tenure per se, but all of this as an incentive to move from a static condition to one in which the desire for economic and social self-improvement can express itself. More specifically, this is bound up with the question of agricultural productivity. To the extent that the reform, at best, leaves the social composition of the village unchanged, to that extent it will be more difficult to deal with the problem of raising agricultural production.

It takes a successful agrarian reform to bring that about. In this connection, Mr. Minister, I should like to point out that the significance of the Japanese agrarian reform, for example, does not lie merely in the fact that so many tenants became owners of so much land. Its real significance lies in the fact that the shift from a bad tenancy system to an ownership system released all kinds of dormant energies inherent in a farmer who owns the land and wishes to make the most of it. The well-known and tested proposition that a farmer who cultivates his own land or a tenant secure on his rented land "can turn sand into gold" is as true in our own day as it was in the seventeenth century when this observation was first made. To be sure, even in Japan or Taiwan after the reforms, the attitude of the government towards farmers and farming had to undergo a decided and positive change in regard to price, credit, and technical assistance policies of every type and description in order to make the agrarian reform meaningful, which is the attainment of agricultural rehabilitation in the widest sense. In saying this, it is not my intention to minimize the redistribution of land factor or better tenurial conditions. On the contrary, they are the preconditions of dealing with a stagnating peasant economy.

In the preceding pages I indicated that the reform has fallen short of its mark. What needs to be considered now are the "whys" of this state of affairs. I have already hinted at the causes and here I intend to spell them out in some detail and with an eye to pointing out the one cause that matters most.

It has happened time and again that an agrarian law was so drawn that it precluded the realization of its well-intentioned purposes. I do not believe that this is true of the Indonesian legislation. One may find fault with this or that provision relating to new rentals (on the high side), permissible land retention (possibly high), and official land prices (too low); but, considering the fact that the legislation had to reflect the political balance of forces on the one hand and socioeconomic forces on the other, the legislation mirrors that situation. I believe, therefore, that the principal agrarian reform provisions are about as good as the ministry could have devised under the circumstances. For the same reason, as well as for a reason I shall mention subsequently, I do not take seriously the criticism that the retention limit is too high. And this is one of the major points upon which the legislative enactment may be questioned.

No matter how good a reform law is, it is no better than its enforcement. The field trip was too brief to enable me to render a carefully considered opinion as to how good or bad the enforcement is, and my view in this matter is impressionistic and possibly hasty. Granting this, I am, nevertheless, inclined to the view that the enforcement of the agrarian reform is not nearly as good as it might be. There is evidence to substantiate this.

I don't have to call your attention to the fact, Mr. Minister, that a good deal of the success of the program depends upon the maximum acreage available for distribution. Yet, my experience in the field, supported by data in the possession of Agraria, leads me to the conclusion that in this important respect the local administration has not done nearly well enough. I have in mind the *downward* revisions of all of the first registrations or estimates of the excess land earmarked for redistribution. This is true of every province. For west, central, and east Java, the first registration called for 57,000, 22,000, and 56,000 hectares of excess land, respectively; the second registration or estimate now officially accepted is 22,000, 11,000, and 22,000 hectares, respectively.

"Estimates" are made to be revised, particularly in instances where the data on land ownership are so lacking in precision as is the case in Indonesian peasant proprietorship. Revisions, therefore, are not surprising. What puzzled one before visiting the countryside is that they were always revisions downward. The trip revealed that the first registration was probably correct; and, if it erred, it was even then on the low side, since an undetermined acreage which should have contributed to the pool of excess land had been divided up by the owners just prior to the enactment of legislation. Naturally, the local administrators cannot be held responsible for that act. Their responsibility for accepting the second registration is something else again. I say this because there is evidence that the difference between the first and second registrations came as a result of further family subdivision of properly excess land, after the enactment of reform legislation. In south Sumatra, for instance, checking the excess acreage registered by an owner is almost impossible; great distances and lack of transportation, if nothing else, preclude that. These conditions do not apply in Java. Besides, the records of the owners and the area that they own are being maintained fairly satisfactorily in connection with the assessment and collection of land tax. It should have been possible to determine from these records the owners who hold land above the ceiling and the surplus land which was to have been acquired from them. Greater official care could have prevented the "disappearance" of some of this land.

It is common knowledge in the field that the retention limit has been subject to much evasion. This could not have been done without the support of local officials, particularly the lurah, and including the local land commissions. It is one of the reasons, though not the most important one, why so very little of the land available for redistribution comes from the average resident owners with somewhat more than the permissible limit. The main source of the excess land is the absentee owners, nobility, and hadjis. For example, in Brelis Regency of central Java, the 1,200 hectares of excess land came from 1,071 owners; but a breakdown of the figures shows that 15 owners of the type mentioned provided from 50 to 75 hectares each, or most of the excess land. The evasions are a common phenomenon practiced by the economic man who naturally doesn't want to divest himself of his property at a compulsory price which bears no relation to the "going" market price. However, this doesn't excuse the enforcers of the law in the field from seeing that the letter and the spirit of the law are carried out with maximum vigilance.

The reference to the price of land as a cause of evasion calls for elaboration. My notebook contains many items on two sets of land prices, official and unofficial, the latter being the market price. In all cases the two diverge widely, namely, official Rp45,000 and unofficial Rp400,000; Rp55,000 and Rp600,000, while in one case the market price was supposedly as high as Rp2 million as against the official price of Rp65,000 per hectare. The built-in resistance to the pricing system and to parting with the land is indeed very great. I raise this question not in order to advocate a price that would suit the owner of the land. In the long run, a tenant has to pay for the land; and no tenant in Indonesia, or for that matter in any part of Asia, could possibly meet the burden of the market price of land. If he could meet that condition, there would be no point in instituting land reform programs. Hence reforms the world over with pricing schemes favoring the tenants, the burden largely shouldered by the owner. Viewed thus, the land price policy in Indonesia makes sense, but only if the difference between the official and unofficial prices is not so wide as to make the former confiscatory. My impression is that the latter is true and that it serves to explain, at least in a measure, one of the difficulties of enforcement.

One of the readily observable facts in the course of our trips was the little information available to the farmers about the basic provisions of the agrarian reform program. Whether a village has excess land for redistribution or not, surely all of them should be involved in the security of tenure part of the program; and its highlights should be placed at the disposal of the farmers on the widest possible basis and in the manner we discussed the other day. Laws tend to be complicated, and a graphic presentation of its ABC's in such a manner that even an illiterate farmer could grasp its essentials is an important part of the enforcement process. This should be a primary concern of the land reform commissions, but my impression is that this has not been the case. This type of education, which is normally part of an intended fundamental change, is mainly conspicuous by its absence.

Part and parcel of the failure to provide

useful information for the benefit of the tenants is the poor performance of the land commissions in promoting the security of tenure program. While I realize that such programs are difficult to carry out, the land commissions could have done much better than the preparation of only 20,000 agreements, many of them of a dubious character. It is true that the land distribution procedures are complicated, but the acreage involved in most villages is too small to keep a commission busily occupied throughout the year. Additionally, most of the villages have no excess land to deal with, and the land commissions could, if they so wished, devote their efforts to a more substantial attainment in creating better tenurial conditions, a problem affecting *all* villages of Java, Bali, and the rice plains of south Sumatra. Part of the explanation is that in all our talks with officials and farmers, *land reform stands for land redistribution above all*, despite the fact that the prevalence of tenancy in Java and Bali demands that the means of living of the tenants and their employment be treated with the utmost concern, protecting them against the overwhelmingly strong position of the landowners. This attitude on the part of the local administration, on top of the admittedly inherent difficulties of bringing about better tenure conditions, explains in large part why this part of the land reform program has yielded poor results.

A few other items gathered in the field may be of some further interest to you, Mr. Minister. What stands out is the slowness in the procedures of determining the amount of excess land: Who is to get the land; the issuing of the certificates of acquisition to the new owners; the large number of public organizations somehow involved in the program; and, consequently, the large volume of paper work. For example, the chief of agriculture inspection, east Java, communicates a tenant's acquisition of a plot of excess land to the following fifteen organizations:

1. The minister of agriculture in Jakarta
2. The central land reform committees in Jakarta
3. The Committee on Supervision and Control over Implementation of Land Reform in Jakarta
4. Committee on Land Reform, First Level, east Java, Surabaja
5. Land Reform Committee, Second Level at the respective locations
6. The relevant district land reform committees
7. The relevant governor, resident, and regent
8. Chief of agricultural control and regional agriculture
9. Chief of inspection of land registration, Malang
10. Chief of inspection of land registration, Surabaja
11. Chief of the Office of Land Registration in Pasaruan
12. Executive officer of land reform gifts, Jakarta
13. Chief of Farmer's and Fisheries Cooperatives Bank, Pasaruan
14. Financial inspector, Bogor
15. Office of Land Taxes, Malang

I leave to you to judge if this widespread proliferation of authority is a means to strengthen the cause of the reform. On the other hand, recalling a talk with a chief of agrarian inspection of one of the provinces, the mere task of communicating with so many agencies is a difficult task. It seems to this observer that the immediate responsibilities of enforcing so difficult a program in the countryside call for the utmost concentration of effort, and anything that tends to detract from the effort or disperse it is not in the best interest of the program.

No one in the field denies the validity of the most common complaint of all, namely, the great time gap between assigning a piece of excess land to a new owner and providing him with a certificate of ownership. A year of waiting is not rare. The process is circular and cumbersome, and it goes something like this: the Village Land Reform Committee, District Regency, Provincial Inspection Agraria, Provincial Committee on Land Reform, First Level, then the bupati with his alleged tendency to take much time to affix his signature, then to the Village Land Reform Committee once again—and to the would-be new owner. This is not to say that we have not seen farmers with the certificates of ownership; all that needs to be underscored here is that getting them is a major exercise and a slow one.

The Village Land Reform Committees, as nothing else, make the difference between implementation of what can be implemented and nonimplementation. As of now, Mr. Minister, they must be credited mostly with nonimplementation, and the reason lies in the composition of a committee made up of the village headman and two representatives of national farm groups. These three, usually landowners, do not necessarily represent the interests of the sharecroppers. The peasants themselves, for whose benefit the government of Indonesia devised the reform, do not participate in its implementation. It is imperative that they do play that role in the light of the following main functions of a committee: (a) determining the amount of excess land in each village and its suitability, (b) drafting the purchase plan and establishing the eligibility of purchasers, (c) deciding cases requiring unusual or special treatment and approving cases of exemption from the purchase provision, and (d) implementing the security of tenure provisions.

All these activities call for intimate knowledge of conditions relating to such basic matters as who owns what land and who rents whose land. Only a committee composed of a number of tenants and owners directly involved in the reform can deal with the relevant issues of implementation. But for a committee to be effective, a number of preconditions are called for. They are as follows: (a) The tenants and the owners on a committee should be popularly elected by their respective groups; (b) whatever the size of a committee, and it can be as large as ten, the tenants must have a majority representation; and (c), if "(b)" is to have meaning and not be submerged by fears of the owner group, the committee must have the fullest backing of all the judicial and police resources of the government. Barring that, there would be no point of creating a committee with emphasis on the tenantry as the principal implementor of the reform.

There may be another way, Mr. Minister, of dealing with enforcement, but I wish to underscore that the existing method is not working at all. The often-heard argument that illiterate and "ignorant" tenants cannot fulfill such functions has little validity. They know where their interests lie; with strong government backing, they would defend them. Experience has shown that this can be the case; it has also demonstrated something else, namely, that the proper work of a land reform committee can give rise to an indispensable type of adult education which no formal schooling can compete with. I refer to the likely emergence of a new leadership, tested on issues vitally affecting the village and which is so badly needed to sustain any other major program of rural rehabilitation your ministry is bound to undertake.

As indicated earlier, in Bali the picture is somewhat brighter on the redistribution side, principally because of the proportionately greater prevalence of princely land within ready reach. From the point of view of security of tenure conditions, Bali doesn't differ from Java; the same difficulties prevail and for the same reasons already pointed out.

South Sumatra, Mr. Minister, is indeed a special case, which you so well summed up during our first meeting in early December. You stated then that the problem in Sumatra is one, to quote you, "of finding people to give land to." Brief though my stay in south Sumatra was, I learned enough to testify to the correctness of your remark.

South Sumatra impressed me as an uncharted sea of land of all types and descriptions and of customs of ownership which have little in common with those of Java. The exception one finds is in the *sawah* plains around Metro, developed by the Javanese immigrants of three decades ago or more. There, as in Java, the holdings are of the order of two hectares, maximum, and much less, mostly one-half hectare in size. There, too, more than half of the land is cultivated by sharecroppers or agricultural laborers; and there is no sign of any application of the provisions to improve the conditions of the sharecroppers. But, in the main, the agricultural economy of south Sumatra is not based on wet paddy fields but on the type of dry land of which there is an abundance, managed by the clans and subdivided among its members as the need arises. Redistribution of land in terms of the agrarian reforms as applied in Java barely enters into the picture. For this reason, south Sumatra's reported 13,000 hectares of excess land (of an estimated million hectares of cultivated land) and 1,200 hectares supposedly already distributed do not reflect the potential of the region. However, this doesn't matter one way or another. The fact is that in conditions of great abundance of land and the existing traditional method of land subdivision by the clan, the land reform program can have but little application in south Sumatra.

In summary, it is clear that, when we talk about reform, we mean reform in Java. That is where the pressure of a huge population on a limited acreage lies. The soft spots of the program in Java have been noted; and suffice it to say here that, taken together, they explain, if in part only, why the program has not done well. With *drive* and *determination* on the part of the local administration, better results might have been attained both on the excess of land side and improved tenurial conditions. I speak of "better" results only rather than of the realization of the principal aims, for it seems to me that even a perfect administrative machine imbued with missionary zeal couldn't bring about all the far-reaching aims of the enacted legislation.

The reason or the "true cause" for this conclusion lies in the all-important fact that there simply is not enough land in Java for a significant land redistribution program. This would hold true even if the maximum permissible retention limit were cut in half. Furthermore, so long as the rural population continues to grow at the current rate, competing for a virtually fixed area of 9 million hectares of cultivated land, the attainment of better tenure conditions is a very difficult task. In short, though a confirmed land reformer, I am of the opinion that the real issue in Java (and Bali) is not land redistribution but population redistribution on the one hand and a breakthrough in agricultural productivity on the other.

There is nothing startling in this thought, Mr. Minister, and I have tried to point this out in our talks. Although not all the important data on land ownership and distribution are available, numerous and detailed village investigations, not to speak of my own observations in the field, all point to the fact that possibly 75 percent of all the rice landholders own less than one-half of a hectare, and possibly 90 percent of all the rice-land owners belong in the category of less than one hectare. On the other side of the scale are the fraction of one percent of all owners who possess more than 5 hectares each. Since peasant proprietorship in Java ranging from 5, 10, 15, and 20 hectares is indeed minute, it must be recognized that land redistribution in Java has a very inadequate base for the current agrarian reform program.

One might construe from all of the above that the reform is not a success and that nothing much more need to be done about it. This impression is subject to serious qualifications if one draws a line between what can and cannot be done in the agrarian reform field of of Indonesia. While there cannot be all success, there need not be a state approaching failure. I believe, for example, that with an abiding interest and effort it would be possible for the local administration to get more excess land for redistribution as well as to speed up the entire process of implementation. It is one thing to recognize the difficulties of improving the terms under which the sharecroppers work somebody else's land, but it is an entirely different thing to display little or no interest in the matter—as has been so amply demonstrated. While it is not possible, therefore, to carry out this part of the program as originally intended, it is possible to do much more than the record of recent date indicates. Hence my conclusion, Mr. Minister, that the immediate issue is not one of overlooking or accepting the reform difficulties as if they were all inescapable but of doing more and better of what is possible within the admittedly serious limitations of too many people on too little land.

Mr. Minister, in the course of this letter, I made repeated references to agricultural productivity. I am aware that you haven't charged me with any duties in that regard, but no interested observer traveling in your countryside or thinking of your food situation now and in the years to come can fail to take note of these facts: that the peasant agricultural economy is in a state of stagnation and, by the same token, agricultural productivity is at a low ebb. It is to this subject that I take the privilege of addressing myself very briefly. There is nothing that I can tell you about these problems that is either revealing or new to you. If I trouble you to look at the obvious once again, it is only in the hope of being of some assistance to you in your effort to place the question of agricultural rehabilitation and productivity where it belongs—in the very center of the national economic issues.

A tradition-bound agricultural peasantry, accustomed to a village subsistence economy, has been both an obstacle to economic development and a partial safeguard against crises. It may well follow from this proposition that the country "can take it," even if agricultural production doesn't show any significant increase in yields. I realize that this is not your view, Mr. Minister, and that you have striven to increase yields in every possible way. Statistically, and with the good fortune of favorable weather conditions and some increase in acreage, rice production between 1957 and 1963 shows a steady if small rise in production. However, production goals of more than 10 million tons for 1962 have not been fulfilled, and the great goal of nearly 16 million tons in 1968—and self-sufficiency—is not likely to be attained. Considering also the sharp increase in population, the per capita intake of all principal foods in Java and Madura has declined from 209 [2] kilograms in 1936–40 to 172 kilograms in 1956–60. The conclusion is warranted that one of the consequences of the traditional—though intensive—system of land use is that food production has not kept pace with the population growth.

The above doesn't imply that the core of the difficulty lies in your traditional land-use system. On the contrary, one is impressed by the skill and ingenuity of your farmers in using the land as they do with the means at their disposal. But the means, as you know, are inadequate to the demands placed upon them; and rice yields, for example, are only slightly bigger than those in a number of Asian countries where an extensive type of cultivation prevails and where soil and rainfall conditions are less favorable than those in Java. I am not speaking of Japan, where rice yields are roughly three times those of Java.

The immediate future of the food problem and its relationship to higher productivity are directly related to our discussion. Despite an increase of rice imports from about 0.9 million tons in 1962 to about 1.2 million tons in 1963, net availabilities of rice in Indonesia declined from just under 10 million tons in 1962 to only about 9 million tons in 1963. The resulting sharp rise of rice prices, as of other foods, is too well known to call for comment. Weather conditions in the current crop year being somewhat unfavorable, one can hardly anticipate a crop of more than 8 million tons, which is likely to raise imports to 1.3 or 1.5 million tons. The drain on the foreign exchange position of the country is quite obvious, for the prospect of spending 125 to 150 million dollars for food imports is little short of one-third of all export earnings. There are other consequences of a social, economic, and political nature which go with food shortages and high prices but they need not be touched upon here. This, in brief, is the current picture.

To some who remember the food shortage of 1961, the present situation doesn't seem serious, the argument being that the current difficulties are mainly induced by unsatisfactory climatic conditions and unusual rat infestation of the fields, both temporary phenomena, and the inflationary pressures which, one hopes, might be overcome sooner rather than later. As I see it, Mr. Minister, this argument is of limited application only. The point is that even if these adverse factors didn't exist the food problem would persist because its origin is not due to temporary causes but to the lasting combination of *normally* low yields, a so far very limited expansion of acreage under food-crops, and of a rising population of 2.5 to 3 million a year. Even on the basis of 80 kilograms of rice per person a year as against the required 90 to 100 kilograms, the annual additional rice requirement in the years immediately ahead is between 200,000 and 250,000 tons. On the basis of this conservative estimate, the country's imports in 1968 or 1969 may well amount to a minimum of 2 million tons—unless agricultural productivity, which is to say overall agricultural rehabilitation, begins to register a decided step forward.

A country doesn't have to be self-sufficient in food if it earns in other ways the means to pay for the food deficit. But even in such

2. Total rice equivalent of cereals and roots.

countries, the food deficit is not necessarily a result of inadequate use of the land. Since your country, Mr. Minister, is not yet in that comfortable position of trading, so to speak, one resource for another, a major attempt to raise the productivity of the land in order to meet the food deficit is the more imperative.

There is another vital reason why this should be so; I have in mind the condition of the peasantry, the vast majority of your country's population. Indonesia is indeed a richly endowed country; but one of the most striking impressions of the rural districts of Java, particularly of central and east Java, of Bali, and of south Sumatra, is that large numbers of your farmers are in difficult economic straits. Small holdings—even if owned—and the sharecropping system are primary causes of difficulties. Additionally, they are greatly accentuated by the fact that the land produces one ton of food where two could be produced. Small-scale agriculture is not by itself an impediment to higher productivity; the cases of Japan and Taiwan are well known. The conclusion is that the higher yields or better use of the land plus better safeguards of the rights of the tenants would indeed serve to improve the economic condition of your farming population as well as to provide more food for the country as a whole.

Last, but not least, yet another consideration enters the picture: the role of peasant agriculture in the national economy. There is a widespread idea carried over from the past that agriculture in Indonesia is plantation agriculture, one of the country's principal claims to wealth and fame. While not denying the great role of this type of agriculture in meeting some of the country's needs, it cannot do what only the great mass of the people—the peasants—must do on their land, for themselves, and for the nation. Though certain advances in peasant agriculture are undeniable, it is fair to say that 60 or 70 percent of the people are engaged in a neglected, retarded, largely subsistence-based agriculture. So long as this continues, with incomes and purchasing power restricted, any progress in other segments of the economy is bound to be adversely affected.

There is ample evidence furnished by other countries of Asia in support of this conclusion. More recently, the great enchantment with industrialization on the part of a number of less developed countries had to be tempered by recognition that their agricultural economies must be strengthened if economic development, as a whole, is to advance at a desirable pace. India is a good example of this, and even a highly industrialized Japan has seen fit to support agriculture to the hilt. Your country, Mr. Minister, need be no exception to this, and your national economy will make progress if it rests on the firm basis of industry, plantation agriculture, *and* peasant agriculture.

From all of the above, Mr. Minister, it follows that the day has indeed come for your government to center its attention not only on imports as an unavoidable measure in a time of food crisis but also to center its attention above all on the overriding propositions that the productivity of peasant agriculture must be measurably increased. There is no other way out of the problem of achieving a sufficiency of food produced domestically, of improving the economic condition of your peasantry, and of the important role it must play in the national economy as a whole if its advance is to be insured. The three elements which are so vital to the welfare of Indonesia can be made to go together, for in effect they are interdependent.

The inevitable question arising from the above is: How does one go about increasing agricultural production? Although I raise the question, I do not intend to answer it; I leave the answer to more technically competent people, at the same time mindful of the fact that your ministry, as ministries the world over, does not lack for lengthy prescriptions of *what* to do in this regard. I will add only that in the generally favorable soil and climatic conditions of Java, your knowledgeable farmers would make the most of the well-known ingredients of agricultural rehabilitation. After all, despite the conservatism of peasant societies of which the Javanese farmers furnish an example, your farmers have successfully incorporated exotic crops into their agriculture without much difficulty. While leaving the answers of what to do to others, I wish to mention in passing one point: the agroeconomic survey your ministry is intent on undertaking. I cannot think of a better beginning than taking stock of your agricultural resources as a case for the planning and implementation of an agricultural productivity program. There is no substitute for such preparatory work, and its value cannot be overestimated.

My purpose in raising the question of higher productivity is not to spell out a relevant program; I am cognizant of your long-standing preoccupation with this problem. My immediate concern is as I indicated earlier: the fullest commitment on the part of your government to treat the condition of peasant agriculture as a prime, national problem. If viewed in this light, an *attitude* can be created which makes the difference between success and failure. The kind of position taken by the government should stand for a number of things and, to begin with, ample budgetary support for agriculture. I was surprised the other day when you told me how small your budget was for what is expected of your ministry. The approach I am advocating means something more than a larger budget; it can and should mean "fighting" for the cause of better agriculture as one would fight a war meant to be won. With such an attitude it should not be difficult, Mr. Minister, to equate the needs for fertilizer, better seed, agricultural experiment work, irrigation facilities, credit, and so forth with the needs for armaments in waging war—and in this instance a protracted war. At best, the modernization of a tradition-ridden agricultural economy, unlike industrial development, is a long drawn-out affair; and the government must be prepared for a long and active commitment. Failing that, the chances of attaining the principal goals are indeed questionable.

It may seem to you that I overstress the role of the government, but my fairly long experience in dealing with somewhat similar problems elsewhere tells me that no major effort in the field of agricultural development can be undertaken and successfully carried out without the kind of commitment I am talking about—and commitment, first of all, at the highest level of government. This *is* the starting point, and I write this in the hope that this starting point will be secured as the precondition for all the tasks that lie ahead of you.

Mr. Minister, I know only too well that in our talks and in this letter I did not prove to be a bearer of cheerful news. On the contrary, you may look upon me as a Cassandra, piling you with problem upon problem. However, I hope you will agree that my views do not reflect the mere whims of the pessimist. Realities as I encountered them are the basis of my appraisal. I may have misconstrued them, a possibility I do not rule out. On the other hand, if they are as they appear to be, you will give my findings, analysis, and suggestions your careful consideration.

Finally, I cannot end this letter without telling you how deeply I appreciate the neverfailing goodwill you have shown me. This has been a source of great encouragement; and, if I have failed to give you the kind of assistance you expected, the fault, I am sure, lies entirely with me.

46. *The Industrialization Bias in Economic Development*

Early in 1964 Ladejinsky participated in a Conference on World Tensions. From a summary of his remarks I have selected for presentation here the five concluding paragraphs. They express very well his views on the unfortunate results of the overemphasis developing countries have generally placed on industrialization as the high road to development and the need to redress the imbalance "so that industry *and* agriculture may constitute, as indeed they must, the base upon which an effective national economy must rest."

THE DISCOURAGING HISTORY of agrarian reform in Asia doesn't speak well for its immediate future. Yet, the prospects are not without a silver lining, mainly because in a rapidly changing world the status quo in the countryside cannot be preserved and is bound to give way. It is the experience of this observer that an increasing number of landlords are cognizant of this and are not of an easy mind about the shape of things to come. In this connection the legislative enactments with all their faults of commission and omissions are not altogether a waste of effort. The mere preoccupation with reform is on the plus side, pointing to the fact that the issue will not just fade away. The question is how to speed up the process in order to preclude a disorderly or revolutionary takeover of the landlords' land and all that a violent act brings in its wake.

For the time being the landless peasantry, discontented though it is, cannot on its own generate sufficient political influence to redress the balance in the countryside. It is not yet a source of authority and a mainspring of change. The change can come about in one of two ways: under the leadership of the Communists, whose expertise in using the land issue for their political ends is well known, or under the leadership presently in power. Since Communism in free Asia is not yet strong enough to seize the initiative, it behooves the existing governments to do what they have failed to do so far, while it is not too late. If this is to be done, and with dispatch, the first order of business is a proper political climate, which only a dominating political group or country can create. Assuming its presence, many of the impediments standing in the way of an effective reform, including landlord opposition, can be overcome. However, this need not imply that a break in the cake of custom and tradition will be attained democratically, peacefully, solely through the due process of law. The creators of the favorable pro-reform climate must be prepared to resort to all the institutional resources of the country and willing, if need be, to act "undemocratically."

Agrarian reform and higher agricultural productivity are closely interrelated. But what is also true is that stagnation of peasant agriculture is a widespread phenomenon in Asia, even where the reform problem doesn't loom large or is nonexistent. Its most telling expression are low yields, and that of the principal foodgrain—rice—is only approximately one-third of those of Japan or Taiwan. In conditions of rapidly rising populations and grave food shortages in a number of countries, this condition is not only keeping the peasantry in poverty but is also becoming a serious and increasing drain on a country's foreign exchange reserves as well as a drawback on a country's overall economic development. Moreover, even in the surplus food-producing countries, the economy of the peasant owner is far from a dynamic one, which is to say one of rising yields, rising purchasing power, and a stimulant to economic development in general.

What makes for improved agriculture is no secret and need not detain us here. One doesn't know of a single ministry of agriculture in the region which hasn't in its files all the necessary prescriptions for a better agriculture. Yet, relatively little has been done, and not only because the conversion of a traditional peasant society to a somewhat more technical base is fraught with major difficulties. More immediately to the point is another roadblock, namely, the heavy bias in favor of industrialization in the typical underdeveloped, agrarian country, as against setting its agrarian house in order. This approach, favored by planners and economists, both Western and Asian, has tended to retard the much-needed agricultural effort. This criticism is not leveled against industrialization as such; its usefulness is all too obvious. What is at issue is *the overemphasis on industrialization in relation to agriculture*. Moreover, this bias is not a mere theoretical concept. It is concretely expressed in terms of the relative allocations of scarce development resources, whether indigenous, borrowed, or in the form of gifts. In largely agrarian Asia this approach is an economic and political fallacy which, if persisted in, is bound to lead—has already led—to serious stresses and strains affecting the progress of developing countries.

The inescapable fact is that, even with Communist China excluded from this picture, 70 to 75 percent of Asia's vast population is peasants. There is no denying the leavening effects of ambitious postwar schemes for industrialization and the material benefits the factory is already bringing to numbers of Asians. Nevertheless, for years to come, industry, however rapidly developed, can make at best only a relatively small dent on the agrarian character of underdeveloped Asia. Agriculture, not industry, will continue to be the pivot of Asia's economic life; to disregard this is to court grave difficulties already apparent in one of the biggest countries of the continent.

In light of the above and while searching for answers to problems of political and economic instability, the time has indeed come to treat the rehabilitation of peasant agricultural economies with surely as much concern and priority as the problems of industrialization have enjoyed. It is a welcome sign of the times that India, for example, is beginning to do just that. But this issue is far from resolved—even in principle—in the greater part of the continent. Admittedly, it is far easier to build the most complex industrial undertaking and make it work than it is to induce the "average" farmers of a country to become innovators.[1] The relative ease in the first instance doesn't argue for the neglect or only inadequate preoccupation with the more difficult task. By and large, peasant agriculture has not received the attention it deserves, with consequences all too apparent. It is easy to blame the traditional character of a peasant economy, but this serves no good purpose either for the peasants or for industry or for a country's general welfare. The fact is that the peasant *can* be an innovator, and there is ample evidence that through underpropitious circumstances he has become one, even if not on a large scale. And the reference here is not to the well-known cases of Japan and Taiwan; pockets of innovation in other Asian countries demonstrate this as well. But to enlarge these pockets so as to encompass a country's entire agriculture, initiative, enthusiasm, and above all appropriate resource allocations must be provided by planners, economists, and national governments to a degree comparable to that accorded to industrialization. Failing this, there is little chance that the line of demarcation between city and village will ever be bridged or that nonagricultural developmental activities will ever attain their ultimate aims. Hence the conclusion that the existing imbalance in the treatment of industry and agriculture must be redressed so that industry *and* agriculture may constitute, as indeed they must, the base upon which an effective national economy must rest.

1. [Many a reader, surely, will be inclined to challenge this statement. At the least it would appear to call for some qualification.]

47. *Land Reform*

This is a seminal paper, not to be missed by anyone interested in the subject. Here are Ladejinsky's views on the meaning and content of agrarian reform, security of tenure and rent reduction, ceilings and land redistribution, methods of land purchase and tenant repayment, implementation, land reform and productivity, the prospects and pressures for reform, and more. The piece complements and at the same time considerably overlaps his "Agrarian Reform in Asia" (CB–94) of a few months earlier. Unable to provide space for both, I have concluded that this one simply could not be omitted.

The paper was prepared for presentation to a Conference on Productivity and Innovation in Agriculture in Underdeveloped Countries held at the Massachusetts Institute of Technology June 29 through August 7, 1964. The proceedings were published under the title *Policies for Promoting Agricultural Development*, edited by David Hapgood (Cambridge, Mass., 1965).

LAND REFORM DEALS WITH the adjustment of a cultivator's relations to the land in a land tenure system. When the system of land tenure in predominantly agricultural countries provides the cultivator with a reasonable reward for his efforts, it stands for economic, social, and political stability in the countryside and very often, and by the same token, in the country as a whole. The obverse is true when the system of landholding denies the cultivator the conditions under which he can secure for himself a reward for his labor commensurate with his role as a producer. In the main and with more recent notable exceptions such as Japan, Taiwan, and Egypt, the latter condition still prevails in many parts of Asia, Latin America, and the Near East. This state of affairs was accepted as "normal" only a few decades ago, but this is no longer the case. The pressure for change in the status quo has been rising, and most countries beset by tenure problems are now preoccupied with the crucial issue of how to change their land tenure systems in greater consonance with the improvement of the cultivator's standard of living, the improvement of the land, and greater agricultural productivity. The issue is the more difficult because, at least in Asia, the scale of farming is very small, particularly so on tenant-operated farms. One to three fragmented acres is often the rule. Although Japan and Taiwan have managed to create a technology and economic and social institutions to suit their conditions, for the greater part of Asia this is still a far cry from reality. More often than not they are "uneconomic farms," uneconomic in the sense that they fall below the subsistence minimum rather than the technical optimum.

The Present Condition

The current search for adjustment of the land tenure systems stems from the circumstances under which vast numbers of farmers live and work somebody else's land. The principal features that characterize the tenants' plight are stagnating agricultural economies; scarce land, yet concentrated in few hands; low yields but high rents; poor farmers but expensive farms; too many people living on too little land and small holdings get smaller under rising pressure of population with no alternative occupations; inadequate tools, indebtedness and usury, malnutrition and illiteracy; keen competition among the peasants seeking scarce land in the hands of the relatively few who have it; absence of any chance for advancement within agriculture; little margin for risk-taking; and subsistence farming with a lack of dynamic or regenerative capacity.

Many of these conditions are due to institutional land arrangements over which the peasant has no control. An exploitative system of tenancy prevails in most countries of the underdeveloped world. Rack renting and insecurity of tenure are its hallmarks, and governments and their judicial bodies have given official sanction to this type of tenant-landlord relationship through the centuries. The farmers have been obliged to pay exorbitant rentals because the compelling need for any kind of employment depresses wages and raises rents. Nor is their contract of tenancy, which is supposed to ensure security of tenure for the tenant for a specified period, often worth the paper it is written on. In most cases the contracts are oral. But whether written or oral, they can be abrogated at the whim of the landlord. The incentive to improve the land and produce more does not exist, nor is there a place for creative technology on a wide scale. To the extent that these conditions preclude a measure of equalization of opportunities, they stifle progressive impulses and tend to underwrite stagnation in agriculture.

Farmers have never been satisfied with this state of affairs. They have often expressed their discontent overtly against persons and governments that they believed were the causes of their distress. But in the main until the end of World War II, in Asia, Latin America, and the Middle East the conservatism and inertia of the farmer and his ingrained feudal subservience to the state and the landlord kept the pot from boiling over. More recently and at least in part under the threat of Communism, the bonds that kept the peasant down are loosening under rising agrarian discontent, both covert and overt. Country after country is struggling with these problems: how to relieve the plight of cultivators working for a pittance, how to revive depressed agricultural economies, how to root the peasant securely and beneficially on the land he cultivates. These problems have close bearing on agricultural productivity and economic development in general.

The Meaning of Agrarian Reform

The answer to the questions posed is broadly known as "agrarian reform." The term conveys many things to many people. To the Communists, agrarian reform is simple enough: It is a means to political power, based on a promise to the peasant of the one thing he wants most—ownership of the landlord's land—in exchange for his badly needed political support. To the non-Communist world, agrarian reform involves such issues as who owns or does not own land, how it is used, who gets what out of the land, the productivity of the land, the rate of economic development, and, additionally, social status and political power.

No single panacea will deal with all these issues effectively. In conditions of rural poverty even redistribution of the land will not suffice unless it is accompanied by the necessary means to work and improve the land. It follows that countries that carry out land redistribution programs must make great efforts to increase agricultural investment, particularly if their rate of population increase is fast. The economic opportunity and psychological incentives that come with possession of the land or security of tenure must go hand in hand with a host of other developmental measures. For this reason agrarian reform in the sense considered here encompasses all or most of the following elements: distribution of land among the landless; security of tenure and fair rents; and better methods of cultivation through technological improvements, adequate credit, cooperative marketing, and other measures. However, it must be stressed that not all these elements are of equal importance. The most important is land ownership. If this is absent, all else may prove ephemeral, including security of tenure and rent reduction—measures extremely difficult to enforce. This explains why in the final analysis the issue is one of land to the landless. This is the real vehicle of security and opportunity upon which a more resourceful economy can be built.

The Content of Agrarian Reform

In most instances the content of enabling reform legislation is twofold: security of tenure and rent reduction, and the redistribution of land among the tenants. In Asia, except Japan and Taiwan, great stress is laid on security of tenure and rent reduction, while in the Middle East and Latin America the stress is on land redistribution and the creation of peasant proprietorship. No country approaches the content of either main measure in the same way, but the meaning of its approach depends upon the answer to this basic question: For whose benefit and for how large a group of beneficiaries is the reform designed? If the reform is indeed one for the benefit of the great majority of the tenants, its specific content and its enforcement will differ substantially from a reform attempting to satisfy both landlord and tenant in the difficult conditions of scarcity of land, land monopoly, pressure of the farm population on the land, high rentals, high land values, and subsistence and unremunerative farming. The Japanese and Taiwanese reforms reflect a thoroughly pro-tenant attitude. Hence they emphasized ownership of land for the majority of tenants through the abolition of absentee ownership altogether and a low permissible retention of land (ceiling) for resident landlords, genuine security of tenure and low rentals for the remaining tenants, arbitrary and low valuation of land prices, easy repayment terms, and a type of enforcement in which the tenants play a major role in order to ensure a minimum of evasion of the principal provisions of the reform legislation. These reforms involved drastic redistribution of property, income, political power, and social status at the expense of the landlords. This was the purpose of the reforms, the basic proposition being that half measures or attempts to satisfy both parties could not bring about conditions under which those who cultivate the land would enjoy the fruit of their labor.

Security of Tenure and Rent Reduction

It is axiomatic that great population pressure on scarce land brings about keen competition for the right to cultivate a plot of land. The result is that rents are high and security of tenure is low. Rents in Asia, for instance, of 50, 60, or 70 percent of the crop are common even when the landlord contributes only the land while all other elements of production are furnished by the tenants. In such circumstances the bargaining power is all on the side of the landlords; they are not obliged and cannot be compelled to give tenants binding security of tenure for a specified period of time. Insecurity of tenure combined with high rents adversely affect agricultural productivity, not to speak of the tenant's welfare. If this situation is to be reversed, a situation must be created that leads to a reasonable ratio between the farmer's share in the effort and costs of production and his share in the crop produced. Coupled with this must be the appreciation of the fact that the margins of subsistence are so narrow that tenants can afford little risk; any innovation in production practices that goes wrong may result in starvation.

Rent reduction and security of tenure is one form of agrarian reform. Existing reforms do not reveal any common standard upon which a reasonable rental may be based. Each country or, in the case of India, each state deals with the problem in its own way. Some maximum rentals are 25 percent of the crop in Japan, 37.5 percent in Taiwan, 50 percent in Nepal, and in India from 16 to 50 percent. The "why" of the range of these rents stems from the natural assumption that the smaller the rent the greater is the benefit to the tenant. They have virtually nothing to do with an apportionment of rent among the various factors that go into the growing of a crop. Rents are of either the "fixed" type, a stated quantity regardless of the harvest, or the crop-sharing type, a percentage of the crop. From the point of view of the tenant, crop sharing makes sense where there is the danger of recurring crop failure, but it has a disincentive effect upon the tenant. Rent regulations under a reform tend, though not always, to effect shifts from crop sharing to fixed rents, thereby providing the tenant with greater incentive. The problem with rent reductions of whatever type is that they are difficult to enforce. Japan and Taiwan have succeeded in enforcing them as they have much else, but they are very special exceptions rather than the rule. This applies particularly in areas such as India and Indonesia where there is fierce competition for any plot of land in order to eke out any kind of living.

Security of tenure is part and parcel of rent reduction, and all reform legislation aims to give stronger protection to a tenant's occupancy rights. Like rents, such measures vary considerably although they have such common prescriptions as written leases, specified periods of occupancy, and compensation for useful improvements.

Tenant Rights on Taiwan

Taiwan, more than any other country concerned with the problem, provides the best tenant occupancy rights. The Taiwanese law provides that farm lease contracts (of six years' duration) shall not be terminated before their expiration except under any one of the following conditions: (a) if the tenant dies without an heir; (b) if the tenant waives his right of cultivation by migrating elsewhere or changing his occupation; (c) if the tenant fails to pay a total of two years' rent. Such safeguards extend to tenants even after the expiration of the lease. The law provides that the landlord cannot take back the leased land for his own cultivation on the expiration of the contract if any one of the following conditions exist: the landlord cannot till the land himself; the landlord's income is sufficient to support his family; or the landlord's action in taking back the land deprives the tenant's family of its subsistence.

For all practical purposes this means that tenants can remain on the land undisturbed even after the expiration of the contract, which in Taiwan virtually ensured the enforcement of the rental provision. Less carefully worded stipulations might have undermined this part of the reform program. In India, for example, the provision that a landlord can take back land for "personal cultivation" has proved to be a serious drawback in enforcing security of tenure and rent provisions.

Land Redistribution

There are regulated tenancies in the United States, England, and even in India that so protect the tenant that he prefers to invest in capital goods to improve the land rather than in buying land. But this is not true in most parts of Asia, the Middle East, and Latin America. What the farmer wants is a piece of land of his own. To the Japanese farmer, one without land is one without a soul. This mystique about land ownership may not be so aptly expressed in other countries, but its validity is there no matter how expressed. Prevailing tenurial conditions suffice to explain it. Where tenants are plentiful and poor and land is scarce, regulated tenancies are almost impossible to enforce. Japan and Taiwan have succeeded, but they are very special exceptions to the rule. Examples to this effect are too numerous to cite here. In one case, that of India, an unwise attempt to regulate tenancies led to still greater insecurities and the driving of tenants to the very bottom of the agricultural ladder to the rank of agricultural laborers.

The arguments are strong for the diffusion of ownership among the tenants as the main purpose of agrarian reform. As we shall see later, enforcement of a land distribution program in Asian conditions is no easy task either. Yet one cannot but agree with a student of the problem that "administratively, there is virtually no question but that a system of owner cultivatorship in which the basic distinction between 'mine' and 'thine' can be marked out on the ground is much simpler to administer than a system of regulated tenancies." This of course assumes that a government, in the first place, devotes itself single-mindedly to the promotion of this kind of a program. There is another reason why agrarian reform should stress ownership. Agricultural practices on most of the would-be new-owner farms need improvement, and this means new investments. Financial institutions, even if they are public, make a distinction between owned land and leased land in making loans. In the first case, land is the security for obtaining investment; but this is not so in regulated tenancies with their continuous shifting of rights and duties between landlord and tenant.

The Land Ceiling

There are reforms the sole purpose of which is security of tenure and reduction of rent (Nepal), but in most countries redistribution of land and the establishment of new owner farmers is the main objective. This aims toward abolition of tenancy altogether. The immediate problem of reform makers then is to secure enough land to endow a given number of tenants with land of their own. The core of it is the "ceiling." The landlord is permitted to retain a certain amount of land (ceiling), the remainder being earmarked for redistribution among the landless. It is the size of the ceiling that determines how far-reaching the program might be, always assuming that the ceiling is enforced.

Since countries have varying amounts of land available and varying scales of tenancy, it is understandable that the ceiling should differ from country to country. More fundamental, however, is the fact that in some countries the ceiling is made to serve the objective of widespread ownership whereas in others the ceiling can become a device to defeat this objective. In Japan and Taiwan the ceiling was so set that it encompassed all the land of the absentee owners and permitted (in Japan) the resident landowner to retain only 2.5 acres. This is an instance of a ceiling cutting to the very bone of the landlord's holding. It cuts less deeply in South Vietnam with a ceiling of 250 acres, apparently deeply in Iran where a landlord is permitted to retain only one village out of many, while in India the great variety of ceilings might have served their purpose except for lack of enforcement and widespread evasion.

Clearly what matters in extending ownership is not only at what level the ceiling is set but also that it not be evaded. The cases of Japan and Taiwan on the one hand and Pakistan on the other are instructive. The low ceiling in the first two countries made it possible to extract a great deal of surplus land for redistribution. Before the reforms 54 percent of Japan's land had been owner operated, and after the reform the figure rose to 92 percent; the respective figures for Taiwan were 60 and 85 percent. West Pakistan is not Japan or Taiwan, but fixing a ceiling of 500 irrigated acres and 1,000 unirrigated acres in a country where the land of two-thirds of the farmers averaged five acres each was not really a measure for land redistribution. Moreover, exceptions and subdivisions of large holdings among members of a family were permitted on the eve of the reform. This, combined with the high ceiling, meant that the land distribution program was divested of any meaning. It appears that, if a ceiling is to provide the landless with land, it must be set at a low level; the acreage retained by the landlord must be fixed retroactively or on a date well in advance of the reform; and, finally, the ceiling must be fixed on the basis of land owned by a household rather than an individual. With a low ceiling, no evasions, and effective implementation, large numbers of tenants can become farmer proprietors.

The Landlord's Holding

Unless a reform is confiscatory—and the ones considered here are not—the question of how much to pay for the land earmarked for redistribution is of great importance. Various countries approach this crucial issue in different ways. Usually the value of the land for compensation purposes has been derived from one or a combination of factors such as valuations shown in land tax records, recorded land revenues, rent collected, landlord's net income, or value of produce. But one thing is common to most of them: Prevailing market prices are seldom accepted as the basis of valuation. If they are considered, it is only as a yardstick for sharp deviation downward. Even where the intent is clearly not to penalize expropriated owners but rather to compensate them adequately for their loss, they are not in the same position as a person under no legal pressure to sell, free to negotiate on the open market for the best available price. This puts the compensation outside the market in the strictest sense. Land prices in underdeveloped and densely populated areas are commonly very high, bearing no relation to the productive value of the land. Low assessments of agricultural land, comparatively low land taxes, and negligence in tax collections have frequently been the main reasons for high market values, particularly in areas with high population pressure on the land. Added to this are such factors as the social and political prestige and influence gained from the possession of land, and a favored tax position compared with that of investors in commerce or industry. The combination of all these elements has made land values unrealistically high, a fact that must be faced by reformers in deciding what to pay for the land to be purchased from the landlords for distribution among tenants.

In Japan the price of land was fixed by capitalizing the annual rent in a manner to ensure a fair profit for the farmer buying the land (taking into consideration that all rice save his own consumption was collected by the government at fixed prices). In practice this meant 40 times the rental value of a rice field and 48 times the rental value of an upland field. These cash rents were quite low compared to a price level augmented by inflation. In Taiwan the government decided that a landlord would get 2.5 times the annual value of all crops on a given piece of land. In Italy assessed valuations shown in the land tax records served as a basis of pricing. A number of Indian states used annual tax revenue or net income from the land as a base. But whatever the existing method of pricing, with rare exceptions the price is fixed below the market price. The fact is that land purchase under a reform is not an ordinary real estate transaction where seller, broker, and buyer meet in a free market. If it were and if tenants were able to pay the "going price," there would be no need for reform. It may be concluded that the price fixed by a government is an arbitrary one, the degree of arbitrariness depending upon how a reformer answers the question already referred to: "For whose benefit is the reform designed?"

Methods of Purchase and Repayment

There is ample evidence that, in order to bring about a land redistribution program, a government must buy the land and resell it to the tenant rather than institute direct landlord-tenant negotiations. This procedure rests on the proposition that, where land is at a premium and the economic and political power of a landlord is strong, the bargaining power of the tenant in relation to the landlord is so weak that an agreement can be reached only on the landlord's terms. The two decades of Japanese paper reforms before World War II amply demonstrate this point. This is true in other countries in Asia where purchase transactions are left in the hands of landlord and tenant. In the long run this means prohibitive prices—and no sales—which explains why landlords are usually compelled to sell a certain amount of land at officially fixed prices.

How a government pays for the distributed land is of great importance to government, landlord, and tenant alike. Experience shows that a government, no matter what the price, is not in a position to pay for the land in cash in one lump sum; Colombia and Venezuela are perhaps the only exceptions to the rule. Payments to the landlord are made mostly in interest-bearing bonds spread over a period of years, with cash seldom exceeding 10 percent of the fixed price of the land. Madras state and the Philippines are exceptions. In the latter a maximum of 50 percent of the compensation may be paid in cash and the remainder in land certificates, except that full payment in cash is made in certain cases of expropriation through the courts. However, very little land has been thus acquired, especially since land acquisitions are based on "fair market value." Payment through bonds, extended over periods of twenty years or more and bearing an interest rate of 3.5 to 5 percent, has eased budgetary difficulties and has served to avoid the inflationary impact of large cash disbursements. In some countries the bonds are negotiable and redeemable in equal annual installments; or the bonds can be used for the purpose of industrial and commercial investment, for the payment of taxes, and such.

Taiwan's case is unique. To avoid the effects of inflation, which resulted in virtual confiscation in Japan, Taiwan tied the price of land to the prices of two principal products of the land and to shares of stock in government-owned industrial undertakings. This meant that 70 percent of the compensation was in the form of commodity bonds payable in twenty semiannual installments over a period of ten years and 30 percent was paid outright in stocks. This method worked well for all parties concerned. The commodity bonds preserved the value of the sales price against fluctuations in the value of the currency for a ten-year period. And an estimated 40 percent of the total compensation found its way into industrial and business investments. This is perhaps the only known case of an agrarian reform that has consciously planned and succeeded in transferring private capital formerly tied up in the land into the general developmental field.

The methods of repayments to the governments by the former tenants vary from country to country, but the underlying principle is that the payments must not constitute an excessive burden on the new owner. A government's annual recovery from the new owner must be less than his former rent. Usually repayments by tenants and payments by governments for the acquired land are extended over a similar period of years. Since all governments are eager to recover their original outlays, they aim to strike an annual balance between what they pay out and what they take in. This does not necessarily work in practice because of changes in the economic conditions of the farm community; a government's decision to charge (or not to charge) to the new owners the administrative costs of a reform; and, above all, the fact that some governments have realized that, if land redistribution is to be viable, heavy investments are needed to improve the land and to create infrastructure to help the new owners during the formative years of the reform. The recovery of these extra costs is not often charged to the new owners directly.

On the whole the record of repayments is good. This must be attributed to the relatively easy terms set by the governments in question. In Italy, for example, a new owner under the reform pays an annual sum equivalent to $19 an acre, on the average, as against $40 to $50 an acre on similar, nonreform land. In Venezuela the new owners amortize in twenty to thirty yearly installments a predetermined proportion of the cost of the land; improvements and the first year's working capital are provided by the government as a gift. In Egypt repayment is in forty annual installments; and, though the farmers are also charged 1.5 percent interest annually and 10 percent for administrative costs, the sum total of annual recovery is considerably below the level of pre-reform rent. It is clear that, even with good recovery, reform imposes budgetary strains on a government. Two facts must be kept in mind, however. One is that an agrarian reform is not an ordinary commercial enterprise but one involving social justice and political stability, an investment in the future of the nation's people. Second, if the assumption is correct, as it seems to be, that agrarian reform does lead to an increase in agricultural productivity, then it is reasonable to say that eventually the government and the country as a whole will more than recoup the investments called for by the initiation of a reform and its support during the early years.

Implementation and the Political Climate

Effective enforcement rather than reform legislation, however well drawn, makes the difference between reform in being and reform on paper. There are too many of the latter kind in the regions under consideration. The agrarian reform movement in Asia anticipated most of those in the Middle East and in Latin America, but this advantage in time did not bring about uniformly good results. With the exception of Japan and Taiwan, the picture of Asia leaves much to be desired. The landlords' property in South Korea has found its way into the hands of the tenants, but the government's failure to come to the financial assistance of the new owners has brought about indifferent results. The civil war in South Vietnam caused that reform to be short-lived. The Burmese tenants are owner farmers now, but government after government has failed to capitalize upon this and develop the rich land resources of the country. The Philippines has four reform failures on its record and is trying now for the fifth time with no greater prospects for success than heretofore. West Pakistan and Indonesia have proclaimed reform but with hardly any implementation, while Nepal, after a dozen years of cogitation, is just barely taking the first step. Finally, there is India, significant and encouraging for what it has attained in unprecedentedly difficult and bewildering conditions and just as significant and discouraging for what it has failed to attain, and for the reasons why.

None of this has anything to do with ignorance of the tenantry's conditions or inability to draft legislation to meet those conditions. Even Nepal, a closed and remote society only yesterday, is skilled enough to write good measures—if it wants to. Surely the same can be said of every country where reform has failed. A number of reasons explain this state of affairs. To begin with there may be insufficient administrative and technical skill to mount a reform; but this is not a crucial issue, and failure to implement a reform cannot be attributed to it. More to the point are inadequate measures or half-measures deliberately drafted so as to retard—if not obstruct altogether—the application and implementation of a reform; absence of leadership among the peasants to propagate the reform idea and exert effective pressure on legislative bodies; disinterestedness—if not overt opposition—on the part of intellectuals and molders of public opinion; and, most important, the built-in opposition to reform by the landlords, whose role in the body politic is out of all proportion to their numerical strength. This combination of factors inhibiting reform is hard to overcome.

Land ownership as the main point of an agrarian reform is difficult to achieve. Examples are legion, and the reason is not far to seek. Land redistribution under agrarian reform is a compulsory measure imposed by a government upon the landowners on economic and legal terms unpalatable to them. In effect this involves a drastic redistribution of property and income at the expense of the landlords. It becomes a revolutionary measure when it passes property, political power, and social status from one group in the society to another. This is a real meaning of an agrarian reform where land redistribution is its central objective. Considering the fact that, in the areas under discussion, legislative assemblies are still dominated by land-propertied classes, it is not difficult to see why both the enactment of appropriate legislation and its enforcement present such formidable problems. Thus land reform, despite its economic implications, commences as an essentially political question involving a most fundamental conflict of interests between the "haves" and the "have-nots."

Who Makes Agrarian Reform?

No matter how ripe a country may be for agrarian reform, it does not follow that an agrarian reform then occurs. Nor does enactment of a reform always meet popular aspirations in terms of social justice and better distribution of rights and opportunities among the landless. The dragging of the feet on the part of governments and legislative assemblies in enacting reforms is notorious. Technical expertise in preparing and administering the necessary legislation is indispensable, but experts do not make reforms. Politicians, and only politicians, make good or poor reforms or do not make them at all. They control the political climate, which determines the will or lack of will to proceed with the task; the specific measures with which the reform is or is not endowed; the care or lack of care with which the enabling legislation is formulated; the preparation or lack of preparation of the pertinent and administrative services; the presence or absence of technical services with their bearing upon the success or failure of the reform; and, most important, the drive or lack of drive behind the enforcement of the provisions of the law. The "positive" makers of reform must represent strong political leadership deeply concerned with the land and its cultivators and capable of translating those concerns into action. There is no substitute for this kind of leadership at the crucial stage of determining a reform's main goal and initiating it. The financial, technical, and administrative needs of a reform can prove to be formidable obstacles to its initiation and enforcement. Nevertheless experience shows—and India provides a number of illustrations—that they are not insurmountable difficulties if the concerted will of political leadership is brought to bear. The built-in landlord opposition, abetted by public servants, can be dealt with successfully if the political leadership is bent on carrying out its goals. This is especially important because the peasantry has not developed a popular political movement of its own capable of effectively representing and advocating its own cause.

The Case of India

India, which has done both much and little, demonstrates how much can be accomplished when will and determination are present and how little can be attained if they are absent. India's reform objectives were twofold: first, the abolition of the zamindari system;[1] second, security of tenure, reduction of rent, and distribution of land to the landless. Despite opposition and administrative and technical problems, the zamindari tenures were virtually abolished. Not all have benefited equally and not all the 20 million cultivators affected have received permanent, heritable, and transferable rights. Nevertheless the effort was a major step toward a reconstruction of Indian agriculture. The measure succeeded because it was politically popular to abolish an agricultural system full of abuses imposed by a foreign power. With the British gone, the government went about drafting the necessary legislation; the zamindars were disestablished despite the opposition of the landlords, lack of financial means, and insufficient administrative and technical staff. Not every "t" was crossed and not every "i" was dotted, but the job was done. On the other hand, in dealing with its indigenous landlord system, India faltered and very often badly, both in content of the measures and in their implementation. The nation is not without its bright spots; but security of tenure, rent reduction, and landlord ceiling provisions have not been successfully implemented on the greater part of India's 80 million acres of tenanted land. In fact, perhaps as many tenants have lost their tenancies as have been made secure on the land through deliberate rather than accidentally faulty legislation. Such states as Bombay and Uttar Pradesh have demonstrated that, given strong and willing leadership, many of the reform problems can be dealt with successfully. But where anti-reform sentiment prevails, as it often does in Indian state legislatures, vague and complicated measures generously seeded with loopholes, delays in legislative enactment, failure to inform the peasants what the law is about, enforcement officers who behave as if reforms are not meant to be enforced, and refusal to enlist the support of the farmer in helping carry out the program—all these become the rule rather than the exception. With variations the same is true of a great many other countries.

1. The zamindari system was a product of early British rule. The zamindar was given the right to collect land taxes and undertook to pay the British administration a fixed revenue. In return he was not only permitted to keep a portion of the revenue but was also recognized as the proprietor of the revenue-bearing land.

The Japanese Example

Japan and Taiwan, on the other hand, proceeded with their reforms with no vacillation or half measures. They developed their own techniques of initiating and enforcing reform. That the American occupation in one case and the loss of mainland China in the other were important elements in fashioning the reforms does not invalidate the proposition that above all else a favorable political climate is the precondition of a sound reform and its implementation. In such an atmosphere even the would-be non-enforcers become enforcers, and technical difficulties are only challenges to be resolved. To be sure, reform has an element of coercion. This is so because the key to the implementation of any reform that transfers property and privilege from one group to another is the degree to which the controlling political forces of a country are willing to support these revolutionary changes and ready to use *all* instruments of government to attain their goals. Government coercion, then, whether practiced or implied, is virtually unavoidable. But so long as this kind of a political wind is conspicuous by its absence, a meaningful reform intended to meet some of the minimum needs of tenants can be neither properly initiated nor carried out.

Even well-drafted measures in a pro-reform atmosphere cannot be easily enforced if the task is left solely to the bureaucracy. There are never enough bureaucrats and reform experts to go around; the experts become experts only in the process of application of the reform. If a reform is to be carried out successfully, it must win the active participation of the people directly affected by the reform—the farmers—who best know the conditions of their rural community. With the exception of Japan and Taiwan, this principle has not been recognized by most countries currently engaged in reform activities.

Japan is the pioneer in entrusting the villagers with the lion's share of the implementation of the reform. A close look at the Japanese experience is very instructive, for it carries a lesson that other countries might well ponder. The device created by Japan for the administration of the reform at the village level is the local land commission made up of five tenants, three landlords, and two owner cultivators—all elected by their respective groups. The preponderance of tenants was a deliberate move based on the theory that, if only one owner cultivator sides with them, a working majority is ensured for the commission. The commissions were entrusted with the actual purchase and sale of land. They had broad powers that they could exercise with a minimum of governmental interference. These included: drafting the purchase plan for each village, determining the suitability of the land to be purchased, establishing the eligibility of purchasers, deciding cases requiring unusual or special treatment, and appraising cases of exemption from the purchase provisions. The Japanese scheme was based on the fact that no staff could have been gathered to deal quickly with the transfer of 30 to 40 million plots of land and that only the local people knew at a glance who was who in the village, who owned what land, who leased land and how much, and so forth. The work of the local commissions, coupled with the successful effort to bring to the attention of the farmers the basic points of the reform, more than justified anticipations. It is questionable if the reform could have been carried out so speedily and in such orderly fashion but for the local land commissions. The duties they performed were an important form of adult education and served to stimulate new leadership. Tenant members who were ill at ease and insecure at the beginning of the exercise were seasoned performers a year or two later. The very creation of the commissions enabled all the adult farm population to vote on a matter of utmost concern to all groups in the village. At the same time the composition of the commissions proclaimed the fact that the tenants' interests were to be protected by the tenants themselves rather than by someone acting on their behalf within the traditional pattern of rural Japan. At the height of the reform approximately 150,000 commissioners received this kind of special leadership training, and half of them were tenants. The prereform contention that only landlords could exercise leadership, make decisions, and administer the community proved baseless.

The Need for Farmer Participation

It may be argued that what Japan could do with a literate farm population is not possible in other countries with less literate farmers. But even illiterate farmers know their own condition and that of others in their community. A judge in court may dispute a farmer's occupancy right because he lacks a written contract; but there is hardly a village where the tenants, like the Japanese tenants, do not know the details of the existing pattern of tenure. This knowledge, if enlisted through some sort of organization (not necessarily the Japanese type), could greatly aid any reform effort. So far, farmer involvement has been conspicuous by its absence unless, as in Bolivia, farmers deal with reform in their own spontaneous fashion. So long as the would-be beneficiaries are treated as mere onlookers, often ignorant of even the ABC's of the reform measure, the reform cannot succeed. The widespread assumption that only white-collar officials can administer a reform is an idea best laid to rest. Reform, if effective, must be carried out at the village level; and the participation of the farmers in an affair that affects them deeply is one of the vital conditions for translating a reform from paper to reality.

Reform and Productivity

Just as insecure and difficult tenure conditions make for low productivity, it is often assumed that agrarian reform with ownership as its objective makes for higher productivity. This is in tune with Arthur Young's statement in the eighteenth century that "ownership is the magic that turns sand into gold." It flows from the idea that investments and productivity are promoted through the incentive of private, individual ownership of land. This assumption may or may not be correct, depending upon the circumstances surrounding the fact of ownership. In Latin America, for example, many a private owner of a hacienda has not proved to be a good producer—and many a small owner has done no better.

While it is easy to postulate that land ownership stimulates productivity, it is extremely difficult to measure its actual effect. While assessment of output is comparatively simple, the contribution of a land distribution program to productivity cannot be separated with any degree of accuracy from that of other factors affecting output. But a close interrelationship between agrarian reform and productivity does exist if agrarian reform stands for ownership of the land or secure tenure and such supporting elements as land improvement, irrigation, credit, marketing, and so on. Without these added inputs, agrarian reform is not necessarily the touchstone to a new owner's increased output and higher standard of living.

The pervading spirit of enthusiasm that comes with reform and the will to make the most of the awaited land of one's own are important incentives to production; they play an indispensable role when the supporting services are there to play their role. The experiences of Japan, Taiwan, Italy, Egypt, and India provide ample evidence that agrarian reform and productivity go hand in hand under these conditions. Therefore, if the anticipations of the beneficiaries of agrarian reform are not to be thwarted, a government promoting a reform must be prepared at least in the initial stages to invest in a variety of other measures that promote agricultural productivity. On the other hand, efforts to promote agricultural productivity where tenancy reforms are long overdue are likely to be adversely affected by the existing tenure system. Unless new inputs come to overburdened tenants as a gift or under exceptionally favorable terms, the chances are that they cannot afford to innovate. The reaction of the tenants to the package program in India is a case in point. It follows that agencies concerned with ways of building a technical and organizational base for raising productivity must be equally concerned with the problem of reforming existing tenures when these constitute a problem.

Social Status and Political Stability

Agrarian reform in its inclusive sense stands for salutary changes in social status and political stability. In Western Europe the changes in tenurial conditions from feudalism to our own day amply demonstrate this view. The changes in Japan since the implementation of the reform tell the same story. The Japanese case carries implications for underdeveloped countries as well. A reform worthy of its name is supposed to strengthen the principle of private property where it was weakest, at the base of the social pyramid. By multiplying the number of independent landowning peasants, there comes into being a middle-of-the-road, stable rural society with enhanced status, rights, and privileges. Government then cannot take the peasantry for granted but must perforce pay close attention to the desires of the countryside. This has happened in Japan to a degree unknown in Japanese history, and it reflects the rise in social and political status of the peasantry due to agrarian reform.

Reform can narrow the traditional differences in the class structure of the village. As the tenants step up, the landlords step down. As the landlord loses much of his affluence, he loses much of his influence. This does not presuppose that the resident landlord is completely displaced or that it is necessary to eliminate him altogether. What is taking place now in the Japanese village is the sharing of power between the old and the new leadership. Both are meeting now on the agricultural committees, cooperative and school boards, and in village offices. They rub elbows dealing with common problems. This is a new and welcome development. Japan's agriculture after the reform still is a marginal enterprise with acre and two-acre farmers. It needs all the available skills and social peace and stability to deal with the reality of 6 million farmers on 15 to 16 million acres of land. This new leadership and the idea of citizenship rights were both foreign to pre-reform Japan. The passing away of the old reality of second-class citizenship is a symbol of the new social order and of greater political maturity and independence. The latter expresses itself in the attitude of the full-fledged citizen toward a government or a party in or out of power. Unlike pre-reform days, voters pick and choose according to their economic interest. The candidate's own proven concern with farm conditions and the record of his party's agricultural policies are decisive. Whatever action a government in power may choose to take with respect to the price of agricultural products, fertilizer prices, scope of land improvement work, and short- or long-term credit funds now affects the distribution of the vote. No longer is the vote delivered en bloc. Clearly, "we support those who support us" has a modern democratic ring and it can be traced to the social and political changes ushered in by the agrarian reform.

So it is in Japan, but it need not be only in Japan, due allowance being made for different conditions from country to country. If an agrarian reform is imbued with the meaning ascribed to it here, then, whether it be India or Egypt, Mexico or Peru, changes in social status and political attitudes in the countryside are inevitable. And a government's response to these changes is likely also to be inevitable if remaining in power requires popular support of the peasantry.

Prospects of Agrarian Reform

In the past two decades "agrarian reform" has come to epitomize the problems and hopes of rural people in Asia, the Middle East, and Latin America. In Asia the issue of reform has long since moved from the talking phase to that of action. The old order in the countryside is under attack—vigorously in very few countries, much less so and with results to match in many more. From the record to date, it might be concluded that only in critical conditions, such as defeat in war for Japan or the escape of the Chinese Nationalist government from China to Taiwan, can far-reaching agrarian reforms be carried out. India disproves this thesis at least in part, as does Egypt in the Middle East. Even in Latin America, legislation has been enacted here and there; and the subject is no longer taboo but has become a burning issue. Admittedly, however, a wide gap exists between the anticipations engendered by the agrarian reform movement and its realizations.

What of the future? The answer lies in a combination of an act of faith and certain realities that are likely to induce a more vigorous prosecution of reform in the days ahead. While village communities seem to carry on as of old, the winds of change blowing from all directions are bound to affect them and the bonds of custom are bound to weaken. In India there are already numerous deviations from the norm, entire districts "going places," reaching out for horizons that only yesterday seemed unattainable. These are still only "islands within," but the drastic changes they presage will at some point touch upon the expectations of all groups of Indian rural society, including the large tenant class. Impatience with present conditions is very likely to come to the fore. Whether in an organized or disorganized form, whether through the due process of law or revolutionary upheaval, the change will come—and not only in India—and an all-inclusive agrarian reform will be part of it. This is the faith part of the argument.

Pressures for Reform

Agrarian reform will continue to command attention and ultimately action for other reasons. The areas we are discussing are agriculturally underdeveloped, particularly in foodstuffs. There is growing recognition that agricultural production must be increased if the standard of living of the producers is to be raised and if capital is to be accumulated for overall economic development. In landlord-ridden countries there is a rising awareness that agricultural progress depends upon a set of incentives that the existing land tenure systems do not provide. This recognition may yet prove to be the augury of the kind of action that makes the difference between reform and "reform." Willingly or unwillingly many countries have been preoccupied with agrarian reform and have drafted reform legislation. Even though its content is faulty and incomplete and its execution poor, the mere existence of the legislation serves notice to the landlords that the issue will not just go away and that better legislation and enforcement may be fashioned at a later date. Evasion of the ceiling, as in India, does not destroy it as a potentially serious attack on the status quo. Landlords recognize this. They know that their tug-of-war with the tenants is far from over, and many landlords would dispose of their land for something less than its market price. They recognize that the old order in the countryside is not what it used to be and that their best days are over. How soon or how late all these elements will be brought to bear in a more drastic effort to redress the position of the landless still is up to the governments in power. If willing and wise, they can speed up the process; barring that, somebody else will take over this long-overdue task, and much more will be at stake than a new rearrangement of income distribution and status in the countryside.

Reform Is Not a Panacea

We must recognize that in overpopulated rural areas an agrarian reform, even with all the supporting services, cannot solve all the problems standing in the way of a better livelihood and greater production. Where the pressure of population on limited land resources is great, agrarian reform can do nothing to change the land-man ratio. With no alternative occupation outside of the village, the rise in the farm population is bound to reduce the size of holdings still further and to increase the number of farms as sons and grandsons take over. These limitations are obvious even in prosperous Japan. While in the United States between 1920 and 1957 the number of farm units was reduced by 22 percent and the working farm population by nearly 40 percent, in Japan during the same period both categories increased by 10 percent despite the country's striking industrial upsurge. The current prosperity of Japanese agriculture cannot hide these harsh realities. Japan's nonagricultural economy has served as a safety valve by providing the farmers with 40 to 50 percent of their income. Clearly this is not possible in developing nations without substantial nonagricultural economies.

Agrarian reform in such conditions is not a final solution of the farmers' or nation's problems. Rather it does away with the worst features of a system that has outlived its usefulness economically, socially, and politically. Reform provides a partial escape from the severe handicaps of a resourceless small-scale agricultural economy pressed upon by a large farm population. It is a means of elevating the human condition. We may arrange for all the material supplies of water, seeds, fertilizers, implements, and credit. But we cannot give the peasant the psychological incentives he needs unless he is secure on the land, particularly on land that belongs to him, for uncertainty in this regard is one of the great disincentives under which a farmer labors. In any attempt to provide a better material and organizational base for the rural sector, the solution of the land tenure issue is the clearing of the ground, the preparatory step for all else, the sum total of which raises agricultural productivity.

V. THE WORLD BANK YEARS, 1964–1975

IN THE LATTER PART OF 1964 Ladejinsky was invited to participate in a major World Bank study of India's economic development effort. This study, a special effort of the Bank, was intended to penetrate to the very fundamentals of India's development problems and prospects and provide the basis for intensified World Bank efforts to assist India in its development efforts for years to come. Such luminaries in the agricultural field as Sir John Crawford and David Hopper, as well as Ladejinsky, were enlisted to participate in the study. The four papers on agrarian reform, community development, the cooperatives, and local administration that Ladejinsky contributed to this study—a book-length volume in themselves—comprise his greatest single piece of work on India and provided the foundation on which all his subsequent work on India rested. This was the beginning of Ladejinsky's association with the Bank, which was to last until his death. Upon the conclusion of this study, he was asked to assist in more conventional Bank missions to Mexico and Iran in 1966. Early in 1967 he was invited to become a member of the Bank's permanent resident mission in India. It was here that he finished out his working years.

These World Bank years were productive of a large body of forty-seven pieces of written work: forty-one for India; two for Iran; one each for Mexico, the Philippines, and Sri Lanka; and one conference statement on land reform.[1] The India pieces include five clearly identifiable Ladejinsky contributions to the resident mission's annual reports; eight inimitable reports of field observations in Punjab, Bihar, Andhra Pradesh, Maharashtra, Bengal, eastern Uttar Pradesh, and refugee camps; and a number of other papers in which he addressed himself to the passing scene and noteworthy issues—special problems and programs affecting the small farmer, credit developments, institutional constraints on farm production, government procurement, renewed efforts at establishing landholding ceilings, the effects of the Green Revolution, and so on. For space considerations and because of a certain degree of overlap, only fifteen papers of this period are presented here, five of them in truncated form.

In a significant sense "Agrarian Reform in India," the first of Ladejinsky's four contributions to the World Bank's 1964–65 study and the first in Part V, represents a watershed in Ladejinsky's arduous and frustrating love affair with India. This had started in 1950 (if not, indeed, with his 1939 paper on "Agricultural Problems of India") when on a

1. See the Chronological Bibliography.

visit he "was privileged to take part in the preparation of the agrarian reform part of the first plan . . . and closely followed its fortunes since then." It continued and developed with his three 1952 studies of land reform in in Madras, Punjab, and Kashmir; his 1954 study of the status of land reform; and his 1963 study of tenurial conditions and the "package" program. Ladejinsky therefore came to the 1964–65 study with an acute sense of the unfinished business of agrarian reform in India—its early postindependence promise, its good intentions, its partial and muffled successes, and its misdirections and frustrations. This major effort of the World Bank, he thought, provided unusual occasion and opportunity to influence the government to make a mighty effort to set things right—hence his lengthy closing remarks under the headings "Suggestions for Action," "Administrative Reform," the "Congress Party and the Peasant," and the paragraphs following. Although the language is measured and objective, one cannot help but recognize that Ladejinsky is making here a supreme and even passionate effort to achieve a breakthrough.

This was not to be. The early successes of the Green Revolution led to a kind of euphoria in the minds of the policymakers and to a period of preoccupation with production goals and efforts; consequently reform objectives receded into the background. Vagaries of the monsoon and production losses brought other periods of preoccupation with import programs, centralized procurement, and price and distribution controls. When interest in reforms was renewed, it took the impracticable form of what Ladejinsky called "The Land Ceiling Flap," rather than concentrating on the security of tenure and rent reforms which in his view were much more practicable roads to reform. Ladejinsky consequently became increasingly stoic over these last years. Convinced that no outside agency or institution could influence significantly the outcome of so fundamentally political an issue in India, he sought publication for his frequent articles in the highly respected and widely read *Economic and Political Weekly* (Bombay), rather than in Western journals, in the hope of influencing the all-important political climate for change within India. But while he never ceased to strive, it can scarcely be said that his hopes remained high or that his satisfactions were many. "The very best thing that happened to me this year or in the past seven years . . .," he wrote in a private letter at the end of 1974, "was two weeks I spent in the Philippines upon Bill Sullivan's [the U.S. ambassador] invitation to tell him if Marcos's agrarian reform is a paper proposition or not. Well, I told him as per the enclosed and minced no words with Marcos in a long session. He seemingly bought what I was selling and I am suddenly something of a 'doer' once again and a 'white-haired boy' all around. . . . For once in many years I have the feeling that a piece of paper can relate to action. A novel experience after seven years of writing papers which have not influenced any policy decisions." The statement was perhaps exaggerated by the euphoria of the moment. By contrast, he clearly denigrated his India contribution. But the statement also conveys a truth. Ladejinsky lived and worked to influence policy decisions and to achieve results, not to produce papers, however elegant and admired. Neither, at his age, was he inclined to take the long view of what effect his work might have in the long run. This only his readers can do.

48. *Agrarian Reform in India*

The introductory notes to Part V have already commented on some aspects of this paper. Further to those remarks, I shall say here only that in my opinion this must rank among the very best of Ladejinsky's written works. It is a great paper. It was originally part of the World Bank's monumental study, "India's Economic Development Effort," dated October 1, 1965 (volume IV, appendix IX).

Introduction

AGRARIAN REFORM IN INDIA is more than a century old. Its history is replete with twists and turns, ranging from the early attempts to protect the cultivator's rights in the land to "the land to the tiller" idea to "cooperativization" of agriculture and back to such fundamentals as security of tenure and controlled rents. The early history can be left to the historians. The stress here is on a few underlying principles of agrarian reform: the conditions which give rise to reform; India's attempts to deal with the problem since independence; the success, failure, and the consequences of these attempts; and the measures that would be required to ensure the implementation of the reforms in the years immediately ahead. Above all, the emphasis is on the fact that the Indian tenurial system is one of the elements that inhibit agricultural productivity. Any consideration of this problem involves such questions as who owns or doesn't own what land, the conditions under which land is held by those who do not own any, how the product is shared, and the effect on investments in agriculture and productivity. Additionally and related to all of this are the important questions of social status and political power in the countryside. For all these reasons, in India where land is a scarce prime resource and pressure on the land continues to mount, the farmers' rights in land are an issue of transcendent importance.

Agrarian reform defined

The term "agrarian reform" is a loose one. Broadly viewed, it is made up of a good many elements. It stands, to begin with, for redistribution of land among the landless, the creation of individual proprietorships, security of tenure, and controlled rents. But no single panacea meets the issue. For even redistribution of the land could not answer all the basic needs of the new owners unless it were accompanied by the necessary means to work and improve the land. With any new economic opportunities and psychological incentives which come with the possession of the land or security of tenure, there must go, hand in hand, a host of other developmental measures. This is illustrated by the underdevelopment of the land of those who have small holdings of their own. Any psychological incentives created by land reform must be put to use for productive purposes by augmenting the resources of the new owners. For this reason agrarian reforms in the sense considered here include also the following elements: favorable financial arrangements for land purchases; better methods of cultivation through technical assistance; land consolidation of fragmented holdings; adequate credit; cooperative marketing facilities; farm price schemes to stimulate agricultural production, and so forth.

Though agrarian reform is a combination of a great many things, not all of them are of equal importance. Important though the other ingredients are, unless those who work the land own it or hold it securely, it could be asserted that all the rest will not have the anticipated results. Hence proprietorship and security of tenure are at the top of the list. It is because not all tenants in India (or elsewhere) are likely to become owners that "security of tenure" becomes a paramount issue. In the sense interpreted here, security of tenure is seen as calling for a set of conditions which tend to stimulate agricultural production, improve the economic conditions of the tenants, and preclude social unrest. The following, then, become the principal elements of "security of tenure": fixity of tenure, fair rentals, and compensation for improvements made by the tenant.

The relationship between the two parties must be in the form of a written agreement, spelling out all the pertinent rights and obligations of the two parties. (This is an important point, for it is probably correct to say that even now the majority of the Indian tenants have no written contracts, carrying on at the whim of the landowners. There is no security of tenure under such an arrangement.) Crucial to the concept of security of tenure is that the period of tenure must be long enough for the tenant to realize the opportunities provided by the leased land, added resources, and reasonable management. The obverse is true of a tenant with a short-term contract, which by itself spells uncertainty and insecurity. A contract can be as long as a lifetime (United Kingdom) or as short as one crop season (India); a five- to six-year period is considered a minimum. The right of renewal is considered to be a necessary part of the contract unless the tenant "abuses" the land and/or unless the owner takes back the land for legitimate self-cultivation. (In India, as we shall point out elsewhere, "self-cultivation" is often a misnomer.) Part of the arrangement of a fairly long lease is that the tenant is entitled to ample notice in advance of the expiration of the contract.

We indicated above that the tenant must be compensated for improvements made with his own resources. Failing that, the tenant will have no incentive to improve the land if the ultimate beneficiary of the unexhausted value of the improvement is the owner or the new tenant. The sharing of the cost of inputs by owner and tenant is therefore thought to be a vital part of a sound tenancy agreement. (The practice of this kind of mutual investment is not, however, prevalent in India.) The resources of the tenant are minute at best, and it would be idle to expect of him to use a volume and quality of inputs commensurate with good farm practices—unless the landlord contributes no less than his share of all the input requisites. It is thought desirable, therefore, to stipulate in the contract that the variable costs of inputs will be shared in proportion to the share of crops received. Therein lie two advantages: an incentive to the tenant to invest and an incentive to the owner to extend his interest in the land beyond mere rent collecting.

Perhaps the most important element in a tenurial arrangement is the amount of rental and how it is paid. It needs no arguing that the level of rents can make the difference between a tenant's incentive to invest or not invest in agricultural production, just as it can in the final analysis determine his standard of living and his ability to climb or not to climb the "agricultural ladder." (With respect to India, rentals before and since the reforms were and are admittedly unduly high, above the level justified by the other factors mentioned in this and the preceding paragraphs. The Indian reformers have recognized this and it explains why the prescribed rents, as distinguished from those actually paid, range in most cases from a high of one-third of the crop to a low of one-sixth.) How the rent is paid is also a matter of considerable importance to the tenant. Crop sharing is seen as making economic sense only where there is a danger of recurring crop failures. Barring such exceptions, it may be said that part of security of tenure should be a fixed, monetary rent. In the long run this would be an incentive to produce more, retaining the increment of additional input in whatever form. Finally, secure tenure is envisaged as providing a system of arbitration to adjust the differences arising in landlord-tenant relationship. Such differences, whether relating to rents, sharing of inputs, crop failures, and so forth, are inevitable; and proper arbitration procedures are required as a basic part of any tenancy improvement program. Taken all together, these are the main elements which serve to create "security of tenure."

Obstacles

Agrarian reforms are difficult to attain. In most cases land redistribution or putting land securely under the control of a nonowner are acts by a government imposing upon the landowners economic and legal terms unpalatable to them. In effect, such policies if carried out are revolutionary acts which pass property and redistribute income, political power, and social status from one group of society to another. To the extent that legislative assemblies are still dominated by land-propertied classes, it is not difficult to see why both the enactment of appropriate legislation and its enforcement present formidable problems. It may be concluded that land reform has not only powerful economic implications but commences essentially as a political question, running head-on into a fundamental conflict of interests between the "haves" and the "have-nots."

The Indian setting

The conflict of interest takes place in a setting of which India is a classic example. It may be summed up as follows: many people on too little land; scarce land, yet concentrated in relatively few hands; negligible capital invested in each unit of land and generally underemployed land; inadequate irrigation facilities and dependence upon mercy of nature; low yields but high rents; poor farmers but expensive farms; small holdings getting smaller under the rising pressure of population with no alternative occupations; inadequate tools, indebtedness and usury, malnutrition and illiteracy; and absence of the agricultural ladder. Probably three-fifths of the cultivating families have little or no margin for innovation and risk taking, and it explains the prevalence of subsistence farming with its lack of dynamics or regenerative capacity. This setting has undergone some changes since independence, but it still holds true to a large degree even after the reforms came into being.

Tenancy as an institution has had a bad name in underdeveloped countries, but it can be and is a sound economic system. Numerous examples can be cited where a cultivator prefers the tenant status, investing his capital in basic inputs of productivity rather than in the purchase of land. In such cases tenancy arrangements have created conditions mutually satisfactory to the tenant and owner. Before the reforms this was not true of India where the bargaining power between landlord and tenant was altogether on the side of the former. That tenancy conditions in India were not mutually satisfactory was alluded to in the preceding paragraph, but more specifically the following may be added. Whether rents were paid in cash or in kind, they exceeded more than 50 percent of the crop; and rents as high as 60 to 70 percent of the crop were no rarities.[1] In most cases the landlord paid only the land revenue, the tenant shouldering all the costs of production. There were also illegal contributions, too many to list, sanctioned by custom so as to remain in the good graces of the landlord and thereby insure the renewal of the lease. In the zamindari areas (about which more later), sub-infeudation or successive grades of intermediaries or tenure holders between the landlord and the actual cultivator also tended to separate the latter from a considerable share of his meager share of the crop.[2] The landlord's participation in the agricultural process was more often than not merely that of a collector of rents and whatever else he chose to impose upon the tenant. Leases were mostly short-time affairs, oral; but, even if written, they did not provide the peasant with security of tenure. Indebtedness on usurious terms was the rule, and neither the cultivator nor the land could prosper. In such a scheme of things, there was no place either for peasant initiative or savings to innovate and improve the land.

The "have-nots" in India are made up of "pure" tenants, tenants with some land of their own, and a vast army of landless agricultural laborers. Just how many there are of each will

1. To cite a few examples: In the Punjab before independence in the prevailing share-cropping areas, the landlord's share could go as high as 80 percent of the crop (H. D. Malaviya, *Land Reforms in India*, p. 164). The Tanjore Tenants and Pannaiyals (agricultural laborer) Ordinance of August 23, 1952, provided for a reduction of rentals by lowering the landlord's share to 60 percent of the crop; prior to that the tenants' share ranged from 15 to 33 percent of the crop; in Hyderabad it was two-thirds to the landlord and one-third to the tenant.

2. Malaviya, *Land Reforms in India*, pp. 103–05.

be discussed in the next paragraph. What should be noted here is the unequal distribution of land ownership, area cultivated, and size of holdings. Thus, 53 percent of the rural households of India own 16 percent of the land, and at the other end of the scale 7 percent of the households own 52 percent of the land. Twenty-two percent of the households own no land at all. The data on area operated indicate the same disparity: 70 percent of the households operate 16 percent of the land and, at the other extreme, 9 percent operate 51 percent of the area.[3] The third criterion also reveals great disparities. Seventy-eight percent of the holdings are up to ten acres, but they cover only 33 percent of the cultivated acreage; at the other extreme, 6 percent of the holdings account for 35 percent of the area. Since most of the big owners find it more profitable to lease out a great deal of their land rather than to cultivate it personally, the institution of tenancy is inevitable.

Accurate data on the number of tenants and the tenancy acreage are difficult to find, and data in general vary from source to source. According to the census of 1961,[4] in the country as a whole the percentage of pure tenants in terms of households and acreage held was 8 and 4 percent, respectively; the respective figures for mixed tenancy were 15 and 18. There is reason to believe that both categories are greatly underestimated. The estimates may not always take into account the common practice of oral agreements; thus tenanted land appears as owner cultivated; the definitions of "tenancy," "ownership," and "cultivation" are not always accurate; some land actually held by tenants is probably included by the respondents as area under "personal cultivation"; there is a rise in unrecorded sharecropping type of tenancy; important, too, is that since the onset of the reforms both landlord and tenant have been reluctant to speak frankly to casual inquirers—such as census takers, among others—about their tenurial arrangements. But even if the census figures are accepted at their face value, in about half of the states of India the land cultivated under pure and mixed tenancy[5] ranges from 31 percent in Bihar to 67 percent in Kerala. According to the Planning Commission, in the late 1950s about 20 percent of the cultivated land was under various forms of tenancy and sharecropping arrangements.

A writer on the subject interpreting data of the National Sample Survey of 1954–55 concluded that "it is clear that the actual proportion of tenant-cultivated area is much larger than any available statistics indicate. If only *all* kinds of tenants were given the ownership of the land cultivated by them with their own hand . . . perhaps half or more of the cultivated land of India could be affected by this process alone."[6] In view of the developments in this field under the reforms, this estimate is undoubtedly on the high side. While nobody seems to be certain about the extent of tenancy in India at the present time, we are inclined to the view expressed by A. M. Khusro. He wrote that "recent researches permit the statement that while the total acreage under open tenancy in India might be about 12 percent of all acreage, the acreage under camouflaged sharecropping tenures of an exploitative type is likely to be at least as much as that under open tenancy."[7] On this basis, about 85 million acres of the total cultivated land of India is tenant cultivated. But whatever the figure, there has been a decline of tenancy in India since the reforms. How much of it (in peasant proprietorship areas) can be traced directly to the positive and negative effects of the reforms no one can say. But as will be pointed out elsewhere, the reduction in tenancy was only partly a result of the shift of tenants to an ownership base. The evidence points to the fact that only a small if undetermined number made the grade and acquired some of the "de-tenanted" land; the majority remained on

3. Planning Commission, "Reports of the Committees of the Panel on Land Reforms" (1959), p. 3, table 4.

4. P. Sharma, *A Study of the Structural and Tenurial Aspects of Rural Economy in the Light of 1961 Census*, table 7.

5. Mixed tenancy includes in part land also owned.

6. Raj Krishna, *Some Aspects of Land Reform and Economic Development in Land Tenure, Industrialization and Social Stability* (The Marquette University Press, 1961), p. 228.

7. A. M. Khusro, "Rural Development," (mimeographed) p. 12.

the land as tenants or lost what little hold they had on the land.

India is a big country with a great variety of rental conditions; but, as we shall see, they have not responded to the provisions of the reform laws. One can argue that, though fixed rents (so much produce or so many rupees per acre) remained unchanged, increased yields and higher prices have succeeded where the laws have failed.[8] As against this, there is sharecropping with its annual and mostly upward changes and what might be called the fixed "moving" rent: As yields increase, the landlord raises the rent accordingly. When these are juxtaposed against Dr. Khusro's argument, one sees no significant change in the rent burden, exceptions notwithstanding. Overall, as a producer of the crop the cultivator continues to share the output in a manner that adds only little to his income, his standard of living, or investment capacity. His rentals are still nearer to 50 percent of the crop, and whatever production inputs he applies are often—though not always—at his expense. The problem is made worse since the majority of tenants cultivate only an acre or two.

The consequences in relation to agricultural production have been described on numerous occasions, all adding up to the tenants' limited capacity to innovate and improve the land. Perhaps the most recent study on the subject deals with one of the most tenant-ridden districts of Madras. We have reference to *Effects of Tenure on Use of Improved Production Practices on Paddy in Thanjavur (Tanjore) District* (February 1964), prepared by the Ford Foundation's Intensive Agricultural District Program (IADP). The burden of the findings is that, while the improved packages of practices could indeed raise the tenant's productivity, his obligations to the landlord are such that it neither pays him nor has he the means to apply the full package of practices. Since he must, perforce, use only a limited part of the available inputs, his increase in yields is correspondingly lower and so is his net gain. Under the circumstances, his burden of risk taking as a function of innovation becomes heavy enough to preclude him from becoming a full participant of the program. We should note that the study assumes that the tenants pay the prescribed rent of 40 percent of the crop. In practice, the rent is closer to 60 than to 40 percent.

Fragmentation and consolidation[9]

At first glance, fragmentation and land consolidation lie outside immediate land reform considerations. Yet, where fragmentation affects so adversely all the cultivators of India, including tenants, it becomes a "legitimate" subject of agrarian reform in its broad sense. Most operating holdings in India are small,[10] but the difficulty of working them is greatly accentuated by their fragmentation to a point that, in the expression of a farmer, "the earth is crumbling under our feet." Fragmentation refers to land scattered throughout the village area in plots separated by land in the possession of others, and it must not be confused with excessive subdivision. The former connotes holdings which, irrespective of their size, are broken up into scattered fragments; the latter, holdings that repeated partition has made dangerously small. It is not often that one finds a holding that is not split up into at least three or four plots. On an all-national basis, the average holding has approximately six plots of 1.1 acres each. In individual cases fragmentation can assume a grotesque form; this was underscored by that mine of information, the Report of the Royal Commission on Agriculture in India. The commission noted that:

> In extreme cases the result is ludicrous: in Ratnagiri, for instance, the size of individual plots is sometimes as small as 1/160th of an acre . . . In the Punjab, fields have been found over a mile long and but a few yards wide, while areas have been brought to notice where fragmentation has been carried so far as effectively to prevent all attempts at cultivation.[11]

8. Ibid., p. 11.

9. The information contained in this section is derived mainly from Sundar Singh, "Consolidation of Holdings," (Planning Commission, 1957).

10. About 63 percent of the holdings are less than 5 acres each, and about 40 percent of these are less than 2.5 acres each.

11. *Report of the Royal Commission on Agriculture in India* (London, 1928), p. 134.

The origin of fragmentation cannot be attributed directly to India's inheritance laws. According to the latter, the land is equally divided among the heirs, usually the sons. In theory, then, if the inherited land consists of three fields to be divided among three heirs, each receives one field. The practice is quite different, each getting one-third of each field. This is done to insure to each one a share in each kind of land, since very often the quality of the land is not uniform. The result is that successive generations descending from a common ancestor inherit not only smaller shares of his land but inherit that land broken up into smaller and smaller scattered plots. Fragmentation, therefore, is not a result of the inheritance laws as such but of the method by which the property is divided among the heirs.

For the country's agricultural economy, whether owner- or tenant-operated, fragmentation is an unmitigated evil for which no advantages can be claimed. Time is wasted and extra expense involved in moving workers, animals, seeds, and fertilizer implements to and from farmstead or from one field to another. Supervision is made more difficult, depredation of animals and birds is harder to control; expenses on water supply, buildings, threshing floors, and so forth, are often much greater; comprehensive irrigation and drainage facilities or other measures of improvement become extremely difficult. The sum total of these disadvantages seriously impede agricultural progress and act as a deterrent to full utilization of the land.

India's answer to fragmentation is consolidation. The term stands for amalgamation and redistribution of the fragmented land so as to reduce the number of plots in the holdings, thus making them more compact. Consolidation aims, therefore, at giving every right-holder a compact area equivalent in value to what he held before in scattered plots. It neither aims at nor results in creating economic holdings where they did not exist before. Consolidation is a difficult process, particularly when the stress is not merely on regrouping of scattered plots into compact ones but also as a means to better land use through more efficient water use, better cropping patterns, and so on. Consolidation is made up of a number of phases, most of them presenting formidable problems. The so-called "preliminary steps" are indeed difficult to take, since they involve securing an agreement among the villagers to undertake consolidation. Voluntary agreements are not easy to secure; what is involved is the conviction on the part of the farmer that in the process of consolidation (correction of records, land measurements, land classification, land valuation, and reassignment of the consolidated land) his interest will be protected. The touchiest of questions is land valuation. While it is a truism to say that plots of land vary greatly in the quality of soil, facilities for irrigation, productivity, and distance from the village, it is a Herculean task to evolve methods of valuation by which different classes of land can be reduced to a comparable basis. Unless valuation of land satisfies the community, consolidation cannot succeed. On the other hand, if correctly valued, the possibilities of disputes and discontent are reduced to a minimum.

Consolidation in India, begun in the middle of the nineteenth century, was very slowly proceeding on a voluntary basis and all villagers had to agree to the scheme. Compulsion was gradually introduced so that consolidation became obligatory for a village if a certain percentage of landowners holding a certain percentage of area had agreed to consolidate. More recently, virtually every state, through legislative enactments, has the authority to undertake consolidation where it deems necessary. This factor and the keen interest displayed by the Planning Commission have brought about considerable progress. Detailed and exact data on consolidation are not readily available, but whatever is available seems to justify the above claim. By the end of the second plan, about 30 million acres had been consolidated. The target for the third plan was 31 million; of this, 15 million acres had been consolidated during the first two years, and 8 million acres were to have been consolidated in 1962–63.[12] We do not know how much of India's 320 million acres of cultivated land is in need of consolidation; but, whatever it is, considering the extremely difficult technical, economic, and hu-

12. Planning Commission, "The Third Plan Mid-Term Appraisal," (November 1963), p. 101.

man problems which accompany consolidation, the figures cited represent a very considerable achievement. Punjab is the star performer, accounting for almost half of the total, with Uttar Pradesh and Madhya Pradesh following in order of importance. Our impression is that the cost of consolidation is not excessive and has never been a serious problem. Neither in talks with the farmers nor in the literature on the subject has this subject been raised as a limiting factor. Moreover, though in theory the cultivator is responsible for the cost, a portion of it is actually met by the state governments. And it is interesting to note that the land revenue is not raised following consolidation, the theory being—and this is only a theory—that the value of the consolidated land has not undergone any significant change.

What happens to a village after consolidation can be seen at a glance from two maps [not included here] which show the layout of the same average village of 974 acres before and after consolidation. In the words of the commentator:

> Even the first look at these maps will show the tremendous change that has been brought about as a result of consolidation. The previous fields, some minute, others large, and all of irregular and haphazard shapes, have been completely replaced by rectangular fields of a uniform size. Similarly, roads have been straightened and properly laid out wherever needed and regular paths have been provided to the holdings of individual owners.[13]

Not in all instances was it possible to convert the scattered plots into single blocks, but the total number of plots in the village has been reduced from 1,398 to 426, or an average of more than 2 acres per plot as against 0.7 of an acre per plot before consolidation. The advantages of consolidation along such lines as operational and administrative efficiency, extension of cultivation, increased productivity, and social gains do not call for much comment. One socioeconomic gain, however, should be noted. Indian villages abound with costly—often bloody—and long drawn-out disputes about boundaries, right of passage for men and beasts, and water courses. Through consolidation much of this has been greatly reduced.

Considering that the average holding in India is still made up of six scattered plots, much remains to be done. Much, too, needs to be done to preclude further fragmentation so as to avoid wherever possible the very complicated consolidation task. What is called for, therefore, is preventive action. States have recognized this by attempting to regulate partitions, transfers, and leases of land fragmented or in danger of being fragmented. In Bombay, for example, the principal legislative measure is the "standard area," or a plot of a size below which profitable cultivation is not possible. For dry crops, the standard area, depending upon the region, runs from 1 to 4 acres; rice, from 0.7 to 1 acre, and grasslands, from 2 to 6 acres. A plot smaller than the standard area is declared a "fragment" and can be transferred only to a landholder of a contiguous piece of land. It cannot be partitioned. A number of other provisions define the eligibility priorities of persons who are to use the fragments. One might note that the priority is given to the tenant of the holding or his heirs. The state of Bihar seeks to prevent the transfer of a fragment to persons other than a co-sharer or one with contiguous land. Yet, this measure is of dubious validity because partition of a fragment is allowed. The one useful provision is that a consolidated holding cannot be transferred in a way that might create fragments. With variations, most other states have enacted somewhat similar legislation of a preventive kind.

Enforcement, as usual, varies from state to state. The measures can be enforced and they are in some states, particularly Bombay. Naturally, there are many evasions caused by administrative, economic, and social problems. When carefully explained, the farmer can readily see the advantages of consolidation or those of preventive action but it may not suffice under certain circumstances. Despite the regulations, an owner of a compact field may have a compelling need to sell off a piece of land informally. There is always a buyer for it because, with holdings small and alternative occupations at a premium, a fragment of land assumes great importance. Consolidated land

13. Ibid.

can be and is occasionally held by co-shares. This in turn may lead to informal partition of the land, which is administratively difficult to enforce. Thus, even wisely designed measures to create compact holdings may run into purely individual problems of the character just mentioned. It is obvious that both consolidation and preventive action is essentially a never-ending process of careful adjustment and change in tune with conditions peculiar to the land arrangement, extent of the acreage, and customs prevailing in a village community. But in the main it may be said that the village has come to recognize the beneficence of these schemes.

Indian Reforms since Independence

Zamindari abolition

Agrarian reform in India falls into two broad categories: (a) zamindari abolition and (b) changes in the tenurial system in the rayotwari or owner proprietorship areas. The zamindari system was a by-product of the British rule and covered about 40 percent of India. Under it, a zamindar, or an intermediary between the British administration and the actual cultivator, was given the right to collect a fixed revenue on behalf of the administration. In return, he was not only permitted to keep one-tenth[14] of the revenue but he was also recognized as the proprietor of the revenue-bearing land. At independence this system covered about half of the cultivated land in former British India and princely states. In time as the population pressure on the land kept rising, the system created some of the worst abuses that can be perpetrated upon a peasantry, including a long chain of noncultivating sublessees, all getting a share of an inflated rent from the same piece of land and the same cultivator.

The following case shows the system's disincentive character and how badly it affected agricultural production. Between the zamindar of a certain estate of 2,000 acres and 360 cultivating tenants, there were four groups of intermediaries, each with its own distinctive nomenclature. The principal zamindar received 8.3 percent of the total rent paid by the tenant. The four individuals of the group directly under the zamindar received a total of 12.5 percent; the 20, 80, and 160 individuals of the three subsequent groups received 20.9, 25.0, and 33.3 percent, respectively. In rupee terms, this meant that, though for revenue purposes, the estate was assessed at 200 rupees, the principal zamindar and the 264 other intermediaries extracted from the tenants 4,800 rupees.[15] To this should be added a long list of illegal dues imposed by the higher groups, everybody protecting his position in relation to the group above him. These tiers of interest were often many times four, but whether four or *eighteen* as the author cites, they all formed "a Jacob's ladder of which each rung is occupied not by an angel but by a tenure-holder, and the topmost by the proprietor."[16] They were all middlemen, rent collectors, in no way interested in agricultural progress, draining the land of its resources. On the other hand, so little was left for the actual cultivators and so insecure was their position that they had neither the incentive nor the means to improve the land and raise agricultural productivity.

The British did not plan it that way, and between 1859 and 1935 they introduced more than a dozen tenancy acts to rid the system of its grosser abuses and absurdities by giving the tenants a measure of protection. They succeeded in some states (Bengal) and failed in many more, always working within the zamindari system, trying to preserve rather than abolish it. It remained for the Indian government after independence to do away with the system altogether as the first order of agrarian reform. Within six years (1951–56) and despite the opposition and administrative problems, the zamindari tenures were virtually abolished. The abolition of the system is important for what it has achieved and also in relation to the slower and less successful progress in dealing with tenancy problems in nonzamindari areas. The zamindari system was the

14. In the temporary settled areas (Uttar Pradesh and Madhya Pradesh), the zamindars retained a much larger proportion.

15. R. Mukerjee, *Land Problems in India*, p. 111.

16. Ibid., p. 98.

weakest enemy to attack because it was imposed by a foreign power which handed out property rights to which neither the British nor the recipients had any claim. Thus, abolition of the system became one of the symbols of freedom from the British rule; and it is not surprising that, as that rule went, the zamindari went along with it.

What of the consequences of the abolition? Some students of the reform minimize or deny altogether its economic benefits while crediting it with social significance. This much can be said. Tenures have been simplified; in the Uttar Pradesh, for instance, out of the chaos of forty types of tenures only three types have emerged. The tenants pay land revenue directly to the government. But since the revenue is equivalent to the rent they used to pay the zamindars, the reform is much criticized.[17] The tenants, the arguments run, have exchanged one landlord for another, the state; and the state benefits financially from the arrangement. The latter is true; in the typical case of Uttar Pradesh, the annual land revenue of 20 crores is considerably above what the state used to collect from the principal intermediaries. The rationale behind this is twofold: (a) Land revenue had not changed in years and is, in fact, low when compared to the rise in prices during the same period; and (b) the government uses part of the revenue as compensation for the intermediaries[18]—a basic provision of the reform—and for developmental purposes.

The critics do not give sufficient weight to these facts: Many of the tenants are securely on the land; they are no longer burdened by a host of illegal exactions, and some have acquired ownership from the state for a moderate purchase price. To be sure, considering the difference in conditions between states and within states, not all have benefited equally; and not all of the estimated 20 million cultivators affected have received permanent, heritable, and transferable rights without strings attached. But overall the effort was undeniably the first and major step—the clearing of the ground toward a reconstruction of Indian agriculture. The abolition provided many of the cultivators with incentive, freedom from fear, freedom from being ordered around, and an enhanced political and social standing which they did not enjoy before. At the same time, recalling our experience in the typical ex-zamindar Aligarh district, one need not tire repeating that small owners or tenants with occupancy rights paying low rentals (land revenue in this case) still meet with difficulties searching for the means to invest and produce more and better crops.

Tenancy reforms

Getting rid of the zamindari system did not put an end to tenancy in India. Even in the ex-zamindari areas, the "home farms" retained by the former middlemen continued to be operated by tenants. Above all, there were millions of tenants in the "rayotwari" areas where owner proprietorship had long predominated. Prior to independence numerous attempts had been made to relieve the conditions of the tenants in the rayotwari areas through some form of land occupancy rights. In the main, however, the practical effects of the measures were to safeguard the rights of the owners while denying occupancy rights to many cultivators. There were numerous exceptions to this, but on the eve of the first five-year plan the great majority of the tenants did not enjoy security of tenure.

In its first five-year plan, begun in 1951, the Planning Commission laid down the general policy for new legislative enactments on tenancy. The goals of that policy centered on the following: (a) security of tenure subject to the right of an owner to resume a limited acreage for self-cultivation, (b) reduction of rents, (c) conversion of nonresumable land under tenancies into ownership, and (d) ceiling on land ownership so that the excess above the ceiling might form a pool of land for redistribution among tenants and farmhands. In addi-

17. Daniel Thorner, *The Agrarian Prospect in India*; and Walter C. Neale, *Economic Change in Rural India: Land Tenure and Reform in Uttar Pradesh 1800–1955*.

18. Compensation is based on the net income of an intermediary at the time of the acquisition of rights by the state. The multiple is higher in cases of lower income groups. In the Uttar Pradesh, compensation varies from 8 to 20 times the net income and in Bihar from 3 to 20 times. In some states compensation is paid in cash, in others in bonds. The total compensation to all the intermediaries is estimated at Rs641 crores, of which Rs225 crores has been paid off. Planning Commission, "Progress of Land Reform" (1963), p. 4, table I.

tion, there were a number of more specific guidelines: The resumed acreage had to be limited only to bona fide cultivators; eviction of tenants had to be avoided wherever possible; rents were not to exceed one-fifth or one-fourth of the crop; and land purchase prices had to be so fixed as not to exceed a new owner's capacity to meet them.

Agrarian reform, like agricultural policy in general, is constitutionally a function of the states; and the Planning Commission's reform policies or guidance are not binding upon the states. However, under the continuous prodding of the Planning Commission, the states have enacted a voluminous body of legislation, each state acting in its own way. Hence there is a great variety in the content of the reforms and manner of enforcement, although the stated principal goals are seemingly the same. The question is: How successful or unsuccessful have these enactments been after nearly fifteen years of effort?

The answer is that, in the main, the reforms have been in serious trouble. This is not to say that significantly useful changes have not taken place. Along with the zamindars, the jagirdars (counterparts of the former in Rajasthan, Hyderabad, and so forth) have gone; and even some of the big landlords of the south are no longer what they used to be. Many tenants acquired lands following the zamindari abolition, and some have acquired *some* land and *some* security of tenure under the tenancy reforms we are about to discuss. But as against all of this there are the delays in enacting reform laws, faulty content of the laws, half-hearted enforcement, and the widespread evictions of tenants traceable to the legislation itself. To this should be added the prevalent failure of rent provisions and the wholesale evasion of the ceiling provisions upon which the new ownership was to rest. Reform was in the air before the enactment of the laws; and this, combined with the long delays in passing a law, made it possible for the landlords to reduce the number of claimants for rights in the land by evicting tenants or shifting them to the status of farmhands. The same held true after the laws went into effect. This doesn't mean that all tenants were thus treated, but there is enough evidence that for large numbers the reforms were far from a blessing.

The eviction movement has been widespread, but data on how many tenants have been affected are not available; what is available relates to the early 1950s and to a few states only. Nevertheless, the practices described below are not only typical of the states mentioned; they illuminate the eviction problem in the countryside as a whole. For the former states of Bombay and Hyderabad, the Planning Commission's Panel on Land Reform[19] gives the following information: in Bombay, between 1948 and 1951, the number of "protected" tenants[20] declined from 1.7 million to 1.3 million, or by 20 percent; in Hyderabad, between 1951 and 1955, the number declined by 57 percent and the area held by them by 59 percent. In a study of the reform movement at Hyderabad, the best of its kind in India to date, we find that of every 100 protected tenants in 1951, by 1954 only 45 percent still held the same status; 12 percent purchased their land, and 2.4 percent had been legally evicted; 22 percent had been illegally evicted, and 17 percent had "voluntarily surrendered" their claims to the land. The author states: "The so-called voluntary surrenders are very often a subtle form of illegal evictions and only a proportion of these surrenders are genuine," and "The implementation of tenancy legislation is a function of the degree of consciousness among the tenantry." The author concludes his findings in these words:

> Surrenders sometimes are genuinely and purely voluntary because many a tenant is completely reconciled to the idea that the landlord, after all, has every right to have his land back, notwithstanding the law . . . Sometimes there have been offers of money by the landlords in return for voluntary surrender of land . . . All too often . . . the landlord is in a stronger position vis-à-vis the tenant and has a greater capacity to rally official (village) support . . . a tenant has often thought it fit that he should surrender the land to maintain good relations, particularly if by this action he can make

19. "Reports of the Committee of the Panel on Land Reforms" (1956), p. 36.

20. Tenants who are supposed to be protected by law against eviction.

> certain that he could continue his lease on another part of the land . . . Many a landlord has often taken advantage of the ignorance of his protected tenant and has asked him to leave on the ground that the period of lease was over! Notwithstanding the fact that a protected tenancy, by definition, is one which is free from the question of time duration, many a tenant has left the land lest he should be involved in litigation with the landlord. Cases have been noticed where tenants have had to leave due to their inability to pay enhanced rents—a device used by many landlords to eject them, notwithstanding the law . . . in plenty of cases force has been used for evictions.[21]

The emphasis in the above is largely on poor or total lack of enforcement, the superior position of the landowner, and the inferior social and economic status of the tenant in the village. Additionally, the laws are very complex and the "bulk of the peasantry find it difficult to understand them. Moreover, after the enactment of a law, it was generally left to the tenants and landlords to take advantage of the provisions of the new legislations and no organized effort was made to make the tenants understand the law and to ensure that they take advantage of it."[22] There is, however, another side to the story, namely, the faulty content of the legislation itself and of some of the questionable underlying principles upon which the reforms are based. The most glaring manifestation of this is the right of the landlord to resume tenanted land for what is euphemistically called "personal cultivation."

On the face of it, the right to resume land for personal cultivation was a sensible policy toward what the Planning Commission called "small and middle" owners. The commission went to great pains defining and redefining the meaning of "personal cultivation," the conditions under which it should apply, setting limits to resumption, and all with an eye to satisfying or reconciling the interests of the owners and the tenants. The provisions on resumption, which in effect meant ejection of tenants from a certain amount of land, varied from state to state. A few examples will suffice: In Assam an owner can resume 33⅓ acres and in Punjab, 30 standard acres,[23] subject to the minimum area to be left with the tenant. In Andhra Pradesh the landlord can resume the entire area. In West Bengal a landlord who held 10 acres or less was permitted to take back the entire area; if he held more than 10 acres, he could resume 10 acres or two-thirds of the area owned, whichever was greater, subject to a maximum of 25 acres. On the other hand, in Uttar Pradesh the landlord couldn't resume any land.

The minute this right became part of law, it opened a Pandora's box of all sorts of problems. In most cases the legislation favored the owners, but even so they went beyond the permissible. This was not difficult to do with policing of the provisions at a minimum. More than that, there was a built-in contradiction between the right of resumption on the one hand and security of tenure on the other. This is so particularly in the case of small and middle landlords because there just is not enough land to meet the goals of both recommendations. It applies to big landlords as well because they can resume any parcel or parcels of land cultivated by a tenant. The Planning

21. A. M. Khusro, *Economic and Social Effects of Jagirdari Abolition and Land Reforms in Hyderabad*, sponsored by the Research Programmes Committee of the Planning Commission of India (Hyderabad: Osmania University, 1958), p. 169.

22. "Reports of the Panel," p. 37.

23. A "standard acre" is the measure against which comparable values can be assigned to different classes of land. This was first evolved in Punjab in the attempt to give uniformity to the amount of land distributed among refugees from Pakistan after the partition of India. For example, if an average acre of irrigated or generally good land yields 20 bushels and an unirrigated and generally poorer land yields only 10 bushels, this meant in practice that a settler was entitled to an acre of irrigated land or two acres of unirrigated. In instances where given land yields more than the standard acre, that land gets a higher valuation than the standard acre. The criteria in some states is not necessarily yield of principal crops; it can be the rate of assessment, the rent rate, or the market value of the land. Currently, the equivalent of standard acre in terms of an ordinary acre in a number of states is as follows: Andhra Pradesh, 2.32; Kerala, depending upon the region, 1.92 to 1.61; Gujarat, 1.58 to 1.23; and Madhya Pradesh, 2.38 to 1.33. Planning Commission, "Progress of Land Reform" (1963), pp. 120–23.

Commission's Panel on Land Reform gives an excellent example of legal loopholes and its adverse effect on the position of the tenants under the law. The panel states:

> Though a restriction was placed in many states on the extent of land which a landlord may resume by ejecting tenants, no provision was made for the demarcation of the resumable area as distinct from the non-resumable area. Thus, though the landlord's right of resumption was limited in extent, he was able to exercise an undue influence over all the tenants which added to his bargaining power and rendered the law ineffective. He could even extort money by threatening to resume land.[24]

We conclude that, while the principle of self-cultivation of a limited area was well-intentioned on the part of the Planning Commission and its refinement in the second plan much to the point, in the context of the real distribution of power—economic, organizational, and political—between the classes affected, it resulted in a chaotic situation of which the landlords have taken full advantage. Despite the examples cited above, data on ejections and of those who remained securely on the land with occupancy rights are practically nonexistent. But one thing can be said with certainty: The right of resumption for "personal cultivation" has weakened the main goal of security of tenure and, doing that, has made the enforcement of reduced rents impossible. The latter is not easy to enforce even given security of tenure; the demand for a piece of land in India is keenly competitive with land and alternative occupations scarce. With much of security of tenure in the rayotwari areas in jeopardy, it is altogether a hopeless task. Resumption of land for personal cultivation served to accentuate the inherent difficulties of that problem. The recommendations of the Planning Commission were not without influence upon the official fixation of rents in the states, although in Andhra Pradesh, Kashmir, and for sharecroppers in West Bengal, the rent may go as high as 50 percent of the crop, Bihar 35 percent, and Madras 40 percent.[25] In other states (Gujarat, Maharashtra, and Rajasthan) the prescribed rent is as low as one-sixth of the gross product, while in still others it accords with plan recommendations. However, whatever the prescribed rents, for reasons just mentioned, in most cases they are more nearly in accord with the customary rents of the pre-reform days.

Converting tenants to owners

The tenancy reforms have done rather poorly in shifting tenants to an ownership base. Some tenants have purchased land, and it would be extremely useful to know just how much land had changed hands, changes directly traced to the land purchase reform provisions. (Surely just as important would be a census on land ownership and tenure in which the states would give most careful consideration to the evasions that have already taken place under the land reforms.) But no data are available to give even an approximately correct idea as to how much land was acquired. But even if available, they would indicate that relatively little land changed hands from landlord to tenant. Recent observation in the field and the official testimony make this clear. The "Review of the First Five-Year Plan" (p. 322) has this to say: "Judged by the amount of land purchased by tenants and the numbers of tenants who have been able to acquire ownership rights, the steps (for making tenants owners) can be said to have produced no significant results." The second plan adds more of the same: "Progress in this direction has been slow." This is partly because the land ceilings, about which more later, are inoperative in India and partly because the resumption scheme limits drastically a tenant's right to purchase a particular piece of land. The reason is simple: A landlord can render that right ineffective by saying that he would resume that particular plot of land for personal cultivation. But there are other and more vital reasons which underlie the weakness of the entire purchase idea as incorporated in the legislation.

24. "Reports of the Panel," p. 37.

25. Forty or 50 percent of the crop is rental for a single crop; if the land is double cropped, the rental in most cases is less than doubled.

From the point of view of the tenant's right of land purchase, the reforms have three serious shortcomings: The price of land is often much too high for the tenant to brave it; the installment payments are spaced within a too limited period of time; and the transaction itself is essentially a matter of landlord-tenant bargaining, which is to say that the outcome is heavily weighted in favor of the landlord. The purchase price has been determined in the following ways: as a multiple of land revenue, occasionally with a wide range, such as 15 to 20 times in Assam, 20 to 200 times in the former state of Bombay; as a multiple of rent; as a portion of market value—three-fourths of market value in Punjab; and as "fixed" value, as in Bihar, with a wide range of from Rs30 to Rs1,050 per acre. But whatever the method of official prescriptions and even if they favor the tenants, realities observed in the field are quite different. The role of agrarian land tribunals, the supposed watchdogs of land transactions, is not nearly as important as might have been anticipated. A landlord is not really compelled by law to sell lands; and, when he does wish to sell, he is seldom constrained by the official provisions. This is the limiting factor so far as the tenant is concerned because, installments aside, few tenants in India can take on such a burden of financial responsibility. It should be added here that three states, Uttar Pradesh, Kerala, and Mysore, have enacted legislation that gives them the right to resume land directly for reallocation to the tenants. Only Uttar Pradesh exercised that right; in Kerala and Mysore that right has not been so far exercised.[26]

In India as in other underdeveloped (and developed) and densely populated areas, land prices are high, bearing no relationship to its productive value. Unrealistically low assessment of agricultural land and comparatively low land revenue are the main reasons for the high market value, particularly in areas with high population pressures on the land. Added to this are such factors as social and political prestige and influence gained from the possession of land as well as a favored tax position as compared with that of alternative investments in commerce or industry. These are facts that must be faced by reformers in approaching the question of land pricing so as not to impose a heavy burden upon the would-be new owners.

Viewed in this light and assuming that the state governments are bent on promoting land ownership among tenants, the price fixed must be considerably below the market price. Land purchase in India under reform conditions cannot be an ordinary real estate transaction where seller, broker, and buyer meet in a free market. If it were and if tenants were able to pay the going price, there would be no need for this part of the reform. Since this is not the case, the price fixed by a government must be an arbitrary one, the degree of its arbitrariness depending upon how a reformer answers this question: "For whose benefit is the measure designed?" Important, too, from the point of view of enforcement of the reform is that the purchase transaction must not be left in the hands of landlord and tenant. The government must purchase the land from the landlord and, in turn, resell it to the tenant. The reason for this is obvious: If left to the landlord and tenant, even a fixed price favoring the tenant is not likely to prevail. Enough has been said about the dominating role of the Indian landlord to labor the point. These are not theoretical musings. Pre-Second World War reform attempts in Japan foundered on just this point. The lessons of that experience eventually led to fixed prices favorable to the tenants as well as to direct government intervention in the buying and selling of land. In India, on the other hand, the legislative enactments are markedly at fault on these all-important issues. Co-joined with indifferent enforcement and the failure of the ceiling measures, the results are as they are: Little land has shifted from hands of the landlords to those of the tenants, or, in the words of the draft outline of the third plan, "action in this direction has been inadequate in most areas."

Ceilings

One of the basic reform objectives was to reduce the number of tenants and increase the number of new owner farmers. Its main tool is the well-tried and tested land "ceiling," a

26. "Progress of Land Reform," pp. 57–58.

device which permits a landlord to retain a certain amount of land (ceiling), the remainder or excess being earmarked for redistribution among the landless. It is the size of the ceiling that determines how far-reaching the program might be, always with the proviso that the ceiling is enforced.

The extreme maldistribution of land in India with nearly a quarter of the rural households owning no land at all and another one-fifth owning less than an acre each provides ample social and economic reasons for the use of the ceiling as a means of redressing this imbalance. The ceiling question gave rise to more debate and arguments than any other reform issue, for it did touch on the raw nerve of tampering with private property rights. But regardless of the arguments for or against ceilings, the Planning Commission laid down the basic policy, concluding that "We are, therefore, in favor of the principle that there should be an upper limit to the amount of land that an individual may hold.[27] The second plan went a step further, concentrating on a number of questions of execution and policy. These questions were:

1. to what land should ceilings apply;
2. the levels at which the ceiling may generally be fixed;
3. what exemptions should be made;
4. steps necessary to prevent *mala fide* transfers;
5. the rate of compensation; and
6. redistribution of land, among those with primary claims.[28]

It would take one too far afield to touch even in general terms upon all or any of the questions just mentioned. What should be said is that in a *formal* way the states followed the recommendation of the Planning Commission, each state determining the ceilings, compensation, exemptions, priority of land allotments, and also taking note of illegal transfers before legislative enactments and how to deal with such transfers.

At one time, a rough estimate attributed to the Planning Commission indicated a possible surplus of about 37 million acres. According to one assessment, "this would be 90 percent of the area required to give any/every landless family a minimum basic holding or 42 percent of the area required to increase sub-basic holdings to basic holdings; or 29 percent of the area required for both these purposes. The basic holding was assumed to be 2.5 acres for one state, 5 acres for seven states, and 10 acres for ten states."[29] Nothing came of it in practice, for the problems associated with legislating and implementing ceilings are indeed most formidable even in conditions of good administration, sound and speedy legislative enactments, and a minimum of opposition. Such conditions, however, are not prevalent in India. Some land vested in state governments was distributed. Not much land was taken from the private owners who did have land in excess of the fixed ceilings. With becoming candor, the Planning Commission in the third five-year plan noted that "on the whole, it would be correct to say that in recent years transfers (illegal) of land have tended to defeat the aims of the legislation for ceilings and to reduce its impact on rural economy." For the moment—and for the foreseeable future—the question of how much land might have been available for redistribution is academic. Holdings have been divided up among members of families so that on the face of it they are *under* the ceiling. Field observations reveal that many didn't bother even with such precautionary steps. There was no reason why they should have. The legislative enactment did not contain the "teeth" to prevent such transfers, as was notably the case in Japan and Taiwan. Enforcement was not a problem; really, there was little to enforce.

Why reforms are subverted

We have noted the reasons why it is so difficult to carry out land reform programs in most countries. It cannot be stressed strongly enough why in India a similar attempt is incomparably more difficult than in most other countries. And yet this is not a justification

27. *First Five-Year Plan*, p. 188.
28. *Second Five-Year Plan*, p. 194.
29. Krishna, *Some Aspects of Land Reform*, p. 229.

for vague and complicated measures generously seeded with loopholes and for the entire array of evasions which separate so many tenants—literally and figuratively—from the land, reduce their production incentives, affect adversely overall production, and on the social and economic scale keep them near the bottom of the village hierarchy.

We have already pointed out that the political milieu of India with its antireformism shared by landowners and many politicians alike has much to do with this state of affairs. It will be pointed out in another section, "The latest phase," that the nature of the village itself contributes to the same end. That part of these notes also ventures to suggest certain corrective actions. But at this juncture we wish to underscore the proposition that the peasants, too, have been unwitting allies of all those who have done so much to subvert some of the most important segments of the reforms.

It is a singular experience for a close observer or student of land reforms in other countries to find many tenants in India surrendering land to owners even though they have a right to remain on it undisturbed. It is equally anomalous to find tenants surrendering their all-important "protected" status in exchange for the uncertainty of finding another piece of land or becoming another member of the rural proletariat. The answers to this puzzle are many and closely intertwined, but one of them stands out: lack of political and social consciousness. Where this is the case, the tenant sacrifices his position of security and status. The obverse is just as true and such cases are not lacking. In trying to single out the basic reason explaining the lack of will to stand up for their rights, one concurs with Dr. Khusro's major finding that, although the peasants hungered for the land or to remain on the land securely on reasonable terms, they neither fought nor agitated for their economic "liberation" in post-independence days. The reforms in India are not a result of popular demand but rather the brainchild of the intellectuals of the Congress Party. This is not a criticism of external intervention; on the contrary, such interveners are indispensable catalysts of a reform movement. Where the criticism comes in is that they stopped short of going one step further—of seeing to it that the peasants shared in the process, which would have been one of the important ways of enhancing their political and social consciousness. And failure to stir them up turned the would-be participants into mere onlookers and, for a good many tenants, with results to match.

Position of landless and farm laborers

The reforms in India aimed at the improvement of the condition not only of the tenant but of the agricultural laborer as well. These workers form the group whose major source of income is agricultural wages, the group which is at the very bottom of the social and economic structure of the village. The increase in population has borne particularly harshly on this section of the population because of the heavy underemployment from which it already suffered. The sheer number of agricultural laborers and the continuous rise in their numbers, more recently augmented by dispossessed tenants, makes any ameliorative effort very difficult to carry out. In 1956–57 out of a total of 66 million rural households, agricultural labor households were estimated at 16.3 million or nearly 25 percent of the total. Of those, 57 percent were landless, while 43 percent held small pieces of land, either owned or leased. Nearly 75 percent of the workers are casual laborers, depending upon a daily wage; the remainder are so-called "attached" laborers, the attachment arrangement ranging from one month to one year. The total number of days of employment of a male laborer is 221, of which 21 days are in nonagricultural activities. Harvesting being the busiest season, it accounts for 25 percent of the man-days worked; plowing and transplanting account for 14 and 7 percent of total man-days, respectively. The earnings and way of life of this village group which consists of from 80 to 90 million persons was pithily summed up as follows: "The agricultural labor households cannot generally make both ends meet and are thus in chronic debt."[30]

These are some of the basic facts; and it is obvious from them that agricultural labor poses grave economic, political, and social problems. The first, second, and third plans have all ad-

30. *Agricultural Labor in India,* vol. I, pp. 43, 46–47.

dressed themselves to the problem of improving its lot. The first plan included proposals for its settlement on land "and protection against ejectment from homesteads." As to the first, "not much progress was made in schemes for land settlement;"[31] as to the ejectment from homesteads, this is a recurring theme in the subsequent plans. Considerable sums of money were allocated for a variety of welfare services for the benefit of this group, but the question of settlement on the land was and is of continuous concern to the Planning Commission. Since above-ceiling land is not available, culturable wasteland and land gathered under the Bhoodan movement through the efforts of Vinoba Bhave have been brought into play. It is not clear just how much of this poor land was distributed nor among how many agricultural laborers. But the amount of land and the number of settled laborers could not have been very significant, judging by the effort provided for in the third plan. Thus, "It is envisioned that during the Third Plan efforts would be made to settle 7 lakh (700,000) families of landless agricultural laborers on about 5 million acres . . ."[32] It is fair to conclude that, though some of these workers may have gained a new footing on some kind of land, the vast majority of this group still carries on as of old. Moreover, since the group grows in size by from 2 to 2.5 million yearly, the immediate concern is not so much betterment as warding off the worsening of the existing conditions.

What of the prospects in the years ahead? Aside from India's long-range and basic problem of the high rate of population growth, the answer to the question depends upon two shorter range factors: the extent of agricultural development as a major source of village employment and the extent of the demand for rural labor to be generated by the country's industrial and commercial expansion. With respect to the first, the answer depends in part on whether agricultural development will be employment oriented or labor displacing through the utilization of modern equipment. As regards the second, the answer is not at all reassuring for years to come.

31. *Third Five-Year Plan*, p. 375.

32. "Progress of Land Reform," p. 26.

With respect to the first factor, it is significant that mechanization or rationalization of agriculture is beginning to take place and is likely to increase with the passage of time. In the short-run, therefore, a reduction rather than an increase in labor input is to be expected. Despite the rise in the rural population, the demand for labor in the harvest season forces wages up sufficiently to justify the introduction of labor-saving machinery. And sometimes not only in consequence of higher wages during the harvesting season; in the Punjab, for example, new farm equipment is often labor saving, that is, intended to displace hand labor. Mechanization, the main purpose of which is to increase agricultural productivity, may have similar results if India's experience is like that of other countries; and a reduction in the demand for the presently already unemployed or underemployed agricultural workers is probably inevitable. The government of Punjab is helping the process along by subsidizing the purchase of farm equipment to the tune of 50 percent of the sales price. This raises a question of policy: Is it wise to take measures leading to the "economization" of labor? It appears to us that, given the current state of all aspects of Indian economic development, any policy which has the effect of increasing the rural labor surplus before urban occupations can provide employment not only for the growing urban labor forces (which it cannot quite do today) but for rural workers also will only add to an unemployment problem already beyond the capacity of the economic and political system to cope with.

Agricultural development of the type indicated cannot result in absorption of any significant volume of rural surplus labor. We shall point out subsequently in some detail that within the foreseeable future industrialization cannot do it either. The answer or partial answer lies in increased employment *within* agriculture and, as a corollary, increased agricultural production. To attain that:

> This requires capital formation and changes in techniques; but not all techniques can be introduced in the context of heavy underemployment. If we selected those techniques which are labor-absorbing rather than labor-displacing in character, and if we used labor

> for capital construction, we meet the requirements of increased productivity as well as of employment, without having to indulge in costly population transfers. Fortunately, agriculture happens to be precisely the sector where, with a small additional amount of fixed capital, a relatively large output can be generated . . . If only organization could be brought to bear upon the situation and surplus labor geared to the activity of further capital formation, we have an elaborate list of capital projects which can be put through.[33]

The list covers such well-known items as irrigation and drainage facilities, single cropping into double cropping, other changes in cropping patterns, reclamation, leveling and bunding of land, more frequent fertilization, weeding, and so forth—all absorbing more labor in the process of more intensive cultivation of the land and all activities the importance of which cannot be overestimated.

The absorption of surplus labor within agriculture will also depend, at least to a degree, upon the opportunities that public works schemes in rural India might readily provide, such as flood control, river embankments, large irrigation and drainage canals, roads, bridges, and so on. These can produce not only employment but essential rural capital, with positive effect upon farm production. In addition, such measures can, as they have done in other countries, stimulate the latent local talent which awaits only the chance and resources to demonstrate what this combination can attain. In the final analysis, what is involved here is the question whether rural India can reach its full growth potential if the wastage of her manpower no longer continues unabated; all this, provided the organizational problems generated by an effort of this kind can be successfully overcome. Another possible source of rural-nonrural employment opportunities lies in what John Lewis calls "The Agro-Industrial Rural-Urban Continuum,"[34] a combination of economic activity that springs in circle from large cities through medium cities to villages.

Whatever the long run holds in store for the great multitudes of the underemployed village proletariat, up to now the improvements have not been significant. The third plan is frank on the subject; it notes that "It is apparent from the experience of the first two Plans that while special schemes in the interest of agricultural laborers are useful, they can touch only the fringe of the problem." The third plan itself provided for programs estimated to cost Rs114 crores for the welfare of the village backward classes, a considerable part of which was to have benefited agricultural workers. Public works projects were to have been the major component of the programs, but in most instances they have barely gotten off the ground. Community development with its agro-industrial plans has done no better. The problem of how to provide an additional 100 days of employment for only 2.5 million persons by the last year of the third plan is still to be resolved. What compounds it is that the persistence of a rigid social structure largely based on caste makes the position of agricultural laborers particularly vulnerable. This is especially true in conditions of lagging rural development. In these circumstances, the writers of the third plan come to this conclusion: "Ultimately, it is by achieving rapid and intensive development in the rural areas as part of the process of economic development for the country as a whole that the landless section of the (village) population can be substantially benefited."[35]

The conclusion of the Planning Commission is quite right, especially when industrialization is considered as a possible means of syphoning off rural unemployment. Dr. Khusro's illuminating *Economic Development with No Population Transfers* throws much light on the subject. In brief, he divides the pool of the available labor supply into two categories: nonagricultural and agricultural. Calculating the *nonagricultural labor supply* based on 2 and 1.75 percent growth of population per annum, he arrives at this conclusion: "If there were no other source of the supply of labor,

33. A. M. Khusro, *Economic Development with No Population Transfers: A Study in Demand for Supply of Labour in the Non-agricultural Sector of the Indian Economy, 1951–76*, Institute of Economic Growth, Delhi, Occasional Paper no. 4 (London: Asia Publishing House, 1962), pp. 41–42.

34. John P. Lewis, *Quiet Crisis in India*, p. 183.

35. *Third Five-Year Plan*, pp. 376, 378.

i.e., if the non-agricultural sector were the only sector in the economy, a state of full employment would be reached by 1975"[36] if the rate of population growth is 2 percent and in 1972 if the rate is 1.75 percent. This would not be bad, but the findings become ominous when the obvious is noted: the agricultural sector with its millions of unemployed and underemployed, confounding matters still further. V. K. R. V. Rao's study on *Population Growth and Its Relation to Employment in India*, referred to by Dr. Khusro, leads to the same conclusions: a speedier reduction of birth rate and agriculture—not industry—as the prime vehicle to absorb its surplus labor.

The lessons to be drawn from all of the above are clear, as the preceding paragraphs demonstrate. But the problem of what to do with the rural underemployed is so crucial that we do not fear laboring the point by way of a summary in Dr. Khusro's words:

> If the demand for labor in the non-agricultural sector is not sufficient to absorb the supply of labor from within that sector for many years to come, then the precepts of the recent theories of development about population transfers have to be modified . . . until a situation of unemployment and underemployment persists in non-agriculture, every man that is transferred from agriculture to non-agriculture merely adds to the pool of unemployed in the latter sector . . . Apart from creating open unemployment and all that it connotes, transfers of population now appear to be positively wasteful, for the cost of such transfers are unnecessarily incurred . . . Two fundamental questions then arise. If the door (in India) is closed, even though temporarily, for any net absorptions of agricultural population in non-agricultural jobs, what is the solution for underemployment in the agricultural sector? Secondly, if population is not transferred and food is not released, how is the non-agricultural sector to get its annual increment of raw material and food for the annually growing labor force of that sector? The answer to the first would seem to be increased employment *within* agriculture rather than in the other sector; the solution to the second is increased productivity of agriculture. Since increased productivity depends on capital formation in agriculture, the obvious solution to the dual problem of low productivity and low employment is to use the underemployed labor in the agricultural sector to build capital *within* that sector, rather than attempt to transfer it to the non-agricultural sector.

This is the crux of the author's summary. It doesn't differ in essentials from the position taken by the Planning Commission, except that Dr. Khusro and, on another occasion, Dr. Rao have furnished chapter and verse to demonstrate that for a long time to come relief to the underemployed agricultural labor can come only from greater development of agriculture itself, the emphasis being on labor-absorbing rather than labor-displacing techniques. It is not at all certain that the new trends in Indian agriculture will be of this character; what little there is points in the opposite direction. Nor is it certain that government policy can effect a change in this trend. If, on the other hand, agricultural development does proceed apace along the line propounded here, the amelioration of the conditions of the farm laborer is possible. But a word of caution is in order. The ameliorative effects can be realized, provided they are not overtaken by further surge in population numbers; and we come back once again to the question of population control. There is another item that must not be overlooked. Assuming the population surge tapers off, the degree to which the condition of farm laborers can be improved will depend upon the loosening of the social structure of the village. If the

36. P. 32 and tables on pp. 37–39. Dr. Khusro adds a footnote to his conclusions, part of which is reproduced here: "Unless the set of assumptions least favorable to our argument is improbable or wrong, the results hold good. And even if they are wrong—i.e., the assumed rates of income growth are too meager or the share of non-agricultural income is too small, or capital-output ratios too low or capital-labor ratios too high, or the assumed rates of population growth are too large—even then, the effect of introducing the correction will be to exhaust the pool of unemployed somewhat sooner. It is also possible that the mistakes in our assumptions will be of a compensating character, the effect of one being countered by that of the other."

line of demarcation between "high" and "low" is not narrowed, the position of agricultural labor will remain precarious. While the effect of more employment cannot be gainsaid, the social status of the "underprivileged" or their "acceptance" as part of the village community is of great importance. So long as betterment continues to be measured along a social ladder which shows little or no resiliency, it is inevitable that even more employment will not preclude the laborer's getting the short end of the sum total of all the measures intended for the reconstruction of the agricultural economy of India.

Cooperative farms

The preceding references to agrarian reform in India relate to individual, private cultivation of land, be it owned or tenanted. But it is of some importance to touch on yet another type of agrarian reform—cooperative joint farming. This should be mentioned not because it has attained a significant degree of acceptance but rather because of (a) the considerable ideological preoccupation in important circles of India with the question of peasant proprietorship versus farm cooperativization and (b) the demonstrated gulf between official intention and the peasants' attitude in regard to that intention.

The rationale behind cooperative farming runs like this: small-scale farming in India is inefficient, whereas large-scale voluntary cooperative farming is efficient. The latter would provide fuller employment for the landless; give substance to the idea of social justice; make better use of available and borrowed resources and create a better technological base; provide a food surplus normally not available from small-scale, individual producers; finally, is a "higher," "socialist" form of social and economic activity. With the country or rather the Congress Party committed, at least in theory, to a socialistic welfare state, the last reason is not among the least important ones propounded in favor of cooperative farming. As to the main features of a cooperative farm, they are as follows: the pooling of the land and centralized management, the retention of proprietory rights in the land, ownership dividends in addition to remuneration for work done, and heavy financial assistance to insure the success of such farms. Clearly, this is not cooperation for specific proposals such as tube wells or land consolidation; it is presumably cooperation for an integrated farm enterprise.

There has been no lack of arguments to demonstrate the fallacy of this type of agrarian reform.[37] It is not our intention to add to the debate but rather to state the official policy, what it led to, why so few become "cooperative" farmers, and what of the future of this type of farming.

While in agriculture the plans concentrated mainly on the reforms already discussed, cooperative farming was not neglected either. The first plan approached the matter rather gently, suggesting that "Small and medium farms in particular should be encouraged and assisted to group themselves into cooperative farming societies."[38] This expression of hope gave way to a rather bold and ambitious view voiced in the second five-year plan: "The main task during the Second Five-Year Plan is to take such essential steps as will provide sound foundations for the development of cooperative farming so that over a period of 10 years or so a substantial proportion of agricultural lands are cultivated on cooperative lines."[39] Sweeping as this statement was, the Indian National Congress, in its Nagpur resolution (January 1959), went the Planning Commission one better. It affirmed without equivocation that "the future agrarian pattern should be that of cooperative joint farming." Despite this, the third plan deviated from this dictum even if asserting that "it is essential to intensify efforts to develop cooperative farming throughout the country and to realize as speedily as possible the objective in the second plan"; at the same time and perhaps mindful of the fact that not much had been accomplished during the first and second plans, its emphasis is much more cautious and matter of

37. N. G. Ranga and P. R. Paruchuri, *The Peasant and Cooperative Farming*; Raj Krishna, "Socialism and Farming," *Modern Statesman* (June 19, 1957); Raj Krishna and others, *Cooperative Farming—Some Critical Reflections* (New Delhi: All India Congress, 1956).

38. *First Five-Year Plan*, p. 194.

39. *Second Five-Year Plan*, p. 194.

fact, while accepting the recommendation of the Working Group on Cooperative Farming to set up 3,200 cooperative pilot projects, or 10 per district. The plan states: "In the main, cooperative farming has to grow out of the success of the general agricultural effort through the community development movement, the progress of cooperation in credit, marketing, distribution and processing, the growth of rural industry, and the fulfillment of the objectives of land reform."[40] So much for official policy. What of the results?

Measured in terms of numbers and size, as of the end of October 1964[41] there was a total of 1,806 "pilot" societies and 1,651 "nonpilot" societies. The first had a total membership of about 32,000 and covered an area of 184,000 acres; the respective figures for the second type were 37,000 and 200,000. The average membership of a pilot society was 18 and of a nonpilot, 22; the average size of the farm was 102 acres as against 120 acres for nonpilot society. The fact that numerically they are far below anything anticipated by the second plan or by the Congress Party is important for what it reveals about the premature expectations and the lack of response on the part of the farmers. But more important than this is what lies behind these statistical bare bones. The official findings and personal interviews with officials of community development are not too illuminating. But the inquirer is not without help; the state governments (and the union government) also farm out all manner of studies to schools of higher learning. Some of the evaluation studies are revealing, and the same may be said of a personal visit to two cooperative farms.

The cooperative farms are indeed voluntary affairs except for certain involuntary influences yet to be touched upon. They are heavily financed, which is the chief attraction to the members of the farms. Very few live up to the principle of "jointness" in work, management, and distribution of income. The character of the membership and why the great majority of the cooperatives came into being shed much light on this point. Contrary to official expectation, in most instances it is not the small or the medium farmers who find their way into these cooperatives. The greater part of the membership is a mixture of a few fairly large owners—often not more than one or two—and a much larger number of agricultural workers, tenants, and ex-tenants who are invited to join for two reasons: (a) to attain the minimum prescribed membership and (b) to provide the requisite labor force. Some who pooled their land are absentee, and often the resident owners do not participate in cultivation of the land either. The agricultural laborers do the work for a stipulated wage, which doesn't exceed the prevailing village wage. Their advantage lies in a longer period of employment, but they are not really members of cooperative farms; they represent the "bogus" membership.

Only part of the owners' land (40 percent in the case of Madras societies), often the poorer part, is assigned to the cooperative. The few exceptions aside, a landowner organizes a cooperative in order to get the rather substantial financial assistance from the government. In the Punjab, for instance, "the eagerness to possess a tractor seems to be the main motivation force for the organization of a society." And it can be other reasons, such as the more readily available inputs or the paying off of old debts. Another reason why one owner or more organize a cooperative is to evade some of the provisions of the tenancy laws. This picture is relieved by the fact that there is a small number of societies organized by convinced "cooperators," and there are also those organized by ex-servicemen, landless, displaced persons, and so forth, on government-owned land; the problem here is that a good deal of it falls into the category of culturable wasteland.

There are a few genuine cooperative societies managed and "owned" by a landowner[42] in the spirit of true cooperation. But the tone is set by the majority. Therefore, the recurring refrain that "no joint farming was in evidence in many of these societies" is not surprising. Some farmers actually work the land individually within the society. Anticipated land improvements have been spotty; the hopes for

40. *Third Five-Year Plan*, p. 209.

41. *The Indian Express* (December 21, 1964).

42. Daniel Thorner, "Progress for Cooperation in Indian Agriculture" (mimeographed).

significant increases in yields have not been realized; the available evidence for the Punjab, Uttar Pradesh, Andhra Pradesh, and Kerala indicate no superior production compared with that on an individual holding. In fact, a declining trend has been reported in the societies of the two first-mentioned states. True, occasionally there has been an economy in animal power and equipment, but it is not clear with what consequences. Poor management and poor allocation of labor tasks had not been conducive to the best utilization of the available inputs or in the creation of new crop patterns. And so, it is "no wonder cooperative farms in general did not achieve spectacular results in terms of yields per acre." Nor could these farms demonstrate any superiority of the much-argued point of "economy of size." In short, the authors of one of the principal evaluation studies concluded that the societies have not attained the advantages of cooperation. With refreshing candor they arrive at yet another conclusion: "The ruling party which is committed to cooperative farming . . . had not mobilized its cadre at the grass-roots levels to support this movement." This plaint has a familiar ring with respect to the entire agrarian reform issue.

In retrospect, one wonders if the supporters of cooperative farming in India have probed deeply enough into its realities, in the context of the Indian village. Hence the apparent assumption that there is hardly any gap between the verbalization of an idea and the deed itself. One may rightly question whether any degree of exhortation can induce large numbers of farmers to opt for this type of farm organization. Only compulsion imposed by a dictatorial regime might have done it on a large scale. As it is, the farmers of India, individualists, and devotees to their pieces of land and their way of doing things simply took no notice of this would-be movement—despite the tempting prospect of generous financial assistance. The lessons of the current experiment are a very poor augury for any shift in attitude by the nonjoiners.

Community development is presently trying to correct the situation by: (a) making cooperative farming an integral part of the Community Block Development Program, (b) by pouring more financial support into the cooperatives, and (c) by urging a number of states to take "effective steps" to accelerate those programs, which is to say to meet the numerical target of pilot societies as laid down under the third plan. To clinch this emphasis on "targetism" rather than quality of performance, the National Cooperative Farming Advisory Board saw fit to recommend that "the Governments of Madras and Kerala should reconsider their decision not to organize any new societies until the working of the existing societies was improved."[43] If, as it appears, numbers count, not content, then the comments in the preceding paragraph should suffice.

Even for a staunch proponent of individual, private ownership of land individually operated, it is to be regretted that the record of "cooperative joint farming" leaves so much to be desired. There is room in India for this type of farming among the landless and the owners of small holdings with next to no other resources to sustain them. Given state reserve land for the first group plus careful technical guidance, financial support, and emphasis on performance rather than numerical targets, the chances are that the results of this kind of experiment might present a brighter picture. The landless in particular might be coming in in much greater numbers as members of their own farm rather than as the coerced ones, for reasons which have little to do with "jointness" or "cooperation." If the political leadership is indeed so convinced of the promise held out by "cooperative joint farming" that it is willing to allocate for this purpose scarce resources, let it be tried among those who have farming skills but no land of their own and among those who have little land but no resources to go with it. While this would not advance the country's agricultural problems nearer to a solution, it might ease the lot of some few with little or no stake in the agricultural scheme of the village; and this is important. Finally, having delineated the scope of cooperative farming, the confusion aroused by the Magpur resolution about cooperativization on the one hand and all the other basic reforms on the other could be very usefully laid to rest.

43. *The Indian Express* (December 21, 1964).

The latest phase

The year 1964 has witnessed an unusual spurt of land reform activity on the part of the Planning Commission and the National Development Council. The cause behind it was the critical report on tenurial conditions in some of the food "package" districts, prepared for the Planning Commission.[44] More specifically, the National Development Council, in its meeting in late December 1963, concluded that "the speedy execution of the program of land reforms was vital for increasing agricultural production and strengthening the rural economy and called upon the state governments to complete implementation of the land reform programs before the end of the third plan." In a meeting of the Land Reforms Implementation Committee of the National Development Council held in June 1964, it was stated that the "loopholes in the legislation rendered implementation ineffective. As a result, provisions for security of tenure and rent were not enforced. This leads to problems of production. The tenants, for lack of security, cannot get necessary supplies and credits; and, in view of high rents they have to pay, they cannot utilize such facilities even if they were available." Two items stand out in these deliberations: (a) the recognition that the problem will not just fade away and (b) the tone of urgency that somehow reform matters must be dealt with if the tenants are to get deeply involved in raising agricultural production on a large acreage of the total cultivated area.

The immediately significant result of the deliberations was the appointment of two teams of officials from the Planning Commission for an on-the-spot reexamination of the agrarian reform problem. Twelve states were covered and reports for the following nine states had been prepared as of the end of 1964: Bihar, Guajarat, Madhya Pradesh, Madras, Maharashtra, Mysore, Punjab, Rajasthan, and Uttar Pradesh. The reports put the record straight with no attempt to soothe, whitewash, or gloss over India's agrarian reform doldrums. Achievements are not overlooked, but the emphasis is on the shortcomings. Nor are the reports short of specific recommendations as to how to raise the quality of legislative enactment and implementation procedures. Whether this latest effort will help to redress the balance at this date is another matter, but we believe that this kind of official enlightenment is a precondition of an attempt to improve on past performance.

The reports reveal once again the success in abolishing intermediaries, progress in land purchases here and there, as well as the slow progress and failures of other aspects of the reform movement; the great variety of tenures and their extreme complexity which would be a source of wonderment, perplexity, and envy even to a hardbitten Philadelphia lawyer; the timidity of the tenants to stand up for their rights, born of fear of the landlords; the absence of a protecting hand from many state legislatures and the character of the village society which doesn't always breed independence of thought and action where and when most needed; the insufficiency of financial and technical assistance if the psychological and economic incentives of a measure of new ownership, of security of tenure, and reduced rents are to be given full play; the omissions and commissions in the legislative enactments which, by themselves, have tended to inhibit implementation; the failure in most cases to inform the intended beneficiaries as to the ABC's of the enactments; the temporary character of some of the laws (Madras); the wide time gap between the enactment of new laws and the initiation of implementation, notably in Bihar with its twelve-year gap; the lack of technical preparation prior to or following the enactment of a piece of reform legislation; and, finally, the successful opposition to the reforms by those bent on thwarting them. The recommendations are pointed and practical in the sense that they can be carried out—provided the states have the effective will to apply themselves to the task with more vigor than in the past and, by way of a comment, provided the peasants are participants in a movement affecting them profoundly.

A number of quotations picked at random throw additional light on the current state of land reform in India and why this issue cannot be neglected. This is so unless one is to assume that: (a) poverty-stricken tenants or poor

44. Wolf Ladejinsky, "Tenurial Conditions and the Package Program."

farmers in general make good producers and that (b) social and political stability can be maintained in rural India for years to come in conditions already noted. In Bihar, for example, on the subject of tenancy, we read the following:

> The tenancies in Bihar are still regulated under the Tenancy Act of 1885 with some modifications. Special provisions for conferment of security of tenure and ownership on tenants have been made in the Ceiling Act, but this provision will apply only to the tenants of owners who are subject to the ceiling. As the number of such owners is not likely to be significant, not many tenants are likely to benefit from the provisions of the Ceiling Act. Practically all the leases in the area are oral and according to the law; therefore, the tenants should have security of tenure. However, it is generally admitted that this law is completely ineffective. A tenant holding land from a raiyat acquires occupancy rights on twelve years continuous possession. Few have acquired the right of occupancy so far.

While the share rent cannot exceed one-fourth of the produce of the land, "in actual practice, the rent payable by the tenant is usually half of the gross produce; and, in some cases where it is fixed in terms of a given quantity of produce or its value, the rent goes up to even about 65 percent of the gross produce." Adding to this the tendency of a landlord to change his tenants with little difficulty and the failure of the ceiling law, the case of Bihar speaks for itself.

The state of Gujarat demonstrated many of the reform handicaps similar to those of Bihar, but of special interest is that part of the legislation which intended to convert tenants into owner cultivators through land purchases. It shows what happens when two unequal parties—landlords and tenants—are presumed to bargain freely in determining the actual price of the land as well as the method of payment. Says the report:

> The implementation of laws relating to tenancy reforms has been slow and unsatisfactory. Further, there is a very large percentage of ineffective purchases which would lead eventually to the ejectment of tenants. The purchase price covered a very wide range from 20 to 200 times the assessment, and the tribunals had to determine the purchase price in different cases. Even though the tenant's capacity to pay is to be taken into account in determining installments, the tribunals have generally fixed the installments only on the basis of the agreements between the tenants and the landlords. In some cases the purchase price is high and the number of installments is so small as to render payment difficult.

Little wonder that "in a very (great) many cases purchases have become ineffective on account of the pressure of the landlords." This in turn leads to the ejectment of tenants altogether. The suggestion of the team that State Mortgage Bank advance purchase loans would not solve the main problem because few tenants, if any, especially in India, are in a position to buy land at a price fixed by the landlord even if the installments were more widely spaced, namely, from the currently prescribed twelve installments in ten years to twenty installments in as many years. In any event, the high price alone couldn't meet the sensible recommendation of the Planning Commission that the total burden of annual payments falling upon the new owner, including installment of the purchase price and the land revenue to the government, should not exceed one-fourth to one-fifth of the gross produce.

Just as Gujarat state represents a case of how *not* to convert tenants into owner cultivators, "Madhya Pradesh presents a classic example of how a good measure of tenancy reform is becoming ineffective due to lack of adequate steps for implementation. The law provides for ownership for all tenants (on that portion of land which the landless cannot resume). In practice, very few have become owners." The law is too complicated—and so are the procedures involving its application—to warrant a detailed comment. But the following quotation from the report should suffice:

> As the right of ownership accrued automatically (to tenants on nonresumable land) it was necessary for the (state) government to initiate steps for effecting mutations in

> favor of the new owners under sections 108 and 110 of the Madhya Pradesh Land Revenue Code. This was not done and it was left to the tenants to make applications for acquisition of ownership. As a result only a few tenants have acquired ownership.

The failure on the part of the state to do its duty cannot be interpreted as other than an antireform action, thus canceling out a major provision intended to benefit the tenants. The failure of the tenants to make the applications is typical of the fear and timidity of an unorganized tenantry to face up to the whims of the landlords on this, as on many other issues, affecting landlord-tenant relationships in India.

In the Punjab where the reforms differ somewhat from those of the other states, the authors conclude: "that the main object of the tenancy reform is to enable a small landowner to have unfettered right over his land, unfettered by any opposing claims of the tenant," and "provisions with regard to maximum rent do not appear to be effective." This doesn't mean that no tenant has acquired proprietory or occupancy right in land; actually, 19,000 tenants out of an estimated 80,000 have acquired such rights in a total of 128,000 acres. Nevertheless, when the realizations are measured against the anticipations, the above conclusions of the investigator of tenancy conditions in the Punjab are warranted.

In the fact of the most bitter opposition to land reform measures in India, Rajasthan did yeoman work in abolishing its powerful intermediaries. Rajasthan also tried to regulate rents, but in this proved much less successful. In fact, Rajasthan exemplified the well-tested proposition that in the absence of security of tenure rents cannot be controlled. "The Rajasthan law seeks to regulate rents (the maximum permissible rent being one-sixth of the gross produce), but it doesn't provide for security of tenure. It is not surprising that, in the absence of any provision for security, the provisions for maximum rent are ineffective. The tenants are in actual fact rack rented . . . paying rent amounting to half of the produce in many cases." Additionally, "Information on some of the important aspects of the implementation of land reforms is either not available or else is so incoherent and inconsistent that little reliance can be placed upon it."

The former Bombay area of western Maharashtra is not without its serious lapses in the application of tenancy provisions. This is true of its principal provisions—conferment of ownership upon tenants in respect of land not subject to resumption. What is obvious is not only failure which the report emphasizes but success as well. By the end of December 1963 out of 1,098,000 cases, purchases became effective in 400,000 cases, while 190,000 tenants couldn't keep up with their payments and turned the land back to the owners. One gleans from the report that one of the causes of the failures is that the prices fixed by the agricultural land tribunals (from a range of 20 to 200 times the assessment) is on the high side. The authors conclude that "the position is disturbing and the object of the land reforms may be defeated if remedial action is not taken." But there is a more vital reason for the failures and the authors recognize it when they say that "if transfer of ownership is placed on a compulsory basis, the landlords would not be in a position to exercise influence on some of the tenants." The importance of the *compulsory* element in land purchasing has been discussed previously and no other comment need be made here. What is surprising and encouraging is that, though the purchase program is very far from completion, 400,000 tenants did acquire 1,054,000 acres of land.

All of the above sums up the current state of India's reforms after more than a decade of legislating and implementations: good in some respects, spotty in others, and altogether poor in many other vital aspects. The most recent effort at fact finding, analysis, and recommendations is one more round on the part of Land Reforms Implementation Committee of the National Development Council to convince the chief ministers of the various states that, for reasons noted before, they must revitalize the reforms and enforce them. Discussions, or rather negotiations, are now on. The pressure upon the chief ministers of the states exerted by the Implementation Committee is seemingly greater than before. But the outcome is in doubt, for only an organized peasantry, conscious of its strength and capable of telling the seekers after their vote that "we support those who support us," could at least in part effect the changes which have so far eluded them. Barring that, one of the important preconditions for improved agricultural production on a large acreage of the cultivated land of India as well as one of the important preconditions for a greater degree of social and political equality will be long in coming.

What Reform Requires

Social character of the village and reform from below

In the preceding paragraph, reference was made to effecting changes "at least in part." The qualifying part of the sentence comes from the notion that, in India, even good reform laws and sympathetic implementors might not quite suffice to carry out the basic tenurial readjustments. The word of caution stems from the character of the Indian village society. The picture is not an encouraging one. To Gandhi the village was an idealized community with eternal verities, the very bedrock of India. But this breaks down upon the evidence of those who cared to look at the village more clinically. To Dr. Ambedkar, the leader of the harijans (untouchables), "What is the village but a sink of localism, a den of ignorance, narrow-mindedness and communalism?"[45] To yet another Indian of vast knowledge and experience, "There are now few values which can be said to be common to the whole (village) community, and certainly there is no common purpose which inspires all sections equally."

The principal features from which the mentioned strictures are derived and of concern to us on grounds of agricultural production and tenurial adjustments may be summarized like this. The touchstones of the village society are certain built-in restrictive and unproductive factors which have a long history of economic, social, and legal relations, peculiarly typical of the Indian village. Part of it is the subject of this paper. But it is more than that. Many peasants have limited economic aspirations and they had come to regard poverty almost as an accepted mode of living. The idea of self-help is not widespread, whereas reliance upon the government to do what they themselves can do is widespread. Whether it be cause or effect, sections of the village suffer from long-standing social and political disabilities in relation to the elite of the village. Relationships, values, and loyalties often transcend economic self-interest and betterment. One of the banes of village life, heavy indebtedness for social purposes (weddings, funerals, and feasts), is just one more facet of it. Despite the new winds which even the village cannot escape, factionalism is rampant and the barriers separating group from group have been only slightly dented. Each group knows only too well its place in society, deeply entrenched as it is in a tradition of inferiority and inequality—conditions accepted on all sides.[46]

Certain religious beliefs, such as the godlike veneration of beasts of burden—though they are mostly underfed and mistreated—have a debilitating effect on agricultural production. There is not much evidence of community goodwill, community loyalty, and of concept of village welfare. "Caste and class distinctions are so acute that the feeling of belonging to the community has not yet come. Self-help, therefore, has amounted to helping certain sections or individuals only."[47] Significant, too, is that those of the village who in recent decades succeeded in acquiring a formal education to qualify for a job in government or in business are lost to the village with hardly a trace, which means the loss of potential new leadership. Finally, since village organization is more social than administrative, villages are hardly

45. Quoted by Hugh Tinker, "The Village in the Framework of Development," in *Administration and Economic Development in India*, Ralph Braibanti and Joseph Spengler, eds. (Durham, N.C.: Duke University Press, 1963), p. 97.

46. Walter C. Neale, *Economic Change in Rural India*, pp. 207–08. *All India Rural Credit Survey, Report of the Committee of Directions*, vol. II, *The General Report* (Bombay, 1954), pp. 277–78. Daniel Thorner, *The Agrarian Prospect in India* (section I, "Agrarian Structure," and particularly p. 11). Kusum Nair, *Blossoms in the Dust* (particularly chapter I).

47. *Kurukshetra* (April 1959), p. 12.

capable of initiating and carrying out needed economic and social changes. And Dr. Hugh Tinker concludes, "so one could go on, demonstrating that village associations are powerful but mainly negative, working against corporate action and initiative."[48]

Though this litany of woes is true in its essentials, this picture is overdrawn in somber colors. It must be tempered by the fact that, while the core of the old society is seemingly intact, there are many deviations from the norm. Many more people, including women, take greater interest in village affairs. Village schools are multiplying rapidly and the children of the "untouchables" do partake in them, suffering no overt discrimination. There is widespread acceptance of modern medicine and a virtual elimination of malaria; very important is the fact that younger people contest for the leadership, though it is still largely in the hands of the greybeards. Whether it be in Punjab in the north or in parts of Andhra Pradesh or still further south in Madras or Kerala, one can find economically developing communities with a spirit of self-confidence formerly lacking. As far back as 1925, one of the great British civil servants of India, Sir Malcolm Darling, in his classic, *The Punjab Peasant in Prosperity and Debt*, noted the change.[49] "Everywhere," he wrote, "the age-long isolation of the village is breaking down and, as intercourse with the outer world is established, a new self-consciousness is dawning and new forces are being released." Anyone who cares to look at the village at close range can see also the much stepped-up momentum of change forty years later, and not only in the Punjab. A number of new developments are in evidence. Universal suffrage has come and is practiced; though this right is not used to full advantage by a goodly number of the village community, the vote is theirs and someday it will be put to better use.

Not all farmers of India are about to turn Calvinist in the sense that the magic of the carrot of "profitability" will quickly undo the heritage of centuries of tradition and the acceptance of the status quo. As against this, it is reasonable to assume that the changes noted and the changes in the claims on the land, buttressed by the application of the *existing* technology, let alone the first outcroppings of a new technology already practiced by some farmers, are bound to induce people to take a few more steps, even if hesitant steps to begin with, in the direction of greater change. One agrees with Darling that "the new light is definitely in conflict with the old," and we believe that this conflict born out of developments already noted will gradually undermine the still-predominating and inhibiting values and attitudes. Under such conditions there would be no place, for example, for the common aberration of "voluntary surrenders" of land, under the reform, *by the tenants in favor of the landlords*.

The new stirrings cannot be denied, and these straws in the wind are encouraging signs. The question is how to speed up the process of change, how to be rid of the "aberrations" which benefit neither the tenant nor the land he cultivates, nor, for that matter, agricultural production as a whole. Walter Neale has a ready answer—effective reforms do not depend upon gifts from on high and they must be fought for by the "lower orders in the hierarchy," taking the responsibilities that go with such effort. He quotes approvingly from Gunnar Myrdal's *Rich Land and Poor*:

> No society has ever substantially reformed itself by a movement from above: by a simple voluntary decision of an upper class, springing from its social conscience, to become equal with the lower classes and to give them free entrance to the class monopolies. Ideals and social conscience do play their very considerable role, which should not be forgotten. But they are weak as self-propelled forces, originating reforms on a large scale—they need the pull of demands being raised and pressed for.
>
> When power has been assembled by those who have grievances, then is the time when ideals and social conscience can become effective.[50]

48. Tinker, "The Village in the Framework of Development," p. 128.

49. (New York and London: Oxford University Press, 1925), p. 263.

50. Gunnar Myrdal, *Rich Land and Poor: The Road to World Prosperity* (New York: Harper and Bros., 1958), p. 71.

Dr. Neale sums it all up in terms of an all-embracing principle in these words: "Land reform does not make new men of peasants—new men make land reforms."[51]

In relation to Indian agrarian reforms as to those of other countries in need of the same, neither Myrdal's nor Neale's prescriptions are wholly valid, though an organized, articulate peasantry bent on satisfying its demand is ineed an essential condition for effecting changes. Denying the government and the elite the role of a catalyst, Myrdal and Neale fly in the face of historical precedent, be it Western or Asian. The French Revolution (and one of its consequences, establishment of peasant proprietorship in France) did not wait upon the masses to make their demands; nor did the English Whigs wait upon their masses to initiate the long series of political and economic reforms beginning in 1832. Perhaps even more to the point and closer to our own times is the case of Japan. Literate Japan was devoid of the constraining forces of a village society à la India; in addition, it had an effective cooperative movement and strong tenant-farmer unions—to mention only these two elements. Yet, it took strong initiative from the top, indigenous and foreign, to come to grips with the agrarian reform issue. With minor variations, the same may be said of Taiwan.

Without equating the two mentioned countries with India, there are, nevertheless, lessons in this for the latter, discussed in subsequent paragraphs. To be sure, the process of change in India is excruciatingly more difficult, since the feature of the poverty-stricken village is, in the main and so far, the absence of a strong tendency to improvement, for reasons already commented upon. One gets the impression that in certain segments of the rural economy of India low output and income tend to perpetuate themselves. Institutions do chain economics to the past and the breaking of the chains is essential to progress. Failing that through internal pressures, there is little chance for agrarian reform and significant and widespread application of new farm practices in general in a country laboring under a system in which a large group of the population is confined by tradition and law to share a meager product.

51. Walter C. Neale, *Economic Change in Rural India* (New Haven: Yale University Press, 1962), p. 258.

Suggestions for action

We come back full circle to where we started: how to break the chains which still tie a *multitude* of farmers to conditions best discarded. One approaches the task with great diffidence. This is particularly true of one who was privileged to take part in the preparation of the agrarian reform, part of the first plan, and who has closely followed its fortunes since then. Moreover, one is aware of the fact that so much of the advice proffered the government of India about what is wrong with this or that facet of its economy sounds like cliche, resembling the endless repetition of an old, familiar, much-played-over record. With these caveats out of the way, the question still remains: What is to be done on the tenurial issue in the time immediately ahead?

In attempting to answer this question, it is well to make clear what is now in India *not* meant by agrarian reform. The problems of land ceiling and, by the same token, of shifting large numbers of tenants to an ownership base are *not* the current issues. This is not the expression of a personal predilection. On the contrary, the failure to enforce land ceilings and institute a working purchase program (from the point of view of the would-be owner) has diverted the agrarian reforms of India from their original course: He who cultivates the land should own the land. But the above-ceiling lands are no more; they have disappeared into the hands of sons and daughters; uncles and aunts; cousins and second cousins; "charitable" institutions; and, hard to believe but true, as in a case known to us in Madras, into a "foundation." Such are the realities and it would be impossible to unscramble what was too blithely and with impunity scrambled up under the very eyes of the state governments. What remains, therefore, of the reform program is its less ambitious though important part: *the enforcement of security of tenure and with it the enforcement of fair rents—and the enforcement of the purchase provisions of nonresumable land.* This, in briefest form, is what the reform is about at the present time.

On the legislative side, this program calls for a number of enactments and amendments. The Planning Commission in its "Memorandum on the Fourth Five-Year Plan," published in October 1964, notes some of the points demanding action. In the words of the Commission they are as follows:

1. replacement of temporary provisions for prevention of ejectment of tenants by comprehensive legislation on security of tenure,
2. restricting the right of resumption for personal cultivation,
3. removal of legal and administrative deficiencies relating to registration of voluntary surrenders of tenancies,
4. conferment of rights of ownership on tenants of nonresumable land, and
5. commutation of rents in kind into cash rents.

To this list a number of items should be added, and item (2) should be drastically revised. It is well known that absentee owners (along with resident owners) and owners whose principal occupation is not agriculture have evicted many tenants. They do not cultivate the land themselves but resort to concealed tenancy arrangements. All such land should be acquired by the government at a fixed price and placed at the disposal of a state land board for resale to the tillers. If a tenant declines to buy the land, the land board may lease it to the tenant at a controlled rent fixed by the state. At the same time, tenants on such lands should be given full occupancy and heritable rights.

Policing of tenancy arrangements has not proved effective either in regard to security of tenure or controlled rents. In view of the social conditions prevailing in the countryside, it is doubtful whether even a perfectly framed law can be enforced; it is recommended, therefore, that the state governments should interpose between the tenant and landowner, be the latter absentee or resident. *The law should be so framed that the tenant pays the controlled rent to the government and the latter reimburses the landowner accordingly.* There is a workable precedent for this approach in the former state of Trivandrum, now part of Kerala. In order to allay in a certain part of the state the rising disputes between the jennis (landowners) and the kudiyans (tenants), the then existing government stepped in and took upon itself the collecting of rent. It paid the landowners the rent after deducting an "administrative fee" of 5 percent. The practice came to an end after the tenants acquired the land upon a reasonable compensation fixed by the government. The moral of this is clear. With the intervention of the government, the rent evasions were eliminated. The resistance of the owners was softened while the bargaining position of the tenants was strengthened. Above all, the right of the tenant to remain on the land so long as he paid the rent became legitimized in practice. Hence the suggestion of adopting the practice of rent collecting by the government in other states. The revenue service might be pressed into this additional job, or the task may be entrusted to a new organization to administer the reforms.

There are many tenants willing to buy what is left of the nonresumable land or any other land that might be available for sale for the benefit of the tenants. Compensation and what form it is to assume enter the picture very importantly. We have stated repeatedly why few, if any, tenants can pay for the land at once (in cash) even if its price is fixed by the government or why they cannot meet yearly installments if payments are too closely spaced or the land is too highly priced. We have also indicated and so have the members of agrarian reform division of the Planning Commission that, unless the landlord is effectively compelled to sell the land subject to purchase, progress will be halting at best. Experience demonstrates that this issue cannot be resolved through voluntary compliance. To this end, *it is recommended that the state governments buy the land directly from the landowners for resale to the tenants.* Many of the landlord-tenant behind-the-scene involvements, which so often prevent a land transaction, would be eliminated if the government intervened in so clear cut a fashion, thereby providing the tenant with a much-needed protective shield. This is the reason why the Japanese and Taiwanese governments placed the entire land purchase scheme on this basis.

If the government adopts this recommendation, how will it pay for such purchases? We may recall that, more than a decade after the zamindari abolition, the government still owed the former intermediaries not quite two-thirds of the agreed-upon purchase value of the land (footnote 21). For reasons explained in the zamindari abolition section of this paper, the same treatment cannot be applied to the landlords considered here; they are not British created, but the "more equal," indigenous brand who must be dealt with with greater circumspection. Whatever the price agreed upon, the state governments will not find it easy to pay promptly and in fairly large sums. They could, of course, dip into the chronically depleted state treasure boxes or resort to the usual payments in long-maturing, low-interest-bearing bonds. We believe that this will not suffice and that the problem can be eased by tapping a financial resource hitherto untouched.

We recommend for the consideration of the Planning Commission *the exchange of stocks of government-owned enterprises in part-payment for land purchased under the reform.* At first glance and in conditions where the government has never practiced such a scheme and owners have never handled a share of stock, this idea may seem farfetched. Yet, Taiwan offers a successful precedent. There, the national government decided to part with some of its public sector industries (cement, paper and pulp, agricultural and forestry development, and industrial mining corporations) to cover 30 percent of the land value purchased by the government from landlords for resale to the tenants. The properties were evaluated, stock issued, and a requisite amount of stock transferred and accepted by the landowners. The two cardinal points of the scheme are that the stock is not overpriced and that the selected industries are well-managed and profitable. In Taiwan the stocks were possibly overpriced, but the industries proved to be extremely profitable; and those who held on to the shares have reaped a return beyond their expectations. It is not without interest that in Taiwan the sellers of land under the reform were much smaller holders than those of India and knew no more about stocks than their Indian counterparts.

Our information about India's large public sector is minimal, but we know that it contains a profitable segment of industries and suggest that the latter might provide a basis for the transfer of shares in payment for purchased land. In relation to India as against the familiar, to us case of Taiwan, we cannot venture outside the general character of the idea. It is in such terms that we aired it before an important official of the Planning Commission. It did not seem forbidding to him and, in fact, he could see its application in other fields, while preserving government control over the public sector. Nothing said here can be interpreted to mean acceptance of the idea. The wide and complex ramifications of the recommendation are recognized and so is the need for a thorough study and analysis of a mass of pertinent information before judgment is rendered. Only the Planning Commission can initiate the appropriate examination along those lines and determine what the public sector could do to further the cause of this aspect of the reform program.

The foundation upon which all other recommendations must rest is the preparation of a basic record of tenancies. Its importance cannot be overestimated. In many states tenancies are on an oral basis, and a tenant cannot assert security of tenure rights unless they are recorded. *Without a written record any and all provisions relating to security of tenure cannot be enforced.* In the Uttar Pradesh a few million records were corrected or newly inscribed in the course of a special drive organized by the state government in connection with the implementation of the Zamindari Abolition and Land Reform Act. The same cannot be said of a sizable part of the country, particularly of Andhra Pradesh, Assam, Bihar, Kerala, Madras, Mysore, and Orissa.[52] Evidently the Uttar Pradesh method, largely based on the determined leadership of Charlan Singh, the then revenue minister, was not or could not be duplicated in many other states. In order to overcome this difficulty, we suggest the following. The patwari[53] cannot carry out so important a task, and here is where the participation of tenants comes into the picture. *A committee should be set up in every village made up of two tenants, one part owner–part tenant, one owner cultivator, and one landlord* to help the

52. "Progress of Land Reform," p. 180.

53. A patwari is the hereditary village official who keeps the land records and the revenue rolls.

patwari and other revenue officials to prepare the record of rights. This suggestion is based on two practical considerations. What the officials don't know or pretend not to know about "who is who" in the village, the three first-mentioned members of the committee do know. The other important consideration is to give the tenants the reality of participation in the implementation of the land reform program.

Deeply involved in every aspect of agrarian reform is the agrarian reform division of the Planning Commission. The one suggestion we offer in this regard is *the urgent need to enlarge its staff in order to expand the scope of its work*. The division is the only one of its kind in the center, for neither the Ministry of Food and Agriculture nor the Ministry of Community Development and Cooperation engages in any agrarian reform work. In size, the staff, including the director, is not a dozen strong. Yet it is called upon to perform a great variety of services ranging from the formulation of major policies to rushing all over the country "putting out fires." From our personal observations, the division is more than overburdened trying to keep up with a whole array of difficult and essential chores connected, especially in recent years, with the implementation of the programs, their evaluation, and the continuous prodding of the states into more vigorous action. The surprising thing is that so few have done so much and so well in difficult circumstances.

While all this is true, it is just as true that the division should probe a good deal more into basic problems if it is to fulfill its unique position as the watchdog of Indian agrarian reforms. We have in mind the need for basic information and analysis that might be of great assistance to the policymaker, administrator, and the student alike. Much is known in a *general* way; much is unknown in a *specific* way.

To cite a few examples, eviction of tenants following the application of the reforms is a common phenomenon, but how many, even if not to the last decimal, remains to be determined. Co-joined with this is what happens to them after eviction. Some tenants have acquired land under the reforms, but the answers to the questions of how many, how much, and in which states are very inadequate. The extent of open tenancy is highly debatable despite the census data. More or less tenanted acres or more or fewer tenants are not academic questions: they have serious bearing on agrarian policy. It is one thing for officials of a well-known state to claim that tenancy is almost nonexistent any longer; it is another matter to find that it is there all the same but in a different and worse guise.

"Underground" tenancy is a post-reform phenomenon and rising fast. There is ample reason for studying it and drawing the necessary policy conclusions. Studies of the effects of tenancy as they relate to landlord, tenant, and land have been made, but they are far from common and are not particularly up-to-date as one might wish. Besides, often the subject enters by indirection, as it were, the emphasis being placed on other aspects of farm management. The economic and social consequences of the reforms implemented and unimplemented are still practically a virgin field; it could stand a good deal of deep plowing and much light would be shed on these issues. The land reform laws are on the statute books of every state, and there is nothing in those laws and their amendments with which at least two individuals of the agrarian reform divisions are not thoroughly conversant. At the same time, a codification and interpretation of the reform laws would be of great value for all concerned with the reform movement in India. The "anti-reform reform legislation" of some states is pregnant with touchy and crucial questions on why reforms in certain states have gone wrong. There is much else in this area of problems that the division could explore on a continuous basis. And there might be also the possibility of enhancing its advisory services to the states in an effort to improve the legislation enactments. A well-staffed special unit, branch, or bureau of the agrarian reform division could and should address itself to issues mentioned and unmentioned as befits the clearing house of reform ideas, policies, and basic information.

Administrative reform

The recent recommendations of the Planning Commission are well-taken; and it is tempting to echo the Planning Commission by saying that they ought to be enforced, as indeed they ought to be. The same should apply to all the recommendations grouped under the heading "Suggestions for Action." But even the best recommendations don't get carried out merely because one wishes them well. To reach their objective they must first of all function within an effective administrative agrarian reform framework. Such a framework India does not have.

There is much poor content in the reform legislation in India, but it can be corrected. The technical problems of administering a program are not insurmountable, and India offers examples of resolving such difficulties. But these are encouraging exceptions which should not blind one to the fact that for all practical purposes India has no administrative mechanism especially designed to implement its reforms. For reasons to be discussed in subsequent paragraphs, India is doing it "on the cheap," relying almost entirely on the existing revenue administration.

The Planning Commission realizes the inadequacy of this instrument. It states rather frankly that "the collection of land revenue and other miscellaneous tasks take so much time of the field staff (of the revenue administration) that they are not able to give adequate attention to the maintenance of land records and the administration of land reform programs." More specifically, in a number of states, "particularly Bihar, Orissa and West Bengal and over large areas of Rajasthan, there was hardly any revenue agency below the district or sub-divisional level. In some other states, such as Assam and parts of Madhya Pradesh, the staff at the village level has been inadequate to discharge the multifarious duties entrusted to it. Although steps have been taken, notably in Bihar, Orissa and West Bengal to build up agencies in the field, much still remains to be done."[54] There are other reasons that explain why the revenue administration is not the vehicle for reform administration.

Consider the administrative structure from the bottom up. At the base stands the patwari, the hereditary village official who keeps the land records and the revenue rolls and is part of the revenue administration. The reform entanglements have added no staff of any kind to assist him. Then comes the revenue inspector, covering from 20 to 40 villages, the subdeputy collectors covering 100 villages, presided over by the deputy collector and collector in charge of an entire district. Only occasionally does one find in the collector's office a special officer for agrarian reform affairs. On the state level there is the revenue minister, and in some states only is there a land reform commissioner assisted by a small staff. By a most generous estimate one might add up 6,000 to 7,000 people not connected with the revenue offices, directly involved in reform administration. But fundamentally agrarian reform is in the hands of the revenue department and the patwaris. And what must be stressed here is not the lack of numerical strength of these administrators but their attitude toward reform.

The patwari is the only official crucial to the reform at the grass roots, but he is a hindrance to rather than a promoter of reform enforcement. Personal observations in other states amply support this comment on the patwari in Uttar Pradesh: "The enforcement of land law in particular faced the other immense difficulty, the character of the patwari or village record keeper."[55] According to a classic description of this ancient village official, he is capable of any form of deceit, corruption, and malpractice.[56] Not all of them are cut of the same cloth, but the author's most charitable comment is: "the methods of dishonesty pursued by the various occupants of this office differ as widely as their habits of thought." With the passage of time his role and attitude have not undergone a significant change. Past and current assessments make this clear. So much for the patwari, the village kingpin and supposed reform implementor.

The other two vital personages are the revenue minister and the collector, for they set the tone, particularly the minister, for the entire revenue establishment. By and large and understandably so, they look upon reform work as

54. "Progress of Land Reforms," pp. 180, 182.

55. Neale, *Economic Changes in Rural India*, p. 201.

56. Sir Cecil Walsh, "Indian Village Crimes," pp. 149–51 (quoted by Neale).

an intrusion into their multifarious activities. That traditionally these activities also involve settlements of rent disputes doesn't detract from the statement. But more germane is the fact that more often than not these individuals, by social caste, are not reform zealots. On the basis of field observations, the same may be said of their principal subordinates. The position taken by many states in view of the reforms serves to compound this negativism.

In fairness, not all patwaris are devoid of all rectitude and not all revenue officers are "negative." After all, the zamindari abolition is a fact and so is the implementation of some of the other new tenancy provisions, spotty though it is. The overriding consideration is that neither the patwaris nor the revenue establishment are instruments for administering so gigantic an enterprise quite alien to their thinking and way of life. We suggest that a *land reform administration* in every state, appropriately staffed by people solely devoting their efforts to the reform work, is a prime need. It will be indicated elsewhere that this is envisaged as only part of the administrative scheme; but for the moment let it be stressed that even this bare minimum doesn't exist. This is the more incongruous when we note that, numerically speaking, community development has 67,000 village level workers as against hardly 6,000 workers devoting all their time to agrarian reform work only.

This contrast is the more striking when we take a leaf out of the Japanese experience. There, the total amount of land transferred from landlord to tenant was no more than 5 to 6 million acres. Yet the enforcement of the program called for nearly 120,000 members of village land commissions made up of tenants, owner cultivators, and landlords as well as upward of 400,000 village volunteers and many thousands of officials and clerks on national, prefectural, and village levels. And this is a country with no patwaris to doctor records and no opposition worth mentioning. The moral of it all is: If India is to proceed with so much that has been left undone, it is indeed time to create an administrative structure for that very purpose. There is a second moral in the Japanese story: There was the *will* in Japan to carry out the enabling legislation, whereas in India and with the exception of the zamindari abolition and only occasionally in tenancy reforms, the attitude has been one of great reluctance.

Congress party and the peasant

One can go on at still greater length pointing out weaknesses and laying down "should be dones" to eliminate them. Instead, we shall rather repeat the one all-embracing soft spot just mentioned: the lack of sufficient *will* to grapple with the problem by those who legislate and those who are charged with the duty to implement reforms. And it is the former rather than the latter with whom the responsibility lies. There is yet another group of people upon whom the primary responsibility devolves. In fact, it predates, if one may put it that way, enactment of legislation and the means of enforcement. This group, too, can affect profoundly both the character of the legislation and enforcement. We speak of the Congress Party.

Not too many years ago it was the leadership of the Congress Party that had the vision, social conscience, and dedication to pioneer the reforms. This fact was of singular importance, for it is political leadership which makes or unmakes reforms. It is political leadership which provides the impetus or lack of impetus which decides between reform and "reform." There is no gainsaying the fact that the economic environment, population pressure on the land, and customary relationships sanctioned by a long history of social and religious traditions exert great influence on what happens to legislation designed to change old institutional molds. But this does not invalidate the main premise—that the content and implementation of agrarian reform are a reflection of a particular political balance of forces in a country. India, of course, is no exception to this; and it is India's political leaders who at one time strove with might and main to give birth to the agrarian reform idea—zamindari and non-zamindari. The party was then the authentic "agrarian reformer." Without going into the reasons why, it must be recorded that the all-important role has been gradually diluted to a point of nonrecognition. It is not in evidence even at ceremonial party gatherings when pro forma tributes are also paid to outworn party slogans. And yet, the party's active allegiance to the reform idea is more important now when early idealism has given way to the painful realities of attempted changes in the old rural order. More concretely, unless the political leaders assume that role once again, the chances of improving the legislative content of the reforms and of implementing them are very poor indeed. By itself, even the best administrative machine cannot meet the task assigned to it. It is understandable why in a review of the *Alliance for Progress*, the writer, noting its slow development, wrote in this wise:

> The execution of agrarian reform, rural community development and an adequately financed and staffed national education program depend ultimately on political leaders . . . and not on experts . . . what the Alliance for Progress needs to be a success in these countries is only the commitment of the political leadership to back up the experts.[57]

Another serious problem related to the one just mentioned is that the people in whose behalf the reforms were designed have never been, as was noted above, a party to the process of reform formulation and enforcement. They are *objects* of the reforms, but never *means* of helping formulate and carry them out. Whether consciously or unconsciously, the Congress Party had never reviewed this distinction with the consideration it deserves. The reliance is upon a reluctant, numerically small bureaucracy to do the job. Reluctant officials and timid farmers are not a good combination to advance the cause of reform. Besides, the exceptions notwithstanding, no bureaucracy, however competent and devoted, as exemplified by the Japanese and Taiwanese cases, can carry out the task without the assistance of the top policymakers and of the peasantry. To illustrate but one point: Even though many Indian villages have records of rights full of holes, the villagers know who is who, who owns what land, who rents from whom and at what rate, and much else. When nudged in that direction, their role is indispensable. There was a time when community development might have played the role of a catalyst in this regard, but it carefully shunned the reform problem for reasons best known only to itself. Nor has there been in recent years any nongovernmental group, in or out of the village, that felt deeply enough about the reform cause to carry the message into countryside. This being the case, it would be appropriate for the Congress Party leadership, which once spark plugged the agrarian reform movement in India, to re-assert its interest in the welfare of its own offspring.

The renewed and active interest can assume a number of forms, reaching out, in the first instance, to the state legislatures and long overdue "tidying up" of enactments. More important than that is the task of forging the missing link—the participation of the tenants in the reform process. This means organizing them by the Congress Party—if need be into "farm lobbies"—so that they may play an active role in all matters relating to agricultural development, including, of course, agrarian reform matters. Reforms in Japan and Taiwan demonstrated that adding this party to the "bargaining table" is the essence of a meaningful reform. Since the initiation of such a movement cannot come from the tenants themselves, it must come from the centers of power just mentioned. This would not constitute an altogether novel experience for the party. The "kisan" (peasant) movement in the zamindari areas in the late nineteenth century was guided by the very people who later on spawned the Congress Party. The latter displayed its great organizational talent in the struggle for independence, in ushering it in during the early perilous years, and in settling down to the tasks of running a new state. It has lacked no talent in soliciting the peasant vote since independence, and it should lack no talent to organize the farmers so that they may articulate and actively advocate their interests. Admittedly, this may not be easy and will take time. The cake of custom is hard to crack, and values and attitudes do not change overnight. But they are not forever. The Communist leadership in Hyderabad in 1948 and the same leadership in Madras witnessed by this observer in 1952 demonstrate that the peasants respond to those who assume the role of their leaders and protectors and act upon it. Under such conditions customary predilections, when pitted against

57. Juan de Onis, *New York Times* (December 15, 1964).

known self-interest, do change even in the Indian village. It is the Congress Party's opportunity to speed up that change.

A move in this direction is a political act, but the entire reform movement in India is not without political considerations, heavily weighted to be sure in favor of the landowners. The union government through the Planning Commission has repeatedly pressed the state governments to get on with their responsibilities. The pressure, however, has fallen short of its mark. The state governments depend upon the union government for nearly 75 percent of their developmental funds, but have the purse strings been used as a "handle" for extracting agrarian reform concessions from some of the recalcitrant state governments? The same political party dominates the center and the states. If the latter continue in their refusal to accept and act on the center's important policy decisions, might it not then be advisable to bring into play the financial resources of the central government? It is recognized that stronger political pressure might provoke political dissensions within the Congress Party as well as give rise to overt discontent in the countryside on the part of the reform opponents. But the alternative to refocusing on the land problem is putting the reform issues, at least in certain states, into a deep freeze to retard agricultural production on 70 or 85 million acres of land and eventually to risk political tensions the country could well do without.

There was a time when, under the care of the Congress Party, agrarian reform was a national issue, in the sense that the principles of tenurial reforms had been accepted as one of its major guiding policy principles. To be against reform was akin to being against independence, for it had been for some years past part of one grand design of the country's development. Is it too late to infuse the party with the old, original meaning, when the welfare of India's submerged was one of its articles of faith? If the voluminous legislative enactments of some of the states give birth to reform measures of questionable value, must not then the leadership of the Congress Party somehow assert itself publicly, reaching out into every nook and corner of those states, carrying the message of people's rights in the land? Would not then the proponents of reform in the countryside take heart and make their contribution to an effective enforcement of what is now known as agrarian reform in India? Would not a good purpose be served if the Congress Party convened and examined the course of the reform movement since independence? It is possible that a deliberation of this kind might provide much of the needed stimulus to push agrarian reform in India off dead center.

Pro-reform and other considerations

The current food difficulties provide a propitious occasion to link them with the reform problem. Greater agricultural productivity and better tenurial arrangements go hand in hand. The discussion of security of tenure, if nothing else, points that way. If one is serious about the first, one must be just as serious about the second. The argument, recently restated by a top official in Punjab, that it is best to leave things as they are so as not to disturb the status quo is neither good argument nor good advice. There is nothing in the agrarian reforms of India as envisaged in this paper that infringes on big landed properties. Most of them are already broken up, worked by tenants, farmhands, or tenants turned farmhands by the very laws which were supposed to prevent much of that. What is at issue is the injection of a touch of dynamism into the use of land by giving the cultivators the incentives that go with security of tenure and reasonable rents as well as such modest purchases of land to which the tenants are entitled under the law. And if, we repeat, there are state governments not prepared to grant these reasonable conditions, at the same time laying aside questions of social justice and equity, then it is up to the centers of power to step into the breach and provide an effective stimulus for the fashioning of the ameliorative measures long overdue.

There is another reason why it is *not* best to leave things alone, as they are, but instead to finish the job expeditiously, salvaging what can be salvaged of the original reform goals. To date, the landowners have done well, while the tenantry has not done well. But it would be correct to say that even the landlord group is not in too happy a frame of mind—their successful evasions notwithstanding. The preoccupation with agrarian reform is bound to continue; there is too much at stake for too many for it to die a natural death. It can't fade away. Deficient though some of the laws are and obstructed though their execution is, their very existence is a promise to the tenant and a threat to the landlord. Even though the cards are still stacked in favor of the landowners, it has been this observer's experience that surely some of them are not of an easy mind—not about the plight of the tenants but about their own future. They know that this is not the last round in their tug of war with the tenants. Some of the provisions most damaging to them (ceilings, occupancy rights, security of tenure, reduced rents, and so forth) are still on the statute books and someday someone may venture to apply them. If the sense of uneasiness is to be allayed on the part of the owners and the sense of despair on the part of the tenants, what better remedy than the enforcement of the minimum program rather than prolongation of the state of uncertainty with all its debilitating effects?

Minimum programs and maximum application

The stress here is on *amelioration* rather than on final solutions, for the minimum program can do no more. This is not to denigrate the beneficence of such measures effectively carried out. It is well to remember that the objects of the ameliorative reforms are small, hand-to-mouth farmers with whom every bit counts. The point can be illustrated simply. At its most rockbottom level—the food level—more or less rent may mean the difference between two meals a day or one meal, a full rice bowl or a half-empty one. It is that basic. And it may mean more, too—just as it can also mean less. It is one thing to be secure in one's tenancy rights; it is altogether a different matter to be at the bottom of the heap as a landless laborer. It is one thing to be a tenant; it is another thing to be an owner or nonresumable land. Granting these most significant distinctions, it is understood that the measure that lies closer to a "solution" is the widespread ownership of the land by those who cultivate it. This has been the guiding principle of the reforms in countries of Asia, in the current reform developments in Iran and in Latin America, and of course in India. But for the many reasons stated, no such drastic changes in the agrarian structure of India will take place in the foreseeable future. One must settle, therefore, for the proverbial "bird in the hand" for the minimum that can be realized as against the maximum which cannot be now attained. Anything that eases the immediate burden of the tenants is an opening wedge of twofold significance: It contributes to the unlimbering of a large part of the agricultural economy and, by the same token, tends to weaken the village setup which smothers initiative in all manner of ways. It assumes, of course, that technical and financial assistance will be there to underwrite the new psychological and other incentives.

Admittedly, past experience has taught that the realization of even the minimum program presents enormous difficulties, especially in India. These stand repeating for their content and for reasons given in the next paragraph. Among the difficulties are the sheer enormity of the task of giving security to millions of tenants on millions of acres of tenanted land; the correction of some of the mischief already committed; administrative decentralization, with each state a law unto itself; the fact that an estimated third of the tenanted land belongs to owners with 5 acres or less; that a quarter of the farm population has hardly any stake in the land at all; the fierce competition for any tillable plot of land on almost any terms; the prevailing lack of peasant initiative and the failure to stimulate it; and the millions of people added annually to the already overcrowded land. This is enough even for some stouthearted, let alone the fainthearted, to seek refuge in "all hope abandon."

These difficulties are restated not only for what they are—and surely not to weaken one's resolve—but rather to call the attention of the Congress Party and of the central government that, if they are to take a more searching look at these problems and to redeem at least part of the pledge to the tenants, it would take *all* instruments of government—due process of law and, when necessary, unceremonious vigor—to deal with them. Above all, it will take *effectively organized will*, without which much else will be written in water. Assuming that this will be the approach, they ought to bring into the open, through all possible media, that anti-reform sentiments are thwarting the expectations aroused by the promised reforms nearly two decades ago. That this is so is an open secret, but it would be useful if they aired them from the housetops. They should proceed with the creation of an administrative organization encompassing all aspects of the agrarian reform programs. They should make the strongest possible bid "to place the peasant," as the late Prime Minister Nehru put it, "in the center of the piece." This is one of the main roads to progress, certainly until the day when an awakened peasantry becomes a source of authority and a mainspring of change.

The Planning Commission is an arm of the union government. As such, it has not failed to try to advance the reforms in the desirable direction. But the commission does not pass laws; only the states have that right, and the commission is not in a position to honor the promissory reform notes issued to the tenantry by the Congress Party before and upon independence. It is for the party which embodies both union and state government to come to the aid of the Planning Commission by matching words with deeds. What is at stake is much more than security of tenure, reasonable rents, and a measure of land ownership—these, of course, but also much else that might generate correctives to social instability and low agricultural productivity. It is recognized that the very attempts of this kind might usher in a disturbing shaking-up and waking-up process. But the risk is worth taking, for the Indian village as presently constituted cannot play anything like the full role in developing its not inconsiderable and still undeveloped resources—even with the help of better and more physical inputs. In the long run the "unsettlement" might prove to be part of India's "silent" revolution, heralded with so much fanfare and so much hope in early days of community development but which died aborning en route from anticipations to realizations.

Closing remarks

It may be presumptuous even for an interested outsider to treat this subject at such length and in such terms. If any excuse need be offered, it is this: It is *not* trite to say that the grave difficulties raised by the reform issues are a challenge to great efforts; failure to bring them to the fore in order to do what *can be done* is to accept the unacceptable, namely, that tens of millions of acres of land will not produce what they are capable of producing and that for millions of cultivators' lives, in Kipling's words, it will continue to be "a long-drawn question between a crop and a crop." All this and much else unspoken but implied fall into the oversentimentalized and dubious village maxim of "let all things old abide." Not everything that is old in Indian villages is worth preserving; surely not at a time when the country must mobilize every resource—including the agrarian reform resource—if it is to grapple successfully with a host of seemingly intractable problems. The agrarian reform is an important part of that resource that must be used to its full capacity. Its implementation and completion should be one of the orders of the day.

Ceilings aside, India provides examples which demonstrate that some aspects of the program can be dealt with. The flaw is that so much of the fulfillment is a partial success only, the path being strewn with many failures. The pervasiveness and stubbornness of the reasons that lie behind them are not minimized, but we are of the opinion that, given the underpinnings already discussed in addition to the indispensable ingredient of *will from the top*, the unfinished agrarian reform business of the possible can be managed even in the excruciatingly difficult Indian conditions. This presupposes no further delay. The long drawer has been in use too long and with lamentable results. Some are apparent to the naked eye, while others, the less apparent, are possibly more dangerous in their political consequences for the future. Taken together, they spell out a slower tempo of rural development, diminishing at the same time the tempo of social change. We agree, therefore, with M. R. Bhide, vice governor of the Reserve Bank of India, that "the (reform) problem is both serious and immediate and unless it is handled firmly and promptly it may well be taken off our hands by persons who are more radical in their thinking and more ruthless in implementing their plans."[58] It is the business of the government of India that that day never comes.

58. From Wolf Ladejinsky's "Tenurial Conditions and the Package Program."

49. *The New Agricultural Strategy and Institutional Factors*

With the advent of the Green Revolution—not yet so termed in this June 1968 paper—Ladejinsky noted the emergence of a "disquieting" emphasis in India on "everything for production," on "biology rather than sociology." Ladejinsky here argues that "leaving out sociology means neglecting such nonmarginal issues as the inadequacies of the agrarian reforms and their adverse economic and social effects, the poor performance of the extension service, the questionable developmental and other activities of the village councils . . . the declining administrative standards, the inadequate performance of the cooperative societies . . . and, finally, the question of subsistence agriculture." These institutional constraints, if uncorrected, would confine the benefits of the new technology to some 10 to 20 percent of the farmers, rather than extending them to most. Because the argument in this paper recapitulates and updates in less space much of the ground covered in other articles (CB–97, 98, and 99), I have elected to present this one.

Introduction

ONE OF THE MOST STRIKING impressions of talks with state officials, officials in Delhi, and with Westerners involved in the agricultural development of India is that everything must center on agricultural production. This is indeed an encouraging sign of the new times, generated by the 1967–68 sharp rise in farm output, attributed in part to the new technology. Needless to say, the writer of this note shares in the hope that at long last an agricultural breakthrough is on the horizon. At the same time, "everything for production" has some disquieting features about it—it leaves out much that should be part of agricultural policy. A seasoned Western agriculturist summed up the new creed of inputs as the "do all" when he stated that what Indian agriculture needs at the moment is "biology rather than sociology."

This is a pregnant statement which, correctly interpreted, stands for the efficacy of agricultural research, massive and coordinated; distribution of inputs concentrated on areas of assured water supply; and, of course, incentive prices. The fly in the ointment is the neglect of "sociology." If by sociology is meant what the armchair social scientist normally pursues, there would be no quarrel with the statement. The difficulty is that in this instance leaving out sociology means neglecting such nonmarginal issues as the inadequacies of the agrarian reforms and their adverse economic and social effects, the poor performance of the extension service, the questionable developmental and other activities of the village councils—the panchayats, the declining administrative standards, the inadequate performance of the cooperative societies (despite their growing role as credit providers and as important marketing agents in a few states), and finally the question of subsistence agriculture.

The improvements in the work of some of the enumerated and neglected parts of the agrarian structure do not involve a choice either of concentrating scarce material resources or spreading them thinly; what is primarily at issue is the better use of political and administrative leadership in order to raise available human resources to higher levels of productivity. Not to grapple with them is to give one pause about the future progress and stability of Indian agriculture. It is quite possible that for a time the disregard of "sociology" will not impede the rise of farm output. This happened in Mexico and it may repeat itself in India. Favored with good monsoons in addition to a sustained new technology, there is no reason why India, like Mexico, should not become self-sufficient in food—and possibly an exporter of rice—in the next few years. The 10 to 20 percent of farmers around whom the new agricultural strategy centers could accomplish that task. But it is just as possible that in the longer run India will also reap a harvest of ever greater polarization between the rich and the poor farmers and with it social and economic discontent that might gravely affect the fragile body politic of the country. It would be foolhardy to assume that rural India is immune, as some contend, to the disruptive pressures of poverty of the many amidst prosperity of the relatively few.

The problems mentioned previously are of long standing, but they have been allowed to come under eclipse to a much greater degree than in the past. The plausible pretext now is that nothing must stand in the way of the current attempts to attain higher rates of production, though they all bear on the same outcome. One's field experience in this regard can be disconcerting.

In the publicized showcase which is Tanjore district of Madras state, officials are ever ready with figures, colored charts, and graphs on some aspects of agricultural production but take no interest in the rather troublesome socioeconomic conditions of the village, which an interested observer can easily detect. And so it is in Andhra Pradesh. There, a most senior official is being very patriotic when he speaks with pride of his five tractors on his broad acres and of his "philanthropy" when he leases an acre or two to a tenant, while in his west Godavari district, another showcase, tenants are dismissed in droves and turned into landless laborers in open contravention of the existing tenure provisions. In the same state, an official in a moment of truth urged this observer not to waste time on official cooperative reports, while confiding to him that if he were "king" he would disestablish most of the 15,000 cooperatives, so badly do they work. Yet the issue is covered over with a wet blanket so far as official action is concerned. In Kosi district of Bihar the new irrigation facilities are turning dry land into gold. But in the same Kosi district 50,000 to 60,000 cases have been wandering from court to court for nearly a decade, all of them concerned with the rights of tenants to the land occupied by them for many years past. Officials make light of the rising tension and occasional "subversive" actions on the ground that this, as much else in Bihar calling for correction, must not interfere with agricultural production, as if its steady progress were only a biological function. The same is true with variations in sleepy and backward Orissa and elsewhere. The general impression is as if the urge for high productivity were antithetical to all the other issues raised. If this is correct, the new farm strategy based on material inputs only may fall short of its potential.

It is axiomatic that, with the availability of irrigation facilities plus credit, fertilizer, better seed, and insecticides, productivity is bound to rise sharply from its present low levels. This on top of well-functioning agrarian institutions would maintain farm productivity at high levels. But extension, cooperatives, agrarian reforms, panchayats, and administrative arrangements do not function well, and some not well at all. To the extent that they are as they are and the new farm strategy accepts them pretty much as the norm, the strategy contains something in the nature of a built-in barrier to the full utilization of productive capacity. This will become evident at later and more sophisticated stages of agricultural development, a point touched upon in subsequent paragraphs.

Another aspect of the same problem applies with much greater force to the largest group of farmers with small holdings. They, unlike the relatively small group of big farm operators, are not direct beneficiaries of the new strategy. Instead of easing their lot through augmented productivity, the performance of extension, cooperatives, and so on often acts upon them as depressants. In conjunction with the small size of their holdings, meager resources, and lowly social position, the outcome often is as if rights, privileges, and opportunities are not meant for them. But it is important to note that from the point of view of productivity there is ample evidence that they could—and some do—equal and exceed yields on large holdings.[1] The transformation from traditional to modern technology on small farms is indeed possible but on the condition that resource assistance, material and human, is measurably increased to fit basic innovation requirements. Failing that, the majority of them are doomed to subsistence farming, while the technological revolution is likely to grow apace on the well-endowed and amply financed farms of the large owners. If, therefore, the small farmers are to make their contribution to productivity, it is not too soon to bring into the open once again the small farm issue, why this class of farmers receives very limited assistance, and to consider anew ways and means of how to shift an increasing number of them to a more productive base. In sum, since in the new strategy there is little room for them, might not measures be taken, particularly with respect to credit for inputs, designed so that they too share in the country's agricultural progress, now so largely geared to the affluent farmers?

Every problem noted above is part of the "institutional" aspect of the agricultural economy. They are all subjects for detailed studies, but this is not contemplated here; the Bell report[2] did that to a very considerable degree four years ago. What is intended rather is to take account of the more recent developments. There will be occasional references to the report as background, but what follows is partly based on field observations and partly on pertinent recent material.

Extension Service

Official view

It is not necessary to extoll here the importance of extension as a means of improving agricultural performance. It is enough to say that in India sixty million families engage in farming and the problem is how to induce them to produce more by weaning them away from those traditional practices which have outlived their usefulness. This is where extension is supposed to come in by demonstrating convincingly how to apply the innovations of agricultural science and technology. The official point of view is that, whatever the difficulties of the recent past, extension is on its way. In the words of the extension commissioner of the Ministry of Food and Agriculture,

> The performance of the Intensive Agricultural District Programme (IADP) . . . has adequately demonstrated the effectiveness of our extension worker provided he is fully supported by a strong administrative set-up, receives adequate technical guidance and assistance from the higher level subject-matter specialists and is able to assure farmers adequate and timely supply of inputs. All these facilities were available in the IADP districts. The results achieved in these areas were extremely encouraging.

This, the argument runs, has led to higher yields, encouraged the government to create the Intensive Agricultural Areas Program in 260 districts, and the expansion "has unmistakably demonstrated the effectiveness of our extension agency."[3]

The village level worker and extension as practiced

Yields have indeed risen in 1967–68 and so has the utilization of new inputs, but the view that the extension service is effective in present circumstances is highly questionable. It is questionable even in IADP districts amply supported by the Ford Foundation, let alone in the districts not so generously financed by the government. It is our contention that in the past agriculture benefited from extension, but its contribution was indirect and limited; the contribution to the new agricultural strategy will

1. "Results of studies made by Farm Management Research Centers in India, analysed recently by Khusro (1964), Krishna (1961) and Long (1961) indicate that gross output per acre averages higher on small farms than on large privately operated farms. Long concludes that gross output per acre was largest on small farms." From K. L. Bachman and R. P. Christensen's chapter on "the Economics of Farm Size," in *Agricultural Development and Economic Growth*, H. M. Southworth and B. F. Johnston, eds. (Ithaca, N.Y.: Cornell University Press, 1967), p. 246.

2. The World Bank's intensive study of "India's Economic Development Effort," October 1965.

3. D. V. Reddy, "Problems of Agricultural Extension and Farm Education," *The Indian Journal of Public Administration* (July–September 1967), p. 583.

be minimal so long as the system continues in its present form. This conclusion rests primarily on the inadequate service of the village level worker (VLW), the central figure of extension, and the great lacuna in agricultural research, without which the VLW has little to contribute.

There are two VLW images—the theoretical one, the paragon of all virtues, and the second one as he is in reality. To the propagandists of extension he is the "friend, philosopher, and guide" of the farmer; "he will go down in history as one of the greatest social inventions of the present era," or, when imagination runs riot, "his sense of duty and service pervades the village in the same way as fragrance in a flower." All of which implies that as a direct link with the farmers the VLW provides them with the kind of services they need. The reality is something else again, for he is not at all what he is painted to be by the image makers of community development. The VLW is relatively young, though many of them have been in service for a decade or so; he is not necessarily of the village; he is in most cases a half-educated young man looking for any job; he is a functionary at the very bottom of the administrative totem pole, working for Rs125 or more per month with little hope of climbing the bureaucratic ladder and supposedly taking care of an area of about 5,000 acres cultivated by 1,000 to 1,200 farmers; above all, his training and activities are of a multipurpose character, a "Mr. Fix It" of sorts, with little knowledge to impart to the farmer or authority born of knowledge to impress him with. There are, of course, exceptions, but the reference is to the majority of the VLW's who number some 70,000. About five years ago an official report dealing, among other things, with the role of the VLW's noted that ". . . at present very few cultivators seek his advice regarding the adoption of improved practices." The situation is worse now, notwithstanding the numerous recommendations to improve the extension service.

On recent trips to the countryside, one of the questions pursued centered on the performance of the VLW. Summing up the impressions of talks with farmers, officials, and extension agents, it can be said that the work of the VLW is not useless but its usefulness is decidedly circumscribed. As in the past, he helps some farmers to prepare the so-called production plan to help them secure a loan from the local cooperative, expedites here and there fertilizer distribution through a cooperative credit line, and puts on a few farm demonstrations. As to his intended primary function as a purveyor of new agricultural techniques, farmers' reactions varied from bemusement to outright statements that the VLW has little to teach them. Another reaction expressed by an official was that the VLW is not serving the farmer or the farm community but serves rather the Block Development Administration, as if the two are not pursuing the same aims. This is reminiscent of the conclusion reached by one of the leading protagonists of community development, of which the extension service is the core. Said this writer, "They (the VLW's) had largely become the errand boys of program administrators rather than catalyzers and servants of village people and organized village groups." Why the extension agents are "boys" and not "men," who enjoy for the most part little respect or confidence of the farmers, becomes evident even on a casual visit to the countryside.

The VLW *in Tanjore*

A talk with a VLW in Tanjore is worth recording. Tanjore is selected because it is the best known, best financed, best staffed, and supposedly with some of the best claims to being a successful IADP district. In the presence of a group of farmers, a VLW was describing his various tasks and telling something about himself. He was 29 years old with nine years service in a number of blocks of the district. He commenced his service at Rs90 a month as base pay and presently earns Rs125 plus Rs90 as dearness allowance. The increase is due to a yearly increment of Rs5 in the past seven years. His prospects for advancement are extremely limited. After nearly a decade of service he hopes some day to become a panchayat extension officer, in which case his base pay would be Rs140 to Rs210 plus certain allowances about which he was not clear. He was ill at ease in the presence of the farmers, and seemed worried over his job because, as he put it, "too much was expected of him." Indeed too much. In describing the character of his tasks, he mentioned agriculture first, but as he kept on enumerating chore after chore, agriculture was not the dominant item in a long list of other activities. The list went as follows:

1. agriculture, including field demonstrations and veterinary work,
2. seed distribution and testing,
3. inspection of farm practices,
4. preparation of farm plans to secure crop loans,
5. receiving and passing on loan applications,
6. collection of loans,
7. school building inspection,
8. enlistment of new cooperative members,
9. family planning with a 1968 target of 750 sterilizations,
10. panchayat work,
11. sales of government bonds,
12. receiving contributions for special affairs arranged by the local administration,
13. maintaining of basic village records incorporated in thirteen registers of one kind or another, and
14. last but not least, written reports on his activities to block headquarters.

How deeply he is involved in the enumerated tasks is not difficult to judge. Even if it is true, as he maintained, that he devotes one week to each of his five villages, it is doubtful that he can expertly perform the multifarious tasks with which he is entrusted. Agriculture is not his main preoccupation; even if knowledgeable and not soldiering on the job, he has other duties to worry about. He was vague about his field demonstration work, noting that the farmers know their jobs. Very much on his mind was his failure to fulfill the sterilization target. At the close of the year, he had helped induce 250 sterilizations against the target of 750. He was bound to hear from higher up about this "failure," and it worried him. There was yet another aspect to his family planning work recorded for our benefit by a local official. With sterilization goes an incentive payment of Rs30 to the man undergoing it. But his "net" is around Rs15, the remainder going as "reward" to the "inducers," of whom the VLW is one. His concern about not attaining the target may be real.

What stands out is how little has changed in the extension work of a VLW in the years past. He may have developed into a good bond salesman, but he is not a well-trained agricultural specialist. Even if he were, he would falter under the weight of so much else. He is not infrequently a garbage can into which a motley refuse of responsibility is dumped. One cannot but sympathize with him in his predicament. Partly because of his own shortcomings but mainly for reasons not of the VLW's making at all, he has been placed in an unenviable position. There is no denying that the extension service makes an indirect contribution to agriculture through a number of tasks already mentioned, while some VLW's do it more directly. One may even go so far as to say that extension had made a notable contribution passing out the word that new inputs are good for agriculture, but it has proved woefully short on how to demonstrate their use to the farmers in a practical way. In the final analysis, the latter makes the difference between a crop and a crop. It can be said therefore that, by and large, the VLW is not a vital innovation agent; this is particularly true when it is no longer necessary to "sell" the farmer the idea that new inputs and productivity go together.

Extension and new technology

Fortunately for Indian agriculture at this stage of development, the importance of the extension service has been exaggerated. How else is one to account for the unmistakable strides registered by the enterprising and some not-so-enterprising farmers? Simply stated, the Ludhiana farmer of the Punjab and his counterparts elsewhere have taken the bit in their mouths and have run with it. A study on the adoption of hybrid maize seed provides supporting evidence. The authors note that:

> Awareness for this innovation was more due to the pioneering effort of some farmers in Mandsaur town. There was hardly any lead given by the government agencies . . . none of these farmers informed that they came to know about the hybrid maize from the village level worker although all the farmers knew him . . . The adoption of the hybrid maize cultivation depended more on

> the self-initiative of the farmers rather than the efforts of the extension agencies of the Government.[4]

This the farmers could do when the new agricultural strategy was barely rearing its head. Now that it is beginning to come into its own, even the progressive farmers of India face the problem of more efficient investment; in this respect they have much to learn about the new technology and much more as the technology grows in sophistication. For this reason, the very considerable expansion of utilization of inputs doesn't mean that they are currently used as they should be, for the information about the relative efficiency of investment in different fields—irrigation, fertilizer use, new strains, pesticides, and so forth—is rather meager. The farmer innovator is not sure what to accept and what to reject among the alternatives which constantly come to the fore as part of a developing economy. The high-yielding varieties involve more than the mere substitution of one kind of seed for another. Their successful introduction will require changes in nearly all components of Indian foodgrain production technology. More specifically, an Indian writes notes the following on the biological complexity of attaining higher rates of output. "The new strategy," he writes,

> Relies on high yielding cereal strains. But these are not miracle seeds which merely need to be sown in order to produce record crops. A great deal more is involved—micro-nutrients, moisture, conservation techniques, bacterial fertilizers, weedicides, plant hormones, scientific cropping patterns and constantly changing plant protection schedules to suit variations in the physiological characteristics of insect pests and diseases.[5]

This being so, Indian agriculture requires tested innovations and a variety of well-operated servicing facilities. Regardless of how willing the farmer may be to accept innovations, an institutional structure composed of well-trained technicians and administrators is critical to his judgment and final decisions and, of course, to the farmer's rate of return. In the view of one close student of Indian agriculture and problems of agricultural development in general, "two specific inputs are particularly likely to have high returns. One is research . . . the other is education, its priority arising from past neglect and because it plays the key role in providing the trained manpower that is the critical complement of all the other critical complements of almost all the other input forms discussed."[6]

Almost everyone conversant with extension in India recognizes that it is outdated and provides no answer to the new requirements. In January of this year at a conference arranged by the University of Agricultural Sciences (Bangalore), speaker after speaker underscored this fact and for many reasons already cited. Even the commissioner of extension mentioned previously conceded on that occasion that the "inadequacy of ability and competence to guide and assist farmers has been increasingly in evidence" among extension workers. The proliferation of a great many VLW training centers, social education training centers, orientation and study centers, Panchayati Raj training centers, extension education institutes, and the National Institute on Community Development—all these and more have fallen far short of what agricultural progress demands. These are largely improvisations to accommodate some interested farmers but chiefly to turn out quickly a mass of new agents for the greatly expanded intensive agricultural development areas. Many, as noted earlier, are familiar with this state of affairs; they described it in detail, have been unstinting in their criticism, and lavish in their recommendations on how to bridge aims and results so lacking in extension. Many of the recommendations are still valid—for example, that the VLW must be "100 percent agricultural"—but implementation has not caught up with them.

4. Faculty for Management in Agriculture and Cooperatives, Indian Institute of Management, *Planning and Implementation in Agriculture. Studies on High Yielding Varieties Programme*, vol. 3, *Hybrid Maize in Madhya Pradesh* (Ahmedabad: 1967), pp. 44 and 51.

5. Ashok Thapar, "Extension Services," *The Times of India* (December 16, 1967).

6. John W. Mellor, "Toward a Theory of Agricultural Development," *Agricultural Development and Economic Growth*, H. M. Southworth and Bruce F. Johnston, eds. (Ithaca, N.Y.: Cornell University Press, 1967), p. 52.

Can the extension service be reformed?

In the face of this record, changes are difficult to effect. If they were to be attempted, the lines of approach are well-known; their realization would be both difficult and time-consuming, but they ought to be set forth at this point. Agricultural development entails "saturating" the farmer with both inputs and knowledge. As to the first point, in the atmosphere of smugness and self-satisfaction about the first outcroppings of the agricultural revolution, it cannot be repeated often enough that readily available fertilizer and so forth is not enough. These inputs, by now conventional in many areas, have to be fed with a continuous flow of knowledge which can come only from the research input and extension as its transmission belt. But agricultural research in India is in its infancy, and "India continues to lead the world with the lowest number of agricultural research workers per 100,000 of the people active in agriculture. The current figure is less than 3 compared with 6 in Pakistan, 60 in Japan, 79 in Taiwan, and 133 in the Netherlands."[7] The policy that suggests itself is self-evident, and all that needs to be added is that the eight agricultural universities of India might follow in the footsteps of the Punjab University and the Agricultural University of the U.P., which are now almost the sole centers of agricultural research and training among the universities.

The writer of these lines is reluctant to fall into the familiar trap of "should be dones," but there is no way of avoiding them and they are restated. If research workers are to play a useful role, they must remain in contact with extension personnel in order to know what is happening in the field and above all to provide them with useful ideas to extend. This in turn presupposes that the VLW is not a mere "errand boy peddling news about new research" but one with specialized knowledge, devoting himself solely to matters agricultural. Additionally, the selected candidates, be they agricultural college graduates or not, must be of a kind who do not look upon extension as a source of any kind of employment in preference to no employment at all. This implies fewer but better people, or the very opposite of the community development approach. The current practice of Rs125 or more is not very much more than the pay of the lowly "chowkidar." If additionally the VLW has little prospect for advancement, the consequences are twofold: He has no incentive to perform and extension has no opportunity to tap better qualified personnel. Finally, something would have to be done about the complicated lines of administrative command which in themselves often inhibit the work of the best extension agents.

This said, it must be borne in mind that the rehabilitation of the existing extension system is an excruciatingly difficult task. As institutions go in India, extension is only a fifteen-year-old stripling, hardly old enough to suffer from the deadweight of a questionable tradition. But it does. What the would-be reformers are faced with is an *establishment* of 70,000 agents; a vast network of hastily established educational institutions spread throughout the country; formidable and complicated administrative relationships which permeate the center, states, districts, blocks, and village panchayats; and, finally, the not inconsiderable and still potent ideological pretensions of community development which fashioned this particular extension system and this particular extension agent. Three years ago the Ministry of Community Development was disestablished and incorporated into the Ministry of Food and Agriculture. This was part of an attempt to rid the Ministry of Community Development and the extension service of some of its grosser aberrations. But so set has the institution become in its ways that one cannot detect any significant change in policies, administration, or style of work. Admittedly, too, and not merely as an afterthought, extension faces the dilemma of quantity versus quality. The latter is the touchstone of the progress of the new technology. If the numbers involved were small, the resolution of this dilemma wouldn't be half so difficult; but how are 70,000 going to acquire quality and become agriculturally knowledgeable?

The preceding may sound like an argument for leaving things alone. Yet to do so is to court trouble because important segments of

7. Ashok Thapar, "Pilots of Rural Change—Agricultural Universities," *Times of India* (March 29, 1968).

Indian agriculture can no longer be sustained by a service dealing, as one put it, in "yesterday's problems and no more." For reasons already stated, a break will have to be made in the very interests of farm productivity. Admittedly, this calls for a de-emphasis of the widely accepted idea that fundamental changes in agricultural practices can be effected through ample supplies of inputs without bothering about the presence, absence, or quality of the institutional "inputs." This is reflected in official programmatic statements on the new strategy, and it is due in good measure to the facile, aggressively optimistic assumption presently in great vogue that the transition from a stagnant to a dynamic agriculture is practically at hand. The evidence is not yet that conclusive in a country with 330 million cultivated acres, only 15 million of which are officially estimated to be under high-yielding varieties in 1968–69.

It must be stressed at the risk of repetition that the farmers, including the enterprising ones, will have to learn a great deal more before lasting attainments are firmly in hand and Indian agriculture becomes less dependent upon the vagaries of the monsoon than it had been in the past. A vastly increased responsibility for the extension service is inevitable. Whatever the impediments, therefore, it should behoove the leaders of the new agricultural strategy to promote an improved input of this kind, just as they are trying to promote more and better irrigation; the production, import, and utilization of fertilizers; high-yielding varieties; pesticides; and so forth. In short, it is high time that the theme song "all for agricultural productivity" should contain a stanza or two on the beneficence of an improved extension system advocated for some time but never realized. Failing that, the contradiction between aims and means may prove costly to the new strategy and to the farmers.

Extension is more than the agents it employs and trains. Training of the farmers to bring them as nearly in tune with the new agricultural developments as possible is part of it. To one official closely involved in the new agricultural strategy this is the crucial part of extension because of lingering doubts that "they [VLWs] would automatically pass on the contents of [their] training to farmers." He has so little confidence in the role of the village extension agent that: "This hope is not much different from Lord Macaulay's hope that the middle class brought under the influence of western education would transmit the fruits of modern knowledge automatically to the masses. Like the Englishman's hope, the hope pinned on the Village Level Workers is also likely to be belied."[8]

This judgment may be too severe, for it demonstrates how little importance he attaches to the future performance of the VLW's. The emphasis, therefore, is on farmer's training and of a kind that "will become meaningful if it is treated as an essential input (along with fertilizers, pesticides and irrigation) of programmes like those of the High-Yielding-Varieties, Multiple Cropping (and) Intensive Cash Crops . . . The training programme should be given the same priority by the organizers of agricultural production as by the trainers and educators." All this is well and good and so are the suggestions that the training programs should aim at "quick and improved agricultural production" by acquiring proper skills, by creating a "connection between the supply of inputs and the imparting of training," by establishing close communications between farmers and "experts of a level higher than the average extension worker," and finally by organizing small groups of farmers "affiliated to institutions where specialized and longer training would be conducted."[9]

Assuming the obvious need of farmer's training and the soundness of the scheme, who is going to infuse it with life? Who is going to train the trainers to train the trainees? Possibly the extension service as part of its inevitable but slow adjustments to the demands for doing things in agriculture on a new footing; possibly as a result of the activities of all the agricultural universities as they acquire new research skills and seek the dissemination of their results. But most of this still lies in the future. For the time being it will be up to the farmers themselves to acquire whatever knowledge they can from the outside while relying on trial and error to solve some of their im-

8. J. C. Mathur, "Farmer's Training is Central to Farm Production," *Yojana* (June 25, 1967), p. 15.

9. Ibid., p. 16.

mediate problems. A costly process but probably the only practical one in the existing circumstances.

Rural Cooperative Credit

Cooperatives in trouble

India is wedded to the proposition that the credit needs of the farmers, as much else, can be best promoted through the medium of credit cooperatives. Events have proved that the high hopes reposed in the cooperatives as the mainstay of institutional credit and as an answer to private moneylending have been exaggerated. Though they are a very important source of credit and the volume of loans keeps on rising, their performance on an all-Indian scale is not getting much better. In the words of the late V. T. Krishnamachari,[10] "The more and more we have expanded the cooperative movement since independence, the less and less confidence it commands in the community." This and much else are common knowledge, but what makes the cooperatives topical again is that their state of disarray is more conspicuous and worrisome, just when new and promising stirrings in Indian agriculture call for an efficient system for the distribution and utilization of credit.

Balance sheet of pluses and minuses

For nearly half a century the Indian credit cooperatives played an insignificant role in rural financing. They came to life only following the All-India Rural Credit Survey completed by the Reserve Bank in 1954. Having concluded that there is no substitute for rural cooperative credit, the Reserve Bank has been pumping into the cooperatives an ever-increasing flow of funds accompanied by a steady stream of recommendations to improve their performance. On the face of it, the liberalized credit policy is a success and the cooperatives are a force to reckon with. A few figures tell the quantitative side of the story. Between 1951 and 1966 the number of credit cooperatives increased from about 100,000 to 192,000 and their membership from 5 to 26 million. Loans advanced by them increased from Rs23 crores to Rs365 crores, and their share of total farm credit from 3 to 25 percent. Paid-up share capital grew from Rs8 to Rs128 crores and deposits from Rs4 to Rs46 crores. Borrowings from the Reserve Bank shot up from Rs5 crores, and the average individual loan increased from Rs44 to Rs314. Nor have they stood still as distributors of farm inputs, aided to be sure by the vast expansion of demand on the one hand and until very recently a virtual monopoly of fertilizer sales on the other. Judged in such global terms and with a network of 23 state (apex) and 348 central (district) cooperative banks, a growing number of noncredit cooperatives and here and there the successful marketing of cash crops, the movement has made much progress. Above all, it provides an important and additional source of credit to millions of farmers at a moderate rate of interest of about 10 percent per year.

This is only one side of the coin, for the cooperatives were and are in a bad way, and not only on account of two drought years in 1965–66 and 1966–67. One can say in 1968 with equal justification what the Bell report said in 1964, namely, that the credit cooperatives are inefficient; their leadership is poor; and that corruption, vested interests, and politicking are common phenomena. The reliance of the cooperatives on outside borrowing, instead of decreasing, is actually on the increase. The percentage of overdues to total outstanding loans is higher now than ever before, and this is true of the states where the cooperatives are most successful. The production-oriented loan policies are in serious difficulties, and concentration of institutional credit in the hands of the well-to-do farmers has deprived many small and medium farmers of access to loans. Thousands of cooperatives are dormant and ready for the mortuary, and many more thousands are veering in the same direction. The extension of the movement and coverage of farmers are quite large on paper, but in reality the members with "live accounts" in cooperative societies form less than one-half of the total membership. Underlying it all is that with

10. Vice chairman of the Planning Commission for over a decade and leader of the National Cooperative Union.

notable exceptions the cooperatives are concerned mainly with quantitative rather than qualitative achievements.

The above charges explain, but only in part, why "only a small fraction of the 191,000 primary cooperative societies could be deemed to be successful in any sense of the term."[11] This in turn raised in some minds the altogether novel question: are credit cooperatives necessary? An example of this is provided by the former chief minister of Haryana state. He is reported to have said that the cooperative movement has gone "so awry in Haryana that if my Viashal (Big) Haryana Party is to fight in the General Elections with a plea that the (cooperative) movement be abandoned, it will sweep the polls."[12] The validity of the chief minister's pronouncement is difficult to judge, but even the avowed friends of institutional credit are deeply concerned about its present and future performance.

High-yielding varieties program and credit

The Bell report took note of the limited effectiveness of cooperative credit in relation to agricultural production at a time when traditional farm practices still blanketed the country. It is useful to inquire into the influence the cooperatives exert on agriculture now that the improved technology is beginning to make headway. What is available by way of evidence is a two-year time span during which new farm inputs were becoming an important factor in production. This is admittedly too limited a period for firm judgments, especially since those were years of severe drought. Granting this, it appears nevertheless that the "Appraisal of the High-Yielding Varieties Program and Implementation of the Crop Loan System," recently completed by the Reserve Bank, carries lessons worth recording. Precisely because these are, as we shall see, not new, they are auguries for the years immediately ahead. Much of the evaluation relates to overdues and the application of the crop loan system; a word about the latter is in order before the two principal bank themes are taken up.

Of all the recommendations to inject new life into institutional credit—and they are legion—the key one of the Action Program for Cooperative Credit is the crop loan idea with its emphasis on productivity. Briefly, its guiding principles are these: Short-term loans are to be based on acreage and anticipated production; the amount of a loan must relate to the estimated outlay on raising the crop; and the loan should be largely in kind, and its recovery should follow immediately upon the harvest, in cooperation with a marketing society. Implicit in this is relaxation of restrictive features such as insistence on mortgageable property as security for most loans, absolute financial limits on individual borrowings, outright refusal or severe limitation of credit to tenants, and so forth. The loans under the new scheme rests on three components, popularly referred to as A, B, and C. Component A stands for the cash portion of the loan to meet household expenditures during the crop season; component B is the in kind portion of the loan, theoretically the largest one, in the form of inputs, principally fertilizers; and component C is another cash portion of the loan for expenses incurred in the utilization of the inputs.

No single standard of operation emerges from the detailed and voluminous findings of the Reserve Bank; this is natural in conditions varying so much from state to state and within states. Yet, viewed in relation to the introduction of high yielding varieties (HYV), the cooperatives "did not, on their part, have a feeling of sufficient involvement in the program" and "the central cooperative banks in certain states did not seem to evince much interest in the program as their representatives did not find a place in the Coordination Committee appointed to implement and review the progress under the program." There are more fundamental reasons for the lack of interest than the one just mentioned, but whatever they may be, the following should be noted: In 1966–67 the Reserve Bank earmarked special funds to stimulate the HYV program, but only 16 percent of them were utilized. The bank's experience with the coops in the nondrought year 1967–68, based on its normal advances, was better

11. P. R. Dubashi, "Cooperative Movement—Strategy for Development," *The Financial Express* (March 22, 1968).

12. "Rural Society Being Ruined by Movement Malpractices," *The Times of India* (November 8, 1967).

but not altogether satisfactory either. With a number of exceptions, the limits of financing by the cooperative banks sanctioned by the Reserve Bank lagged behind anticipations. In part, the Reserve Bank fell "victim" to the unfulfilled targets of HYV estimates prepared by the states; but there is more to it than that, even when the difficult drought years are taken into consideration. According to the bank:

> The reasons for shortfall in the lifting of credit were natural to conditions when the implementation of targeted programmes could not make expected headway because of lack of coordination among the various services, lack of adequate publicity, lack of provision of scientific advice, deficiencies and delays in making arrangements for credit, fertilizers and improved seeds, non-association of cooperative credit agency in an adequate manner with the program, and so forth and so on. Responsibility, however, lies squarely on the cooperatives for having failed to make necessary supervision arrangements to tackle overdues on the one hand and to improve policies so as to make them production oriented and to remove the bottlenecks in providing adequate credit for production on the other.

Overdues

Much of the quoted statement, made in late 1967, has the familiar ring of an old and much-played gramophone record, but it doesn't touch on the reasons why the credit cooperatives behave as they do. Nevertheless, the "should have been dones" are useful reminders of things gone wrong, and the crucial one among them is the failure to reduce the overdues which have in recent years risen to about 30 percent of loans outstanding. These restrict the flow of credit and have, not surprisingly, in the bank's view, "a crippling effect on agricultural production." The Reserve Bank singles out overdues as the worst manifestation of the credit system's operations. Overdues are an old story except that they have been rising—and not only in years of bad crops. According to the bank's findings, practically all states have districts with an "alarming" number of credit societies with overdues of more than 50 percent; many of them have defaulted altogether and are no longer eligible for credit. The examples of the cooperative movement, Maharashtra and Gujarat, are no exceptions and the same goes for the progressive state of Madras. This observer visited a block in the well-known Tanjore district in late 1967, a good crop year. There, twenty-five out of the thirty-seven credit cooperatives were in total default and had to be shut down; but, even before that action, only 15 percent of the membership borrowed from them. The principal cooperative officer of Tanjore prefers to discuss his operations in terms of the aura surrounding the district as a great "success story," but when presented with unpalatable evidence he admits that many other blocks tell the same story. The bank made a detailed examination of Andhra Pradesh which reveals conditions just as bad, if in a different form. In nine of the state's twenty-five central cooperative banks, overdues in 1966–67 ranged from a "low" of 55 to a high of 75 percent, which is a measure both of overdues and of the viability of the primary societies. Not unrelated to this is that out of Andhra's 15,381 cooperatives, only 831 and 367 of them respectively had full-time and part-time secretaries.

The causes leading to overdues were explained in considerable detail in the Bell report. Here it is to be noted that, while the official figures on overdues are bad enough, they do not reflect the real position of nonrepayment. Many cooperatives indulge in considerable "window dressing" to put their best foot forward on the eve of a new fiscal year. It is a widespread and normal practice among apex and central cooperative banks and primary cooperative societies to resort to all manner of artificial "adjustments" to cover up their inadequacies in respect to deposits, loans, recoveries, and their overall financial position.

Credit cooperatives have been notoriously lenient in collecting overdues in good as well as bad crop years. Preliminary soundings about the 1967–68 recoveries, an outstandingly good crop year, are not encouraging, according to officials of the Reserve Bank. Repayment of loans to moneylenders and other private sources is a priority obligation not enjoyed by the cooperatives. The decades-old philosophy that credit cooperatives are social welfare rather than business institutions and that government funds make their wheels go round has not been lost on the borrowers, and the prevailing attitude is that government loans can wait. The same attitude, supported by political consideration, extends to the cooperative leadership and personnel and expresses itself in a growing reluctance to press for prompt repayment. This explains the lament that "the members of the panchayats, of the primaries, and the boards of directors of the banks seem to be coming under the influence of thinking which considers energetic action for timely recovery as something undesirable, being likely to be unpopular."[13]

This comment goes back to 1964 or 1965, but nothing has changed since then; it is officially attributed to the fact that "in most cases the overdues may be said to have risen from the falling off in the standards of good cooperative conduct."[14] All such official statements are followed by exhortations to make good the loans through cooperative marketing of farmers' products, thereby insuring the collection of loans at the source. Cooperative marketings have never accounted for more than 5 percent of the outstanding loans in any given year. But there is more to it than the inadequacy of the marketing cooperatives. The often bruited-about assumption that all members of cooperatives (as farmers of India in general) have saleable surpluses of food grains is not warranted because the majority of the farmers have little or nothing beyond what they retain for their own consumption. It is true, therefore, as official circles admit, that, perversity of the borrowers aside, many of them cannot repay their loans even if they wanted to. Meanwhile, no stigma is attached to a defaulter who is part of a goodly company in the largest number of village communities. The upshot of it all is that poor recoveries have proved to be a major inhibiting factor in the way of expansion of cooperative credit.

Crop loan system and cooperative credit

The crop loan system was intended to provide the badly needed production-oriented credit chiefly in the form of fertilizer by eliminating features restricting the flow of credit from the coops to the farms. Save for Maharashtra and Gujarat, the system is not working well six or seven years after its inauguration. The explananation is threefold. In the first place, the in kind portion of the loan, which was expected to tie credit to production, is meeting with little success; "it has to be recognized that the linking of distribution of inputs to cooperative credit has not made much progress."[15] In the second place, the cooperatives continue to grant loans on a basis which is in total variance with the crop loan idea, about which more later. In the third place, one must keep in mind the mistaken idea already touched upon—that the cooperative membership produces crops mainly for sale. Regardless of the valid or invalid criteria underlying the crop loan system, having examined the application of the in kind portion of loans as part of the HYV program, the bank concluded that "district studies in all the states reveal that much remains to be done to implement this provision."

From the point of view of input needs financing and the magnitude of the task, when agriculture may well be poised for a significant breakthrough, the Reserve Bank's appraisal cannot be questioned. More revealing of in kind versus cash credit, looking ahead to the crucial 1970–71 season, is the following:

> While, for the package areas the portion of the short-term cooperative credit advanced in kind can be said to be on an average slightly more than 15 percent of the total short-term credit provided, for the particular areas in which high yielding variety crops are grown, the corresponding figure is about 40 to 50 percent. For the country as a whole, however, the percentages may be said broadly to be 15 percent in kind and 85 percent in cash, though precise information is not available. With the insistence now placed by the Reserve Bank and the Central Government on the crop loan system, . . . it is possible to hope that about one-third of the total short-term cooperative credit for

13. Reserve Bank of India, *Report of the Informal Group on Distributional Arrangements for Agricultural Credit* (1966), p. 39.

14. Ibid., p. 46.

15. "Report of the Fertilizer Credit Committee of the Fertilizer Association of India" (1968), p. 131.

> production for the country as a whole will, by 1970–71, be advanced in the shape of fertilizers.[16]

This is probably a realistic estimate based on a sound appreciation of how far the cooperatives can go in dispensing credit in kind, and it is also a comment on the future utilization of fertilizer worth remembering. But perhaps the single most important factor is that the borrowers prefer cash loans and an undetermined but large number of cooperatives go along with this preference, advancing what should be the in kind portion of the loan in cash, which on numerous occasions means granting the entire loan in cash.

The penchant for cash loans doesn't mean that the peasants did not want fertilizers and other inputs. In fact, many of them turned around and acquired them, often from the cooperatives; but in doing this they frequently had other considerations in mind not related to production, such as household expenditures, repayment or partial repayment of loans to noninstitutional lenders, and so forth. In sum, this meant that the *size* of the package of inputs was being reduced. This was especially true of the vast number of small owners and tenants who do not fare well at all as borrowers on the ground of lack of creditworthiness. We already mentioned that the bulk of institutional credit goes to the well-to-do cultivators, and the measure of it is pointed up by the Reserve Bank. It appears, then, that the 41 percent of the cultivators whose asset holdings are less than Rs2,500 are getting only over 9 percent of the credit, as compared with the 17 percent who are better off with assets of Rs10,000 and above, who account for 56 percent of the credit. Since the richer farmers are no better repayers than the poor farmers, they share the greater responsibility for impeding the more rapid circulation of cooperative resources.

The bank findings leave the impression that, while much is yet to be done to effect a link between the in kind portion of the loan and production, improvements are bound to take place; and Maharashtra and Gujarat, with their long record of outstanding top leadership, are cited as prime evidence. Moreover, the vast increase in utilization of inputs in 1967–68, most of it distributed through the cooperatives, is an even more telling case in point—regardless of whether or not this development is an outcome of that particular provision of the crop loan system. One must also assume that, with the spread of the new farm practices, bigger owners with a marketable surplus may opt in increasing numbers for the in kind portion of the loan. But even if so, future projections indicate that the problem of solving production credit is not assured. This stems from the fact that the cooperatives have not yet mastered another basic provision of the crop loan system.

It will be recalled that, under that system, loans are supposed to be advanced to farmers not on the basis of creditworthiness judged in the context of tangible security they could offer for the loan but on the basis for raising the crop and their repaying capacity from the expected crop. The merits and demerits of the system having been considered; on balance this has been official policy for a number of years, the purpose being the removal of restrictions on the flow of credit to those who need it most. In the main, the policy has been a failure. What with rising overdues, the insistence of the coops on tangible security has reached the point of absurdity, although the loan limits in some states have been greatly increased from those of three or four years ago. In certain districts of Andhra Pradesh, for example, cooperatives "insist on simple mortgage deeds for loans up to Rs100. For loans above this amount, registered mortgage deeds are required to be given"—similarly and with variations in a number of districts of other states.

From the bank's point of view, the test of creditworthiness as practiced by the coops contradicts the main principle of the crop loan system and it discriminates severely against small owners and particularly tenants who have no tangible property to offer as security. The latter, to be sure, receive loans but they are more in the nature of subsidies than credit. The predilection of the cooperatives to play "safe" by imposing security restrictions is understandable, for even in these circumstances overdues are increasing. But reducing the credit flow in order to improve recoveries is not a way out

16. Ibid.

either, placing the bank, the cooperatives, and the borrowers in a difficult situation for reasons already discussed. The Reserve Bank's appraisal doesn't draw from this an overall conclusion, but those of the "Report of the Informal Group" also sponsored by the bank, may be quoted here:

> By and large, the picture which emerges is one of inadequate implementation of the crop loan system in the country as a whole, with the size of loan being restricted in many cases by absolute quantitative limits and being determined by such considerations as the value of assets owned which could result in over-financing in some cases or underfinancing in others with reference to the actual need for production credit.[17]

Another important conclusion of the appraisal is that the inadequate work of the coops cannot be attributed to lack of financial resources. If resources were doubled, the problem posed by their failure to absorb properly what had been available to them would remain unresolved. In the final analysis the main shortcomings of the coops and the membership they serve converge on their absorptive capacity problem. There is no gainsaying the benefits of a continuous expansion of institutional credit; but, if its improper distribution and utilization goes on unabated, if they fail to tie, firmly in most cases, credit to inputs, inputs to output, and output to storage and marketing—all these cannot help but constrain the great potential of Indian agricultural resources. There are other reasons for the existence of these constraints, not all of the cooperatives' making, and more about them later.

Commercial Banks and Rural Credit

Why commercial banks and their past record

Because they perform as they do and because even under sounder conditions of granting credit the cooperatives cannot supply the rising rural needs engendered by the new agricultural programs, it became evident for some time that another source of credit must be added to what is available. The draft outline of the fourth plan (August 1966) took cognizance of this. Thus, while "The Cooperative credit structure will of course be the main agency for provision of rural credit . . . the credit needs of agriculture will be met by a variety of credit institutions."[18]

This departure from a long-established policy gave rise in the past year to heated debates as to who might provide the additional source of credit, and all indications are that the commercial banks are to be that source, a source still largely untapped. One of the sticks the government uses to get the banks in line and help create a multi-institutional approach is the threat of "social control," which is at best tighter Reserve Bank supervision of credit allocation by commercial banks "in keeping with the new national priorities," loosening the ties with favorite industrial and commercial enterprises or restrictions in regard to loans to firms in which bank directors are interested, and certain changes in top management. But the real objective behind the move is to compel the commercial banks to divert a fair amount of their resources to the agricultural sector. The alternative to "social control" is the threat of outright nationalization of the banks. Much of the criticism against the banks derives from the alleged fact that the total number of their accounts is small and that out of an estimated 700,000[19] accounts not more than 600 draw 50 percent of all the credit. Ideological hangovers of the Congress Party may have something to do with this pressure, but one of the main charges is that commercial banks extend very little credit to agriculture. This is correct, and in the opinion of the cooperative specialist what little agricultural credit is provided by them "is nothing but an eyewash to blur the view of the critics."

The resources of the commercial banks have risen sharply over the years, as evidenced by

17. Ibid., p. 41.

18. "Fourth Five-Year Plan—A Draft Outline," p. 140.

19. Actually closer to a million, but many borrowers have two or three accounts at the same bank. The figures cited have not been checked and are considered only approximate.

the amount of loans advanced: from Rs585 crores in 1951 to Rs2,347 crores in 1966. This is not reflected in their relationship to agriculture. In 1951 commercial bank loans to agriculture totalled Rs12 crores, or 2 percent of the total loans advanced by them; in 1966 the comparable figures were only Rs5 crores and 0.2 percent. Now as in the past this credit is given mostly to plantations where the risk is more predictable, but the decline in recent years indicates that even they are not creditworthy or sufficiently remunerative to banks enjoying booming profits in a lenders' market. Most banks consider agriculture too volatile, risky, and costly, and the mobilization of rural deposits not worth their time and effort. Besides, the unenviable financial record of the cooperatives was more than sufficient to keep the commercial banks out of agriculture. This is the way it came about: "The provision of credit had been left either to the traditional money-lenders (with full regard for profit and none for welfare) or to the cooperative organization (which is nearly all welfare and not cost or profit-conscious and has no surplus to meet the welfare from)."[20]

Cooperatives fear competition

The reluctance of the banks to engage in agricultural financing is matched by the opposition of the cooperatives to "outsiders" who aspire to provide credit to farmers, even if government sponsored. Competence or incompetence aside, cooperative leadership assumes that, with the exception of moneylenders, institutional rural credit is a monopoly not to be infringed upon by others and competition is to be avoided. The entry of others is not seen as an expansion of total credit supply that might in the process help the cooperatives but as a reduction in their business and in the business of cooperative banks. A few instances illustrate the point. The coops have been opposing the setting up of central government and state-supported agricultural finance corporations in areas where the cooperative credit structure is very weak, for the purpose of providing special credit for seed farms, minor irrigation, mechanization, construction of warehouses, transportation of agricultural products, and so forth. The cooperatives look upon them as another "eyewash" scheme on the ground that the sponsors allegedly do not understand farm financing, that the contemplated resources were inadequate, that the cooperation contemplated no direct risk taking, and that the majority of small farmers could not benefit from the corporations anyway. So far the coops succeeded in preventing any action since the scheme was first mentioned in 1964.

The cooperatives reacted similarly to taccavi loans. Every state government had been giving such loans at lower interest rates than those charged by the cooperatives; they were originally distributed to farmers only through state agricultural offices. Upon the insistence of the cooperatives that such handling of credit is unfair and not in their best interest, several state governments started routing the taccavi loans through the cooperatives. In fact, so vehement was the opposition to any other agency entering the agricultural field that the former minister of food and agriculture, M. C. Subramaniam, spoke as follows:

> I would beg of my friends to realise that, while cooperation is a good way to render service to the farmers, other institutions should not be ruled out. The "Establishment" in credit circles was quite worked up when the suggestion was made that the Food Corporation should enter the field of credit for the agriculturists. When the Food Corporation proposes to enter the field, we are told, it would cut across the philosophy that cooperatives are the only institutions which should extend credit.[21]

If the vested interests of institutional credit could display such a disposition toward the government's own schemes, their negative attitude toward commercial banks as potential suppliers of rural credit is not surprising. According to existing regulations, relations between cooperatives and banks are circumscribed to a point of almost nonexistence as

20. A. M. Khusro, "Banking and Agriculture," *A Symposium on Banking* (April 29, 1968), p. 2.

21. O. P. Karla, "Nationalization of Rural Banks," *Indian Express* (December 19, 1967).

borrowers from the banks, or having accounts with them, or learning something useful from their practices. In all of this the cooperative system maintains a stance of "splendid isolation," which in turn is derived from a long history of idealization of the cooperatives as something of a "cure-all" rather than from the economic services they render to the farmers.

Yet, the cooperatives can no longer offer blanket opposition to the government's pressing of the commercial banks into the rural credit field and get away with it. If there is, therefore, no avoiding the banks, the contention is that their role must be only an indirect one. This was spelled by the chairman of the Gujarat State Cooperative Conference in late January of this year. He stated that the banks should not extend credit to farmers directly; instead, they ought to lend money to the cooperatives or subscribe for debentures of the cooperative land mortgage banks. In some such manner commercial banks and cooperatives would create a partnership, but of a kind that would preclude direct involvement of the banks with the farmers. Avoidance of "direct involvement" is a recurring theme of the cooperative leadership. The explanation, as noted earlier, lies in the fear of competition. The above referred-to chairman of the Gujarat State Cooperative Conference let the cat out of the bag when he reportedly spoke to this effect: "Mr. Patel expressed his apprehension that if the commercial banks started competition the discipline in the cooperative organizations would be adversely affected."[22] Mr. Patel is no exception.

According to an acidly penned major survey of the cooperatives of Madras, published in early 1968, "Farmers are increasingly going to private money-lenders instead of cooperatives offering low interest rates. The reasons could be the farmers' inability to give security, non-acceptance of farmers as members, delay in processing applications or denial of loans to tenants."[23] This is familiar stuff, if worth repeating. Of greater immediate interest is that, however badly the coops work, they must be protected against competition because "Once the commercial banks are permitted to enter the field," the report notes, "the cooperative movement, already weak, will be rendered weaker . . . such *onslaughts* [emphasis added] on the cooperatives should be properly countered only through proper service to farmers." The expressed hope is also familiar and the means to translate it into reality are just as familiar—and so far largely ineffective. Because of the latter, this much is clear: Competition is one thing cooperatives cannot stand whether in respect to credit, distribution of inputs, or marketing of agricultural products—despite the fact that they are supposedly closer to the villagers and have decades of experience behind them.

22. "Banks asked to give Farm Loans through Co-ops," *Financial Express* (January 28, 1968).

23. Excerpt from the report quoted in "Inability of Co-op Banks to Utilize Funds Deplored," *Financial Express* (March 5, 1968).

Prospects of commercial bank credit

As of the moment, it is evident that the commercial banks are beginning to inch into the rural credit business, both indirectly and directly. So far and bending under pressure, among which is a provision that no bank can open a new urban branch unless it opened also two rural branches, they have invested Rs18 crores in debentures of the Central Land Mortgage Bank. Additionally, they have created a new institution, the Agricultural Finance Corporation Limited, which is considering the financing of a number of development schemes to the tune of over Rs10 crores. The corporation has an authorized capital of Rs100 crores, of which Rs10 crores had been issued and Rs5 crores had been paid up and available for lending. The corporation is expected to select commercial banks as agents, the choice depending upon the strength of a particular bank's concentration in a selected area. A few of the banks with branches in small centers, which are rural or semirural, have actually gone into financing of tractors and other farm inputs. In Rajasthan, for example, the maximum of such loans can be as high as Rs30,000 with an interest rate of 9 percent per year and repayment in four years on a quarterly or semiannual basis. The same bank, the Rajasthan branch of the Dena Bank, is also prepared to extend short-term loans up to a maximum of Rs5,000 for the purchase of fertilizer, seed, and insecticides. The State Bank of Patiala has more recently begun to provide more credit, while the State Bank of India is ready to underwrite some of the credit required by the hybrid seed project in the Terai.

The one pioneer in the field is the Mysore state-based Syndicate Bank. Though only eleventh among commercial banks in overall resources, in the past few years it has accumulated 3,000 active farm accounts, ranging from Rs500 to Rs15,000 and a record of repayments of 96 percent. Its "pigmy deposits" scheme permits a farmer to put into his saving account as little as 25 paise at a time, but the net result has been the mopping up of a good deal of rural deposits where the bank operates. One of the innovations is that the bank doesn't always stick to conventional banking practices. In a personal interview, the president of the Syndicate Bank did not disclaim the need for safe security, but he practices a "mix" of security in the usual sense where possible and/or whether the farm has a growth potential combined with the farmer's repaying integrity. The theory tested by the bank is that, if the lending policy leads to greater production and absorptive capacity of the borrower, it would also lead to a greater repayment capacity.

Implicit in this approach are the bank's economic "feasibility" studies about the economy of the farm in question and the farmer's condition; granting of loans mostly in kind—be it a purchase of a tractor, a pump set, fertilizer, and so forth; supervision of credit so that a loan earmarked for production items is spent accordingly; and technical advice by the bank's staff. The bank then is both a provider of loans as well as a source of agricultural extension work. It may be concluded that, but for the fact that the Syndicate Bank doesn't engage in marketing of agricultural products, it practices something akin to the integrated cooperative scheme (ICS) recommended by the Rural Credit Survey many years ago but not always practiced by the cooperatives. The services rendered by the bank are expensive and the cost per farm loan is higher than the cost of an industrial and commercial loan. Yet the bank is making money though at a lower rate of profit, has few overdues, and is expanding agricultural financing with an eye of reducing costs and in anticipation that the new technology will be affecting a rising number of cultivators, thus creating more business for the bank.

Time will tell how widespread the rural credit operations of the commercial banks will become; but, if they do take a leaf out of the Syndicate Bank, chances are that they will be in business on a fairly significant scale. At first glance, the widely dispersed lending operation, illiteracy of the farmers, uncertainty of agricultural production, and legal and institutional obstacles in collecting overdues stand in the way of farm financing through commercial banks. On the other hand, a number of conditions favor them. The principal one is that agriculture is becoming production oriented and demand for credit is bound to grow. Whatever the volume of business and the new methods of expanding credit among the cultivators, they will draw a sizable clientele among the large owners of unencumbered and mortgageable properties.

In many instances, the interest rate of the banks and the cooperatives are approximately the same, but even if the former charged another 2 percent, the differential would be too small to deter one from dealing with the banks. From the borrower's point of view there is considerable advantage of not getting involved with the time-consuming cooperative procedures in granting a loan. Finally, some farmers might borrow from the banks and the cooperatives, and the latter are justifiably concerned lest the former enjoy repayment priority. With this in mind, in a recent talk with the registrar of cooperatives of the Punjab state, he argued strongly that such dual borrowings ought to be prohibited as far as the banks are concerned. The rationale is that in the absence of proper coordination such schemes may lead to duplication or multiplication of institutional agencies in some parts of the country where the cultivator borrower may be overcredited; in that event, overcrediting the farmer is as dangerous as not granting him any credit at all. But above all and not unreasonably, the cooperatives dread the prospect of being left with fewer big borrowers and more of the small farmers with poor security standing and a poor record of payment. If this comes to pass, the problem will not rest with the banks, for the presumption is that in the main they will deal with small farmers only to the extent of meeting minimum political considerations.

Since problems and solutions have often a way of passing each other by, not all of the above might turn out as suggested. But if they do and the banks secure a fair share of the rural credit business, the question that arises is whether or not this would induce the cooperatives to greater efficiency. If the answer is in the negative, the dire forebodings of the coops may come true. Regardless of how the cooperatives feel about the intrusion, the participation of the commercial banks in rural credit, even if on a selective basis, is a welcome step in the right direction; and the move is just another sign of the weakness of the cooperatives, the rise of agricultural productivity, and the energizing effect of the threat of "social control."

Assessment of the Cooperative Movement

Neither the commercial banks nor any new source of farm credit can displace the cooperatives or even significantly reduce the scope of their activities. The new sources, if and when they come, promise only indirect and limited answers to coop disabilities. Virtually all official reports on cooperatives are of a critical nature, but they all end on the note that they must continue and must succeed. "If cooperation fails, there will fail the best hope of rural India," or "Cooperation has failed, but Cooperation must succeed"[24] are *the* guidelines of any official attempts to improve cooperative credit system. The latest in the series, the very outspoken Madras report, brands as "shameful" the fact that after sixty-three years in being, only 30 percent of the members of the coops borrow, but there is no hint of doing away with them. Whatever their weaknesses and however vociferous the grumblings of the membership, they are too much a part of the rural scene; and the vested interests controlling them and their supporters in the central and state governments are sufficiently strong to insure their existence. Realistically speaking, therefore, the question, "are cooperatives necessary?" cannot be answered negatively.

Clearly, the credit cooperatives are here to stay, but the question of how to rid them of their shortcomings remains unanswered. The plethora of recommendations for remedial action made over the years have proved of little avail. The Bell report makes ample references to these measures and they need not be repeated. What should be stressed, however, is that with the spread of the new technology the cooperatives will be faced with more formidable tasks than ever before. "In 1970–71 the total cooperative short-term loans are envisaged on the order of Rs6,500 million to Rs7,000 million, as against Rs3,000 million in 1965–66."[25] This includes Rs2,700 million for fertilizer alone.

Funds to enlarge their capital base to meet these requirements are likely to be found; whether the coops will be better prepared for the additional responsibilities is doubtful. The same committee just referred to states that in arriving at the estimated figures "we have made the assumption that all possible steps at all levels would be taken to improve the performance of the cooperative credit structure not only quantitatively but also qualitatively in the next few years."[26] Judging by past experience, the assumption of a more durable and better functioning movement may prove premature. In the customary fashion of the years gone by, a restatement of the well-known recommendations will undoubtedly follow, but the issue as far as we are concerned is not their content but why they fared badly. Why, then, have the recommendations—many of them sound ones—gone by the board? Are the shortcomings of the cooperative movement remediable or are they inherent in its structure to a point of no return? In other words, is the assortment of cooperative weaknesses and the educational and technical backwardness by which they are accentuated a subsidiary ailment or the main disease itself?

The answers to these questions are many. Some are rooted in the origins or philosophy of the movement; some derive from the size of the holdings, tenure systems, quality of land

24. *All-India Rural Credit Survey*, pp. 250 and 372, respectively.

25. "Report of the Fertilizer Credit Committee," p. 222.

26. Ibid, p. 234.

and rainfall conditions, type of population served, and resulting income; some, and just as importantly, are rooted in the social environment of rural India and some in the fact that in recent years the cooperatives have also become political institutions, an arena of contention among political parties. These cause categories are not of the cooperatives' making; but the structural and functional arrangements, insofar as they relate, respectively, to the organization and working of the cooperative credit societies and banks, are of their own creation for which they cannot shed responsibility. In much of this, interpretation must deal with many intangibles not subject to quantification. And it should be added that in not all instances do the categories fall in neat, segregated compartments without overlapping and affecting each other. With these caveats out of the way, the "whys" of the cooperative credit movement are considered in some detail in the subsequent paragraphs.

Historical and environmental factors

Cooperations as an idea and credit cooperatives as an institution were introduced by the British in response to the poverty of the mass of the small cultivators. From the very inception, however, and down through the decades, this purpose did not stick because "From character and repaying capacity, judged by local knowledge and kept under review by local vigilance, cooperative credit, at a very early stage, gravitated to ownership and landed security."[27] By virtue of being affluent and politically influential, the rich rather than the poor farmers came to dominate the coops and enjoy the lion's share of the benefits from cheap credit. This cannot be viewed in isolation from another historical condition, namely, that the coops did not commence nor have they developed at any time since as a popular movement in which the membership is "involved" and cooperatives are cared for by the members as their very own. This apartness which still bedevils them harks back to their origins when they were being created, rigidly administered, supervised, and financed by just another government department. The fostering of self-government and self-reliance was minimal. When looked at from the point of view as to who dominates the local cooperatives now, it is clear that what may have been a period of trial and error gradually evolved more or less into a norm in postindependent India. Another early precedent essentially gone unchanged is the idea that the credit cooperatives are largely relief rather than business institutions, that their own resource participation is small, government providing most of it. The 20 percent share of the working capital owned by the coops in the late 1960s is a case in point. Taken together, the historical conditioning carried over with no significant amendations into our days explains at least *partly* why sixty-three years after their existence quantitative as against qualitative factors and original purposes compared with actual practices and ends they serve diverge so widely from one another.

Important as these are in shaping the cooperative structure, more vital still are the factors which determine the cultivators economic activity and social background against which the activity takes place. The reference is to the Indian subsistence farmers who constitute the majority of all farmers and whose income and consumption expenditures lag behind production. Whatever institutional credit they do get is helpful, but it is not sufficient to make a nonviable enterprise viable. With all their faults, credit cooperatives cannot be held solely or mainly responsible for this state of affairs. Credit for productive purpose is self-financing only—in the sense that repayments are made—if the farmer produces a surplus over and above his costs and consumption needs. In the case of subsistence farmers, it is not "self-financing." In recent years nearly 50 percent of their total borrowings was for "household expenditures." This would not constitute a serious impediment if the size of their holdings and productivity left a marketable surplus large enough to cover all elements of production, including "household expenditures" as a legitimate cost of production or the wage equivalent of family labor utilized in cultivation. This, however, is not the normal condition.

Against the background of the 1967–68 excellent crop and high hopes that an agricul-

27. *All-Union Rural Credit Survey*, vol. II, (Bombay, 1954), p. 259.

tural breakthrough is practically at hand, there is a tendency to overlook the seamy side of India's agricultural structure. But it is there just the same and is not about to fade away. Thus:

> Farm management data of recent years lend support to the belief that under average conditions, a farm of less than 7.5 acres size—which incidentally is the average size of farms in India—can neither provide adequate family consumption, nor support a pair of bullocks nor adequately employ family labor or bullock labor of a single pair, nor, indeed, deliver surpluses for purposes of industrialization. Such a farm is quite seriously disabled and if full value were inputed to human and bullock labor, it turns out to be a non-profit making venture.[28]

More to the point is the case of farms averaging 5 acres, or what the above-quoted Dr. Khusro calls "the limit of economic feasibility under Indian conditions of today." The area occupied by this group is only an estimated 12 percent of the total farm acreage. "But to say this," Dr. Khusro remarks, "is not to forget the problem of 56 percent of farm families and a mere 12 percent of acreage and facing disabilities of small consumption, small employment and small surplus."[29] These facts go far toward explaining why, apart from the reasons noted in previously, millions of cultivators do not meet their loan obligations; it points also to an important cause of the strain imposed on the cooperatives for reasons other than their poor judgment in making loans, poor administration, or the like. Since the reference is to farmer owners, it is all too clear how much greater the strain is on the cooperatives in extending loans to the many tenants who work somebody else's land. Any assessment of the cooperatives must, therefore, take into consideration the state of the country's agricultural economy in general and the position in it of a vast group of badly situated farmers. In a real sense and while not explaining everything, the cooperatives are to a large extent victims of a situation not of their making. They cannot revise the inequitable land structure of India, and only occasionally and *temporarily* can they alleviate its worst features. The following written sixteen years ago with no special reference to India is germane to an important "why" of Indian credit cooperatives:

> The fact that the cooperative credit movement has made such limited progress . . . suggests that the causes of failure lie deeper than administrative defects . . . They lie . . . in the chronic insufficiency of the farmers' income and consequent tendency of consumption to outrun production. In countries of subsistence agriculture, the mobilization of savings is impossible, if savings do not exist . . . the problem of credit cannot be solved in terms of reform of the credit mechanism alone, and the extension of credit facilities to small farmers through the cooperative principle can only be successful if it is applied as part of a general programme of agricultural re-organization.[30]

These observations illustrate that not all the reasons usually given to explain the inadequacy of the coops can be placed at their door. Besides, considerations on the part of other policymaking agencies enter into the picture. But it would be a mistake to conclude that the structure of the agricultural economy is the sole explanation of the generally accepted thesis that coops are not nearly as good as they should be. After all, by far the greater part of the land of India is not in subsistence agriculture, yet the records of the coops in financing farmers with such holdings is far from a success story. Part of the difficulty affecting small and big farmers alike is the enormous effort that has gone into the expansion of the cooperative network in order to meet quantitative targets and only here and there—and upon reminders—a side-glance to quality standards. The exceptions merely prove that a reasonably sound administrative mechanism with a staff which is less ill-paid, less ill-trained, and more adequate is not always out of reach.

28. A. N. Khusro, "Rural Development," unpublished paper, p. 13.

29. Ibid.

30. *All-India Rural Credit Survey*, vol. II, p. 254, quoted from International Conference on Agricultural and Cooperative Credit. "Selected Readings in Agricultural Credit" (1952), pp. 72–75.

Village leadership or vested interests

In any attempt to augment the exception category, note must be taken not only of efficiency factors but also of two additional and very significant "whys": the small but powerful interests which rule the local cooperatives as if they were their private property and the undesirable interference in the affairs of the cooperatives by the national political parties. We are not repeating the obvious: the usual recommendations that these realities must fade away. It needs to be stressed that they are "flesh of the flesh and bone of the bone" of the cooperative movement and that it leads to such abuses as the exploitation of the coops for the benefit of the very few, outright corruption, and the accentuation of the already existing wide gap of the privileged and the nonprivileged. Therein lie some of the explanations why intended reforms of institutional credit have so largely failed.

The items on the "interests" and politicians do not lend themselves to statistical proof, but the validity of the nonstatistical evidence cannot be swept under the rug. The evidence stretches throughout the history of cooperatives and is incorporated in official reports, books, articles, and eyewitness accounts. This observer, with no special mission to investigate their work, listened to farmers on two occasions with fascination as they were describing the "considerations" to secure a loan. One of the farmers who paid for the "privilege" but did not get the loan elicited no more sympathy than a chorus of guffaws from a large gathering of cultivators. The tale of piling insult upon injury was not as unique to the gathering as it was to us; it appeared like a minor and acceptable abuse to live with, lubricating, if not always, this particular aspect of agricultural development. A variation on such episodes and other malpractices and shady deals derives its significance not from the action of an impecunious clerk but rather from the behavior of those whom the All-India Survey stamps as the "upper 10 percent," or in the felicitous expression of one writer, the local "all-in-all."[31]

The role of the "vested interests," a common nomenclature in cooperative circles, is as old as the system itself, except that they have expanded with the expansion of the system. Who they are and what they do is well known. The "not uncommon inclination of members," the survey authors write, "to treat the society as a close preserve of one class of persons in the village may be traced to the character of the village leadership, to the bias towards ownership of land and to influence of caste." As if to make certain that there is no mistaking who these people are, the following is worth recording: "It is only in India one finds . . . millowners, *rentier* landlords and traders being the leaders of cooperative organizations and yet we hear not a whisper from any quarter against this entry of inimical elements in the cooperative body."[32] There are also the village headman, panchayat leaders, local moneylenders, representatives of the highest caste, and revenue officials. Their power is enhanced because some of them are connected with the seats of administrative power as appointed or elected officers. The survey was prepared in the very early 1950s, but the "all-in-all" are stronger in the 1960s; they have successfully "extended their hand" on the course of the agrarian reforms, have grown richer, the cooperative business has vastly expanded and so have the stakes in it.

The survey makes clear who controls the local credit societies; the findings a decade or more later do not differ in any essentials. In 1963 after a wide-ranging trip through the country, Daniel Thorner wrote that "It almost seemed as if a man couldn't be considered important unless he had in his grip an agricultural cooperative whose funds he was diverting to support his own private business venture." The state of Mysore "furnishes the example *par excellence* of leading families who completely control village affairs, including the cooperatives." And so in state after state, and he concludes that "The little people cannot sit as equals with the important people . . . It is doubtful that they work together on a cooperative basis."[33] The most pointed comment on

31. Daniel Thorner, *Agricultural Cooperatives in India—A Field Report* (New Delhi: Asia Publishing House, 1964), p. 6.

32. *All-India Credit Survey*, pp. 263, 265.

33. Thorner, *Agricultural Cooperatives in India*, pp. 5–7.

the economic consequences and the so far resigned acceptance of the situation is a saying among the Uttar Pradesh peasants, which, freely translated, stands for the familiar "To him that hath much, much shall be given." In late 1967 in a series of articles by a leading corporation of India under the general title "Co-ops with the Lid Off,"[34] one finds yet another confirmation of the same story. The theme is wider than the one dealt with here, but one thing is clear: The cooperatives continue to be a playground for the village leadership described.

All this is true, but it must be qualified. The village is under tied control of the few; but, as we shall point out elsewhere in connection with panchayat elections, the village is no longer silent and the villagers are picking and choosing. The new agrobiological and political happenings have their economic and social effects and voices of dissent are heard. Their significance should not be exaggerated, but the very promise of the new technology being realized by a minority of cultivators is provoking some of the majority to speak out about their wrongs as well as their rights—more about the former, to be sure, than the latter. The following episode, one of the many, is not without interest. "It was a fascinating sight," a reporter writes, "at a public meeting in the vicinity of Nagarjunasagar where the Andhra revenue minister, V. B. Baju, was distributing titles for uncultivated land to landless laborers at an impressive ceremony. Suddenly a woman stood up. Her face was covered with a veil, but her voice was firm. 'What is the use of this ceremony,' she shouted, 'when the land is useless and every little thing we need we have to bribe your officials'." This, appearing in an article under the suggestive title of "Many-Splendoured March of Farm Revolution" is no more than a single strand of that revolution, and the woman's outburst will not put an end to bribery or to the distribution of "useless" land. And yet in any assessment of the current scene, particularly in its relation to the new technology and hopes aroused, that woman's dashing of cold water on the big "do" is an encouraging aspect of an otherwise trying state of affairs.

34. *Times of India* (November 6, 8, 14, and 17).

Village polarization and obstruction of reforms

Two lessons can be drawn from the preceding. The first one is that, whereas cooperatives imply union of common interests and effort, there is nothing in their records that points in that direction. The social and economic separatism of the Indian village is an old story, but the cooperatives as catalysts in bringing the community closer together misfired. Led by the local leadership, they have actually drawn sharp lines of demarcation between cultivator and cultivator, caste and caste, and generally speaking between the weak and the strong. In sum, the irony of it is that in these circumstances it fell to the lot of the cooperatives to further the polarization of the village to an unprecedented degree. What the consequences of this might be as time marches on we leave to the Indian astrologers, but it can be said that the vast expansion of institutional credit has serious chinks in its armor; the "vested interests" have managed to subvert its benefits, just as they subverted much else that didn't suit their purpose.

The second lesson is that they have so far succeeded in precluding any meaningful cooperative credit reforms where such are technically possible. It is very tempting to let the survey speak for the record, for what it had said on this crucial point in the early 1950s has particular relevance in the late 1960s. In a section on "village leadership and administration" we read that "This close conformity of association and interests between the subordinate officials of Government and the more powerful elements in the village is a matter to be borne in mind as of great significance in explaining the failure of implementation of the policies and directives, cooperative and other, emanating from higher levels of administration." Even when measures reaching the village are translated into action, they are not necessarily lasting, "especially when they involve some disadvantage to the more powerful in the village." And so the judgment: "Acting in concert with these, the subordinate official, whose function takes him to the village, creates for the benefit of superior officers what might be called the illusion of implementation woven round the reality of non-compliance." But the burden of "change around changelessness" is always attributed to the few wheeler-dealers of the cooperatives; while "The *status quo* and the non-compliance are often achieved conjointly" with the subordinate agencies of government, the prime movers are "the leading elements of the village."[35] So much for this group, their use of the cooperatives for their particular interests, and by the same token the disadvantaged of the great mass of cultivators who get the short end of cooperative credit. It all sums up in one of the most important "whys" of the nature and performance of the credit cooperatives.

Party politics and cooperatives

The role of the political parties in the local credit societies and credit banks is closely related to that exercised by the stronger elements in the village. In fact, socially and economically the politicians and the "strong" are often undistinguishable from one another, partly because the leadership of the coops is made up of many politicians and partly because the interests of the two are closely interrelated. With the rising proliferation of political activity since independence, the cooperatives couldn't escape becoming an arena for political involvement or, as some prefer it, hotbeds of political strife. The prevailing view of those who shape cooperative ideology is twofold: (a) The politicization of the cooperatives is interfering with the normal operations of basic purposes of a cooperative organization; and (b) it deflects them from the principle of political neutrality, thereby introducing an element of divisiveness and conflict. And everything in the history of the cooperatives tells us that they could do very well without such additional constraints. But there is yet another weighty reason for this negative attitude, which is not dissimilar from the one already discussed about the vested interests in relation to the cooperatives.

It was natural to expect that politics, in the sense of aims pursued by political parties, would extend to the cooperatives. And so it did, and local politicians aspiring to higher office and MLA's (members of the legislative assemblies) are found in commanding positions throughout the cooperative structure. In certain states the MLA's are in sufficient presence in important jobs as to term that condition "emmelaitis."[36] The thrust of this development, however, is not on the democratization of India's body politic but on something more immediately tempting, namely, the fact that cooperatives are big business with millions of borrowers and large and constantly expanding government resources at their command. To "capture" a cooperative, especially the big ones, or a cooperative bank opens vast opportunities of a varied kind. They become centers of political power, raise the leader's prestige in the eyes of the membership; and it enables him to build a following to influence election results. On the economic side it opens the opportunity of favors and special favors and gains as distributors of patronage through loans, contracts, and jobs. The evidence is that the politicians, whether in direct or indirect leadership, and the usually influential members of the coops have not neglected these opportunities. The question is: What good does it do to the cooperatives as an institution called upon to serve the legitimate and best interests of the membership?

One should note first that political involvements are not universally detrimental to fortunes of the coops; there are many cooperative organizations which prospered under a politically involved and dynamic leadership. Maharashtra offers many such examples. Secondly, it is virtually in the nature of a cooperative organization to serve as a communication channel for political ideas from on-high down to the villages. For this reason, "some degree of politics seem to be inherent in the cooperative structure in the context of the wider political system. It has to be accepted as a fact of life."[37] Where the rub comes in is in the difficulty and failure to draw a line between what one might call "legitimate" and "excessive" involvement. In India the relationship of politics and the

35. *All-India Credit Survey*, p. 277.

36. Thorner, *Agricultural Cooperatives in India*, p. 11.

37. B. S. Baviscar, "Cooperatives and Politics," *Economic and Political Weekly* (March 23, 1968), p. 495.

cooperatives is often so intimate that the latter become political footballs. Some have gone through that experience (Kerala) with the change of political party leadership. The same holds for individual or groups of coops, and favorites today are on the losing end tomorrow. The cooperative as a unit, or the apex or central banks for that matter, suffer when decisionmaking is motivated by political rather than economic considerations. And these are not rarities. Apart from this, it is well to reiterate the politically inspired gains of the insiders. They accentuate the already existing wide gap between various groups in the community; and, on the top of the so-called "normal" activities of the people of standing and influence, the two cannot but have a detrimental effect on the healthy growth of the cooperative movement.

Conclusion

Such in brief are the principal environmental factors which go a long way to explain why the cooperatives behave as they do. They are by far more difficult to deal with than the so-called "normal" technical cooperative problems. Over the years the latter have been subjected to continuous investigations and an endless stream of recommendations how to cope with them. In the main, the medicine did not take and laments about the state of the cooperative system are as widespread as ever.

The large expansion of membership, financial resources, and loans outstanding are undeniably beneficial to many farmers and are signs of growth. Yet though this paper is not a detailed examination of the system, enough has been said to them that, barring certain exceptions, the signs of growth do not indicate that the cooperatives have finally found the road to health and prosperity. It is appropriate to quote the following sentence from the survey: "Indeed, it might not be wholly impermissible to detect a certain degree of truth in a description we have come across of cooperative in this country 'as a plant held in position with both hands by government since its roots refuse to enter the soil'." Fifteen years later the sheer weight of institutional credit argues against so pessimistic a view, but it cannot be overlooked that, if shorn of continuous pump priming, the majority of the cooperatives would collapse like a house of cards.

No such upheaval is going to take place. What with the new technology in being, more funds will be provided from governmental sources regardless of whether the system works better or not. Competition may or may not smooth out the worst features of credit distribution. One thing is certain: The largest share of the credit will be given to those farmers who are the principal participants in the new technology. They may not utilize it most efficiently for reasons already stated, but the institutional credit will gravitate to them because the new agricultural policy staked its goals on the skills and resources of these very farmers. If cooperative credit should prove insufficient, they are in a position to tap other sources to a greater degree than ever before. The conclusion may be drawn that, as to credit availability, the new technology and the farmers it primarily serves will not be affected adversely to any significant degree.

This is only one side of the story. The other and more difficult one can best be stated by raising the following question: What about the great mass of the farmers who are not part of the new strategy or technology but are capable of raising productivity on their small holdings, who are in greater need of credit than the well-to-do and yet are getting little of it or none at all? To pose this question is not to imply that small farmers have not taken advantage of the new inputs or that they will not do so in the future. They recognize that some of the new varieties and what goes with them spell the difference between subsistence and better living. It is not surprising, therefore, that many have taken them up, but many more —and possibly the majority of them—have not. A recent bit of evidence throws much light on the subject and it is worth recording. "Many villagers said to me," writes an enthusiastic reporter of the new changes afoot, "that they were turning not only green with jealousy for the big landlords and farmers but also red with anger against the stepmotherly treatment meted out to them."[38] This attitude is not an incen-

38. M. B. Lal, "Many-Splendoured March of Farm Revolution," *Statesman* (April 10, 1968).

tive for keener search of inputs, for the channels supposedly open to them are often blocked. According to the same reporter, a farmer, evidently speaking for the multitude, had this to say: "The big farmer gets fertilizer first and we have often to buy it from the black market. He can get any amount of credit in time; we have to wait and wait. Water and electricity are his for the asking. We never seem to get them on time." After a swing through the country the reporter concluded that "The intensity of the feeling varies but never the substance of the complaint." He also leaves the impression, supported by our own observation, that in an atmosphere of rising expectations engendered by the new technology, the complaints can no longer be neglected as in the times past.

None of the above detracts from the obvious merits of the new strategy and its aims. The only thing in question is the short end of accommodations received by the majority of the small farmers and some receiving none. And yet these farmers, largely engaged in subsistence farming, constitute the central piece of the environmental or socioeconomic setting of the village. After all is said and not done about this deeply rooted cause of cooperative difficulties, a more direct assistance approach to this group of farmers can no longer be delayed and for reasons of productivity rather than social equity.

It is highly questionable if the caste system or the existing local leadership can be knocked on the head for at least a generation or more; but it is possible to make a few-acre farms more productive, a condition already attained by some of them. The evidence from the Indian experience demonstrates that, not only are small farms productive in the sense that large farms are; but "In respect to the ratio of output to paid input, the small farm turns out to be more productive than the large farm, and in respect to output per acre, the small farms appear to be even more productive."[39] This assumes, of course, that all things are being approximately equal, in addition to the well-known fact that more labor is used per unit of cultivated area on small farms than on large farms. But above all, the proposition of all things being approximately equal assumes, among other factors, that the small farmer has the resources or credit for the requisite inputs. Since this is not within reach of the majority, it constitutes the crux of the matter when subsistence farming is related to the new strategy or, more specifically, to cooperative credit.

The purpose behind this question raised previously is not so much to seek and immediately find a solution as to stress its urgency. Already the structure of land distribution as well as the distribution of assets in the rural areas is such as to make it extremely difficult for the poorer farmers to maintain their position relative to the top farmers. The scales would now be tilted still further against them. Under such conditions the new technology would be producing not only "kulak farmers" but also a crop of farmers no longer accepting their lot as something nature ordained. As against this prospect, there is little in the contemplated plans of adding new sources of credit that holds out a promise to the commonality of farmers of reasonably priced credit in reasonable volume. What promise there is lies not in future credit activities of the commercial banks, the existing land mortgage banks, or in the proposed Agricultural Development Corporation to be financed by AID. All these are added props for the "strong" of the farm community rather than for the "weak."

That something is amiss in the policy of "everything for agricultural productivity" is gaining recognition. One can cite many a published comment to this effect, but the telling straw in the wind is that the sense of unease about the policy has reached the highest level of government. Jagjivan Ram, union minister of food and agriculture was reported (April 14) to be "worrying" about the new agricultural strategy because it tends to create "a gap between farmer and farmer . . . imbalanced and uneven development of the countryside" by depriving small farmers of facilities that the rich agriculturists enjoyed. "The minister said," the report continued, "that his new approach would continue to be selective but would be based on development of areas 'rather than on the basis of individuals as hitherto'. This ap-

39. Bachman and Christensen, "The Economics of Farm Size," p. 246. The entire chapter bears on this point.

proach would cover both rich and small farmers alike." If the minister is to be taken at his word, though he didn't spell out his approach, it is obvious that to narrow the gap the subsistence farmers surely need something of the assistance currently enjoyed by the "progressive" farmers. If the cooperatives are to remain so largely the preserve of the latter group, a device or agency or a new approach must be created for that specific purpose. The idea that not everything can be done at once on account of shortage of resources is not altogether valid, partly because most of the big institutional credit beneficiaries can find credit outside the cooperatives. Moreover, the needs of the two groups are quite distinct with emphasis on production credit in one instance and consumption or "household expenditures" on the other. This argues for a separate treatment and a separate arrangement. But perhaps more to the point is the record of the Indian cooperative movement which makes clear that the rich and poor farmers cannot sup at the same cooperative table—or any table for that matter.

It is just possible that Mr. Ram's remarks are an augury of new developments; but, until it is known what they are, one cannot be so naive as to believe that Indian agricultural planners in the center, in the states, or the cooperative leadership are prepared to single out at this juncture subsistence farming as a subject for special consideration. If they don't, the answer to the problem is not in taccavi loans, never a significant element in farm credit, or in the agricultural credit corporations which have so far failed to get off the ground. So long as the new agricultural strategy excludes subsistence farming from its purview, this type of agriculture will persist with no benefit to farmers, cooperatives, or to the country as a whole. The preponderance of the costly private sources of credit and the weakness of the cooperative credit institutions in financing them are not just an accident, but a part of the subsistence economy itself.

We noted earlier that an undetermined number of small farmers have managed to benefit from the new inputs. It follows that the idea that one must start from scratch doesn't apply here. The task is one of widening the breach—if only on a selective area basis—thereby strengthening the position of cooperative credit or of any new instrumentality for purveying credit to small farmers in order to help them shift more of the resources for purposes other than self-consumption. And if it is true as the new strategy proclaims that the rich farmers cannot maximize production on their own, how much truer this is in the case of subsistence farmers? If it is true, as indeed it is, that rich farmers treat cooperatives with scant respect, using them for all they are worth, laggardism among subsistence farmers is more readily understandable and must be treated as another unpleasant fact of small-scale farming which could demonstrate its productivity capacity on a wider scale, given the opportunity. This is in relation to the often heard argument that they are not creditworthy or trustworthy. So possibly they are and so they will remain until the initiation of a significant change in the character of their existing economies. It should be the business of government, therefore, to spare some effort and some resources to help break the circle of low productivity, indebtedness, and poverty among the multitude of Indian farmers.

From all this there emerges no neat a, b, c, d-like scheme as to how the government should do this or that to help change the character of subsistence agriculture or how to recreate the cooperatives into sounder institutions to meet the demands of the new technology. Nor is there any intention to correct this seeming omission in these last few paragraphs. The reader accustomed to a string of "recommendations" following a lengthy piece of paper is in for a disappointment, but there are reasons for this omission. The purpose of this exercise was mainly to explain the principal "whys" of the failure of recommendations to rehabilitate the cooperatives and why small-scale agriculture is entitled to a larger share of input—credit to raise its productivity. Also, it would have been presumptuous on our part to do much more than that because cooperative leaders, farm officials, and academic students of the system have stated repeatedly what this is all about—including recommendations numerous enough to stretch from Delhi to Bombay. The problems of subsistence agriculture have not yet been subjected to the same searching analysis as have the cooperative ailments, but this is an obstacle that can be overcome. India has the professional competence to analyze and prescribe curative measures as well as to point the way to implementation. What the country lacks is something else, and on this we shall wind up this part of the paper.

India appears to be on the verge of significant innovations in agriculture, but India is far from prepared for the organizational, cultural, social, and political changes that must accompany the fullest realization of the technological changes. To move from unpreparedness to preparedness presupposes a stable, progress-oriented government with power, recognition of what the people want, and determination and professional competence for sustained agricultural progress. When this theory-reality is measured against India since independence, the conclusion is that, while much of it is in force, one vital element is missing. India has a stable government, changes in state governments notwithstanding, and India is progress oriented; the center and the states have been exercising power; government knows that the people crave for a better living; and India has a pool of professional competence which should be the envy of many other developing countries. What India lacks is the crucial ingredient of will and determination without which no reform can be carried out. The preference, therefore, is for action meeting least resistance, which is to say, for a stance that eschews hard choices and avoids stepping on a good many economic and political toes. This, more than the lack of resources, tells a good deal about the cow question, the state of land revenue, grain procurement, extension service, cooperatives, subsistence farming, the doldrums of the agrarian reforms, and the condition of the panchayats. Generally speaking, it explains why recommendations and realization so often diverge.

So much of what is being said involves painful breaks with customary attitude and beliefs, ways of living, doing things, and profitmaking. They are painful in any society when shifts in privileges are the order of the day. They are particularly so in agrarian India, which over many millennia has created so stratified a society along economic, social, and religious lines that the winds of change are only beginning to make themselves felt. In these circumstances and at a time when political fragmentation is at its most pronounced since independence, the exercise of power in relation to the issues discussed is admittedly more circumscribed; and determination or political nerve, the most formidable engine of economic progress, is a rare commodity. This is as of now, but it is not the end of the story.

Lurking in the background is Gandhi's admonition that "It should be the business of every government of India to dry every tear of every Indian eye." Grandiloquent and perhaps too vague for some tastes, but this is *the* true, if overgeneralized, prescription for Indian ills. In a real sense Gandhi's dictum has not lain dormant these many years—not in the eyes of those who remember the India of 1948 and know something about the India of 1968. But it may be taken for granted that no thoughtful and responsible Indian in or out of government would dispute that by far the hardest task of reconciling gaping differences of wealth and power in the countryside still lies ahead. If it is to be translated with reasonable success and one hopes in an orderly fashion, willy-nilly there can be no avoiding of "unceremonious vigor" or political determination. When that day comes and on the assumption that the agricultural technology is being sustained, *Indian*-made recommendations and *Indian* forms of implementation will be finding their marks in evolving a better extension service, soundly managed credit cooperatives, more meaningful agrarian reform programs, village panchayats which are developmental not in name only, and much else besides.

Admittedly, concrete action with regard to any of these spells a lot of unpleasantness for the politicians of the union and state governments. India is no exception in this respect, but neither in India nor elsewhere can this stand in the way much longer of meeting economic and social necessities. One can think of no better overall prescription than the merger of India's existing technical skills with a strong dash of political discipline. In this combination lies the key to much that needs to be done, and only then will the "should be dones" acquire a significance so largely lacking now.

50. *Punjab Field Trip*

Late in February 1969 Ladejinsky made a brief field trip in the Punjab, the heartland of the Green Revolution. This piece, dated April 2, reports his observations on the progress, characteristics, and consequences of the Green Revolution in its early phases. The article, later published in the *Economic and Political Weekly* (Bombay) on June 28, 1969, was widely noted. Amid the euphoria created by spectacular production increases, it called attention to the other side of the coin—soaring land values and rents, tenant exploitation and displacement, and growing inequalities of income.

When published in the *Economic and Political Weekly*, the title was changed to "Green Revolution in Punjab—A Field Trip." This article was subsequently reprinted by the Agriculture Development Council, Inc. of New York in June 1976.

Introduction

THE WRITER OF THIS NOTE visited Punjab between February 23 and 29 for the purpose of observing the conditions of the wheat crop, farm profitability, procurement prices, land values, land rents, and the effect of the new technology on the absorption or employment of the rural landless. Time being very short, we avoided local officials and spent no time collecting official statistics. Instead, the subsequent paragraphs are based on contacts with farmers in the field or in villages. The sample is not overly large and no claim is made for unchallenged accuracy of the conclusions, particularly in regard to the last point. Nevertheless, we believe that the pattern which emerges reflects a set of existing conditions.

Productivity, Prosperity, and Polarization

Rainfall in the Punjab has not been as plentiful, or as ideally spaced as a year ago. The saving feature are the tube wells which dot the fields as far as the eye can see; but even their owners are as one that they, too, will be adversely affected to the extent of 5 to 10 percent decline in yields per acre. Yet, preliminary estimates point to a considerably greater output than the record crop of 3.4 million tons in 1967–68. Two developments explain this. The first is an increase in the wheat acreage caused by a sharp fall of potato prices and the consequent conversion of a considerable part of this acreage to wheat. The second and more important explanation is the continued upsurge of the new technology, as evidenced by the following figures. In 1966–67 the high-yielding varieties covered 4 percent of the wheat acreage, the year following 35 percent; in 1968–69 the estimated coverage is 60 percent. This will more than compensate for the potential loss attributed to low precipitation.

One of the outstanding impressions carried away from talks with farmers who fully adopted the new package of practices is not so much the increase in wheat yields from 16 to 36 maunds[1] per acre (average for Ludhiana) or 50 to 60 maunds on many individual farms, but that they have pretty much reached a plateau or, as one of them expressed it, a state of "stagnation." Mr. Singh, one of the most successful farmers of Ludhiana district, thought that the stepping up of investment from the current Rs200 to 250 per acre would be counterproductive unless new varieties are developed with a capacity to sustain larger ap-

1. One maund equals 82.5 pounds.

plications of fertilizer. He and his opposite numbers look forward to a new "miracle" variety, commonly referred to by them as the "triple dwarf." Wheat breeders are rather skeptical about its potential yielding capacity, and the prospects just ahead may prove disappointing to the eager innovators. They will have to be satisfied with what has been attained; and this, as we shall point out, should not be hard to take.

The farmer in question has sixty acres of tube-well-irrigated land and all the modern equipment a highly successful owner operator can acquire in present-day India. His principal crop is wheat and he manages to double crop all his land with subsidiary crops, although the intensity of cultivation for Punjab as a whole is 135 percent. The rate of utilization of high-yielding varieties on his farm is worth recording and it is as follows: 5 percent of the acreage in 1965–66, 30 percent in 1966–67, 65 percent in 1967–68, and 85 percent in 1968–69. There will be no complete coverage unless a variety is developed with color and taste to suit the consumption preference of the producers. Mr. Singh is one of the rare farmers among even the large holders who keeps detailed accounts of all his operations and has no reluctance to exhibit them. Every conceivable item relating to the cost of production is there, including amortization of everything that is amortizable, and not excluding food for seventeen members of his family, three permanent workers, and food for fifteen to twenty daily laborers during the peak of the season. The last item, return to management, sums up the 1967–68 season: It shows a net income of Rs1,600 per acre. The figure is made up of sales of seed which commands a price of Rs180 per quintal as against Rs75 for ordinary grain. But had he devoted all his land to grain only, his net income per double-cropped acre would have amounted to Rs800 to 1,000. The extra profit from seed production will come to an end in another year or two, for too many farmers, he explained to us, are engaging in seed multiplication on their own. On the whole, it was clear that he and many like him—even if with much smaller acreage—have taken advantage of the new technology and profited accordingly.

In calculating his net income or return to management, no reference is made to taxation. This is not an oversight. With a land tax of Rs2 or so per acre and no income tax, the omission is understandable. As with this farmer, so with the rest. A farmer richer still told us that he had written to the state government suggesting a scheme of progressive taxation. Such a rara avis we haven't encountered before or since, and we still wonder whether he sought release from a surfeit of tax-free money or was merely pulling the leg of an inquisitive visitor. The fact is that the well-being of Punjab is made possible to a considerable degree by an elaborate infrastructure in the form of irrigation projects, power development, agricultural research, and seed development involving large amounts of state and center expenditures. And yet the principal beneficiaries are virtually innocent of taxation to sustain the ever-rising revenue needs of the donors. What can be said for Punjab is that it represents only another instance of practically no direct taxation typical of all the states of India. Clearly, the much fought over "no taxation without representation" doesn't apply in rural India in general or in Punjab in particular. The well-off farmers are so heavily represented in state legislatures that the historic adage has been reversed to read "no taxation despite representation."

Aside from looking forward to a still more productive new wheat variety and the complaint about the absence of combines to expedite the harvest, Mr. Singh's one immediate concern is a new weed, almost indistinguishable from the wheat, which has infested his fields this year—and grows as fast as it is being pulled out. He looks upon it as a not unexpected side effect of new varieties. So far the weed disease is not common and the geneticists of Punjab University have been alerted to it. What concerns them more is the shape of wheat prices now that the harvesting season is within sight. He and other progressive farmers know that the total wheat output and marketable surplus will be larger than last year and the chances of a drop in price, whether officially procured or not, are real. Coupled with this is the uncertainty about the size of the buffer stock and therefore the volume of the surplus the central government will be willing to procure. At the same time, there is the feeling that regardless of the procurement prices and quota fixed in Delhi, their state government will not let them down on both scores. As to the procurement price, they would like Rs100 per quintal; they know this is out of the question and will be content with last year's price of Rs75 per quintal. If so, they will market their surplus with even greater alacrity than in the 1967–68 season.

All in all, one is impressed with the sturdy, self-confident, purposeful, innovation-minded group of farmers who in a short span of years have succeeded in translating promise into reality. So much so in fact that an occasional though rare sight of a camel pulling a traditional plow or the more frequent sight of a Persian wheel doing its rounds appear as anachronisms from a bygone age. Despite these apparitions, what is immediately striking is that for the owner class of 10 to 20 and more irrigated acres of land, making money is a new way of life. The process of self-enrichment is there for all to see; and, for those who remember the rural Punjab of Malcolm Darling's classic, *The Punjab Peasant in Prosperity and Debt*, observing the more recent changes is an exhilarating experience. We are not referring to the big farmers only. The cultivator in mind with nine owned acres of land, a tube well of his own, a ripening thick stand of wheat promising an abundant crop, and no debts and no borrowing—so he assured us—is the more telling symbol of what the Green Revolution stands for than the kulak of 50 acres who struck it very rich indeed. Webster defines a peasant as one "tilling the soil as small landowner or as laborer . . . a person of low social status especially when comparatively uneducated and uncouth." Neither in appearance nor speech nor judging by the transformation of his nine acres is he any longer a peasant, and this is one of the attributes of the Green Revolution. But lest one wax too lyrical, it is well to remember that the growing polarization in agriculture between the rich and the poor is also an integral part of the Punjab scene. Part of this story is told in a latter section of this note, and it suffices merely to mention here that the welcome drive for productivity and betterment doesn't obscure the fact that there remain the small farmers, the tenants, and the landless agricultural workers who do mar the picture of achievement and contentment in India's premier state of rural development.

Land Values, Land Rents, and Tenancy

Land values in Punjab have been rising over the years, but it is the steady penetration of the Green Revolution that has pushed them up to new peaks. There is not enough land changing hands to turn Punjab into a boom and bust real estate operation; but, concerning what little is being bought and sold, the farmers' refrain is the same: Today's land prices are going to be higher tomorrow. Land sold five years ago at Rs2,000 or 3,000 or 5,000 per acre sells presently at Rs10,000 or 12,000 or 15,000 per acre. Instances are not lacking of prices doubling within the last two years. Farmland close to a highway is particularly high; but any owner of productive land, regardless of its location, is in a sellers' market. The profitability of farming explains it and with it the changing attitude of using savings and making investments with an eye to the main chance. This doesn't mean that traditional investments are obsolete. In the debate on the union budget, Morarji Desai is reported to have said that some rich farmers buy gold "not by the tola[2] but the pounds."[3] This authoritative source is not disputed; but it is equally true that the major investments go into the re-equipment of the existing holding to fit the current phase of agricultural innovations and into land acquisition, if such is for sale. This, as one farmer put it, is the "real gold" and the source of security. Hence the temptation of the well-off farmers to invest in more land that successful farming and tax-free income can buy. What has given land transactions a risky speculative tinge here and there is the appearance on the scene of "gentlemen farmers." They are mostly urbanites who know little or nothing about farming but have a great deal of "unemployed" rupees, mostly acquired through undeclared earnings. They look upon investment in agriculture as a means of legal-

2. One tola equals 11.5 grams.
3. *Hindustan Times* (March 21, 1969).

izing illegal resources; as a tax haven, which it is; and as a source of earning tax-free supplementary incomes. In the circumstances, land prices are of no great moment to them. They are unfair competitors in a limited land market, and there is no love lost between the would-be farmers and the genuine farmers, with the latter predicting that the former will come to grief. In the meantime land values are going up and up and evidently with little thought as to whether productivity and prices will sustain them.

From rising land values to the enhancement of rents is hardly a stone's throw. But this in itself is not the whole story, and a bit of background is in order. Prior to the enactment of the Punjab Security of Land Tenures Act of 1953,[4] rentals paid by the estimated nearly 583,000 tenants and owner tenants or tenant owners was roughly 50 percent of the gross crop, with the owner contributing little to the making of the crop. The act provided for a maximum rental not exceeding one-third of the crop. But this provision, like security of tenure, ceilings on holdings, and so forth, has been so emasculated that the very title of the act is one of the misnomers of the agrarian reforms of India. The result was twofold: large-scale ejection of tenants by owners on the ground of land resumption for self-cultivation and the persistence of traditional rental arrangements on the remaining tenanted holdings. No data on ejections are available, but the following tells something of their magnitude, though the data are probably only approximations. Thus we read:

> The law regarding security of tenure appears to be on the whole somewhat illusory. This is supported by the fact that the number of tenants in existence in the Punjab area at the time when the Punjab Security of Land Tenure Act was amended in 1955 has greatly decreased. From the information supplied by the State Government on March 30, 1955, it appears that the total number of tenants including tenants-cum-owners was 583,400 . . . From the information supplied by the Punjab Government recently (1964), it appears that the number of tenants who were holding land at the commencement of the Act and are still holding land as tenants is 80,250. It seems difficult to account for the large difference between these two figures.[5]

Actually the difficulty is not very great; it is an open secret that the difference between 583,400 and 80,520 is largely due to out-and-out ejectments or the transfer of tenants to the status of sharecroppers or agricultural laborers. So much for the background and tenancy conditions accelerated by the so-called reform of the tenurial system.

The new technology in Punjab has demonstrated how very profitable farming can be; and, coupled with high land values with the absence of any effective, restrictive rent measures, the upward revision of land rents was inevitable. And so it is, rents having moved from a 50 to a 70 percent share of the crop, or from Rs500 to Rs750 per acre where the arrangement is in cash. Landowners and tenants readily testified to this change. Considering that a good acre of irrigated wheat land on a well managed farm yields now 35 to 40 or more maunds, in theory the tenant should receive more maunds per acre than when he was paying 50 percent of the crop. This presupposes, however, that he uses the full package of inputs in a manner applied by the innovating cultivators. Owner farmers with sizable holdings who also lease in land are in a position to do just that. On the other hand, interviews with "pure" tenants do not lead to the same conclusion. For them, sharing the crop on a 70–30 basis is not a boon. Rising rents in consequence of rising land values may be an economically valid and unavoidable process but a hardship nevertheless for those who cannot extract from the land a return large enough to meet living requirements over and above the rental obligations and inputs which must be largely their own.

Finally, some tenants receive the short end of the stick because of the familiar owners' fears of claims to occupancy right. This was apparent in the case of an owner of 50 lush

4. Punjab then included what is now the state of Haryana.

5. Planning Commission, *Implementation of Land Reforms* (New Delhi, August 1966), p. 116.

acres under vegetables on the outskirts of Ludhiana city. His is a showcase with five to six crops to the acre, an operation of which the most skillful Chinese or Japanese farmer would approve. In addition to his own land, he leases 20 acres under wheat but, as he explained to us, not on the 70–30 basis. He farms his land on something less than the traditional rent because he is creditworthy; because the lessor knows that he can get the land back any season he wants to; and because he doesn't want to deal with ordinary run-of-the-mill tenants lest they make claims to rights in the land, even if they be only imaginary rights. Our "tenant" in turn sublets the land to real tenants, providing at the same time all the equipment and inputs. When the crop is in, he deducts, in kind, all the costs connected with producing it, the remainder being shared 70–30. The answer to the question as to what the 30 percent share meant to the tenants in terms of produce or cash did not come through. What did come through was this: "If my 'partners'," as he called them, "don't like my terms, I can always find others to take their place." Here he spoke the truth. Official information speaks of only one-seventh the number of tenants Punjab had a decade and a half ago when they cultivated approximately one-fifth of the cultivated land. It says nothing about the displaced or about the tenants now disguised as sharecroppers or agricultural workers, mostly the latter. In the field this becomes clear, though it cannot be measured statistically. This is why the big owner "tenant" knew what he was talking about when he said there was no lack of others willing to become his "partners."

The sample relating to rents and tenancy was not large enough to assert without equivocation that the conditions described hold good for all of the Punjab, but the consistency of what we observed in the districts of Ludhiana and Jullundur leads us to the conclusion that it may well be fairly representative.

New Technology and Employment

The new technology is characterized by frequent application of water; fertilizer; insecticides and weeding; double cropping; bigger crops; and a larger volume of transportation, marketing, and food processing. All these are more labor demanding on primary, secondary, and tertiary levels and therefore the prevailing belief that the new farm practices would relieve the acute problem of underemployment in rural areas—and even beyond. The very same practices, however, also give rise to such labor-saving devices as tractors, rotary ploughs, harrows, and tillers to go with them, electrically-operated tube wells, sprayers, and threshers—to mention only modern equipment already present on some Indian farms and particularly in the Punjab. So far, in discussions relating to the Green Revolution, the stress is on its labor-requiring aspects while giving little play to the widespread experience of other countries that as agricultural technology grows in sophistication it leads to less employment of labor. The probable reasons for current emphasis on the one as against the other are the early stage of the package of practices and the prevalence of the small-sized holding in India. But be that as it may, it is not our intention to argue the point of more or less employment at this or that juncture of technological development but to record our observations of the current rural labor situation in Punjab and to draw some tentative conclusions. Before we touch upon this, it is well to begin with the prevailing farm wages.

Wage Scales and Duration of Employment

We have inquired into this matter on every possible occasion, and from the testimony of employers and employees it is clear that daily farm wages in the past five years or so have moved up from Rs2 to 3 to Rs4 to 5. In both instances one must add the food and tea provided by the employers, which bears a cost of about Rs2. There are significant exceptions. On two occasions in the district of Jullundur, more typically agricultural than Ludhiana, reported cash wages were as low as Rs 3 to 3.50. On the other hand, in the very crucial harvest season, wages may go up as much as Rs2, making a total daily wage for a period of a month of Rs7 to 8 a day—not counting the food portion. Permanently employed laborers are not doing as well. Their monthly wage is around

Rs100 to 110, exclusive of food. Their advantage lies in the year-round employment. Reports gathered in Delhi that wages have shot up to Rs10 to 15 and more a day are without foundation. The same applies to industrial wages. As a check against our information from farm sources, we sampled a few industrial undertakings employing thirty to fifty laborers. Aside from the important consideration that the labor there is on a more or less permanent basis, wages are not as high as in agriculture. The Texan Industries of the small Goraya City (Jullundur district) produces axles for cars, employs forty workers, and draws upon labor from the surrounding rural communities. Unskilled labor is paid Rs3 per day, graduating to Rs6 for the most skilled—and no food is offered as part of the wage arrangement. Similarly in two other enterprises of about the same size. In all these wages have gone up somewhat, but the real explanation why they lag behind farm wages is the availability of labor throughout most of the year. The owner of Texan Industries, who prides himself on his trademark "Texan," looks like a Texan and thinks that in his free and easy manner he is a transplanted Texan, recounted for our benefit his daily experience of refusing employment to people looking for jobs. An encounter with a large group of landless laborers digging up sand and gravel and carting it to the railway tracks revealed wages that compared well with those of unskilled labor in industry. Their pay for heavy labor, closely supervised by four straw bosses, was Rs3.50 compared with Rs2.50 about two years ago.

With reference to the farm labor situation in the areas visited, the conclusion is that, with the spread of the new technology, the demand for labor has increased and that the increase in wages is partly a reflection of that demand and partly a reflection of the steady rise in the cost of living during the five-year period. The net gain in monetary terms over and beyond the increased cost of living cannot be very large. Even if it is a rupee or two, it is not as important as the answer to these two questions: How much greater is the demand for farm labor or, more particularly, how many more days are they employed on the farm now compared with the pre-Green Revolution days; and is there a shortage of farm labor in Punjab, as is often heard in various circles? With regard to the first question, the estimate ranges from 150 days per year to a high of 180 days. If it is the latter, it would appear that they are employed from 30 to 40 days more than was customary in the past. Any such estimates are only rough approximations, because, surprising though it may seem, by and large neither among the employers nor among the employees is there anything like a consensus on just how many days they do work on the farm. The progressive farmer already referred to with sixty owned double-cropped acres has a clearer idea on the subject. With tractors and other labor-saving devices, he economizes on labor; but this is more than offset by double cropping, weeding, and his eagerness to harvest the crops as quickly as possible to avoid shattering of a bumper crop and to expedite its threshing before the rains set in. The upshot is that he employs, depending upon the season, from 10 to 20 percent more laborers, in addition to the three permanent ones he normally employs.

As to the question whether the new farm practices have given rise to labor shortages, our answer is that they have not. To an owner increased demand for labor and higher wages may appear as conditions resulting from a labor shortage, but when pressed for specifics they know better. The big farmer near Ludhiana city, who employs at any given time fifty to sixty casual laborers to keep his factory-like vegetable farm going, went so far as to say that 30 to 40 percent of Punjab's labor is imported from Uttar Pradesh. It is possible that this is the case in border areas, albeit on a smaller scale; but upon closer examination neither he nor any other owner could furnish any evidence in support of the statement. In his own case not only does he import no labor, but he manages to secure some Punjabi for a song; a group of women we watched preparing cauliflowers for the market received nothing more than the discarded leaves, used by them as fodder for their animals. Getting something for nothing in the very environs of the bustling with activity, fast-growing, commercial and industrial city of Ludhiana is indeed an entrepreneurial feat. It is doubtful, however, if this particular sense of achievement would have been his if he had to contend with a critical supply of casual labor. In short, in answer to the much-repeated question, "Do you experience a labor shortage?" the replies have been in the negative. Our sample being rather small, we wouldn't swear on the Bible that this is typical of Punjab without exception, but according to our information no field has remained unattended even during the harvesting season due to lack of labor. The same is true of industrial labor in Punjab's numerous small-scale industries, which all depend upon its supply from the rural communities. In the light of this, the much talked about labor shortage supposedly created by the new technology essentially stands for a greater demand for labor, particularly during harvesting, and hence the higher wages caused in part only by this condition.

Employment Prospects

It is apparent from the foregoing that, at this stage of Punjab's rural development, the demand for casual labor has increased despite its 14,000 tractors[6] and many more of threshers. By virtue of this alone, the agricultural laborer is somewhat better off, and particularly in a state where agricultural technology is spreading alongside numerous small industrial establishments creating their own demand for labor. One cannot gainsay the importance of this combination on top of the demand generated by the very prosperous farmers and nonfarmers for employment other than strictly agricultural. As against this, it should be noted that even in the present-day Punjab there is a sufficiently large reserve of labor, the supply of which becomes "critical" only during a period of a month or two at most. Underemployment still exists, although facts and figures are difficult to come by. Whatever they are now, this observer left Punjab with the strong impression that underemployment there is bound to grow apace. This melancholy thought would jar some observers who have all but concluded that the casual worker has been well taken care of. And yet it is not a matter of pessimism versus optimism when this question is posed: What of the future for the landless workers when the process of mechanization moves into a higher stage and becomes as irreversible as the Green Revolution which fathered its current stage? The question is raised to highlight some of the thinking among the very successful farmers.

If this development comes to pass, the demand for farm labor in Punjab would inevitably go into a sharp decline, even if the overall growth of the population were arrested at the present level and even if the same applied to the tendency of rounding out the bigger holdings at the expense of the very small and uneconomic ones. Since neither is likely, especially the first, Punjab will be faced with the problem of what to do with a rising army of rural underemployed and all its consequences. There is firm ground for this assertion. The leading farm innovators of the state are no longer satisfied with the equipment they have, although admittedly it represents a great advance over what they had not many years ago. To be sure, they are not given to the flights of fancy of a farmer we know who cultivates 300 acres of irrigated wheat land. Not ten years ago he was in the bullock stage. Presently there is not a single bullock on his big spread; and the seven tractors (two of them spares), the rest of the equipment to match, the eleven tube wells, and practically no taxes enable him to net Rs1,000 per acre. Formerly, he single cropped 50 acres with the aid of twelve workers; now he double crops 250 acres with the aid of nearly a hundred workers. Keeping his vast irrigation network shipshape absorbs a good deal of the labor. He thinks he can better himself, and he is in search of the latest labor-saving devices. Aside from looking around for the conventional yet more efficient tractors and specialized implements, his thoughts about the future of his farm are touching the science fiction stage. He dreams of the day, induced by an article in a Soviet magazine, when he will perch atop a tower and guide the work of his tractors by remote control. Practical, rich, and restless farmer that he is, he went to the trouble and expense of erecting strategically located utility poles as a preliminary step on the road to his magic tower. Useless to him now, they stand there, thirty feet high, just in case the dream comes true.

Most well-to-do Punjab farmers are more

6. The total number of tractors in all of India is estimated at 70,000 to 90,000.

modest in their expectations, but economy of operations and returns are uppermost in their minds. Not surprisingly, they would like more modern and more powerful tractors and specialized equipment to go with them, and some would like to exchange their crudely made and none-too-efficient threshers for a variety of a combine that would harvest, thresh, fill, and tie the sacks ready for shipment. This is not universal, but the desire is there. If this is realizable in the Punjab, there would come into being the modern agricultural version of the "trinity"—the tractor, the combine, and electric power—which in other countries has displaced hired labor to a vanishing point. Precisely the same is not going to happen on the irrigated lands of Punjab with its demand for hand labor compared with what did happen in the dry farming areas of the United States and in other countries. But the difference is only one of degree; and labor elimination, particularly during the peak seasons, is bound to take place.

Discussions among the watchers of the Green Revolution as to whether government policy could or should retard the rate of mechanization is academic, certainly for the present. According to A. P. Shinde, union minister of state for food and agriculture, the government has decided to meet some of the shortage of agricultural implements by importing during the current financial year 15,000 tractors, 4,000 power tillers, 10,000 discs, 75 combines, and a number of other items. Additionally, firms are being licensed for the manufacture of power tillers.[7] The numbers may not seem large in view of the pent-up demand, but the move is clearly in the direction of intensifying mechanization. The advocates of a "go-slow" policy may not have had the last word; but one would be bold to predict that, in a contest—if it should come to that—between the further refinement of technology on the one hand and social policy on the other, the latter would win. Moreover, as development feeds upon development, so labor-saving devices feed upon the preceding generation of the same. Freezing the level of mechanization can no more be contemplated than freezing any of the new practices, which in turn depend upon more efficient, labor-saving equipment. Much of this wouldn't apply in anything like the same measure to the principal rice areas of India, and the "whys" need not detain us here. But talking about Punjab and keeping in mind most of the northern wheat belt of India, a reversal of the current demand for more employment created by the present stage of the new technology is inevitable. This development is not around the corner just yet, but it is in the making all the same.

7. *The Economic Times* (March 29, 1969) and *The Indian Express* (March 29, 1969).

Summary

Such are the findings of a brief trip through two of the principal districts of Punjab. For reasons pointed out in the first paragraph, we recognize that the random sampling carried out on the fly and hastily jotted-down bits of information are a method that could be improved upon. We recognize, too, that loquacious as a farmer can be about matters that touch him most, not everything he shares is necessarily all he cares to share; nor for that matter does everything he says filter down through the interpreter in a manner that leaves nothing to be desired. And this is true of the many-sided Mr. Sharma, one of the [World] Bank office drivers who manfully doubled in brass on this as on another occasion.

With these caveats out of the way, we believe that the topics dealt with and the conclusions drawn could stand the test of more rigorous field observations. Meanwhile, in summary, the conclusions are as follows: The presence of the Green Revolution is very much in evidence—and so are its consequences; Punjab's crop prospects are better than ever, and procurement prices are likely to be sustained. Owner farmers with irrigated land are making money hand over fist; and the bigger the farm, the more they make. Of the "burden" of taxation, there is none to speak of. Land values are spiraling; land rents are going up; and the condition of the tenants is not better—if not worse. The demand for casual labor has increased and so have wages, and the landless laborer is somewhat better off than in the past. For the moment there appears to be no shortage of labor in agriculture or industry, but our contention that Punjab imports no labor from Uttar Pradesh could stand closer scrutiny and we intend to re-examine this point at an early opportunity. Underemployment has not been eliminated, except at the peak seasons; with the further sophistication of mechanization, Punjab will be confronted with an ever-rising number of underemployed. This is not about to take place, but the prospects are there nevertheless. One more conclusion is to be noted, and it deserves separate treatment.

Regardless of whether an observer probes deeply or ever so lightly into the Punjab scene, the signal impression he is bound to carry away is one of the air of prosperity that permeates the state. Those bent on measuring everything quantitatively can feast on the 60 percent coverage of wheat land with high-yielding varieties, the doubling of yields per unit of land, the five- to sixfold increase in fertilizer consumption between 1962–63 and 1967–68, the 45,000 tube wells, the steady increase in the number of tractors, the waiting list of would-be tractor purchasers that stretches from here to there, the 5,000 threshers reportedly sold this past year, and so forth. This and much else not statistically measurable already remarked on elsewhere testify to Punjab's rural muscle. All this is true, but it is not without a fly in the ointment. Growth and prosperity cannot hide the fact that the new agricultural policy which has done a yeoman's job in generating them is also the indirect cause of a widening of the gap between the rich and the poor. Precisely because the Green Revolution has found its widest application in Punjab, the probability is that, relatively speaking, the gap is greater there than in any other part of rural India. The assumption that the lot of the landless worker has undergone a change for the better because he earns Rs2 a day more or is employed a few weeks more is correct, but it doesn't measure up against the well-to-do farmers who are better off still and that rich farmers are getting very much richer. Nor can one leave out of account the rising insecurity of tenants, reduced in number though they are, and that the overall conditions of tenants and sharecroppers are anything but improving.

Punjab is not without its small farmers and some of the problems this category entails. There are, to begin with, an awful lot of them and, numerically speaking, Punjab is no different from any other state of India. According to a combination of data relating to 1955 and 1961, farmers operating holdings less than 7.5 acres accounted for roughly 73 percent of the total number of farm households and just as roughly for 20 percent of the land. We barely sampled them, but we assume that their condition is better than the generality of their counterparts in dry areas or in areas where irrigation opportunities are not as favorable as in Punjab. But in relation to the growing gap in income distribution or levels of living, the situation is not as reassuring as it might appear when one looks at the innovation indexes of Punjab. According to an estimate by one of the keenest students of the new agricultural strategy, not more than 10 percent of the 73 percent of the small farmers are in one way or another involved in the innovation process. The explanation is not in a lack of interest to partake of it but in the restraints imposed by the lack of resources and the failure of the cooperative credit system to help them breach this formidable barrier to any significant degree. How badly small farmers fare at the hands of the credit cooperatives has already been described.[8] The rising voices in favor of supplying small farmers with more credit may eventually lead to a desirable change, but in the meantime it can be said that even in the up-and-coming Punjab inadequate participation in the new technology is another sign of the increasing economic polarization.

Neither self-enrichment—a very encouraging development in itself—nor the less promising outcroppings rooted in the peculiarly Indian rural conditions are arguments against the new technology or the urgency to strengthen it in every possible way. There is a need to expand its coverage in order not only to narrow disparities in income but also to curb the sociopolitical manifestations of the rural affluent, which often tend to thwart the developmental process through their influence in state

8. Delhi Office, "A Note on Small Farmers" (February 13, 1969); "A Note on Distributional Problems" (March 15, 1969); and "The New Agricultural Strategy and Institutional Factors" (particularly the section on rural cooperative credit) (1968).

legislatures and Parliament as well. Cases in point are the scrapping of land revenue in several states, the resistance to the introduction of fair rates for irrigation and power, the opposition to raising land tax rates which are totally out of line with mounting land prices, productivity and agricultural prices, and the general refusal to consider a revision of rural taxation in consonance with the changing conditions in agriculture. For all of this and much else already noted, no government can accept the existing situation overly long even if it be defended on the ground that in certain circumstances this is the price or penalty an early stage of economic growth exacts. It is therefore a welcome sign that the growing gap in income distribution in the countryside and its disturbing political connotations are becoming subjects of increasing concern on the part of the union government of India. Without detailing the move to do something about it[9] or the plethora of statements to the same effect, the aim is not to circumscribe the scope of the new technology but to extend it beyond its present unstated but nevertheless real limits. The same goes for Punjab. Despite the state's innovative character, the unquestioned progress it enjoys, and the still brighter promise it holds, there are chinks in the armor of Punjab's prosperity. Not to take notice of them would be to overlook some of the significant, if indirect, consequences of the Green Revolution.

As already indicated, these consequences can be dealt with, although not in terms of final solutions. The scope of the new technology can be widened, especially if the bias against institutional change gives way to recognition of the interdependence between technology and socioeconomic reform. Cutting across all these considerations is the highly significant factor that Indian agriculture is increasingly amenable to renewal because a rising number of farmers wish it that way. The potentials of this mental attitude, even if not often supported by resources to bridge the gap between desire and fulfillment, cannot be overstated. It derives from the proposition that, at long last, in India too it is being demonstrated that ideas have the remarkable property of being powerful instruments of change and mutually self-reinforcing. This may sound "old hat" in a highly developed country, but it is a novel and most important tool among large groups of Indian farmers, particularly when the outlet for new aspirations is helped along. Helping it along is the issue not only in Punjab but in backward Bihar as well, where we recently visited. What with a level of literacy not more than 15 percent and other problems piled upon problems, Bihar is not Punjab; but there, too, not many farmers still need to be persuaded that the new technology is good for them. But what is in inadequate supply are the "outlets," except for the relatively few. In this regard the two states are not altogether dissimilar, although in most other vital characteristics they are poles apart. The moral of it all for Punjab (as of Bihar) is this: If India is to take full advantage of the sharply awakened anticipations for betterment among all manner of farmers, it is time to re-examine the current agricultural strategy and chart its course for the years immediately ahead.

This may well be provided by the All-India Agricultural Commission about to be created. While all its terms of reference have not yet been spelled out, according to Jagjivan Ram, the union minister of food and agriculture, the commission "would examine the present conditions of agriculture and rural economy and formulate directions under which the infrastructure for agriculture should be developed to keep up the tempo of growth." Expanding on the theme, the minister noted that the time had come "to take steps for all-round development of the agricultural economy" and not only in areas "of relatively larger production potential." In the same vein was his reference to the "imbalances" between small and big farmers. In his view these had existed for a long time, but "they had been accentuated by the intensive agricultural strategy followed over the past few years." Whether the "imbalances" examined by the commission will cover the whole gamut of problems touched upon in this paper and whether the prescriptions and their implementation will culminate in more widespread increases in production, better income distribution, and peace in the countryside cannot be foretold at the moment. The stresses and strains

9. See Delhi Office, "A Note on Small Farmers."

generated by the contending rural economic and political interests will undoubtedly cast their shadow on the ultimate results of the contemplated effort. In the circumstances, one can only express the hope and the wish that the agricultural commission will see fit to deal with all the pertinent issues as an integral part of the future development of Indian agriculture.

51. *Bihar Field Trip*

If the Punjab field trip was a vivid sketch of the advent and early effects of the Green Revolution, this paper on Bihar is a full-blown portrait of the same, in a state noted for tenurial relations so exploitative that frequent and bloody outbursts in the countryside have been common. Ladejinsky points up these conditions when he digs up and exposes an official state government instruction calling for the sabotage by field officers of tenurial rights legislation. The picture presented is an extremely dour one. But "All said and done, it is not the fault of the new technology that the credit service doesn't serve those for whom it was originally intended; that the extension service exists largely in name only; that tenurial legislation is deliberately miscarried; or that wage scales are hardly sufficient to keep body and soul together. All these are primarily man-made iniquities which, if corrected, could fuse a measure of economic or social justice with economic necessity, thereby adding another essential dimension to the Green Revolution." This is a powerful and moving piece.

This paper, dated July 29, 1969, was based on a field trip made in late April. It was later published by the *Economic and Political Weekly* (Bombay), September 27, 1969, under the title "Green Revolution in Bihar—The Kosi Area." The article was reprinted by the Agricultural Development Council, Inc. of New York in June 1976.

Introduction

IN LATE APRIL OF THIS YEAR the writer of this note visited Purnea and Saharsa districts of Bihar state. The two, and particularly Purnea, are well known for two reasons. The first stems from the fact that the two districts fall within Bihar's most important irrigation development project, the Kosi project. Because of this, their agricultural economies are subject to drastic improvements where canal water is available, and signs of it are not far to seek. The second reason is on the less happy side of things. They are known for extremes in the size of landholdings; generally prevailing low levels of farm practices and production; chronic food deficits and occasional famines; the prevalence of small farmers; the high rate of sharecropping under rack-rent conditions; the ejection of tenants under the threat of agrarian reforms and the growing insecurity of the sharecroppers as the potential of the Kosi irrigation scheme rises; and finally, though wages have improved, the large and rising contingent of landless still receive the short end of the economic stick. We looked into these problems, without neglecting technological change, in the Kosi area. In fact, we begin our observations with this development.

The New Technology

The new technology is no longer a hallmark of only Punjab, western Uttar Pradesh, and a few other parts of the country. By and large, Bihar still conjures up an image of agricultural primitivism and unrelieved poverty. But it is no longer all that sweeping in parts of Purnea or Saharsa, for there too there exist fairly large islands of change where water and other resources are available. This we encountered as soon as we landed at Purnea's small air strip. On the edge of it and practically undistinguished from it, a Massey-Ferguson was plowing up a field. We agreed with the development commissioner of Purnea who greeted us upon arrival that we would not have to go far afield to look for the new technology.

It was our experience in Punjab that the sample of cultivators engaged in the new practices contained a doctor turned part-time farmer. In Purnea, within steps of the airstrip, our first "case" also proved to be a doctor, an ex-doctor turned full farmer, and small farmer at that. Our find was in his early sixties, spry and wide awake, who six years ago in the midst of a busy practice set out to change his way of life. He bought eight acres of land and built himself a neat but simple shack where he lives and from which he farms his land. Unlike the pretentious Punjab doctor farmer, he is not doing it "for my country"; farming is his new way of life because, as he explained to us in his precise English, "if you want to die don't take poison—go to Purnea." Ministering to the sick must have gotten on his nerves, but in moving from the discomforts and bustle of even so small a city as Purnea, he didn't exchange activity for merely reading holy scripture or contemplating his navel. On the contrary, he is a very busy farmer, as progressive minded as the big innovator farmer elsewhere. He dug a tube well on which he spent Rs6,500 and assured for the farm an adequate water supply. Wheat is his principal crop; he buys high-yielding seed varieties at Rs250 a quintal and the amount of fertilizer he uses is just what the recommended package prescribes but very few apply: 33 kilograms nitrogen, 25 kilograms phosphorous pentoxide, and 20 kilograms potassium oxide. He virtually double crops his land, plowing being done by tractor at Rs25 an acre, threshing by machine and harvesting by hired hands, all at a cost of one-eighth of the crop. As among the best wheat farmers, his yield is 45 to 50 maunds[1] to the acre and his "net return to management" is the by now familiar Rs800 to 1,000 an acre. He has done well in yet another way. He paid Rs500 an acre for the land and he could sell it now at Rs6,000 an acre. And if ever the airstrip is enlarged, he will find himself with a valuable "commercial" property only a real estate developer could dream of—all in all, just an aspect of the new technology that came to Purnea.

This is reminiscent of the Punjab story, particularly in relation to farmers with irrigated land. But one point needs emphasizing, though it is obvious by now. The doctor farmer, the cloth merchant with his four-acre hobby, the lieutenant colonel with his 16.5 newly purchased acres, the gasoline station owner who is happily new in the game with 25 acres acquired two years ago, and the numerous old-time farmers we encountered during the trip were able to shift from traditional to modern farming because they could muster the resources needed to effect the change. And these are considerable even for a small farmer; for one with 7 acres must have a pair of bullocks worth Rs2,000, a tube well demanding about Rs6,500, and inputs worth Rs1,500—or a total of Rs10,000. Once these basics are provided, the new practices are no respector of the size of holding; and our Purnea experience doesn't differ in essentials from that in Ludhiana. One can derive a larger or smaller income depending upon the acreage farmed, but water, high-yielding varieties, and insecticides if applied within proper norms are as surely productive on small as on large holdings. While this much has been settled, the fact remains that the overwhelming majority of the small holders in Purnea—as in Punjab—do not have the resources to utilize the new package of practices. We shall elaborate on this theme in another section of this note.

Wheat

Purnea is overwhelmingly a paddy district, but wheat, not paddy, is the pivot around which the new technology revolves. Strange as it may appear to the reader—and it surely appeared strange to us when we ran into the phenomenon—it was in Purnea *city* that we first came across something of a "wheat craze." The local

1. One maund equals 82.5 pounds.

officialdom—from the development commissioner, magistrate, and on down the bureaucratic line, so it seemed—were all growing high-yielding variety wheat in their back or front yards. Our visit coincided with the harvesting and threshing season, and it was a common sight to spot the city farmers on their half-acre or quarter-acre plots helping gather in the crop. Only the flower beds have suffered, but the local magistrate and his wife supervising the threshing—to cite one instance—have done very well. Their crop is not big, the holding being not quite half an acre; but, with the help of the garden hose, effective varieties, and fertilizer, their yield is a record 30 maunds and it will be all sold as high-grade seed at Rs250 a quintal instead of as grain at Rs75 a quintal. Not all these new-fangled "farmers" invested equally in their miniature farms, producing mostly wheat for their own consumption; but most important is the fact that the new technology has made it both possible and, from their point of view, very much worth their while. Nor does it matter if this spurt of activity on the part of all manner of bureaucrats is only a passing vogue. For the time being, they turned the late Prime Minister Shastri's motto "let there be a kitchen garden" into "let there be a wheat plot," thereby indicating the nature of the change, even if only on a small scale. But it is in the countryside that one sees its full scope; there, the new kind of wheat and some of the practices that go with it are firmly entrenched.

Looking at the crop pattern of Purnea district from 1964–65 through 1968–69, it is obvious that wheat is becoming a major crop. Whereas the rice acreage has hovered around 1.1 million and jute has declined from 300,000 to a little more than 200,000 acres, the acreage under wheat has increased from 56,000 to 220,000, practically on a par with jute and next to rice; the fourfold increase has been effected within the last three years. Even more remarkable is the changeover in Saharsa district: from 8,000 acres of wheat in 1965–66 to 130,000 in 1968–69. These global figures speak loud about a basic transformation in the district, but it is not without interest to trace the quick reaction of the farmers to what they considered was good for them. Thus, the farmers of Kalaiye village (Purnea), with a total of 1,855 acres, have taken to the new wheat technology with the greatest alacrity: from 15 acres to 995 acres under high-yielding varieties between 1965–66 and 1968–69. Given a fairly assured water supply, other villages reacted similarly, if not so spectacularly. Worthy of note, too, is that in these particular villages the literacy rate was undoubtedly below the notoriously low state average of 15 percent; but evidently the idea of better practices was quickly assimilated just the same. This is not to extol illiteracy as a conveyor of useful information, but it is just possible that, at an early stage of technical agricultural revolution, literacy is not as crucial a variable as one might have been led to believe.

Paddy

We noted that the Green Revolution in Kosi is centered around wheat, not paddy. The "miracle rice" is well known among the farmers, but it takes little searching to find that in Kosi area, where paddy is king, the crop is in a state of flux. This is not to say that it can't make up its mind which way to go; the direction is clear, but getting there will take time. In Purnea, with a million acres under paddy, the concern is not, as in wheat, to expand acreage but rather to substitute traditional varieties yielding 10 to 15 maunds an acre with taichung or IR8 yielding double or triple that, or with improved local varieties. Farmers readily concede that under certain conditions they outperform local varieties, but for good and sufficient reasons they hesitate to plunge, as so many of them have done in the case of wheat. They know that the high yields of the "miracle" varieties are having difficulty in stabilizing themselves; 40 or more maunds to the acre one season and only 10 the next is not an uncommon experience.

Nobody knows how much "miracle" paddy there is in Purnea, and we found no one to venture an estimate. One can only surmise that, since it hadn't caught on like fire and the general attitude is one of caution, the acreage is small. In one large package block with 32,000 acres in paddy, the high-yielding area was estimated at only 400 acres in contrast to 95 percent for new wheat. Not all villages visited show so low a proportion of new to traditional paddy, while in some the increase has been very steep—from 5 acres in one year to 350 acres the next, or from nothing in 1966–67 to 200 acres in 1967–68. But in 1968–69 one finds instances of a drop-off in acreage, notably of the taichung type. Farmers distinguished between taichung and IR8 and prefer the latter because its yields are higher, it tastes better (although it suffers by comparison with local varieties), and the extraction rate is 80 percent compared with taichung's 62 percent. But the real reluctance to use taichung is due to its greater proneness to disease in the rainy season and its resistance to insecticides. This variety had something of an earlier start than IR8, and the richest farmer of Dogachi village (Purnea) was so impressed with it after the first season that he named his rather spacious new house the "Taichung House." This gesture proved to be an ill-chosen tribute to the variety. With the silting and closure of the canals in the summer and sole dependence upon the rainy season, taichung is only a memory as far as he is concerned. Only the name of the house reminds him of the unsuccessful experiment. IR8 is doing better, but it wouldn't do at this stage to say that it has really turned the corner.

It is altogether too early to judge the future of the high-yielding paddy varieties. Our impression is that many, if not most, farmers are still experimenting with them whereas wheat is now firmly rooted. Kosi is not singular in this regard, for this holds for new wheat versus new paddy in state after state. The new paddy, unlike wheat, is a much more sophisticated plant which responds favorably only to a specific set of conditions. On their own, the farmers of Purnea, as elsewhere, cannot fashion them. The same can be said of the Kosi irrigation scheme even if all the difficulties connected with proper water delivery were eliminated. The problem is a national one and only sustained research of the highest order, with adjustments for regional and local characteristics, can develop a new paddy technology. In the meantime, there is no indication in Purnea that the mixed bag of successes and failures have in any way lessened the appeal of the miracle rice. The imagination of too many farmers has already been stirred by what it offers under appropriate conditions. For the moment some farmers burned their fingers and taichung is down; IR8 is evidently up, though still on a small acreage. Indigenous dwarf varieties now released—jaya and padma—will undoubtedly be tested soon by the most enterprising of Kosi farmers. At the same time, some local improved varieties are on the increase. In summary, measured quantitatively, there is no Green Revolution for paddy yet, but some of its ingredients are already there in terms of the existing trials and errors and the prevalent aspirations to expand paddy output. These are no mean achievements in themselves.

Irrigation

The rapid rate of assimilation of the new wheat culture and, to a degree, the new paddy culture reflects above all the expansion of the water supply provided by the Kosi irrigation scheme and so far to a very small degree only by the appearance of tube wells. Under existing rainfall conditions a farmer could count on only one crop, whereas under Kosi a three-crop pattern—rabi, kharif, and summer—is being adopted. Since the Kosi irrigation scheme is the central piece in this radical transformation, a word about Kosi—what it is now and its future prospects—is in order.

Without going into many details, suffice it to say that if the scheme is completed in 1973 or 1974 as planned, it might accomplish the following: a gross command area of 2.3 million acres, out of which the culturable command area is estimated at 1.5 million acres and a potentially irrigated crop area of 1.8 million acres. On this basis, about 40 percent of the cultivable land of Purnea and as much as 80 percent of that of Saharsa district would be benefited by the Kosi irrigation scheme. Whether the scheme will be completed on time or not—its financing is meeting with increasing difficulties—presently Kosi is officially providing water for 650,000 acres. Unofficially it is not quite so; but the changes the project has already effected cannot be overestimated, and a trip in any direction encompassed by the scheme makes this clear. But in relation to what the scheme has set out to attain now and in the future, there is likely to be a gap between promise and fulfillment. It is increasingly recognized by the administrators of Kosi and farmers alike that its effective coverage will ultimately fall short of its mark, just as it is falling short of the acreage it is supposed to cover now. Part of the explanation lies in the silting problem which has already reduced Kosi's water delivery program. While no information is available on just how large an area is involved, it is a safe surmise that the effective coverage is something less than 650,000 acres or, to put it differently, that the area commanded doesn't receive fully assured water supplies throughout the year. It doesn't if only because the canals are closed for desilting in the summertime (about three months) and in October before the rabi season commences. No effective solution of the problem of how to desilt flood waters originating in the Himalayas is on the horizon; and the interruptions, inadequacy, and insecurity of the Kosi water supply will continue within the foreseeable future. This said, Kosi with all its problems is the most important source of irrigation for the two districts, for there are areas in Purnea and Saharsa where next to the uncertain rainfall there is no other water facility. Jaladgarh village (Purnea) with its 1,300 acres is a case in point. The farmers there have water on their brains, and understandably so. The village has only eight tube wells thinly stretched over 100 acres, and this is where modernized agriculture is centered and, of necessity, at a reduced tempo: better but not up-to-standard quality seed, only slightly more than a third of the prescribed fertilizer dosage, a token of pesticides, and so on. The conclusion is obvious and needs no comment. What does call for a comment is the prospect of tube well development.

The preceding paragraph might leave the impression that there are next to no tube wells in these districts. There are, finished and unfinished, but one hesitates to cite a firm figure because the number depends upon the person giving the information or the brochure one consults. One can only hazard a guess of a few thousand. That many more thousands of tube wells are badly needed as a supplementary source of water to Kosi is recognized by all concerned. Kosi's silting problem only accentuates the need, but so far the efforts to this end have not been too successful. Package block "Purnea East" may serve as an illustration. The tube well target for 1967–68 was 453; applications filed, 194; and tube wells dug, some of which energized, 92. One of the reasons for this state of affairs is the assumption by many farmers that somehow Kosi would provide more water than it ever promised. But the fundamental obstacle is the lack of resources and the difficulty of securing the credit of Rs6,000 or so needed to finance the construction of one. The new breed of farmers of the kind mentioned have no resource problem; they latch on to the new technological wave with cash or creditworthiness good enough to borrow what they need if they need it. The same is true of the large owners, but little of this applies to the bulk of the farmers.

They must do their financing through borrowing, but for many even to obtain the minimum which a small tube well requires entails almost insuperable difficulties. One of the conditions to secure electrification of the well is an advance of Rs2,000 per power connection to the state electricity board, which is set off against the electricity charges. This was enough to discourage many applicants, and the provision was abolished this spring. But another important provision has remained unchanged. When Bihar State Development Bank advances a loan, it does so upon the condition that the borrower mortgages his land as security, land worth twice the amount of the loan. With an eye to ensuring repayment, the provision may be valid; but this type of security, and particularly the doubling of the security, automatically excludes all tenants—and Purnea and Saharsa have many of them—and most small farmers. It leads also to self-exclusion of not a few of those who have sufficient land to mortgage but don't wish to have it encumbered for the life of the loan, which is seven years. On top of that, the bank takes a conservative position in determining land values. In the circumstances, only the select few are privileged to secure such a loan. This very cautious approach, good to ensure prompt repayment, has little to recommend itself on developmental grounds. It helps explain why even in the favored package blocks the gaps are so wide between targets, the number of applications filed, the applications granted, and the tube wells constructed. It appears to us as a restrictive approach to tube well financing, the more so since a tube well program is designed to advance the new technology in areas where its acceptance is ripe and secure water availability is its motive power.

Such handicaps notwithstanding, Purnea has just set a target of 8,000 tube wells in twenty selected blocks during the coming five years, at a cost of Rs8 crores. Considering the need, the program is a modest one. It is also likely that groundwater reserves are ample for such development. Its implementation, therefore, may well be attained and an additional 80,000 or more acres brought under irrigation, though not necessarily within the time projected. Since the average cost per tube well is estimated at Rs10,000, it is safe to say that the principal beneficiaries will be the large landholders with owned or borrowed resources. This is not surprising; it is characteristic of the new pattern of technology in Purnea as in Ludhiana, and the same with regard to all other investments in inputs which distinguish the role of the few and the many in the current phase of rural modernization.

Fertilizer

It is not necessary to labor the point that, next to water and improved seed varieties, chemical fertilizers are a prime ingredient of the new package of practices. Our first case, the doctor farmer has demonstrated that, aside from his initial investment in a tube well, the major part of his expenditures is for fertilizer because he uses the prescribed dosages. He is not an exception but he is not typical of most farmers of Purnea either. Talks with farmers reveal that while fertilizer effectiveness is conceded on all sides, some use none and most of the small and not so small farmers use, on average, relatively little. An investigation of 1,500 Purnea farmers, with holdings ranging in size from less than one to 20 acres, concluded that "only 18.5 percent of the total requirement was fulfilled."[2] The nonfulfillment is not in the midst of supply shortages but of ample availability. The reluctance to use more fertilizer is due to an insufficient water supply, but the basic deterrent even in the case of those who have water is lack of credit. This will be discussed in subsequent paragraphs, but it should be noted that, despite the presence of restrictive factors, fertilizer consumption is very much on the increase. Coming back to the package block Purnea East, the utilization of fertilizer between 1965–66 and the incomplete year of 1968–69 jumped from 62 to 900 tons. This nearly fifteen-times increase doesn't mean of course that the farmers are applying the prescribed norm. If they did, the consumption of fertilizer might have been closer to 2,700 tons. But the trend is very much upward, and the results are similar in the Kosi area as a whole. In the period of 1965–66 and 1967–68, consumption of fertilizer has risen from 2,500 to 24,000 tons, and, if the official estimate for 1968–69 is correct, the volume should be up to 44,000 tons, or approximately the same rate of increase as Purnea East.

If one considers the fact that only three years ago the Kosi area was on a par with some of the most backward sections of rural India, the remarkable thing is not the shortfall from the norm but the volume of fertilizers actually used. It is a tribute to the pervasiveness of the new technology, as it is to the farmers who recognized it for what it is. Nevertheless, the much-involved development commissioner of Purnea lays stress on the failure of the farmers to take up by a considerable margin the volume of fertilizers available in the district. Aside from disinterestedness on the part of some farmers which may be attributed to backwardness, there are a number of other factors which retard the adoption of the new technology.[3] But of all the retarding factors, the prevailing view is that inadequate credit is the principal explanation of the shortfall in fertilizer consumption. This underscores the role of an agrarian institution in stimulating or retarding technological change. In Kosi area and in Bihar generally existing institutions, and more immediately credit facilities, tend to retard such change.

2. "Problems of Small Farmers of Kosi Area (Purnea and Saharsa Districts)" (Bihar: Patna, 1969), p. 13.

3. Fragmentation of holding, land status, inadequate and untimely supply of inputs, inadequate and untimely irrigation, unsuitability of land, lack of credit facilities, and other factors like storage and marketing.

Credit

A sample study prepared by the office of the development commissioner of Purnea, covering Purnea and Saharsa districts for the year 1967–68, show [p. 46] the following sources of short-term credit:

	Percentage		
Credit sources	*Purnea*	*Saharsa*	*Consolidated*
Government (taccavi loans)	15.5	18.2	17.8
Cooperatives	15.0	12.2	13.0
Moneylenders	67.3	69.1	67.6
Commercial banks	2.2	0.5	1.6
Total	100.0	100.0	100.0

Commercial bank credit is too negligible to be worth discussing, and its much talked about future role need not detain us here. If all the bank loans were spent on the purchase of fertilizer, which is not the case, the volume could not have been of any significance. Farmers like taccavi loans; they are provided by the state governments at lower interest rates than the 8 to 9 percent charged by the cooperatives. What most farmers don't like is their extremely uneven distribution. Assuming that the sample is reasonably correct, farmers in the category of 17.5 to 20 acres received an average of Rs200 per family, compared with Rs26 for farmers in the category of 2.5 to 5 acres, or Rs28 in the 5 to 7.5 acre category [table 16]. All these loans are essentially handouts, not often repaid, and it is not surprising that they are so small. Only the larger landholders probably use some of these loans for fertilizer, thus contributing to the skewness of the rate of application as between groups of farmers. As to the volume purchased, it cannot be very large, partly because the total of the taccavi is relatively small and partly because even the larger recipients of such loans have more than developmental problems to worry about.

Cooperative credit was supposed to have been the principal source of rural financing, particularly for the poorer farmers. This is not so for India generally, despite the vast expansion of credit in recent years, and Bihar is no exception. As a matter of fact, Bihar is one of the weakest states in this regard, and the Kosi area reflects the disarray of the credit societies. Compared with the all-India 26 to 28 percent of all rural credit provided by the cooperatives, the figure for Purnea and Saharsa is only about 15 percent. It bespeaks a number of things: that only a third of the farm families are members of cooperatives; that less than half of them receive credit; that vast numbers of small farmers receive the short-end of credit even if their needs are most pressing, and the size of a cooperative loan ranges from a high of Rs195 to as low as Rs12. Interestingly enough, the highest figure applies to landholders in the category of 10 to 12.5 acres. This is a departure from the norm, which is to say that the farmers with the largest holdings, in this instance from 15 to 20 acres, might have been expected to control the cooperatives and receive credit accordingly. It may well be an exception to prove the rule. Cooperative credit, which accounts for 15 percent of the total and on an individual basis considerably less than Rs100, cannot but restrict consumption of fertilizer, except for the few farmers high on the credit-receiving scale. The following sheds additional light on this problem in Kosi and beyond:

> While for the package areas the portion of short-term cooperative credit advanced in kind can be said to be an average slightly more than 15 percent of the total, short-term credit provided for the particular areas in which the high-yielding variety crops are grown the corresponding figure is about 40–50 percent. For the country as a whole, however, the percentages may be said broadly to be 15 percent in kind and 85 percent in cash, though precise information is not available.[4]

Purnea and Saharsa have some better-financed package blocks but their cooperatives are not exceptional, and even they do not provide 40 to 50 percent of the credit in the form of fertilizer and other inputs. It is a safe conjecture that, on the average, the farmers who receive cooperative credit spend no more than 15 percent of it for fertilizer. If so, this type of credit, and in the amount received, makes only a limited contribution to the consumption of

4. "Report of the Fertilizer Credit Committee of the Fertilizer Association of India" (1968), p. 131.

fertilizer; but this is not to say that the richer farmers depend only on credit for the purchases of their inputs. In the case of the underprivileged small farmers of Kosi area, the situation is worse; according to the study by the office of the development commissioner, "the cooperatives and government agency (taccavi loans) between them hardly contribute 10 to 15 percent of the requirements of the small farmers," and this doesn't mean only input requirements.

The lion's share of credit in Kosi, approximately two-thirds of it, is furnished by moneylenders. In making loans they draw little distinction between the rich and poor farmers, and a good deal of the total finds its way to farmers with 10 acres and less. Even farmers "below one acre" average around Rs55, whereas this group hardly ever receives cooperative or taccavi loans. For the greater part loans furnished by the moneylenders range from Rs100 to Rs300. Only farmers in the highest category, 15 to 20 acres, find little need to deal with moneylenders. Considering that the moneylenders provide so large a pool of credit compared with the other types, it is doubtful that such loans are used for fertilizer, and chiefly because of the high interest rate. Getting loans from the moneylenders is so common that individually or in groups farmers are not reluctant to tell what they pay for it. The usual response in Kosi is not interest rate in terms of a percentage but in so many annas to the rupee. With but few exceptions the going rate of interest is 12 annas to the rupee or 75 percent a year, and sometimes only for the growing season of particular crop—a much higher annual rate. This compares with the highest we encountered in the most backward sections of rural India. The corollary of this is that, if a farmer uses such loans for the purchase of fertilizers, he pays in effect at least 70 to 80 percent more than the normal price. It is questionable if many farmers would use such high-cost fertilizer. Additionally, many small farmers avail themselves of such loans mainly because of immediate cash requirements for household expenditures which have little to do with inputs. In the conditions described, one cannot expect loans of this type to play a significant role in the consumption of fertilizers.

The conclusion is that the credit facilities of Kosi are not conducive to the optimum utilization of fertilizer and of other inputs in consonance with the requirements created by the greater availability of water and new practices. The prevalence of this condition is attested by the already mentioned shortfall in the offtake of fertilizer of which there is no longer a shortage in Kosi. Precise data on the role of credit in the lifting of the roughly 30 percent of total requirements are not available, but it cannot be very significant. The question that arises is how and who among the Kosi farmers bought the 24,000 tons in 1967–68 or the estimated 40,000 tons in 1968–69? The plausible answer is that much of it was bought for cash, and this in turn suggests that the principal buyers are the bigger farmers who are in a position to command the necessary funds. The unevenness of per acre utilization readily observable in the field bears on the same point. This, of course, is not typical only of Kosi; the same is true to a considerable degree in Punjab, where the bulk of the small farmers are either not participating in the new package of practices, or are participating to a small extent only. In Kosi as elsewhere this demonstrates the sharp economic stratification of the village community and its bearing on the Green Revolution. The availability of credit according to need rather than influence could reduce this form of polarization, and it goes a long way to explain the stress of the "Report of the Fertilizer Credit Committee" on more credit more evenly distributed among the present and future consumers of fertilizers. Barring that, consumption will undoubtedly continue to rise along with the increasing number of participants, but the problem of unequal application of fertilizer and, by the same token, that of failing to lift the needed and growing availability of fertilizer, will remain unresolved, retarding at the same time the spread of the new technology.

Land Values

The new technology, coupled with the anticipation of increased water availability, has sent land values up. They are not, to be sure, in the Punjab category, but they are rising just the same and not only for irrigated land Kosi water and the beginnings of tube well development have "turned sand into gold." While this is having deplorable effects on the tenant class, about which a great deal later, land prices have shot up from Rs1,000 or less to Rs5,000 to 6,000 and more per acre where, along with water, the profitability of the new inputs has become apparent. Kosi abounds in such examples, and the lieutenant colonel already mentioned illustrates an aspect of this development. In September of 1968 he was lucky enough to buy 16.5 acres of what was then unproductive land for a total of Rs22,000. He added a tube well at a cost of Rs6,500, spent another Rs2,000 for leveling the land, and inputs worth Rs175 per acre. Unbelievable as it may seem, nine months later he harvested 10 acres in wheat, returning 40 maunds to the acre. Listening to him on the threshing table, he was full of future plans and hopes of purchasing another 10 acres to "round out" his holding. He was not modest about his venture and understandably so; his total investment of some Rs30,000 of not a year ago has a ready resale value of Rs72,000. Whether fortune will smile on him again if he succeeds in acquiring another 10 acres is another matter; water availability or the prospect of getting it has, on the one hand, raised land values and given urgency to "round out" land properties; but on the other hand, the market availability of such promising land has virtually dried up. What little land is sold falls into the general category of "distress" sales—death in the family, mortgage foreclosure, and so forth. Land once neglected is all too precious now to part with.

The values of nonirrigated land and downright poor land with precipitation so scanty that a good deal of it is not in cultivation are also rising. Traveling between Purnea and Tejpur, the location of the Kosi barrage, one can't but notice a huge expanse of land, estimated at a million acres, most of it dry, more nearly scratched than cultivated. In response to questions about land prices in that predominantly empty wilderness, farmers volunteered that much of the land is in the hands of large proprietors and not for sale; what is for sale has gone up from between Rs300 and 500 to between Rs1,000 and 1,500—and that small holders are also reluctant to sell. The reasons given appeared to this observer highly speculative: that Kosi will somehow extend its water command so as to include at least part of the land in question; that conditions for digging tube wells will ease; that large-scale government tube wells will come into the picture, though admittedly there are very few in the area; or, finally, that perhaps a brand new irrigation scheme will revive the land. But underneath it all is the none-too-well articulated yet hopeful thought that the agricultural transformation now taking place in parts of Purnea must eventually encompass the less favored areas. At a crossroads situated in something resembling blighted land, the group of farmers knew that "over there" better things were taking place; and it was not surprising to hear a farmer speak with derision about the 8 to 10 maunds of wheat or rice that he extracts from an acre of land, a yield that reduces a farmer to debt and penury. The voicing of discontent is, of course, a by-product of the Green Revolution which has so far left him and multitude of others untouched, except in a psychological sense. This in itself is a change and departure from the customary attitude, although we cannot judge how widespread it is. If widespread it may work for good or ill, depending upon the presence or absence of a suitable release for desires no longer easily contained. Not many miles further up the road we chanced to meet a tenant who cast some doubts on the imminence of changing traditions. Contrary to our usual experience, he proved a noncooperative witness about his own condition. He did give us the reason and it can be best roughly translated in these words: "To discuss my circumstances is to complain, and I do not want to dig my own grave." Chronologically, this belongs in the section on tenancy and we leave the subject in abeyance. In the meantime, however, and reverting to the subject at hand, land values are moving up even in that neck of the woods.

Wage Scales and Labor Conditions

The first round of the new package of practices caused land values to rise, but the situation is not quite the same for the wages of landless agricultural workers or farmers with little land who are also in the market as casual laborers, though wages too have risen. The reasons are twofold. First, while the precise size of the labor force in Kosi is not available, an estimate for Bihar as a whole a decade ago suggests that it accounts for about a quarter or slightly more of the total number of farm households. This explains why it is generally accepted in Kosi that those seeking farm employment are in great surplus. Second, the new technology, still representing only a small percentage of all the cultivated land of Purnea, is not extensive enough to give these groups the kind of bargaining power their counterparts in Punjab have during the peak seasons. What is in their favor on the limited area affected by the Green Revolution is the three-crop pattern with its greater demand for labor and the shortage of labor-saving devices. After all, Purnea can boast of only 250 tractors and of very little other modern farm equipment. In these conditions, the general run of wages on irrigated land in the busy season is Rs3 to 4, more 3 than 4, plus food, for a 10 to 12 hour working day. At least one owner informed us that his hired hands worked during the harvesting season from 4 A.M. to 7 P.M., with a two hour break, for Rs3 plus food. This may not be typical, but information from other sources suggest that it is not exceptional either. Clearly, Purnea is not Ludhiana where wages are approximately twice as large; and the same holds for farm wage scales on nonirrigated land and for remunerations received by permanent or attached laborers.

Farm wages in Purnea have not stood still just as they haven't throughout the country. A shortage of labor has nothing to do with this upward movement, for there is none; on no occasion have big landowners complained about any labor shortage during any season of the year. But the steady rise in the cost of living has done for Purnea what a similar development has done for other rural and nonrural sections of the country. The upshot is that in the Kosi area farm wages during the nonpeak seasons, if translated into cash, have risen from one rupee to Rs2, plus one or two meals. This is for male labor; female labor rates are a rupee or slightly more plus food. Not all payments are in cash, especially for harvesting and threshing, where custom dictates that each of these operations be paid at the rate of one-sixteenth of the crop. This practice, which is something of a food price insurance, is no longer favored by owners of irrigated, high-yielding land. An owner was patiently explaining to us, as if it needed explaining, that it was to his advantage to pay Rs3 or 4 rather than the usual share of a much bigger crop. The impression carried away is that real wages have remained pretty much where they were five years ago and that the increase in monetary terms is essentially a reflection of the rise in the cost of living and the very limited scope of alternative occupations. The striking feature of the Kosi area is that one can travel miles on end without ever coming across a smoke stack or a big or small industrial establishment. The nonagricultural work consists mainly of odd jobs like carrying loads, repairing or building houses, driving carts, sometimes canal digging, and so on. Since the scope of these types of occupation is restricted, the conditions of a rapidly growing labor force could not have changed much from the patterns evolved in the past, the exception being the pockets of technological change and their additional labor requirements. The absence of labor migration in any significant degree to areas outside of Kosi and the still prevailing old agricultural practices explain the long stretches of rural labor unemployment. Occasionally an owner tends to endow the casuals with full employment, forgetting that January, February, June, October, and November are very slack months. Laborers questioned privately present a more realistic picture: For the greater part they are employed from 150 to 180 days per year, not counting 30 to 40 days of nonagricultural employment.

The preceding is even more germane to the conditions of work for permanent labor, which constitutes about one-fourth of the total labor force seeking employment. The system of permanent or attached labor is widely practiced among the bigger owners, and this is but one of the differences between Kosi and Ludhiana. In Punjab a permanent laborer generally receives Rs120 per month, including food and often clothing. In Purnea and Saharsa one is brought up short when the usual answer is anywhere from Rs15 to 30 per month plus food and clothing. The shock over, a more complicated arrangement emerges. A farmer with 120 acres has twelve permanent laborers, using casual labor only during the busy seasons. He pays them Rs25 a month but he also leases out to them small plots on the usual tenurial terms, which are anything but liberal in Purnea. It is from this that they draw their additional compensation, but it was also carefully explained to us that the amount of land leased out is small enough so as not to interfere with the much larger labor requirements of the employer. If the permanent laborers do as well as the casuals by the combination of Rs15 to 30 plus the extras, they also have the other advantage of "permanency." But the system is showing signs of weakening. Continuous tie-in with land allotments or crop sharing with the *same* employees is creating apprehension in the minds of the landowners, for according to certain land reform provisions this practice might infringe on their proprietory rights if ever the provisions come to be taken seriously.

On the whole, the labor picture in Kosi is not a pretty one. Observing the workers in the village and in the field and mentally calculating their take-home pay, the size of their families, the food expenditures which amount to about 75 percent of their total income, and the knowledge that two-thirds of them are in debt, some nagging thoughts come to mind. First of all the never-ending wonderment about the poverty in which a not insignificant segment of the rural population lives and above all the *patience* with which it is endured. Close familiarity with this sort of human condition is just as depressing as if it were seen for the first time. For the same reason it is more revealing in its implications, and hence the not altogether idle question—what if the cup begins to run over? Second, with alternative occupations in the Kosi area being what they are, and even if the number of employables remains stationary, an unlikely condition—their lot can't be relieved unless the coverage of the Green Revolution expands sharply. Finally, the much debated question of disparity of incomes between the rich and the poor, whether for reasons of the new technology or the existing rural structure, calls for no elaboration. It was there before the advent of the irrigation facilities and there is more of the same now, its extent being determined by the ownership pattern, the size of holdings, the water they command, and the farm practices pursued.

Small Farmers and the New Technology

The discussion so far has confined itself to Kosi farmers as a group and to agricultural laborers. An occasional reference to "small farmers" and to tenants was made mainly in passing, with a promise of more later on. The "later on" is here and the subsequent paragraphs are an attempt to deliver as it were on the "promissory note." It should be noted at the outset that, while the two groups claiming our attention are of the Kosi area, they are not limited by geography. With variations, they are typical of Bihar as a whole, just as they are typical of parts of practically all other states of India. What they represent and what issues they raise are important for two reasons: They are slowing down, even if unwittingly, the rate of growth of the new technology; and, if their condition undergoes no improvement, they could just as possibly turn to raising hell as raising crops. This would not be in the Indian rural tradition, we are told, but the Green Revolution isn't either and so is much else in India and elsewhere until overtaken by new pressures or new ideas. These gloomy forebodings rest mainly on the limited land base of the great majority of cultivators of Bihar, the pattern of land distribution, the unenviable state of farm labor, and the somewhat similar condition of the tenants or sharecroppers.

Growing land pressure and small holdings are no novelty in India and even less so in Bihar. In 1951 the average size of a holding was 4 acres; in 1961 it was only 3 acres (fragmented in about a dozen plots) as against an all-India average of 7.5 acres. This evidence of land hunger becomes clearer still when the land distribution pattern is examined; the following figures sum up the situation:[5]

	Percentage	
Size of holdings (acres)	*Number*	*Area*
0 to 1	31.0	3.6
1 to 2.5	25.5	10.3
2.5 to 5.0	20.5	17.9
5.0 to 10.0	14.3	24.2
10.0 to 25.0	7.1	24.9
25.0 to 50.0	1.1	8.6
Above 50.0	0.5	10.5

5. From Kedarnath Prasad's *The Economics of a Backward Region in a Backward Economy* (Bombay: Scientific Book Agency, 1967), table 81, p. 160.

Those familiar with the character of land distribution in India will not be aghast that 77 and 23 percent of the holdings and holders owned 31.8 and 68.2 percent of the land respectively. More serious from the point of view of the future fortunes of the Green Revolution is that, if the category of small farmers consisted only of those with 5 acres and less, they represent over two-thirds of all the households and 32 percent of the land. Not all the farmers with 5 acres or less are necessarily small farmers—if they have water—just as not all the farmers with two or three times the acreage are big farmers—if they have no assured water supply. But exact quantification of "who is who" aside, one can see at a glance the enormity of the task that lies ahead of the Green Revolution.

For the specifics of the case or the circumstances in which the "forgotten majority" composed of certain categories of owners, tenants, or sharecroppers carry on, we turn again to the sample studies carried out in Purnea and Saharsa districts in 1967–68. Taking Purnea as an example with its tested groups of farmers from one acre and less up to 20 acres, the following picture emerges. The average size of land per farm family member is 0.65 acres, ranging from a low one-tenth of an acre to a high of 1.58 acres. The cultivable holdings of 5 acres and less account for 52 percent of the investigated households, and 74 percent possess holdings of below 7.5 acres. All of them cultivate most of the land, and the rate of intensity of cultivation has gone up in the past two to three years from 28 to 43 percent. This is the result of the gradual expansion of the Kosi irrigation scheme which provides them with an estimated 25 percent of their water requirements. For the rest they depend upon the uncertain rainfall because all other water sources —tank, state tube wells, private tube wells, and surface wells energized or not—provide less than 2 percent of their water needs. What with 58 percent of the farmers classified as "backward," 28 percent "average," and only 14 percent "progressive," the water problem is probably not only a reflection of being mostly resource-poor. And yet this is the principal cause which induces the wide gap between the prevailing desire for more and better inputs and using them.

We remarked earlier that in the Kosi area the remarkable phenomenon is not underutilization of inputs but the degree of their utilization. Nevertheless, it cannot be overlooked that the percentage of households in Purnea *not* using high-yielding varieties is 68; *not* using locally improved seed is 80; *not* using chemical fertilizers is 71; *not* using pesticides is 80; *not* using improved implements is 91; and the percentage of households *not* using improved methods of cultivation is 81. Backwardness and shortage of water are not the only explanation of these shortfalls; lack of resources to invest in better practices and the failure of credit facilities to fill the gap are contributing to the same end. The result is that, according to the sample study, the average income per acre of traditionally cultivated land is Rs180, or not more than a fourth or a third of the income per acre enjoyed by a fully practicing "Green Revolutionary." It takes little imagination to surmise that the Green Revolution cannot prosper if the overwhelming majority of the farmers are left largely to their own devices. We say "largely" because the "progressives" are assumed to participate in the new technology, and probably some of the "average" category have nibbled at its edges. The disconcerting part of it is that many of the outsiders are where they are while recognizing the technology's advantages and that most physical inputs other than water are in ample supply. Water is indeed the bottleneck, and only 24 percent of the total cultivated land of the sample offers opportunities for improved cultivation. Yet there are many recorded cases of farmers with Kosi irrigated land but otherwise with limited resources to buy fertilizer, pesticides, better equipment, and so forth who are only peripherally involved in the Green Revolution.

If we labor the resources point or if not sufficient stress is placed on "backwardness" or on such handicaps as small size of holdings, excessive land fragmentation, irregular delivery of water, untimely though available supply of fertilizer and other inputs, and the scanty contribution of the credit facilities, it is only because of the indelible impression of the field trip that there are too many farmers in Purnea with a little land and with plenty of nothing of much else. A sympathetic writer on the subject bemoans the "overwhelming numbers pressing heavily on land and [that] recurring cycles of flood and drought offer a part explanation for Bihar's backwardness." He contends, however, that "A more serious drag on progress . . . is the deadweight of a tradition-bound, caste-ridden society, undistinguished by any marked spirit of initiative and self help."[6] This is a familiar and, to a degree, a valid charge. Yet it takes no account of the fact that in a very short time ideas of change and betterment have also touched Purnea, not only in the back and front yards of the city wheat growers or in the fields of the new breed of farmers. For reasons already stated, while only a minority of the farmers of Purnea partakes of the new technology, its significance and the promise it carries for the future cannot be overestimated. The conclusion is that, regardless of the negative effects of the deadweight of tradition and other inhibiting factors, the countervailing power of the Kosi irrigation scheme, for example, and the package of new practices it is giving rise to are watering down tradition and turning the mind of the big *and* small farmers toward change. The crucial difference between the two is not so much a matter of "the spirit being willing" but in having or not having the wherewithal to enlarge a holding, to dig a well, to hire on occasion a tractor for plowing, to buy fertilizer, or to be judged creditworthy or not.

On a more practical level, there is ample evidence that a farmer with 3 or 3.5 acres is viable if he has an assured water supply and the new inputs to go with it. The few who are in that position double and triple crop and have a net income of Rs800 or more per acre. Even a farmer with 2 acres who triple crops can maintain his family at a reasonable subsistence level. If on the other hand, "the farmer (with water) is not progressive enough and adopts high-yielding varieties to a limited extent," the size of the farm has to be 5 to 6 acres for him to be viable. Finally, a nonirrigated farm between 11 and 23 acres can yield the cultivator a fairly good living depending on his "progressiveness," the quality of the soil, and other agro-climatic conditions. These findings relate to Purnea and Saharsa districts; and they are on the encouraging side provided, of course, that government policy and action can nudge them closer to availability of water, credit, and other essential inputs. If left alone with desires thwarted and growing disparities all around, and with a multitude of farmers "on the outside looking in," so-to-speak, a setting will eventually emerge where customary restraints may prove a very thin reed to lean on. India is not without a few cases where restraints all but melted away under violent agrarian outbreaks. Who will be bold enough to assert that in Bihar's Purnea frustration and discontent will stop short of such outlets? And let it be stressed that these references are to farmer owners, however small their holdings may be. What, then, about tenants or sharecroppers who either have no land of their own or have so little of it that they mostly work on someone else's land? The remainder of this note is devoted to this question.

Tenurial Conditions, Reform and Consequences

If the account of the small farmers and farm laborers is not a pretty story, the sharecroppers' story in Kosi area is surely not prettier. The conditions imposed upon them by the owners are probably among the worst in the country. The new agricultural policy with its emphasis on greater productivity hardly applies to them. They remain where they have been all along, except that psychologically they are worse off than before the advent of the new practices and added rewards; neither are they better off because of the reform legislation. The crux of the matter is that in Bihar such essentials as security of tenure, regulation of rents, and ceilings on land ownership are in the deepest of doldrums, perpetuating a land structure as inequitous as it is inefficient. So long as this prevails and supporting rural institutions generally render them little assistance in the context of technological changes that are taking place on other men's fields, greater disparities in the levels of production and income are inevitable. Inevitable, too, are the grievances; they help swell the chorus of discontent already felt among the small farmers and farm laborers.

6. Ibid., p. 48.

In sum, Kosi area is a good, if unfortunate, example of this state of affairs.

The farmers of Purnea, Saharsa, or of Bihar in general are divided into three groups, according to "status." These are "raiyats" or owner farmers; "under-raiyats" or tenants generally holding land on cash or share rents and with certain recognizable rights in the land they cultivate; finally, the "bataidars" or sharecroppers, paying rent as the term indicates and working the land on oral leases. It is the sharecroppers who are Kosi's biggest and most troublesome problem because, unlike the farmhands who have no legal claims to the land they work, the sharecroppers are entitled to such claims under the Tenancy Act of Bihar. The owners deny them these rights on the ground that they are not recorded in the record-of-rights, and this is what the tenancy measures are in large part all about. According to the Purnea sample study, which included 1,444 households in thirty-one selected villages, the number of sharecropper households represented 43 percent of the total. Of these, only 6.4 percent had no land of their own, while the majority had some land either as owner or tenant but not sufficient for subsistence living. Just as important as the number of households involved is that the acreage they cultivate on a sharecropping basis is 23 percent of the total, or 26 percent if the land rented by the tenants is added. What the latter indicates is that the tenants represent not quite one percent of the households. This has not always been so, but under the threat of the reform measures owners have managed to "convert" tenants through a variety of questionable methods into insecure sharecroppers. But this system stands for something worse than high rent or that the owner contributes nothing but the land towards the making of the crop. As sharecroppers, their names, the names of the owners, the plots they lease from them, and rents paid are not inscribed in the record-of-rights. Not to be recorded is to have no legal standing in claiming occupancy or security of tenure rights in courts of law. Many social and economic consequences flow from this, all of them comprising grave and keenly felt issues, which so far haven't been remedied by a number of tenancy reform enactments stretching from 1885 into recent years.

An observer in Purnea is not surprised by owners who view reform measures as if they were not meant to be enforced and utilize every loophole to their own advantage. It is a common attitude shared by owners in India involved with tenants. Nor is it unique that with the Kosi irrigation scheme and technological changes under way, the once-unproductive and neglected land is becoming a prized asset. Not unlike the bigger owner innovators of Punjab, their counterparts of Purnea are eager to make the most of the changed situation induced by rising land values and production. One way of achieving it, they contend, is few entanglements with sharecroppers and more hired labor. Relevant to this is a conversation with a big Purnea landlord. He first informed us that he owned 16 acres of land but corrected himself under the good-humored prodding of a crowd of farmers that he had failed to mention another 484 acres. The lapse of memory might have had something to do with the ceiling on landholdings and its maximum permissible limit of 60 acres; but, on the other hand, no owner bows his head in shame on account of ceiling evasion. He already irrigates a fifth of his land and plans much more of the same in the next few years. As he looks ahead, he finds his present arrangement with the sharecroppers a hindrance. His main reason is the modernization of his land, which he maintains his sharecroppers cannot undertake. Speaking as one economic man and, mistakenly, to another economic man and moving away from the crowd, he tried to explain to us in fluent English that the new agricultural policy calls for large resources and unencumbered management, measures he could apply only with hired labor. This is a position we encountered in other parts of Purnea, yet the environment is against complete separation of the tenants from any hold on land. There has already been too much bad blood spilt in the process of downgrading tenants to the lowly status of unprotected sharecroppers by preventing them from being recorded as tenants. Any attempt at yet another round to augment still further the ranks of farmhands carries no promise of success, but rather of much turmoil that might set Purnea on its ear. The local scene leaves one with the impression that the big owners renting out land will be very lucky indeed if they can maintain a semblance of the existing arrangements. The reasons are plain to see as we turn to the conditions presently governing the leasing of land.

On the merits or demerits of tenancy as a form of land usage, there are reasonable differences of opinion but there are virtually none about sharecropping as practiced in Bihar. Even its supporters would admit, if pressed, that the system is good neither for efficient production nor for the well-being of the sharecroppers. This is brought out very clearly by the case study already referred to and prepared by local officials with no specially known bias in favor of sharecroppers. We can do no better, therefore, than let the summary of the study speak for itself:[7]

> The land-owners do not allow the share-croppers to cultivate the same land from year to year for the fear that they may lay claim over the land. In irrigated areas the land-owner generally gets the irrigated lands cultivated by his own men or hired labour and settles unirrigated or less fertile lands with the share-croppers. Big cultivators who have large areas of irrigated land have no option but to get part of the irrigated land cultivated by share-croppers. In such cases, also, the land-owners generally change the share-croppers every year. The harvested crop is kept mostly in the "Khalihan" of the land-owner till the division of the produce. Though according to law the land-owner is entitled to ¼th of the produce only, in actual practice, the produce is divided half and half between the land-owner and the share-cropper.
>
> In a few cases the share-croppers who were recorded as under-raiyats in the last Survey and Settlement Operations in the district of Purnea have been dispossessed on paper either through ex parte title suits or rents suits though they are still continuing to cultivate the land and divide the produce on 50:50 basis with the land-owners. All the share-croppers who have been examined have invariably stated that the insecurity of their tenure is the biggest handicap debarring them from adopting the new technology.
>
> The study reveals that barely 5 per cent of the share-croppers have used high yielding varieties of seeds, fertilizers and insecticides. The majority of the land-owners do not share the additional expenditure on irrigation, seeds, fertilizers and inputs. Through the efforts of the Extension Machinery some progressive land-owners have started sharing 50 per cent of the cost of high-yielding varieties of seeds, fertilizers, insecticides and irrigation.
>
> As the share-croppers are not recorded and their legal rights over the lands cultivated by them are not recognized by the land-owners they do not get inputs or loans. Hardly 5 per cent of these are members of Co-operative Societies and their share capital contribution is very low. They get only nominal amount as loan from the Society. Since the land-owner goes on changing the land settled with any particular share-cropper from year to year, the different agencies granting loans like the Revenue Department or Agriculture Department or even the Co-operative Societies hesitates to advance loan to them. The share-cropper has to depend mainly on the land-owner or the local Mahajan for credit. The land-owner generally supplies seed but takes back double the amount at the time of the division of the produce.
>
> The study has also revealed that most of the share-croppers belong either to the Scheduled Castes, Scheduled Tribes or Backward Classes. They are socially backward. They supplement their income from share-cropping by working as part-time laborers either in the fields of the land-owners or other raiyats of the area. The women get part-time employment during transplantation or harvesting of paddy and other crops, jute retting, etc. Some women also work as maid servants in the houses of the landlords or other well-to-do families. The children also work in the field or tend cattle. Even with all these the family somehow ekes out a miserable existence.
>
> To sum up only a small percentage (about 5 per cent) of the share-croppers have adopted the new technology to a

7. "Problems of Small Farmers in Kosi," pp. 49–50.

limited extent. The major handicaps against their adoption of the new technology are (i) the insecurity of their tenures, (ii) lack of credit facilities and non-availability of inputs, and (iii) social backwardness.

Where and how have the tenurial measures gone so wrong as to perpetuate so obsolete a system of owner-tenant relationship? Since this is not a study of agrarian reform in Bihar in all its aspects, the comments will be limited to a few essential items. The first one of a general nature is that the entire array of legislative measures suffers from acts of commission and omission, the sum of which is that in the vast majority of cases the goals are yet to be realized. A few examples illustrate the point.

Under the Bihar Tenancy Act of 1885 a tenant is entitled to the right of occupancy if he has held a plot of land continuously for a period of twelve years. The owners defeated the provision by changing the tenants. Under the same act, nonoccupancy tenants holding land on written leases are liable to eviction on the expiry of the lease, while those holding land on oral leases cannot be evicted. Since virtually all of them hold land on oral leases, they are, in law, protected except on ground of nonpayment of rent or misuse of the land. In practice, however, the measure is completely ineffective. So long as the act ties security of tenure rights to *continuous* occupation of a particular plot or plots of land, the owner evades the provision by rotating the sharecropper. Even if the sharecropper cultivates the same land for a stated period of time, oral leasees cannot prove the existence of continuous tenancy in order to acquire occupancy rights. Worse still is the ill-defined provision that permits substantial owners holding more than the ceiling limit to resume land for personal cultivation even from tenants who have security of tenure under the tenancy act.[8] Such an owner can resume half of the tenanted area of one acre and a maximum of 5 acres. Further, a tenant can become an owner of such land upon payment of suitable compensation. As the ceiling provision remains ineffective, neither provision proved to be of any benefit to the tenants. The principal landowners in Purnea are the Takurs and the Brahmins, castes which to this day do not touch the plow. When they resume land it is not for personal cultivation but it leads instead to eviction of tenants, to so-called "voluntary surrenders" of land by tenants, and to the conversion of tenants into sharecroppers or agricultural laborers. At best, the right of resumption of any piece of land has served the owners well as a threat to keep the tenants in line concerning security of tenure or ownership questions. The conclusion of the Planning Commission that "at present few under-raiyats (tenants) feel secure" in Bihar points to the fact that the foundation upon which the agrarian reforms were to have rested is a goal yet to be attained by the vast majority of Bihar tenants, including those of the Kosi area. The Planning Commission's statement was made in February of 1965, and nothing has changed since then.

With no security of tenure and no bargaining power, there can be no enforcement of the officially prescribed rental not exceeding 25 percent of the produce. We know of no case where this provision of the act (1961) is applied. Generally, rent averages from one-half to two-thirds of the gross produce, all the cost of production and equipment being furnished by the tenants. In a few cases rent is being paid as a fixed quantity, and any increase in yields leads, more often than not, to a rise in the fixed rent. In the end, the rental provision could not possibly be enforced even if the competition for leasing land was not as keen as long as the sharecroppers are denied security rights. Land ceilings or limitations on the size of holdings in order to create a pool of land for the conferment of ownership upon tenants have fared equally badly. The limitation varies from 20 to 60 acres, depending upon the quality of the land. In the Kosi area where many holdings run to 1,000 acres or more, a fairly large surplus of land might have been expected. But this is not so because the ceilings apply to the land held by a person and not to the aggregate area held by the family. In fact, every person was permitted to transfer land up to a ceiling area to near and distant members of the family within a period of six months from the commencement of the act. There has been a dispersion of land but on paper only, and in the process the anticipated surplus acreage has

8. Fixation of Ceiling Area and Acquisition of Surplus Land Act, 1961.

virtually disappeared. In the light of this summary and the situation in present-day Kosi, it is understandable why the conditions of the share tenants are as bad as they are, that they are in no position to advance the cause of the Green Revolution, or that the latter is of no help to them either.

If a break is to be made in the direction of security of tenure, more reasonable rents, and restoring tenants to the land from which they had been illegally evicted, the first and immediate order of business is to try once again to establish documentary records that so and so worked so and so's land. Only such written evidence might persuade the courts and none too enthusiastic officials that share tenants exist and that they are entitled to certain rights in land. This attempt failed, but not because of technical difficulties or the common excuse of lack of funds. A more telling obstacle lies in an official's remark that the registration of tenants and giving them their due "is bound to upset the social order of things." This remark was not altogether out of order because the first special drive to recognize their legal existence resulted in Purnea in owner-tenant disturbances, including a number of murders on both sides. The drive was called off, the Pandora's box was shut tight, presumably in the hope of insuring social order. The instruction issued by the Bihar revenue secretary on August 12, 1964, which put an end to this major reform measure, is well worth quoting:

> Reports have been received that in spite of clear instructions some field officers and staff have started recording under-raiyats (tenants). This should stop at once. Until further orders no work relating to recording of under-raiyats should be taken up during the Drive period. Even preliminary work relating to collection of data about possession of under-raiyats . . . should be kept in abeyance . . . Circle Inspectors and Karamacharis (village accountants) should specially be warned to follow these instructions carefully. Persons found violating these instructions will be seriously dealt with. Treat most urgent.[9]

9. Government of Bihar, Revenue Department of Land Reform Section, Revenue Secretary (Patna, August 12, 1964).

Another instruction was issued a month later with the ominous opening reading: "Reports have been received about evictions of under-raiyats and other agrarian disturbances." This was to make doubly sure that the one remaining measure for redressing the grievances of the tenants was terminated, as indeed it was and remains to this day. In late 1967 the union minister of food and agriculture was moved to remind the chief minister of Bihar that "it is essential that a special drive should be organized for preparation of records of the under-raiyats and bataidars," but with no visible effect. On a recent visit to the capital of Bihar we could elicit little more than that the lack of funds precludes action. And it should be added that the act of recording doesn't always guarantee the anticipated benefits. Far from it. In Purnea, for example, a cadastral survey carried out about a decade ago recorded tenant holdings; but it led to the above-mentioned disturbances and, in addition, to 40,000 to 50,000 appeals filed in the courts by owners contesting such entries. These cases have been wandering from court to court there these many years, while in the meantime the tenants are denied security rights.

On the face of it and as already indicated, this is not a pretty picture. On all basic legislative counts and from the point of view of agricultural productivity, the situation is discouraging. What offers a glimmer of hope is that 15,000 share tenants, of whom 12,000 from Purnea, were courageous enough to file applications with the collectors of the Revenue Department for restoration of their tenancies. That the "seasoned" attitude of the officialdom rejected the bulk of them (11,000) doesn't minimize the significance of their appeal action. To this must be added the 40,000 to 50,000 above-mentioned cases of those who evidently refused meekly to accept their fate and thereby compelled the owners to carry their displeasure to the courts. Who knows how many thousands more would have asserted their rights if the government of Bihar had evinced any sympathy for their cause? One thing can be stated affirmatively: The ground in Purnea is fertile for a change with or without the due process of law, and at least in part because of the new stirrings currently generated by the Green Revolution. A determined leadership, however small in number, could make a good deal of political capital by advocating nothing more radical than the rock-bottom program of security of tenure and more reasonable rents. Considering that land redistribution has never really been the intent of Bihar's reform legislation, one questions the position of the government of Bihar in smothering the minimum reform under the pretext of maintaining peace in the countryside. It is doubtful, too, if the owners were thereby well served. Nothing really has changed except for the hasty papering over of agrarian tensions. The new status quo, which is worse than before the enactment of tenurial measures, can hardly endure; after all, it rests on nothing more substantial than an official fiat against rocking the boat. Even though the cards are presently stacked against the tenants, another round in their tug of war with the owners is likely. When the confrontation comes, the chances are more than even that a price will be paid for a posture of no concession and no compromise. As to the tenants, they have nothing to lose.

If strife is the eventual outcome, is there anything to prevent it and who is going to do the preventing? The answer is simple enough—the by now well-known causes obstructing good legislation and good implementation should be removed. The difficulty with this suggestion is that it has been made repeatedly and unsuccessfully by the Planning Commission, the union minister of food and agriculture, and by other competent observers. Furthermore, this inaction cannot be attributed to a lack of experts, lack of pertinent data, or lack of funds. These, to be sure, are not in abundant supply but they are not crucial to the failure of the reforms. What is crucial is that in Bihar the leading politicians have not really accepted the idea of tenurial changes as a means to improve the conditions of the tenantry. And yet only they are the authentic makers or unmakers of reform. In Bihar as elsewhere they can provide the impetus or lack of impetus or draw the line between reform and "reform" by endowing or failing to endow sound economic measures with the indispensable political support. In Purnea as in Bihar as a whole, the necessary precondition doesn't exist, at least as of this writing. It is not lacking during the election campaigns when a public posture of concern about the downtrodden is useful, but it is evidently deemed less useful once in power; and mainly for this reason the reforms stagnate. If, therefore, an answer is sought to the question posed at the opening of this paragraph, the answer can be provided only by the political decisionmakers of Bihar. They know what the situation is, and they know what needs to be done to reverse it.

Summary

Until only yesterday, so to speak, Purnea and Saharsa were very backward rural districts. They are less so now with the Kosi irrigation scheme gradually coming into its own. Basic technological changes are taking place on holdings with an assured water supply and farmers with resources to acquire other inputs. Tube wells which contribute the same end are in their infancy; but the trend, slow though it is, is in the desirable direction. The expansion of the wheat acreage and the rise in yields have been dramatic. Although not nearly as advanced technologically as Punjab, in its essentials and on a more limited scale the Purnea experience is a repetition of the Punjab story. The high-yielding paddy varieties have not done nearly as well, but improvements may be anticipated following the adjustments they require. Aside from the "hard" evidence that the Green Revolution is under way, the idea is fairly widespread among a good many nonparticipating farmers that it provides a method of getting away from meager crops, low income, and poor living. While the majority of farmers in Purnea and Saharsa are still rated as "backward," the impression is that those who recognize the implications of the changes are not in a minority. The relatively quick change in mental attitude, induced not by the ineffectual extension service but by what the eye can see and the ear can hear, is in itself a significant result of the new technology in action.

The Green Revolution is limited in scope both as to acreage and participants, and this is likely to remain so even when the irrigation scheme is completed and tube wells enjoy vast expansion. A goodly number of cultivators will continue to have water problems and will remain with insufficient resources to enable them to partake of the new technology. The existing institutional credit arrangements with their well-known bias in favor of the big owners, the generally poor state of the cooperative credit societies, the predominant role of the moneylenders providing loans at usurious interest rates, and the lowly and insecure position of the tenantry—all these preclude the participation in the new package of practices of the vast majority of the cultivators. Recognizing the promise, yet with no access to it, creates its own grave social and economic problems, some of which are readily observable. Chief among them is the disparity in levels of production and income. Unlike Punjab, the Green Revolution in Kosi area has still a long way to go, and the polarization on this account is still limited. What is worse is that Purnea and Saharsa districts are burdened with an accumulation of problems, the sum of which is sufficiently serious to give one pause about the consequences of being merely an onlooker at better production, higher income, and better living. The inevitable tensions that this breeds are heightened by the facts that the new technology applies equally beneficially on small as well as large holdings and that the overwhelming majority of the cultivators are in the former category. Unless agricultural policy widens the scope of the Green Revolution, higher farm output and rising incomes are bound to be restricted to a relatively small number of farmers.

The new technology in Kosi area, whether in being or only anticipated, has pushed up land values very sharply, particularly where productivity is rising. Contrary to the experience of Punjab, rents have not risen; they are already so exorbitant that there is no room for further enhancement. But for reasons already stated, the sharecroppers neither gain materially from the Green Revolution nor are more secure on the plots they cultivate. They are, if anything, worse off because as ownership of improved land is coming to be prized very highly, so is the mounting determination of owners not to permit the tenants to share in the rights of the land they cultivate. Their preference is to be rid of them. No clean sweep of this nature will take place; there are too many tenants to deal with successfully, to mention but one reason. Yet it cannot be denied that one of the consequences of the Green Revolution is the weakening still further of the tenants' already very tenuous hold on the land. Despite this, the owners fear that some day certain pertinent tenurial provisions might be resurrected and their rights circumscribed. This is in itself a sufficient cause to opt for the hired hands whenever possible. Wages have doubled in the past few years, but very little of it can be attributed to the new technology; the principal cause is the rise in the cost of living. The net gain over and above this is insignificant; and, coupled with the lack of alternative occupations, the conditions of farm labor compare only with the very worst prevailing among farm labor in India. There is nothing in the offing by way of remedy through higher wages. Even if such were officially prescribed, who would enforce them? It has been rightly observed that "Those who dodge the rent laws with impunity will also evade the wage legislation. The real question is not how to raise the wages but how to put teeth into any new law that is passed."

This comment applies perfectly, if unfortunately, to the years-long attempts to provide the tenants of Kosi area with security of tenure, fair rents, and a modest measure of land ownership. Though the owners will not succeed in dispossessing the sharecroppers, turning them into hired hands as they have succeeded in shifting tenants into the status of sharecroppers, they have nevertheless managed to deprive them of the elemental rights to which they are entitled under the reform legislation. This being so, most of the 25 percent of the cultivated land of Kosi so held can undergo no significant change in modernization and productivity. By the same token, the economic position of the sharecroppers will remain at their current low levels. And there is nothing in the picture to indicate any deviation from the existing conditions, certainly not through the already subverted due process of law insofar as tenurial measures are concerned. Whether this state of affairs can endure as the Green Revolution extends to all irrigated land, widening the gap between the well-off and the poor, is problematic. The record of the stresses and strains between owner and tenants in Kosi is not an augury for a peaceful and purposeful development of the area. Chronic poverty does lead to political instability, which can in turn upset the foundation upon which the precarious economic life of Kosi rests.

The small farmers of Purnea and Saharsa districts are better off than the sharecroppers or hired farm labor, and for the good reason that they own the land. Yet enough evidence has been marshaled to show that as producers they fall short of the desirable, and shorter still of the full participation in the Green Revolution that many would like. It may be recalled that hybrids prosper only on irrigated land. On this score, the prospect of irrigating 40 percent of Purnea's cultivated land and 80 percent of Saharsa's should be encouraging for the widespread adoption of the new technology. This might be so provided that in addition to water the would-be beneficiaries have resources for all other inputs. And this is what the multitudes of the small farmers are short of even when they have water. This may not be the sole reason why nearly four-fifths of them do not use new inputs, but it is an important reason. In relation to modernization of agriculture, at least some of them can be likened to tenants who receive land under a reform and nothing else to go with it. Willy-nilly, many of them will be "selected out" from the purview of the Green Revolution.

The administrators of Kosi are keenly aware of this state of affairs; and, to shore up the position of the small farmers, they estimate that, in terms of short-term loans, Purnea will require Rs180 million and Saharsa Rs127 million. Instead of the conventional sources of institutional credit, a corporation such as the Agriculture Development Corporation "or a subsidiary branch of the Agro-Industries Corporation may be set up in each of the districts . . . for advancing long-term, medium term, and short-term loans" and for a great variety of other services ranging from storage, marketing, processing, and up to and including acceptance of deposits "to mop up the surplus capital in the agricultural sector."

If such an agency is established, it is questionable whether it could provide financing on so large a scale. Purnea and Saharsa are not the only claimants on scarce resources. One can, however, visualize some expansion of credit over the past because the Kosi irrigation scheme has become Bihar's symbol of technological advancement and its claims enjoy a certain priority. It is likely, too, that, under the energetic and far-sighted leadership of the development commissioner of Purnea, what can be done to assist these farmers toward agricultural modernization will be done. But one suspects that, in the main, the Green Revolution in Kosi will proceed in its usual selective way: high production and high income for the full participants and low production and low income for the noninvolved. The inevitable spillover will be there, it is hoped, among the farmers who somehow managed to add inputs to Kosi water or to a tube well. As for the dryland farmers, they will have to shift for themselves in the more or less accustomed fashion. A large-scale publicly assisted tube well program could raise the number of participants as it did in Punjab and with a signal increase in productivity. But looking at the Green Revolution and the demands it imposes on the one hand and the variegated group of claimants for its favor on the other, the benefits are bound to be unevenly distributed in Kosi just as they are in Punjab or wherever the new technology is practiced.

As a technological phenomenon, the Green Revolution at this stage cannot be faulted, for Kosi already demonstrates salutory changes in productivity and crop patterns. For the same reason, one must accept the need to choose between neglecting development in one area in order to encourage a more rapid rate of growth in another more suitable one. What doesn't fare well in this logic of development is the notion of economic justice. The best that can be said in justification is that, in a resource-poor country, economic necessity and economic justice do not often ride in tandem. But the story could read differently if the time comes when the Green Revolution is no longer only a combination of a set of economic factors in which the human condition has no place or is diluted beyond recognition. This point is made because, within the limitations of the Kosi rural setting, for example, and precisely because of the presence of the Green Revolution, not a little could be done to relieve the lot of the small farmers, the plight of the sharecroppers, and the subhuman condition of the rural proletariat. All said and done, it is not the fault of the new technology that the credit service doesn't serve those for whom it was originally intended, that the extension service exists largely in name only, that tenurial legislation is deliberately miscarried, or that wage scales are hardly sufficient to keep soul and body together. All these are primarily man-made institutional inequities which, if corrected, could fuse a measure of economic or social justice with economic necessity, thereby adding another essential dimension to the Green Revolution. That it will not resolve all or most of the problems of the disadvantaged is not crucial, simply because egalitarianism in rural India is not even good as a slogan; what is important is that it *can* mitigate them, and this is a minimum program worth striving for. If the Green Revolution is to broaden its scope while maintaining peaceful progress in the countryside, achieving the minimum program along with the technological changes should be the first order of business in Kosi as elsewhere. If this sounds like just another "should be done," the answer is that we know of no other remedy in sight. The consequences of leaving things alone are as evident as they are disturbing, and any hope to reverse or minimize their baleful effects lies in a determined effort to pursue the essentials of rural stability and productivity.

52. Agriculture–Some Broader Considerations

When each year the consortium of aid-giving nations to India meets under the chairmanship of the World Bank, one of the basic documents for consideration is the annual report prepared by the World Bank's resident mission in India. This was Ladejinsky's contribution to the annual report, "Economic Situation and Prospects of India," dated April 21, 1970. The audience differed from that of many other of Ladejinsky's reports from New Delhi, which were circulated among Bank staff only, and this helps account for both an overlap in subject matter and a difference in tone.

The "broader considerations" treated here are the social implications of technological change, as distinct from the productive and economic aspects treated elsewhere in the report. These social implications are dealt with under four major subheadings—the Green Revolution, land reform, cooperative credit, and rural poverty programs. The same major topics recur in his contributions to the annual reports in succeeding years.

Some Social Implications of Technical Change

WHATEVER THE DIFFICULT PROBLEMS on the agricultural research side, they do not reflect adversely on the technical prescriptions of the new strategy. The following, too, may be usefully kept in mind: "maintaining the historical trend in foodgrain production means a sharply increased rate of growth in productivity per hectare to offset the fact that net cropped area has not increased much in the 1960s and that in the 1950s the increase in area accounted for nearly one-half of the increase in output."[1] The recent rate of growth, therefore, is no mean achievement, even if it does signify that the job is more difficult than sometimes portrayed. The technological knowledge for a more rapid rate of growth is better understood now than at any previous time, and the same holds for

1. Martin M. Abel, *Agriculture in India in the 1970's* (The Ford Foundation, February 1970). See also R. S. Minhas, "Fourth Plan: Objectives and Policy Frame," *Commerce Pamphlet 20-21* (September 1969).

rice as well; the illusions that it can be achieved quickly have been largely dissipated. Even the growing social stirrings in the village may work to the same end. And this is not to overlook the presence of all the promising elements previously noted. But if all this is to bear fruit, the well-known "must be dones" need to be stressed once again. These are, among others, the high priorities in research and water supply and water management, political commitments relevant to social and economic rural problems, and consideration of the acute regional imbalances. That Punjab, Haryana, Western Uttar Pradesh, and pockets of Tamil Nadu, Andhra, and Maharashtra have recently developed an investment capacity sufficient to sustain a self-generating growth rate does not mean that the rest of the country is in anything like an equally comfortable position. Moreover, as the country moves from subsistence to commercial agriculture, the risks as well as the gains are multiplied. In a technical sense this calls for making fewer mistakes, which in turn depends upon the transmission of useful information to millions of farmers from the relatively few scientists and technicians as well as an extension service of a much higher caliber than presently available.

If not all that needs be done can or will be done, careful selectivity and concentration on a few strategic areas might suffice in the circumstances. The Small Farmers' Development Agency now in its early stages of formation is an example. Three-fourths of India's cultivated acreage is not irrigated; the gradual introduction of better practices into the dry areas and what that might take is one of the numerous instances of what might be done. In general, further advances can be accomplished depending how public policy, particularly public policy in the states, deals with the technical, economic, social, and organizational problems which the new strategy has thrust upon the farmers in different farm-size groups, functioning under greatly varying conditions. Genuinely dealt with, the Green Revolution might come into its own not only as a better technical tool but also as a tool in the hands of a much greater number of cultivators. And the latter is crucial, for the larger the coverage of the Green Revolution, the narrower the disparities among cultivators clamoring for shares in India's agricultural new deal. To this extent and barring monsoon failure, self-sufficiency in foodgrains could become a realistic target for the 1970s, even if not in 1971–72 as is assumed in some quarters.

As noted earlier, the advances in Indian agriculture have taken place not in a vacuum but in a given framework of resources, inputs, level of technical knowledge, and organization. As a general proposition and in practice, the soundness of this approach cannot be contested. The results, though very uneven, do point to the renewal of Indian agriculture. The fact that rice, coarse grains, and cash crops have not come up to anticipation presents serious difficulties but they are not insuperable. The muscle of the country's agriculture as already developed shows what can be accomplished, though at a much greater concentration of effort, beginning with the yet-unimproved varieties, the more complicated water problems, and on through a host of other technical problems. Yet, without minimizing the importance of attaining food self-sufficiency and raising the output of cash crops, this is not the sole issue facing the country as it proceeds in that direction. The more difficult and vital questions calling for answers are these: Who among the farmers is largely responsible for the additional output; how is the additional income being distributed; and how can rising productivity be integrated with the welfare of the farm community as a whole. Put another way, the troublesome issue that is beginning to make itself felt is the quality of the Green Revolution.

The Green Revolution is mainly confined to land with an assured water supply, larger holdings, and farmers with resources (both owned and borrowed) sufficiently large to take care of all other inputs. Such farmers, including those partially participating in the new package of practices, are relatively few; even in a technologically advanced state like Punjab, their total number probably doesn't exceed 10 to 20 percent of all the cultivators. But such is the recent shift in attitude generated by the innovations that the great majority of the cultivators who are not practicing them for lack of resources also recognize that they offer a method of getting away from meager crops, low income, and poor living. Recognizing the promise but not sharing in it creates its own social and economic problems, some of which are readily observable. Chief among them is the widening gap between the recently self-enriched and the mass of the poor farmers. Thus, directly or indirectly the Green Revolution has pushed to the fore rather suddenly and dramatically the nagging question of rural income distribution. The inevitable tensions that this breeds are heightened by the fact that the new technology applies equally beneficially on small as well as large holdings and that the overwhelming majority of the cultivators are in the former category. The distributional and other awkward implications of a progressive agriculture are well understood by the government of India, although much less so by the governments of individual Indian states. Out of deep concern with these problems, the government understands equally well that, in order to raise production, narrow down economic polarization, and insure peace in the countryside, agricultural policy must widen the scope of the Green Revolution.

While the bigger farmers are producing the surpluses and profiting handsomely from the improved package of practices, the same cannot be said about many other farmers, including, of course, the small farmers and particularly the tenants, sharecroppers, and the landless agricultural laborers. The traditional handicaps of the small farmers aside, the new strategy has exacerbated the notoriously troublesome tenurial problems. Where the new farm practices are in vogue, land values have risen three-, four-, or fivefold; and unrestricted control over such land by the owners has never been more prized. As a consequence not only have rents risen to as high as 70 percent of the crop in some places, but security of tenure and other rights in land the tenant might claim are in greater jeopardy now. Where not too long ago the principal issues between owners and tenants were the size of the rent and the degree to which the tenants or sharecroppers could remain on the land undisturbed, now Green Revolution land is so profitable that owners would like to get rid of the tenants altogether to resume the land for self-cultivation and to opt for the plentiful supply of hired labor which has no claims on the land whatsoever. There are too many tenants or sharecroppers to deal with them with impunity without courting trouble, but there is evidence that the long-run process of shifting tenants to the rank of sharecroppers and landless is being accelerated as more of the bigger owners become participants of the Green Revolution. In such circumstances, security of tenure, fair rents, and modest measures of land ownership provided by the reform laws are less attainable than in the recent past. In the end, one other consequence of the Green Revolution is that it has brought into sharper focus the vexatious tenurial issues while at the same time further weakening the tenants' uncertain hold on the land they cultivate.

The new type of agriculture has important consequences for labor. These are discussed in this report in the chapter on employment [omitted]. It is in fact difficult to foretell now what the net impact will be of all the changes which are going on. Certainly the new technology demands and enables farmers to afford mechanical assistance, but it also creates demand for labor. Thus, while one should question the view that the Green Revolution will necessarily help to solve the employment problem, if one comes to the conclusion that the outlook is for an overcrowded, low-wage farm labor market as the 1970s unfold, this too can hardly be blamed on the new technology.

There are other imbalances long part and parcel of the rural scene. The new technology has undoubtedly served to tilt some to a critical point, and the biased distribution of all kinds of resources in favor of the more affluent cultivators is one of them. But after all is said and done, the imbalance or, to put it more accurately, the pervasive rural poverty must be attributed to the sum total of all the social, religious, economic, and political arrangements which govern the village and which are in turn mirrored in the character of the Green Revolution. In such conditions on top of scarce national resources, willy-nilly agricultural policy tends to encourage a more rapid rate of growth in a promising area at the expense of a less promising one. Economic justice, then, almost inevitably takes a back seat. This said, there is no necessary immutability in current agricultural policies. The strains they impose and the discontent they feed could be eased if the time comes when modernization of agriculture is a combination of technical factors and of the long-awaited man-made adjustments in the institutional agricultural framework. When the question is raised, therefore, where the Green Revolution stands in relation to the nontechnical items of production, the following may be claimed with reasonable certainty: It is not the fault of the new technology that the credit cooperatives do not serve well those for whom it was primarily intended, that the extension service continues to languish, that the panchayats are essentially political rather than developmental bodies, that security of tenure is still a prime concern of many tenants, that rentals are exorbitant, that ceilings on farm land are notional, or that farm labor wage rates in many areas barely keep up with the cost of living. To repeat, the Green Revolution did not spawn these conditions, whereas various acts of commission and omission have surely a share in the debilitating forms they assumed.

The perennial question how to deal with these problems is upon the country once again. As there is no recipe for "instant development," dramatic and fast solutions are not in the picture. On the other hand, some of the conditions for amelioration are riper than in the past, precisely because social consciousness and political rights have properly made people more impatient and more demanding. In relation to agriculture, this is a consequence that extends beyond the Green Revolution. More specifically, it is reflected in the stage India has reached, well summed up in these words: "The real divide in India today is between traditionalism and modernity, revivalism and progress, conservators and innovators, the superstitious and the scientific and rational."[2] As for India as a whole, so for the countryside, and with the possibility that (a) the new forces will prevail, (b) "modernity" in agriculture stands for a good deal more than the further upgrading of the well-to-do farmers, and (c) "practical circumstances would compel the necessary corrections."[3]

If this is indeed so, then—reverting to the main theme—the scope of the new technology could be considerably enlarged. In that event all its parts would mesh in a more liberating sphere of greater promise and better performance. Admittedly, preceding this phase are political considerations and decisions which are studded with some big "ifs." If their elimination depended only on the apprehension of the government of India about rural institutional disabilities, the current political climate could usher in a desirable breakthrough. But it is the state governments which have the sole responsibility for matters agricultural, and it is they who have so far displayed little interest in dealing with the touchy subject of rural inequities. If they continue in the same vein, the full potential of the Green Revolution will not be realized even if India becomes self-sufficient in food. In present-day India with the rising clamor for more widely shared benefits of rural growth, achieving food self-sufficiency in the present manner is much less than enough. The underprivileged small farmers, tenants, sharecroppers, and the landless can be left to fend for themselves only at the price of overt discontent, which no longer lies in the realm of theoretical assumptions of interested reformers. This, as remarked earlier, is better appreciated now by the government of India than in the past. How much longer will the state governments find inaction politically profitable and at what price—to this question there is no ready answer.

It is in this connection that three particular aspects of the agricultural scene may be discussed because of their importance to the improvement of India's agricultural society—improvements not only in production and employment but also in the quality of rural life. These are the situations of the small farmer, farm credit, and land reform.

The Small Farmer Program

Who is a "small farmer" and "smallness" in farming are not easy to identify, and there are about as many definitions as there are economists trying to define them. To cite but one example: A farmer in India with 3 to 4 acres of irrigated land is possibly better off than one with 10 to 15 acres of dry land. Perhaps the most meaningful definition is that "a small farm is one which does not provide a net farm

2. B. G. Verghese, "An Agenda for India," *The Hindustan Times* (January 18, 1970).

3. Ibid.

income sufficient for the subsistence of the farm family."[4] The size of this group of farmers is estimated at approximately two-thirds of all the farm households, and they account for about 20 percent of the cultivated land. Whatever the differences in the size of holdings, all owned, partly tenanted or all tenanted, quality of land, cropping pattern, and so forth, the group as a whole is characterized by the following features: badly fragmented holdings of sometimes more but mostly less than 5 acres; little irrigation or none at all; insecurity of tenure; low productivity and low income; poor implements; insufficient access to credit; malnutrition to the point of affecting productive efficiency; underemployment; and, not surprisingly, the average per acre amount of capital expenditures a mere Rs75 as against Rs716 among big farmers.[5] Most of these farmers have hardly any margin left above subsistence living, and saving for investment is almost precluded. The corollary of this is that the majority of them are bypassed by the new technology and their assets are in a state of gradual deterioration. In order to provide for basic needs, they must compete for outside employment for a sizable portion of their income. The sum total of all these inadequacies constitutes the central problem of small farms and small farmers.

For a long time this issue has been in limbo except for a great deal of theoretical consideration in academic circles and decidedly tinged with pessimism about shifting it to a practical plane. Now, though, the government of India is in the midst of doing something about it. "Distributive" justice and the urgency to allay discontent are some of the considerations. But the basic factor that makes action possible now is the presence of the new technology. While it widened the gap between the rich and poor cultivators, it also demonstrated that, given requisite inputs, its beneficial effects can be applied with equal success on small holdings as well. It is the combination of all these elements that has brought about the creation of the Small Farmer Development Agency with the following aims in mind: promotion of small farm interests before official bodies; ensuring the availability of inputs and services to these farmers; and helping design production programs to fit the specific needs of a group of farmers in a given geographic area while tailoring credit and supplies to its requirements in the light of improved farm practices.

Financially and organizationally, the scheme could not serve all of the 40 million or so farm families in need of assistance, particularly the farmers with less than 2.5 acres of dry land. The immediate stress, therefore, is primarily—though not exclusively—on the "potentially viable small farmers" whose problems are more manageable. At present, the agency's goal is 45 pilot projects or separate agencies spread throughout the country. The selection is based on criteria favoring higher productivity and essential assistance provided by the projects and the states where the projects are located. Since the aim is to cover approximately 30,000 to 50,000 farm families per project, the scheme contemplates the rehabilitation of 1.5 to 2 million families by the end of the fourth five-year plan. On the financial side, a total of Rs670 million or Rs15 million per project has been allocated. In relation to small farmer pilot projects, the contribution of the center is expected to be considerably augmented by those of the state governments. This on the top of the anticipated institutional credit and credit from commercial banks, the center's Rs15 million per scheme is likely to be increased four- or fivefold.

Getting the program off the ground is proving to be time-consuming, but projects in Purnea (Bihar), Darjeeling (West Bengal), and Chindwara (Madhya Pradesh) have been initiated, while a number are in preparation in seven other states. For all practical purposes they are operated by the states. The chairman of a local project is the collector of the district or the development commissioner of the area. All other active participants on the management side are the district officers of the Agriculture and Animal Husbandry Department of the state and the heads of local credit cooperatives. There are two representatives from the center—one from the Planning Commission and the other from the Ministry of Agriculture. As part of the administrative setup, the projects

4. V. M. Dandekar, foreword to the *Seminar on Problems of Small Farmers* (Indian Society of Agricultural Economics, 1968), p. ix.

5. "All-India Rural Debt and Investment Survey," *Reserve Bank of India Bulletin* (June 1965).

do not intend to do what other appropriate organizations are already doing; but direct participation is not excluded, especially in the area of water supply, which is the most important input no matter in which district the program will operate. Another anticipated direct function of the agency relates both to the ownership and leasing out on a no-profit-no-loss basis of a variety of equipment ranging from tractors to sprayers and to servicing of equipment owned by farmers.

It is recognized that credit for both consumption and production purposes of the small farmers is the centerpiece of the effort, but emphasis will be on the latter. To this end, each agency has obligated itself to assist the credit cooperatives with special funds and also to hire additional staff "to ensure that the financing is undertaken as an operation of supervised credit." This is a departure from previous practices and a significant one if implemented. Annual short-term loans of Rs500 and one long-term loan of Rs1,500 during the five-year period is part of the scheme. Compared with the past, these will be generous advances, for few farmers in this category were known to receive such accommodations. If the new approach succeeds in covering 1.5 million or more farm families, it is estimated that the total amount of such loans in the last year of the plan may exceed Rs1 billion, or roughly 20 percent of all the coop credit in 1968–69.

In addition, the union government has formulated another scheme to render assistance to submarginal farmers and landless farm laborers. The former fall in the category of holdings up to 2.5 acres and the latter are those who earn more than 50 percent of their income from agricultural labor. The scheme calls for forty projects and outlay of Rs175 million, and it expects to cover approximately 1.6 million individuals out of an estimated total of 52 million. More significant than the projects just mentioned is the anticipation that they will induce the states to mount similar programs of their own.

Such, in brief, are the problems of the small farmers and the mechanism to deal with them. It is much too early to venture a judgment how this effort will be translated into action and implemented. The difficulties facing the agency are enormous. The task of picking and choosing the "proper" small farmers is enough to give one pause. The effective gearing of material and human assistance belongs in the same category. Besides, many a small farmer is a tenant farmer or a sharecropper, and how troublesome a problem this is was well expressed by the guiding spirit of the Small Farmer Development Agency, B. Venkatappiah:

> There are certain prerequisites which are even more fundamental than, say, the infrastructure of credit, marketing, storage and communications. These have references to tenants, sub-tenants and more especially sharecroppers and are concerned with such basic requirements as the existence of land records and the implementation of land reforms. It is being increasingly borne in on those concerned with the formulation of these projects in different parts of India that, where the potential beneficiaries of land reforms are concerned, the Small Farmers' Development Agency will be virtually inoperative in various sectors, including the important one of credit, unless a number of measures are first undertaken in the project area in regard to land records and land reforms.[6]

To these formidable barriers one can add a string of others, such as the uncertainties about the expertise rendered by the experts; the lackluster performance of the credit cooperatives, particularly in the face of added responsibilities; the same about the extension service; how unequivocal the state governments will be in supporting the program; and so on.

To touch on these items is to note that the transformation of the small farmers into nonsubsistence farmers is bound to have all manner of snags. But however great the handicaps, they in no way argue for the dismissal of the program out of hand or for taking the position that subsistence farming is beyond help because "it is futile to search for the solution of subsistence agriculture which is not also a solution of general underdevelopment. Correspondingly, that which is a solution for general

6. B. Venkatappiah, Planning Commission, address before 29th Annual Conference of the Indian Society of Agricultural Economics, December 30, 1969, p. 25.

economic underdevelopment will also be a solution for subsistence agriculture."[7] The plausibility of the argument, as the difficulties mentioned, hold only so long as the government has done nothing to counter them. This speaks for the past. What is new and significant is that, for the first time in the history of the country's agriculture, this particular class of farmers has been singled out with a view to rehabilitation. Whether at the end of the fourth five-year plan the targets are attained is not all that important. More immediately important is that these farmers will no longer be neglected. There is just a chance that the testing process may yield useful results, even though the program, to start with, deals with farmers who live and work in hostile conditions and surroundings and are "backward."

As against these constraints, there are the lessons of the new practices with their proven transferability of experience and response to economic incentives, which apply to all types of farmers in propitious circumstances. On the plus side, too, are at least some of the lessons learned from the administrative arrangement and drive exemplified by the implementation of the new strategy. Without minimizing the tenurial drawbacks, the agency which is directly involved in forcing the issue into the open may yet prove something of a catalyst in effecting desirable changes. Finally, it would be foolhardy to expect the impossible in the sense that all problems of the small farmers can be resolved satisfactorily and within a five-year period. Partial response to some of them may spell the difference between a firm beginning and failure. It is recognized that much of the above is speculative; what is not speculative is that the small farmer sector, even if selectively determined, has been placed on the agenda of India's priorities. This is in itself an advance. The program therefore is eminently worth trying, without waiting either for improvements of the condition of the village poor via the long drawn-out process of an overall economic development of India or for a sharp decline of population pressure on the land.

7. M. L. Dantwala, "Problems of Subsistence Farm Economy: The Indian Case," *Seminar on Problems of Small Farmers* (Indian Society of Agricultural Economics, June 1968), p. 6.

Credit

The Bank's 1968 economic report touched on the essentials of rural credit provided by the cooperatives. The intervening year shows no change in their rather inadequate performance, although from the point of view of loans advanced they have continued to forge ahead. However, perhaps more significant for rural credit in 1969 than the expansion of short and medium-term credit by Rs600 million was the publication of the monumental and searching "Report of the All-India Rural Credit Review Committee." Apart from its thorough analysis of the current state of institutional credit and the future credit requirements, the report contains a number of conclusions, both familiar and new. Chief among the latter is that the credit cooperatives can no longer remain the sole mainstay of institutional credit and the only answer to private moneylending. This sharp break with a traditional policy is due not only to the manner in which the cooperatives have been discharging their tasks but mainly because of the significant changes in the rural economy and the anticipated huge credit demands, which the cooperatives could not possibly meet.

The image of the credit cooperatives does not fare well in the committee's report. Judged quantitatively, they are a landmark in the countryside and it could not do without them. Between 1951–52 and 1968–69 their share in total credit used by farmers increased from 3 to between 33 and 35 percent, membership from 5 to 28 million, number of societies from 100,000 to 172,000, and short- and medium-term loans from 140 to 4,900 million in 1969. On the plus side, too, is the long-term credit provided by the land development banks which increased from Rs1,700 million in 1966 to 3,620 million in 1969. The value of farm inputs distributed by the cooperatives rose from about Rs360 million in 1960–61 to Rs2,500 million in 1968–69; of these fertilizer accounted for Rs2,000 million, or 60 percent of the total consumption of fertilizer in the country. Lastly, the Agricultural Refinance Corporation in the past couple of years has improved its stand as a financier of long-range agricultural projects with credit of Rs790 million in 1968–69 as against an average of Rs100 million during the first four years of its existence, beginning in 1963.

These accomplishments cannot be underestimated, and the credit cooperatives remain the main source of credit to the farmer borrowers or the aspiring borrowers. But the global figures cannot hide the fact that qualitatively the cooperative credit system is replete with major shortcomings. Despite the fact that many more farmers are benefiting from them, not more than 40 percent of the membership borrows from the coops; the tie-in between credit and production is rather weak; distribution of credit is often below reasonable standards of timeliness, adequacy, and dependability; cooperative-owned capital is 27 percent of their working capital, and dependence upon funds from the Reserve Bank and other official financial institutions goes on; mobilization of deposits is far short of the feasible; and overdues account for one-third of the outstanding loans, causing the credit line to be choked up and credit restricted. The central cooperative banks do no better than the primary cooperative societies; out of a total of 344 central cooperative banks, as many as 67 accumulated overdues exceeding 50 percent of the loans granted; and an estimated third of these banks have no reason for existence. Worthy of note, too, is that the more prosperous states are as derelict in repayment as are the poorer states. Many credit cooperatives are such in name only. While their number has been reduced, with an eye to weeding out the weak ones, from 212,000 in 1961–62 to 172,000 in 1967–68, another 50,000 are "sick" or slated for the mortuary. Finally, the discrimination against small farmers and tenants goes on unchecked. According to an official source:

> The traditional emphasis on linking of credit to security offered by the borrower in the form of land and other tangible assets, exclusion of small farmers from the membership of the cooperatives . . . absence of tenancy records and prevalence of the system of oral tenancies are among the major factors that have led to the denial of adequate credit to small farmers.[8]

8. Fourth Five-Year Plan, "Cooperation and Community Development," p. 13.

Much of the above is attributed to poor management and poor leadership of the credit cooperatives and cooperative banks. All these are familiar charges. What is less well known is the political character of the credit cooperatives and its negative effects on credit disbursement. In the words of the committee's report:

> this often implies that individuals not belonging to the particular political parties or factions are virtually out of the scope of cooperative credit. Another disturbing factor is what amounts to political interference in the working of cooperative institutions such as central cooperative banks. For reasons such as this, again, the assumption that the cooperatives could meet the credit demands of all creditworthy cultivators has proved untrue or only partially true. Even where technically all could be served, politically only some have benefited.

The strictures against the existing cooperative credit system come at a time when new developments are calling for a vast increase in rural financing. These are the Green Revolution with its augmented requirements, a rising trend towards more long-term credit, and the requirements of huge numbers of small farmers who have little or no credit standing with the cooperatives. In the light of this, the committee estimates that the short-term requirements of the farmers in 1973–74, both in kind and in cash, will amount to about Rs20 billion or nearly fivefold compared with current advances. Since this type of loan can be utilized most advantageously only if supported by medium- and long-term loans for basic improvements of the agricultural economy, such requirements be-1969–70 and 1973–74 are estimated at Rs5 and Rs15 billion, respectively. These anticipated huge demands for credit and how to meet them constitute the core of all the committee's deliberations, and its recommendations are tied in with the problem of mobilizing additional resources and how to disburse them properly.

Part of the effort to gather in such resources is expected from the land development banks with Rs7 billion, the Agricultural Refinance Corporation with Rs2 billion, and the commercial banks with Rs2 to 4 billion during the fourth five-year plan. Most of the remainder will depend on the cooperatives, and the immediate concern of the Review Committee is to correct all that has gone wrong with the existing cooperative credit structure while at the same time opening the rural credit field to new institutions. The recommendations about cooperatives are essentially a reiteration of the familiar theme that it is time they grew up to their obligation of better distribution of short-term credit. Similar injunctions in the past were not productive, and it is far from certain whether the committee's newly laid down policy for new institutions to compete with the cooperatives will yield much better results. But whatever the outcome, it appears certain that the cooperatives will no longer continue as the sole purveyor of institutional credit. The financial institutions, which are to help effect this change while providing more badly needed resources, are the Reserve Bank once again, a newly established Rural Electrification Corporation, the land development banks, an expanded Agricultural Refinance Corporation, and the commercial banks—only recently involved as an important source of rural credit.

The movement of commercial banks into the rural areas was an inevitable development with or without nationalization, but until 1968 their role as providers of rural credit was negligible. The shift in their attitude followed the government's pre-nationalization policy of "social control" of the banks, which meant tighter Reserve Bank supervision of their credit allocations "in keeping with the new national priorities." The objective was to compel the commercial banks to divert more of their resources to the agricultural sector, considering the high priority in national planning enjoyed by agriculture and, of course, the profitability of the new technology and the large scope for tapping rural deposits and for reinvestment through credit advances for all manner of inputs.

In addition to the sharp increase in rural lending by the commercial banks, there was also set up in April 1968 a joint Agricultural Finance Corporation with an authorized capital of Rs1,000 million. The objective of the corporation is to finance every conceivable and well-conceived agricultural project through the selection of commercial banks as agents, the choice depending upon the strength of a particular bank's concentration on a given area. Direct bank financing of a multitude of individual farmers is still very much in the future, and emphasis will continue to be on direct and indirect financing of large projects and on a sizable clientele among the large owners of unencumbered and mortgageable properties. On this basis alone and concurrent with the experience gained in lending to agriculture, the prospects are that the commercial banks will continue to extend their operations on a significant scale.

Neither the commercial banks nor any other source of farm credit can displace the cooperatives. The new sources promise only indirect and limited answers to the coop disabilities. Whatever their weaknesses and however vociferous the grumbling of the membership, they are too much a part of the rural scene; the long-lived notion of their indispensability, the variety of interests controlling them, and their supporters in the central and state governments are sufficiently strong to insure their existence. Besides, they do render a useful service to those who do borrow from them. For the reasons stated, even the well-conceived recommendations in addition to the competition incentive offered by other institutions, may not suffice to effect drastic improvements in their service. Without decrying the pressure to shed their old ways, the possible solution lies primarily in the emerging new agricultural setting and different mental attitudes that are beginning to come into their own. Of sheer necessity, therefore, the management of the cooperatives and the distribution of credit will have to adjust to the new situation. This is when a recommendation to correct past failures can become effective. In the meantime this much is certain: Regardless of how well or poorly the cooperatives will do in the future and whether or not the projected Rs40 billion of requirements are met, a great deal more money will be poured into rural credit in the next five years. The "climate" of expanding agricultural production favors it. The quality of credit disbursement is another matter, and its lopsidedness and inequities may well persist. But the funds will be expanded just the same, assuming, of course, that the scope of the new technology is expanded way beyond its present limits.

Land Reforms

This old and vexing problem was in the Indian picture in 1969 as in the preceding years—except much more so. The past year did not stand out as one of improvements in security of tenure, fair rentals, firmly recorded rights to the land, and so on. Nevertheless, it was a year of soul searching and breast beating, of patience wearing thin, of tensions and outbreaks of violence in the countryside, and of cases of forcible occupations of land or crops by tenants and landless in the majority of the states of India. On the positive side, the year was also notable for the conference of the state chief ministers, sponsored by the prime minister for the sole purpose of impressing upon them that it was later than they thought and urging upon them an "integrated" land reform program for the current plan. It was the year, too, when the Home Ministry, not usually given to involvement with such issues, plunged in to pinpoint the causes of agrarian unrest and urge in no uncertain terms remedial measures to allay fears of worse unrest.

All these worrisome trends are of long standing, but they acquired special relevance as a result of adverse consequences of the new technology in exacerbating social tensions in the village for the reasons already given. If to this is added the dimension of all other tenurial disincentives, then one must recognize that the inequalities in rural India are likely to increase rather sharply. It is the combination of the old and the new disabilities of farm groups accounting for about 25 percent of the total farm population and roughly 20 percent of the cultivated land which is at the bottom of the overt discontent. It is the same combination that refocused attention in 1969 on such questions as these: how to relieve the plight of cultivators working mostly for a pittance, how to revive their stagnating economies, and how to root them securely and beneficially on the land they cultivate.

It would be pleasant to record that the 1969 preoccupations have yielded some tangible results, but nothing very much went beyond verbal affirmations. Although a number of states are contemplating tenurial improvements, the evidence is far from clear as to the character of the would-be legislative enactments and their enforcement. On the whole, for the time being agrarian reforms tell the same familiar story of many a year past and just as often recounted; but then as now its essentials are worth repeating once again.

If judged by the agrarian reform debates, one can easily assume that the entire reform movement in India is a failure. This is not the case. As pointed out in last year's report, the zamindari system[9] was abolished and 20 million cultivators and 40 percent of the cultivated land of India came into direct relationship with the state, and for all practical purposes most of these cultivators are now owners of the land they cultivate. The elimination of this system with its absurdities and injustices was not difficult to accomplish because it was imposed by a colonial government which handed out property rights to which neither the British nor most of the recipients had any claim. A vastly more difficult problem presented itself in protecting the minimal, if basic, rights of the tenants in the non-zamindari, ryotwari areas where owner proprietorship predominated. The government of India did decide to provide them with security of tenure, reduction of rents, and the conferment of ownership through the familiar ceiling device of limiting the size of holdings and distributing the surplus. Under the guidance and continuous prodding of the Planning Commission, the states with their sole jurisdiction in agricultural matters have enacted a voluminous body of legislation designed to meet these goals.

These efforts were not without their benefits; some 3 million tenants and sharecroppers have become owners of an estimated 7 million acres of land, while here and there farmers who continued to work someone else's land came to benefit from the legislative provisions. The difficulty is that it is only "here and there," for, in the main and aside from the formidable administrative obstacles to implement the reforms, they have been shot through with acts of commission and omission which often negate the reforming intentions of successive five-year plans. In Kerala, West Bengal, and Tamil Nadu, legislation has been enacted or is

9. See *Economic Situation and Prospects of India* (April 18, 1969), footnote p. 29.

about to be enacted with an eye to plugging some loopholes; but how the new measures will be enforced remains to be seen. Despite these new efforts and without going into too many details, the implementation of the reforms in most—not all—states is in serious difficulty. As against the noted positive side, there are the wholesale evasions of ceiling provisions upon which an increase in peasant ownership was to rest, widespread eviction of tenants on the so-called ground of "voluntary surrenders" of land by tenants, and the failure of the rental provisions due to lack of security of tenure or the right to remain on the land undisturbed barring some exceptional conditions. One of the many glaring loopholes in the legislative enactments is the seemingly reasonable but ill-defined right of a landlord to resume tenanted land for what is euphemistically called "personal cultivation." This has led to extensive evictions or to "voluntary surrenders" of land. The Green Revolution and the value it places on land clear of any claims by any individual other than the owner is yet another invitation for tenant displacement. Such developments are not productive of the kinds of incentives among tenants that stand for good farming. But apart from this and with the recognition that new technology can potentially produce all the food India needs without recourse to tenanted land, there are nevertheless the pervasive questions of social injustice, inequity, and political instability which come in the wake of reforms that fall far short of their goals.

It was the realization that the land reforms during the past two decades have not measured up to anticipations that led to the conference of the state chief ministers in November 1969. The prime minister left nothing unsaid to stress the urgency of the meeting. In her view, "The warning of the times is that, unless the Green Revolution is accompanied by a revolution based on social justice, I am afraid the Green Revolution may not remain green." Mrs. Gandhi wanted the chief ministers to "act now when there is still time and hope" to implement the reforms properly as part of the new agricultural strategy, because "No single program so intimately affects so many millions of our people as land reforms." Proceeding from the general to the specific, the prime minister pointed out the well-known ABC's of what should be done, noting that the problem lies in the lack of effective implementation. Important, too, was the emphasis on reforms effectively carried out along with access to credit and other resources so that the newly obtained rights in the land could be productively used.

If the chief ministers were to commit themselves in earnest to the cause espoused by the prime minister, the following would have to be attended to: The preparation of a record of rights of ownership as well as tenancy, for without it no claim could be legally established and no tenant could attain security of tenure; written agreements instead of the prevailing oral agreements, rentals clearly defined, and receipts for rents made mandatory; as has been provided in some states, a tenant should not be liable for eviction for nonpayment of rent when arrears of rent are being recovered from the produce and other assets of the tenant; the right of resumption should be extinguished at least until "personal cultivation" is so defined as not to infringe on tenancy rights; "voluntary surrenders" should be so regulated that the government or local authority can settle other tenants on surrendered land if such takes place; while the ineffective ceiling provisions are practically beyond redemption, they should be reexamined primarily with an eye to preventing further concentration of land in relatively few hands; in any new rights gained by a tenant, access to water where available should be part of it, particularly as a means of utilizing new farm practices; finally, the creation of the type of village body in which the representation of tenants could become directly involved as an aid to the implementation of a reform.

These are not novel measures; all are mainly concerned with security of tenure rather than distribution of land. The failure to enforce the ceiling provisions to create surplus land for distribution has been so general that any corrective measures are probably futile. The recent (late February 1970) announcement of the Uttar Pradesh government and that of Tamil Nadu that they are contemplating the lowering of the ceilings is hardly the panacea it appears at first glance, even if such measures are enacted. One must note not only the evasions, litigations, and harassment produced by the ceiling measures in the past but also the distinct possibility of diverting attention from the fundamental problem of how to achieve security of tenure. For these reasons, security of tenure rather than redistribution of land is the main issue of reforming what has remained unreformed in the agrarian structure of India.

The enforcement of any of the proposals depends above all upon the political will of a party in power. Absence of this crucial factor has been a major drawback in the past, and the conference of the chief ministers veered in the same direction. On the all-important question of preparing an up-to-date record of rights, the "utmost priority" has been recommended but no time limit set. The prime minister's call for a commitment "not only to broad objectives but to meeting them by a certain definite date" has been largely ignored. A very useful suggestion of separate judicial machinery to deal with cases arising from implementation has not fared any better. The same may be said of measures to protect tenants from eviction, fair rentals, and prevention of land resumption. Looking back, therefore, at the results of the conference, they are reminiscent of the familiar pattern of representatives of states subscribing in "principle" to much of what the reform proponents have to offer but shying away from translating the principles into something concrete. And it is the states which have the power to legislate, administer, and enforce land reform programs.

The agrarian reform prospects for the 1970s are certainly not bright, for only politicians strategically placed make or unmake agrarian reforms. This is not to understate the fact that, especially in a country like India, the economic environment, population pressure on the land, and customary relationships of a long history of caste and religious traditions exert great influence on what kind of legislation is written to break institutional molds and what happens to it after enactment. They point, too, to the back-breaking task of renewing the old farm structure. But the fact that changes in that structure are inevitable does not invalidate the main premise—that the content and implementation of agrarian reforms are a reflection of a particular balance of forces in a country. In India a balance conducive to reforms has been largely absent all along, and the immediate future holds no great promise of reversing that trend. In fact, the political hurdles may be greater following the split of the Congress Party. Reforms which were unattainable in the Congress heyday of undisputed authority are much more difficult of attainment now when the support of the political leadership of this or that state, the bargaining involved, and the price exacted become crucial to the governing power. Countering this logic of political development is that not many years back agrarian reform was not nearly the subject of political debate it is now. The widespread surrender of land by tenants used to evoke hardly a ripple of official concern—nowadays it is very likely to ensure a fight. It cannot be overlooked, therefore, that the more recently pronounced reform stand of the ruling Congress Party can be a deterrent against bargaining reforms away from political considerations.

Despite the mixed setting with the preponderance of negative factors, it is reasonable to conclude that the issue of reform will not remain in the doldrums. The tide of unrest is bound to rise and with it agitation and organization of the discontented by political parties, both old and new, and not only the parties of the extreme left. In these circumstances, what states' legislatures normally refuse to enact and enforce under the due process of law, they may well be compelled to grant under duress. Tamil Nadu, for example, may be a case in point. Prior to the 1969 outbreaks of violence and loss of life, Tamil Nadu was reluctant to revise its tenurial enactments in line with the suggestions of the Planning Commission; recent indications are that it might, at long last, fall in line. Rajasthan might be another straw in the wind judging by the overt agrarian unrest in Ganganagan district and the chances of allaying it. There, opposition parties have organized groups of underprivileged farmers to make sure that the land about to be irrigated by the new Rajasthan canal is not auctioned off to the highest bidders but redistributed among the underprivileged at fixed prices. Under this pressure and in the hope of bringing the agitation to an end, the government of Rajasthan was compelled to set up a committee to conduct an on-the-spot probe into the existing rules for the allotment of land to poor farmers. The outcome of the committee's deliberations remains to be seen, but the point to stress is that even the concession of looking into the matter would not have taken place but for the agitation and organization of the farmers by a variety of political parties. None of these cases nor the land and harvest grabbing in West Bengal, Kerala, and similar instances in a few other states are models of agrarian reformism. They are nothing of the kind, but it takes little imagination to see that the stage is gradually being set for a determined group to make political capital by organizing the village poor and pressing their demands outside an orderly legislative process. It is for this reason that the Home Ministry envisions an "explosion" in the countryside if these problems are not dealt with vigorously and without delay. "Explosion" may be long in coming or in its literal sense it may not come at all. But the prospect of mounting instability and violence cannot be excluded—unless the states come to realize that theirs are the opportunities of enforcing beneficially the modest, nonrevolutionary programs which have always been the core of the Indian agrarian reform movement.

53. A Note on Agricultural Reforms

The interesting aspects of this August 1970 paper, despite a certain redundancy of earlier ones, lie in its background and apparent motivations. Within a short time after Robert McNamara's advent as president, significant changes in insight and objectives had come to the World Bank. In his address to a Columbia University Conference on International Economic Development in February 1970, McNamara had stated that "However important an increase in GNP may be as a necessary condition of development, it is not a sufficient condition." Limiting attention to an expanding GNP and overlooking "the other economic, social, and moral dimensions of the modernizing process . . . can only lead to greater political, social, and economic disequilibrium . . . If our investments are to meet this wider goal . . . we and other investors need to add . . . a new dimension of social concern."

Statements like these, and the new directions in Bank policy and activities to which they pointed, were a welcome sign to Ladejinsky that he was no longer swimming against the stream. At the same time they probably suggested to him that the Bank's staff might well benefit from a timely summary note that recapitulated what agrarian reform in India was all about and what a reasonable "minimum program" would be. But he also added a caution against undue expectations of how influential a role the Bank might attempt to play in such an outcome. The basic story, by now familiar, need not be repeated; only Ladejinsky's concluding summary is presented.

* * *

WHATEVER THE FUTURE MAY BRING, it is generally agreed that India is confronted in the countryside with a situation in which desirable agrarian reforms cannot be brought about either by persuasion or legislation. Twenty years of blowing mostly cold provide sufficient evidence on both scores. Political organization among the have-nots and the exercise of their voting power can be crucial. But this is a long time off. One comes back, therefore, to the principal point made in this paper, that the state governments must turn a new leaf in dealing with agrarian reforms. That new leaf is direct action and the mobilization of forces and resources of a kind with which wars are fought. At this juncture this is seemingly the only road to progress until the day when the peasantry becomes a source of authority and a mainspring of change. But in whatever way it comes, come it must simply because rural India cannot prosper in a state of turmoil. The Green Revolution, yes, for it carries the promise of an eventual sufficiency of food for the people of India. No one can minimize this achievement when it finally comes to pass, as it is likely to; but it should be noted that it can come through the efforts of something considerably less than the majority of the cultivators. In this connection, the warning of Mr. Chavan, the former home minister, that unless the land reforms are speedily implemented the Green Revolution might turn into a red one goes beyond rhetoric. Rising productivity, which is the hallmark of the Green Revolution, doesn't necessarily stand for the alleviation of the plight of the underprivileged ruralists. The negative aspects of the Green Revolution, due to the special Indian setting in which it operates, could be mitigated by implementing the minimum reform program. If done, agricultural production would receive an added impetus, but not the least would be the fusing of a measure of social justice with economic necessity. And after all is said and done, this is the real goal of an agrarian reform. And this is the goal India is yet to achieve.

Has the Bank a role to play?

Unsolicited, one is tempted to raise this question: Is there anything the World Bank could do by study or exhortation to help advance the cause of agrarian reform in India? The answer is "No!" Precedence alone argues for this position. FAO, the Ford Foundation, and experts singly and in groups, domestic and foreign, have tried their hand directly or indirectly and with no success. Researching Indian reform problems in depth is not now a useful preoccupation. It has all been done over and over again, and it only went to prove that experts and solicitous advice don't make reforms. For the same reason we firmly believe that, with all due recognition of the Bank's standing in India, the same would happen if the government of India accepted and the Bank launched a "reform project" to restate what the situation is and what might be done to alleviate it. The reluctance to involve the Bank stems from this overriding consideration: If any issue of development and social change is beholden to politics, agrarian reform in India is surely that. As such, it is an area where angels fear to tread, and this should apply to the Bank as well. It appears, therefore, that, while it is indeed fitting for the Bank to be informed about an issue which is tied in with the future of the country's stability and development, anything that goes beyond that will yield no useful results. In the final analysis—and without attempting to discover new Americas—it is for the Indians to settle this purely Indian issue, which spells nothing less than passionate adherence to property rights on the one hand and an attack on the same on the other hand. In this kind of a struggle the better part of wisdom is for the Bank to stay out of it.

54. *Field Trip to Eastern Uttar Pradesh*

In late April 1971 Ladejinsky visited eastern Uttar Pradesh. His report, dated May 6, sheds much light on problems of water, mechanization, the consolidation of fragmented holdings, tenancy, and so on in an area recently touched by the new agricultural technology.

I VISITED EASTERN UTTAR PRADESH in late April to acquaint myself with a reputedly backward, poor, overpopulated, and caste-ridden part of the state. The population is 93 percent rural as against 87 percent for Uttar Pradesh as a whole. Industrial development—large or small scale—is virtually nonexistent; except for a few sugar mills and brick kilns, industrial activity is largely confined to traditional cottage and household units. Ten years ago (more recent data are not available) the per capita income of the state was Rs261 (national average Rs330), while that of eastern Uttar Pradesh was Rs172. Severe shortage of irrigation facilities, recurring floods, small holdings, and farm practices which inevitably meant low productivity largely accounted for this state of affairs. In the past five or six years the agricultural economy has been undergoing promising changes; agricultural production and productivity have been rising and the conditions of some farm groups show signs of improvement. On the whole, though, the region is only beginning to find its way out of what was (and still is) essentially a subsistence rural economy.

In the course of a six-day jeep tour one collects not only aching bones but also certain impressions about current farming conditions and the thinking of the farm community. The paragraphs that follow are indeed impressions only, but after visits to a dozen villages and talks with numerous farmers in the field they are worth recording. In summary they are as follows: (a) Regardless of the size of holdings, lack of water is uppermost in the minds of the farmers. While they are supposedly sitting on a large reservoir of underground water and despite the sharp increase in the number of tube wells, the water, for the greater part, still is underground. (b) The Green Revolution is well known in eastern Uttar Pradesh and farm practices are gradually changing, but of mechanized agriculture there is virtually none. (c) The larger owners fear a further reduction of the land ceiling and are taking measures to meet that threat by another round of land transfers to near and far relations. (d) Tenancy is on the decline but not because tenants are being converted into owner proprietors. (e) Land consolidation is taking place on a fairly large scale, but apart from its beneficial aspect it has one serious negative aspect, which is the displacement of tenants as part of the process. (f) The recent abolition of land revenue on 6 acres and less is not favored by the farmers, big or small. (g) Finally, the English-speaking harijan (untouchable) farmer spoke for many harijans, small cultivators, and landless when he concluded his tale of woe in these words; "Sir, I am very poor."

Wherever we went, the principal item which concerned all farmers was shortage of water. All other serious problems were clearly of secondary priority to the bigger and smaller farmers, and the same applies to the landless agricultural laborers from the point of view of the duration of employment. It is a cliché to say that the reliance on the monsoon is risky; but, when talking in the field to two small farmers jointly renting three-fourths of an acre under cane, the water issue assumes grave proportions. Since they had only an inadequate masonry well, the monsoon would have to be bountiful for them to earn a gross Rs175 apiece. This is not an extreme case, extending as it does to better-off farmers who so far depend primarily on the annual vagaries of the monsoon.

We are not in a position to deal even in rough terms with the main ramifications of the water problem of eastern Uttar Pradesh. It would take much time, effort, and specialized knowledge to do that. But something can be said that points to the magnitude of what bothers the farmer community. The following is one indication that bears on the above:

> Mobilization of all the available resources for irrigation to the maximum extent and as expeditiously as possible is the kingpin in the strategy of accelerated development in this region. The Team has come to the conclusion that by 1971 effective irrigation will have to be extended to 72 percent of the cultivated acre in Azamgarth, 73 percent in Jaunpur, 75 percent in Ghazipur and 80 percent in Deoria.[1]

It should be added that the same team placed the principal reliance on private minor irrigation works, with strong emphasis on tube wells, whether diesel or electric. The six reasons given by the team to stress this preference as against major types of irrigation are well known and need not be repeated here.

Much has been achieved. The tube wells are not dotting the landscape à la Punjab; spotting one from a moving vehicle is not easy, and

1. "Report of Joint Study Team," (Uttar Pradesh [Eastern Districts], January 1964), p. 76, paragraph 4.

some of the villages we stopped in had none. Nevertheless, this is a case where the eye can mislead the casual observer, for the figures tell a different story. In the district of Jaunpur between 1965–66 and 1969–70, the total number of tube wells increased from 509 to 5,595, in Varanasi from 546 to 7,011, in Ghazipur from 237 to 4,076, in Azamgarh from 424 to 7,463, and in Deoria from 2,283 to 9,205. Electric connections are a problem because of a power shortage, and in the first four districts roughly half the tube wells are electrified and in the last one only 25 percent of the total. The progress is unmistakable, but water deficiency is there just the same, and if the situation is to be materially improved their numbers must increase manifold. If it is assumed that a tube well irrigates 10 acres and if this capacity is related to the sown acreage of the above-mentioned districts in the order given, the tube wells provide water for only 9, 8, 7, 7, and 8 percent of their respective acreages. Other irrigation facilities take up some of the slack, but it may be concluded that, while most farmers of the region suffer from a land shortage, there is apparently more land than water. The state tube wells are a story in themselves but by all accounts not a success story, and farmers want tube wells.

The villages are small, but not so much in terms of farm families as in acreage. A village possessing anything above 600 to 700 acres is exceptional; but, of what land there is, distribution is, as usual, highly skewed. Taking the eastern region as a whole, owned holdings of less than one acre accounted (1961 census) for 22 percent of the total; 38 percent are in the category between 1 and less than 3 acres; 18 percent between 3 and less than 5 acres; and 22 percent owned 5 acres or more. What the figures don't reveal is that a relatively small group of farmers with 10 to 30 acres own most of the land and the rest are "mini" farmers, an undetermined number with so little land that for all practical purposes they can be classified with the landless. Holdings are badly fragmented. About the village of Sakaldia (Varanasi district) our notebook reads as follows: a holding of 40 acres in 15 to 20 plots, 14 acres in 10 plots, 4 acres in 7 plots, while the biggest holder with 45 acres testified to 60 to 61 plots. Its depressing effects on agricultural production are obvious; but as we shall point out elsewhere this condition has induced a good deal of consolidation effort, and this is one of the brighter developments observed.

Tenancy is on the decline but for rather special reasons. For all practical purposes the agrarian reform laws do not apply in eastern Uttar Pradesh, and a case can be easily made that they have worsened the position of the tenantry through the familiar "voluntary" surrenders, outright ejectment, high rentals, and total disregard of the ceiling provisions. Though the ceilings are not enforced, the larger owners are disturbed by the reports of lowering them. As a precautionary measure, a new round of land transfers is taking place in anticipation of such legislation. According to two "tasildars," in their "tasils" (district subdivisions) hundreds of owners have already legalized such transfers in the guise of sales. How widespread this movement is in other parts of the country remains to be seen; but, if eastern Uttar Pradesh is any indication of what is in store, lowered ceilings will not result in any significant acreage slated for redistribution.

Tenancy conditions, in what remains of tenancy, are what might be expected and occasionally a bit worse. Rents are rising, which is not surprising in conditions of land hunger and increased productivity. On better irrigated land the fifty-fifty sharing formula is in effect a net of 25 percent of the crop to the tenant because the cost of the inputs are often deducted from the tenants' share. Higher rental is not the worst the tenant faces. Under the impact of the new and more profitable farm practices, the trend is away from tenancy and towards self-cultivation primarily with permanent or "attached" labor and casual laborers during the peak seasons. In both instances, little cash changes hands. In the first instance the owner turns over half an acre to two acres of land to the attached laborer in exchange for carrying out whatever chores he is assigned in the course of the year. The land reverts to the owner when the arrangement lapses. An acceptance of a loan may lead to the same condition, the loan—not the land—being the consideration. The attached laborer receives all of his food while on the owner's farm. Where casual or daily laborers are involved, the usual practice is to provide breakfast and a kilogram or two of whatever the main product of the season happens to be, the amount depending on whether he works half a day or a full day. In the harvesting season the remuneration ranges from one-sixteenth to as low as one-thirtieth part of the crop. Wages have moved up in eastern Uttar Pradesh as elsewhere, but if converted to cash they do not exceed Rs2 a day. The food thus earned during four to six months of the year doesn't suffer for more than 3 to 4 months. In most instances, significant improvement in wheat and rice yields has not yet created new alternatives for farm laborers. They continue to be faced with the problem of large supplies of surplus laborers. Two practices help to supplement such meager incomes: remittances from members of the families (not only of the poorest) who have migrated to other states in search of employment, or employment in nearby brick kilns or elsewhere in Uttar Pradesh itself. A case in point was the rickshaw driver who in the nick of time saved the day for us when our jeep ran out of gasoline. He proved to be an owner of an acre and a half of land plying this trade to augment his income. The poverty of the underprivileged is self-evident, but it is mitigated where the sons or heads of the families have pulled up stakes and found employment outside of the villages.

That all or most farmers need credit is axiomatic, but it is equally true that the credit cooperatives in Uttar Pradesh are among the country's worst performers. This is surely true of eastern Uttar Pradesh. Had we visited many times twelve villages, the conclusion would be the same. They are either defunct altogether, lead a hand-to-mouth existence, and even the one good society we ran into—Rs80,000 advanced and only Rs8,000 overdue—did not prove to be as painted for the benefit of the visitors. Only a third of the farm families were members and fewer still received loans. About 90 percent of the advances were in cash with no concern for how the credit was used, and curiously enough the farmer owners are charged a 9 percent interest rate as against 12 percent paid by the landless. Presumably the degree of creditworthiness had something to do with the different rate. The chairman of the cooperative society and his co-leader, obviously the "all-in-all" of the village, were extolling its progress, but there were gaps in the account. The principal gap filled in for us privately was that the chairman was both the main consumer and source of credit, relending it at 30 percent or more. The strong impression carried away is the apathy of the farmers about credit societies as if they had no real function to perform. The reasons are too well known to call for elaboration. At least one farmer dismissed the entire cooperative movement by saying that "the embezzlers are in charge of the coops." This is not a new charge, even if not always phrased in such terms; but evidently the vested interests in this particular eastern Uttar Pradesh village behaved just as the rest of them in multitudes of ill-performing credit cooperative societies in other parts of the country. The branch of a nationalized bank six miles away did not come in for much praise either, the main point being that it concerned itself only with loans to rich farmers.

In conditions sketched above one wouldn't expect the Green Revolution to be a factor of any significance. And yet, many farmers actually practice it by using fertilizer and improved seed varieties. In a region where none was used only a few short years ago, it is remarkable how widely this use has proliferated and the line the farmers draw between the relative utility of one type of fertilizer compared with another. For a variety of reasons the volume used varies greatly, but a farmer's self-answering question—"How can you raise a good crop without them?"—tells a good deal about the changing practices. Very significant is the volume used, *given water availability*. In this regard, two facts emerge from a sample study dealing with four categories of farmers, ranging from an average of 16, 7, 4, and 2 acres. The first is that a number of them use the prescribed dosage (nitrogen-phosphorous-potassium 36:18:18 kilograms per acre on wheat or rice); some few use more, and even the two-acre farmers deviate but little from the prescribed mix. The second item is worth noting, even if not a new one, is that the size of holding is at least neutral from the point of view of yields attained. Thus, the farmers in the category averaging 2 acres produced 40 maunds of wheat and 37 maunds of paddy per acre, compared with 44 and 49 in the farmer category averaging 16 acres. Considering that the latter applied much more fertilizer per acre, the small farmers hold their own rather well in proportionately similar resource conditions. They evidently try harder.

Apart from this sample taken in some of the villages we visited, the average wheat yields on irrigated land is 30 maunds per acre; this compares well with the former 15 to 16 (without fertilizer and better seed), not to speak of the 8 to 10 maunds per acre on dry land farming. Approximately the same holds for paddy. The cropping pattern too shows some signs of change, and the introduction of potatoes as a cash crop has been very profitable and is expanding. This region is not Punjab or western Uttar Pradesh, and it would be idle to draw comparisons between them. But this much should be noted: Many farmers with little technical assistance from outside know how to mix or stretch their limited resources gainfully.

The net effect of the contact with the Green Revolution has certainly benefited the larger landholders and, to a degree, the smaller holders as well who are in a position to buy fertilizer, improved seed varieties, and make the best use of what water they do have. The farmer with 40 irrigated acres who has just built himself a Rs100,000 "pucca" house is the great exception, but a few have gone from thatch to tile roofs and a few from tile roofs to new and better houses. It is fair to say that the front-seaters in any peasant meeting have not done badly in eastern Uttar Pradesh, and some of the less well-placed farmers are not without gains either. The picture of improved conditions must not be overdrawn; all the inhibiting factors mentioned earlier continue to affect adversely the very small cultivators, tenants, and landless.

A feature no can can miss, however brief the visit, is the apparent absence of mechanical equipment normally associated with the new technology. One is tempted to say, for instance, that there is not a tractor in the region. This is not true, of course, or else why would there be an agency selling tractors in the city of Benares? We are merely saying that a tractor is a great rarity; there was only one in the villages we contacted. The visit to a reputedly most progressive farmer of eastern Uttar Pradesh revealed that on his block of 45 acres of choice land only the thresher was a concession to mechanization. The owner innovator saw no point in using more labor-saving devices in order to operate an already very profitable farm with four well-fed pairs of bullocks. They provide him with compost which he successfully mixes with chemical fertilizers, and he is in a position to maintain a proper time schedule for his field operations. This is not to say that some farmers have no tractors or haven't applied for them, but the tractor is not yet part of the changes taking place. Almost the same goes for the power tiller. A very unhappy farmer sold his bullocks, borrowed some money, and bought one of domestic make for Rs8,000. And that was a big mistake. To listen to him, everything went wrong with it and the replacement wasn't any better. This useless symbol of mechanization will give way to bullocks once again; and the immediate prospects are that, within a considerable radius of this particular fiasco, no farmer will venture to buy one. Threshers are just beginning to make their appearance; but the prevailing sight, this being the threshing season, are bullocks making their endless rounds on the threshing floor. The rich farmer with the Rs100,000 "pucca" house has not been touched by the mechanical phase of the new technology; his participation is limited to the sharing of a thresher with another farmer. The change is mainly in the improvement of the plow—from the customary one with a thin iron spike to an all-iron plow. The shift is anything but universal; one village did boast of ten such plows, and the best one can say that it is the beginning of a trend.

Pressed as to why the bigger farmers have not gone much beyond the iron plow stage, the answers are about as follows: cheap and abundant labor; fragmentation of the land; initial heavy outlay; absence of repair facilities in neighboring towns, and, as always, uncertainty about the water supply. Any one of the reasons or a combination of them may explain the reluctance to mechanize the farms of 20 or more acres. Our traveling companion, Srinath Singh, a native of the area with much knowledge of its rural economy, adds another reason: arrested entrepreneurship. An irrigation facility is a must; a modest investment in fertilizer and in high-yielding varieties doesn't involve much risk-taking, whereas mechanical equipment demands a considerable investment and an array of changes for which they are not yet prepared or which they don't consider vital. The conclusion he draws is that for some time traditional agriculture in eastern Uttar Pradesh will continue but with strong admixtures of the Green Revolution in the form of tube wells, fertilizer, and seed. What may tilt this relationship in favor of mechanization is the consolidation of holdings currently very much in evidence.

Previously we cited a sample of the extreme degree of land subdivision or fragmentation. It is an old story not peculiar to eastern Uttar Pradesh; but the administrators of the region are combating it by the time-consuming, costly effort to consolidate the fragments into more viable blocks of land. Five years ago the total land consolidated in the region was an estimated 700,000 acres, and by 1976 another 2 million acres are to be treated similarly. The answer to the question of how much has been achieved to date is that the job will be completed in another 4 to 5 years. Looking at individual village cases, the maps with neatly drawn enlarged holdings, the wrangle among the farmers about exchanging plots—and this is the core of consolidation—one is left with the impression that the anticipated completion of the job is on the optimistic side. The court cases challenging the decisions of the consolidators are enough to give one pause. Nevertheless, the results we looked at or those about to take place are or will be an improvement over the former totally uncontrolled scatteration of plots. The majority of the farmers will not have all their land in one piece, but the expectation is that a goodly number will and that the plots of many others will be reduced in number. A farmer with 11 acres and 64 fragments now reduced to three plots may not be typical, but an attempt to eliminate extreme subdivision is under way. It is understood also that the best consolidation effects are good only for a generation. The reference is to the inheritance laws, especially in cases where the joint family tradition of cultivating the land as a unit breaks down. In such event, the process of parcellation enters the picture once again.

Whatever the future, more immediately consolidation in relation to agricultural productivity should mean greater scope for tube well development. Officials and farmers are agreed that, with the new incentives provided by fertilizer and better seed, there will be a greater urge to dig tube wells. The tube well development in the past five years is an encouraging sign of the prospects that lie ahead. And yet the consolidation process is not without its fly in the ointment as far as the tenants are concerned. Testimony on this score from various sources is clear—consolidation is an occasion to eliminate tenancies. The process begins, at least with some owners, once a village has been earmarked for consolidation. While it is true that, even before consolidation the tenants had no firmly assigned plots beyond a crop season or could be displaced on the ground of "self-cultivation," the reshuffling of the plots and the presumed rise in the value of the entire holding offers the owner an added incentive to be rid of his tenants. In this regard consolidation and its economic consequences are not dissimilar from those of the Green Revolution in general with its enhanced land prices, rents, and ultimate ejectment from the land in many cases. The consolidators are alive to this by-product, but they can only point to the agrarian reforms which failed to achieve their goals and which are not within their competence anyway.

Effective May 1, 1971, all holdings in Uttar Pradesh of 6.25 acres and less are relieved of paying land revenue. It is an outgrowth of a promise made by a variety of political parties on the eve of the election to woo the farm vote. Now that this is one promise kept, how do the farmers feel about it? The consensus seems to be that they don't favor it for a number of reasons. The last revenue settlement in eastern Uttar Pradesh took place in 1882 and the per acre revenue varies from Rs2 to Rs6. Considering the changes in agricultural productivity, particularly in more recent years, for the individual farmer the amount involved is too small to be considered a boon. Since the revenue is levied on owners, there is no reason for them to worry that their title to the land will be in any way jeopardized without the revenue receipts. But they are concerned and would feel more at ease by paying the land revenue. Its abolition in these categories has not been accompanied by an upward revision of revenue for farmers above that acreage. The decline of land revenue for the state as a whole is estimated from Rs21 crores to 9 crores; in a district with which we are familiar, the respective figures are Rs45 lakhs and 13 lakhs. The farmers believe that the government of Uttar Pradesh will try to make up this loss through other imposts, ultimately nullifying their small net gain. Finally and more importantly, the argument against abolition is that, if the government is in earnest about the lot of farmers, why no greater assistance in furthering irrigation facilities, more electric power, more schools, more effective credit facilities to the very farm groups relieved of paying land revenue, more and better village approaches from the highway, cheaper water rates, lower fertilizer prices, and so forth? When all these reasons are put together, it is not surprising why the politics of land revenue abolition don't sit well with the beneficiaries.

As far as the writer of these observations is concerned and in relation to the World Bank, one recommendation emerges from all of the above: an examination of the water problem of eastern Uttar Pradesh with an eye to a tube well project. Whether the bigger farms are mechanized or not is not an issue; quite a few of them are doing well without the benefit of the tractor, power tiller, or even a thresher. But for the entire landowning farm community, including the agricultural laborers in search of employment of longer duration, a rise in productivity is of prime importance; and the main stumbling block is the shortage of water. If it is true that eastern Uttar Pradesh sits atop a large supply of still untapped underground water, further tube well possibilities cannot be excluded. Whether, despite the consolidation efforts, land fragmentation and the small-sized holdings are an insuperable obstacle should be examined in line with this overall suggestion. The encouraging fact is that in the peculiarly difficult conditions of the region, 32,000 tube wells have been installed in the past five years. We presume that sound economic reasons generated this development. It may well be, therefore, that a Bank mission would find conditions suitable for a tube well project as such or as part of a broader agricultural credit project.

55. Agriculture–Problems and Needs

This was Ladejinsky's contribution to the World Bank's annual report, "Economic Situation and Prospects of India," May 11, 1971. The problems discussed are for the most part familiar, but they are examined in a changing, increasingly emotion-laden scene. The Green Revolution has moved forward, but so have its side effects. Mechanization has become a cause for concern and debate; a rural works crash employment program is in preparation; the Small Farmers Development Agency has got its program started; in the absence of progress on agrarian reform, the Communist and other leftist parties joined to organize a national "land grab," which was tried and failed. All these and more enter into Ladejinsky's analytic review. He includes a powerful quotation from Jayaprakash Narayan on the state of social injustice in the Indian countryside and offers some telling observations on appropriate goals for external assistance. "Foreign lenders," he argues, "should seek out projects which combine reasonable production potentials with a concern for social justice."

To reduce redundancies between this 1971 annual report and those for 1970, 1972, and 1973, only the section on land reform is included here; in it Ladejinsky clearly formulates his famous "minimum" program.

THIS OLD AND VEXING PROBLEM [of land reform] is familiar to readers of previous reports on India. The past year has not been notable for steps towards resolving such issues as ceilings on land ownership (except for a few new legislative enactments with results yet unknown), security of tenure, and fair rentals or recorded rights of tenants and sharecroppers to remain on the land undisturbed. In 1970 the union government made yet another attempt to induce the state chief ministers to move the unfinished reform business off dead center, but with far from satisfactory results.

The year 1970 will be remembered in Indian agrarian history as the year of the land grab. Though the result was a fiasco, it highlighted the fact that the land reform issue is not about to fade away. The tide of unrest is bound to rise, and even the politicians normally opposed to reforms may yet be compelled under the threat of circumstances beyond their control to grant what they have long denied to the underprivileged under due process of law. This may not lead to any significant redistribution of land, but it could result in a fairer distribution of agricultural income through a number of measures which constitute a feasible "minimum program."

In our past reports the following issues have been examined at considerable length: the need for reforms, the efforts to that end spanning two decades or more, the important achievement in the abolition of intermediary (zamindari) tenures, the far from successful tenancy reform measures affecting cultivators on about 20 percent of the cultivated land of the country, the peculiarly difficult Indian rural setting which partly explains the character the reforms have assumed, and finally the insistence of the government of India—as distinct from the state governments—that the errors of commission and omission of the reform legislation and of its poor implementation must be corrected. We will not go over this familiar ground, but four items may usefully be discussed: (a) a minimum reform program, (b) the land ceiling and the conference of the state chief ministers, (c) the land grab movement, and (d) implementation prerequisites and prospects.

Granting that reforms in India may yet see better days, what should be the character of a *minimum* program to which the states of India should address themselves? Experience here and elsewhere suggests an action program as outlined in the following paragraphs.

First and foremost is the preparation of a record of tenancies without which no tenant's claim could be legally established and no tenant can obtain security of tenure. The importance of this for the securing of credit has been underscored above. Written lease argeements instead of oral ones and receipts for rents should be made mandatory. Crop share rents are extremely difficult to regulate and should be replaced by cash rents; to facilitate this, either the rents should be fixed as a multiple of the land revenue or else the state governments should divide land in each area into a few broad categories and determine the average produce of each such category of land and the average price at which it may be converted into cash; as a rule of thumb, no rental should exceed a third of the crop, particularly where the tenant is responsible for all the inputs.

The essential purpose of tenancy reform is to give security of tenure and to regulate the rent tenants pay. One without the other makes no sense, and the two measures must go hand in hand. The Indian land-man ratio being what it is, no land at a rental below 50 percent or more of the crop can be secured without the force of law behind it. The tenant will not be able to protect himself without the assurance that, barring misuse of the land or nonpayment of rent, he can remain on the land undisturbed and with a right of renewal of the lease. This is the basic part of "security of tenure" and is crucial to the entire scheme of land reform legislation. To make this effective, the most debilitating part of the land reform in India, the ill-defined right of an owner to "resume" or take back the tenant's holding for "personal cultivation," should be abolished or permitted only in exceptional cases defined by law. In practice hardly any tenants voluntarily surrender land; they do so only under the threat of total separation from the land, even as an agricultural laborer. Such surrenders, if they are not fictitious, should be so regulated that other tenants are settled on the land; in all other instances, which is the majority of them, they should not be permitted except when the land is required for some specific nonagricultural purpose. Most of the surrenders are past history, and restoring the tenantry to their formerly cultivated land is about as difficult as undoing the ceiling provisions mischief discussed below.

Nevertheless, no program of this kind can overlook it when a demonstrated wrong has been done or agreement breached. Access to water where available should also be part of any revised tenancy legislation, especially as a means of utilizing the new farm practices.

If such measures are enacted and enforced, the familiar discrimination by the credit cooperatives against the tenants would undergo a drastic change. It is important that tenants be given a sense of participation in the implementation of a land reform program. With that in mind, and as in Japan and Taiwan, village reform committees should be created with a suitable representation of tenants, owner lessors, and owner cultivators. What the officials do not know about "who is who" in the village the committee would surely know. Even a literate tenant would lose his way in the maze of the provisions of the enacted legislation and amendments piled upon amendments; this explains why most of them have only a hazy notion of what the reform laws mean or do not mean. The owners have no such problem. We suggest, therefore, that the basic provisions of the laws pertaining to landlord-tenant relations be reduced to a sheet of paper and, like the Magna Carta, be nailed to the door of every village office. Many a tenant would find out what the law is about, and under propitious circumstances he might put the knowledge to good use.

Perhaps the most difficult part of any reform effort is a ceiling on land ownership to provide a pool of surplus land for the establishment of owner proprietors. Virtually every state reform enactment contains a limitation (ceiling) on the permissible retention, but it is equally true that the intent of the ceilings has been widely evaded. The ceilings have generated an estimated 2.4 million acres of surplus land of which 1.6 million acres have been taken over by the state governments, but only approximately 1.2 million acres have been distributed. The numerous exemptions (twenty-six kinds in Tamil Nadu)[1] and the division of the land among members of the owners' families and close and distant relatives have combined to make certain that their own divided holdings were each *under* the ceiling. In the light of the overall performance, the third five-year plan concluded with moderation that "On the whole, it would be correct to say that in recent years the transfers of land (to relatives) have tended to defeat the aims of the legislation on ceilings and to reduce its impact on the rural economy." The drafters of the legislation could have anticipated this result when they elected to permit the application of a retention limit not to the family as a unit but to individuals comprising it, while at the same time handing out exemptions right and left.

This brief account of past history is relevant to yet another land reform conference of the state chief ministers (September 26–27, 1970) sponsored by the prime minister. Coming as it did not long after the nationwide land grab movement and the recognition that a similar conference in November 1969 was anything but a success story, something more tangible was expected this time. The immediate outcome belied these expectations, but not for lack of exhortation on the part of the prime minister. With an allusion to land grabs, Mrs. Gandhi noted that "political parties may organize, even exploit, rural discontent but they do not create it. The time has come to face the facts." She pointedly asked the chief ministers: Why is social discontent in the countryside on the increase? Why is it erupting in violence with greater frequency? In her answer which should have set the tone for the coming deliberations, she stated:

> The land reform measures implemented have failed to match the legitimate expectations which were first fostered among millions of cultivators during the national movement . . . In short, we have yet to create institutional conditions which would enable small farmers, tenants, and landless laborers to share in the agricultural New Deal.

It would be pleasant to report that these sentiments were shared by the assembly. Reducing the ceiling limit was the main point of discussion, imparting to it an importance far beyond its deserts; but most of the chief ministers rejected such a proposal outright. This preoccupation with the most intractable issue of all

1. V. G. Prasad Rao, "Land Ceiling in Tamil Nadu," *Times of India* (October 8, 1970).

diverted attention from practically all other aspects of land reform which economically and politically truly matter at this juncture. What concerned the chief ministers were higher prices for agricultural products, more credit, and more of other inputs to raise production. They were making their stand on the basic assumptions underlying the new agricultural policy, one of them being no disturbance in land relations. The upshot was the classic shelving device: The Central Land Reforms Committee was to "examine in depth the difficulties" brought out during the conference.

Nevertheless, a number of states have since enacted new and more meaningful legislation by lowering the retention limit and by tying it to the family holding as a whole. These states are Kerala (Communist), Tamil Nadu (notorious for the plight of its tenantry and landless laborers), and West Bengal (under president's rule and overflowing with rural strife). If actually implemented, a fairly large number of cultivators would get some land but not enough to create anything resembling viable holdings. In Kerala and Tamil Nadu "all rural households will have some land to cultivate but . . . between 40 and 45 percent will have just half an acre each."[2] If similarly reduced ceilings were applied nationwide, approximately the same proportion of cultivators in the various states would have holdings ranging from 0.5 acres to 2.5 acres, the only exception being Rajasthan with 5 acres and much of it poor land. Though this exercise was not precise, Dandekar and Rath conclude that there is not enough land to raise the existing smaller holdings above these levels and that to provide land for the landless would vastly increase the number of uneconomic nonviable holdings.

The validity of this argument is recognized, but in a country with great land hunger the line of demarcation between a viable and nonviable holding is not as important as it appears at first glance. Even half an acre, let alone an acre or two of reasonably good land, spells the difference between abject poverty and something approaching security. Nevertheless, we are not suggesting major emphasis on new ceilings at the present time. For one thing, the new land recipients are likely to get the worst land. The state governments would have to finance the conversion of such land into more productive units. This is unlikely to come about; but, if it does not, such holdings would eventually go back where they came from. In view of state government finances, the obligation of paying for the land, which the states would have to undertake in the first instance, would have to be at the expense of other development expenditure unless it were financed, as it well could be, by taxing agricultural income. But not the least of the problems is the enforcement of such ceiling measures. The past record cannot be disregarded by even the staunchest supporters of land redistribution in India. It is very doubtful that what has been so thoroughly scrambled up with impunity can be unscrambled, especially at a time when the role played by the bigger owners in the new technology is deemed important. Finally, it would be highly presumptuous to assume that all or even most of the land claimants would indeed share in some land or, as a corollary of that, that lowered ceilings would do much to contain rural unrest. This said, the accumulation of excessively large holdings in a few parts of India is a threat to the social order and should be tackled on a selective basis by the states concerned.

In retrospect, 1970 will not be remembered for the ceiling controversy but for the nationwide so-called land grab movement or, as its initiators prefer to call it, the "de-grabbing" movement. Measured in land seized and distributed, the entire affair can be dismissed as a failure, even if the claim of having occupied and distributed 30,000 acres is correct; but as a reflection of the forces of unrest, it demonstrated that a better prepared movement in future could lead to drastically different results. The phenomenon of land grabbing is not new in Indian history. Unlike the modern land grabbers, those of a bygone age went about their business by taking the land rather than indulging in agitations, flag waving, and symbolic land occupations. But one doesn't need to invoke history to note that the movement discussed here can be carried out successfully in modern times; all one has to do is look two or three years back at West Bengal, where un-

2. Dandekar and Rath, *Poverty in India* (Ford Foundation, December 19, 1970), p. 102.

der the United Front (Communist) regime 100,000 acres or more were forcibly seized and distributed, though at the price of considerable loss of life on both sides. Kerala's pre-1970 land grab experience is a peculiarly Keralan story, but it also points to the fact that the rural poor can be persuaded to take the law into their own hands and can get away with it.[3] Nor is the *idea* a novel one that some day the land question of modern India might be settled in such manner. On this point, an exchange between Gandhi and Louis Fisher, his biographer, is instructive. Asked Fisher: "What would happen in Free India?" Replied Gandhi: "The peasants would take the land. We wouldn't have to tell them to take it. They would take it."[4] As with all good prophets, it didn't happen in his lifetime; but the "takings" in recent years, however small, hark back to Gandhi's prophecy.

West Bengal has been very much in the limelight for some years, with the rest of the country claiming only an occasional headline. There was a particularly big one from Tamil Nadu, where forty-two allegedly recalcitrant landless were burned alive. Official sources recorded disturbances in the states of Assam, Andhra Pradesh, Bihar, Gujarat, Punjab, Rajasthan, Tamil Nadu, and Uttar Pradesh, including a large number of forcible land occupations in 1969. But a more concerted and militant move was decided upon at the meeting of the National Council of the CPI (Communist Party of India) on May 15, 1970. Supported by the two socialist parties and other left-wing organizations, a joint action committee was formed to initiate the first phase of the "struggle-for-land day" on July 1, followed by the second and third phases, all ending in late September. Three reasons explain this redoubled effort: the established precedent in the preceding years, the need for a political "vote grab" device to mobilize the peasantry under their respective party banners in the coming national elections, and discontent with the slow pace of the land reforms.

Judging by the proclaimed intent of the "grabbers" or "liberators" of land, they were setting out to undo all the injustices which have cropped up in the wake of the land reform programs. In reality, the scope of the activities was limited and more symbolic than real. The brunt of the attack was on certain prominent individuals in agriculture, business, and public office with known large holdings who circumvented the ceiling provisions. In addition, it centered on large holdings vested in the state since the abolition of the zamindaries—which had been either distributed among the wrong people or not distributed at all. Government wasteland, tribal lands, Bhoodan lands (gifts by rich owners to the V. Bhave movements but allegedly not distributed among the landless), homestead lands, and temple lands were also included.

The movement created a great stir throughout the country and was not without a good deal of sympathy on the part of many of the onlookers. One of its weaknesses lay precisely in the fact that it could boast of more onlookers than participants. A fervent partisan of the movement bemoaned the readiness of the landed rich to protect their interests as against the indifferent part played by the village poor in support of their cause. "Even in the villages," he writes, ". . . there is much misgiving regarding the land struggle. The left parties have not been able to allay these misgivings."[5] This is part of the explanation of why the immediate results were meager.

Other reasons are not far to seek. Though the leftist groups allied themselves to a cause close to the hearts of the village poor, their battalions were not there at the showdown to lend tangible support. They were conspicuous by their absence partly because the Communists and their allies jumped the gun too soon and partly because their enthusiasm was evaporated, as in the case of Rajasthan, by a notification of the government that landless people forcibly occupying government land (one of the main issues there) would be disqualified from future allotments. By far the most debilitating effect on the success of the land

3. T. K. Commen, "A Close Look at Land Grab—Allepey's Experience," *Times of India* (September 19, 1970).

4. Louis Fisher, *A Week with Gandhi* (Bombay: International Book House Ltd., 1944), p. 43.

5. Bauke Behari Das, "Lessons of Land Movement," *Indian Left Review* (February 1971), p. 19.

seizure move was the fact that practically all state governments took it very seriously. It was a remarkable display of single-mindedness and zeal about a land issue such as has never been demonstrated about the mass evasions and wholesale subversion of basically modest and useful reform measures. An estimated 20,000 political activists were placed under arrest; law and order were upheld; and the land grab movement became only a memory—or so it seemed.

It may be too soon for a final judgment on the episode. The three political parties dominating the land grab have managed to establish a degree of identity of purpose which may serve them better should there be a next time around. Much more important, the movement has succeeded in focusing public attention on the land hunger among the landless; more so in fact than any official and unofficial declarations could do. It pinpointed an inordinate accumulation of land in a few private hands contrary to the law and spirit of the reform enactments, and it brought into the open the entire question of public land vested in the state governments and how poorly such land is being allocated to those in greatest need. To some close observers of the movement, its main achievement lies in the "waking up" of state governments and of the political parties which heretofore took a very conservative view of landlord-tenant relations. The apparent signs of greater concern about fairer and speedier allocation of public land to the landless and the easement—often to the point of abatement—of eviction procedures against the illegal encroachment on land prior to the land grab could be treated as signs of the "waking up."

One cannot predict whether this is a desirable portent of changes in the making. The prime minister's sweeping election victory may well prove to be far more important in this regard. We hazard this assumption because India's only option as a democratic welfare-minded state is to alleviate the conditions of those who have been promised and are entitled to certain rights in the land they cultivate. Therefore, the prime minister's repeated stress that this is the only course for India to pursue can be disregarded only at the risk of the very interests anti-reformism is supposed to protect.

The difficulties in the way are too well known to need elaboration. Suffice it to say that no land reform is an amicable affair; wherever reforms have succeeded, governments have had to use all the power within their command to overcome the obstacles. India can be no exception to this rule, and strong pressures and clear-cut measures from the top will have to be applied to attain the desired goals. This is the only road to progress until the day when the peasantry itself becomes a source of authority and a mainspring of change. The minimum reform program outlined in the preceding pages is not a revolutionary program; it requires only the legitimation of customary peasant rights and the removal of abuses which are in fact an outgrowth of legislation often obstructively drafted and cynically executed. Rightly or wrongly, the land redistribution question is muted, and the landowners and the politicians supporting them would be wise to accept what impinges upon their economic interests least. If so, the instabilities and the tensions would be minimized, if not eliminated. And so would the fear of an "explosion." Agricultural productivity would receive an added impetus with a more equitable expansion of the new technology. Not least, one could anticipate among such reform-affected farmers the fusing of a measure of social justice with economic improvement. After all is said and done, this *is* the goal of an agrarian reform worthy of its name. And this is the goal India should pursue.

56. *Refugees in West Bengal*

Toward the end of August 1971, the number of refugees from East Pakistan who had fled to India had reached an estimated 8 million, almost all of them Hindus, and about 6 million of them in West Bengal. The World Bank's resident mission in New Delhi had gone to West Bengal to study the situation at firsthand and assess how this new responsibility so suddenly thrust upon India would affect its economy and needs for additional external assistance. Ladejinsky was a member of this party. In this letter of August 28 addressed to I. P. M. Cargill, then director of the South Asia department of the World Bank, Ladejinsky reported his personal observations and assessment of the situation. Among the plethora of reporting by the news and other communications media at the time, it is hard to recall anything approaching this in significant fact-finding, insight, and analysis.

Ladejinsky's plea that the lack of precedent or other conventional reasons not be permitted to stand in the way of emergency assistance was not necessary. The India aid consortium had already met to consider the problem, and a sizable volume of emergency aid was quickly made available through the United Nations and supplemented by bilateral sources with the active support and assistance of the Bank. But Ladejinsky's reporting of the magnitude and urgency of the need and of the admirable and efficient handling of the problem by the Indian authorities was undoubtedly of considerable help to the Bank's management in formulating a constructive response to the emergency.

I AM JUST BACK FROM CALCUTTA where Mac[1] and his entire staff were wetting their feet—literally so on two occasions—in connection with the impact of the refugee problem on West Bengal and India in general. You will get a detailed report on the subject; in the meantime I am giving you my impressions, some of which may not fit in the overall body of the report. What I have to say will be no great news to you, but here it is for whatever it is worth.

If seeing is believing, it takes but a fleeting glance to recognize that the refugees are in one hell of a fix and ditto for West Bengal and the Indian government. In saying this, I am not thinking so much of the present, which is grim enough, but of the worse shape of things to come in the not too distant future. But to begin with a few figures and who the refugees are. Just how many refugees there are is debatable, but the official estimate as of August 19 is just a touch over 8 million. If it is a million less, the problem is only marginally less bad. Approximately 6 million of the total are in West Bengal, and of these an estimated 1.5 million are reportedly with relatives and friends. The bulk of the refugees are in 1,100 camps, with 750 in West Bengal. The overall figure is not one pulled out of a hat; it is based, first of all, on the inoculation card which is a must for every refugee who crosses the border, and this is followed by the even more important ration card issued to every refugee. It is this double-check system that enables the commissioner for rehabilitation of West Bengal to cite from a little book the daily number of arrivals and the total. The imperative of a ration card and shelter would seem to give the numbers a good deal of reliability. Under these circumstances, there cannot be much slippage at the border, and if the total is wrong it could also err on the low side. If 8 million is a

1. [Orville McDiarmid, then chief of the World Bank's resident mission in New Delhi.]

fairly correct figure, it is not the end of the story. The time when the daily influx ranged from 70,000 to 100,000 is over. In late July and early August the number had declined sharply to about 26,000, mainly because of extensive floods; but more recently the daily intake had been averaging around 40,000. This, of course, is the most disturbing part of the whole affair as it means no end of the inflow immediately in sight, let alone a reverse flow. As the monsoon begins to draw to an end, it is not unlikely that by the end of October or early November India will have to cope with about 10 million refugees. Not without interest is the fact that the more recent arrivals originate more and more from the outlying districts, 100 to 200 miles from the borders. The presumption is that the border districts have been largely cleaned out of the Hindu population.

Who the refugees are bears significantly on the issue at hand. I used to think that, since the opposition to West Pakistan engulfed the vast majority of the East Pakistanis, the exodus would reflect it. This is not the case and it is demonstrated by the fact that the official estimates refer to only 7.5 percent Muslims and 92.5 percent Hindus. The figure could be closer to 5 percent—*if* that many. In the three camps we visited one has to look for a Muslim—and not find one—as was our experience. Certainly in the Salt Lake camp, near Calcutta, with an estimated 150,000 "inmates," as our guide referred to them, we didn't stumble on any group of Muslims. There may have been some, and other camps surely have some; but the inflow is essentially that of East Pakistan Hindus. There is enough circumstantial evidence to say that the occupying powers of East Pakistan are "cooperating" in the exodus not only by making the lives of Hindus intolerable but also by fomenting the forces of communal antagonism. But whether they are "cooperating" or not, if 10 million refugees is to be the figure, which is approximately 80 percent of the Hindu population of East Pakistan, one must assume that the core of the West Pakistan policy is to turn East Pakistan into a Hindu-free state or another version of a *Juden rein* Nazi Germany. It is this that makes so improbable the repatriation of the refugees in the foreseeable future—if ever—and it is this prospect that bedevils the government of India.

As to the motives of the escape, inclusive of refugees with properties left behind, they are known, do not need much laboring, and they are cited only as a part of the picture. The answers to the question why they left East Pakistan are all the same: They did it in fear of their lives. This explanation is not given in terms of unsubstantiated fears but rather in the face of murder and pillage visited upon their relatives or friends. As we listened to a number of cases, and admittedly a small sample, "guilt by association" has emerged as one motive for decamping. When an English-speaking Hindu publisher and book dealer had learned that two of his politically involved friends (not Awami League activists) were shot during the night, he had no choice but to flee for his life the morning after. The other motive is the wholesale depredation of entire Hindu communities, or selective parts of such, and the belief that the same would befall them unless they cleared out. Finally, regardless of the immediate causes and the degree of embellishment with which this or that account may be painted, I was reminded of an old Russian saying that "no one runs away from a good life." A young farmer with 13 acres of land is a case in point. As farmers go in East Pakistan, he was among the substantial ones and deeply attached to his property. Yet he, too, felt he had no choice but to leave and, not without interest, with the deed to his land safely brought along. It is questionable if the proof of ownership will serve its purpose; he feared that his land had been already occupied by the Muslims. Just as questionable, though he did not allude to this, is his prospect of getting back to his village. A mere glance at a group of refugees in their new surroundings and the inevitable thought of what is in store for them even under the best the government of India could provide are good enough explanations of the strong pressure they were under to escape from an intolerable situation created by the military occupation of East Bengal.

Geographically, the refugees are dispersed in eight states. However, West Bengal and Tripura claim a total of 7.4 million of which 6.1 million are in the former. Dispersal of refugees among other states is small. No state of India really welcomes them. When the GOI decided (June 6) to revise an earlier stand of not allowing dispersal, most states pleaded one excuse or another for not accepting refugees. The chances are that West Bengal above all will have to live with the refugee problem so long as their return to East Bengal is only a pipe dream. At this point it would be fitting to say something about the burdens shouldered by West Bengal and the government of India, but I prefer to give the pride of place to the forthcoming report. Instead, I would rather say a word or two about the refugee conditions in the camps and about measures the government employs to improve them as well as note some of the things the government cannot do for fear of upsetting altogether the delicate and troublesome economic and political balance of forces of West Bengal.

My impression upon meeting the evacuees is that they don't seem to be people who have suffered great trials in their track to the Indian border. They are silent, cheerless, and idle; but for the most part they are not emaciated-looking, starving people. The food ration, about which more later, does apply in most cases. Encountering them outside the camps one might take no more notice of them than of any other Bengalis. This said, it is equally true that they are not free Bengalis moving about at will; they are in camps constrained by a regimen that condition entails. Few of them have brought along any wordly goods and only small amounts of money or none at all. Rings, watches, and ornaments are possessed by few; most of them were removed from their persons by unfriendly Muslims or Majlis workers. The huts are a picture of plenty of nothing when measured by such basics as a change of clothing, total lack of bedding and bedsteads, and an inadequate supply of cooking utensils. If any improvement in these and other respects is to take place, it can come only from the resources of the government of India and the international relief organizations. For reasons I shall mention shortly, the able-bodied refugees cannot be gainfully employed and their dependence upon outside assistance is total. This is a situation pregnant with dire consequences if the refugee problem should become a permanent one.

A word about living conditions in the camps. Small though our sample was, there is enough corroborative evidence to make this much clear: Almost everything is still makeshift. This is reflected in the poor conditions of a good many shelter facilities; the inadequate water supply; the hospitals few in number; the great shortage of doctors, nurses, and medical supplies; the absence of school facilities; inadequate supply of nourishing food for the young; and the same may be said of a variety of other items to sustain bare living. In monsoon conditions, and this year's monsoon in West Bengal is the most plentiful in twenty years, the seamy side of camp life is magnified manyfold. The monsoon has turned the countryside soothingly green, but to the multitudes of the refugees it is a cruel experience. The reference is not to one of the big camps reportedly 70 percent under water. In the two smaller camps we looked into on a rainy day we only dirtied our shoes; the refugees, on the other hand, who sleep on the wet or damp ground with nothing between them and the ground and no blanket to cover them will reap more pneumonia, tuberculosis, bronchitis, dysentery, and other assorted diseases. Mid-October, when the monsoon gives up its last, is looked upon as the day of deliverance. The same mid-October heralds the coming of the winter season, and how well the refugees will stand it will depend upon the timely arrival of 6 to 7 million blankets.

This litany of woe is jaded old stuff to many from the outside looking in, and I am not reciting it merely to evoke greater sympathy for those who for no fault of their own must live with catastrophe. The point I wish to make is that, despite all the shortcomings affecting the life of the refugees, the Indian government and the state government of West Bengal have performed remarkably well. What stands out in the huge Salt Lake camp, for example, is not the makeshift arrangements but that there are arrangements, however inadequate; that shelter huts were and are being built as the influx increases; that the building material is somehow procured and delivered; that the lanes between the sheds have been laid out in reasonably good fashion; that wells are being dug although, observing the long water queue, not yet in sufficient numbers; that the rations are bought, transported, and distributed; that there is a hospital and other medical facilities at least to cope with emergency cases; that, whatever the shortages, there is recognition that they have to be whittled down; that beyond it all there is an organizational mechanism that somehow sees to it that supplies of the voluntary agencies as well as those procured internally are channeled to ultimate users. The prime responsibility rests with the government of India and its constituent parts; and it deserves high credit for achieving so much, overnight so to speak, and in conditions of indescribable complexity. A purist or a Bank feasibility study team could find much fault with every aspect just touched upon; but, to the extent that the issue is the *survival* of 7 or 8 million people in the midst of limited human and material resources, he is a bold man who thinks he could do much better.

The GOI is not without high marks for keeping the lid tight on communal riots in the wake of the mounting refugee tide. It is a "non-development" worthy of many a useful development. History of Muslim and Hindu communities tearing themselves to pieces in 1951, 1961, and 1964 or the less gruesome affairs in between or since then should be invoked to note the *total* absence of such episodes since April of this year. In the light of the past experience induced by less provocation than more recent events might have brought about, this is novel and encouraging. The forbearance of the government, of the Indian press, and of the populace in general must be underscored for yet another reason: the decidedly pro-West Pakistan position taken by most Muslim leaders and Muslim press in India. Regardless of the pent-up emotions which give rise to such support, it goes against the grain of everything India stands for in relation to the events of recent months. Nevertheless, while the tensions and the desire to "settle accounts" are there, they have not been permitted to translate themselves into the familiar forms of pillage and murder.

The government of India has undertaken to provide the refugees with shelter and food, but clothing only in hard cases. The ration is made up of 300 grams of rice, 100 grams of wheat, 50 grams of dahl, 150 grams of vegetables, and bits of salt and spices. The total daily cost of feeding a person is one rupee and all other expenditures, including one-tenth of a rupee in cash for every adult and one-twentieth of a rupee for every child, raises the total expenditures to Rs2.5 a day. Wheat is procured from the adequate stocks of the Food Corporation and rice from the same source, plus some foreign contributions now beginning to arrive; this explains why prices of these commodities have remained stable. The prices of supplementary foods, particularly legumes, which are locally procured, have risen steeply. While the ration is considered adequate except for the wheat component to which the refugees are not used, supplementary food for children is a major problem; the estimate that malnutrition affects 40 percent of the children may err on the low side. Since the government has not specifically obligated itself to provide it, agencies such as UNICEF, WHO, CARE, Red Cross, and OXFAM are the most important sources for narrowing the gap.

The refugees cannot make any contribution to that end nor to many another sharply felt need of a material sort. They simply have no earning power and no prospect of reversing this condition; they are admitted only as foreigners and as such, in the official view, are not entitled to any employment. To make sure that this condition holds, there is a kind of an invisible yet functioning sanitary cordon surrounding the camps. Military commandants of the camps see to that. It is understandable then why in the city of Calcutta one doesn't encounter refugees and not only because they are indistinguishable from the local population. This is the more remarkable when one stops to think that Salt Lake camp with its 150,000 inmates is only 15 miles away from the center of Calcutta. The possible but limited exception to this state of affairs is provided by some of the 1.5 million or so refugees who live with their friends and relations. Those of them who succeeded in shedding their refugee identity may be competing for jobs. Very few well-educated, professional people have secured jobs as East Pakistanis, but in the main the total number of employed among the refugees outside the camps can't be large; besides, the labor market is very thin. The strain on them, as on their friends and relatives, must be very severe.

I believe that the restriction on their employment has little to do with the "foreigner" classification. Politics aside (if that is possible), the record of the government of India regarding the refugees is one of great compassion, amply proved by deed. It might not have imposed that restriction or imposed it only perfunctorily, but for the fact that they are mostly in the unemployment-ridden West Bengal where the industrial scene leaves much to be desired and law-and-order problems are at their worst. Second, in certain areas the refugees exceed the number of local population. To let the former loose in search of employment would be to invite political and economic upheaval. With variations, this would hold for West Bengal in general. Possibly to underscore that the refugees are not entitled to employment, even the construction of the camps is carried on by outside labor in most instances. Here and there they have been given access to chores but to chores involving hardly any skills.

Despite the many impediments, some of the camp refugees have secured temporary employment, although not in industry; the strongly organized trade unions of West Bengal preclude that. It is easier in agriculture, but those who ventured may not do so again. Where the going daily wages are Rs2 to 4, the refugees are willing to work for one rupee. Such distress wages have not gone unnoticed, and the complaints raised by the local agricultural laborers are likely to put an end to such competition. In short, for the reasons indicated the refugees are essentially an idle, unemployed lot of people and grave problems are in store for all concerned. This is increasingly recognized by the officialdom as the idea of early repatriation begins to fade away. It is less clear if this prospect is shared by the refugees; currently their choice apparently lies between self-deception or the realization that life has nothing or little to offer. When the "moment of truth" dawns all around, which is to say that no political solution is possible calling for their return home, India's problem will become excruciatingly more difficult and urgent than it appears today. The financial and administrative burdens India is shouldering would seem much less grave then than now. The current status quo based on charity could in no way provide any contentment for millions of idling people cooped up in camps with no hope of escaping beyond their confines. While better rations and overall better amenities are badly needed and welcome, they will not prevent the rise of unrest and rebelliousness. This is bound to occur among a people bereft of any illusions about returning home or any likelihood of a reasonable integration within the Indian community. One cannot but sympathize with the dilemma India is likely to be confronted with: to contain the refugees in the camps by force of arms or throw the gates wide open and court an explosion in Calcutta and in other cities, which "tremors," as one put it, "would even shake the fair garden city of Delhi."

Miraculous shifts in policy on the part of West Pakistan may prove this prognosis fallacious. On the other hand, if it is indeed beyond miracles and only the familiar political stances continue to prevail, India will find itself pretty much in the position outlined. The only way for India to escape the choice of either evil is to enable the refugees to go in security where they came from. If there is a pilot who can navigate safely between this Scylla and Charybdis, I wish he would come forward. In his absence, and no return to the homeland a fact beyond dispute, this much is warranted: It is inconceivable that so great a number of refugees can be kept in the camps indefinitely with no right to forage for jobs. I suspect that this is well understood in official circles, and only time can tell which way the government might turn. Here Mac reminds me that India can absorb 1.5 percent increase in population (the 8 million refugees) as it does more than this every year without great distress. This should enable it to face up to reality and disperse the refugees as soon as, and here comes the catch, the move is *politically feasible*. In that event, the problem would be one of providing India with additional investment resources required to employ them. So runs Mac's argument. In the long run, Mac's suggestion may well have to be acted upon, but only in the long run. Eschewing for the moment all the economic, political, and linguistic difficulties dispersal would run into, the fact of the matter is that so far there are no indications that the government is thinking along such lines; and for the time being Hobson's choice is about the only option open to India.

So much for some of the things observed and impressions carried away and interpreted. I now turn to something else even at the risk that you might consign me to perdition.

As I mentioned at the outset, Mac and the other members of staff are preparing a paper on India's extra financial needs induced by the refugee avalanche. At the same time, as you know, the GOI is planning to seek an IDA credit to help cover part of them. I am sure that the paper will provide sufficient fuel to justify, in your mind, the granting of such a credit. In the meantime, I have been hearing noises about precedents preventing IDA from such an undertaking. If this is so, let us not forget that the situation in India is itself without precedent and I do not believe that the precedence question is relevant.

Banker and economist that I am not, I am tempted nevertheless to put in my two cents worth to this end, and not necessarily on economic grounds. I assume that the Bank is not a "regular" Bank, and I do so not only for the sake of *my* argument. It is an institution dedicated to development only insofar as the end result is betterment of man's condition. This is why the Bank is justifiably involved in such seemingly noneconomic projects as education and population control. In the final analysis the ultimate purpose in either case is to improve the quality of the individuals as well as of the community as a whole. I submit that, apart from assisting India in a financial emergency, the Bank can treat the "refugee loan" similarly. The benefits of the projects mentioned are long drawn-out affairs, as they must be, whereas the life of the refugees calls for immediately enhanced encouragement to which the Bank can contribute its share.

One of the lasting impressions of a camp visit is that any amount of aid measured against the needs is like a drop in the bucket. This is not to minimize the fact that every bit helps, and an IDA credit judiciously spent belongs in that category. Another way of putting it would be to say that such a credit would obviate the necessity of reducing developmental outlays by that much. With the cost of caring for the refugees now running at an annual rate of nearly $1 billion compared with no more than $200 million of refugee aid now in the offing, a large cut in developmental outlays seems inevitable; this should be prevented to the extent possible. My apologies for introducing an economic argument, and I am back on the theme that the credit in question is not of the conventional type and that it must be handled with a strong dash of noneconomic realities. I cannot tell you what approach the Bank should employ in this instance or how to skin this particular kind of cat. I leave this to the Bank.

Peter, the preceding may strike you as just another species of my "irritating evangelism," and you may consider me the naivest of naive for bringing up some arguments which, admittedly, do not add up to two plus two making four in the accustomed fashion. This I will grant you, but you must grant me that the very reason for having refugee camps of the magnitude one finds in India is exceptional even in this year of our Lord, 1971. And the responsibility to help sustain this mass of humanity is not only India's or of the international organizations I referred to earlier. It is also mine and yours and that of the Bank as a whole. The judgment, therefore, about granting the "refugee loan" cannot avoid such considerations. If the articles of the Bank's incorporation don't provide for the kind of credit that presumably departs from the more normal attributes of a Bank project loan, they should be rewritten or a new article added in tune with the conditions imposed upon India for no fault of its own. Finally, even if the loan be considered only as a gesture of goodwill, I am eager to see the Bank make that gesture.

57. *Agriculture–Issues and Programs*

This paper was Ladejinsky's contribution to the World Bank's annual report, "Economic Situation and Prospects of India," May 10, 1972. By this time Ladejinsky had come to dread the necessity for the preparation of this annual contribution. He had already had his say on the problems and issues he regarded as fundamental. At this stage, he felt, the choice was to be either trivial or repetitive, and he would not deal with trivia. He need not have been so concerned. The fundamentals with which he was so deeply concerned had become, due in significant degree to his own earlier efforts, the subject of a continuous debate in India in academic and research circles, in the intellectual media, within the government, and among politicians. Year by year ad hoc national commissions issued new reports bearing on one or another of the basic problems, and new programs and policies were initiated by the government to deal with them. These developments, together with passing experience, enabled Ladejinsky to return, year after year, to the fundamental problems and issues which concerned him, investing them with a surprising freshness and even a heightened interest.

Thus, in the present paper, he reports once again on various aspects of the Green Revolution. Rural poverty this time is viewed in the context of a battery of rural development programs which have been developed around the nucleus of the Small Farmers Development Agency and the crash program for rural employment previously encountered. Cooperative credit is again scrutinized not only in the light of the latest data but also in terms of the proposal for an alternative scheme for agricultural credit advocated by the National Commission for Agriculture. He examines the new guidelines announced for landholding ceilings and points out the essential error of this emphasis. (The treatment of the new ceiling program here accounts for the omission of his published October 13, 1971, article [CB–119] on the same subject.)

One further word is appropriate about the battery of rural development or rural anti-poverty programs Ladejinsky describes and assesses in this paper. Although he is obviously wary of them and they certainly do not represent his preferred means for dealing with rural poverty, he nevertheless—being less sanguine about the likelihood for adoption of his preferred means—strongly supports them. "Finally, despite the problems which beset all the programs, there can be no turning back; they do represent areas of great promise aimed at the right things by attempting to build employment, better standards of living, and social justice into economic growth."

Space considerations have required the deletion of Ladejinsky's treatment of the issue of farm taxation.

The Agricultural Scene

IN THE "YEAR OF THE REFUGEES AND WAR," Indian development problems are more numerous than usual, but food shortages do not claim the headlines. Early in 1971 India was afflicted with damaging drought and devastating floods, but such are the new trends in the agricultural economy that 1971–72 registered the further rise in agricultural production previously described. So depending on one's definition, and with the present crop prospect and the buffer stock well above the fourth plan targets, it may be said that India is on the threshold of self-sufficiency. In view of the situation in the not too distant past, this is by far India's major economic achievement in the fourth plan.

Critics often point out that the increase in agricultural production is limited to a few crops and that the Green Revolution has a "narrow foundation." This "narrow foundation" is part of the strictures leveled against the new agricultural strategy. And yet, though this and other problems are still to be mastered, the preoccupation with them is greater than ever. In the light of rice research under way now, the next few years promise significant results. Nor is it too soon to display a similar concern about the vital commercial crops which leave much to be desired. The fact that the passage from traditional to modernized agriculture cannot be effected as if by magic is no reason why these crops should continue to stagnate in the sixth year after the advent of the Green Revolution. It explains the often expressed view that "the revolution we have witnessed has been a wheat and price revolution, but not yet one of the widespread agricultural and rural development." This implied critique must be tempered, however, if we take a larger view than the time horizon of the fourth plan. During the period 1964–65 through 1971–72 aggregate food production increased from 89 to 112 million tons, or a 10 percent per capita increase. But for the new agricultural strategy, India would have to import a minimum of about 8 to 10 million tons of wheat yearly at a cost of $600 to 800 million. Wheat accounts mainly for the rise in output, and there are reasons to believe that its production will continue to rise, even if at a lower rate. If so, at current consumer prices—and barring an act of God—a surplus above internal demand cannot be excluded. This in turn would pose a question of how to dispose of the same, especially if present support prices remain unchanged.

The overall increase of 23 million tons of grain between 1964–65 and 1971–72 has not come by osmosis. Favorable climatic conditions have played their very important role, but it is equally true that numerous farmers have taken advantage of the new technological trinity—assured water supply, Mexican "dwarf" wheat varieties, and chemical fertilizers—the basics of what is popularly known as the "Green Revolution." With wheat yields rising sharply in the entire northern tier of India at a compound rate of 7.5 percent a year, the success in wheat provides a window on the new technology which is being extended beyond wheat. The chief beneficiaries are the large farmers with ample resources. Since the new agricultural practices are neutral to scale (apart from the possible economics of capital intensive mechanization which are yet to be established), a number of small farmers are also involved. More recently, along with the original Green Revolution, there is developing a "mini" Green Revolution which is a mixture of traditional and modern practices and is applicable to more than just wheat. Such developments are anything but common; but, where they do take place, as in parts of eastern Uttar Pradesh for example, they make the difference between a poverty-stricken life and a reasonably secure one.

Thus the prevailing psychological urge is to share in betterment even if the means to realize it are still not there. Generally speaking, the climate for improvement is good. Therein lies not only the precondition for rising productivity but also for profound changes in the pattern of life and thinking in the Indian countryside. In the areas where it has gone farthest, in Punjab and Haryana, the process of modernization is no longer confined only to techniques of cultivation; there, it is seeping through into other walks of rural life, stretching from changes in the consumption pattern down through better living conditions and in a variety of other ways, including the first signs of a break with rural squalor. That economic values are gaining over traditional considerations is observable in the mode of community affairs, the urban-oriented behavior, the rise of individualism among the farmers, and the fact that farming as an occupation in those states is considered highly prestigious. By the same token, the Green Revolution is affecting the old established relationships in rural society. To be sure, not all of these changes are beneficial;

but the loosening of the bonds of a formerly stagnant economy and society and the reaching out for new horizons carry a significance of their own that cannot be overestimated.

In the past year there appeared a spate of articles with such titles as "How Green is the Green Revolution?" The argument of the skeptics rests on the proposition that "nature's bounty" is responsible for the upsurge of agricultural production and productivity. Favorable climatic conditions continue to make the difference between a good and a poor crop, but they don't explain the character of the gradual shift from traditional to modernized agriculture where it is taking place. Nor is rising output the best measurement of the changing scene. More to the point as a sign of ambition and endeavor is the veritable upsurge in the utilization of inputs. For the recent past this is examined in chapter 3 [omitted], but longer-term trends are also relevant. Fertilizer consumption ($N+P_2+K_2O$) increased from 306,000 metric tons in 1960–61 to 2,350,000 metric tons in 1970–71. During the same period the number of electric and diesel pump sets increased from 421,000 to over 2.4 million, the number of tube wells from 90,000 to 460,000, and tractors from 31,000 to 140,000. Since most of these inputs, and supplementary ones, materialized in the past five years, two conclusions stand out: "The first one is that the combination of factors cited here, in addition to favorable prices, has raised output beyond what would have been possible in favorable climatic conditions but without the attributes of the new technology; the second one is the willingness to invest, experiment, and take risks, and this in itself is a break with practices of the pre-Green Revolution days.

In the light of these accomplishments, a keen student of the country's agricultural scene was moved to say:

> Whatever else may be wrong with India in this decade—population explosions, limited success on the family planning front, massive unemployment, competitive communal tensions arising therefrom and educational malaise—the food problems will no longer pester us in this decade in any serious manner. *One can perhaps stick one's neck out and assert that India's food problem has been solved for good.* [Author's italics.][1]

From such an unconditional assertion to euphoria about the future contribution of the Green Revolution is but one step. Thus, the author goes so far as to state categorically that

> The green revolution will, therefore, give us now not a great deal of quantitative increases in grain but a qualitative improvement—a diversification of diet patterns away from foodgrains and in favor of processed foods, vegetables, fruits, milk, eggs and meat products, and a changing cropping pattern, changing away from foodgrains towards commercial crops like jute, cotton and oilseeds.[2]

One hopes that this anticipated shape of things will become a living reality, but the problems standing in its way are formidable.

Constraints—environmental and socioeconomic

India is indeed only at the beginning of the modernization process. The crop production indexes of selected districts and regions and sharp investment differentials between "progressive" and "nonprogressive" farmers in the same categories (small, medium, and large) tell the story at a glance.[3] This is part of what is increasingly spoken of as "regional differences." In one knowledgeable view, "the decade of 1961–71 seems to have accentuated regional differences . . . backward areas seem to have gone still more backward by comparison."[4] Even the areas under high-yielding varieties are characterized by wide variations in the quantitative and qualitative means of production. The upgrading process must proceed slowly even under the most propitious circum-

1. A. M. Khusro, "Agriculture as Business," *The Illustrated Weekly of India* (February 13, 1972), p. 19.
2. Ibid.
3. K. William Easter and Shrinath Singh, *The Importance of Regional Differences in Agricultural Development,* paper presented at the Annual Conference of the Indian Society of Agricultural Statistics, New Delhi, December 1971.
4. Asok Mitra, "Aspects of Levels of Development," *Yojana* (February 6, 1972), p. 110.

stances, and these are not present in India. The fact is that three-fourths of India's cultivated acreage is under "dry" farming, depending entirely upon the vicissitudes of the monsoon. This is well understood in governmental circles, and it explains why vast sections of the country have not been touched by transformation and equally vast parts can only boast of "small islands within." It is the more surprising, therefore, that, though the Green Revolution is far from green all over, between 1968–69 and 1971–72 the acreage under the new practices has increased from 9 to 18 million hectares or by 15 percent of the cultivated food grain area. But it may be a long time before this ratio is significantly increased. Not the least of the problems is that "land is going out of agriculture every day at an alarming rate due to the spread of industry, roads, houses, salinity and alkalinity and various forms of erosion. At the same time, population is growing at 2.25 percent per year, thereby making imperative the production of more and more food from less and less land."[5]

Apart from environmental and other limitations, the contention propounded in these reports on previous occasions that the Green Revolution is primarily "selective" remains valid. It was natural for the new technology to concentrate in regions with optimum conditions, especially with assured irrigation facilities or reliable rainfall. The new agricultural strategy calls for large investments, and it was in the nature of such requirements that rich or relatively well-off farmers were the first to take advantage of the new opportunities and profit from them. While smaller farmers have not been bypassed altogether, the majority of them, and particularly the tenants and sharecroppers, are often barred from active participation in the new agricultural trends for lack of resources. The preconditions inherent in the new agricultural strategy have raised these already existing issues into sharper focus, at the same time adding brand new ones of their own.

The issue of "selectivity" aside, there are other developments, direct and indirect, attributed to the new technology. Mechanization of agriculture is one of the best known. During the past five years mechanization has been rapidly expanding, and the ongoing debate is whether it serves a useful purpose in a country where rural unemployment and underemployment is rampant. There is no denying that the new practices are highly intensive and labor demanding; but at the same time these practices also give rise to labor-saving devices such as tractors, rotary ploughs, harrows and tillers, sprayers and threshers—to mention modern equipment most of which, other than combine threshers, is already on the farms of many practitioners of the new technology. Labor-absorbing and labor-displacing practices meet each other, and the question is how the two balance off in the light of this novel experience.

The answer to this question depends upon carefully gathered empirical evidence which is only beginning to make its appearance. The World Bank's arrangement with appropriate institutions to study the problem might help shed light on this question. For the time being, however, the answer seems to be that at this stage of mechanization—and barring the use of combine threshers—along with the rise in real wages, the demand for labor has also increased. The immediate "victims" of tractor power are the bullocks. A 30-acre farm employs four pairs of bullocks; with the introduction of the tractor 60 to 75 percent of them are displaced, but this is not the case of the four men behind the bullocks. They are more than gradually absorbed by the labor-demanding new practices, especially on the biological side which is land augmenting, or in secondary or tertiary occupations, which are also an attribute of the new technology. The findings of a study of farm management data for one Punjab district concludes on this note:

> "the immediate impact of tractor-use is the displacement of farm labor to the extent of about 15 percent and bullocks to the extent of about 60 percent. . . . When the dynamic effects of tractorization such as increase in cropping intensity and yield per acre are taken into account, output is found to increase by about 28 percent and employment by about 7 percent . . . it would appear, therefore, that the employment potential

5. M. S. Swaminathan, "Indian Agriculture Tomorrow," *Yojana* (February 6, 1972), p. 113.

from farm tractorization for the economy as a whole is considerable."[6]

The observable, if unquantifiable, experience in other parts of the hotbed of the Green Revolution, the Punjab, seems to point in the same direction. Under the impact of mechanization, the tractors leading the parade, the radical changes in the overall intensive cropping patterns have acquired a dynamism all of their own. As such, it has been labor absorbing rather than labor displacing. The sharp rise in output and farm income has favorably affected employment in such related fields as irrigation, road building, transportation, marketing, warehousing, processing of agricultural products, and in manufacturing of a large array of farm inputs. The same rise in income, coupled with the changes in the consumption pattern and other improved living conditions, is beginning to give more employment to masons, carpenters, blacksmiths, weavers, potters, utensil makers, tailors, and so on. While no data are available to measure these developments, it would seem that this rhythm of activity calls for more rather than less employment.

Nevertheless, the issue is far from settled and only in part on account of the prevailing feeling that the available studies on mechanization are not the last word on the subject. Besides, there are other bits of evidence contrary to the one cited in the same Punjab and also in West Pakistan and the Philippines. Even a confirmed supporter of tractorization as a labor-absorbing tool notes that "it is not certain whether this absorption will be full or partial."[7] These lingering doubts stem primarily from the apprehension of labor displacement following the spread of mechanization in scope and sophistication. The first appearance in the Punjab of 75 combine harvesters is a case in point. Harvesting-threshing is the most labor-demanding and the highest-paid work. But according to a study in preparation by the Institute of Economic Growth at Delhi University, a modern combine thresher displaces 30 percent of the "normal" labor force and reduces its wage intake by 40 percent; but, unlike the tractor, it is only labor saving and not land augmenting. At the moment the number of combines is too small to be overly concerned, but more are likely to come; for, in the areas of concentration of the new technology, mechanization feeds on mechanization. The incentive is coming from the gradual mastering of high-yielding varieties, demand for increased production, assured farm prices, and rising wage rates—particularly during peak seasons. This is why a student of mechanization went so far as to say that "in pockets of high agricultural growth . . . one should expect a relationship between the rate of mechanization on the one hand, and wage rates . . . on the other, to be in the same direction as in a developed economy."[8] "The same direction" in this instance doesn't, of course, mean levels of mechanization similar to those attained in the United States or in any number of Western European countries, especially in regard to displacement of labor. If nothing else, labor required for the irrigation and water management so crucial to Indian agriculture will preclude that kind of erosion in labor demand.

Whatever the future trend, one can discern a growing opposition to a free-for-all mechanization process. Until about a year ago, the official policy was heavily weighted in favor of "the more mechanization the better" through the subsidization of farm machinery and credit availability for the same on easy terms. Questions of employment were not to the fore. More recently, this policy is being tempered with respect to the tractor; and, while nothing specifically is said about the combine thresher, the chances are it will be subject to severe restrictions. The imposition of a variety of duties which greatly increase the cost of an imported tractor is not only a measure in favor of the domestic tractor industry but also a sign of a "go slow" policy, at least until such time as all pertinent issues bearing on tractorization are sorted out. More important are the often recurring references to a "selective" mechaniza-

6. C. H. Hanumantha Rao, "The Impact of Tractorization on Farm Equipment," memo to the National Commission on Agriculture.

7. A. M. Khusro, "Structural Changes in Indian Agriculture," mimeographed, p. 24.

8. C. H. Hanumantha Rao, "Farm Mechanization in a Labor-Abundant Economy," *Economic and Political Weekly* (annual number, 1972).

tion policy. It is quite likely that the National Commission on Agriculture will lay down new guidelines for a mechanization policy.

In typical Indian rural conditions of land ownership, resources, and marketed surpluses, income inequality is a normal state of affairs. Since it takes a minimum of Rs10,000 to 12,000 to reequip a 7 to 10-acre holding—minus a tractor and appliances to go with it—the tractor is not normally within the reach of a medium-sized farmer, let alone of the smaller holder, unless they can secure cooperative credit. More often than not they can get only insufficient credit and, on many occasions, none at all, for the distribution of credit in most Indian villages is biased in favor of the affluent. Many would-be innovators can be likened to tenants who receive land under a reform but not much else. Not a few are excluded from the purview of the Green Revolution altogether or participate in a limited way at best. This is not an argument against modernization, but it tends to make the rich farmer richer and the poor relatively poorer.

Data to measure income distribution among the principal farm groups are not available. However, some measure of income inequality is indicated by the fact that 62 percent of the cultivators farm less than 5 acres. Households of landless farm laborers as a percentage of total households—with exception of Kashmir where it is only 2.5 percent—range from a low of 12 percent (Rajasthan) to a high of 35 percent (Andhra Pradesh). About 20 percent of the land is farmed by tenants, but in ten states tenancy is this percentage or higher.[9] It should be noted, too, that, when incomes are low and savings limited, as is true for the majority of small farmers, a good deal of their food must be purchased in the slack season in conditions often of rising prices. This is on top of unequal access to factors of production. These alone, apart from the effects of the structural configuration of land owned or operated, raises the degree of intra-agricultural class income inequality.

Surplus-producing farmers (particularly those in the wheat regions) are the major beneficiaries of the future shift in incomes in favor of agriculture as well as within the agricultural sector activated by the price support policy. On the other hand, the burdens of change are likely to fall on those who gain least from it. This is not an argument against the new technology. It relates only to the need to make available resource input allocations with an eye to raising the number of participants in the Green Revolution. Unless appropriate measures are taken, the Green Revolution is likely to exacerbate income disparities. An economist not given to alarms and a believer that the Green Revolution "has done enough to solve India's grain problem during the 1970's," is nevertheless alarmed. "If orderly redistribution of surplus is not quickly brought about," he writes, "the situation can easily blow up and there are already straws in the wind."[10]

A problem within the context of the new technology is the position of the tenants and sharecroppers who overtly and covertly cultivate about 20 percent of the land. The Green Revolution has not formally imposed new restrictions upon these farmers; in practice, however, it weakens their hold on the land and their already unenviable economic position. For a variety of reasons, their displacement into the ranks of agricultural laborers is likely to be speeded up and their capacity to make a living impaired accordingly.

The new package of practices and good support prices, on top of no land or agricultural income tax to speak of, make modern farming in India an enticing proposition for farmers with large holdings ecologically well-placed. It is understandable why well-off farmers strive to invest in more land or lease more land. Good as this is for productivity, from the point of view of the tenants it is anything but a boon. The first consequence of rising land values is that rents have risen above the traditional (although illegal under the reforms) 50–50 share of the crop. This may be unavoidable in conditions of land hunger, hardly any security of tenure, and evasion of rent controls; but hardships increase for those who cannot extract from the land a return large enough to meet living requirements over and above the rental

9. Ranjit Sen, "On Rural Poverty," *Economic and Political Weekly* (December 25, 1971).

10. A. M. Khusro, "Agriculture as Business."

obligations and inputs which are often at their own cost. But the real issue is not higher rentals but the desire of the owners to be rid of the tenants altogether and resume the land for cultivation with modern equipment and hired labor, all with the view of unencumbering the land from any claims on it. The ostensible rationale for all of this is that the current farm policy calls for efficient production, larger and compact holdings, and centralized management in the hands of a single owner; tenancy and sharecropping, the argument runs, have no place in it. Nor is it unreasonable to assume that the same consequence is in store for some of the mini-owners.

How far this technological side effect is being translated into reality is another matter. Statistical evidence one way or another is inadequate, but observations on the spot indicate that separation from the land under different guises is not rare. Wholesale dispossession is not likely, partly because a great deal of it had already taken place before the Green Revolution and partly because this would mean courting trouble for the big owner. But the practice and threat of turning tenants and sharecroppers into agricultural landless laborers are there. Once again, the new agricultural policy did not plan it this way; but it takes place so long as the ineffective security of tenure provisions cannot prevent such developments. The upshot of it all is that meaningful tenancy reforms which were difficult to enforce before the Green Revolution have become much more so since its advent.

Rural poverty

We have discussed income inequalities above, but the issue that beclouds the entire agricultural scene is rural poverty. Agricultural growth is a goal all are agreed upon, but in the peculiarly difficult Indian agricultural structure even self-sufficiency on higher consumption and qualitative levels is not necessarily an anti-poverty remedy for millions of disadvantaged. Rural mass poverty in India is an old story; and the industrial growth of the country, the new agricultural developments, the growing number of the rich in industry and commerce, the emerging middle class, and the rising expectations that followed upon promises of better days—all these are adding to the frustrations of the poor. The have-nots know that "over there," near and distant, the rural haves are enjoying considerable gains. All this has affected the thinking of some of the economic and political elite within and outside the government. These groups are beginning to recognize that development divorced from the question of poverty is a threat to the stability of the country, and the issue is how to invest development programs with an anti-poverty orientation.

The enormity of the problem becomes obvious when viewed against the dimension of the rural poor. They are made up by the landless, labor households with some land, and a number of owners in the group holding less than 5 acres. The study group on the welfare of the weaker sections of the village community reported in 1961 that 80 percent of the rural households had annual incomes less than Rs800, and of these 45 to 50 percent had incomes less than Rs500.[11] At such income levels, many live below the poverty line. Just how many is estimated by R. S. Minhas[12] (table below). It appears that between the mid-1960s and 1967–68 there was a slow but steady decline in the proportion of people below the poverty line and in absolute numbers as well. But even so, "after 20 years of economic development, between two-fifths to one-half of the rural people of India today are living in abject poverty," in round figures, between 154 to 210 million people. Another estimate disputes the decline in numbers and the total is given as 250 million.[13] A breakdown by states points to roughly the same story.[14] Whichever set of figures is nearest correct—and probably not one of the three is the last word on the subject—the magnitude of the problem is obvious.

11. E. P. W. Da Costa, "A Portrait of Poverty in India," in *Challenge of Poverty in India*, A. J. Fonseca, ed. (Bombay: Vicas Publishers, 1971), p. 50.

12. R. S. Minhas, "Mass Poverty and Strategy of Rural Development in India," Washington, D.C.: World Bank, Economic Development Institute, (March 1971, mimeographed).

13. Pranab K. Bardhan, "On the Minimum Level of Living and the Rural Poor," *Indian Economic Review* (April 1970).

14. Ranjit Sen, "On Rural Poverty," table 1.

Year	*Below Rs240 a year at 1960–61 prices*		*Below Rs200 a year at 1960–61 prices*	
	Percent	*Millions*	*Percent*	*Millions*
1956–57	65.0	215	52.4	173
1957–58	63.2	212	50.2	169
1960–61	59.4	211	46.0	164
1961–62	56.4	206	43.6	159
1963–64	57.8	221	44.2	169
1964–65	51.6	202	39.3	154
1967–68	50.6	210	37.1	154

The computations of the table and guesses about the post-1968 period may support the following conclusions:[15] (a) Between the mid-1950s and 1967–68 the absolute number of people below the poverty line did not undergo any clearly discernible change. (b) Their numbers seem to fall in good harvest years but shoot up in bad crop years.

The emphasis of the fourth plan on "the common man, the weaker sections, and the less privileged," the political slogan to "abolish poverty," and the more recent outcroppings of a similar sort are welcome signs of great apprehension about the problem. Stepped-up industrial development is not likely to make a dent in the number of the poor, and only the rise in the productive capacity of millions of small farmers and agricultural laborers through the rejuvenation of the agricultural economy as a whole can make a difference in their standard of living. Failing that, in Minhas's view, there is the additional danger that the water-seed-fertilizer-tractor revolution, "which has not delivered much yet" might "get stuck at the periphery."

We do not agree with the stricture that it "has not delivered much yet." We are in agreement, however, that an "integrated" or multipurpose program of agricultural development involving the less endowed ecological regions is the order of the day, just as the agricultural strategy of 5 or 6 years ago was the order for its day. A new strategy is both inevitable and overdue. Almost every advance carries some penalty, but in no way does this diminish the contribution of the new technology, despite the imbalances which flow from it. As an instrument of greater productivity it has proved its worth, and what is needed is more technologically involved farmers rather than fewer. This is the way to take the country forward rather than getting it stuck at the periphery of the Green Revolution. Fruitful biological research that might open the dry land areas to better cropping is, of course, the essence of the process. But this, as with other forms of technological innovation, will not suffice if the farmers of those regions are to benefit from them fully. As one put it and with application to India as well, "most of the problems of Southeast Asia will not be solved by straight line manure furrowing . . . but by other sorts of inputs."[16]

The "other sorts of inputs" as part of an expanded technology are familiar. Apart from the already mentioned vastly expanded research activities, they are: a rural credit service to serve those who need it most; an extension service worthy of its name; panchayats (village organizations) which are indeed developmental bodies; security of tenure; rents which are not usurious; ceilings on agricultural land which are not paper propositions; consolidation of holdings; farm wage scales in the less endowed states which are relevant to the human condition; uplifting the position of small farmers and tenants by alleviating restrictive practices such as difficult access to credit because of the tenurial handicap; a vastly expanded rural works program to lift employment among rural unemployed beyond "digging holes and filling them up," now in an experimental stage; and the essentials for much of the above, the augmentation of national resources via greater revenues from the well-to-do rural sector, which currently contributes little by way of direct taxation.

Rural programs

In the 1960s India had unsuccessful experiences with a variety of rural aid programs. The failure then is not necessarily an excuse for not

15. For details, see R. S. Minhas, "Rural Poverty, Land Redistribution and Development Strategy: Facts and Policy," *Indian Economic Review* (April 1970), pp. 97–128.

16. Manning Nash, *Agricultural Revolution in Southeast Asia*, vol. 2, report of the Second SEADAE International Conference on Development in Southeast Asia, New York, June 24–26, 1969, p. 53.

trying once again in the more propitious circumstances of the 1970s, especially when the number of the rural poor is so great and likely to grow and when they have the capacity to produce. For reasons stated elsewhere, the fruition of most of the programs lies in the future; but it is encouraging that the government has been diverting funds for schemes which directly relate to the improvement of levels of living of the poor.

These programs have been operated under such titles "Rural Works Program" (RWP), "Small Farmers Development Agency" (SFDA), and "Marginal Farmers and Landless Laborers Scheme" (MFAL). The RWP has concentrated on fifty-four drought-prone areas. The government of India provides Rs20 million to each area for four years to finance roads, irrigation, and soil conservation. Where the work benefits individual farmers, they are granted a subsidy of 25 percent and a loan for the remainder of the costs of the work done. The SFDA and MFAL programs are essentially agricultural credit schemes. The former was discussed in detail in last year's report. The two agencies involved use their funds to provide risk subsidies to the cooperative credit institutions and other financing institutions. They also assist small and marginal farmers by granting subsidies up to 25 percent and 33.3 percent of the investments for which they borrow. In addition to finance, these organizations may perform a variety of services for the betterment of their clients.

The programs initiated in the past year are the "Crash Scheme for Rural Employment" (CSRE) and the "Dry Farming Scheme" (DFS). Under the first one the union government allots Rs1,250,000 annually to each of 350 districts to finance durable and productive labor intensive rural projects such as village roads and minor irrigation. About 80 percent of the cost is to be spent on wages, providing employment for 1,000 workers at a daily wage of Rs3 and not more than Rs100 a month for an average employment of ten months in the year. Local bodies participate in the selection of projects which are subject to approval in New Delhi. The Dry Farming Scheme (integrated dry land agricultural development scheme) is a pilot and demonstration program designed to test the application of the latest research findings and inputs in some twenty-one selected areas of some 12,000 acres each. For each of the areas Rs10 million has been allotted for the plan period. In all, anticipated expenditures in the plan period for former and recent programs make up the not inconsiderable sum of Rs3,800 million.[17]

There is ample scope for expansion of the rural works just outlined. This doesn't imply that, even if greatly expanded, they would constitute a sufficiently comprehensive program for overcoming rural poverty in India. Nor are they viewed that way by the government of India. While the programs are an entity in themselves, their success cannot be divorced from an "integrated program" which stands for a package of agrarian reforms, consolidation of holdings, effective rural credit for all sectors of the rural economy, research, and the widening of the scope of the new technology. Nevertheless, the problems of rural poverty are so urgent and the immediate need to earn a wage or improve the productive facilities of the small farmers is so great that the importance of the programs by themselves cannot be gainsaid. Besides, these essentially pilot projects, if successful, could be the forerunners of a much wider scope of similar activities, the sum total of which could just possibly make a dent in the poverty line where they are carried out successfully.

With respect to the plan's twenty-four dry farming projects at a total anticipated expenditure of Rs200 million, an outlay of Rs20 million was provided in 1970–71 but the expenditure

17. With no reference to the overall figure or parts of the same mentioned in the preceding paragraphs, the following is from the new budget: "the budgetary provision for the small farmers development agency is being doubled from Rs60 million to Rs120 million next year and for marginal farmers and agricultural laborers from Rs30 million to Rs60 million. Programmes for dry farming development, rural works in drought-prone areas and the crash programme for rural employment are being continued with a total provision next year of Rs720 million. It is our hope that in the light of the experience already gained and the assessments recently made in consultation with the State Governments, it would be possible next year to utilize in full the provision that is now being made." As of now, the crux of the matter is the proper utilization of the funds rather than the quantum of funds available.

actually incurred was only Rs4.5 million.[18] The "Mid-Term Appraisal's" comment is that "this program . . . has not made adequate headway so far." This is not surprising, mainly because of the newness of the scheme. Since the beginning of the drought-prone area programs (1971–72), an expenditure of nearly Rs180 million has been incurred out of a contemplated total outlay of Rs1.25 billion through 1973–74. Minor irrigation and soil conservation are the principal ingredients of the scheme. The actual implementation of the crash scheme for rural employment (CSRE) started only in October 1971, but expenditures incurred already amount to Rs400 million. At this rate of expenditure the intended Rs1 billion through 1973–74 might fall short of requirements. The labor component is expected to account for 70 percent of the expenditures. Field trip observations of the operations of a couple of these programs by a World Bank mission in late 1971 are encouraging. Expenditures for these programs are an important measure of progress, but their durability is just as important and only time can tell the effectiveness of the schemes as a whole.

More or less the same applies to the earlier programs such as the Small Farmers Development Agency (SFDA) and the Marginal Farmers and Landless Laborers Scheme (MFAL). Statistically speaking, a good deal has been done. More than a million small farmers and over half a million marginal farmers have been identified; Rs104 million have been advanced to them as short-, medium-, and long-term loans; and they had been able to take up 16,500 minor irrigation works and a sizable number of livestock and poultry projects. Yet they are meeting with difficulties. The "Fourth Plan Mid-Term Appraisal"[19] sheds considerable light on the current status. Briefly, the situation is something like this. The intended forty-six projects of SFDA with a total provision of Rs675 million are finally in operation, but "the expenditure during the first two years of the Plan was only Rs30 million.[20] Thus the real work of implementation on the ground is expected to be taken in hand during 1971–72." In the case of MFAL and the Rs475 million allocated, the expenditure up to the end of the year 1970–71 was only Rs10 million.[21] According to the "Mid-Term Appraisal," "It is obvious that, even more than the programme of small and potentially viable farmers, the programme of marginal farmers and agricultural labor will require a significant acceleration in the remaining Plan period if the program is to utilize the Plan provisions and to have requisite impact in terms of the objectives that were visualized."

The reasons for the delays are problems such as these: the elapsed time for sanctioning a project, time lag between sanctioning a project and staffing it, failure of some of the states to provide their share of expenditure, and the most crucial of all handicaps is the "stagnant or inadequate" flow of institutional credit to small farmers. This particular problem continues unabated. The official comment is that "the problem of re-orienting to loaning procedures and policies in favor of the less privileged . . . is proving doubly difficult." Added to this is the fact that "the commercial banks have not generally been able to do much in several SFDA areas." A smooth flow of credit is the very essence of the successful implementation of these programs, and it is understandable why "in the remaining period of the Plan one of the principal tasks is to re-examine some of the assumptions behind the SFDA programme with regard to institutional credit and take such measures as will help increase the flow of credit into the hands of small farmers."[22]

The "Mid-Term Appraisal" reflects the content of official project reviews of SFDA programs. The total number of such reviews being only three, it need not be presumed that they are typical of all programs. With this caveat it is nevertheless of interest to summarize the findings of the most recent review of one program carried out in late 1971. Its major con-

18. Government of India, "Economic Survey, 1971–72," p. 23.

19. Government of India, Planning Commission (December 1971).

20. According to a more recent paper prepared by the Planning Commission, funds released for the SFDA as of late February 1972 were Rs74 million and for the MFAL, Rs38 million.

21. Planning Commission.

22. Ibid.

clusion is that for reasons of physical, financial, and operational constraints, by the end of 1973–74 only 10,000 farmers, or 20 percent of those identified as small farmers, are likely to achieve viability through the SFDA. The remaining 40,000 may succeed in increasing their incomes if the SFDA succeeds in providing them with productive credit under adequate supervision, but "viability," as distinguished from "increased" income, will not be within their reach. The reasons for this anticipated outcome are many, but the two principal ones are the poor state of the institutional credit facilities and the problems of identifying the small farmers. Without going into details, all cooperative credit–serving institutions, from the central cooperative bank down to the primary cooperative societies, are in a bad financial way in the district where this particular program of the SFDA is in operation. About 80 percent of the cooperatives are in the "C" and 17 percent in the "D" class, the two lowest performance categories. About 80 percent of the societies are in default to the central cooperative bank, whereas the owned funds and deposits of the latter were almost totally locked up in overdues. As of the date of the investigation, the credit furnished by the commercial banks has been negligible; but in effect they are expected to meet at least one-third of the total financial needs of these farmers. Minor irrigation is the main accelerator of any SFDA program, followed by the adoption of modern farm technology and the use of improved inputs. Both are costly and both depend upon the adequate and timely supply of credit, but "during 1970–71 the credit institutions were hardly prepared to discharge the task assigned to them."

A year and a half after the establishment of the project, 12,000 farmers from 569 villages had been identified; the process of identifying all of the 50,000 is expected to be completed by the end of 1973–74. "Who is who" is of prime importance, for it forms the basis for extending the benefits and determines a farmer's eligibility for different forms of assistance. But the task of identification is proving to be more difficult than anticipated. The preparatory work based on "farm production plans" from revenue records and personal inquiries by village-level workers proved to be deficient "in requisite competence," and many of the records had no basis in fact. Big farmers divided up their land in order to fall within the category of "small farmer"; farmers with holdings in other villages have not been identified and some few with incomes way above the average also succeeded in claiming small farmer status. Above all, many tenants were excluded altogether because of the prevalence of oral leases and hence the inability to prove the obvious—their status as tenants—and credit eligibility. Additionally, the agency had not drawn a distinction between irrigated and unirrigated land, it is quite likely that it has enrolled a sizable number of farmers who are already viable. The findings conclude that "the mechanical application of the parameters in terms of size of holdings, without reference to income, is a serious limitation in the procedure adopted by the agency.

The reported situation in Nalganda district of Andhra Pradesh is much better. The account is incomplete, but it would seem that in the matter of selecting likely candidates for the SFDA and the MFAL, much greater care has been exercised, particularly from the point of view of good credit use and its timely repayment. What else the agency of this district has done or not done is still a subject for investigation. "But Nalganda," the author of the article writes, "is only one of the 90-odd districts covered by these programs. In most of the others no comparable effort has been made."[23]

One final point relates to the state of minor irrigation. It had not received the attention it deserves from the agency and mainly from the credit institutions. A repetition of the same in 1971–72 "might create serious problems for the Agency at a later date when it finds that, in the absence of adequate minor irrigation facilities, the bulk of the small farmers, though identified and otherwise assisted, have not attained viability. In other words, the basic objectives of the SFDA scheme would get defeated." Even with remedies suggested in a number of directions, the difficulties revealed in getting under way cannot be eradicated with great dispatch. It is clear from the foregoing

23. Ashok Thapar, "The Neglected Small Farmer—What Can Be Done to Help Him," *Times of India* (March 20, 1972).

that the apprehension expressed in the Planning Commission's "Mid-Term Appraisal" about the present condition of the SFDA is well founded. On the plus side is the fact that in assessing the programs no attempt is made to condone, overlook, or minimize the shortcomings that have cropped up in the course of their operations. Effective remedial measures for the tenants will depend upon appropriate legal measures at the state level; the question of credit and weak credit institutions represents an equally serious problem which only time, skill, and ample funds can correct. The recommendations of the National Agricultural Commission (see Cooperative Credit) and their implementation could be the right shot in the arm. Finally, despite the problems which beset all the programs, there can be no turning back; they do represent areas of great promise aimed at the right things by attempting to build employment, better standards of living, and social justice into economic growth.

Apart from all the programs already described, the recommendations of the Bhagvati Committee on Unemployment should be noted here.[24] The report of the committee issued on February 11 of this year urges an additional Rs7,400 million in the fourth plan to support a crash program to create *four million new jobs over the next two years.* The new funds, if available, are to step up the allocations for such familiar projects as rural electrification, road building, minor irrigation, education, rural water supply, and so forth, and for the inclusion of a new Rs2,300 million rural housing program in the fourth plan. The raising of such large funds and the ways and means of utilizing them present serious problems of their own, the more so in this case because the report has little to offer financially or organizationally by way of mounting an effort of this magnitude in so brief a time span. The best that can be said at the moment is that the committee's recommendations reassert the urgency of providing jobs for the unemployed or underemployed in the shortest possible time with the rural areas as the main focus.

Reverting to the schemes under discussion, it may be pointed out that the problems encountered are to a large degree the end results of something that goes beyond resources, planning, preparation of projects, execution, maintenance, and so on. These are all technical problems which can be overcome, provided it is recognized that the future of the schemes will be decided at the village level. And the village level, or rather the arrangements under which "our" farmers live and work, is not easy to contend with from the point of view of vigorous prosecution of the schemes. The question of credit is a case in point. Its flow in their direction is "stagnant or inadequate," not always because the credit cooperative till is exhausted but also because the "other" farmers who run village affairs can intrude into the schemes not meant for them to the detriment of those for whom they are intended. What credit they do get is not necessarily related to their needs or their entitlement under the schemes but often to what the others decide they should or shouldn't get. There are other interferences between material inputs and outputs which only most special care can neutralize. This is illustrated by one of the most perceptive examinations of a variety of small farmer schemes undertaken in Maharashtra before SFDA came officially into being. Says the author of the study:

> This class of citizens enjoys little or no economic and social status in the village community. He is never approached by any of the extension agents. Most of the time he works on the land of the rich and influential farmers. Often he is unemployed most of the time in the year. In any meeting in the village either he is not seen or is seen generally sitting at the fringe of the group. This is the micro-unit that must not be lost sight of. Unless this micro-unit is correctly identified and most appropriate plans for its development are prepared the entire efforts of the agency would be fruitless.[25]

24. Government of India, Committee on Unemployment, "Interim Report on Short-Term Measures for Employment" (February 1972).

25. V. R. Gaikwad, *Small Farmers: State Policy and Programme Implementation* (Hyderabad: National Institute of Community Development, 1971), p. 64.

The author of the quoted statement is not denigrating the numerous schemes currently sponsored by the government but uses his findings as a warning against the oversimplification of what goes into the making of a successful project. Precisely because so much is at stake due to the ongoing process of polarization between rural rich and the rural poor, the power exercised by the privileged "gatekeepers" of the village community must be watered down by challenging that power through economic opportunities made available to the underprivileged in the form of the programs discussed here. But the lessons of the past failures must be kept in the forefront because the programs deal with the weaker sections of the rural population with very special political, legal, technical, and administrative problems that their condition implies. Nothing said here is an argument against the programs. Not to venture is to admit defeat, a condition that cannot be contemplated even in the face of fear of failure. Nor must one be discouraged by their slower growth payoffs than the primarily growth-oriented projects which have been common thus far. But even if these schemes necessarily involved some degree of "trade-off" between growth and economic and social development broadly conceived, there can be little doubt that economic growth thus achieved would be more lasting and, in human terms, more desirable.

Cooperative Credit

The statistical picture

Demand for institutional credit continues to grow and loans to farmers have been on the increase. Firm figures for 1970–71 are not available, but in 1969–70 cooperative disbursements of short- and medium-term loans amounted to Rs5.4 billion or Rs380 million more than in the preceding year. Since by the end of the plan (1973–74) the target is Rs7.5 billion, its realization will call for additional mobilization of considerable resources. On the noncredit activity side—sales of farm inputs, distribution of consumer goods, and marketing of agricultural produce—the cooperatives have not done badly either. Now that marketing cooperatives are assuming a greater role as procurement agents of the Food Corporation, their turnover is bound to rise accordingly. On the qualitative operational side of the cooperatives, there is the elimination of 6,000 nonviable cooperative societies, reducing them to a total of 162,000. Reorganization of the cooperatives with an eye of greater efficiency is reported from a number of states. Punjab and Haryana expect to convert most of the existing societies into viable units by the end of the plan. In addition to cooperative credit, the contribution of the commercial banks has increased sharply from Rs1.84 billion in June 1970 to Rs2.4 billion in June 1971. Clearly, in a very short time the banks have become a major force in rural credit. As regards long-term credit, the Land Development Banks have made significant progress. Between 1960–61 and 1968–69, the loans advanced by these banks to agriculture increased from Rs120 million to Rs1.48 billion, and in 1969–70 the volume of new loans stood at Rs1.55 billion. As for the Agricultural Refinance Corporation with its stress on minor irrigation, land improvement, storage facilities, farm mechanization, and so forth, it continued to make progress. Up to 1968–69 ARC had sanctioned 233 schemes involving financial assistance of Rs1.82 billion, of which it committed Rs1.56 billion. As of August 31, 1971, the corporation had 489 schemes with a total outlay of Rs3.14 billion, including its own commitment of Rs2.7 billion.

Strength and weakness

These statistical bare bones speak of continuous quantitative expansion of rural credit and of qualitative improvement as well, or attempts in that direction. To this may be added what has already been remarked on previous occasions, namely, that there are many well-performing societies and that importance of the cooperatives in providing 30 percent of the rural credit at a low cost cannot be overestimated. In doing this, they have helped to rationalize distribution patterns, increase purchasing power, and improve standards of living; and they have certainly contributed to modernization of agriculture. In a word, millions of farmers have benefited from them and rural India without this landmark is hard to visualize. On the other side of the ledger, and a very important side at that, are the shortcomings which are growing in scope as the volume of credit keeps on expanding. The same source of information which shows the numerical progress of the cooperatives tells a less cheerful story about the quality of a good deal of their performance. This is not new nor is it new that they have not been improving with age. Problems of "revitalization" are decades old, and the reports and recommendations to that effect have been standard fare for just as long. Despite these efforts, in the year 1969–70 their problems are too great to be ignored. The Reserve Bank of India, the mentor and perennial financial prop of the cooperatives, notes the following:

> Although quantitatively there has been some progress in the working of the societies, a great deal of effort is still required in the matter of revitalization of primary credit structure. Firstly, as against the total number of societies at 162,700 the number of dormant societies was 19,000 and these covered 49,703 villages. Secondly, the number of societies working without profit or at a loss was 49,910, i.e., about 30 percent of the total. Thirdly, out of 150,636 societies audited during the year, only 22,094 societies were classified as A or B. As many as 128,542 societies were classified as C, D or E . . . and these constitute 85 percent of the societies audited during the year.[26]

Since C, D, and E classifications denote poorly operating coops, perhaps a better way of putting it is that only 15 percent of all cooperatives meet the high standard of the Reserve Bank. A comparison with similar data for 1962–63 shows that in that year the A and B cooperatives numbered 33,000 as against 22,000 in 1969–70. In line with this and other shortcomings—all of old standing—the number of dormant societies is probably considerably larger than the official figure given. According to the official definition, an active society can be one "which has issued even one loan, but has undertaken some other activity during the year . . . We have reason to believe that several of the societies which have been classified as active or non-dormant have in fact been doing only token business and cannot be really considered 'active'."[27] It may be presumed that a goodly number of societies in the D and E categories are ready for the mortuary. Although the total membership is close to 30 million, only 39 percent are borrowers, while in the late 1950s the percentage of borrowers was closer to 60. Not without significance is that the idea of thrift and savings as an important factor in the promotional work of the cooperatives has fallen into disuse. The emphasis is entirely on credit, relying for the greater part on borrowed funds to the exclusion of self-reliance, thereby limiting growth of cooperative credit.

Overdues

In the final analysis, an important test for judging whether the credit system is working efficiently is whether loans are recovered on due dates. This is a measure of the strength of the cooperatives as well as of the attitude of the borrowers toward the cooperatives. It is the failure in recoveries that mainly accounts for the stagnation or recession in recent years in most states. In short, overdues have grown apace along with the expansion of credit. Between 1960–61 and 1969–70 those of the primary societies have gone up from 20 to 38 percent; this in turn was reflected in the overdues to district and central cooperative banks which have risen from 12 to 29 percent. According to the Planning Commission's "Mid-Term Appraisal," "in some states the position is very alarming indeed . . . there has been a steady deterioration over the last several years and the same trend was maintained during 1969–70." The global percentage figures tell only part of the story. In Assam it is as high as 82 percent, Orissa 63 percent, Kashmir 60 percent, and Bihar 50 percent; even in the most prosperous of agricultural states, Punjab, overdues are at 38 percent; and in that cradle of

26. Reserve Bank of India, "Credit Societies," *Statistical Statements Relating to the Cooperative Movement of India, 1969–70*, pp. iii–iv.

27. *Report of the All-India Rural Credit Review Committee* (1969), p. 169.

cooperativism, Maharashtra, overdues are at 54 percent and as a single state it accounted for one-fifth of the overdues incurred by all the states. The "Mid-Term Appraisal" notes that "an even more disturbing picture is obtained if these figures are disaggregated in terms of individual central cooperative banks." Paradoxical, too, is the fact that the habitual defaulters are not the smallest and poorest but the biggest and richest borrowers.

Nonpayment is often a result of an "act of God," and it is inevitable in some circumstances. It is not inevitable in a number of states enjoying in the past five years unprecedented prosperity. In such conditions nonrepayment is only a reflection of the inadequate performance of the cooperatives. Added to this is the apprehension of large numbers of members that due to poor recoveries from others, a society may not be able to give them a fresh loan. This in turn is one of the reasons for the failure to repay the loans promptly and causing overdues to mount. It weakened the capacity of the cooperatives to advance fresh loans even to creditworthy farmer members, and this had resulted in relatively small numbers of borrower members.

Cooperative membership

The membership of the cooperatives is as varied as the social and economic structure of the village and just as unequally represented. One finds in it rich and middle-class landowners, proportionately fewer tenants, and still fewer agricultural laborers and artisans; nor is the membership lacking in traders, rice millers, moneylenders, and local politicians. On its face it is a mix of all the elements of the community, but obtaining any credit and the volume of credit received are related to the lines of economic stratification. The wealthier groups take the fullest advantage of what the cooperatives have to offer; they are the "insiders" and they control the cooperatives. Whatever the idealized notion of "cooperation," "village democracy," and "egalitarianism," the fact is that the village imposes its own pattern upon the cooperatives. Part of this pattern is that the cooperatives do not belong to the mass of the membership but to powerful factions who use or abuse them for their own advantage. This is compounded by yet another reality that the cooperatives also "belong" to the government. Government control and direction of the movement is a historical legacy. Although many efforts have been made since independence to strengthen the nonofficial voluntary character of the movement, it is still primarily the Department of Cooperation which decides how funds are used, terms of membership, who are to be given loans on what basis, in what amounts, and so forth. There may be good reasons for the tight governmental control; but in the eyes of many of their members the cooperatives are primarily instruments of the government, and the overdues are at least partly a reaction to this condition.

Credit distribution

All this is relevant to the fact that rural India has entered an era of agricultural development when the question of how institutional credit is distributed is of primary importance to millions upon millions of the poorer farmers. How much credit they get must be linked from now on with the major attempt of the government of India to raise their standard of living. The answer to the question just raised doesn't readily lend itself to precise statistical analysis, but enough information is available to conclude that credit is distributed in a manner that favors the well-to-do as against the underprivileged section of the village.

In last year's report we quoted the Haryana Minister of Cooperation to this effect: "It is now a widely known and accepted fact that the smaller cultivators and other weaker sections of the rural community have been totally ignored in the field of cooperation." It is perhaps an exaggeration to say that they are "totally ignored," but they have not done well either in sharing available credit. If the available data roughly reflect the credit distribution pattern, a few conclusions may be drawn. According to the Reserve Bank, in 1969–70 tenant cultivators, agricultural laborers, and "others" received credit amounting to Rs240 million, which is 4 to 6 percent of the total credit disbursed. Since the tenants alone cultivate an estimated 15 to 20 percent of the total acreage, their share is manifestly below what they might be entitled to under a more equitable system of distribution. Assuming that they cultivate only 15 percent of the land or 50 million acres and that they received all of the Rs240 million, the inference is that large numbers of them received no cooperative credit at all or very little of it. As for credit distribution among the farmers in general, ample statistical evidence is provided by the monumental and authoritative *Report of the All-India Credit Review Committee* published in 1969.

It would take one too far afield to explain how the committee arrived at its conclusions. For this reason, a few quotations will have to suffice. Thus:

> It doesn't, therefore, appear so much that the bigger cultivators are getting undue large loans in absolute terms as that a large proportion of the small cultivators get either no cooperative credit at all or such inadequate credit that even in relative terms, they are worse served than the bigger cultivators . . . the principle of open membership is not always effective and several cooperatives operate as a closed shop for the benefit of one particular economic group or caste or faction. Secondly, the repaying capacity of the small cultivator is called into question, and loans [are] often ruled out on that ground. Thirdly, in the distribution of the limited funds available, it is the small farmers who get left out . . . the tenants are discriminated against and, irrespective of the crop raised by them and their acreage, their entitlement to credit is restricted to only Rs250 or thereabout. Insistence on the security of mortgage of land, again, operates against the provision of adequate short-term credit to the tenants. Finally, in spite of the provision for appeal to the Registrar of Cooperative Societies, even membership—let alone credit—is being denied to cultivators belonging to certain classes, factions or castes. All these have had the total effect of restricting the access of the small cultivators to cooperative credit. *The substantial expansion in cooperative credit witnessed in certain areas, therefore, represents, in effect, more an increase in the amounts borrowed by a limited number of members rather than a widening of the area over which the benefit of credit is spread.*[28]

Finally, "the small cultivators are indeed handicapped in having access to cooperative credit."[29]

The Interim Report on Credit Services for Small and Marginal Farmers and Agricultural Laborers, prepared by the National Commission on Agriculture (December 1971) affirms the above findings. The nationalized commercial banks have done no better because "even their operations excluded the bulk of the small and marginal farmers, as they also tended to finance agriculture according to the prevailing criteria and methods."[30] Seven or more reasons are cited to explain this state of affairs, which are in effect the strongest indictment of the cooperative system on the grounds of poor management; poor leadership; disregard of all the principal recommendations of more equitable distribution of credit; credit monopoly enjoyed by the "all and all" in the villages; divergence between accepted principles and practices as demonstrated, for example, by the neglect of the crop loan system in favor of acreage held and land mortgage as security; the questionable creditworthiness of the cooperatives themselves; and the multiplicity of agencies through which a farmer must seek credit and put it to use (see attached on the flow of credit) [omitted]. While extolling the benefits the system has rendered regardless how credit is distributed, "the present structure of the organization has to be modified to give a better bias towards the small and marginal farmers and agricultural laborers."[31] It inveighs against a dual agricultural economy growing up under the aegis of the new technology and pleads for the incorporation of its underprivileged part "within the compass of the upsurge of agricultural modernization." Hence the proposal for a new credit system.

Integrated agricultural credit service

Both the *Report of the All-India Rural Credit Review Committee* and the *Interim Report of*

28. Italics added. All the quotations are from p. 174 of the committee's report.
29. Ibid., p. 131.
30. *Interim Report*, pp. 13–14.
31. Ibid., p. 10.

the National Commission on Agriculture have not limited themselves to a mere analysis of the present-day strength and weakness of the cooperative credit system. In the first-mentioned report the immediate and practical outcome was the attempt to remove the anti-small farmer bias by creating the Small Farmer Development Agency and focusing national attention and effort to combat this problem. The fact that SFDA does not have smooth sailing only tends to underscore how deeply rooted cooperative malpractices have become with the passage of time. In the *Interim Report* and apart from advocating "mixed," specialized farming as the most promising pattern for small farming, the main stress is laid on the creation of an Integrated Agricultural Credit Service. "Considering," the report notes, "that there are over 20 million operational holdings in the country with less than 2.5 acres, the gap in the coverage and the magnitude of the problem assume staggering proportions and point out the inadequacy of the present institutional system in ushering in the structural changes necessary to bring the small farmers as sharers of benefits that accrue from agricultural growth."[32] More specifically, one of the principal recommendations is to do away with the multiplicity of credit agencies so that the borrower will have only *one* contact point for obtaining his credit *and* input requirements. Equally important is to make certain that the farmers in question who cultivate nearly one-fifth of the land should get "at least 40 percent of the total agricultural credit."[33] To a degree, this tilt in allocations is based on the proposition that many of the big farmers are capable of financing their requirements from their own resources. Special attention is to be paid to the cooperatives in the eighty-eight districts where the SFDA and MFAL programs are being implemented to ensure that they are immediately reorganized into strong units capable of canalizing credit to the farmers in the selected areas.

How this is to be accomplished is presented in considerable detail but it need not be repeated here, except for a few main points. By way of a preamble, the agricultural credit service will have the following constituent parts:

1. Farmers' Service Societies—one for each tehsil/block or any other viable unit of convenient size, with as many branches as are required in the area;
2. A union of these societies at the district level, and functional district organizations for specific commodities; and
3. Lead Bank of the district assuming leadership in the matter of organizing integrated agricultural credit service.[34]

In organizing the proposed framework, a number of considerations are kept in view. Apart from the injunction that no small farmer should be debarred from membership in the cooperatives, the report stresses that:

1. Farmers will deal with a single agency for their short-, medium-, and long-term credit requirements as well as their demands for inputs and services. That agency should be their own organization, viz., Farmers' Service Society.
2. The Lead Bank will assume the responsibility for integrating credit with supply of inputs and services.
3. The Farmers' Service Societies will be responsible for organizing, on commercial lines, the supply of inputs and services, either directly or under contract arrangements with other public agencies or private firms.[35]

Additionally, a Farmers' Service Society is expected to assume a variety of functions, eight in number, indeed turning it into a multipurpose cooperative, talked about in the past but never tried. As an aid to this and not only with an eye to employing a full-time paid managerial and technical staff to support the operations, the basic cooperative law is to be amended in a number of ways. But what comes through in everything pertaining to the proposed integrated agricultural credit service is the theme that the credit and other services to the small farmers, whether members of SFDA

32. Ibid., p. 20.
33. Ibid.
34. Ibid., p. 26.
35. Ibid., pp. 26 and 28.

or MFAL or not, must be given top priority. The Reserve Bank and a great many committees and commissions have dealt with this problem, but the *Interim Report* discussed here takes the *Report of the All-India Rural Credit Review Committee* a step further by restating and prescribing for the needs of all farmers who for a variety of reasons have least benefited from the cooperative system. One of the suggestions is to keep out bigger farmers, the reasons being that (a) "they would exploit the small farmers" and (b) "that with the expansion of commercial banking into rural areas, this class of entrepreneurs can have access to bank credit." The scheme concentrates an institutional credit arrangement for the small farmers and evidently assumes that the existing cooperative systems will go on with or without improvements. This presupposes a dual cooperative system, although the report makes no reference to it.

Comment

Credit cooperatives have functioned primarily as moneylending institutions, and their interest in the overall economic activity of their members has been peripheral. In the case of the small farmers, even availability of credit wouldn't be altogether crucial without providing them at the same time with many other services. It is recognized, therefore, that credit cannot stand by itself but should represent such other areas of activities which make up the essentials of farmers' needs. This recognition is, in fact, the rationale of the multipurpose societies or what the *Interim Report* terms "integrated agricultural credit service." The idea is not a new one; the lasting benefits to the farmers of this type of societies have been stressed for nearly two decades past, but only a few sugar cane cooperatives have practiced it. If such cooperatives can be created and the poorer farmers provided with a single contact point for obtaining credit, inputs, and assistance in processing and marketing their products, it would signal a major change in the cooperative movement. It would demonstrate for the first time a close linkage between means and ends.

The credit medicine prescribed for the small farmers should apply to the cooperatives in general. Decades-old way of lending and the inimical pressures upon the cooperative system cannot be cured very quickly by multipurpose credit societies, but the advantages of a version of a supermarket credit scheme to the borrower—whether big or small—are obvious. Apart from the benefits of shopping under one roof, the cooperatives could exert better control over lending, utilization, and recovery of loans. The difficulties of a major shift of this kind, particularly considering the vested interest entrenched in all the constituent parts of the country's institutional credit operations, is recognized. At the same time, there is considerable disenchantment with the gradual accumulation of lending agencies, each one having been created to meet this or that problem the primary societies couldn't resolve. To the concerned cooperators who have been watching the expansion of credit along with the expansion of problems, the primary societies and district and central cooperative banks as they are now have outlined their usefulness. In their view, therefore, the multipurpose idea has much relevance for a new deal in cooperative credit if overdues, the political character of many of the cooperatives, and the role of the wealthy clientele in shaping them are to be dealt with more effectively. For all of these reasons at some point soon, the argument runs, the existing system must undergo a drastic change if it is to serve well the needs of a changing agricultural economy.

The real question is whether this scheme will succeed this time whereas somewhat similar proposals by the Reserve Bank of India and other involved bodies have never quite reached the trial stage of significance. Considering the endless array of recommendations to improve the work of the cooperatives which went unheeded over the years, one may be tempted to dismiss the new attempts as just another paper proposition. To take that position is to assume that no improvement in the cooperative system can be contemplated and, more particularly, that the Small Farmers Development Agency programs are likely to fail for want of "integrated credit." Viewed optimistically, the new schemes will be tried on a pilot basis to begin with, and the chances are that through trial and error innovations will be introduced. There is no inherent reason why the cooperatives so dependent upon government assistance couldn't, for instance, insure that all the small farmers are eligible for membership, that credit be based on crop production rather than on security of land, that it is linked with marketing, that inputs are provided, and that the cooperatives are there to serve the overall needs of agricultural advancement and better standards of living. And as for the small farmers, so for the generality of farmers. For all too long these basics have been regarded as unrealizable tall orders. This time around one likes to think that they will prove more manageable—if the recommendations of the *Interim Report* are not to be another exercise in futility and the interests of the farmers in question are to be secured. But for the reasons touched upon, there is more at stake than credit for the small farmers. Left to its present devices, even many of the beneficiaries of the system run the risk of ultimately being choked off from a ready source of credit.

Ceilings on Land Ownership

The past year left the unfinished business of agrarian reforms where it has been in the years past—in the long drawer. But one issue, ceilings on land ownership, or the acreage an owner can retain, came in for a good deal of public discussion. It is this that constitutes the subject of the following paragraphs.

New ceiling approach

On August 3, 1971, the Central Land Reforms Committee announced new guidelines for reducing the acreage an owner may keep. Their purpose is to widen the scope of land redistribution among the tenants, sharecroppers, and the landless. The guidelines to effect this are as follows: (a) nationwide, no owner can keep more than 10 to 18 acres of irrigated, double-cropped land or 54 acres of dry land; (b) the permissible ceiling must apply to the family as a unit rather than as in the past, in most states, to individual members of the family; (c) these limits, especially on dry land, can be exceeded depending upon soil and climatic conditions; (d) existing exemptions in favor of mechanized farms, well-managed farms in general, and so forth should be withdrawn; finally, (e) exemptions in favor of plantation crops shall be re-examined in consultation with ministries concerned and state governments. The recommendations make no reference to compensation, but the Minister of State for Agriculture informed the Parliament that he favored "reasonable" compensation but not the market value of the surplus land. This is not surprising in the light of past treatment of surplus land and constitutional amendments relating to land pricing under the reforms. From the point of view of the promoters of the new scheme, it is an improvement over the existing one on two counts: the maximization of surplus acreage for distribution and its application on a family basis only. The latter is intended to eliminate illegal transfers to all and sundry, thus helping ceiling goals as has been the case until now.

On the face of it, if the primary aim of land reform is the narrowing of rural income inequalities and the easing of rural tensions, the new proposals are a step in the right direction. On the other hand, the results of ceiling legislation in the past decade and more and the political realities still extant argue against optimistic expectations. It cannot be excluded, therefore, that the considerable effort invested in formulating the new approach and inducing the states to formally scrap the old legislation and enact appropriate new legislation will not attain the stated objectives. This does not denigrate the issue involved, which is serious enough. Yet too great a priority is attached to land ceilings in the absence of a similar concern about reform measures of immediately more vital promise such as a record of tenurial rights, security of tenure, and reasonable rentals. These constitute the "minimum program" discussed in detail in the World Bank's last year's annual report. It is this consideration that leads to the conclusion that, presently, land ceilings as a part of land reform in India, while necessary, appear as something less than a fruitful preoccupation. This is particularly so if past experience with the attempts to enforce the ceiling provisions is anything like a prelude to the future.

Ceiling program in retrospect

At least in theory, ceiling legislation covered the entire country. The permissible retentions were most generous regardless how and to whom they applied. In Gujarat, for example, the retention varied from 19 acres of irrigated land to 132 acres of other types of land, per family; in Mysore from 27 to 216; and even in Kerala, with small holders predominating, the range was between 12 and 15 acres. The same repeated itself in states where the ceiling applied on individual basis. In Bihar, for example, each member of a family was permitted to hold from 20 to 60 acres; in Andhra Pradesh from 27 to 324 and, where a household exceeded five members, additional land was allowed at the rate of 6 to 72 acres per member. And so it went from state to state.

Permissible retentions of this magnitude were in themselves a deterrent to the program's success. This was compounded by legal and illegal land transfers both in anticipation and after the enactment of the ceiling laws. On top of that were numerous exemptions, of which Tamil Nadu could boast of as many as 26, Uttar Pradesh 20, Madhya Pradesh 19, Maharashtra 11, and fewer exemptions in other states. It is not surprising that there was little land to distribute. The outstanding exception is Kashmir where special political circumstances created a situation where the declared surplus land (180,000 acres) was actually distributed. Not so in other states. Between the early 1960s and the end of 1970, the states of Bihar, Mysore, Kerala, and Orissa have not contributed a single acre of surplus land. In all of Andhra Pradesh only 1,400 acres have been taken over —and none distributed. The big state of Tamil Nadu contributed so little of declared and distributed surplus that its performance is only marginally less than that of the noncontributors. Summing it up for India as a whole, by the end of 1970 the "declared surplus" was only 2.4 million acres and "area distributed" just half of that, or 0.3 percent of the total cultivated land of India.

Looking back and in relation of any new attempt to *reduce* the permissible ceiling, it must be kept in mind that the entire ceiling episode was one of evasion by commission and omission. With deliberately poorly drawn enactments, holdings had been fictitiously divided up among close and distant relations as to make them appear *under* the ceiling, while many owners did not bother even with such precautionary measures. Lacking a single punitive clause in the entire mass of legislation—with the exception of the ineffective one in Maharashtra—there was no reason why they should have. It made no difference whether the ceiling applied on an individual or family basis; there was no concern about the bona fides of the owners subject to ceilings or for that matter administrative arrangements worth mentioning. Enforcement was not a problem; there was little to enforce. In sum, while officially and in principle the states accepted the ceilings programs, they rejected them in practice—and with results one might have expected.

Can a new program succeed?

This experience raises a fundamental question: Are new ceilings necessary? Lack of space forbids even a summary of the pros and cons of the ceiling controversy, but our answer is in the affirmative. India's poor record to date is no proof that the ceiling idea is too fanciful to be useful. A glance at the country's agricultural structure, with its inequalities and multitudes pressing on the land on the one hand and the narrow industrial base on the other, reveals conditions in which any ownership of land or for a tenant to remain undisturbed on the land are minimum securities the underprivileged can hope for. This, apart from ideology, is the principal economic justification of the ceiling. The question, therefore, is not whether a ceiling program is necessary but whether, in the experience of the past decade, it can be carried out.

In the light of the same experience, what is there, then, in the current picture that would be conducive to a right about-face as suggested by the Central Land Reforms Committee, with permissible retentions made more restrictive and exemptions eliminated? Little encouraging to speak of, even though in recent months ceilings have been reduced in Assam, Bihar, Kerala, Tamil Nadu, and West Bengal. Not until these measures are implemented is there any reason to assume that this is in earnest. Moreover, developments of a familiar sort tell a different story. In early 1970 on a visit in eastern Uttar Pradesh, the "taluka" offices were overworked selling stamps for registering new land deeds or fictitious sales in the wake of rumors that the ceilings might be lowered. More recent events in Maharashtra tell the same story. None of this is surprising. It is part of a well-established custom, and the likelihood is that there are probably almost as many similar Maharashtras as there are states—an exception or two notwithstanding. A positive acceptance of any new ceiling program would call for the abrogation of all the so-called legal and illegal land transfers and the acknowledgements on the part of the state legislatures and administrators at all levels that for most parts the ceiling programs were essentially make-believe affairs. Under whatever other guise this may be dressed up, it is highly questionable that as of now or in the immediate future ceiling programs of the kind recently proposed can be enforced—even if the appropriate legislation is enacted.

Conclusion

For reasons stated, a ceiling program is essential. Any meaningful reform without it is a misnomer, while its effective implementation is one of the main reasons of the few reforms in Asia which have succeeded. But considering the past history of this measure and its future prospects, major emphasis on ceilings now is not warranted. This is in line with the numerous problems obstructing its implementation, the recognition that what has been so thoroughly scrambled up with impunity cannot be unscrambled, the unwillingness of state legislatures to get on with the unfinished business of agrarian reforms in general, and the possibility of impairing the position of the big owners in the new technology. Apart from all these considerations, the *new stress on ceilings could very well be at the expense of the more immediately promising concentration on the minimum program of recorded rights of tenants or sharecroppers, security of tenure or their rights to remain on the land undisturbed, fair land rentals, nonexploitative farm wages, the enlarged scope of land consolidation, and a further expansion of the Green Revolution into environmentally more deficient areas.*

Admittedly, the obstacles that stand in the way of a ceiling program apply also to the still unresolved tenurial measures. The latter, however, do not carry the same "bite" the former does and are much more feasible for implementation. Additionally, enough was said in last year's World Bank report why this "minimum" deserves top priority and it need not be repeated here. If the state political decisionmakers are serious about it, they should take to heart Prime Minister Gandhi's injunction that "it is time to face the facts." This is the understanding that there is no contradiction between the steady progress of the new technology and those elementary and persistently denied rights of the underprivileged peasantry. This holds good in a country like India in quest for more production, more equitable distribution of farm income, and rural stability. But the only reason we underplay the land redistribution question at the moment is that the climate might not yield positive results within a relatively short time. And it is for this very reason that preoccupation with it now might also detract from, if not smother altogether, the effort to get on with the long overdue minimum program.

* * *

58. *Drought in Maharashtra: Not in a Hundred Years*

This was Ladejinsky's report to the Bank dated January 12, 1973, on the devastating effects of the drought in Maharashtra. The dramatic subtitle makes further introduction unnecessary. This paper was subsequently published by the *Economic and Political Weekly*, Bombay, February 17, 1973, under the same title.

Introduction

IN THE ANNALS of Indian agriculture, 1972 will be remembered as another "Year of the Drought." In view of Bengal in 1943, it is by far not the worst recorded, yet it is bad enough in large areas of the country to create landmarks of misery all of their own. Varying in degree of intensity, there are Maharashtra; Andhra Pradesh; Gujarat; Rajasthan; and large pockets of Mysore, Orissa, Madhya Pradesh, and parts of Uttar Pradesh. Countrywide, the final summing up of the visitations that drought brings in its wake upon stricken states, including the lucky escapees, must be left to the future historian. At the same time, something must be said about its scope and aftermath before the drought becomes just another episode in the long chain of similar episodes.

Looking at the afflicted states as a whole, Maharashtra may claim the dubious honor of being the worst-affected state. The drought there "is in the center of the piece," as Nehru was fond of saying, referring, to be sure, to an altogether different matter. Particularly in the past two months reams of copy have been devoted to its crop failure, painting it as the worst in a hundred years. In the circumstances there was nothing to do but visit some of the affected areas. This we did in late November to early December, and what follows is an account of conditions observed and of discussions with farmers and local officials involved in the disaster. And let it be stated at the outset that, while officially the situation is still called one of "scarcity," in reality it is a disaster of unprecedented dimensions, especially if one is to assume that for Maharashtra one hundred years is a sufficient span of time within which to judge a crop failure as precedented or unprecedented. And there is one more thing to note. This is an occasion when the writer intends to keep his emotions in leash. He labors under the impression that a "bleeding heart's credibility may suffer from dramatizing the incontrovertible cruelty of nature—no rain and no crops; individual tragedies galore; short rations and no little starvation; a multitude of farmers seemingly without a hope in the world; a good deal of listlessly wandering skin and bone that may be termed livestock by courtesy only; and a horizon which upon closer examination is essentially a wasteland, exceptions notwithstanding. It is best, therefore, not to play on the heartstrings of the reader. Instead, the facts and figures will speak for themselves. Gathered on the fly, so to speak, or from hastily compiled documents, some of them may be disputed. However, whether overestimated or underestimated, they do reflect the overall dimension of the drought calamity. With these caveats out of the way, we shall begin with the principal consequences of the drought as it affected the Maharashtra state as a whole and then proceed with the main object of these observations: drought conditions in the Poona, Sholapur, and Ahmednagar districts.

Maharashtra State

The reason the Maharashtra drought is so much in the limelight is not only because of its extreme severity but because it is the third in succession. The remarkable thing is how relatively little had been made of the first two; not less significant is that the lessons of the preceding years have not been fully taken to heart. This apart, historically the struggle of the Maharashtra farmer has been one of a quest for water. Nature has been stingy in providing him with a sufficiency of rainfall, and neither did it help him in many of the operations reaching out for this indispensable resource. Not only has it deprived him of adequate rainfall; for the greater part it has also put a layer of hard rock between him and an undependable subterranean flow. This helps to explain to a large degree why the rate of irrigation in Maharashtra is the second lowest in India and partly why the irrigated area during the past decade has risen only marginally—from 6.5 to 8.3 percent. Regardless of the reasons, it is obvious that rural Maharashtra depends very largely on the monsoon for irrigation and, as we shall presently see, for drinking water as well.

The year 1960–61 was the best recorded crop year; but between then and 1970–71 a number of relatively minor droughts has reduced food production by a reported 14 percent despite a rise in cultivated area, improved varieties, and other supporting measures. The corresponding increase in all-India foodgrain production was 35 percent. A similarly unhappy position prevails in respect to most commercial crops. Few irrigation facilities and the monsoon playing truant so often are reflected in the decline of farm income. The per capita income of the state of Maharashtra at constant prices, in the past decade, increased by 3 percent compared with an average per capita income increase of 13 percent for all of India. In this growth, the primary sector (agriculture, animal husbandry, fisheries), employing 70 percent of the population, showed a decline of 16 percent. This is not because the Maharashtran farmer is conservative and indolent but because he is so largely dependent on the water gods who have been bypassing him these many years.

In Maharashtra the subnormality of the monsoon has become more a rule than an exception, and all measures to combat it have been either insufficient or nonproductive for a variety of reasons. The years 1970–71 to 1972–73 tell the story of the persistence of this subnormality and its consequences. According to official estimates, out of a total of some 35,000 villages, 23,000 villages in twenty-one districts (out of a total of twenty-six districts) in 1970–71 and 15,000 villages in twenty districts in 1971–72 were affected by water scarcity. The current year is apparently the "banner" year; officially, twenty-five districts are facing or likely to face crop failures in larger and smaller degree. The number of villages involved is not yet known, but it would exceed that of 1970–71. Statewide, the numbers of people subject to the drought in one way or another during the mentioned years are officially estimated at 19, 12, and 20 million, respectively, out of a total rural population of 35 million. These figures, as well as the number of affected districts, may well be on the high side, but even smaller magnitudes would not change the picture to any significant degree.

The explanation for this state of affairs, which may be somewhat exaggerated for Maharashtra state as a whole, lies above all in the poor monsoon, or regionally in its almost total failure, on the eve of planting the kharif (summer) and rabi (winter) crops. In the Sholapur district, the rainfall between June and November was only 3 inches as compared with a normal 20 inches. With variations similar statistics can be cited for a number of other districts. Sowing operations were held up in large tracts of the state; and, in some places where sowing had been done, the seed was wasted. Even the better water-protected rice areas of the state have not escaped all the ravages of the drought. This is not to say that all crops failed—some survived well, some were badly damaged, while in the three districts mentioned earlier and with the exception of sugarcane practically all other crops failed utterly. In the light of all of this, the kharif crop, which in Maharashtra accounts for about two-thirds of the total agricultural output, presents the following picture:

	Percent of normal
Jowar	45
Groundnut	30
Bajra	20
Cotton	60
Rice	50

If these figures are not eventually revised upward, in terms of money value the kharif production this year is estimated at Rs315 crores against a normal crop value of Rs574, or a fall of 46 percent. The total normal grain output of the state is around 6.5 to 7 million tons; provisionally, this year's anticipated output is put at about half of that. This is explained by the fact that the prospects of the rabi crop are equally bleak, since the anticipated rains either did not come or were below normal. Whatever the size of the income generated in a non-drought year by the kharif and rabi crops, including animal husbandry, this year it cannot be expected to be much more than half of that. The consequences are widespread, and we shall touch upon them in dealing with the three districts. Suffice it to say here that one of the consequences of the sharp decline in income is the suspension or remission of land revenue taxes and the nonrecovery of loans by the cooperative and banking institutions and government. One of the worst-hit districts is Sholapur; although it is probably not representative of the state as a whole, yet the recovery position to date is more than a straw in the wind. Thus, if we read the official statistics correctly, up to the middle of 1972 less than one percent of the land revenue was collected; of the irrigation assessments, 4.5 percent; of the education cess, 2.7 percent. Recovery of the land bunding assessment is near zero, and the same is true of repayments on two special loans. On a field trip in Sholapur, Poona, and Ahmednagar, it is crystal clear that what is true of mid-1972 will be equally true at the end of the fiscal year 1972–73. Allied with this is the obvious sharp rise in indebtedness, leading to a vicious circle from which extrication is difficult, if not impossible.

For Maharashtra as a whole, two other consequences of the crop disaster must be mentioned at least by way of summary; they will be discussed at greater length as part of our observations in the districts. These are the shortage of drinking water and of fodder. The available figures for drinking water are anything but precise and must be taken with caution. Yet, however exaggerated they may be, they should be mentioned as an indication of the gravity of the situation. The lack of drinking water is more debilitating than is an inadequate food ration. This was attested to on all sides. Desertion of villages, of which there are not a few cases, is attributed to the first rather than to the latter. A writer notes that even in normal years an estimated 12,000 villages must search for water within a radius of one mile or more. This year, the number of villages with an inadequate water supply is much larger. In the Gandhi Centenary Year (1971) the state government had promised that the above-mentioned 12,000 villages would be provided with drinking water, but in fact only 15 percent had been supplied so far.[1] To a lesser degree, urban centers have not escaped this problem. In short, it is reported that in the greater part of the state, surface water has dried up and that underground water can be reached only through high-speed boring at 200 feet or more.

Cattle are in still greater jeopardy; they suffer from serious shortages of both water and fodder. But the stress here is on the latter, for along with three successive crop failures, fodder has also come a cropper. Fodder prices are mounting far beyond the capacity of the farmers to sustain them, since the price of a bundle of fodder is four to six times higher than in a normal period. Above all, fodder is in very short supply and is only marginally available in the worst drought areas. No wonder that in the affected regions no farmer buys cattle; all are trying to sell them to the slaughterhouses and this, as we shall point out later, for a song. Many cattle perish, and it is perhaps not a great exaggeration to say that, of the 15 million livestock population in the badly hit areas, about half are expected to perish of starvation or to be consigned to the slaughterhouses. An examination of this problem in the three districts supports this contention. So much for the sketchy overall picture of drought-stricken Maharashtra.

The Three Districts

General impressions

Poona city, the first city on the way to the countryside, doesn't prepare one for the drought

1. Pannalal Surana, "Who Cares About the Drought?" *Economic and Political Weekly* (November 18, 1972), p. 2269.

of Maharashtra as a whole or, for that matter, for the calamitous conditions prevailing in the greater part of Poona district and its two other neighboring districts. On the contrary, the modern "Blue Diamond" hotel speaks of the remarkable industrial upsurge in the past two decades within the periphery of the city. As usual, the countryside and the city are worlds apart, although willy-nilly the economic distress in the rural areas is bound to affect—and is already affecting—all kinds of traders and craftsmen who also depend for their living on the economic condition of the rural population. But the grim struggle for life comes into the picture only some ten miles away from the well-irrigated green belt adjoining Poona city. From there on and hundreds of criss-crossed miles later the enormity of little or no food, little or no fodder, and little or no drinking water is inescapable. Once beyond the Poona green belt, the fields are deserted of man and cattle. Having missed the devastated fields of Bihar in 1965–66, the sight of these fields leaves one shaken by the perversity of nature. The striking feature of what one sees on both sides of the road is this: Where the rabi crop should have been 2 to 3 feet high, it can be measured in very few inches, mere blades in mute evidence of the catastrophe. The kharif crop was almost a total failure, unless a 70 to 80 percent loss can be described in milder terms. In a great many instances there was no crop at all to harvest, while the luckier farmers harvested a fifth or a tenth of the normally notoriously low yields. Mindful of that experience, they approached the rabi season very cautiously; about half of the land was not planted at all, and in the other half most of the seed was wasted. And let us stress that we are speaking not of pockets of failure but, with the exception of sugarcane along canal-irrigated land, of the overwhelming part of the cultivated 6 million acres of the three districts.

It is their misfortune that they lie in the so-called "rain shadow area" of the state where the rainfall is the least. True to form in particularly bad years, the "rain shadow area" may have produced much shadow but only a few showers in September when the rabi planting was under way—and nothing more since. The overall statistics of kharif production cited earlier do not apply to this region. During and after the trip it has not been easy to assimilate the fact of fields turned to wasteland, which neither man nor beast can use to advantage. Innumerable examples can be cited of the immediate effect of the rainfall playing truant, but one of them will suffice. Here is an instance of a big owner of 70 acres of land, of which 20 are irrigated in a normal year. He has three surface wells, two of which have been fitted with electric motors. But this year the wells are dry and for the time being the equipment is a useless piece of machinery. His rabi sowing is hardly off the ground and his kharif crop consists mostly of bajra stalks with not enough grain "even for a sparrow to feed on." The great majority of the cultivators have only 3 to 5 acre holdings, and their staying power is not worth talking about. We do not mention the farm laborers because they, like the normally well-off and not so well-off smaller cultivators, find themselves in the same predicament but mainly because they prefer the customary field work to the jobs they are doing now.

If an interested observer wishes to talk to farmers, it is best to look for them neither in the fields nor in the villages. This is not intended to leave the reader with the impression that most villages are deserted; only a few are on account of lack of drinking water and lack of employment. With no work in the field, there is no work in the villages either. The cultivators, the agricultural laborers (who, incidentally, comprise a large contingent), and the few craftsmen—all these are employed or are seeking employment on relief projects. Farmers do not always tell all; but there is enough evidence that, barring exceptions, they have already disposed or are in process of disposing of such marketable possessions as jewelry and ornaments, other finery, and brass utensils, inherited or acquired in better days, for the purchase of food or fodder. After land their next most precious possession is the livestock, and they are disposing of that at an increasing rate lest it perish for lack of fodder. At this point and going only a few months back, not to speak of the hardships of 1970–71 and 1971–72, relief work employment is their only safeguard against utter starvation.

Relief projects

The government of Maharashtra and the collectors of the districts recognize what was so well summed up in the following statement: "The drought problem is human, economic and political, and essentially exposes the stark reality of poverty."[2] The only recourse to mitigate these problems is through public relief projects, and any criticism leveled against them is at the risk of total underestimation of the lifesaving role they play. In all of Maharashtra there were 6,600 "scarcity" relief projects and 4,000 "normal departmental" projects, or a total of 10,600, as of November 21, 1972. In the districts of Ahmednagar, Poona, and Sholapur, the respective figures were 642 and 117, 460 and 38, and 743 and 102. These projects naturally fall into such main categories as soil conservation; minor irrigation works; excavation of canals for major, medium, and minor irrigation projects; percolation tanks; contour bunding as part of soil conservation; road bunding; and currently a great many stone-breaking projects. An official list runs into twenty-two items, including the principal ones just mentioned. They need not be cited here, but, in brief, they run the gamut from priority projects included under the state or national plans, to priority for incompleted works undertaken during the two previous drought years, and on to measures relating to the fodder problem, the employment of students, and concessional examination fees, and finally to schemes for the distribution of "nutritive food." The latest (January 7) is a "relief gift" announced by Prime Minister Gandhi during her tour of one of the Maharashtra drought districts. This is the center's decision to build the Konkan railway and three other railway projects, all with an eye to creating permanent assets and providing sustained employment. With the number of drought-hit people seeking work anticipated at 4 million in March–April, one cannot but welcome public undertakings of this magnitude as part of the relief work program. All in all, almost nothing has been overlooked that can produce employment; much has been accomplished in terms of work completed, the number of people employed, and deaths from widespread starvation prevented.

It is not possible to quantify the work accomplished in the first half of 1972–73; neither are we in a position to judge the quality of this work or the overall progress made along these lines in the visited districts. Yet it is obvious that by far for the greater part the projects are not in the nature of digging holes and filling them in. During the past two years ending October 1972, the official figures for Maharashtra state are impressive: the area covered under soil conservation is 1.8 million acres; the number of completed irrigation and percolation tanks is 383; the road earthwork completed is 43,000 kilometers, of which roads at a "safe stage" span 34,000 kilometers. Even if the official figures err on the high side, we have seen enough that is very reassuring. Much else not mentioned here must also have been done; and this can be seen, if not measured, in the districts, particularly in the field of canal excavation. But for the drought and the imperative need to provide as many jobs as possible, the canal projects would not be as far advanced as they are now. A case in point is Kukdi Canal in the Ahmednagar district, which is reported to be five years ahead of schedule. Whether five years or less, on at least two occasions when visiting a relief project, the collector of Sholapur pointed with pride to work which would not otherwise have commenced until three years later. To a degree, the same is true of the huge Bhima irrigation project initiated in 1964 to irrigate 400,000 acres at an estimated cost of Rs60 crores. A visitor is inclined to accept the assurance of the chief engineer that, under the impact of the drought, the work is proceeding at a faster pace, immediately creating more employment. The same goes for a number of less ambitious irrigation schemes, some of which were late in coming despite the ample warnings of recent years. This is where the lessons of the past should have been given greater consideration, and this is where preventive therapy leaves much to be desired.

Not all of the 2,100 projects of Poona, Sholapur, and Ahmednagar are productive in terms of lasting benefits; some of the earthwork for road building or road repairs and the numerous "metal projects," which stand for stone

2. "Drought a Trigger For Economic Opportunity," *Hindustan Times*, vol. 49, no. 346.

breaking, are of questionable value. A well-known writer on rural affairs pays unstinting tribute to the projects, but he also introduces a caveat worth quoting:

> The real problem, however, lies ahead. Whereas the number requiring employment can be expected to double or triple, the scope for increased earthworks is limited. In the hard-core drought-zone districts most of the easier tank sites have been used and a major part of the bunding program has been completed . . . The additional employment will largely have to be provided by quarrying and metal breaking. Unless this scheme is implemented with great care, it could very easily result in the production of huge quantities of useless rubble at unwanted sites.[3]

The relief projects do not fall into the category of "integrated" rural development, so dear to the heart of the planner. Granting the beneficial effect of projects conceived as part of a wider and continuing program of economic opportunity, the fact is that the projects in being are keenly short of supervisory technical staff and considerable organization effort; and many of them are not carefully thought out. Visiting any of the bigger projects, one is surprised that so much is being done, though for the reasons mentioned, not always with the care and forethought that would imply quality and integration. Besides, there is another telling reason why, in conditions of a galloping crisis, what should be done cannot be done. The collectors who shoulder the burden of the ongoing projects and are pressed to improvise new crash programs soundly view them as job generators above all else. The collector of Sholapur said something to the visitor with which all other collectors would heartily agree: "You see those boulders in the mountains or wherever they may be—I'll break every one of them in order to keep the people employed." Many observers are particularly critical about this relief measure. Thus, "metal-breaking was once a form of hard labor for convicts under rigorous imprisonment"; and, to clinch the argument, "the British used it as a means of massive employment in the famine of 1875." The argument further runs that "if adequate measures had been taken, the tragedy wouldn't have reached present proportions," and, by extension, stone-breaking would have been only a memory.[4] The validity of this contention is discussed elsewhere; but, given the existing situation, it is difficult to quarrel with a collector's under-pressure approach.

It cannot be stressed often enough that the vast majority of the rural community in these districts cannot and will not be able to make a living from agriculture, surely not until the kharif crop of June–July 1973 and just possibly not till September 1974. The reasoning behind this lies in the prevalent pessimism among farmers and local officials about monsoon prospects in 1973. If the latter variant (September 1974) is to be obviated, the 1973 monsoon would have to be exceptionally good to bring parched soil alive after three successive droughts. This subject came in for much discussion; and, to our knowledge, there are no bettors on a bountiful nature in the coming year. This is crucial to the entire question of the scope and number of relief works which will ultimately determine the number of people provided with jobs as a means of survival.

Relief labor force

In recent months the number of project employed has been growing by leaps and bounds. In eleven readings between June 1, 1972, and November 15, 1972, average daily attendance has increased from 271,000 to 1,700,000. By December 21, the figure was over 2 million. The official estimates through March 1973 range between 3.5 and 4 million. This is for Maharashtra state. The story of the three districts is seemingly grimmer. If Sholapur is taken as an example, it appears that, out of a total rural population of 1,600,000 and a much smaller labor force, about 320,000 are employed on relief projects now; judging by the daily clamor for jobs, the projected figure for

3. Ashok Thapar, "Drought in Maharashtra," *Times of India* (December 7, 1972).

4. M. J. Akbar, "They Must not Die," *The Illustrated Weekly of India* (January 7, 1973).

April 1973 is around 500,000. Since there is no particular reason to question this figure, the local authorities estimate that 70 to 80 percent of the adult population will seek such employment. Ahmednagar presents roughly the same picture. In late November–early December the relief project employed 210,000 and the sugar refineries employed an additional 70,000. With the crushing season over in late February–early March, relief projects of one kind or another will have to take care of them as well. This apart, the collector emphasized that not all looking for jobs are getting them. By the early months of 1973, the collector envisaged an employment roll of 350,000 to 400,000, or between 60 to 70 percent of the adult population. To date, young people of 18 and below are not project employable; but there are persistent demands that the age limit be liberalized, allowing minors in the 14 to 18 age group to engage in relief work. If it does come to that, the numbers will grow apace. What stands out from the preceding and is easily confirmed visiting any of the larger projects is that two or three members of one family—and sometimes four—live off the projects. This is one of the major reasons why the collectors are constantly preoccupied with devising more projects to keep the pot boiling. One can't spend any time in the company of a collector without realizing that in the back of his mind is the problem of maintaining law and order, for which he is solely responsible. This political consideration cannot be divorced from the human and economic problems which reduce, in effect, to feeding the potential starving. Any lapse in employment, however short lived, must be carefully watched, for it could lead to the very circumstances the collectors are bent on preventing.

Wages and food rations

What then are the conditions of these employed which have so far prevented starvation of the bulk of the rural community, though serious undernourishment is prevalent? Most of them work directly for the management of the projects on a piece-work basis. But the system must contend with this limitation: no male worker can earn more than Rs2.5 per day, and no female more than Rs2. There is a growing demand for equal payment to men and women, but the basic complaints are about the wages actually earned and the delays in making payments on the due date. In practice, random samples in the course of brief stopovers reveal that the maximum rates are earned only by some, while an undetermined but large number of males earn much less than Rs2.5 and women less than Rs2; their wages can be as low as Rs1.50 and Rs1. This much is clear: the maximum is largely window dressing. There are, of course, exceptions. At a stone-breaking project, villagers boasted of Rs3 a day, though employed by a private contractor. As might be expected, in many of the projects created overnight, charges are not lacking about dubious practices of the supervisors in making out the pay slips, mostly for short paying them on the work completed. Additionally, certain groups of workers have to furnish their own hammers, picks, and crowbars and the rock blasters have to buy their own dynamite. There is a disgruntlement on both accounts. An official tried to regale us with a story that a project-employed family has done so well as to purchase some *tolas* of gold.[5] Actually, whatever the wage rate and whenever it is collected, it is used to purchase the food ration from a fair-price shop or in the open market.

The ration itself consists of 8 (rarely 9) kilograms of food per month, made up of 5 kilograms of wheat, 2 kilograms of rice, and 1 kilogram of jowar. In addition, a family must buy sugar, salt, cooking oil and pulses, chillies, vegetables, tea and kerosene, and meet other minimum, if basic, nonfood needs. To get an approximate idea of earnings in relation to the food they can purchase, as distinguished from the caloric intake, we assume a maximum of 26 working days per month, an average wage of Rs1.50, and three employed members per family (two males and one female). On this basis the monthly income per family in round figures comes to Rs117. The cost of the basic ration per family of five at official prices is about Rs35 and that of the supplementary food probably as much, since most of these items are expensive and are bought in the open market; as we shall point out elsewhere, a great

5. One tola = 11.66 grams.

part of the basic ration also comes from the costly open market. Theoretically, this would leave a family with Rs47 for all nonfood requirements. This would not be bad except for a number of considerations which seriously dilute our calculations.

In reality, twenty-six working days per month is a standard not attained by the majority; occasionally they have to wait between projects and sometimes they are too exhausted to appear on the job day after day. The latter is a factor emphasized by the chief minister in his plea with the central government for a free supply of high protein food at a cost of Rs7.5 per month per employee. He noted the lack of it in these words: "This has resulted in deterioration of their health and has further affected their physical capacity to work with the result that their output doesn't enable them to earn the subsistence rates of wages allowed at the relief works." The 8 kilograms ration is supposed to be supplied by the fair price shops at concessional prices; but this, too, is not common practice. It is an open secret that these shops are frequently short of supplies and that a good deal of the ration is brought at much higher, open market prices. Five members per family is far from the rule, and the wages earned often serve to sustain more than five. Nor can we ignore altogether the costly fodder expenditures by the many farmer workers who still maintain cattle. Clearly, the calculated Rs47 for "extras" is too high. The conflicting information as between the collectors and the farmer workers do not permit an exact answer; but it seems that not much, if anything, is left for nonfood spending. Though employed, they are below the poverty line and there is no room for spare cash and tales such as buying gold. Incidentally, the 50,000 Indian refugees from West Pakistan, following the Indo-Pakistan wars of a year ago, are doing better than the project employees of Maharashtra. If the published reports are at all correct, each family receives Rs150 per month, not counting tents, woolen blankets, jersies, and educational and medical services; and the forty-two camps in which they live are normally located in areas where water facilities are available.

Related to the food problem are two other questions of prime importance: the quality of the basic ration and its volume compared with the one officially prescribed. We touched on the latter in general terms, but something more specific needs to be said about it. In the company of a high official of the Maharashtra government, we looked at a lunch pail at one of the project sites, which was meant for six obviously related workers engaged in the hard job of blasting rock. The pail contained six *bhakris*, or a crude version of the better known *chapati*. Nothing else. The eye, whether tutored or untutored, could see there was enough to fill the stomachs of only two or, at the maximum, three workers. Neither the official nor this visitor felt like exchanging comments on its meagerness to avoid the obvious embarrassment that this particular ration was tantamount to undernourishment or semistarvation. This was not an exceptional case, for it can be said without equivocation that the ration as finally translated into food is deficient in calories and nutritive value. The then much talked-about drive to supplement the pay with a free supply of high protein food was understandable; and, though something like that had existed during the previous two drought years, in 1972 it was yet to make its appearance. In early 1973 such supplementary food is likely to come into the picture. The main complaint of the employed and the local authorities both is that the ration actually received is much smaller than the prescribed 8 kilograms. A newspaper headline reads: "Maharashtra's Food Needs Fully Met."[6] If so, this constituted a sudden and welcome change from the situation as it existed in early December.

If the information gathered is correct, the procedure for ordering foodgrains is something like this: A collector prepares his requirements on the basis of the official ration, but there is allegedly a wide gap between requirements, allotments, and the amount of food actually supplied. For example, Sholapur requirements were 12,000 tons of grain per month, the allotment was close to 8,000 tons, but the volume furnished was 6,000 tons. Ahmednagar's requirements of 15,000 tons dwindled to an allotment of 8,500 tons, while actual receipts in November were not quite 7,500 tons and

6. *The Statesman* (December 16, 1972).

Poona's 11,500 tons turned into 7,500 tons. The official explanation for much of this lies in transportation bottlenecks. Perhaps so, but only in part; the more likely explanation is the overall food shortage, the dwindling of the buffer stocks, and the stricter rationing among states, including Maharashtra.

The 8 kilograms per adult per month are a puzzle we haven't resolved. Putting two and two together, this basic ration means no more than 930 calories per day. If the requirements versus deliveries are as the collectors claim, it can't be much more than half of that promised. If for the sake of argument it is still 900 or so calories per day, and the supplementary food amounts to another 500 to 600 calories; the sum total is 1,400 or 1,500 calories, admittedly a totally inadequate caloric intake for a manual laborer. If the would-be improved extra food finally materializes, the theoretical grand total might reach 1,700 to 1,800 calories. This, too, is insufficient; and how insufficient becomes apparent when in India a male of 21 years and over requires a minimum caloric intake of 1,800, and additional 900 calories if he works six hours a day as these laborers do. The real difficulty as far as caloric intake is concerned lies in the assumption that the basic ration of 930 calories is indeed received; this is doubtful, for we do not believe that the collectors played games with us when they expounded on their problem of requirements versus deliveries. On top of this are the irregular deliveries; fair-price shops with short supplies; collectors constantly on the phone to Bombay with dire warnings of supplies at dangerously low levels; and, to repeat, many of the recipients doing unaccustomedly hard work. The real question, therefore, is not "scarcity" or "undernourishment"; this much nobody denies. The question is whether those same people we are talking about and their families, or a goodly number of them, are not subject to acute undernourishment or semistarvation, a term which is officially taboo.

"Gratuitous relief"

As regards relief, the government of Maharashtra is not totally unmindful of the old, infirm, disabled, and all others, including children, who cannot be employed. Hence the "gratuitous relief" is in the form of very limited cash doles. The dole rates per day are as follows:

Adult dependents	Rs1.00
Nonworking children (over 12 years but under 14 years)	Rs0.65
Children over 7 years (but under 12 years)	Rs0.45
Children under 7 years	Rs0.35

This by no means involves any significant coverage even of the hardest hit districts judging by the fact that by November 15, 1972, only 17,000 people were under this scheme in all of Maharashtra. This is a highly selective coverage and how selective may be judged by the following: In the admittedly very badly hit three districts we are discussing, with a total population of 5.5 million, dole-recipients numbered 6,200, 3,400, and a mere 2,000 in Sholapur. Since the latter has 955 villages, the number of dole recipients on a per village basis is obviously very low. The inevitable conclusion is that as a relief measure it is only a drop in the bucket, and whatever sustenance comes their way is derived almost entirely from those members of the family employed on relief projects.

Equality, relief expenditures

For all the official concern about socialism and equality in India and the measures supposedly leading to it, equality can be found only on the project sites generated by the drought. The latter, like a plague, is *the* effective leveler recognizing neither rich nor poor, farmer owner nor landless, "untouchable" nor "touchable." The sharp socioeconomic distinctions of the village in normal days disappear as all these groups compete for the Rs2.5 or Rs2 or less per day. This is the unforgettable experience of the venture into Maharashtra, and it can be easily tested on any of the projects. This we did by merely asking the assembled to state how much land they owned. The number of landless being large, the frequency of the replies that "I am landless" is not surprising. What is surprising and what immediately points to the immensity of the drought disaster are replies such as these: "I own 3 acres, 5 acres, 10 acres, 20 acres" and so forth. What crowned it all on one occasion was the reluctant admission of a laborer, prodded by the assembled, that he owned 80 acres of land. For him to break stones was a socially degrading and altogether traumatic experience. Nevertheless, there he was—and others like him, if with less land, who suffered during the previous two droughts but were finally overcome by the current one, the worst within memory. Listening to the account of the big owner, he had fared no better on his vast acreage than had the small owner on his holding. When we suggested that he might have sold off some of his land to spare himself the trouble of this type of manual labor, it was greeted with mild amusement all around. This was like underscoring our ignorance in not knowing that no one is either willing or can afford to buy land in prevailing conditions. To the city buyer such poorly endowed land is not a good investment even as a tax shelter.

Strange as it may seem at first glance, considering the usual social and economic conditions of the agricultural laborers, they are probably not much worse off now, except for the type of work many of them are currently engaged in and dislike. This is not difficult to understand. In the marginal rural areas represented by the three districts, the normal period of their agricultural employment is about 150 days per year or less. Moreover, while employment was not geared only to adult males, employment generally was restricted. Wages were abysmally low, and only what came in kind as supplementary wages enabled them to keep body and soul together. Much of this has been voiced by the landless on the project sites. Now, however, they enjoy certain advantages, tangible and intangible. Unlike the owners, they are not plagued by the psychological distress of being reduced to the lowest status of manual laborers; this has been their condition all along. Class distinction is not a criterion for project employment. So long as they are employed, they earn, pay for the rations, and eat and are undereating in roughly the same fashion as do most owners similarly employed. This is the nature of equality but not as a result of social justice practices or better income distribution in the wake of buoyant agricultural and industrial economies. Only poverty underscored by a disastrous drought has seen to that, and come a good crop the same old inequalities will reassert themselves.

So long as there are boulders to break, canals to dig, roads to be built, and other projects to provide employment and food rations, short though they may be, the situation can be controlled. All this on the proviso that the source of the mounting funds will not dry up. Whether this will or will not happen depends primarily on the willingness and the ability of the union government of India to supply the funds. Maharashtra couldn't do it on its own, especially considering the sums involved. By mid-1972 central government teams, after a survey of the situation, agreed to a relief grant of Rs20 crores. By November, the government of Maharashtra raised its demands by another Rs40 crores; on December 5 and looking beyond 1972, its total stated relief requests upon the central government till the end of March 1973 have risen to Rs150 crores, of which Rs74 crores have been granted, while the Central Relief Committee of Maharashtra has estimated the needed amount at about Rs300 crores. Granting that a hard-pressed state government may have exaggerated its requirements, it is nevertheless true that, in the light of existing and anticipated support needs, ever-rising drought relief funds seem to be the order of the day. Surely so, if the multitude of the formerly relatively better off and the endemic poor are not to be decimated by a crisis of unprecedented virulence.

Cattle, Fodder, and Drinking Water

The consequences of the drought have many ramifications for jobs, food, and survival. These have already been discussed, but two more remain to be related: shortages of fodder and drinking water. One cannot spend five minutes in the stricken region without becoming aware that both man and beast are profoundly affected by these two problems. It is pointless to judge which is the more serious; both are equally grave in their own way, and what follows is a detailed discussion of the scope of these problems and their consequences.

Disappearing cattle

In all three districts the cattle or the fodder problem is critical. These districts have not a touch about them of Punjab, Haryana, or western Uttar Pradesh where mechanization is supplanting animal power. A tractor is a rarity; in all of Maharashtra their number exceeds only 2,000; and of the more than 2 million plows, wooden ones account for three-fourths of the total. Here cattle and plow are the mainstays of all farm operations; and, as the cattle go, so goes the power to produce a crop. The point is that an undetermined but large part of the livestock population may vanish via starvation or throwaway sales to the slaughterhouses. Sholapur district may be cited as an example of the prevailing fodder situation, although Poona and Ahmednagar districts would do just as well. In late October the meager supplies were estimated at 25 percent of requirements, and these will not last much beyond December. During our visit in early December the fodder alarm bells were ringing all over the districts. Import prospects are dim; even if the government of Maharashtra had made an effort in that direction, it is questionable if it would have succeeded; the neighboring states with greater availability have virtually declared a ban on exports. Neighboring districts are in a precarious condition, and districts within Maharashtra with some fodder to spare are reluctant to part with it.

Rising fodder prices are as expected; and the anomaly is that, while sugarcane sells at Rs150 a ton, fodder sells at Rs300. More to the point is that 100 bundles of fodder sell at Rs120, and a good head of cattle requires at least 5 bundles for daily consumption at a cost of Rs6 per day. Few farmers can afford to spend that much—not when a working adult male earns Rs2 or less per day, most of it spent on buying food. Even if three in the family are engaged in such work, the total intake would fall short of feeding a pair of bullocks if they have to buy the fodder in the open market. None of this means that a traveler doesn't encounter numerous farmers carting loads of this and that pulled by cattle, especially government-provided grain supplies or sugarcane. These are the fortunate ones. To begin with, the irrigated sugarcane crop is only about 10 percent down; and, traditionally, numerous farmers with cattle are employed in the sugar refineries during the crushing season, lasting three to four months. Feeding the cattle is part of the compensation, but by February–March they will have to return to their villages and add to the ranks clamoring for jobs and searching for the unobtainable fodder.

Generally speaking and despite exceptions, the drought effect on cattle is about as follows: fodder is running out, keeping cattle is an increasing burden; hundreds of them are perishing daily; and the creation of special camps for communal feeding has become a much talked-about issue, the solution of which is very "iffy." As so much else about the drought and the 1973 outlook, the really difficult situation lies in the immediate future. The same holds for the cattle situation. For example, in Sholapur the total number of "useful" cattle (bullocks, cows, and buffaloes) is nearly half a million. By the end of December about 80 percent of the required fodder would have had to come from government sources, indigenous or imported. Assuming that such sources will be found, which is not likely, the government would have to spend Rs2,250,000 per day or a total of about Rs540 million until the next kharif crop comes—if it does. The collector all but threw his hands up at this prospect, not on account of the vast sum involved but because of the unavailability of fodder. Said he: "The problem is gigantic, and it is highly doubtful if so much fodder can be made available from any source."

Fortunately for the government exchequer and unfortunately for the farmers, this potential burden is being gradually eased by the abnormal attrition in the livestock population and by distress sales to the butchers. The story that a farmer sold in Bombay a pair of bullocks for a bunch of bananas is probably apocryphal, and yet it is not without a point. According to the evidence, the picture is not one of exchanging cattle for bananas but of abnormally low prices that would buy little food or little fodder. For instance, farmer X on one of the project sites informed the visitors that he had bought a bullock three years back for Rs500 and recently sold it for Rs75 for fear it would perish before much longer. Our notebook abounds in such purchase and sale figures as Rs600 and

Rs80; Rs900 (for a pair) and Rs120; Rs400 and Rs55; and, incredible as it may sound, Rs350 and Rs20. A visit to a weekly cattle market told the same story, if not worse. The influx was so large that the many-acred market expanse was flooded with cattle in anticipation of buyers. The sales were few; the butcher buyers were few in number, and feeble animals were selling for as little as Rs10 to 15. In the two other districts the situation was about the same.

These were clearly distress sales at giveaway prices. Not one of the sales was to another farmer; they were all consigned to slaughter-houses. Naturally, no farmer we talked to was happy about this arithmetic. Apart from the immediate loss, a farmer touched on the crux of the matter to the visible approval of the gathering when he delivered himself of this observation: "If the rains do come next kharif season, what shall we use for motive power?" This is a thought that bedevils the farmers; but, considering the circumstances, they are compelled to do the most expedient thing, fully recognizing that eight months later they will be just as heavily indebted and with no credit to speak of to enable them to buy a useful pair of bullocks. We encountered cynics arguing that the sales and the dying out of so many cattle is a blessing in disguise, practically a Malthusian solution to their excessive numbers. "The fewer the better" is based on the proposition that approximately half the livestock population is useless, merely competing with the farm community for food. Whether so or not, immediately the farmer is the loser and in the future he will be deprived of his motive power, however inefficient it may seem to the non-farmer.

The same cynics recognize that the better livestock must be saved before they become enfeebled and share the fate of those already doomed. This explains the hue and cry heard in all the visited districts about the urgency of establishing special camps for communal feeding in forest-grazing areas or other suitable locations. Without going into the conflicting figures on forest-grazing possibilities that emanate from a variety of sources, the chief minister of Maharashtra stated on December 21 that 150 camps had been established in the forest area of Vidarbha region where an estimated 200,000 cattle could be accommodated. This means that in the three districts where forests are conspicuous by their absence, cattle would have to be transported over long distances. This prospect has created a subsidiary issue: If the districts were to come forward with a maximum of 100,000 head of cattle, who was going to pay the Rs60 per small wagon transporting five animals? More important is the concern of the local administration that many "useless" cattle not worth preserving might find their way into the camps, whereas the farmers are reluctant to send their better stock to faraway camps over which they have no control. There are other related problems, some of which do not bode well for this scheme as a major way out of the difficulty.

In fairness, whether the forest camps will materialize or not, officials are deeply preoccupied with this problem of how to preserve as much of the livestock population as possible. Nevertheless, we left with the impression, perhaps a mistaken one, that as of the end of 1972 the contribution of the forest camps cannot be a significant one. The greater promise, though of relatively short duration, lies in the relatively nearby camps provided by the sugar refineries. We visited this type of camp in Ahmednagar where 2,000 animals were munching away on green fodder. This lot of cattle will be preserved. There are a number of similar camps, and the thirteen sugar refineries of the district are likely to take care of about 50,000 cattle during the crushing season; the search for the rare commodity elsewhere will then commence once again. We conclude, therefore, that the "save cattle" campaign can preserve in good condition only a relatively small part of the animal population—unless the forest-grazing idea is activated and the difficulties eliminated so that the farmers will readily agree to long-distance transportation. In that event, the prevailing talk that half of the livestock population may die of starvation or be consigned to the slaughterhouses might hopefully prove exaggerated. On the other hand, the drought being neutral, if the current state of affairs extends into the coming promiseless months, the districts in question will be sharply denuded of poor as well as of good stock. The above-quoted farmer may indeed have a devil of a time looking for cattle to pull his plow if the rains do come.

Drinking water

What fodder, or its shortage, is to cattle, the shortage of drinking water is to humans. This is one of the principal headaches of all concerned. On all occasions this problem stuck out as the sorest of thumbs. Malnutrition is bad enough, but inadequate drinking water and not infrequently the outright lack of it is a basic cause of at least some farmers pulling up stakes. And for a farmer to desert his village, his land, and such way of life as he has been accustomed to is an act borne of desperation. In the field it doesn't take much searching to become aware of the extent of this visitation. All one has to do is look into the wells. While we didn't look into all of the 70,000 wells in one of the districts, those we looked into were mostly as dry as a bone. There are the usual exceptions and long queues where women spend hours to get a bucket of water of doubtful purity—and not always getting it. With the aid of tankers, mostly donated by the oil companies, water is transported across considerable distances, primarily to the relief works, where it is rationed as if it were gold. This, incidentally, is where we saw our first tanker in operation. There, too, the water leaves much to be desired; but in distress of this magnitude, beggars cannot indeed be choosers. In early December with relatively cool weather, the situation was as described; worse to come will be the time when it begins to warm up, particularly during the months of March, April, and May. Measures like transportation of water by tankers and bullock carts and revitalization of some of the wells have been initiated, but so far the impact of these are not commensurate with the size of the problem. What these districts need —and Maharashtra as a whole—is high-speed drilling rigs which can dig a well to a depth of 200 feet or so in one day and, with luck, find water. With the help of WHO/UNICEF, during the past seventeen months, 430 bore wells have been drilled and nearly 80 percent of these wells have proved successful.

In his statement of December 21 the chief minister spoke of twenty-one rigs ordered from the United States, making a total of fifty-one rigs from domestic and foreign sources. As against this anticipation, in early December Maharashtra accounted for only fourteen rigs in operation. Of these, thirteen are capable of drilling bore wells of 4-inch diameter to the depth of 200 feet; the remaining one is capable of drilling a 6-inch diameter to a depth of 400 feet. The results of this last rig are especially gratifying. In the rocky geological conditions prevalent in Maharashtra, this is what induces the search for this type of equipment. In the meantime the drought-afflicted villages of the three districts have to make do with two and occasionally three rigs. The additional equipment will be of inestimable value, but for the time being it would be foolhardy not to recognize the manifestly inadequate supply of this type of equipment if we consider the following:[7]

> Out of nearly 35,665 inhabited villages in the State, the expectation is that at the beginning of the Fifth Plan, there would still be 19,000 villages needing attention. This would include villages with no adequate source of water within acceptable distance, villages which are susceptible to endemic water borne diseases and villages located in saline hilly tracts. In villages where bore well is the source, it is necessary that one successful bore be fitted with a hand-pump and another with a power-pump. In villages where the bore is not likely to yield or does not actually yield adequate water, it is proposed to undertake water supply schemes through the Public Health Engineering Organisation of the Irrigation and Power Department. In such cases, the water is lifted from a nearby perennial river or a stream is dammed across a seasonal flow source. In the difficult villages, piped water supply is proposed to be undertaken leaving the internal distribution by way of house connections to be provided by the Village Panchayats through their own funds if they so choose. While the exact financial implications of the programs cannot be indicated, it is expected that the total financial outlay

7. "Approach of Maharashtra State to the Fifth Five-Year Plan," (September 1972), pp. 12–13.

in respect of the rural water supply scheme during the Fifth Five Year Plan will be approximately Rs100 crores.

All of the above deals with drinking water. Whether it is exactly 19,000 villages or Rs100 crores more or less or whether the drinking water upon the completion of the fifth five-year plan will reach the consumer precisely as described or whether the Irrigation and Power Department will perform as expected—all these are issues which need not be quibbled about at the moment. The villages of the districts provide supporting evidence for what one might view as an alarming statement. There is additional evidence of a different sort. We stated at the outset that individual drought tragedies will be eschewed. Nevertheless, they cannot be altogether avoided. Upon our return to Bombay from the districts, we encountered the unusual that others have also been encountering lately, namely, a group of farmer beggars. Answers followed questions, the primary burden of which was not only the search for work but also to quench the thirst. As one of them put it: "We are used to hunger; we are not expecting to get food-paying jobs in Bombay, but we are still getting water here." The taps in the hotels still run full, but the water supply of Bombay itself is below normal; and the refugee farmers, as beggars or manual laborers, living in open encampments or under makeshift tents, may not have escaped the water pinch or other deprivations.

It would be a mistake, however, to leave the impression that the district administrators are not doing their best to ease this problem. We have been repeatedly impressed with their spoken and unspoken concern. But the facilities at their disposal are negligible. Having failed to heed the warnings of the past years, the drinking water situation, as so much else about the drought and how to deal with it, may be summed up, at least in part, as follows. Rural Maharashtra and certainly our districts are like a city on fire, the fire brigades trying to put out fires beyond their capacity in innumerable nooks and corners. It helps to explain without condoning why the actual expenditure for drinking water relief in the three districts was so small that we hesitate to cite it. In the midst of all the known and unknown, only one thing can be said with fair certainty: Early 1973 will not yield much drinking water; it will probably yield more misery before the cup runneth over.

Refugees

We mentioned refugees in passing, but something more should be said in view of the plethora of dramatic reports on the subject. Figures on the number of refugees are hard to come by. Depending upon the source, there are few, not so few, and the possibility of a great many of them not many months hence. A news item of December 20 reports 2,500 in Poona city and 1,500 in Bombay according to the information of the Maharashtra government, but estimates of the Municipal Corporation of Greater Bombay speak of 25,000. To cap it all, the chief of the Poona division asserted that if the rains continue unkind, the number of refugees from his six districts may eventually reach a high of 1,250,000. Only this much is obvious: There are refugees, their number is on the increase, and their prospects for employment in the cities are not promising. In search for any work to avoid begging, there have been cases of wages driven down, which doesn't sit well with the manual laborers of Bombay on whose territory the refugees encroach. Whether in smaller or greater numbers, they remind one of the East Pakistani refugees but with this difference: The latter were provided with the basics of food, water, and shelter. The native ones are on their own and at this early stage depend mainly on private, not always well-financed, charitable organizations and such funds as are provided by the Municipal Corporation of Bombay. If the prediction of the chief of the Poona division comes to pass, only a national effort could see them through an altogether novel experience. There are reports that Bombay authorities are trying to arrest the trickle of refugees lest it swell into a flood. The main inducement to attain this are promises of jobs where they came from. The obvious reasons to stop the refugee movement are the additional strains and stresses this would create in an already overcrowded city like Bombay. Whether the coming months will lend themselves to a remedy of this kind is a matter no one can foretell. But regardless of the number and future developments, at this juncture we are reminded of an old Russion saying that nobody runs away from a good life. This is surely true of the refugees.

Scarcity or superscarcity

Throughout, the operative word was "scarcity." Maharashtra, other drought states, and the central government have preferred to drop the word famine "as a thing of the past," substituting "scarcity" for it. They are right, for famine means starvation and death on a large scale. This is not the present case, whereas the Bengal famine of 1943 meant just that. Nevertheless, the word "scarcity" calls for delineation since it is used for a very wide range of phenomena. Scarcity may mean that crops are adversely affected and that the yield is less than 50 percent of normal expectations. Yet, it may also mean what has happened in these three districts: intense hardship where there are hardly any crops, a totally inadequate supply of fodder, and a serious shortage of drinking water. It is scarcity in the latter sense that has hit the rain-shadow areas of Maharashtra for the third successive year. The debates in the Lok Sabha on a variant of this very subject are worth mentioning. The opposition's claim that the reported deaths in some scarcity areas were due to "starvation" was countered by the government with the argument that malnutrition *and* disease did them in. This argument partakes more of semantics than substance. The fact is that since the Bengal horror of 1943 which claimed some three million lives, the specter of classical famine and death has disappeared from India. Better distribution and timely inputs are responsible for that, and the successful solution of the Bihar food crisis in 1965–66 was an example. Crop failure nowadays manifests itself more pointedly by depriving large numbers of people of sufficient purchasing power to buy the grain which might be available elsewhere. This is what has happened in the three districts; hence the need for relief works. Once they are created in sufficient numbers, people are employed, wages are paid, and food rations are available—the worst of a starvation threat is over, although malnutrition as distinct from starvation persists. Therefore, whether it should properly be called "starvation" or "malnutrition" or only "scarcity" is not a useful preoccupation. What is important is that "scarcity" as it is known in the region discussed here leads to debilitation and fatalities through illness or epidemics which in themselves are a product of food shortages. Finally, whatever the label, any of these designations are not only a drought phenomenon but also a manifestation of the underlying poverty problem. We need hardly recall that even in normal times most of the people in the three districts manage only a hand-to-mouth existence.

Will peasants rebel?

Considering the prevailing conditions in the three districts, it is a source of wonderment that the afflicted take it without overt protests, let alone without resorting to violence. There was such a case with four persons killed when under the leadership of one of the Communist parties a large group of demonstrators stormed an official's office protesting against sharply rising prices. A few cases of "misbehavior" of a milder sort are not unknown; this was illustrated by a recent riot (January 5) in Chandrapur district where relief workers went on a rampage demanding a flat payment of Rs2.5 per day per man and Rs2 per woman. Apart from such instances, perhaps the only real protest is registered by those who take the extreme step of deserting their villages to search for jobs elsewhere. On the whole, one is mystified by the lack of widespread militancy on the part of the low wage employed and especially on the part of the many who are not employed at all. This observer couldn't help but cogitate about the causes explaining it, spinning out all kinds of theories about age-old customs, culture, religion, caste, and subservience. But in order not to get bogged down in the complexities of each mentioned item and recalling something of Indian history, let us just say that the mystery resides in a traditional rural nonmilitancy. This is not to overlook the fact that causes for violent outbreaks have not been lacking both before and after independence. Thus, in 1895 Maharashtra was the center of the well-known Deccan riots, the cause being fantastically usurious money rates, default, ensuing loss of land, and pauperization—hence totally unexpected riots.

However, although traditionalism persists, it

is undeniable that times have also changed. The government of Maharashtra receives practically no revenue from the stricken villages—and nothing much is said or done about it. Apart from the fact that you can't get blood out of a turnip, no local administrator can afford to trifle with the mood of a depressed, hungry, and angry community—except at grave risk. The rural cooperatives are in shambles, for it is no longer a question of 30 or 40 percent overdues but one of no loan repayment whatsoever. Ditto for the moneylender, but neither the coops nor the moneylenders nor the government would dare to press their demands to the point of taking over the land or indulge in even less drastic measures. There are other factors, one of them a negative one, which also tend to keep unrest within bounds. Unlike urban labor, there is practically no organized farm movement to speak of to press for higher wages, more ample rations, better-stocked fair-price shops, and lower food prices. Finally, the jobs of many are at marginal reward levels; but they do help to blunt the cutting edge of discontent.

Yet the picture is mixed and one of unease, which is another aspect of changing times with their greater political consciousness and promises of better days ahead. On project sites one is repeatedly struck by a sense of misery rather than a prevailing violence in the air, as one observer contends. This is not to argue with the writer of the most graphic of articles on the subject to date.[8] We would add only that, if misery topped by anger could be harvested, the local authorities would have to deal with something much more potent than a mere sense of unease. The collectors are very much alive to this prospect; we haven't met one who didn't express his fear along these lines in the looming, crucially difficult months of 1973. A collector has enough chores to fill a legal-size sheet of paper, but as we observed him in action he is no longer the collector even as of two or three years ago. What he applies himself to above all is the maintenance of law and order, not so much via police force as through the creation of projects yielding employment, and beating the drums for a greater availability of food supplies. It is the combination of all these elements which probably explains why the ample combustible material for violent action has so far not been ignited. Having said this, it is equally true that given a repetition of 1972 in 1973, we wouldn't bet a paisa that all the cited explanations would again suffice to keep Poona, Ahmednagar, and Sholapur in hand.

Conclusion

Little remains to be said by way of conclusion, but, if custom dictates one, a few items may be touched upon, mostly in repetition. First and foremost, the drought in Maharashtra does not lend itself to exaggeration. As Gertrude Stein would have said, a disaster is a disaster is a disaster is a disaster. "Scarcity" is not the word for it, even if only 15 million out of a rural population of 35 million—as against the official 20 million—depend to a larger or smaller extent on the government for their sustenance. The scale of the assistance effort of the Maharashtra government is undeniable as is that of the central government. One may cavil about the amounts allocated, the inadequate food deliveries, and so on; but surely at this point neither the state nor the central governments can be charged with negligence or disinterestedness. Considering with what the district collectors have to make do, their performance on the firing line has been altogether praiseworthy. A number of the projects they initiated or inherited will be of lasting importance. The often heard outside complaint about too many stone-breaking projects has little merit; chipping stone six hours a day is a backbreaking job, but it does stand between starvation and survival. What stands out is that the collectors correctly perceived that what matters most at the moment is not integrated agricultural development for which there is neither the time nor the technical-organizational arrangements but feeding the otherwise starving with the limited resources at their command. The year 1972 might have turned into an old-fashioned famine but for the measures taken. It might have led to a breakdown of law and order. That this has not happened so far is a measure of

8. M. J. Akbar, op. cit.

the farmers' essential political nonmilitancy and of the local administration's preventive assistance programs. This much on the plus side in the midst of extreme privations.

On the negative side are meager wages, meager food rations, meager caloric intake bordering in many cases on semistarvation, and the woeful fodder and drinking water problems. Much of this has resulted from an act of God, yet Maharashtra has had ample warning since 1961–62 (and before); some of the drought misfortunes could have been minimized if only the lessons of the past had not gone unheeded. As in the case of Andhra Pradesh, Maharashtra practices shock therapy now in lieu of the preventive therapy that should have been undertaken a good while back. For all the kilometers of roads built, percolation tanks repaired, irrigation facilities initiated, and soil conservation accomplished, it can be argued that the reputedly well-administered state of Maharashtra was late in recognizing the trouble it has been storing up for itself and, above all, for its people. The altogether too small supply of badly needed high-powered boring rigs points this up. There is hardly any excuse for the whole of Maharashtra to have had only fourteen of them and to be only now anxiously looking for another fifty or more. With adequate foresight this need might have been met years ago. Equally true is the failure to judge in good time the scope of the unprecedented crop failure and to make the state's demands on the central government accordingly. Maharashtra has behaved in this respect not unlike officials who look late in the day for grain imports in high-priced markets depleted of their surpluses. The same may be said about the permissive attitude of Maharashtra toward the inadequate buffer stock deliveries. Only relatively recently, when a rapidly deepening food crisis in the districts was there for all to see, did Maharashtra firmly stand up for an increasing flow of food supplies and for funds more in consonance with its rising needs.

The tragedy of the afflicted millions is that all this is not water over a dam, for the first half of 1973 is breeding social and economic distress and tensions possibly more virulent than those of 1972. Neither is it certain that the rest of 1973 will yield anything better. Banking on the continued quiescence of the poor and degraded is not the better part of wisdom. It is, of course, possible that a good all-India crop and a resurgent industrial economy in 1973–74 could make it easier to fill the current wide deficiencies with which the districts are studded. If that is not in the cards, an explosive situation will continue in the making so long as semistarvation and farmer refugees migrating to the cities continue to be facts of life. If, therefore, the lid is to be kept on and while hesitating to proffer advice, it should be the business of the Maharashtra government to tell its people that the essentials for a reasonably adequate survival will be provided where they are needed, come what may. This would be a promise worth keeping. For all too long promises have been a game played with impunity; but it is not unlikely that the next time around, poor rations, understocked or empty fair price shops, or inadequate labor rewards will lead to the very upheavals the collectors are so eager to avoid. The traditionally nonmilitant farmers may take a leaf out of the Deccan riots and settle accounts in their own way.

To revert to the drought in India as a whole, its treatment leaves a good deal to be desired. From late summer on it was no secret that the drought was a widespread calamity gravely affecting farmer and nonfarmer. And yet for a time there was hesitation and a lack of candor in coming to grips with it as a national issue. The implied excuse that panic had to be avoided did not prevent rapidly rising prices and a squeeze on all consumers, above all on the poor. An adequacy of food imports should have been assured when their effectiveness would have been greatest—and at a lesser cost. The picture may well change next year. The upcoming rabi crop is said to be promising; the cycle of four good monsoon years and one poor one is likely to be in for a renewal, and 1973–74 may demonstrate once again India's unquestioned agricultural potential. But none of this can erase for some time the memory of this year's experience or that of the Maharashtra's year of troubles.

Something else should be added. The suffering and degradation of millions in Maharashtra and elsewhere due to scarcity or "superscarcity" cannot be ascribed entirely to nature's caprice. Such occurrences are a familiar Indian story, but to blame it all on acts of God while treating famines as an "outmoded concept" is a contradiction that flies in the face of reality. That reality, in the light of what it takes to combat droughts five minutes before or five minutes after twelve, raises the question of developmental policies and priorities. Maharashtra and other states are good examples of what we have in mind. By the time the state is done with the 1972–73 abomination, it will probably have spent a good share of what it might have taken to create a protective irrigation system in the traditionally afflicted areas. If such funds had been used for this purpose throughout the past decade, they would have made a salutory difference.

Apart from what we have said thus far, we have no recommendations to make. Maharashtra is a well-governed state not lacking in talented administrators and capable technical experts. They know what has gone wrong and why. It is idle, therefore, to urge upon them to be "innovative," "integrative," to do "the right thing at the right time," and so forth. In fact, we could recite a dozen "should be dones," but this would be neither a balm to this concerned observer nor an aid to the overworked officials who have written them out in the first place. On the face of it, there is seemingly nothing the matter with them—if only timely implementation could be brought into play. This old hat is known to all and sundry, and there is no point breaking into open doors. If a recommendation is to be made, it is the obvious one: It is time that the lessons of the past droughts be assimilated and acted upon in terms of resource allocation to minimize their scope and consequences. And with this go a couple of questions which relate not only to Maharashtra: Is it not time to give thought to the questionable usefulness of crop farming in admittedly difficult rainfed areas? Would it not be more beneficial, for example, if a large part of the three districts or similarly affected areas in other states were converted mainly to livestock grazing demanding smaller investments but arguably more useful than what goes under the name of "irrigation" in drought-prone areas? Though not a novel idea, much needs to be done before answers are fitted to such questions. Whether they will be forthcoming or not, it is well to keep in mind the "normal" situation: very low yields on the one hand and the constant threat of a monsoon failure on the other. Joined with this is the fact that the vast sums spent on relief in the past were for the most part only temporary palliatives incapable of neutralizing or minimizing the onslaught of a new drought.

Lasting improvements or changes in whatever form lie in the future, at best. Immediately and in regard to Maharashtra, we conclude on nothing more cheerful than is dictated by the observations recorded here. Facts and figures are "stubborn things," as a famous political leader asserted; and Maharashtra, particularly the three districts, abounds in them. It is doubtful if more refined evidence gathered at greater leisure would yield anything more encouraging about the current state of affairs. As to the shape of things to come, the picture is both clear and obscure. No one can speak with certainty about the kharif crop of 1973, but the five or six months preceding planting will be more difficult than in the preceding year unless all tribulations are swept away through extraordinary relief measures. The chances of a good 1973 rabi crop are no better or no worse than those of the kharif crop. Both are unknowns and are entitled to nothing better than a fifty-fifty quantification. In sum, the government of Maharashtra cannot exclude the possibility of facing more of the same as the new year gets under way. After all, there is no guarantee that Indra, the God of rain, will not choose to play truant for the fourth time in a row; and it is against this that the government of Maharashtra must be prepared.

59. The Rural Scene

Ladejinsky's contribution to the World Bank's 1973 report on India's economic situation and prospects reviewed the nationwide impact of the previous year's drought and assessed the long-term agricultural outlook, in addition to dealing with the themes pursued in his previous annual reports. Omitted here is his final updating of the progress and failings of the rural poverty programs, already dealt with in some detail in his 1972 report.

The Drought

IN THE ANNALS of Indian agriculture, the year 1972–73 will be remembered primarily for its crop failure and for the new round of land ceilings legislation, the latest in a series of efforts to deal with the most difficult issues of agrarian reform. But overall, where 70 percent of the people depend for their living on farming, it is the decline of the kharif (summer) crop by 9 to 10 million tons which looms largest in assessing the agricultural scene. The principal sufferers are obviously the farmers of the stricken areas, but the repercussions are much wider. The nearly 20 percent increase in foodgrain prices between January 1972 and January 1973 has had a depressing effect on the economy in general and more immediately on the consumers, particularly the overwhelming majority of the poor. In a country where 40 to 50 percent of the people live below an austerely defined poverty line and where 50 to 60 percent of the family budget is devoted to food, the deleterious consequences of the drought need not be labored. This apart, the drought has given rise to much unsubstantiated speculation about the inadequate potential of the agricultural economy of the country and, more especially, about the "failure" of the Green Revolution. All that need be said here by way of a preface is that, while the tragedy of the afflicted is real beyond doubt, taking the long view the gloom emanating from the drought on these two accounts is not warranted. The fact is that the Indian agricultural economy has not been stagnating, and there is reason to believe that it will make further strides in the future. These, then, are the main themes of this year's account of the agricultural scene. Other issues, notably rural poverty and rural credit, are omitted. In these regards and unless the increased volume of credit be considered the touchstone of performance, no significant improvements worth reporting have taken place in the course of the year; and last year's detailed examination of these problems should suffice for the time being.

Only a year ago, in conditions well below those of the record crop already attained, India still seemed on the threshold of food self-sufficiency. Judging by the rising trend of grain production—an overall increase of 19 million tons between 1964–65 and 1970–71—there was reason to believe that the country was beginning to emerge from the era of food shortages and food imports. The officially pronounced policy in early 1972 of no concessional food imports and the flight of unofficial fancy that the country might even export grain in the near future were part of the picture suggesting that India was turning the corner. The only variable omitted in all these anticipations was the behavior of the monsoon. This year, after three favorable ones and two only moderately so, it has played truant in many parts of the country, bringing in its wake severe food shortages, sharply mounting food prices throughout the country, and grave privations.

The statistical and interpretative picture of agricultural production in India is presented in the previous chapter [omitted]; but by way of summary of the impact of the drought it suffices to say that this year's was one of the worst in recent memory, the monsoon rains having failed over a large part of the country. The precise number of people affected is difficult to estimate, but conservatively speaking tens of millions of people have been subjected to very inadequate food rations. To this must be added shortages of drinking water and shortages of fodder, which in turn consigned a large number of the cattle population to the slaughterhouses or death from starvation. This calamity was not as countrywide as some observers claim, but it made its deepest inroads in the states of Maharashtra, Rajasthan, Andhra Pradesh, Gujarat, large pockets of Mysore, Madhya Pradesh, and segments of Uttar Pradesh. In the first-mentioned state alone, an estimated 20 million out of a rural population of 35 million suffered from the drought and will remember it for years to come.

Space does not permit even a fleeting description of the intense hardships borne by the stricken rural communities, and yet the 1972 drought did not develop into the classical famine with its accompanying horrors of mass mortality. This is explained by a number of factors, the principal one being the 9 million tons of buffer stocks. These made the difference between starvation and survival, even if in the midst of privation. Next in order of importance were the massive relief works generating employment. Once they are created in sufficient numbers and people employed, wages are paid and some form of food rations are provided by the buffer stock; the worst of starvation is averted, although debilitating malnutrition persists. This is India's case in 1972. Apart from the traumatic experience of farmers barely keeping body and soul together tending relief works instead of planting and harvesting crops, the material loss of failed crops and general pauperization of the drought-affected communities is enormous; and the farm income of some of the states will have been cut by half, more or less. The government of India has by now incurred emergency expenditures in one form or another to the extent of Rs4.4 billion, a burden it can ill afford. The burden would have been much greater but for very stringent control over the purse strings. This is exclusive of emergency food imports which are likely to absorb some U.S.$200 million. And there is yet another cost—the sharp depletion of the original reserve stocks before the procurement of the new rabi (winter) crop commences in May 1973. In the light of the government takeover of the wholesale trade of wheat and rice, the size of the rabi crop, that of the 1973–74 crop, the volume procured, and the quality of distribution will be crucial. The successful test will come if as a result of the trade takeover the government of India manages to procure 8 million tons of wheat as against 5 million tons a year ago. But as of the moment, India is likely to wind up its agricultural year ending in June with a total estimated foodgrain crop of close to 100 million tons and a reserve stock of about 4 million tons including imports—instead of 9 million tons in the preceding year. For a country like India with its unpredictable climatic vicissitudes and rising demands due to population growth, this is not a satisfactory position unless both the winter and summer crops of 1973 are good indeed. In that event, the current food gloom will have dissipated.

It is not far-fetched to hope that the latter might come to pass, and the 1972 visitation may be just another familiar passing phase typical of Indian agriculture. The cycle of four or five reasonably good monsoon years followed by a poor one will probably be in for a renewal, and 1973–74 may demonstrate once again India's unquestionable agricultural potential. But the suffering and degradations of millions reduced to meager relief work wages and meager food rations cannot be entirely attributed to nature's caprice. Such acts of God are an old Indian story, but their pernicious effects might have been tempered through timely preventive measures rather than through shock treatment after the event. As it is, the record shows that large sums are spent on drought relief without creating a significant residual to minimize the ill effects of the next drought. Mindful of this experience and the dire need to augment production, the government of India is not banking now only on a good monsoon. As an emergency measure it is providing the state governments with Rs1.5 billion to create additional minor irrigation facilities on some four million hectares of land. In addition, Rs1 billion of short-term input loans have been earmarked for this emergency. In this manner it is anticipated that some of the lost ground in crop production will be retrieved in 1973 while at the same time additional assets of a permanent nature will be created, thereby reducing the proneness of some of the affected areas to similar situations in the future. Much of this work is under way, but only time will tell the effectiveness of these measures induced by the drought.

Illusory Drought Consequences

In the midst of rampant pessimism generated by the drought, two of its by-products are the tendency to minimize the past achievements of the country's agricultural economy and, along with it, to treat the Green Revolution as if it were a nonevent. Neither position is valid. Precisely because 1972 was a "Year of Trouble," a brief look at the past record on both scores is in order. The fact that the overwhelming majority of the rural population still lives in abject poverty should not be taken to mean that what has been accomplished is in any sense insignificant. Even in conditions of traditional agriculture, between 1949–50 and 1964–65 the grain output of India increased from 55 to 89 million tons. That half of it was a result of expanded acreage and only half from higher productivity does not invalidate the achievement. Clearly, what stands out is that the rural economy was not one of stagnation even long before the advent of the new technology. The more so with its advent, though admittedly productivity has a long, long way to go before agriculture ceases to be a gamble on the monsoons.

The subject of the Green Revolution has been repeatedly discussed in this space on previous occasions; but, in the midst of decrying the effectiveness of the new agricultural strategy, a few indicators may usefully be cited once again. The outstanding one is that in 1970–71 India attained a record crop of 108 million tons; while wheat made the principal contribution, the new practices were making themselves felt in higher rice yields as well. Since this resulted from a combination of good climatic conditions and modern technology, a better measure of the contribution of the technology is the veritable revolution in the utilization of inputs. Fertilizer consumption ($N+P+K_2O$) increased from 306,000 metric tons in 1961 to 2,260,000 tons in 1970–71; during the same period the number of electric and diesel pump sets increased from 421,000 to over 2.4 million; the number of tube wells from 19,000 to 480,000, of tractors from 31,000 to 117,000, and the area under improved varieties from 1.9 million to 15.4 million hectares. All these helped to raise output beyond what could have been achieved even in favorable climatic conditions, lifting the trough of total grain production by approximately 20 million tons even in a poor year like 1972. To the extent that the Green Revolution makes for stability via irrigation and for higher yields via its other techniques, its role is indispensable in any attempt to modernize Indian agriculture. One should remember also that, as already indicated, the process of modernization has spread to other walks of life; and in at least a few regions the Green Revolution has grown into something bigger—a more complete rural revolution.

Much of this has tended to be forgotten in the traumatic events of 1972, and it is not surprising that the voices of the detractors of the Green Revolution have swelled into a chorus. At this juncture it is important to note not what the new technology failed to do but what it did accomplish in a brief span of time. To begin with and very importantly, there is the lion's share of the 9 million tons of buffer stocks which came from the authentic Green Revolution areas. Without this contribution India would have reaped starvation on an unprecedented scale. The procurement campaign of the summer crop now under way has fallen far short of the target. On the other hand, Punjab and Haryana, the two states where the new technology made its deepest penetration, are the best performers, accounting for nearly 50 percent of what has been procured. Significantly enough, they did it mostly in rice, which countrywide is in short supply this year. This is worth stressing because it points to a major and useful shift in the cropping pattern of these states, due primarily to the new farm practices. Punjab and Haryana have also had to contend with their climatic problems and power shortages, but their investments in irrigation and in other modern practices have sustained them rather well in an unfavorable season. All this is in relation to the immediate food crisis, and these facts of life and other contributions noted earlier cannot be overlooked. Moreover, the denigration of the new technology on the ground that it is uneven, selective, income-disparity prone, and so forth has nothing to do with the modern practices as such. The new techniques as techniques cannot be held responsible for a variety of imbalances such as poor credit, minimal extension service and inadequate farm labor wages, or for the fact that generally speaking economic necessity and social justice do not often ride in tandem. These are essentially man-made issues of long standing. More to the point is that in the course of a very few years the new technology has become the turning point for what agriculture has already achieved and for what it may attain in the future.

Green Revolution's Gainers and Losers

Accepting, therefore, that there is a technological breakthrough in certain areas of Indian agriculture, the question may be raised as to how beneficial this has proved to be so far to the big farmers, small farmers, tenants, and farm laborers. Numbers cannot be assigned either to the size of these groups or to their gains or losses in income terms. Nevertheless, within the areas affected by farm innovations, their respective positions can be ascertained if only in general terms. The point of departure for an assessment of this kind is the admittedly existing dualism in Indian agriculture: the small part of it (10 to 15 or more percent) that is irrigated, prospering, and progressive and the remainder that is dry, poor, and stagnant or what one writer calls the "two rural Indias."[1] Both are conditioned by geography, nature of land ownership, size of holdings, costly investments imposed by the new technology, and much else already touched upon in a preceding paragraph. The predominance of "dry farming" explains why vast parts of the country have not been touched by the Green Revolution. In that sea of farmsteads, unequal income distribution is not nearly as burning an issue as in the smaller areas with assured irrigation potential where big farmers with owned or borrowed capital have taken advantage of the combination of tube wells, improved seed, fertilizer, and tractors. In the process they struck it rich beyond their expectations; the rich became richer. And not only the rich have benefited. In states like Punjab and Haryana, numerous small farmers, though by no means most of them, with owned holdings of 5 acres and more of good land have also benefited from the adoption of new practices. This said, there is no doubt that the gains of the Green Revolution are distributed differently with different categories of farmers, putting the small entrepreneurs at a disadvantage.

In the very same hub of rural transformation, the meek and the humble among the farm owners, mostly holders of 5 acres and less, have been more or less bypassed. In theory the new technology is "neutral to scale," which means that it can be applied to any size of farm; and yields are a function only of inputs. In practice this does not often happen, the constraint being mainly the lack of resources. For all the expansion of cooperative credit in recent years, its distribution has always been weighted in favor of the well-to-do rather than the weaker sections of the farm community. Summing up the gainers or losers among all the farm owners in areas where the new practices predominate, the situation is about like this: (a) More or less all innovators have enjoyed considerable gains in real income; (b) within the innovating groups, proportionately larger gains have accrued to the bigger farmers with larger land and other resources; (c) the income disparities have greatly widened between the innovator and noninnovators; (d) even in the backward agricultural states with islands of improved technology, farmers have substantially benefited; and (e) unequal access to credit is the principal reason why in typical regions of agricultural modernization so many small farmers have fallen by the wayside.

As to whether in the decidedly innovative areas the landless farm laborers are gainers or losers, the picture *at this stage of technology* is tilted in their favor, even if only slightly. The new type of agriculture is labor intensive, employing more labor over a longer period of time in the production of a larger output.

1. Ashok Thapar, "The Challenge to India," *Times of India* (August 15, 1972).

While the cost of living has risen sharply, so have wages, with farm labor gaining some ground. But the technology is also potentially labor displacing, notably with the introduction of the combine thresher. With no mechanization policy to avert an unfavorable impact on rural employment in regions discussed here, in the years ahead the trend will shift from time-saving to labor-saving devices; and the number of displaced laborers can only increase. This movement may slacken its pace from time to time, as is the situation right now, but possibly sooner than later events might well begin to justify Nehru's lament against farm mechanization as a threat to the welfare of the landless farm hands. Assuming that as of now farm labor has not suffered from the impact of the Green Revolution, the outlook however is for an overcrowded, competitive low wage farm market regardless of the scope of the new technology.

Whether farm labor has gained, stood still, or, as others contend, lost some ground, the very important saving grace is the indirect multiplier effects on employment attributed to agricultural modernization. The new demand had created new opportunities for goods and services ranging from small-scale engineering industries and on through expanded activities among blacksmiths, carpenters, masons, brick makers, leather workers, utensil makers, transporters, wholesalers, retailers, and many others. This is the shock wave impact of increased production and the prosperity it generates. That many of the beneficiaries are nonfarmers only speaks in favor of the broader implications of the Green Revolution. In any assessment, therefore, of whether labor has gained or lost under the aegis of the new technology, these newly created employment opportunities must be taken into consideration.

If there is any segment of the rural population that has clearly lost out in the innovative process, it is the tenants or sharecroppers. This is particularly true in areas where agricultural transformation is a potent force. There, land values have risen sharply and so have rentals—from 50 percent to as high as 70 percent of the crop. Besides, unrestricted land control has never been more prized, and security of tenure and other rights in land a tenant might claim have been perceptibly weakened. The owners would like to get rid of tenants and resume the land for self-cultivation, making use of the plentiful supply of hired labor combined with a new array of modern equipment. The old practice, if not always successful, whereby tenants are reduced to sharecroppers and eventually to landless workers is being accelerated as more of the bigger owners shift to the new technology. If anything, they are trying to lease or buy more land to insure the fullest utilization of tractors, pump sets, power threshers, reapers, and so forth. The tenants, therefore, are the losers of the transformation from traditional to modern agriculture.

Agricultural Prospects

In the mid-1960s the authors of *Famine 1975!*[2] argued that India is doomed to starvation. Her agricultural policies being allegedly what they were, from the standpoint of food sufficiency she was past redemption. So much so, the authors contended that "If other more deserving countries are to be 'saved' (through food aid imports), India must be 'sacrificed'."[3] This dire prediction was followed by the introduction of the new package of farm practices and their well-known attendant consequences. If in 1972, however, the optimism of recent years has given way to concern once again, the explanation lies in the time lag to capitalize on scientific advances in a number of important foodgrain and cash crops and in the drought or in the obviously great impact which climatic conditions still have on the productivity of unirrigated semiarid areas which in better years contribute over 40 percent of the country's food supply. While all this does not minimize the progress made thus far, the eminent scientist Dr. D. M. Swaminathan calls attention to the fact that "it would be self-deception to believe that a scientific breakthrough will *automatically* (emphasis added) result in a production revolution and that a production revolution in agriculture will simultaneously result in

2. William Paddock and Paul Paddock, *Famine 1975!* (London: Weidenfeld and Nicolson, 1968).
3. Ibid., p. 218.

a prosperity advance."[4] In his view it would take concerted efforts in many directions (scientific, socioeconomic, educational, and "hard" political decisions) to bring that about. Meantime and in the light of the tasks ahead, the promising lessons of recent experience are the watering-down of the myths about the Indian peasant's inertia and fatalism on the one hand and that of technological stagnation on the other.

The size of the task of feeding the rapidly growing population of India is indeed immense. The plan targets for 1973–74 and 1978–79 are 115 million and 140 million tons respectively. Looking farther ahead, the relationship between population growth and food needs tells its own target stories, varying with the estimates presented below. The 1971 census revealed a population of 548 million, 80 percent of it living in rural areas. Total population is estimated to reach 657 million by 1981, 703 million by 1985, and 864 million by the year 2000. These estimates are conservative in the sense that the annual rates of growth are assumed to be 1.55 and 1.30 percent, respectively, during 1981–91 and 1991–2000. Since so favorable an impact of family planning programs is conjectural at best, by the year 2000 India may have to feed a population of 900 million to one billion. As against this, there are numbers of foodgrain requirement projections based on a variety of assumptions. According to the Planning Commission, by 1981 India would need 168 million tons of cereals and 19 million tons of pulses. The cereal estimates of the National Council of Applied Economic Research (NCAER) are lower by 30 million tons. Finally, there are the Working Group on Demand and Supply projections of the National Commission on Agriculture with an estimated requirement of 228 million tons of grain and pulses by A.D. 2000. Regardless of the acts of commission and omission in all the estimates, agriculture clearly has a long way to go to meet requirements of such dimensions. Additionally, milk and meat production and the fish catch must be sharply raised.

The foodgrain and cash crop expectations are based on a total cropped area of 164 million hectares, or a sown area of 139 million hectares, and a cropping intensity of 118 percent. The real hope lies in the irrigated acreage. As of recent date, the net and gross irrigated areas were 30 and 37 million hectares, respectively. Out of the net irrigated area, 12, 4, 11, and 3 million hectares were irrigated by canals, tanks, wells and tube wells, and other methods, respectively. The potential irrigated area is estimated by the Irrigation Commission at 82 million hectares, but the more immediate prospect is a total of 49 million hectares if all the projects now in various stages of execution are completed. The increased water availability presupposes vastly improved water management, soil research and conditions, and successful genetic manipulations of the yield potentials of various crops.

Despite the formidable enterprise that lies ahead, Dr. Swaminathan is optimistic about meeting the food needs of India. He is aware of the still prevailing low yields in grain production (except wheat), animal husbandry, and fisheries; he is equally aware of the gaps between what has been accomplished in the major crops in the experimental stations and in the field. Nevertheless, he concludes that "The targets are not frightening, considering the irrigated area we have and scope which exists for moisture conservation and integrated land-use planning." He does not discount the ever-present technical and scientific problems; but with regard to India's problem of problems, the dry land problem, the encouraging signs lie in the preliminary work of the All-India–Coordinated Dry Land Farming Research Project; the small-scale but successful Indo-Canadian dry farming project at Hayatnagar near Hyderabad; the Central Arid Research Institute at Jodhpur, Rajasthan, concerned with management of desert soils; and perhaps above all the newly created International Crop Research Institute for the Semi-Arid Tropics (ICRISAT) at Hyderabad. The thrust of all these is one of effective water harvesting and moisture conservation among small and marginal farmers, most of them subject to well-known handicaps apart from being "dry." Part of the effort is the development of new cropping strategies and the exploitation of "the resilience which tropical and subtropical agriculture . . . offers with regard to crop substitu-

4. M. S. Swaminathan, "Population and Food Supply," *Yojana* (January 26, 1973).

tion." All this entails a generation of work, but Dr. Swaminathan's conclusion is that, as far as scientific application is concerned, it "offers cheer and hope for striking a favorable balance between population growth and food supply at least till the end of this century, by which time we can expect a greater stability in population expansion."[5]

In trying to assess this optimistic picture, one must willy-nilly oscillate between favorable and unfavorable factors, hoping of course that the former will prevail. The following pages are an attempt to do just that. To deal with the negative side first, the outlined potential is viewed against the elements which tend to hinder and slow down the implementation of a desirable scheme. The optimistic anticipations have been predicated on appropriate policies and their implementation at all political, economic, and organizational levels, and most particularly at the district level, where the fate of any rural undertaking is decided. Past experience cannot exclude the well-known "implementation gap" even if the purpose and general design of a particular measure might have left little to be wished for. A number of agricultural endeavors like credit distribution, agrarian reforms, small farmer development agencies, drought-prone areas, and so forth point, among other things, to this problem. This gap cannot be eliminated overnight and hence the caveat about executing with dispatch the multifarious and admittedly difficult tasks that lie between anticipation and realization. The hindrances to implementation are particularly pronounced at the village level where the aim is not only increased production but balanced development as well and where the need for people's active participation in decision-making is indispensable. The record on this score is particularly poor though with many exceptions to the contrary. The village panchayats or local development bodies, having become largely political entities controlled by the well-to-do, whether for economic or social-caste reasons, have tended to minimize the developmental and other needs of the weaker sections, beginning with production and on through education, health, nutrition, family planning, water supply, roads, electrification, and the like. The success of a good many programs therefore depends upon the revitalization of the panchayats in the sense that responsibility for carrying out local programs involves initiative and action at the grass-roots level. The latter cannot be effected by fiat, for in good measure their success depends on the elimination of cultural barriers replete with emotional and psychological snags. A close observer of the rural scene can see welcome signs of change, but they are not sufficiently widespread to overcome these barriers and their persistence alone may well inhibit implementation even in the presence of such reliable factors as physical inputs and much else to go with them.

Yet, there is another side to the picture on the face of which the outlook need not be as agonizing as it can be made out to be. Renovative policies and actions to maximize productivity and employment in agriculture and in activities allied to it are not lacking. They have been differently cataloged by different people, but the critically important elements are these: (a) widening the range of economic activities to which the new and upcoming techniques are applicable; (b) development of rural industries and agriculture-allied activities such as livestock breeding, dairying, and forestry; (c) improving the infrastructure facilities for credit, extension, tenurial conditions, and marketing; and (d) underlying most of this, more effective and quicker exploitation of the country's water resources. An aid to these is the favorable position of the fifth plan toward agriculture. The latter is assured top priority with an outlay of Rs70 billion, or double that of the fourth plan. The promise of supplementary special funds should work to the same end. As to the emphasis on particular sectors of the agricultural economy, the "Approach to the Fifth Five Year Plan 1974–79" notes that, while the strategy which fathered the Green Revolution "will continue to receive attention" . . . it will be necessary to place stress on agricultural programs in rainfed and other less promising areas." Useful though this may be, general propositions of this sort and generous funding do not always make for effective programs and only performance will tell the story. There are, however, a number of other factors, which also favor higher productivity and greater rural employment.

5. Ibid.

Science may well play, eventually, the role ascribed to it. What is often in the experimental station is not always within the grasp of the farmer, but technical innovations are now more closely related to the ultimate beneficiaries than they were not many years back. The new technology reflects it. The times when the peasantry had to be persuaded to shift to a higher productivity plateau have, for the greater part, changed. The lesson derived from the Green Revolution that better farming stands for better living is now known among the "outsiders" as well. This year's drought experience cannot but lead to a greater appreciation that vulnerable farm areas must be made less so. The measures to modernize the rural economy are not novel; in one way or another most of them, though with gaping holes, are either dealt with or are under active consideration. The country is not lacking in capable, ambitious, and energetic bureaucrats who may yet find the organizational keys to open many a door. And once again, with a stress on greater productivity only, if the additional 15 to 20 million irrigated hectares which are expected to enter the picture upon the completion of projects already undertaken are indeed realized, the volume of output is bound to increase significantly. This is not to speak of an expected overall rise in production from a rural economy in general acquiring muscle. None of this implies meeting all the targets or the dogma of self-sufficiency. The latter in particular need not be treated as sacrosanct so long as agriculture and the national economy as a whole are buoyant.[6] Last but not least, with the notable exception of agrarian reforms where slogans set the tone, most of the other objectives noted above are conceptually realistic; if they fail of successful resolution, it will be only because science, administrative-organizational arrangements, and policies to unleash rural energies fall below the mark. This cannot be excluded, but the time is fast approaching when failure will be inadmissible. The reasons are twofold. The future, like the past and present, holds out hardly any promise that the country's industrial and commercial development can absorb significant numbers of the village poor. Only within agriculture with heightened productivity, greater employment, and improved tenurial relations will one have to look for signs of rural betterment.[7] Failure, therefore, can be contemplated only at the risk of worse things to come; in that event, all or most of the favorable assumptions recited thus far would be writ in water.

Summing up, in an India which is changing, however slowly, it cannot be assumed that the debilitating elements are there to stay forevermore regardless of where they make themselves felt. Given the emergence of countervailing factors on top of all that already stands for progressive agriculture, it is not inconceivable that India may be able to feed its growing numbers—not always, to be sure, without the aid of imports—while making at the same time some dent in the mass poverty prevailing in the rural sector. Some of the favorable conditions cited may seem to the reader too intangible, more like an act of faith than well-supported evidence. This is conceded, but to one familiar with the Indian rural scene, especially with its compelling needs and aspirations which simply cannot be denied much longer, the economist's facts and figures must also make room for a measure of faith in a timely intervention of benevolent political, economic, and social forces. The immensity of the task of feeding India's rapidly rising population and ameliorating the conditions of the poor may indeed seem forbidding in a country with so complex and varied an agricultural economy and rural society, but who is there to say that the prospect is not feasible if the will is there?

Many of the observations recounted thus far lie in the future and are admittedly speculative. Less so is the case of two important issues,

6. In connection with higher production, we assume that it would not take all the farmers to achieve the objective and that social justice and equitable income distribution may not necessarily be one of its outcomes. This matter has been treated in preceding economic reports.

7. "At the present pace of industrialization, any mass-scale transfer of the labor force from agriculture to non-agricultural sectors is ruled out. The growing labor force in agriculture has to be provided with fuller employment within agriculture. A redistributive land reform derives its basic rationale from this consideration." "Approach to the Fifth Plan, 1974/79," p. 5.

now, as in the recent past, much in the limelight. One is the new land ceiling round; the other is the old and familiar Small Farmers' Development Agency and other allied schemes dealing with the rehabilitation of one of the poorer sections of the farm community.

New Land Ceiling Round

Not every subject dealt with in last year's annual agricultural review is repeated, but the new land ceiling round is one of the two that cannot be ignored. Until early 1972, ceilings on land ownership for redistribution among the landless had been a dead issue because all legislative enactments in the past decade or more were failures. But ever since the Congress Party's victories in 1971 and 1972, coupled with the promises of land to the needy, the ceiling issue has been coming to a head once again. After months of the bitterest political controversy in the history of Indian agrarian reform, the battle was resolved on July 22, 1972, when national guidelines of a new ceiling program were ushered in. In brief, the main provisions are these: (a) An owner can retain a maximum of 18 acres of double-cropped perennially irrigated land or 27 acres of single-cropped land or 54 acres of all other types of land; (b) the ceiling applies to the family as a unit of five rather than, as in the past, to individual holders; (c) "major" children (over 18 years) are treated as separate units, and each one of them is entitled to hold land not exceeding the governing ceiling for a family of five; (d) January 24, 1971, is the effective (retroactive) date for the purpose of ceiling implementation; finally, (e) compensation for the surplus land to be distributed "should be fixed (by the states) well below the market value of the property so that it is within the paying capacity of the new allottees . . . and in a manner that it will be no financial burden on the central and state governments."

To date most states have enacted new ceiling legislation, but it is highly questionable if they will be or can be implemented in a way that would result in a large pool of newly surplus land earmarked for distribution. An examination of the guidelines outlined above and the absence of an administrative organization to implement the provisions lead to this conclusion. On the face of it, the guidelines for the new ceiling round are better than what had preceded them for the following reasons: The drastic curtailment of permissible land retention, acceptance of the family unit as a base for ceiling determination, curtailment of illegal land transactions by retrospective application of the ceiling, and the low price of land to the intended recipients. For all this and for other considerations, the effort appears to be on the plus side. But there is another, less promising, side to the story; a few examples will suffice.

The retrospective date of January 24, 1971, is not a serious deterrent to illegal land transactions. In practice, the land illegally transferred cannot be recovered. The effectiveness of a family as the holding unit has been gravely eroded by splitting the family unit into two parts, both of which are entitled to separate ceiling retentions; the net result will be a decline in potential land availability. Compensation for the acquired land might, in effect, border on confiscation; and as economic men the owners with surplus land can be expected to behave accordingly—that is, resist the implementation of the ceiling reform in every possible way. They are great practitioners of this art, as evidenced by the failure of all previous ceiling enactments. The most basic provision of the new legislation deals with "perennially irrigated land or irrigated land capable of growing two crops." The controlling word is "perennial," which means water all the year round or at least ten months of the year. It implies a sufficiency of water for two crops, and this is not common even in the best-irrigated state of Punjab. It will take some proving that each irrigated acre is perennially irrigated, a feat very difficult to perform. But even if the enumerated handicaps did not exist, implementation would be very difficult if not impossible because of the absence of an organizational structure and an administrative set-up. This was raised by way of a question at the very highest level of government, but there is so far no evidence of any answer to meet it. The chances are that, as in the past, responsibility for the brunt of the effort will continue to rest with the revenue departments. Their primary responsibility is to collect revenue rather than attend to very complex reform chores. More difficult is the situation at local administrative levels where the reform must be enforced if it is to be enforced at all. Apart from political and caste considerations which motivate local officials in their essentially anti-reform attitudes, there is the additional and well-known drawback once again called to attention at the highest level of government, namely, that many states do not have the basic land records and the collateral information necessary for the implementation of an effective land ceiling policy. This explains the anachronism that, while state after state continues to enact new ceiling laws, hardly any of them has taken the trouble to put their records in order and to find out how much surplus land might be available for distribution.

Much else could be added to this list of problems standing in the way of well-formulated laws and enforcement. For the moment, let it be noted that the new ceiling round will in the end be judged not by the levels at which ceilings are fixed but by how much or little land they will generate for redistribution. The range of permissible retention currently prescribed is at most half of that of the former ceiling limits. The presumption is that it should yield a much greater surplus than the actual distribution of a million-plus acres of poor land during the first ceiling round, as against a variously estimated anticipation ranging between 37 to 40 million acres. This presumption, however, has little validity, for the lessons of the past experience amply demonstrate that such estimates and fulfillment are worlds apart. There are occasional state news items which suggest rather low surplus land expectations. This is neither surprising nor encouraging. If these straws in the wind are indicative of a trend and leaving out of consideration waste land vested in the states, the surplus is not likely to be many times more than that secured under the past ceiling enactments. Moreover, in most instances, the land that finds its way into the hands of future beneficiaries will be relatively poor marginal land.

None of the above inveighs against a ceiling program. Any meaningful reform without it is a misnomer, while its presence should be one of the main guarantees of the success of an agrarian reform in its broadest sense. The difficulty with the new ceiling round is the agonizing negativism implicit in it; ditto when the ceiling idea assumes the aspect of a cure-all without the requisite underpinnings in sight. In these circumstances, an exaggerated priority is attached to the land ceiling in the absence of equal concern about such fundamentals as the absence of the recorded rights of tenants, the largely absent security of tenure, high rentals, the very questionable practice of land resumption for "personal cultivation," sharecropping and its invidious effects, exploitative farm-labor wages in most instances, and a number of other items adversely affecting the weaker rural sections. All these are vital matters and are part of a minimum agrarian reform which should be part and parcel of a ceiling program. The separation of the two, as if they were not interrelated, does not auger well for either one and certainly not for the unfinished business relating to the issues just raised. The prospect therefore is for diminishing the ranks of tenants and sharecroppers and adding to the ranks of agricultural laborers, a condition to be avoided at all costs.

* * *

60. *Agrarian Reform à la Punjab*

Toward the end of June 1973, Indian newspapers broke the story of a scandal in the distribution of state-owned lands in the Punjab, where many officials grabbed, at low prices, land intended for the poor. This is Ladejinsky's vivid report of September 17 to the World Bank of that incident. ". . . if there is any lesson to be learned from this episode," he wrote, "it is the same old lesson: The implementation of agrarian reforms in India cannot be carried out so long as many in authoritative positions are essentially antireformist in outlook and practice." "For one fairly familiar with the shenanigans that go into writing and yet unwriting land reform programs, the Punjab caper need shock no one; it is just a 'wrinkle,' a grotesque one to be sure, in the long chain of evasions which all lead to the familiar end . . . another unwelcome reminder of the enduring malaise of Indian agrarian reformism." Could a commentary be more succinct, more telling, than this!

Introduction

IN THE HISTORY OF AGRARIAN reform in India, taking away land from the poor farmers by the rich farmers is an old story carried over into postindependence reform days. The notorious "voluntary surrenders" of land by tenants in favor of their owners, squeezing the tenants off the land as a by-product of land consolidation or loss of any hold on land by the poor through a variety of other dubious measures—all these and much else have not helped the reforms to prosper. But it remained for Punjab (and probably not only Punjab) to add another wrinkle, namely, grabbing *state-owned* land intended for the landless and harijans—the poorest of the poor—by prominent politicians and government officials at low prices. How such land passed into such hands improperly is still to be determined, but the fact itself is not disputed. When the news broke in late June, the press greeted it with such eye-catching headlines as "Exposure of Land Grab in Punjab," "Illegal Land Deals in Punjab: Top Public Men involved," "Many Punjab VIPs accused of Illegal Land Grab," "Punjab Ministers, Speaker Indicted for Land Grab," "When Thieves Fall Out," "Search for More Land-Grabbers," and so on. This was not a storm in a teacup or an event blown out of proportion by the press. Land was indeed grabbed by a number of high and mighty through collusive transactions, misuse of official power, and intimidation of the poor, and connived in by those who are supposed to uphold the law. This note tells the story in some detail; and, if there is any lesson to be learned from this episode, it is the same old lesson: The implementation of agrarian reforms in India cannot be carried out so long as many in authoritative positions are essentially antireformist in outlook and practice.

Historical Background

It is difficult to say which of the Indian states deserves mention as the worst reform performer. With very few exceptions they were equally remiss in this respect, but a bit of history should make clear why Punjab is one of the top recalcitrants and why an opportunity to grab land earmarked for the poor becomes Punjab.

Prior to the enactment of the Punjab Security of Land Farmers Act of 1953, the number of tenants or semitenants was estimated at 583,000. The act provided for a number of measures to strengthen the position of the tenantry, including a reduction in rent from fifty-fifty to a maximum of one-third of the crop.

But this provision, like security of tenure, ceilings on holdings, and so forth had been so emasculated that the very title of the act is one of the misnomers of the agrarian reform of India. The result was twofold: large-scale ejection of tenants by owners on the ground of land resumption for self-cultivation and the persistence of traditional rental arrangements on the remaining tenanted holdings. No data on ejections are available, but the following tells something of its magnitude, though the data are probably only approximations. Thus we read:

> The law regarding security of tenure appears to be on the whole somewhat illusory. This is supported by the fact that the number of tenants in existence in the Punjab area at the time when the Punjab Security of Land Tenure Act was amended in 1955 has greatly decreased. From the information supplied by the State Government on March 30, 1955, it appears that the total number of tenants including tenants-cum-owners was 583,400 . . . From the information supplied by the Punjab Government recently (1964) it appears that the total number of tenants who were holding land at the commencement of the act and are still holding land as tenants is 80,250. It seems difficult to account for the large difference between those two figures.[1]

Actually, it is not so difficult. It is an open secret that the difference between 563,400 and 80,250 is largely due to out-and-out ejections, "voluntary" surrenders, or the transfer of tenants to the status of sharecroppers or agricultural laborers. Since 1964, particularly since the advance of the Green Revolution, the situation is nothing if not worse; what little tenanted land remains, rentals have gone up to 70 to 80 percent in favor of the owners and security of tenure is but pious hope. This is the background before and after the agricultural technological advances of Punjab, and it is not surprising that in this kind of a reform climate the episode about to be recounted can find a congenial place there.

1. Planning Commission, "Implementation of Land Reforms" (August 1966), p. 118.

The Evacuee Land Episode

The genesis of the evacuee land episode goes back to the partition of the country and the creation of Pakistan. After the rehabilitation of the refugees from Pakistan on land left behind by the Punjabi Muslim refugees, the government of India inherited a surplus of evacuee land. On June 3, 1961, the union government transferred this land of varied quality to the state government of Punjab. The acreage and price of such land were as follows: 80,000 acres at Rs495 per acre, 100,000 acres as Rs100 per acre, and 111,000 acres at Rs5 per acre. This came to be known as the Package Deal of 1961; but prior to that on February 2, 1960, the central government also transferred to Punjab 47,000 acres of "inferior evacuee land" at Rs5 per acre and 36,000 acres of a slightly better grade at Rs100 per acre—in all, 374,000 acres. In transferring this land at prices considerably below their market value, it was understood that it would be eventually distributed among the generally landless and landless harijans.

For almost four years the Punjab government had made no move about the newly acquired land. When it finally bestirred itself into action in 1964, the land was not distributed among the two mentioned groups; instead, an undetermined acreage was finding its way into the hands of "unauthorized" persons. This did not go unobserved and, as complaints about such land transfers kept on mounting, on July 4, 1972, the governor of Punjab, with the approval of the government of Punjab, issued an executive order appointing an Enquiry Committee "to probe into the setting up of sizable agricultural farms on evacuee land by officers, their relatives, and other influential public men. The immediate result of that order was the setting up of the Harchard Singh Committee under the chairmanship of Harchard Singh, representing the Congress Party in the state assembly. Of the six other members, all of the state assembly, four represented Congress, one was the leader of the opposition, and one member of the Communist Party (India). The committee's terms of reference were: to find out how the land was acquired—whether through auction, illegal deals, or by unauthorized occupation—the extent and value of the acquisition, the role of the officials concerned, losses sustained by the state, and so forth. The committee did not have an easy time delving into these problems, and the results of the inquiry are far from complete. This is explained by the fact that the only useful data were provided by the rehabilitation department of the Punjab Kisan Sabha (peasant organization); the other sources tapped by the committee remained totally unresponsive. The public notice failed to evince any interest.[2] A paltry thirty complaints, including two oral complaints, were received, out of which hardly eight were relevant. Similarly, the legislators and most of the political organizations remained indifferent.[3] This may explain why the committee has not heard "the officers, public men, and others against whom complaints had been received." It also served as a basis for the accusation leveled against the committee later on that it engaged in an ex parte investigation. Notwithstanding these limitations, the committee gathered enough information to justify its effort. When after much delay its findings were made public on June 26 of this year, it created a political storm in Punjab. In a congress-ruled state this affair added no laurels to the party, and, as will be pointed out elsewhere, at least one eminent member of congress turned into a source of considerable embarrassment for the congress high command. It also provided extra slingfuls of arrows for all those who view with profound cynism the activities of the ruling elite.

Two reasons explain the political storm and scandal. The first one is the evidence of flagrant misuse of authority and probity in disposing of the evacuee land. The second and immediately attention-calling reason is the all-star cast of characters who participated in these questionable deals. The committee named such dramatis personae as the state agricultural minister, number two cabinet member, and a spoken-of candidate for chief ministership; the assembly speaker; the minister of development; the parliamentary secretary; Congress MPs; members and former members of the Punjab State Legislative Assembly; and a former Akala party minister. Even the most radical spokesmen for the downtrodden could not withstand temptation, for prominent among the accused is a leader of the Communist Party (Marxist), Punjab education secretary, some senior police officials, and several provincial civil service officers. In every instance cited land was acquired in the names of a retinue of relatives, obliging tenants, and occasionally in the names of equally obliging landless or harijans. These bogus participants were to cover up the fact that this reform in reverse was indulged in by members of the very elite of Punjab society, individuals in commanding political positions, lawmakers, upholders of law, or molders of public opinion. Coupled with this is that these pillars of society came into landed properties not meant for them not only cheaply but mostly with money borrowed from government agencies on favorable terms.

The committee examined 126 cases of which 61 were given a clean bill of health, mainly because the individuals involved purchased land up to the prescribed limit of 5 standard[4] or 10 ordinary acres and whose total holdings fell short of 17.5 acres; this was the limit adopted by the committee for a "sizable" farm. Among the buyers there were some scheduled castes (harijans) government employees who were permitted to purchase land at "restricted auctions." The latter entitled the purchaser to borrow money from the revenue department, repaying it on an installment basis. As a matter of fact, this privilege was enjoyed by the rich purchasers as well. The number of scheduled caste employee buyers was small, but they comported themselves no better than the elite buyers. They, too, followed the crowd, buying land in several names, real or fictitious; and their farms were also "mostly one-man shows... Nearly all of them have also become real landlords and even behave as such, with the result that the local population is more sick of them and their behavior.[5]

2. The committee invited complaints from the public and political organizations through a notice published three times in all leading English, Urdu, Hindi, and Punjabi newspapers of the state after an interval of 15 days.

3. "The Harchard Singh Committee Report," p. 15.

4. A "standard acre" is the measure against which comparable values can be assigned to different classes of land with different deficiencies and productivities.

5. "The Harchard Singh Committee Report," p. 18.

The committee's attention centered, however, not so much on these buyers as on the prominent members of the officialdom and politicians with appetites for large holdings at low prices. A few cases illustrate the point. Mr. A and his relatives acquired 250 acres at a cost of Rs44,590, the current value of that land being Rs998,000; Mr. B and his relatives acquired 200 acres at a cost of Rs25,310, current value of that land being Rs716,000; Mr. C and his relatives acquired 190 acres at a cost of Rs37,318, current value estimated at Rs752,000; and to cite a relatively minor acquisition, Mr. D and his relatives acquired 62 acres at Rs15,000, current estimated value being Rs250,000. The extent to which relatives come into the picture is shown in one of the cases examined by the committee; it is possibly an extreme but nevertheless illuminating variation on the same theme. The real buyer and owner, a member of the Punjab civil service, acquired 13 acres; but at the same time he also acquired land in the names of his wife, father, brother, sister-in-law, son of sister-in-law, brother-in-law, son of brother-in-law, mother-in-law, servant, aunt, father-in-law, or sister-in-law, and, finally, another sister-in-law. Altogether he bought in his name and in the name of twelve others 129 acres at a cost of Rs60,000 at the current price estimated at Rs515,000. This case has many ramifications too complicated to unravel even for a Philadelphia lawyer. This much, however, is worth mentioning, as just another example of what a difficult time agrarian reformism has in Punjab—and in other states. Only with one exception do the purchasers reside on the land. For the management of this land, the owner organized a cooperative society, "The Partap Cooperative Farm Society Ltd.," presided over by him. This society is pure fiction; it couldn't be legally registered for the simple reason that its membership was made up of a motley crowd of nonfarm resident shopkeepers, government employees, and whatnot. The best that can be said for it is that it is probably no worse than many other so-called farm cooperative societies created solely to evade the tenancy and ceiling provisions of the agrarian reform.

This land, as the land in all other instances similarly acquired, was of poor grade and had risen in value partly because the original price was very low, partly due to improvements, and perhaps above all due to the general rise of land values throughout Punjab. The issue, however, is not the increase in the value of such properties between then (1964–66) and six to seven years later. The issue is that the purchasers had no right to buy such "sizeable" acreage and that the state government of Punjab disregarded the original purpose of the evacuee land. The method of acquiring the land was largely through open auction, although it didn't differ in essence from the "restricted auction." Every auction was subject to a "reserve" price fixed by the government of the last two years' average, but as the committee pointed out "the 'reserve' prices fixed by the government were very low. These prices were fixed by them as far back as 1961–62 and have hitherto remained unchanged despite the sharp rise in the prices of land all over the state." In theory, the poor harijan could compete for the auctioned land as well as the well-off politicians and the well-placed officials. Predictably, this was not the practice because:

> The above sales suffered from numerous defects (many of a criminal nature) . . . These persons, men with means as they are, had an edge over the poor tillers, who couldn't mobilize requisite resources to compete with them. Even otherwise, they had the goodwill and patronage of the auctioning officers on their side. As a result, the poor and needy persons, depending upon agriculture alone, had been deprived of the opportunity of purchasing these lands and such actual landless tillers of land as were occupying the same, of course unauthorizedly, were uprooted. This was a tragic outcome of the policy of open auctions. It frustrated the cherished ideology of "land for landless tillers."[6]

The acquisitions, as noted earlier, were made largely under a system adopted in 1964. The above quotation makes quite clear what was wrong with the sales. The obvious, too, should be added: the familiar pattern of ill-considered governmental policy, administrative bungling, and misuse of authority and influence by politicians and civil servants. For ex-

6. Ibid., pp. 18 and 71.

ample, the basic instruction relating to the sale of land makes no reference to the original intention to turn such land over for the benefit of the landless. This was a crucial omission which hardly calls for comment. Nor did the instruction clarify that only the head of a family could purchase land, debarring all immediate family members, distant relatives, and all others willing or induced to help augment the purchases of the head of the family; hence the subversion of the 5-acre limit. Bidding by proxy was not permitted but practiced just the same; auctions were not competitive, which is to say that they were generally rigged. Many of the acquisitioners also violated the provision that they could not resell, transfer, or mortgage the purchased land till the final payment of loans, with interest, or till the expiry of ten years from the date of purchase, whichever was later. Actually, a number of them disregarded this provision, disposed of the land at exorbitant prices, and in the process "make huge profits by making meager or no investment at all." Not all of them struck it as rich as the father of an important politician. He is reported to have bought 500 acres of evacuee land at Rs25,000 and subsequently sold a third of it at Rs300,000. Not unexpectedly, the process of purchase and sale was often accompanied by the eviction of tenants or small farmers who were squatters on the land. One will never know the number of such dispossessed, but in some districts the eviction complaints "were of a serious nature."

Apart from the acquiring of land through auctions, some of the evacuee land was occupied by persons without authorization to do so. The occupants were of two types: the landless squatting on a few acres and people of high official and public standing "who had occupied large chunks of land and had set up semimechanized farms." Of ten such cases listed by the committee, the unauthorized occupation ranges from a high of 296 to a low of 26 acres, and most of them paying the government neither rent nor revenue. The committee was particularly ired by the occupants of the 296 acres cultivating the land for the past six to seven years "without payment of even a single penny to the government." As in the case of auction buying, unauthorized occupation on a fairly large scale is concealed in the name of numerous individuals. In the instance discussed here, they numbered twenty-six. The actual occupiers were two, one of them superintendent of police; the names of all other alleged occupants entered into the revenue records "were fake." This particular occupation has led to a great deal of opposition on the part of the local population and a few murders. It is for this reason that the cultivation of this land partakes of a demonstration of strong-arm measures. Thus, "At the time of inspection 176 acres of land were under wheat. The committee was told that A and B bring about 100 persons armed with deadly weapons for sowing the crop every year so that there should be no opposition from the local population. Similarly, they bring this many men for harvesting the crop." It is this picture that moved the committee to this comment:

> These are the officers and public men who have taken the law in their own hands; misused their authority and influence and have occupied large areas by terrorising the local population, forcibly evicting the existing tenants and occupants, in collusion with police and field revenues staff. The local population is so scared of them, particularly of the officers, that they were not willing to tell the truth to the members of the Committee when they inspected the farms set up by some of them. But it should not be difficult for the Government to find out the correct position, provided probe in each case is entrusted to an upright and nonsparing officer.[7]

The patwaris, or village chiefs, abetted this or any other category of land grabbers by falsifying the village records. It is the committee's considered view that "the most unscrupulous accomplice" of illegal land acquisitions was the village patwari. This is a very familiar comment, for what holds for Punjab is equally true of the dubious activities of many patwaris in other states. In the light of the above, the words of the committee that "Most of the officers and public men had exploited these inherent defects to the best of their advantage" are not surprising. Instructions from the reha-

7. Ibid., pp. 24–25.

bilitation department issued in September 1966 to eliminate some of the defects came too late, for the land acquisitions boom was tapering off. Additionally, the committee makes it quite clear that the instructions would have been of no avail because they could have been easily evaded with the help of local authorities. This is, of course, not new; nor are the kind of deals exposed by the committee uniquely Punjabi. Given the structure of Indian society, particularly in the rural areas, the rich and influential can get away with a great deal of conceivable misdemeanor.

Committee Recommendations

Proceeding from the general to the particular, the committee's limited probe leaves no doubt about "a great deal of conceivable misdemeanor" committed in Punjab in disposing of the evacuee land. This in turn raises three questions: Will the accused individuals be compelled to part with their ill-gotten properties? Will the officials who connived in the irregular procedures be penalized for their conduct? Third, will the committee's effort as a whole have a salutory effect on legislative enactments bearing on land problems, enforcement, and the general conduct of politicians and high and low officials inevitably involved in such issues? Recent press news items provide a political reminder that the committee report may not fare well and similarly the answers one might have anticipated to the raised questions. But before these matters are touched upon, it might be well to summarize the tentative answers to the questions in conjunction with the principal committee recommendations and the nature of the committee itself.

As to the first question, the committee already rendered its judgment by recommending that the government of Punjab should set aside many of the sales or in a number of instances the defective purchases be brought to the attention of the government for further consideration. It also recommended disciplinary measures for the individuals who participated in assisting the owners in acquiring those properties. But the committee is not a judicial body, and it is a reasonable presumption that the recommendation which hurt the owners most will be tested in courts of law. At best these will be protracted affairs, and one shouldn't bet his bottom dollar that the courts would necessarily uphold the committee's recommendations. The committee is cognizant of this and recommends instead the framing of suitable legislation "which could bring the past cases within its ambit." The answer to this could be "yes" or "no," the reason being that there are indications that political considerations are already inducing the government of Punjab to deal with these recommendations in its own far less drastic way than the committee would. As to the second question, the answer is anything but clear. Although a couple of officials have elected to resign since the publication of the report, it doesn't follow that others similarly involved will act accordingly or that their professional standing will be seriously affected. If nothing else, severe penalties could very well upset the customary norm of a goodly number of public officials doing business with the public. Besides, the unexpected political exoneration of three important personages accused by the committee as land grabbers would appear to favor the relatively subordinate officials in a final accounting. For these reasons, the recommendation that the government of Punjab should "evolve a suitable quasi-judicial hierarchy with statutory powers" to deal with the offending officials doesn't seem to carry much force. The answer to the third question hinges on the validity of the answers to the first two. It is true that the immediate effect of the release of the committee report created a sense of profound unease among all manner of politicians, important officials, and the "neo-feudalists," as the report refers to them. But little more than a month later it begins to look as if they might somehow weather the storm, and political and professional conduct hardly changed. Of course, the government of Punjab professes to pursue action against tainted officials and nonofficials to the "logical end," but its outcome is one of uncertainty.

In the language of the committee, "the government should become sadder and wiser from the mistakes committed in the past." This reference is above all with an eye to the disposal of the remaining evacuee land, which the committee estimates at 150,000 acres or so. This figure is questionable because the committee dealt with only a limited number of individuals who bought land at various auctions or occupied land without authorization. The total acreage thus sold or occupied is yet to be determined, following which the remainder can be flushed out. But the main point of it all is that a fairly sizable acreage is available for distribution and that only the poorest landless should be entitled to it. For this purpose, the committee recommends abolition of the auction system, distribution by allotment, at the same time reducing the 5 standard or 10 ordinary allocation rate so as to benefit a greater number of applicants depending on agriculture for their living. To the same end, the committee would like to see all the unauthorized land occupations by officials and rich and influential persons retrieved, thereby augmenting the land pool earmarked for distribution. The same should apply to the land that might be retrieved through the "de-grabbing" process. Mindful of the importance of correct record of rights in land, the obvious recommendation is to correct the wrongs committed by the patwaris. The committee takes note of the fact that there are many more cases of illegal acquisition of land than it had been able to uncover. It therefore urges the government of Punjab "to take suitable steps to evolve a proper machinery to detect all such cases and to deal with the same mercilessly." Some of the recommendations call for legislative enactments and some for direct action of the government of Punjab. The government is reportedly setting a "suitable machinery" to "process" the committee's report and for "follow-up action"; what this might yield remains to be seen. Running ahead of the fast developing final aspect of the episode, it is not certain that the government of Punjab has set its heart on a degrabbing process. It is not excluded that, under the impact of events described here, the remainder of the evacuee land will be distributed among the needy; but what had taken place so far is surely not the first demonstration of Punjab's unconcern about land for the poor.

Politics and the Land Grab

This skepticism is based on fairly solid political evidence. Among the land grabbers cited by the committee are the minister of agriculture, the parliamentary secretary, and the general secretary of the Punjab congress. All three resigned their offices immediately upon the publication of the committee report. Since then the chief minister exonerated them of any wrong-doing without subjecting their cases to a judicial review; in fact, he did so almost before the ink of the report was dry. Some observers contend that he acted upon higher advice on the alleged ground that no person should be penalized for taking advantage of the flaws in a land policy. Apocryphal or not, rehabilitation was granted and the land the committee claims they acquired improperly will probably remain with them. Another likely consequence is that, if the principals are given a clean chit, why should the book be thrown at some of the subordinate officials who were doing only the bidding of their superiors?

But it is the speaker of the assembly, accused by the committee of buying 90 acres of evacuee land in the name of his relatives and occupying without authorization an additional 10 acres of land, who became the real politico-dramatic element in this affair. He was neither willing to resign pleading not guilty, nor was the chief minister evidently in a position or willing to give him a clean bill of health. It remained for the congress high command, including the prime minister, the congress president, the foreign minister—most prominent in the inner circles in Punjab congress—and the home minister to settle the case of the speaker. The union minister of justice is reported to have examined the case, found the speaker's land purchases both improper and illegal, so reported to the high command, whereupon the home minister advised the speaker to resign for the good of congress unity and as a demonstration of high standard of public conduct. However, this was not to be, the speaker insisting on exoneration. And so to ease the interparty factional pressures and recriminations within the Punjab congress and to refurbish the party's tarnished image, a truly Solomon-like solution emerged: simultaneous exoneration *and* resignation. When asked of the home minister why the speaker should resign if he is exonerated, which is to say not guilty, the answer was that no legal questions were involved but only propriety;

and, as if to clear up matters altogether, he further stated that "certain things are for reasons not mentioned." Regardless of what lies behind this cryptic statement, the worst of the political crisis has been patched up. The committee report, limited though it is in scope as far as finding culprits is concerned, badly unnerved the leading landed members of the congress and Akali (opposition) parties. They are breathing easier now. Whatever the future political consequences, the immediate fallout of the political turmoil created by the land grab revelations can be dealt with at a more leisurely pace.

At this point it no longer matters whether the speaker has truly gotten a clean chit in exchange for his resignation or whether the three other participants cleared by the chief minister was a political act, an act of friendship or influence along the corridors of power, or an honestly considered view that they were not guilty as charged. What matters is the ambiguous position of the committee, particularly with respect either to de-grabbing of the land or administrative decisions the government might take in all cases the committee referred to it. The assertion by the home minister that the solution of the speaker's problem did not amount to a censure of the Harchard Singh Committee is not altogether reassuring. The fact is that the hostility of many members of the state assembly toward the committee is considerable, and many a noninvolved influential Punjabi probably shares this attitude. Nevertheless, politics and hostile attitudes aside, the committee served a good purpose in that the remainder of the evacuee land will be handled with greater care and circumspection. Whether its recommendations about setting aside many of the auction sales will be accepted by the government is less certain. True, this is said in the face of the chief minister's statement that his government was trying to determine whether to enact appropriate legislation or promulgate executive ordinances to take back the land which had allegedly been "grabbed." In the meantime, an official committee has been appointed to examine this question. Reassuringly, but it would be surprising if otherwise, the chief minister asserted that "neither the Center nor the Congress High Command had barred the State Government from taking steps to take back the land."[8] The final outcome is yet to be gauged, for promise and performance in respect to land issues have seldom ridden in tandem. If they do this time, they will put an end to the strong contention in certain circles that the effort of the Harchard Singh Committee is just another nonevent.

Conclusion

The results of the committee's probe raise yet another question: Is Punjab really unique in disposing of state-owned land that leaves so much to be desired? In the wake of the Punjab events, the land reform division of the Ministry of Agriculture has sent out a round robin inquiry to all states about the handling of their state-owned land. To date no reply has been received, and if and when received no state is likely to admit to any improprieties in disposing of it. Unofficially, land grabbing has been revealed in Uttar Pradesh, although not exactly as in Punjab. There the harijans received some land vested in the state, but the largest share of it is reported to have gravitated into the hands of those who needed it least.[9] Regardless what other states reveal or don't reveal, one cannot but agree with the following:

> Every state Government must, in the context of the Punjab Committee's report, put the following questions to themselves and find the true answers. How many influential persons in each state, including politicians, officials and businessmen and others, have bought big chunks of land over the years? How many of them have employed illegal or unfair methods for the purpose? To what extent have loopholes in the land ceiling laws been taken advantage of, obviously with the connivance of the bureaucracy? Was black money involved in any of these land deals, and if so, how much in each

8. "Legal Position about Land Grab under Study," *The Sunday Standard* (September 8, 1973).

9. S. C. Scala, "UP Harijans Deprived of Land Reform Benefits," *The Times of India* (September 13, 1973).

> state? Even if the technical requirements of the law of the land were satisfied, how many of such transactions conform to the norms of propriety? The last question is important, for, as the Punjab Committee has pointed out, the impropriety involved in politicians and bureaucrats purchasing land to grow rich at the cost of the poor cannot be overlooked, even if nothing illegal is found in the transactions.[10]

These questions generated by the Punjab episode have bearing on the general problem of agrarian reform in and out of Punjab. It has been pointed out earlier that Punjab is not known for its proreform proclivities. Essentially the same holds for the greater part of the country. Right now, under the pressure of the new ceiling round, Punjab like all other states is also in the act of fashioning some sort of a ceiling reform. It is predictable for reasons pointed out elsewhere[11] that substantial redistribution of land cannot be anticipated. This is the immediate prospect; and, if the causes are to be repeated, they stem from the character of the state assemblies, the activist lawmakers, the enforcers of the legislative enactments, and the elusive "political will." How then is an agrarian reform to take place, a reform however mild, to suit the palate of those who could share in minimal yet basic rights in their land, not to speak of actually parting with some? We have no answer to offer; only a sovereign state bent on carrying out its promises can provide a satisfactory answer.

For one fairly familiar with the shenanigans that go into writing and yet unwriting land reform programs, the Punjab caper need shock no one; it is just a "wrinkle," a grotesque one to be sure, in the long chain of evasions which all lead to the familiar end. Just the same, it is another unwelcome reminder of the enduring malaise of Indian agrarian reformism. But in the longer view, is the outlook indeed one of unrelieved gloom as summed up by the Urdu poet Momin—"What shall I do, oh God! Nothing is of any avail"? It need not be that desperate if one assumes that some remedial measures must inevitably come into their own for at least two reasons: (a) It would be a serious error of political judgment if the country's leadership thought that going back on the promises already made to the rural masses were possible, and (b) one can argue that nothing is forever even in timeless India. But for the moment, and with all conceivable allowances for past reform setbacks, there is little to cheer about either.

10. "Land Grab," *National Herald* (June 1973).
11. Wolf Ladejinsky, "New Ceiling Round and Implementation Prospects," *Economic and Political Weekly* (September 30, 1972).

61. Agrarian Reform in the Philippines

On the request of the U.S. ambassador to the Philippines, William Sullivan, President McNamara of the World Bank arranged for Ladejinsky to visit the Philippines and study the progress of their land reform. Ladejinsky's observations were reported in this letter of October 7, 1974, to Secretary Conrado F. Estrella of the Department of Agrarian Reform. One noteworthy feature is Ladejinsky's pragmatic recommendation that landlord opposition, arising partly out of opposition to the prescribed mode of payment for expropriated lands, be reduced by "sweetening" the terms of payment. "Unceremonious vigor," one of his favorite phrases in reference to implementation of agrarian reform, is here modified to "unceremonious vigor combined with conciliation." Ladejinsky was gratified by the positive reactions to his views expressed in a personal interview by President Marcos.

I AM ABOUT TO END MY BRIEF VISIT to your country, in the course of which I tried to familiarize myself with the state of the Philippine agrarian reform inaugurated by President Marcos. The purpose of this letter is to share with you the impressions I gathered about the progress of the reform. Considering the magnitude of the program, I have far from probed deeply into its various ramifications. In the circumstances, you will treat my remarks accordingly. Parenthetically, before I proceed with my observations, it may interest you to know that I have read President Marcos' letter to Mr. McNamara of March 6. I was particularly interested in the president's remarks about agrarian reform. In view of Mr. McNamara's own concern about the subject, you will not mind, I am sure, if I share with him my letter to you.

The president's reform proclamation of October 21, 1972, your own valued remarks on the subject, and your government's reform project with USAID—all these provide ample and well-reasoned justification for the abolition of the existing tenurial system and the shifting of the majority of the tenants to an ownership base. I am, therefore, in accord with the goal as expounded in the Presidential Decree No. 27, assuming, of course, that the means of effective implementation are at hand. Apart from this, Secretary Estrella, I am impressed with an important aspect of the program you and I discussed as far back as 1969. I refer to the fact that this time the tenant will be getting not only the land but indispensable assistance, like credit, seeds, fertilizer, and other inputs, as well as marketing facilities that assure him a better price for his produce. When these resources and organizations that go with them make themselves felt countrywide, it would distinguish the Philippine reform from many another well-intentioned but not successful reforms. By not adding these all-important ingredients, the best land distribution scheme is usually written in water. In short, one cannot take exception to the conceptual basis of your reform. I am well aware, too, that you have been able to carry out a great deal of work in identifying tenant tillers and owners, mapping out relevant holdings, issuing LTC's (certificates of land title), while at the same time grappling with the complicated question of determination of land values. All of this leads me to the view that the errors of commission and omission in an undertaking of this size coupled with so short a time span are inevitable. However, given political and administrative will, they can be corrected without detracting from the main aims of the reform.

Turning to implementation of the reform, I assume it will cover all the tenanted land, the exception being the maximum retention in the category of 7 hectares and less. If this is the case, it would, I understand, involve approximately 38,000 owners of 780,000 hectares of land for the benefit of 418,000 tenants. I need hardly remind you that this implies a drastic transformation of tenurial relations and fulfillment of the president's reform pledge. By the same token, it would have been natural to anticipate all manner of pitfalls standing in the way. This is not to say, as some recent observers have contended, that the reform is in danger of disintegration. On the contrary, while the problems are slowing down the implementation process, you are not facing the kind of alarmist prospects just mentioned; the reform has gone too far along to entertain such prospects. I shall touch on the less cheerful side of the picture, but the accomplishments directly related to reform are not far to seek. When, for example, one visits a barrio where all tenants received their allotted land, fixed in price, and are beginning to receive support items, the accomplishment is clear cut. It was significant to listen to tenants not eligible for land distribution who feel protected on the land they cultivate by virtue of the fact that the reform is undermining the position of the landowners. Quite naturally, most of them would like to own lands; but, since a goodly number of tenants on the land of the smaller owners will continue as such, this is in itself an important by-product of the reform. Last but not the least, though the local administration needs bolstering up, it is promoting the reform and its ancillary activities with dedication. At the earlier stages, the excessive zeal of the local structure you have built has led it into difficulties; but, as between an inert and active group of reform participants, the choice is obvious.

All this is gratifying to an interested observer, but the picture would be out of focus if I failed to take note of the stumbling blocks well-known to you which have cropped up in the course of the implementation process. This is not surprising in a reform which compels an owner to divest himself of a good deal of property, often all his property, social status, and political power, on terms he considers disadvantageous. It explains why one doesn't have to travel far afield to be told that the affected owners of the current reform phase (more than 24 hectares) are against the letter and spirit of the reform. Their attitude stems from two propositions: They don't care to part with the land; and, if part they must, they don't like the terms of payment for the land imposed upon them by one of the principal reform provisions. I refer here to the 10 percent cash, the remainder payable in fifteen years, 6 percent interest-bearing bonds, nonadjustable for inflationary conditions. I shall have occasion to deal with this matter elsewhere, but here I should like to call your attention to this obvious fact: If the owners of more than 24 hectares have nothing good to say about the reform, which is quite understandable, the owners from 24 down to 7 hectares, who are less well-endowed economically than the higher categories, will like it even less. In saying this, I am not shedding tears about the fortune's turn of either group. Nor is it a plea to let the owners of 24 down to 7 hectares go scot-free. This is mainly to underscore that they will generate even greater reform opposition. For this the government has to be prepared, whereas it hasn't been quite prepared to deal with the owners of 24 hectares and more. For this reason, Mr. Secretary, the responsibility for eliminating the persistence of obstacles lies partly in your corner and mainly on the highest level of government. I'll have more on the subject later; but let me only add here that, if the obstructions are to be overcome, the time is indeed ripe for the kind of firmness or unceremonial vigor coupled with conciliation that only the president could generate by laying down a few basic guidelines meant to be enforced in the strictest sense of the word. I hope that the specific of this will become apparent as I touch on some of the shortcomings which currently beset the reform.

As to shortcomings, I disagree with the views of some observers that, because of the discrepancies between LTC's issued and actually delivered to the tenants, the land distribution program has slowed down to a trickle. I need not repeat the argument in the rejoinder provided by your office except to say that it was very much to the point. And yet, there is an LTC problem which on at least three occasions of personal experience loomed large.

In one region comprising six provinces the number of certificates prepared as of August 31, 1974, was nearly 21,000, but the number actually distributed was 13,000. The remainder is "pending." The "why" of the pending is a combination of all too quick surveys which leave much to be desired, of owners who succeeded in concealing their true land positions, of owners contesting many certificates on grounds of "partitions" allegedly carried out prior to October 21, 1972, "urbanization" claims, shift of rice and corn fields to sugarcane production, and various other claims to evade reform provisions. The chief of one province is burdened with 40 percent of undelivered certificates partly for the same reasons and partly (25 percent) due to misspelling of names and neglecting middle initials. Carelessness, probably due to eagerness of the local staff to show rapid progress, is obvious in yet another case. In the area visited, out of 3,000 certificates, 500 had to be returned on account of misspelled names, 200 instances of owners' "contestations," and in all nearly a thousand certificates looking for a home. Taking the countryside as a whole, out of an estimated 215,000 LTC's, probably 40 to 60 percent have reached their rightful destination, covering from 150,000 to 200,000 hectares. Since, incidentally, it appears that my findings relating to nondelivery of LTC's to the ultimate beneficiaries depart considerably from other estimates, further investigation of this key element in the program would seem to me to be of the highest priority.

The figures presented suffer from a great lack of precision, but it is generally agreed that they are below the goal. Even so, they reflect an accomplishment in the midst of difficult circumstances. At the same time one cannot overlook the fact that a very large number of tenants are hanging in the air, and they will remain in that suspended state until adjustments have been made; so far, this is proving to be a longer drawn-out affair than should be tolerated. The past technical difficulties are in the process of being corrected, albeit slowly. As to future difficulties which the implementation of the last phase of the reform with its greater number of tenants might bring about, the center must convey to the local administration that the issue is not one of speed and reporting accomplishments in double-quick time. Above all, greater care, even if more time consuming, is essential for the good of the program. The subsidiary consideration is that in the eyes of critics of the reform any problem is apt to be judged as evidence of nonfulfillment. This can be interpreted as such only if one assumes that no rectification will take place, but the causes for such an assumption must be avoided at all cost.

Technical errors, as extensive as they are, are serious enough when certificates of ownership are being held up; but much more serious are the nontechnical difficulties which are at the root of the problem. On this score and based on talks with local officials, one gets the feeling of an air of uncertainty extending from the center downward and vice versa. Here, I wish to call your attention to an impression which, if true, only your office can correct. I have in mind the practice of contesting owners carrying their cases to Manila in search of relief. This tends to delay the resolution of certificate problems while at the same time tying the hands of the regional directors. Additionally, the evidence is that it takes more than Memo Circular No. 2-A, modified by Memo Circular No. 8, to prevent the falsifications many owners resort to. A more effective solution might be to give the regional directors who are in a good position to know whether the claims are valid or not the authority to deal with such matters expeditiously, on their cognizance, and without further recourse to the center. I know of two instances when a regional director so acted, but his files are bulging with many more similar cases yet to be acted upon. I don't know how other directors comport themselves in a situation just described, but I am struck by the lack of clarity in this regard. Any ambiguity of this sort should be cleared up with the responsibility vested in regional directors, as suggested. Failing that, or a better measure the center may devise, the resolution of all falsifications and, in consequence, the delivery of a large backlog of certificates to tenants will be long in coming.

Whatever the ultimate decision on your part, it cannot avoid a fairly strong element of *force majeure*, and not only because you operate under decrees conceived under martial law. You would have to resort to forceful action even under the due process of law, for it is in the nature of successful agrarian reforms that stepping on the toes of those who oppose them is an unavoidable precondition. I call your attention to the obvious despite the penalties "proposed for the agrarian reform violations," and they are quite stiff. I suspect that they are only proposals, and of this the guilty owners are certain. After all, there is probably not a case of a landowner sent to jail for violations which is not in dispute by appropriate authorities. I perceive in this a contradiction between the new society as envisaged by martial law and the ease with which the owners treat their obligations under the new society's reform program. This is apparent not only in illegal subdivisions of large holdings and other evasions but very strikingly in the issue of land valuation.

No matter whether one visits a regional provincial, district, team office, or barrio, the question of land valuation stares one in the face. How vital this question is to the owners and tenants calls for no elaboration. And yet the process proceeds at a snail's pace, and often it does not proceed at all. As things stand now, land valuation is in the doldrums. Leaving aside the question whether two-and-a-half times production from a given plot of land is precisely what a successful reform requires, the reality is that most owners and tenants cannot agree on the productivity of the land. Of course, since both parties have been with the land too long not to know its productivity, particularly its productivity in the three years prior to October 1972, the difficulty lies in the fact that the owners want too much for the land and that the tenants would like to pay considerably less. The attempt on your part, Mr. Secretary, to bring about a closer meeting of the minds through the creation of a thousand or more barrio valuation committees has so far led to no tangible results. The composition of the committees favors the tenants, but the trouble is that the great majority of the owners are avoiding them as if they were poison. Repeated invitation letters from your local officials to attend the barrio committee meetings are mainly disregarded. The suggestion of one of your directors to bring the committees to the owners is not practical either, if for no other reason than the owners' refusal to elect two members to represent them on a committee.

This is a clear case of boycotting the committees, which goes back to something more fundamental than the price of the land, namely the method of payment for the land. This will be discussed subsequently; but, to sum up the current state of affairs, let me note that as of September 23, 1974, the planning service of the DAR (Department of Agrarian Reform) confirmed only 140 owner applications signifying valuation agreements. This number is infinitesimally small compared to expectations based on the number of certificates actually delivered to tenants. The argument of local officials that failure to fix a mutually agreed-upon price is of no great moment so long as the tenant has the certificate of ownership is of no great validity. If tenants do not know what their financial obligations are to the owners, many of them do not consider themselves owners of the land. An investigation in one locality revealed that 44 percent of the tenants who can boast of land certificate titles still look upon themselves as leaseholders; they are not reform beneficiaries because the rentals they continue to pay are higher than the anticipated annual land amortization payments.

Apart from the evidence provided by the planning service of the DAR that land price fixing is making little or no progress, the Land Bank charged with the responsibility to pay for the land points to the same frustration. One common complaint is that the excessive documentation demanded by the bank slows down payments. The bank doesn't deny the charge, defending itself on the ground that it must be very careful in handling public money. Over-documentation is demonstrated by the fact that the cases of nearly half of payment applicants are "pending" for this reason. More to the point, however, is whereas the Land Bank expected and geared itself to an avalanche of applications, the avalanche has not materialized. For reasons already indicated, it is not as surprising as it is disturbing that as of the week ending September 20 the bank had in its portfolio a total of only ninety-five applicants, of which forty-one are pending; and the lands actually paid for are a mere 3,100 hectares or hardly worthy of comparison with acreage already covered by the distributed certificates. It may be noted, too, that according to the bank 90 percent of the land it has so far dealt with was heavily mortgaged. The interpretation offered by the bank is that the owners in question had so little equity in the land that they did not mind parting with it.

In sum, the problem of land valuation cannot be permitted to go on at the same pace, and the measures taken so far have led to nothing substantial. If this issue is to be shifted off dead center, the application of muscle seems to be in order. But I am not advocating jails or fines. I doubt either would be productive or stringently applied for reasons I need not go into or that the instrumentality and authority exist in the countryside for strong-arm measures. Above all, none of this is necessary. You have an easier and more effective way out while applying a firm stance of a different sort.

Excepting a few cases your administration will have to arbitrate, land productivity in the areas of your concern is no mystery; one cannot find an owner or a tenant who does not know what it is, and I have learned enough during my field trips to know that your local administrators are not ignorant about it either. Armed with the information, the first step to break this logjam is stoppage of rentals, which tenants continue to pay laggard owners. The continuation of such payments has undoubtedly played into the hands of uncooperative owners. Certainly the move would be hard on them, but there is no reason why recalcitrants should be rewarded. The next step is an unchangeable cutoff date for the determination of land values, with authority vested in regional directors to fix values in case of failure to reach agreement by that date. This may result in some instances in over or underpayment, but the risk is probably no greater than what emerges from the haggling between owners and tenants. If all these measures are strictly adhered to and enforced, the valuation process could be speeded up. The suggestion may appear too drastic, but it seems to me that these or any other practical moves the DAR might devise are long overdue.

The crux of most of the difficulties bedeviling the country's agrarian reform is the mode of land payment decreed by the government. As seen in the field, this is the principal cause of the extreme reluctance of owners to negotiate land price fixation. While owners in categories of more than 24 hectares do not wish to part with their land, they would dispose of it for these reasons: (a) the realization that their day as owners of rice and corn land is, willy-nilly, nearing an end; (b) the relatively few negotiated land prices are in a range of ₱5,000 to 8,000, which while below the market price are nevertheless reasonable even from the owners' point of view; and (c) a better payment arrangement. Ten percent cash and the remainder in bonds is not attractive, the cash portion being too small and the bonds being of uncertain value over a fifteen-year period. Having been involved in a reform which proved to be needlessly confiscatory, I recognize the merit of your government's arrangement. However, something should be said on the other side. Despite all the glaring impediments generated by the owners, the reform is a stick as far as they are concerned. But by itself the stick lacks effectiveness when measured against reform accomplishments. This being the case, it may be necessary to supplement the stick with a carrot in the form of more cash and a clear-cut clarification that the bonds stand for something more than what presently strikes the eye of an owner.

Speaking first about the bonds, I dare say the majority of owners are not aware that there is a considerable element of conversion or "monetization" available to bondholders. The Land Bank makes it clear to a visitor that a portion of the bonds can be converted into investment or investment-participation in enterprises approved by the government. The Board of Investments (BOI) has listed these opportunities, and they come large and small. It should be the task of the Land Bank and DAR to make certain that there is the fullest appreciation among owners of this convertibility. There is another more familiar field of investment which may provide scope for monetizing binds. I have in mind the availability of public reclaimable lands secured against a portion of the bonds. This suggests the development of agroindustrial complexes of purely agricultural enterprises on the vast reclaimable areas of Mindanao, Palawan, and other outlying islands; this can be done on an individual basis or through a corporation with state-landlord partnership. True enough, the agrarian reform experience may cause ex-owners to shy away from land investments; but appropriate official assurances may prove tempting. On the whole, it may be envisaged that perhaps one-fourth of the total value of the bonds could be used eventually as collateral for approved investments of one kind or another. For the remainder to be paid in nonconvertible bonds, surely in Philippine conditions a more realistic interest rate than the 6 percent, tax free to be sure, would seem a more reasonable treatment.

A partial utilization of the bonds could stimulate greater acceptance of the reform, but the more immediate "carrot" inducement is raising the cash portion. I am not prepared to follow other observers who insist that the cash portion should be raised to 50 percent. As "sweetener," I can suggest only an increase from 10 to 25 percent. This would not meet all or most of what the owners are after, but it would be a practical move in the right direction. The suggested figure is not altogether out of thin air. Assuming 430,000 purchased hectares under the category of above 24 hectares at an average of ₱7,000 per hectare, the 10 percent cash payment would amount to ₱300 million; 25 percent would raise it to ₱750 million. I think the country could stand the strain of unanticipated expenditure of the magnitude in order to complete the reform.

Governments have a way of finding all kinds of resources for all kinds of necessary "visible" programs, but less so for programs of social significance. Even recognizing this, I do not question such commitments or the problems of diverting funds from programs already in being. If, however, the Philippine agrarian reform is, as widely proclaimed internally and externally, the cornerstone of the new society, anything within reason that assists its implementation should not be dismissed out of hand on the ground of lack of funds. In 1975 the government of the Philippines anticipates a budgetary outlay of roughly ₱7,500 million for a variety of public programs. If so, it is not easy to accept the proposition voiced in certain circles here that none of these can be trimmed for the sake of speeding up the reform implementation or new funds secured for this purpose. Finally, I would not have brought up the subject of an additional cash outlay but for my concern about the current state of the reform and the all too easygoing manner of dealing with reform obstructions. If the latter cannot be eliminated through effective enforcement, one might as well be prepared to deal with the principal source of evasion part of the way by liberalizing the payment formula. So long as these persist and stronger measures are not or cannot be applied with good effect, the unexercised stick might as well be supplemented with a carrot. Actually, this is a tested practice in the best of reform climates and could be significant at this particular juncture here.

In my field trips and in Manila, I have been struck by the fact that land distribution is the sole preoccupation of the Department of Agrarian Reform. The Presidential Decree No. 27 imposed that condition. Nevertheless, the Land Reform Code of 1963, which substituted the vastly superior leasehold provision for the inferior status of share tenancy, has been ignored in the midst of the concern about land distribution. The neglect of the leasehold operations is a basis policy misjudgment not only because shifting to leasehold improves the status of the tenant, reduces social tensions, and stimulates agricultural productivity. There is also another, more compelling reason. Even if land distribution is extended down to 7 hectares, approximately 56 percent of the tenants will still remain in that condition. Exactly how many of them are leaseholders as distinguished from share tenants is not known, but it may be assumed that more than half of them are in the latter category. Moreover, if land distribution does not cover the 24- to 7-hectare category, a possibility which cannot be excluded at the moment, approximately 73 percent of the tenants will not be covered by Decree No. 27. Considering all these factors, it is quite apparent that sole preoccupation with land distribution is something that the Department of Agrarian Reform should begin to reconsider at the earliest moment. Nonetheless, this renewed concern for leasehold should not detract from the firm prosecution of the land transfer program.

I noted above that the small segment of your local administration with which I have been in contact appears to be dedicated and eager to go on with the task. And yet, morale could stand improvement. There is a need of clearer and steady direction and firm delegation of responsibility. Some of your administrators are people of vast experience with intimate knowledge of local conditions, but the authority due them is not always there. On such basic issues as combating evasions, their authority is limited and obscure and, hence, ineffective. When one of your best regional directors handled two evaders with firmness, it impressed me as something of an event whereas it should have been looked upon as a normal chore in the course of duty. As you well know, I part company with those critics who interpret the certificate title problem as prime evidence of the reform's disintegration. But the fact remains that correctable as this problem is, given the will to do so, it could have been largely avoided if greater stress had been laid on qualitative rather than quantitative performance. I hope that the lesson of this experience is not lost when the next reform phase comes around.

In regards to the evaders, my position is as already stated, namely, unceremonious vigor combined with conciliation. This does not suggest a blanket reprieve for subverters of the reform. In all such instances, administrators of the reform should be fully prepared to exercise stringent economic penalties and worse if necessary, since "punitive" decrees which are merely intentional are ineffective. Apart from all this, one of the danger signals of inadequate performance is the seeming inclination by too many to treat the reform as if it were "old hat" or with no particular respect. Here, I do not wish to be misunderstood. I am not implying that the reform is not a living institution nor that it cannot overcome its present trials nor that, above all, many owners do not know full well that their tactics of evasion will not avail them much in the longer run. Though this is true, the aura surrounding the early days of the reform's inauguration is somewhat tarnished; and it is best to make certain that the aroused expectations of the tenants do not go begging.

The preceding leads me to the conclusion that the time has come for a thorough and candid reexamination of the reform's accomplishments, failures or shortcomings, and what needs to be done to reinvigorate it. In a broad sense and particularly in relation to reinvigoration, the man to do it is President Marcos; and the occasion for it is the second anniversary of the reform on October 21, 1974. A restatement why the reform is of vital importance to the country, its goals, the encountered difficulties, specific guidelines for overcoming them, and commitment to land distribution from 24 down to 7 hectares—all these could very well clear the air and set the future course for action. In the experience gained and hopefully assimilated and henceforth acted upon without fear or favor, lies the hope, Secretary Estrella, that under your leadership the institution of private ownership of land will undergo the kind of change the agrarian reform as originally proclaimed by President Marcos intended to accomplish.

62. *Food Shortage in West Bengal–Crisis or Chronic?*

In this powerful paper dated December 16, 1974, Ladejinsky reported his observations after a visit to the area during the previous month. The paper was subsequently published by *World Development*, February–March 1976, under the same title.

THIS NOTE DEALS WITH the food distress in rural West Bengal following a brief visit to the state in early November in the wake of drought and flood devastation. There, as elsewhere, widespread hoarding and black marketing or failure of the procurement drives, on top of depletion of buffer stocks and inadequate imports, and the inexorable population spiral are part of the picture. In human terms, it all adds up to under-eating on the part of many, outright hunger among considerable numbers, reported death from starvation or prolonged malnutrition in numbers which are both overstated and understated depending upon the source of information, occasional suicides, desertion of children, greater crowding of cities, and sale of land by small holders to meet their immediate food needs. With the exception of the certainty of rising population numbers, the cited consequences cannot be quantified; the best one can say is that there is a large element of truth in all of them. It is unquestionable, however, that in the light of total food availability millions must contend with very short rations. The statement of the relief minister of West Bengal that 15 million people out of a total population of 45 million are in a state of semistarvation is what made the state a center of attention in a year when a number of other states have not escaped somewhat similar predicaments either.

One of the scarcity features of West Bengal, as of other states with large urban centers, is the sharp difference between food conditions in the latter and in the hinterland. The urban centers leave one with the impression that, while belt-tightening has been the order of the day, in the main they are provided with basic rations chiefly because the food distribution policy dictates that they and industrial centers have first priority on available supplies provided from the central pool, controlled by the central government. This explains why a casual visitor to a city like Calcutta might be left with no thought of special food problems, while in reality the state of West Bengal has achieved great notoriety as one of the worst afflicted by devastating floods and drought. Our observations support both propositions; it is not in Calcutta but in the countryside where one runs into critical food problems. But it must be stated at the outset that those who write about West Bengal under such headlines as "1943 Being Re-Enacted" exaggerate the current difficulties, which are grave enough. In that year an estimated 2.5 to 3 million people perished from starvation; in 1974 a few hundreds and possibly a couple of thousand shared the same fate. This comparison, however, in no way minimizes the fact that the people of many parts of rural West Bengal have been undergoing extreme privation on a scale not experienced in many a year. Moreover, underconsumption is a perennial feature of the state but harsher and more visible in 1974. What follows is an attempt to pinpoint some of the principal aspects of this state of affairs.

West Bengal is a deficit state even in good years, the deficit ranging from 1.5 to 2 million tons. The gap is mostly covered from the central food pool; this supply is utilized to feed the distribution system of Calcutta as well as one or two smaller urban enclaves. In politically volatile Calcutta, to keep it fed is the main concern of no matter which party is in power. Part of the distribution policy is that the countryside with its 50 or more percent of the people under the poverty line is considered "self-sufficient" in food, the fallacious theory being that the village provides for its own. Besides, underconsumption (some call it hunger) is a life-long experience of this group of people between a crop and a crop—even when the crops are good. And there certainly has been no reason to worry about the middle and upper rural groups who profit also in years of shortage when prices are high and procurement is permissive. In short, West Bengal has managed under these arrangements and preconceptions so long as Calcutta didn't suffer from a lack of basic rations at subsidized prices.

In 1974 the situation has undergone a change; the idea that the rural poor are cared for by the village no longer holds even in official circles. To be sure, Calcutta continues to do reasonably well though on reduced rations, such as 2.5 kilograms per person per week, at around Rs1.50 per kilogram compared with open market prices of Rs3.5 to 4. This type of statutory (obligatory) rationing (SR) covers 10 to 11 million urbanites, mostly in Calcutta. In this year of widespread food shortages, which involves West Bengal as well, monthly allotments from Delhi of approximately 120,000 to 130,000 tons have not been easy to come by; but Calcutta being Calcutta, numerous personal interviews reveal that the rationing system is functioning reasonably well, even if the total distribution volume is slightly less than in the preceding year. When the relief minister spoke of millions in search of food and barely getting any, he spoke not of Calcutta but of the distressed conditions in five of the fourteen rural districts of the state, some of which suffered from floods and some from drought. As will be pointed out presently, intentionally or not, he spoke also of the lack of food in districts which have not been afflicted by nature's ravages.

No official would venture an estimate about the damage sustained in the five districts and the number of people affected. This is not surprising in view of the contradictory official assessments and judgments about everything important—size of summer crop already harvested; expected size of the most important November crop; damage done by floods, cyclones, and drought; number of people in the countryside for whom handout food programs have been launched in order to avoid large scale starvation; and so forth. On the other hand, unofficial estimates by interested observers point to an overall loss of crops of only 5 percent; and yet one cannnot impugn the food minister's claim of a "huge shortage of foodgrains" from which the people outside Calcutta suffer. A visit to a couple of districts, one damaged and one *undamaged*, supports the minister's contention. According to him, the shortage is not so much a result of nature's folly as of the hoarding of a possible million tons of rice. This was precipitated by the undetermined crop losses, rice smuggling into Calcutta, and sharply rising prices in and outside West Bengal. In a private conversation with the food minister, he made it clear that the "soft" procurement drive last year led to hoarding and that the subsequent dehoarding campaigns have been no more successful than the procurement drive. What the minister understandably did not say is that the political party to which he belonged favored neither strong procurement nor dehoarding, for these would have violated the political connection with the important district constituents—the well-to-do farmers. This explains why, out of a total production of 6 million tons of rice and a marketable surplus of about one million tons, the state government had decided to procure only 500,000 tons. In fact, it collected only 160,000 tons. This was inevitable as the government was depending on the "goodwill" of the rice-mill owners while the surplus producers were depending on the government's unwillingness and incapacity to extract the grain. Thus, although the millers' quota in the total target was 360,000 tons, they delivered only 40,000 tons.

Regardless of the top priority enjoyed by Calcutta and the political orientation of the government of West Bengal relative to the countryside, in 1974 it couldn't neglect altogether the rural hungry. No one will ever know the exact number of people who fall in that category, but it undoubtedly is large even if the minister's figure already referred to is exaggerated. What is not questionable is the meagerness of the assistance despite an alleged expenditure of Rs15 crores for all forms of relief and an officially claimed monthly distribution of 25,000 tons of grain among landless laborers, holders of plots too small to be self-sufficient, and poorer smalltown people. Much of this goes under the name of "modified rationing," which means nothing more than 500 grams per person per week. There are other relief programs for the flood and drought victims launched on September 15 to feed daily some 650,000 people, of whom an estimated 250,000 get free food from "gruel kitchens" and the remainder receive meals from "cheap canteens" at 25 paise per meal. Responsible officials tell an inquiring visitor that this mix of programs, difficult to disentangle one from another and with evidence that some of it exists only on paper, covers a total of 2.3 million people. These figures are suspect, demonstrated if nothing else by the useless chore of pressing on officials to reconcile furnished data. The hastily set-up rural works programs are utterly inadequate. Only one person per family is allowed to work, and he receives 750 grams of wheat and cash of Rs1.25 per day. Not only are these projects few in number, but their capacity to provide employment is pitifully low. Officials admit privately that some of these works last no longer than a week, partly for lack of resources and partly due to administrative problems. In such circumstances there are neither additional earnings to speak of nor the much talked-about asset-building operations as the core of work programs.

The drought in Purulia and Bankura districts and floods in the districts north of the Ganges have induced unexpected programs for which the administration was not prepared. What little is done is done on the fly and grain is not available to keep various pipelines going outside of Calcutta. Bearing on the same point is the prevalent and unsupportable view that the food scarcity in the rural districts is a passing phenomenon which will be corrected by the promising November harvest of the winter crop. This is questionable, however good the crop turns out to be. This judgment, elaborated in subsequent paragraphs, is based on three visits in the countryside and information relevant to the "normal" condition of the rural poor. It is apparent in almost any rural district that, while floods and droughts create particularly grave food distress, districts not afflicted at all by the perversity of nature are affected only to a lesser degree. It is also apparent that only sharply increased food availability accompanied by a sharp fall in prices and a rise in rural employment over longer periods of time and at better wages could bring a real measure of relief for those who are destitute year in and year out.

Bankura district is a drought victim, and the anguish of its hungry has been amply described and publicized. Suffice it to say here that consequences summed up in the first paragraph fully apply there; the "large element of truth" applies here as the whole truth. Assertions of mortality from starvation are easy to come by, but nothing that might conjure up an image of famine like that in 1943. And yet one cannot escape the visible suffering among the rural poor who bear the brunt of skyrocketing food prices, paltry wages, and little employment at any wage. Even in Bankura rice is available in the market, but at prices only few can afford. Relief has been stepped up both from Calcutta and by some well-to-do individuals and charitable organizations who in effect maintain the gruel kitchens which offer a daily meal but not to everybody in need. There are not enough kitchens to go around, and according to a relief worker they cater to only a relatively small portion of the people in want. The common thread which underlies the food situation in stricken Bankura as elsewhere in rural West Bengal is the total inadequacy of grain for distribution and the limited scope of the emergency relief programs. West Bengal authorities are fully cognizant of the dire consequences that lurk in the wake of great numbers of underfed people, as recently told by the chief minister to Sir John Crawford. But with the food at its disposal they don't dare to cut rations to the bone in a tense city like Calcutta for the sake of providing something more substantial for the hinterland. It is not necessary to spend much time in Calcutta and Bankura to see the official policy treatment of one as against the other.

In Bankura the subject of mortality from starvation of perhaps 100 to 200 is not hidden under a bushel. Politicians and relief workers speak about it, but a visiting observer has no way of ascertaining the real magnitude of the ultimate price hunger exacts. Flat denials of such occurrences by the central government and, in this instance, by the administration of West Bengal cannot be taken seriously. District officials are cautious in reporting to Calcutta cases of this type of mortality. The investigations they launch and the findings are more often than not attributed to causes allegedly not directly relevant to hunger, as if gastroenteritis were a disease caused by overeating. The official approach to this issue has given rise to a terminological controversy waged in the press. It is mainly about whether scarcity is famine or whether malnutrition is starvation. The minister of state for agriculture of the union government saw fit to lay down the dictum that scarcity is not famine, except in instances of 100 percent failure and conditions of "absolute" scarcity. As to malnutrition versus starvation, the government draws a fine line of distinction between the two, evidently on the theory that the former is less embarrassing than the latter. The fact is that many people in Bengal are without adequate food and a few of them are dying for the lack of it. Whatever the terminology, those who die do so inch by inch, finally destroyed by hunger.

Bankura with its heartbreaks is attributed to nature's perversity, but in a large measure this is equally true in districts with no visitations of nature. The 24-parganas district with its two enclaves, Diamond Harbor and Kakdwip, illustrate what we have in mind. The spring and summer crops were good and the ripening winter paddy, the major crop, had all the making of a promising one. Had the traveler not been alerted that there, too, gruel kitchens are the lifeblood of a great many people, he would have passed them by. Forewarned, he is at centers of food distribution, one of them handing out potfuls of gruel to hundreds of people queuing up in line. In all, the Diamond Harbor kitchen is supposed to feed 4,000 people of a block made up of thirty-two villages but feeds probably no more than 1,000. As one watches the line, one is struck by the predominance of women and children, card in one hand and metal containers in the other. The relative absence of men is explained by their reluctance to appear "begging" for food when their womenfolk and children can do the same just as well, or absence in Calcutta in quest for employment. The gruel is a mixture of coarse grains, a bit of rice, vegetables and a dash of oil, and lots of water. For most of the recipients this is the only meal of the day. If one is to believe the elicited statements from a group of eight, it would appear that three do scavenging to find additional food while five wait for the portion of gruel the day following. They all fare very badly, and a visit to a couple of shacks with nothing edible in sight confirms this impression. A chance encounter with a group from the countryside who pulled up stakes and reached Calcutta looking for employment and food shows them seemingly worse off, if that is possible. On the other hand, it is idle to measure the degree of distress between one and the other; both are victims of circumstances already noted and those to be mentioned elsewhere.

The gruel kitchen operates from early morning till noon, and all those who queued up received their allotment—some an extra ladleful and some less. But on the particular occasion described here, the closing of the kitchen was not the end of the story. Not many minutes later pandemonium broke loose. A group of about forty appeared on the scene, cards in hand, clamoring for food. An examination of the cards showed that some were repeaters but that the greater number were latecomers. Relief workers explained that this was not a rare occurrence. When ingredients are available, a new batch is cooked and distributed; when it is not, the would-be recipients are out of luck. The repeaters are problems, resolved on the spot, favorably, if a repeater makes a convincing case for an extra portion.

There are many similarly nature-unaffected villages in West Bengal and badly in need of gruel kitchens, cheap canteens, modified rationing, and other forms of relief from hunger. Since nature has been kind, why the relief urgency? The 1974 familiar answers are overall food shortages, poor procurement, and high prices. But the more specific normal causes are the presence of a large and growing number of landless agricultural laborers, shift from payment of wages in kind to cash as desired by the owners, meager wages, prevalence of single cropping, and short duration of employment—and all this leading to abysmally low purchasing power. With the exception of this year's abnormally high prices, all other factors operate in bad and good crop years alike.

The number of landless in West Bengal is sharply rising. The population of West Bengal is about 45 million, and three out of every four live in the villages. This relationship had not changed between 1961 and 1971, although migration to Calcutta goes on at all times. On the other hand, the number of owner cultivators during the decade declined from nearly 5 to 4 million while the number of landless increased significantly as part of the general rate of population growth, sale of land by marginal farmers, and outright eviction of sharecroppers. The combination of these factors leads to the bottom rung of the agricultural ladder. The all-India percentage of cultivators and agricultural laborers in the male working force was 67 percent in 1971. The percentage of agricultural laborers in the male working force was 15 in 1951 and had increased to 21 by 1971. In West Bengal, however, the percentage was 20 in 1961 and 33 in 1971, the highest in the country. In some districts of the state the proportion doubled within the decade. Along with this development there has been a renewed concentration of land holdings, agrarian reforms notwithstanding. Up-to-date data for the state are not available but in a division of 24 Parganas district, next door to Calcutta, one finds data such as these: 51 percent of the farmers owned land up to one acre, or 5 percent of the total; one percent held 29 percent of the land. Or, while 95 percent of the farmers held only half of the land, 5 percent owned the balance. This is not exceptional and the impact of this ownership pattern is felt in production; in the evasion of land reforms; and, of course, on the condition of the landless. And so it has on the sharecroppers who cultivate 40 percent of the land.

The average family size of landless laborers is between five and six, and wage earners per family is less than two. That living conditions border on near starvation is supported by other factors. Most of the cultivated land of the state is single cropped. On paper 27 percent of cultivable land has assured irrigation, but where water sources have been created the wastage in distribution due to seepage has been very great. Even in areas of perennial water estimated at 3 million acres, the average cropping intensity is less than one and a half crops per acre per year. This alone imposes short employment seasons, the total number of working months ranging under four and not exceeding five. Labor is in surplus and has none of the bargaining power exercised by its counterpart in Punjab. When they are employed, they work in peak seasons from dawn to dusk at Rs2.5 or less. The same Punjab laborer earns in peak seasons closer to Rs8 to 10 per day, particularly during harvest time. In off-peak seasons the Bengali's daily wage is as low as Rs1.50, and the old custom of meals in addition to the wage has been reduced to a minimum. The Rs5 minimum wage is not enforced. The increasing trend of eliminating the traditional practice of payment in kind for work performed during harvesting and threshing is taking its toll. On irrigated, high-yielding land the practice of payment in kind has virtually disappeared and clearly to the advantage of the owners who get higher yields and higher prices. As for the landless, they lost what amounted to something like food insurance for part of the year. The stomach of every interviewed tenant was thoroughly aware of this development, but there is nothing they can do to reverse the trend which is countrywide. It might be noted that neither the central nor state governments have taken the slightest interest in the matter. Directly related to this is that opportunities for gainful employment outside the agricultural sphere are virtually nonexistent.

The shacks they live in, the clothes they wear, the food they eat, and their general appearance—all these speak of miniscule income which inevitably cannot meet their minimum consumption expenditures. Precisely what the income is was difficult to ascertain from those we questioned, but obviously they had plenty of nothing. Nevertheless, figuring generously, their average yearly earnings cannot exceed Rs200 to Rs300. In the late 1960s a Labor Department study of the living conditions of the landless in a number of districts concluded that their yearly earnings were as low as Rs103 and as high as Rs137. This was far below the then estimated expenditure requirements of Rs240 per year to stay just above the poverty line. When 1974 prices are compared with those of 1961, the Rs240 minimum must be doubled. In conditions of extreme food deficiencies of today or shortages in normal days, agricultural laborers of the area in question are worse off than a decade ago. The department comment on its findings ran as follows: "When this very poor daily income is evaluated in the context of the average size of the family in the district under survey, it really baffles reasoning how they continue to live."

The bafflement of survival is the greater in 1974, for this time even with increased cash earnings they are not only priced out of the open food market where rice sells at Rs3.5 to 4 per kilogram, but for most of them the same would be true if they were covered by statutory rationing at Rs1.50 per kilogram. The landless, therefore, are worst hit, and only to a lesser degree the same may be said of marginal farmers and sharecroppers. Local authorities estimate that, in the Diamond Harbor block mentioned earlier, these groups represent about 70 percent of the population and they often experience acute scarcity of basic foodstuffs. And this is in an area where the crops this year have been reasonably good. In a real sense, therefore, the landless and semilandless are entitled to relief in good as well as bad years, for they are the authentic people of poverty. They are not only poor but have no political pull despite their numbers and no organization whatever. Presently, no political party, including the extreme left, is capable or willing to undertake such a chore. The days of the united front government of four or five years ago dominated by the Communists are gone. The 300,000 acres or so of land "grabbed" by Communist activists have since been returned to their owners, and many of the grabbers while their time away in jails. In this political topsy-turvy it is not surprising that the landless and the sharecroppers are left high and dry. Nor is it surprising that in the face of chronic distress and deprivation they are afraid to alienate those who provide them with work at least for part of the year.

One of the consequences of the rural conditions in Bengal in 1974 is the spurt in sales of land by the small and marginal farmers, thereby increasing the size of the landless group. Such sales are mostly "distress" sales. Although land values have been sharply rising in recent years and more so under the inflation impact, sales of such land command a very low price; if occasional reports and discussions in the field are to be taken at their face value, a piece of land normally worth Rs1,000 sells for half that price or less. The badly affected Cooch Behar district (West Bengal) illustrates the situation. June, July, and August, normally dull months for land sales, registered six thousand of them in one subdivision of the district; the total sales for the year will be largest on record. It is also estimated that 80 percent of the sales do not exceed one acre and that these fall in the "distress sale" category. In most cases the unwritten commitment on the part of purchasers of these lands to return the same when the original owners have repaid the sale price with interest is not adhered to. There are legal barriers against distress sales of land, but there seems to be no difficulty to circumvent the law. Many of the sales can be traced to indebtedness incurred in cash or in kind, and all of it at a monthly interest charge of 6 or more percent. The landless and the sharecroppers are also debt ridden either to landowners or moneylenders, who are often one and the same person. In the areas visited, indebtedness ranges from Rs300 to Rs600. Repayment is often in the form of labor performed on the fields of the lenders, but being indebted is a permanent condition of landless. The fact is that neither the landless nor the sharecroppers have any access to the organized money market—cooperative credit societies, nationalized banks, and so forth. They have no alternative but to borrow grain from the rural rich during the lean agricultural seasons when prices are normally high. All this is well known and is mentioned here merely to stress that it is part of their life-style in years when food problems do not claim the headlines but are present just the same as far as a large segment of rural Bengal is concerned.

Migration into Calcutta from the countryside is an old story, but this year is likely to be a record year. To an unaccustomed eye it is difficult to judge whether a larger number of people appear to be living on the streets and sidewalks of Calcutta. But a study prepared by the Calcutta Metropolitan Development Authority shows they are coming in larger numbers. A major finding of the study is that pavement dwellers are Bengali-speaking, Hindu agricultural laborers whose livelihood in the villages had, for a wide range of reasons, collapsed, and that only out of despair do these people venture into Calcutta. In 1971 an estimated 48,000 people migrated from the countryside into the city center encompassing 3 million people. In 1973 an additional 175,000 people found their way there. The current year is expected to show a further increase. These are the poorest of the poor, worse off than the slum dwellers. A sample of 10,000, 98 percent of them born outside of Calcutta, classifies 56 percent as "very poor" with yearly income not exceeding Rs80; 32 percent are "middle class," or Rs80 to 140; and 12 percent are "well off" with income from Rs140 and above. The last two classifications must have been done with tongue in cheek, but they are as symbolic of the life they escaped from as they are of the life-style Calcutta offers them. Recent attempts to seal off Calcutta from migrants have not been successful. A visit to the city's two railway terminals shows that rural destitutes are arriving daily in considerable numbers. They have nothing to sustain them except begging for food and drinking water. After the end of the day they collect the leftover vegetables from the market, eat them half-cooked, which in turn reduces their digestive capacity and gives rise to the widespread epidemic of gastroenteritis and diarrhea. These migrants know nothing about gruel kitchens, cheap canteens, and other relief measures the government of West Bengal is promoting in the rural areas. It takes little effort to establish that these new additions to the growing population of Calcutta are landless, small holders, and sharecroppers who already sold their cattle, plows, and even household articles before trekking to Calcutta or other cities. For the city administration they present a very difficult problem because, even if any significant welfare program to better their lot were within its competence, it would attract more rural people in distress. This is an unacceptable prospect, but so is their condition in the city. To resolve this dilemma, the authorities would like to send them back where they come from. This is acceptable in theory but not in practice, and the human flow from countryside to the city will go on unabated. And so it will be so long as rural West Bengal has little to offer a vast group of its inhabitants.

In the midst of acute food shortages, a visitor is startled by the profusion of grain, vegetables, and fruit in the markets and stalls of Calcutta, albeit at a stiff price. Most of the grain is smuggled in, and thousands of landless and other underprivileged are heavily involved in it as agents of the surplus producers who short-change the state on their procurement obligations. Of plans to prevent smuggling there is no end, but it goes on just the same by bicycle, train, or boat, yielding hired smugglers an income of Rs10 to 15 per day. The connection between the owners of rice and the landless who work their land or those who tenant or sharecrop a large portion of it and now help them smuggle out the rice is not surprising, and not only because the latter find themselves "gainfully" employed at a critical time. It is part-and-parcel of the close ties between the well-to-do farmers, the political party in power, the rice mills, the government purchase agents, and all those who work or rent their land. The sum total of this relationship led to the failure of the procurement and delivering drives and to smuggling as its principal by-product. In fact, it goes a long way to explain the agrarian and political arrangements of rural West Bengal.

With the harvesting of the exceptionally good November crop, the food situation will ease, at least for a while, so much so that according to a reported government declaration "Relief Work Can Close—The Emergency Is Over." If procurement at long last should miraculously yield the anticipated 500,000 tons, it would work to the same end. The same if the ad hoc created various forms of food assistance are not dismantled but expanded and institutionalized. All this, at best, would provide only a barest minimum to keep body and soul together among the poorest. This is the condition because, while in 1974 the food situation was notoriously out of hand, for the landless and nearlandless it is normally bad without an "assist" from nature's havoc. This is a perennial problem, and the fact that it becomes a public issue only in a year of acute scarcities cannot gloss over the yearly shortages which victimize the poor. In either case, these are not met with an intent that would make a significant difference to the needy. For reasons already noted, there is politically much more at stake in Calcutta than in the rest of West Bengal. But this is only part of the explanation and by far not the primary one. The basic reason is that it is questionable whether a state government other than the present one would do better so long as agricultural production remains at levels where yearly deficits are the norm and the agricultural rural structure remains unchanged.

The commonly prevailing view is that West Bengal has no business of being a food-deficit state since its soils are generally good and the state "floats on water." It conjures up a favorable setting for the biological-chemical technologies—irrigation, fertilizers, pesticides, and so forth associated with high-yielding varieties and such benefits as land augmentation, labor intensification, and rising output. The application of this formula in West Bengal is limited, but to the extent of its penetration, the state's agricultural economy didn't stand still. Irrigation facilities, fertilizer use, and grain production have risen and wheat is becoming an important crop. On the other hand, when the state's physical attributes are compared with its needs, the picture is one of deficiencies in every field of agricultural activity. A high official in Calcutta made no bones about it, assigning the blame to the state's preoccupation over time with industrial rather than agricultural priorities.

There are other, weightier reasons already alluded to which have thwarted development and are responsible for the permanency of the lean months and food shortages among the disinherited. A couple of other examples may be added; one of them is the much talked about "integrated rural development." In West Bengal this is little in evidence. Administratively, too, there are special problems arising from a number of uncoordinated ministries all concerned with agriculture, directly or indirectly. One of the ministers rightly bemoaned to an inquiring visitor his remoteness from, and unfamiliarity with, the activities of the other ministers. That something was seriously amiss on the rural scene received once again official recognition recently. On December 14, on the concluding day of congress workers' deliberations conclave, the minister of finance delivered himself of some observations worth mentioning, although not exceptional. He was reported to have said that, unless 500,000 tons of rice were procured this year, the public distribution system would fail. "Either we procure or we perish," he cautioned. In view of West Bengal's record of successfully milking the central pool while procuring poorly, the consequences of the admonition may have been exaggerated; but the need to procure that volume cannot be gainsaid; less threatening, judging by past experience, is the finger of accusation leveled at the well-to-do farmers. In respect to the rural economy, the minister stated that agricultural development, land reforms, minor irrigation, and cottage and small industries should be given greater importance in West Bengal's plans. Sufficient attention, he said, had not been paid to basic problems of agriculture.[1] The minister's strictures are well taken, but injecting new life into the agricultural economy will not be easy and will take great tenacity of purpose and some doing. What will not fit the bill is the recently devised comprehensive development project (CADP) which is talked

1. *The Sunday Statesman* (December 16, 1974).

about in official circles as a major item of rural reconstruction.[2] But taking the most extravagantly favorable view of all the endeavors, public and private, it is assumed that West Bengal could eventually fill the 1.5 to 2 million tons gap. Only in that event will the food situation ease and some of its beneficial effects will devolve upon rural poor as well.

Whether the poor eat better is less a function of increased production than of purchasing power they command, but the latter will remain at low levels or hardly a part of the "income stream" if the agrarian structure continues to be dominated by a handful of wealthy farmers. Land reform—redistributive as well as protective—suggests itself as an obvious antidote against inequality of land and other income-earning assets. This presumably would strike at the root cause of inequalities and their perpetuation. In West Bengal, particularly, this is easier said than done. The pockets of high concentration of ownership are there; the 5 percent who own 50 percent of the land could stand a thinning-out process, although the poorest of the poor, the landless, with no rights in land, would hardly come in for any shares. This approach, however, is mostly academic; the political climate for any redistributive scheme is not "right," and the ceiling legislation and its ultimate effect on land redistribution among the needy can be dismissed as of marginal value, as illustrated by the following development. The area slated for redistribution in West Bengal under the ceiling laws is 452,000 acres. In the past two years 58,000 acres have been acquired but only 5,000 acres were actually distributed. The remainder is under dispute in law courts. A more rewarding way of helping one category of the poor—though not solving their problems—is to give the sharecroppers or tenants complete security of tenure, as distinguished from distributive reforms. This would mean enforcement of the already existing legal rentals, sharing of inputs, recorded rights in the land, an end to evictions, and delivery of promised access to institutional financing of their agricultural operations. As for the landless, the only likely means to improve their food purchasing power are threefold: an extended employment period; wages in kind rather than cash; and, if cash, a binding minimum wage. One is tempted to suggest a drastic scaling down of indebtedness—by fiat if necessary—for it is in part the consequence of short employment and meager wages. But such a proposal is futile unless the mentioned preconditions are enforced.

These are the familiar "should be dones," and they are paraded once again with great reluctance. Experience over the years has shown that these well-known remedial measures are poles apart from what the owners are willing to accept, what the state government is prepared to impose, or the would-be beneficiaries can enforce on their own. The only possible exception is an increase in double cropping with its attendant rise in employment, provided it is not diluted by the rapidly growing number of ruralists looking for jobs. This is not to disparage the special agencies created to assist the small and marginal farmers and here and

2. CADP has been about two years in the making, and the testing of it is some time off; but when in operation it is supposed to be very innovative, the gist of it being compulsory consolidation of holdings and collective utilization of irrigation facilities on holdings within a project block of 10,000 acres or more and the eventual application of the "cooperative principle." Personal interviews with the developers of the idea did not shed much light on how Bengal's agriculture will be reconstructed except in rather far-fetched generalities. Administrative arrangements, sources of "seed money" and just plain money, and an infrastructure in its various forms are still hanging in the air. The only prepared project (about one block in Naida district) offers a glimmer of what CADP hopes to achieve. The area in question, made up of 10,000 cultivated acres, 4,600 owner cultivators and sharecroppers, and 2,714 landless laborers, might offer employment of 275 days per year at an average rate of Rs5 per day to 60 percent of the labor force. There is nothing firm about any of this, partly because everything is still only on paper and partly because the infrastructure is very poor. As the project report puts it: "If the present type of socioeconomic institutions characterized by increased landlessness of peasants, the extremely high rate of sharecroppers' rent, the very high rate of interest charged by the moneylenders (which is practically the only source of credit for the overwhelming majority of farmers) continue, then even the meager physical infrastructure that can be provided by the state cannot be fully utilized by the people." This untried model is far therefore from the sine qua non of Bengal's rural development. The qualms of this observer have been brushed aside; instead, the virtues of this would-be panacea are extolled and with the clearly expressed hope that somehow the World Bank would see fit to lend its support to it.

there the landless through special employment programs. But emerging with the memory of "normal" and abnormal conditions prevailing in rural West Bengal, what can one do but repeat these truisms at the sight of the hungry in good and bad crops years alike. What is one to say at the sight of gruel kitchens which deaden hunger more than they feed? Or at the sight of works programs which are caricatures hardly relieving anybody? Or modified rationing of 500 grams of grain per person per week? All this and more of the same at the crest of a food emergency.

At this writing, the inadequate relief measures are taking a holiday and normalcy is taking over with the benefits of land ownership, credit, and technology sopped up by the well-to-do farmers. For the recently troubled government of West Bengal, this is a time for a sigh of relief; but what it pretends not to know or doesn't care or is incapable of recognizing is a reality that cannot be wished away—that underconsumption and low purchasing power or no purchasing power to speak of for vast numbers is a part of that very normalcy. In the existing framework of West Bengal's land limitations and how it is owned and used and the bulging numbers pressing against it, perpetuation of poverty is virtually unavoidable. And yet acceptance of this state of affairs as if preordained from on high cannot be tolerated. It is an affront to the multitudes suffering from it, and it should be no less so to all those who speak of the abolition of poverty and of democracy, egalitarianism, socialism, and the general uplift of the downtrodden. Giving expression to all these is indeed preferable than passing them over in silence, but with the proviso that the *deed* is not "out of station,"[3] as is frequently the case.

If all these worthy avowals are to acquire meaning, the practical approach is not to mouth the resilience of the poor. That they "can take it" can no longer be treated as an attribute of national hardihood. This has served for all too long both as a source of lassitude for those who should know better and as a denial of basic rights for those who need them most. The fact is that the poor are the Achilles' heel of West Bengal and of all that eats at its vitals. The cure, partial though it may be, lies in the immediate concentration on the elementary—a stitch here and a stitch there on the canvas of the protective reforms to mitigate the worst conditions under which they live and work in normal times as in times of natural calamities.

3. The common expression "out of station" refers to officials out of the town in which they serve.

APPENDIXES

A. The Atcheson-Fearey Memorandum

This is the first of two basic official documents which led to the post-World War II land reform in Japan. On October 26, 1945, George Atcheson, Jr., the acting political adviser to General Douglas MacArthur, transmitted to the general a memorandum "prepared by a member of my staff, in consultation with Far Eastern experts of the U.S. Department of Agriculture, prior to his departure from Washington." The memorandum, entitled "Japan: Agrarian Reform," was prepared by Robert Fearey of the U.S. State Department on the basis of Wolf Ladejinsky's professional input. Copies of the memorandum and transmittal were sent to the State Department on the same date. It was General MacArthur's acceptance and decision to act upon this memorandum which led to the land reform. Ladejinsky's own account of why MacArthur responded positively to this recommendation is presented in his "From a Landlord to a Land Reformer" (II–16).

OFFICE OF THE UNITED STATES
POLITICAL ADVISER

Tokyo, October 26, 1945

CONFIDENTIAL

SUBJECT: Agrarian Reform in Japan

The Honorable
The Secretary of State
Washington, D.C.

Sir:

I have the honor to enclose a copy of a paper, "Japan: Agrarian Reform," and a copy of a memorandum of October 26, 1945, transmitting the paper to General MacArthur.

It appears to us that the problem of agrarian reform is one of far-reaching importance and deserves careful attention in the current program for the democratization of Japanese economy.

Respectfully yours,

George Atcheson, Jr.

OFFICE OF THE UNITED STATES
POLITICAL ADVISER

Tokyo, October 26, 1945

MEMORANDUM FOR: Supreme Commander and Chief of Staff

SUBJECT: Agrarian Reform in Japan

There is attached, as of possible interest, a paper, "Japan: Agrarian Reform," prepared by a member of my staff, in consultation with Far Eastern experts of the U.S. Department of Agriculture, prior to his departure from Washington.

Perpetually depressed conditions in agriculture have enabled the military during recent decades to pose as the champions of the farming class, defending its interests against the privileged classes and arguing with considerable success that the only real solution of the farmer's plight lies in overseas expansion. Continuation of the present conditions may in future be expected to leave the farming class a fertile field for military propaganda.

The ruling and propertied classes have always strongly opposed the initiation of basic agrarian reforms and, judging by the small attention which the Higashi-kuni and Shidehara governments appear to have given the matter, will probably continue to do so. Although the Socialist and Communist Parties have announced intention to carry out agrarian reforms, the changes proposed seem ill-considered and inadequate, and occupy a subordinate position in their respective platforms.

We would suggest as a possible course of procedure that the Japanese Government be instructed to initiate studies of the problem, and to draw up, in consultation with the appropriate sections of your Headquarters, a comprehensive reform program which, after receiving your approval, would be subjected to public scrutiny before presentation to the Diet. Implementation of the reforms should not, of course, be permitted to interfere with the production of foodstuffs during the present period of acute shortages.

George Atcheson, Jr.

The Japanese Agrarian Problem

INDUSTRIALIZATION IN JAPAN did not lead, as it did in England and certain other countries, to the withdrawal from cultivation of a large proportion of the cultivated area and the relegation of agriculture to a position of relatively minor importance in the economy. During the last seventy years the Japanese farm population has increased 25 percent; improved farming techniques have been developed; and the cultivated area has been painfully expanded with the result that the annual farm output during recent decades, far from declining, has been from two to three times greater than during the feudal era. Japanese agriculture has continued to supply the bulk of the country's food requirements notwithstanding the more than twofold increase in the population and at the same time has been the source of the country's leading export item, raw silk.

The rewards to agriculture, however, have not been commensurate with its contribution. The farming population has been permitted to retain only a small proportion of its expanded output for its own consumption or to exchange for the products of industry. The greater part has been taken from it in the form of high rent, interest, and tax charges and used for the benefit of industry and trade and the furtherance of the military program. As a consequence, although living standards in the farming districts have improved materially since feudal times, the improvement has been much less than proportionate with the increase in the average farmer's annual product; and standards are still extremely low.

Nevertheless, until about twenty-five years ago, the farming population remained reasonably content with the gains which it had made. In the interwar period, however, not only were no further gains recorded but there was a definite deterioration in the farmer's economic status. The farm debt increased eightfold, and each year additional freeholders lost their land through foreclosure and had to accept a tenant status. As it became clear that the government planned no important reforms in their behalf and that the future held forth only the prospect of still greater poverty and distress, the long-suffering farming class showed signs of increasing discontent. Dissatisfaction reached its peak during the early and middle 1930s, subsiding thereafter only because of the developing war crisis.

The unsatisfactory state of Japanese agriculture can be traced to two fundamental causes—natural limitations and handicaps, and the government's discrimination against agriculture in favor of industry and trade. While no less than nine separate elements in the farmer's plight are listed and described in the following pages, it will be found that they all derive from these two basic causes.

"Too many men on too little land"

The most important single cause of the unsatisfactory conditions prevailing in Japanese agriculture is overcrowding on the land. About 5.5 million farm households till only 15 million acres, or an average of only 2.7 acres per household. This latter figure is to be compared (leaving out of account differing degrees of fertility) with an estimated 3.0 acres per farm household in China, 3.6 acres in Korea, 10 acres in the United Kingdom, 47 in the United States, and 80 in Canada. Many Japanese farm families, however, cultivate much less than 2.7 acres. In 1938, 33.8 percent tilled less than 5 tan (1.2 acres) and 66.5 percent tilled less than 1 cho (2.5 acres).

Alleviation of this condition cannot be expected through expansion of the arable land area as a consequence of war casualties or barring revolutionary advances in the science of fertilization through increased yields per acre. During the last seventy years it has been possible to increase the area of cultivated land by only 25 percent, and that only with the greatest effort and expense. Moreover, all of this expansion was accomplished before 1921 when the peak of 15,101,000 acres was reached. Since that time total acreage has remained at a slightly lower level, varying between 14.4 and 15 million acres. War casualties cannot be expected to improve the situation, the total working population having risen substantially since 1940; and yields per acre are already among the highest in the world.

The only real solution to the problem would be the absorption of literally millions of farm workers in industrial occupations. For the foreseeable future, however, an expansion of Japanese industry sufficient to absorb this number of workers must be ruled out of practical consideration. The most that can possibly be hoped for is that industry will be able to provide employment for the greater part of the annual increment in the laboring force without thought of effecting a reduction in the farming population.

The wide extent and unsatisfactory conditions of tenancy

Next to overcrowding on the land and a direct consequence of such overcrowding, widespread tenancy under conditions highly unfavorable to the tenant is the second most important evil plaguing the farming class.

After the abolition of feudalism in 1871, only about 20 percent of the cultivated area was tilled by the tenants; but this proportion rapidly increased. The small size of the individual peasant holdings, making profitable operation difficult if not impossible, and a heavy burden of taxation produced a mounting indebtedness which caused many farmers to lose their land. The lack of alternative employment opportunities in industry or commerce for those thus dispossessed and the fact that landowners with more land than they could themselves cultivate could earn a larger return by renting to tenants than by operating with hired labor were important factors which favored the growth of tenancy. By the end of the last decade, 28 percent of the farm households in Japan rented all of the land they tilled, and an additional 43 percent rented part. Thus 71 percent of the total number of farm households suffered the disadvantages of tenancy in some degree. Of the total cultivated area, 46 percent was rented.

The rental for a one-crop field is about 50 percent of the crop, and for a two-crop field about 60 percent. All expenses connected with the raising of the crop—seed, farmhouse, sheds and implements, and all taxes and assessments except the land tax—are normally the tenant's own responsibility and must be deducted from his share of the annual product. He receives nothing from the landlord save the land and usually is not even compensated for necessary improvements which he makes upon his holding. Moreover, contracts between landlord and tenant are so loose and for such short periods and the demand for land is so great that a landlord who is for any reason dissatisfied with his tenant can evict him with relative ease. While custom and more recently the resistance of organized tenant groups have exercised an increasing restraint on this privilege, insecurity of tenure remains a prime source of discontent among tenants.

The tenants' difficulties, it should be realized, are attributable not so much to an exceptionally high rate of rent, for in fact the rate is not so very much higher than in most Western countries,[1] as it is to the small size of his holding. An English tenant farmer with 12 acres or an American with 47 acres, average in those countries, can deliver 40 to 50 percent of his crop to the landlord and still have enough for his own needs. The Japanese tenant, however, after turning over half the output of his tiny one-and-a-half or 2 acre plot often does not have enough left even for the barest subsistence. Investigations have shown that 40 percent of the tenants have to purchase rice before the year is out (with income from subsidiary occupations or with borrowed funds), and 20 percent must start their purchases within six months after the harvest. The intense competition for the land by the excessive number of farmers and the lack of alternative employment opportunities in other fields, however, leave Japanese tenants no choice but to pay the rent demanded.

The rent which weighs so heavily on the

1. Tenant cotton and tobacco growers in the United States generally turn over half their crop to the landlord and grain farmers from 40 to 50 percent. While the landlord bears most of the cost of cotton and tobacco production—seed, fertilizer, and so forth—grain farmers generally meet these costs themselves.

tenant, primarily, as explained, because of the small size of his holding, affords the landlord only a moderate return on his investment. Although the absolute amount of rent and income obtained per acre is large due chiefly to the high crop yields, the rate of return is small, about 4 percent,[2] because of the exceptionally high prices at which farming land sells in Japan. In 1938 an acre of good paddy field cost 2,620 yen (US$748) and upland 1,680 yen (US$479).[3] (In this country an acre of good Iowa farmland cost about $80 in 1938 and in the more densely settled states such as New Jersey or New York up to $130.)[4] It is not surprising that returns computed on the basis of such highly inflated land values should be low.

It should be noted that landlords who bought their land thirty or forty years ago (when population pressure was less intense and prices, accordingly, were lower) and receive the prevailing rate of rent earn a very large return on their investment. By and large, however, the tenant's plight cannot be attributed to excessive landlord profits. There has in fact been a noticeable tendency in recent years on the part of the landlords to move their capital out of cultivated land into other investment fields, notably industry, commerce, and urban real estate, where the return has been higher.

A large volume of indebtedness and high interest rates

Prior to the First World War the total farm debt in Japan was estimated at 750 million yen, or 135 yen per farming household. After that time the increasing disparity between farm income and expenditure led to a rapidly mounting indebtedness so that in 1937 the debt totaled more than 6 billion yen, or over 1,000 yen per household. Fifty-six percent of the debt was at that time (1937) held by moneylenders, 14 percent by the Hypothec Bank and the Agricultural and Industrial Bank, 13 percent by private banks, and 17 percent by other financial institutions.

Although interest rates are high for all types of borrowers in Japan because of the chronic shortage of capital, they have been highest by an appreciable margin in agriculture, averaging between 12 and 15 percent per year and frequently totaling as much as 20 or 30 percent. The annual charge on the farm debt consumed almost a third of the net yearly farm income during the period 1931–35; and, in later years when farm incomes were higher, it consumed between 18 and 20 percent.

Discriminatory fiscal and tariff policies

Notwithstanding the much lower per capita productivity and average level of income in agriculture than in industry, a farmer's taxes, as the following table indicates, have been substantially higher than those of a merchant or an industrialist with the same income.

At the same time the farming class has received much less than its proportionate share in the expenditure of public funds, farm aid being only a fraction of the assistance afforded industry and trade. Also, farm products have been denied the tariff protection accorded many leading manufacturing industries. Imported food products have been admitted free or at very low rates of duty.

Widely fluctuating prices for rice and silk

Widely and sharply fluctuating prices for rice and raw silk, which have together comprised 65 percent of the total value of farm production, have been another factor contributing to the unhappy lot of the farmer. In the case of rice, an inelastic demand in conjunction with large variations in the size of the crop, and the difficulty of controlling rice imports from the colonies have been the principal causes of price fluctuations. As for silk, approximately 80 percent of the raw silk output has been sold in the United States and Europe where, being a luxury product, prices have varied widely in

2. The landlord's return averaged 4.4 percent for paddy fields and 3.7 percent for upland fields during the prewar decade. (Shiroshi Nasu, *Aspects of Japanese Agriculture* [1941], p. 130.)

3. The yen had an average value of 28.5 cents during 1938.

4. These American figures might more properly be compared with the Japanese yen figures quoted above than with their American equivalents, because a yen had approximately the same utility value to a Japanese that a dollar has to an American.

RELATIVE BURDEN OF DIRECT TAXES ON LANDOWNERS AND BUSINESS PROPRIETORS IN JAPAN—1929

	Per capita taxes (in yen)					
	Paid by farm landowners			*Paid by business proprietors*		
Annual income	*Total*	*National*	*Local*	*Total*	*National*	*Local*
1,200	270	68	202	126	44	82
2,000	530	144	286	230	88	142
3,000	874	259	615	366	157	207
5,000	1,395	457	938	701	316	385
10,000	3,485	1,187	2,298	1,603	814	789
30,000	12,097	4,956	7,141	6,819	3,737	3,081
100,000	53,225	22,168	31,057	30,058	17,883	12,175

Source: Hioye Ouchi, "Tax Burden on Salaried Men and Farmers as Revealed by the Official Survey of their Livings," XIX Session de L'Institut International de Statistique (Tokyo, 1930), p. 9.

accordance with business conditions. Government efforts to stabilize the price of rice have met with at least some success, but in the case of silk they have been almost completely unavailing.

The competition of cheap colonial rice in the Japanese home market

Rice adapted to Japanese taste is produced in Korea and Formosa at approximately 25 percent lower cost than in Japan. Imports of rice, duty free, from these colonies to supply the growing deficiency in Japan's domestic supplies during the last two decades have destroyed the monopolistic position formerly enjoyed by domestic producers and have exerted a depressing effect on prices. Because of the ready availability of rice from Korea and Formosa, the Japanese farmer, although suffering the full effect of a price drop during years of good crops, is almost never compensated by higher prices during short years. With the exception of the rice year 1939–40 when poor crops were harvested both in Japan and in the colonies, colonial rice has always been available to fill the market.

A declining income from sericulture

Japanese silk exports declined from \$361 million in 1929 to \$79 million in 1933. From 1934 to 1941 they remained in the neighborhood of \$110 million. In part the decline was due to depressed business conditions in foreign markets. An even more important cause, however, was the almost complete substitution after 1935 of rayon for natural silk in the manufacture of broad goods and beginning in 1939 the rapid replacement of silk by nylon in women's full-fashioned hosiery. The burden of this loss was borne mainly by the Japanese farming class, which in former times had obtained approximately 20 percent of its income from sericulture. If, as is expected, the substitution of synthetic for natural silk continues after the war, this important source of cash income will be even further reduced.

The high price of fertilizers

Artificial fertilizers are an essential element in farm production in a country like Japan where the soil has been depleted by centuries of intensive cultivation. Recent experiments representing an average of results from soils all over Japan show that in the absence of artificial fertilizers of any kind yields are 48 percent below those of plots fully fertilized. Naturally it is of great importance to farmers that they should be able to obtain fertilizers at reasonable prices. The artificial fertilizer industry, however, has set high, monopolistic prices which it has maintained with little regard for fluctuations in the prices of farm products. Government efforts since 1936 to reduce fertilizer prices by bringing the industry under some measure of public control have not been successful.

Exploitation of farmers by middlemen in the marketing of their crops

Middlemen, speculators, and petty traders have long used their stronger bargaining position relative to the individual peasant to extract excessive commissions for their services. Farm cooperatives, which dispose of member's crops, provide common warehousing facilities, and furnish other joint services, have done much to overcome this evil but have not eliminated it. The poorest farmers who need the protection afforded by membership in a cooperative most are frequently prevented by their poverty from becoming members and so remain at the mercy of unscrupulous dealers.

The combined effect of these difficulties, as already indicated, is a very low standard of living in the farming districts. This standard, it is true, is materially higher than in feudal times in Japan or than in most other Oriental countries today. It must be remembered, however, that Japanese farmers lived under the worst possible conditions in the feudal period, and over most of the Orient such conditions still prevail. Moreover, since the general plane of living in Japan is higher than in other Oriental countries, the Japanese having raised themselves to a position in many respects closer to Western than to Oriental norms, the consumption levels of the Japanese farming class are not fully comparable with those of most other Far Eastern farming populations. The more significant comparison is with the living standards of other sections of the Japanese population which are materially better than the farmer's.

The unsatisfactory state of Japanese agriculture has been like a cancer in the economic and political life of the country. The large farming class has remained a perpetually discontented element in the population and a principal support of the military program in which it has seen virtually the only hope of a solution of its economic ills. At the same time low farm incomes and the small total purchasing power of the farming districts have meant a restricted domestic market for Japanese manufacturing industry, contributing to dissatisfaction there. It is difficult to see how reasonably prosperous conditions and peaceful tendencies can prevail in Japan until conditions in agriculture have been markedly improved.

Measures of Agrarian Reform

We have seen that the principal source of the Japanese farmer's difficulties is an insufficient area of land to till. Some five-and-a-half million farm households must obtain a livelihood from only 15 million acres. Unfortunately, however, there is no prospect that this basic handicap can be overcome, either through expansion of the farm acreage or output as a consequence of war casualties or, at least in the foreseeable future, through the movement of the surplus farming population into other occupations. It can only be accepted and attention directed to the solution of other evils plaguing the farming class.

These evils, however, are both numerous and important. While many are a consequence of the basic difficulty of "too many men on too little land," they are not an inevitable consequence and, it is believed, may be overcome while that handicap still remains. The most important are (a) widespread tenancy under conditions highly unfavorable to the tenants, (b) a heavy burden of farm indebtedness combined with high rates of interest on farm loans, and (c) government fiscal policies which discriminate against agriculture in favor of industry and trade.

Widespread tenancy under conditions highly unfavorable to the tenants

Twenty-eight percent of the farming households in Japan rent all the land they till, and an additional 43 percent rent part. The rented area comprises 46 percent of the arable acreage. Rentals are extremely high, considering the small size of the average holding; and the tenant must bear all expenses of the farm except the land tax. Usually he is not even compensated for necessary repairs and improvements he makes on the property. Living standards, while low for all classes of Japanese farmers, are particularly low for the tenants who, in addition, can be evicted by the landlord with relative ease.

IMPROVEMENT OF THE CONDITIONS OF TENANCY. The tenancy problem may be approached either with the idea of improving the conditions of tenancy or of accomplishing the complete emancipation of the tenant class. Considering the former alternative first, the most important handicap under which the tenant operates, the high rent, might be alleviated by setting a ceiling on farm rents, limiting them by law to, let us say, 25 percent instead of the prevailing 50 to 60 percent. Also, the landlords' right to evict their tenants might be made subject to stricter legal provision, affording the latter greater security in their right to till the land. Regulations requiring the landlords to compensate their tenants for needed improvements and to bear a larger proportion of the costs of farm production (such as seeds, fertilizers, tools, and taxes other than the land tax) would be ineffective if the landlord were still able to raise rents and offset any added costs to himself. But in combination with a rent ceiling they might be of help.

This program suffers from several important disadvantages, however. Even under present conditions most landlords earn only a moderate return on their investment;[5] and, while the absolute amount of the rent per acre is quite large due to the high yields, it is by no means exorbitant. The suggested reforms, therefore, would so reduce the landlords' return (if they did not entirely eliminate it) that many landlords would decide either to cultivate the land themselves or to sell it. In both cases this decision would result in the eviction of the tenants already tilling the land. This would clearly follow if the landlord decided to cultivate the land himself and would be scarcely less certain to occur if he undertook to sell it, as few tenants would possess the necessary funds to purchase their land. They might be able to buy over a long period of time, say thirty or forty years; but then during that entire period they would be obtaining little or no benefit from the reforms. Annual payments toward the purchase of the land would offset most of the gain from the reduction in rent and shifting of part of the costs of production to the landlord. Also the landlords would probably choose to sell to buyers who already had the purchase price available. Thus many tenants, far from benefiting from the reforms, would be driven from their land and forced to adopt the even less enviable status of hired farm workers. A situation would be likely to develop similar to that in Hungary, Poland, and other Eastern European countries; much of the land now being farmed by tenants would be purchased by large owners who would convert them into big estates farmed by hired workers under professional managers.

A further disadvantage of plans which seek merely to improve the conditions of tenancy would be that, even with a rent reduction of as much as 50 percent and other significant reforms, conditions for those tenants who were not evicted would still be highly unfavorable. The average tenant's holding is so small that a rent only half as large as at present still would not leave him an adequate income. Finally, since there would have been no fundamental change in landlord-tenant relationships (for the tenants who were not evicted), the landlords, because of the continuing scarcity of land and multiplicity of farmers seeking land to till, would still possess every advantage in dealing with their tenants and would probably succeed in gradually restoring the former conditions of tenancy.

EXPROPRIATION OF ALL RENTED FARMLAND FOR THE BENEFIT OF THE TENANTS, COMPENSATION TO BE PROVIDED THE LANDLORDS IN AN AMOUNT EQUAL TO THE PRODUCTIVE VALUE OF THE LAND EXPROPRIATED. There are of course a large number of plans by which the complete emancipation of the tenant class might be effected, of which only one is considered here. Under this plan all rented farmland in Japan would be expropriated by the government and turned over to the tenants in full title. The landlords would receive in compensation government bonds in an amount equal to the productive value (capitalized value of the average net annual product) of the land expropriated. Such compensation would not be paid until stable financial conditions have been restored and bonds of enduring value could be delivered. If this should not be for some time after the transfer had been effected, landlords without other means of support might have to be assisted.

MARKET VALUE VERSUS PRODUCTIVE VALUE. Undoubtedly the landlord class in submitting

5. See above, under "Wide extent and unsatisfactory conditions of tenancy."

to the expropriation of their rented land would demand compensation on the basis of the inflated market value of their holdings. By the end of the last decade, however, the market price for farmland had been driven by the intense demand for such land to a point 40 to 50 percent higher than its productive value (capitalized earning power). The only valid measure of an earning asset's worth is its earning power. Demands for compensation in excess of the capitalized value of the annual net product of the land expropriated should not be entertained.

THE ABILITY OF JAPANESE FINANCES TO BEAR THE BURDEN OF COMPENSATION. It is believed that compensation computed on this basis would not impose too heavy a burden on Japan's postwar government finances. In 1939 there were 6.8 million acres of rented farmland in Japan with a total productive value of approximately 8 billion yen:

LAND TILLED BY TENANTS, 1939

Land type	*Acres*	*Average market value per acre (yen)*	*Estimated average productive value per acre (85 percent of market value) (yen)*	*Total productive value (million yen)*
Irrigated fields	4,071,000	2,620	1,440	5,862
Dry fields	2,745,000	1,680	920	2,424

Interest of 4 percent on 8 billion yen worth of bonds, assuming a value for the yen approximately the same as in 1939, would amount to 320 million yen a year, or only about 15 percent of the average annual budgetary expenses of Japan's national government during the period 1931–36. At that time the tax rate in Japan was not excessive and an increase of 320 million yen in the tax load could have been borne without difficulty, as was subsequently shown. Naturally, the fact that the debt would be owed within Japan would greatly simplify the problem of its support; service on the bonds would involve no more than a transfer of wealth from one section of the population to another. It might be noted that the United States expects to support a national debt ten times greater than in 1937 with a national income which even in the most optimistic view will not be more than twice as large as at that time.

A heavy burden of farm indebtedness and high rates of interest on farm loans

While the emancipation of the tenant class would be the principal feature of a program of agrarian reform, it would not be sufficient by itself to accomplish the desired purposes. The basic handicap of an insufficient area of land to till and many other evils would still remain so that, if this step were not accompanied by others, the newly established freeholders would still not have achieved a satisfactory economic status and would probably soon lose their land again. Also, nothing would have been done to assist those farmers who at present own their farms, many of whom, borne down by taxes and debt, have been little better off than the tenants.

The farming debt totaled approximately 6 billion yen in 1937. During the period 1931–35 the annual charge on this debt consumed almost 30 percent of the average net yearly farm income; and, in later years when farm incomes were substantially higher, it took from 18 to 20 percent. There is reason to believe, however, that the debt has been considerably reduced since that time. Exceptionally high farm incomes, the lack of spending outlets, and currency inflation are believed to have resulted in the liquidation of a large proportion of the debt. Depending upon its size, part or all of the current debt might be assumed by the government at such time as Japanese finances may be able to take on this comparatively small additional burden.

Reduction or elimination of the existing debt would be of only temporary benefit if steps are not taken at the same time to ensure reasonable interest rates on new loans. Rates have been high for all types of borrowers in Japan, but they have been highest by an appreciable margin in agriculture, averaging from 12 to 15 percent per year and frequently totaling as much as 20 or 30 percent. The principal cause of these high rates has been a large proportion of bad debts combined with the inability of farmers to provide adequate security. The improved credit standing of the farmers as a consequence of the reforms herein proposed should do much to remove this difficulty but would need to be supplemented by other measures to ensure adequate supplies of loan capital to farmers at reasonable rates. A government farm loan agency might be set up and official support accorded private cooperative credit associations, which already exist in Japan in fairly large numbers. Because most farm loans are self-liquidating over a short period, the chronic shortage of capital in Japan should not prevent the establishment of reasonable rates.

Discriminatory fiscal policies

We have seen that owners of farmland in Japan bear a disproportionately large share of the nation's tax burden. Beside the national land tax, they must pay prefectural and town or village surtaxes and numerous local fees and taxes. With the elimination from the national budget of military expenditures of all kinds, it will be possible to reduce taxes for all sections of the population; but in addition steps should be taken to bring about a more equitable distribution of the tax burden. This will require revision of existing prefectural and local as well as national tax laws. Also, in the expenditure of public funds, the needs of agriculture should receive equal consideration with industry and trade, since they have not in the past.

Other fields for reform

Measures regarding tenancy, indebtedness, and fiscal policy would constitute the principal features of a program of agrarian reform during the occupation period. Simply by themselves they would result in a marked improvement in the farmer's status. But even with this improvement the farmer's position, considering the large number of farmers who must obtain a livelihood from so small an arable area, is likely to be precarious. Consequently attention should be directed also to the solution of other difficulties from which the farming class has suffered.

FLUCTUATING PRICES FOR RICE AND SILK. One such difficulty has been widely and sharply fluctuating prices for rice and silk, which together comprise 65 percent of the value of farm production. In the past, efforts to stabilize the price of rice have suffered from a lack of funds, the difficulty of controlling rice imports from the colonies, and an insufficient determination on the part of the government to make a success of the control program. With the elimination of all military expenditures from the national budget, funds for this purpose should be more plentiful. Also, imports from the colonies probably can be more easily controlled when these areas are no longer an integral part of the Japanese economic system and the interests of Japanese nationals engaged in Korean and Formosan rice production no longer have to be protected. Rice purchases from these areas can be regulated by the government so as to offset crop variations in Japan and maintain stable prices.

In the case of silk the problem is more difficult. Silk prices vary with business conditions in overseas markets, which, of course, are beyond Japan's power to control. Also raw silk costs so much to store because of the insurance that plans to stabilize prices by controlling the release of stocks from storage are not practicable. Probably all that can be attempted is a system of relief payments to sericulturalists in years of low prices from funds obtained from a tax on silk in good years.

THE COMPETITION OF KOREAN AND FORMOSAN RICE. Japanese farmers would benefit greatly from a tariff on Korean and Formosan rice, which in the past has entered Japan duty free. On the other hand such a tariff, through its effect on food prices in Japan, would injure the country's export industries. Higher food prices would necessitate higher wages in industry, which in turn would reduce Japan's competitive advantage in international trade. Any reduction in trade would be to the disadvantage of the entire country, agriculture included. It would therefore seem that attention should be directed to other methods of farm aid less obstructive of the interests of industry and trade and of Japanese economy as a whole.

A DECLINING INCOME FROM SERICULTURE. The expected further decline in the demand for natural silk after the war is bound to have a serious effect on farm incomes. In recent years the farming class has derived from 12 to 16 percent of its income from sericulture, which, considering how close to a subsistence level this class lives, has often meant the difference between just getting by and acute privation. There has already been some adjustment to the loss of this business, chiefly through reduction of the mulberry acreage in favor of other crops. It is important, however, that the farm population find other seasonal and part-time employment in which the whole family can engage in partial replacement of the silk industry. Certain small- and medium-scale industries normally found only in cities and towns have already proved adaptable to productive conditions in the rural areas. With study and investigation, the number of these industries might be considerably increased, providing the farmer with the subsidiary employment he requires.

THE HIGH PRICE OF FERTILIZERS. Japanese farmers, who must use large quantities of artificial fertilizers, have been greatly handicapped by the high prices which they have had to pay for these fertilizers. Approximately 10 percent of the farmer's gross income finds its way into the pockets of the fertilizer monopoly every year. It would seem that this industry either should be taken over by the government as an industry intimately affected with the public interest; or output, prices, and sales should be strictly controlled so as to assure farmers adequate supplies of fertilizers at lowest reasonable prices. Joint purchasing arrangements have enabled some farmers to supply their fertilizer needs at lower cost, but this is not an adequate solution of the problem.

EXPLOITATION OF FARMERS BY MIDDLEMEN IN THE MARKETING OF THEIR CROPS. Finally, steps should be taken to assist and protect the farmer in the marketing of his crops. In the past, middlemen, speculators and petty traders have been able to take advantage of the farmers' urgent need for funds to extract excessive commission for their services. Farm cooperatives have been fairly successful in overcoming this evil, but their membership often does not include the poorest farmers who need their protection most. The cooperatives should be strengthened in every way possible and their membership expanded to include the less well-to-do farmers. At the same time, the activities of private traders should be subjected to closer regulation.

Of the above measures, tenancy reform is of predominant importance. Also, it would appear that the second of the proposed solutions of this problem, calling for the complete emancipation of the tenant class, is much to be preferred over the first for the reasons already stated and because any other reforms which might be introduced would have only a limited effectiveness if the tenants had not previously been emancipated. As long as they were still tenants, the 28 percent of the total number of farming households which rent all their land and, to a lesser extent, the 43 percent who rent a part would derive little benefit from a reduction in interest rates, for example, or from a more equitable distribution of the tax burden, or lower fertilizer prices. The increase in net income obtained as a consequence of these reforms would be taken by the landlords, who would raise rents in the same proportion that incomes were increased. A rent ceiling, for the reasons earlier stated, would likely result in a mass eviction of tenants, the landlords tilling the land themselves or with hired labor. As long as the demand for farming land continues greatly to exceed the available supply, and the landlords, as a result, are able to establish the terms of tenancy practically at will, they will be able to transfer to themselves the greater part of the benefits of whatever reforms are instituted.

B. SCAP Directive 411 on Rural Land Reform

This is the order issued by General MacArthur's office on December 9, 1945, instructing the Japanese imperial government to submit by March 15, 1946, a program of rural land reform which would "insure that those who till the soil of Japan shall have more equal opportunity to enjoy the fruits of their labor." In terse military style, it enumerates the "more malevolent" of the "pernicious ills which have long blighted the agrarian structure" and instructs the Japanese government to incorporate in its land reform program proposals not only basic measures for dealing with the enumerated ills but also "provisions for reasonable protection of former tenants against reversion to tenancy status." Interestingly, this far-reaching order is signed by an assistant adjutant general with the rank of colonel.

MEMORANDUM FOR: IMPERIAL JAPANESE GOVERNMENT
THROUGH: Central Liaison Office, Tokyo
SUBJECT: Rural Land Reform

1. In order that the Imperial Japanese Government shall remove economic obstacles to the revival and strengthening of democratic tendencies, establish respect for the dignity of man, and destroy the economic bondage which has enslaved the Japanese farmer to centuries of feudal oppression, the Japanese Imperial Government is directed to take measures to insure that those who till the soil of Japan shall have more equal opportunity to enjoy the fruits of their labor.

2. The purpose of this order is to exterminate those pernicious ills which have long blighted the agrarian structure of a land where almost half the total population is engaged in husbandry. The more malevolent of these ills include:

a. *Intense overcrowding of land.* Almost half the farm households in Japan till less than one and one-half acres each.
b. *Widespread tenancy under conditions highly unfavorable to tenants.* More than three-fourths of the farmers in Japan are either partially or totally tenants, paying rentals amounting to half or more of their annual crops.
c. *A heavy burden of farm indebtedness combined with high rates of interest on farm loans.* Farm indebtedness persists so that less than half the total farm households are able to support themselves on their agriculture income.
d. *Government fiscal policies which discriminate against agriculture in favor of industry and trade.* Interest rates and direct taxes on agriculture are more oppressive than those in commerce and industry.
e. *Authoritative government control over farmers and farm organizations without regard for farmer interests.* Arbitrary crop quotas established by disinterested control associations often restrict the farmer in the cultivation of crops for his own needs or economic advancement. Emancipation of the Japanese farmer cannot begin until such basic farm evils are uprooted and destroyed.

3. The Japanese Imperial Government is therefore ordered to submit to this Headquarters on or before 15 March 1946 a program of rural land reform. This program shall contain plans for:

a. Transfer of land ownership from absentee land owners to land operators.
b. Provisions for purchase of farm lands from non-operating owners at equitable rates.
c. Provisions for tenant purchase of land at an-

nual installments commensurate with tenant income.

d. Provisions for reasonable protection of former tenants against reversion to tenancy status. Such necessary safeguards should include: (1) Access to long- and short-term farm credit at reasonable interest rates; (2) measures to protect the farmer against exploitation by processors and distributors; (3) measures to stabilize prices of agricultural produce; (4) plans for the diffusion of technical and other information of assistance to the agrarian population; and (5) a program to foster and encourage an agricultural co-operative movement free of domination by non-agrarian interests and dedicated to the economic and cultural advancement of the Japanese farmer.

e. The Japanese Imperial Government is requested to submit in addition to the above such other proposals it deems necessary to guarantee to agriculture a share of the national income commensurate with its contribution.

FOR THE SUPREME COMMANDER:

H. W. Allen
Colonel, A.G.D.,
Asst. Adjutant General

Chronological Bibliography of Wolf Ladejinsky

Papers included in this publication or included in part are designated by an asterisk. The number of pages shown for unpublished papers is the actual number for the single-spaced papers and the estimated single-spaced equivalent for all others.

*1. "Collectivization of Agriculture in the Soviet Union." *Political Science Quarterly* (March–June 1934), pp. 1–43, 207–52.

2. "Soviet Harvesting and Procurement Measures." *Foreign Crops and Markets* (July 15, 1935), pp. 52–57.

3. "Agricultural Conditions in the Soviet Union." *Foreign Crops and Markets* (September 30, 1935), pp. 465–70.

4. "Agriculture in Manchuria—Possibilities for Expansion." *Foreign Agriculture* (April 1937), pp. 157–82.

5. "Farm Tenancy and Japanese Agriculture." *Foreign Agriculture* (September 1937), pp. 425–46.

6. "The Japanese Cotton-Textile Industry and American Cotton." *Foreign Agriculture* (December 1937), pp. 589–618.

7. "Soviet State Farms." *Political Science Quarterly* (March–June 1938), pp. 60–82, 207–32.

*8. "Soviet State Grain Farms." *Foreign Agriculture* (October 1938), pp. 439–54.

9. "Agrarian Unrest in Japan." *Foreign Affairs* (January 1939), pp. 426–33.

*10. "Agricultural Problems of India." *Foreign Agriculture* (August 1939), pp. 321–46.

*11. "Japan's Agricultural Crisis." *Journal of Farm Economics* (August 1939), pp. 614–31.

*12. "Chosen's Agriculture and Its Problems." *Foreign Agriculture* (February 1940), pp. 95–122.

13. "Japan's Food Self-Sufficiency." *Foreign Agriculture* (June 1940), pp. 355–76.

*14. "Agriculture of the Netherlands Indies." *Foreign Agriculture* (September 1940), pp. 511–74.

15. "Agriculture in British Malaya." *Foreign Agriculture* (March 1941), pp. 103–25.

*16. "Agricultural Policies of British Malaya." *Foreign Agriculture* (April 1941), pp. 159–64.

17. "Manchurian Agriculture Under Japanese Control." *Foreign Agriculture* (August 1941), pp. 309–40.

18. "The Japanese Silk Industry Faces a New Crisis." *Foreign Agriculture* (December 1941), pp. 515–34.

19. ——— with Fred J. Rossiter. "Food Situation in Far Eastern and Southeastern Asia." *Foreign Agriculture* (April 1942), pp. 147–64.

20. "Thailand's Agricultural Economy." *Foreign Agriculture* (May 1942), pp. 165–84.

21. "The Food Supply of India." *Foreign Agriculture* (July 1942), pp. 265–81.

22. "Australia's Agricultural Resources." *Foreign Agriculture* (January 1943), pp. 3–24.

23. "Food Situation in Asia." *Annals of The American Academy of Political and Social Science* (January 1943), pp. 91–93.

24. "Island Agriculture in the South Pacific." *Foreign Agriculture* (August 1943), pp. 178–84.

25. "Agriculture in Ceylon." *Foreign Agriculture* (January 1944), pp. 3–20.

26. "Agriculture in Japan—Pre-War." *Foreign Agriculture* (September 1945), pp. 130–42.

27. "Landlord vs. Tenant in Japan." *Foreign Agriculture* (June 1947), pp. 83–88; (July–August 1947), pp. 121–28.

*28. "Farm Tenancy in Japan." Report no. 79. Tokyo: Supreme Commander for the Allied Powers (SCAP), Natural Resources Section, June 25, 1947, pp. 1–54.

29. ——— with William M. Gilmartin. "The Promise of Agrarian Reform in Japan." *Foreign Affairs* (January 1948), pp. 312–24.

*30. "Trial Balance in Japan." *Foreign Affairs* (October 1948), pp. 105–16.

31. "The Balance Sheet." (Alternative version to no. 30, unpublished, 1948), 17 pages.

32. ——— with Warren H. Leonard and Mark B. Williamson. "Prospects for Japanese Agriculture." *Foreign Agriculture* (November 1948), pp. 240–45.

33. ———with Darrell Berrigan. "Japan's Communists Lose a Battle." *Saturday Evening Post* (January 8, 1949), 4 pages.

34. "Land Reform Progress in Japan." *Foreign Agriculture* (February 1949), pp. 38–41.

*35. "The Rent Reduction Program in Taiwan: Field Observations." General Report no. 1. Joint Commission on Rural Reconstruction, September 9–22, 1949, pp. 166–80.

*36. "Land Commissions in Japan." September 1949.

*37. "Field Trip in Szechwan: Tenant Conditions and Rent Reduction Program." General Report no. 1. Joint Commission on Rural Reconstruction, October 13–20, 1949, pp. 148–64.

38. "Land Reform in Formosa." *Foreign Agriculture* (June 1950), pp. 130–35.

*39. "Too Late to Save Asia?" *Saturday Review of Literature* (July 22, 1950), 5 pages.

*40. "Rural Reconstruction under the China Aid Act." *Foreign Agriculture* (August 1950), pp. 167–74.

*41. "Observations on Rural Conditions in Taiwan." Joint Commission on Rural Reconstruction, June 1951, 30 pages.

*42. "From a Landlord to a Land Reformer." Informal address at private dinner party in Taiwan, June 1, 1951, 4 pages.

*43. "The Plow Outbids the Sword in Asia." *Country Gentleman* (June 1951), pp. 65–68.

44. "Comments on Keiki Owada's paper, 'Land Reform in Japan'." Conference on World Land Tenure Problems, University of Wisconsin. University of Wisconsin, Papers, 1951, pp. 330–35. Also published under the title "Land Reform in Japan: A Comment." In *Land Tenure*. Edited by Parsons, Penn, and Raup. Madison: University of Wisconsin Press, 1956, pp. 224–29.

45. "Agriculture," *Japan*. Hugh Borton, editor. Ithaca, New York: Cornell University Press, July 1951, pp. 46–63.

46. "Japan's Land Reform." *Foreign Agriculture* (September 1951), pp. 187–89.

47. "Consolidation of Japanese Land Reform Legislation." U.S. State Department Dispatch, September 6, 1951, 7 pages.

48. "Report on Field Trip to Nagano Prefecture." U.S. State Department Dispatch, October 12, 1951, 23 pages.

49. "The Occupation and Japanese Agriculture, October 1945–April 1952." U.S. State Department Dispatch, June 3, 1952, 20 pages.

*50. "Field Observations in the Punjab." U.S. State Department Dispatch, October 10, 1952, 11 pages.

*51. "Land Reform Observations in Madras." U.S. State Department Dispatch, November 6, 1952, 25 pages.

*52. "Land Reform Observations in Kashmir." U.S. State Department Dispatch, November 25, 1952, 17 pages.

*53. "Comments on Land Reform in India." Ford Foundation Conference on Land Tenure, December 8–9, 1952, transcript, pp. 45–72.

*54. "A Comment on the Report of the Ford Foundation Conference on Land Tenure, 1952." Letter, March 23, 1953, 5 pages.

55. "Field Trip to Kyushu." U.S. State Department Dispatch, July 1953, 29 pages.

56. "Japanese Food Supply, 1953–1954." U.S. State Department Dispatch, November 30, 1953, 13 pages.

57. "Food Policy Council Established." U.S. State Department Dispatch, January 12, 1954, 6 pages.

*58. "The Status of the Land Reform Program in India." U.S. State Department, August 4, 1954, 18 pages.

*59. "Advancing Human Welfare." Ford Foundation, November 1954, 13 pages.

60. "Group Meeting of Big Landlords of South Vietnam." U.S. State Department, March 18, 1955, 4 pages.

*61. "Field Trip Observations in Central Vietnam, April 2, 1955." U.S. State Department Dispatch, May 23, 1955, 18 pages.

*62. "Field Trip in South Vietnam, April 1955." U.S. State Department Dispatch, April 29, 1955, 14 pages.

*63. "A Visit with President Ngo Dinh Diem." U.S. State Department Dispatch, June 15, 1955, 7 pages.

*64. "South Vietnam Revisited, July 16, 1955." U.S. State Department Dispatch, August 29, 1955, 32 pages.

65. "The Cai San Refugee Resettlement Project." April 28, 1956, 4 pages.

66. "Developments in Cai San." June 4, 1956, 4 pages.

*67. "U.S. Aid for Land Reform in Vietnam." July 10, 1956, 4 pages.

*68. "Making the Proposed Land Reform Redistribution Program More Practicable." October 9, 1956, 7 pages.

69. "Cai San and Agrarian Reform." October 18, 1956, 2 pages.

*70. "Towards a More Effective U.S. Aid Program in Vietnam." November 15, 1956, 5 pages.

71. "President Diem—An Informal Report." Unpublished manuscript, undated (probably July–August 1956), 12 pages.

72. "Agrarian Reform in Japan." A statement to the Fourth International Catholic Congress on Rural Life, April 1–6, 1957, 13 pages.

*72a. "Agrarian Revolution in Asia." A one-page separate statement to the Fourth International Catholic Congress on Rural Life, April 1–6, 1957.

73. "Carrot and Stick in Rural China." *Foreign Affairs* (October 1957), pp. 94–104.

*74. "Agrarian Revolution in Japan." *Foreign Affairs* (October 1959), pp. 95–109.

75. "Vietnam: The First Five Years." *Reporter* (December 24, 1959), 4 pages.

*76. "Rural Communes in China." May 1960, 10 pages. Published in somewhat abridged form in *Reporter* (October 27, 1960) under title "More than Mao Can Chew."

*77. "Self-Description/Appraisal." June 15, 1960, 3 pages.

*78. "Exploration of Job in Nepal." Ford Foundation, November 21, 1960, 7 pages.

*79. "Nepal's Five Year Plan." Ford Foundation, November 28, 1960, 4 pages.

*80. "Land Reform in Indonesia." New York: The Council on Economic and Cultural Affairs, January 24, 1961, 5 pages.

81. Letter to Rishikesh Shaha, Minister of Finance of Nepal. Ford Foundation, June 30, 1961, 6 pages.

*82. "Agrarian Reform in the Republic of Vietnam, 1961." *Problems of Freedom.* East Lansing: Michigan State University, 1961, pp. 153–75. Also published as "Agrarian Reform in Free Vietnam." Stanford, California: Stanford Research Institute, 1968. B17–42.

*83. "Corporate Farm Management for Japan?" June 7, 1961, 3 pages.

*84. "Tenurial Conditions in Nepal." Ford Foundation, February 15, 1962, 11 pages.

85. "Why a Ford Foundation Program in Nepal." Ford Foundation, March 15, 1962, 7 pages.

86. "Nepal's Development Intentions and Prospects." Ford Foundation, October 1962, 13 pages.

*87. "India After the China Border Clash." Ford Foundation, November 14, 1962, 2 pages.

*88. "Visit to the Philippines." Ford Foundation, January 17, 1963, 12 pages.

*89. "Agrarian Reform in Nepal." Ford Foundation, March 16, 1963, 4 pages.

90. "Visit to Indonesia." Ford Foundation, April 22, 1963, 9 pages.

*91. "Tenurial Conditions and the Package Program (in India)." Ford Foundation, 1963, 69 pages. Also published under same title by the Planning Commission, Government of India, 1965, pp. 9–59.

*92. "Land Reform in Indonesia." Ford Foundation, February 27, 1964, 16 pages.

*93. "The Industrialization Bias in Economic Development." Excerpt from statement to Conference on World Tensions, February [?] 1964, 6 pages.

94. "Agrarian Reform in Asia." *Foreign Affairs* (April 1964), pp. 445–60.

*95. "Land Reform." *Policies for Promoting Agricultural Development.* Report on a conference held at M.I.T., June 29–August 7, 1964. Edited by David Hapgood. Cambridge, Massachusetts, 1965, pp. 295–321. Also reprinted in *Seven Articles on Land Reform.* Asian Studies Papers, Reprint Series no. 2. East Lansing: Michigan State University, 1964–65.

*96. "Agrarian Reform in India." *India's Economic Development Effort,* vol. 4, appendix 9. International Bank for Reconstruction and Development (World Bank), October 1, 1965, 58 pages.

97. "Community Development and Agricultural Production." *India's Economic Development Effort,* vol. 4, appendix 10. World Bank, October 1, 1965, 39 pages.

98. "Cooperatives." *India's Economic Development Effort,* vol. 5, appendix 11. World Bank, October 1, 1965, 43 pages.

99. "Administration, Panchayats and Agricultural Development." *India's Economic Development Effort,* vol. 5, appendix 12. World Bank, October 1, 1965, 52 pages.

100. "Traditional Agriculture and the Ejido." *Current Economic Position and Prospects of Mexico,* vol. 4, annex 7, part 2. World Bank, October 26, 1966, 55 pages.

101. "Land Reform in Iran." World Bank, December 31, 1966, 25 pages.

102. "Land Revenue and the Abolition of the Land Tax." World Bank, 1968, 9 pages.

*103. "The New Agricultural Strategy and Institutional Factors." World Bank, June 1968, 40 pages.

104. "A Note on Small Farmers." World Bank, February 13, 1969, 14 pages.

105. "A Note on Institutional Problems." World Bank, March 31, 1969, 13 pages.

*106. "Punjab Field Trip." World Bank, April 2, 1969, 15 pages. Published in *Economic and Political Weekly* [June 28, 1969] under the title "Green Revolution in Punjab—A Field Trip."

*107. "Bihar Field Trip." World Bank, July 29, 1969, 30 pages. Published in *Economic and Political Weekly* (September 27, 1969) under the title "Green Revolution in Bihar—The Kosi Area."

108. "Report of The All-India Rural Credit Review Committee." World Bank, October 6, 1969, 5 pages.

109. "Note on Iran's Agricultural Cooperatives and Agro-Industries." World Bank, January 1970, 34 pages.

*110. "Agriculture—Some Broader Considerations." *Economic Situation and Prospects of India,* vol. 1, ch. 5. World Bank, April 21, 1970, pp. 37–53.

111. "Ironies of India's Green Revolution." *Foreign Affairs* (July 1970), pp. 758–68.

112. "Rural Credit and The Nationalized Banks." World Bank, August 19, 1970, 7 pages.

*113. "A Note on Agricultural Reforms." World Bank, August 1970, 10 pages.

114. "Agrarian Reform in Asia, The Green Revolution and Its Reform Effects." 28th International Congress of Orientalists, Canberra, January 6–12, 1971, 22 pages.

115. "The Green Revolution and Rural Reforms." *Financial Times* (January 25, 1971), pp. 18–19.

*116. "Field Trip in Eastern Uttar Pradesh." World Bank, May 6, 1971, 7 pages.

*117. "Agriculture—Problems and Needs." *Economic Situation and Prospects of India,* vol. 1, ch. 3. World Bank, May 11, 1971, pp. 51–80.

*118. "Refugees in West Bengal." World Bank, August 28, 1971, 8 pages.

119. "Land Ceilings and Land Reform in India." World Bank, October 13, 1971, 13 pages. Published in *Economic and Political Weekly*, annual number (February 1972) under the same title.

*120. "Agriculture—Issues and Programs." *Economic Situation and Prospects of India*, vol. 1, ch. 4. World Bank, May 10, 1972, pp. 84–120.

121. "The Land Ceiling Flap." World Bank, May 31, 1972, 16 pages.

122. "New Ceiling Round and Implementation Prospects." *Economic and Political Weekly* (September 30, 1972), pp. 125–32.

123. "A Non-trip to Andhra Pradesh." World Bank, October 5, 1972, 6 pages.

124. "The Raj Committee Rural Taxation Report." World Bank, November 15, 1972, 8 pages.

*125. "Drought in Maharashtra: Not in A Hundred Years." World Bank, January 12, 1973, 25 pages. Published in *Economic and Political Weekly* (February 17, 1973) pp. 383–96 under the same title.

*126. "The Rural Scene." *Economic Situation and Prospects of India*, vol. 1, ch. 5. World Bank, May 8, 1973, pp. 40–62.

127. "Agrarian Reform in India." A Statement to a World Bank Land Reform Seminar. World Bank, May 1973, 7 pages.

128. "The Failure of the Wheat Trade Takeover." World Bank, August 13, 1973, 20 pages.

*129. "Agrarian Reform à la Punjab." World Bank, September 14, 1973, 11 pages.

130. "No Rice Takeover." World Bank, October 13, 1973, 11 pages.

131. "How Green is the Indian Green Revolution?" World Bank, December 15, 1973, 19 pages. Published in *Economic and Political Weekly*, Review of Agriculture, vol. 8 (December 29, 1973) pp. A133–44 under the same title.

132. "A Note on Small Farmers." World Bank, April 30, 1974, 5 pages.

133. "The Rural Scene." *Economic Situation and Prospects of India*, vol. 1, ch. 4. World Bank, May 7, 1974, pp. 103–33.

134. "The Emergency Agricultural Production Program, 1972–1973 (Failure of a Program). World Bank, September 5, 1974, 10 pages.

135. "1974 Procurement Kharif (Summer) Crop Prices." World Bank, September 21, 1974, 7 pages.

*136. "Agrarian Reform in the Philippines." World Bank, October 7, 1974, 8 pages.

137. "Wheat Procurement in India in 1974 and Related Matters." December 16, 1974, 30 pages. Published in *World Development* (February–March 1975), pp. 91–111 under the same title.

*138. "Food Shortage in West Bengal—Crisis or Chronic?" World Bank, December 16, 1974, 13 pages. Published in *World Development* (February 1976), under the same title.

139. "Agricultural Production and Constraints." World Bank, February 13, 1975, 13 pages. Published in *World Development*, vol. 4, no. 1 (1976).

140. "Agricultural Prices Commission, National Agricultural Commission and Wheat Procurement." World Bank, March 31, 1975, 8 pages.

141. "The Land Reform Program in Sri Lanka." World Bank, April 17, 1975, 12 pages.

142. "Fertilizer Consumption in Punjab." World Bank, April 28, 1975, 3 pages.

Depository Libraries for the Ladejinsky Papers

Louis J. Walinsky's original compilation of the papers of Wolf Ladejinsky included 97 of the items listed in the bibliography. A microfiche of this draft of the manuscript has been deposited in the following libraries, which are listed alphabetically by country.

Australia

Australian National University Library
P.O. Box 4
Canberra

Commonwealth Scientific and Industrial Research Organization Library
314 Albert Street
Melbourne 3002

University of Melbourne Library
Parkville N2, Victoria

Burma

University of Rangoon Library
Rangoon

Canada

Canada Department of Agriculture Library
Sir John Carling Building
Ottawa

England

Bodleian Library
Oxford University
Oxford

The British Library
Great Russell Street
Bloomsbury
London WC1

Institute of Development Studies
University of Sussex
Brighton, Sussex BN1 9RE

School of Agriculture Library
Cambridge University
Downing Street
Cambridge

France

Bibliothèque de l'Unesco
Place de Fontenoy
Paris 7

Organisation de Coopération et de Développement Economiques, Bibliothèque
2 Rue Andre Pascal
Paris 16

Germany, Federal Republic of

Bibliothek des Instituts für Weltwirtschaft und Zentralbibliothek der Wirtschaftswissenschaften in der Bundesrepublik Deutschland
Postfach 809
23 Kiel-Wik

HWWA-Institut für Wirtschaftsforschung-Hamburg
Neuer Jungfernstieg 21
2 Hamburg 36

India

Allahabad Agricultural Institute Library
P.O. Agricultural Institute
Allahabad, Uttar Pradesh

Calcutta University Library
Asutosh Building
Calcutta 12

Government Agricultural University Library
Ludhiana
Punjab

Indian Agricultural Research Institute Library
Pusa, New Delhi 12

Osmania University Library
Hyderabad 7

University of Bombay Library
Mayo Road
Bombay 32

Indonesia

Perpustakaan Negara
c/o Perwakilan Departemen P.D.
dan K Kalimantan Tengah
Palangkaraya

Iran

Melli (National) Library
Teheran

Israel

Jewish National and University Library
Hebrew University Campus
Jerusalem

Ministry of Agricultural Central Library
Hakiria
Tel Aviv

National and University Institute
of Agriculture Library
P.O. Box 12
Rehovoth

Italy

Biblioteca David Lubin
Food and Agriculture Organization
Viale delle Terme di Caracalla
00100 Rome

Biblioteca del Ministero dell'Agricoltura e
Foreste
Via XX Settembre 20
00187 Rome

Japan

Hiroshima University Library
Higashi-senda-machi
Hiroshima-shi

Hitotsubashi University Library
185 Kunitachimachi
Kitatamagun
Tokyo

Institute of Developing Economies Library
42 Ichigaya-Hommura-cho
Shinjuku-ku
Tokyo 162

Kyoto University Library
Yoshida Hommachi, Sakyoku
Kyoto

Nagoya University Library
I Furo-cho, Chikusa-ku
Nagoya-shi

National Diet Library
1-10, 1-chome, Nagata-cho
Chiyoda-ku
Tokyo

Okayama University Library
Tsushima,
Okayama-shi

Osaka City University Library
Sugimotocho 459
Sumiyoshi-ku
Osaka

Tokyo Agricultural University Library
4-461 Setagaya
Setagaya-ku
Tokyo

Tokyo University of Agriculture
and Technology Library
P.O.D. Musahi-Fuchu
Tokyo

Tokyo University Library
Hongo 7-chome, Bunkyo-ku
Tokyo

Yamaguchi University Library
Kameyama
Yamaguchi-shi

Korea

Korean Development Institute
P.O. Box 113
Cheongryang
Seoul

National Central Library
Sogong-dong, Jung-gu
Seoul

Seoul National University Library
31 Tongsung-dong, Chongno-ku
Seoul

Malaysia

University of Malaya Library
Kuala Lumpur

Netherlands

Koninklijk Institut voor de Tropen
Central Library, Royal Tropical Institute
Centrale Boekerij, Mauritskade 63
Amsterdam

Library of the Agricultural University
General Foulkesweg Ia,
Wageningen

New Zealand

National Library of New Zealand
Private Bag
Wellington

Philippines

University of the Philippines Library
Diliman, Quezon City

Singapore

University of Singapore Library
Bukit Timah Road

Taiwan (Republic of China)

National Taiwan University Library
Roosevelt Road
Taipei

Taiwan Provincial College of
Agricultural Library
250 Kuo-Kuang Road
Nan-chu, Taichung

Thailand

Chulalongkorn University Library
Phya Thai Road
Bangkok

United States

Library of Congress
1st and East Capitol Streets
Washington, D.C. 20540

National Agricultural Library
U.S. Department of Agriculture
14th and Independence Avenue SW
Washington, D.C. 20250

University of California Library
Davis, California 95616

University of Wisconsin Libraries
Madison, Wisconsin 53706

Vietnam

Bibliothèque Nationale de la République
Démocratique du Viet-Nam
31 rue Trang-Thi
Hanoi

INDEX

Numbers in italics refer to entries in the Chronological Bibliography (CB), pp. 580–85.

Abdullah, S. M., land reform and: benefits, 188; implementation, 179, 196, 200, 210; policies, 180, 181, 183, 184, 185
Abolition of the Big Landed Estates Act of 1950 (India), 180–81, 183–85, 186
Acre, standard: definition of, 379, 544n
Action Program for Cooperative Credit (India), 414
Agra Tenancy Act (India), 33–34
Agrarian Law of 1960 (Indonesia), 342
Agrarian reform. *See* Land reform
Agreements. *See* Leases
Agricultural administration, *CB–99*
Agricultural Co-operative Association (Japan), 79, 82
Agricultural Credit Agency (South Vietnam), 308
Agricultural Finance Corporation Limited (India), 420, 470
Agricultural and Industrial Bank (Japan), 42
Agricultural Land Adjustment Act of 1938 (Japan), 84
Agricultural Land Adjustment Law Enforcement Order (Japan), 111
Agricultural production, 364; in China, 115–16; in Indonesia, 59, 349–52; in Japan, 24, 45, 55, 570; in Netherlands Indies, 58–59; in South Vietnam, 231, 266; in United States, 288. *See also specific crops*
—in India, 36, 371–73; credit for, 338–39; development of, 384–87, 564–65; drought and, 515, 532, 533, 534; emphasis on, 405–07; IADP and, 373, 407; improvement of, 373, 539; problems of, 24, 30–39; program for, 500–05; prospects for, 536–40; prosperity of, 432–35, 440; technology and, 406, 432–34, 495–96
—large-scale: in British Malaya, 62, 63, 65; in Netherlands Indies, 60; in South Vietnam by French, 245, 256–57, 301
—in Nepal: implementation problems of, 291–95; objectives of 334; political implications and, 291, 294–95; resource utilization and, 294
—small-scale: in British Malaya, 62, 63–64; in India, 452–54, 465–68, 509–10; in Korea, 49; in Soviet Union, 23, 24–27
Agricultural Refinance Corporation (ARC) (India), 468–69, 505
Aid, external, 13–16, 206, 207, *CB–59, 67, 70*; by India to Nepal, 293; to Netherlands Indies, 59
—by United States, 197; to China, 136, 141; to Latin America, 136; to Nepal, 293; to South Vietnam, 250–51, 264–67, 268–70, 273–79; to Taiwan, 141, 145
Akbar, M. J., 519n, 529n
All-India Congress Committee, 188
All-India National Commission on Agriculture, 441, 508–09
All-India Rural Credit Review Committee, 508–09
All Jammu and Kashmir National Conference (India), 180
Allee, Ralph, 297, 298
Anarchy. *See* Conflict, political
Andhra Pradesh (India), 163, 379; Communism in, 132, 401; credit cooperatives in, 415; crop loans in, 417; legislation in, 337; rent in, 371n, 380; rural programs in, 503;

sociological neglect in, 406; tenancy in, 157, 378; village voice in, 426
Annam. *See* Central Vietnam
Arbitration. *See* Litigation
ARC (Agricultural Refinance Corporation) (India), 468–69, 505
Asia, *CB–19, 23, 39, 43, 72a, 94, 114*
Assassinations in Japan, 44, 85
Atcheson-Fearey memorandum, 5–6, 569–78
Australia (and South Pacific Islands), *CB–22, 24*
Autonomous Rice Control Law of 1936 (Japan), 44

Bachman, K. L., 407n, 429n
Bali (Indonesia): land reform in, 340–41, 342, 343, 347, 348; lease contracts, 343–44; tenancy status in, 344, 351
Bangladesh. *See* Refugees in India from East Pakistan
Banks, land development: in India, 413, 414–15, 421, 469–70, 505
Bao Dai, 215, 253–54, 302, 307
Beg, M. M. A., 179, 181, 182–83, 184, 185, 186
Bell report (India), 413, 414, 415
Bengal Tenancy Act of 1859 (India), 33
Benson, Ezra Taft, 3n, 215
Benton, Thomas Hart, 154, 287
Better Living Societies (India), 35
Bhagvati Committee on Unemployment (India), 504
Bihar (India): agricultural production in, 406; cooperatives in, 448–49; credit in, 448–49, 461; double occupations in, 443; fertilizer in, 447, 448–49; Green Revolution and, 442, 453, 459–60, 461–62; irrigation in, 442, 445–47, 453, 454; jute production in, 444; labor in, 450–51, 452; land ceilings in, 457–58; land holding size in, 453; land reform legislation in, 194, 337, 457, 458–59; land transfer in, 375; land values in, 449–50; landlords in, 377, 457; litigation in, 458; moneylenders in, 449; occupancy rights in, 192, 457; rent in, 380, 457; rice production in, 444–45; sharecroppers in, 455, 456–57; small farmers in, 452–54; taccavi loans in, 448; technology in, 441, 442–43, 452–54; tube wells in, 446, 447; wages in, 450, 451–52; wheat in, 443–44
—tenancy in, 191, 391; conditions of, 454–55, 460–61; restoration of, 458; security of, 457; status of, 455, 461
Bihar Tenancy Act of 1885 (India), 457
Binh Xuyen, 239, 240, 241, 244, 253–54
Bolshevik Party (Soviet Union), 148–49
Bombay Banking Enquiry Committee (India), 32
Botanic Gardens (Singapore), 64
Bowles, Chester, 6, 20–21, 154–55, 198
British Malaya, *CB–15, 16. See also specific entries*
Brunner, Edmund de Schweintz, 51n, 52n, 54n
Buck, Lossing, 74n, 89
Buffaloes. *See* Cattle
Buffer stocks in India, 533, 534
Bullocks. *See* Cattle
Bureau of Food (Taiwan), 139
Bureau of Statistics (Japan), 43–44

CADP (comprehensive development project) (India), 564–65
Cao Dai: American aid and, 250–51; conflict with, 239, 240, 244, 246; control by, 230, 236, 241, 248, 249–50, 251; French and, 242; land reform policies of, 246; President Diem and, 241, 253; tenant conditions under, 249; tenant-landlord relations under, 250; villages, 247, 249–50. *See also* Tac, Pham-Cong
Capitalism in Soviet Union, 27
Cattle: in India, 37, 479, 516, 524–26; in South Vietnam, 219, 258, 265; in Thailand, 265
Ceiling Act (India), 391
Central Board of Irrigation (India), 37
Central Land Reform Committee (India), 511
Central Vietnam, *CB–61*; buffaloes in, 219; characteristics of, 219–22; condition of, 218–19; credit in, 221; debt in, 221; education in, 221–22; land ownership in, 219–20, 222, 224, 226–27; landlords in, 220, 223–25; marketplaces in, 221; moneylenders in, 221; national army and, 228; political conflicts in, 228–29; political tasks in, 227–28; population pressure on land in, 219; rent reduction in, 217, 222–23, 224; resettlement in, 223; taxes in, 220–21; tenancy records in, 218, 220; tenant status in, 220
—land reform program in: attitude toward, 223, 224; goal of, 228; limitations of, 226–27; politics of, 225–26; provisions of, 222–23
Ceylon. *See* Sri Lanka

Chang, Chun, 118, 148
Chen, Tsi Liu, 120, 122, 127
Cheng, Chen, 17, 20, 142; Communism and, 132; land reform implementation by, 144, 145, 151; rent reduction and, 105, 107
Chiang, Mon-lin, 95, 114, 136, 137, 148; China aid program costs and, 141; land reform and, 134; rent reduction and, 123
China, People's Republic of. *See* Kwangsi; Kwantung; Manchuria; Szechwan; *specific entries*
China, Republic of. *See* Taiwan
China Aid Act of 1948, 134, 136
Chinese Association of Land Reform, 148
Chosen (Korea), *CB–12. See also specific entries*
Chosen Agricultural Lands Ordinance of 1934 (Korea), 53
Chosen Tenants Arbitration Ordinance of 1932 (Korea), 53
Christensen, R. P., 407n, 429n
Cochin-China. *See* South Vietnam
Cocoons. *See* Prices (in Japan, for silk)
Collective bargaining. *See* Tenant unions
Collectivization: of agriculture, *CB–1, 4, 8, 73, 76*; in China, 131, 287, 295; Communism and, 209, 287; in Soviet Union, 23, 24–27, 28–29, 287, 289. *See also* Cooperatives
Colonialism. *See* Occupation, colonial
Columbia University Conference on International Economic Development, 474
Comindari, 192
Commission on Social Policy (Japan), 83
Commission on Tenancy (Japan), 83
Communism, 130–35, 150, 275; agrarian revolution and, 279–81, 300; in Burma, 132, 288; in China, 117, 129; collectivization and, 209, 287; human welfare and, 204, 205–06, 208–09; in Indonesia, 132; land ownership and, 151; land reform and, 12, 13–14, 356; landlords and, 280; in Nepal, 292, 295; in North Vietnam, 276, 299, 310; in Pakistan, 132; in Philippines, 326; political goals of, 129, 353; in Soviet Union, 24–29, 148, 280, 287
—in India: influence of, 131, 132–33, 288, 485; land grab and, 485; land ownership and, 168–69; land reform and, 165–66, 167–68; rent and, 169–70, 195; uprisings of, 132
—in Japan, 132, 149, 151–52, 153–54; failure of, 288; strength of, 286; threat of, 281
—in South Vietnam, 254, 288; attitude toward, 225, 226, 270; control of, 302; influence of, 304; opposition of, 299, 310, 312; taxes and, 302; threat of, 278. *See also* Binh Xuyen; Hoa Hao; Viet Minh
Community development, *CB–97*
Compensation, landlord, 359; in India, 183–85, 187–88, 370, 377, 575–76; in South Vietnam, 274, 308
Comprehensive development project (CADP) (India), 564–65
Conciliation of Tenancy Disputes Bill of 1924 (Japan), 46, 83
Conference of the state chief ministers (India), 483–84
Conference on Productivity and Innovation in Agriculture in Underdeveloped Countries, 315, 354
Conference on Social Development and Welfare in Vietnam (1959), 299
Conference on World Tensions, 315, 352
Conferences: on economic development, 315, 354, 474; by Ford Foundation, 67, 189, 198; in India, 180, 420, 483–84; on land tenure, 189, 198–201; on rural life, 279; on welfare, 299; on world tensions, 315, 352
Conflict, political: between China and India, 324; between China and Japan, 47, 48; in India, 132, 175, 196, 324, 528–29; in Nepal, 291, 292, 294–95. *See also* Communism; Revolution, agrarian
—in South Vietnam, 234, 237–38, 253–54, 275–76, 295; with Cao Dai, 239, 240, 244, 246; with Hoa Hao, 239, 241–42, 254; with Viet Minh, 228–29
Congress Party (India): election of, 166; land reform and, 193, 196, 202, 212; nonparticipation of, 163, 167; role of, 400–02, 404
Congress Working Committee (India), 188
Contagious Diseases of Animals Act (India), 37
Contracts. *See* Leases
Cooperative Societies Department (British Malaya), 64–65
Cooperatives: agricultural, *CB–98, 109*; in British Malaya, 64–65; in Japan, 46–48, 79, 82; in Soviet Union, 23, 24–27, 28–29; in Taiwan, 106. *See also* Credit cooperatives
—in India, 34–37; assessment of, 422–28; crop program and, 414–15; environmental factors of, 423–24; future role of, 428–31; historical factors of, 423–24; legislation for, 387; marketing by, 416; membership of, 507; politics and, 427–28; problems of, 413–14;

reform obstruction and, 426–27; village interests and, 425–26
Corporate farms in Japan, 312–14
Cotton production in Korea, 57
Crash Scheme for Rural Employment (CSRE) (India), 501, 502
Crawford, John, 367, 560
Credit, 467; agricultural, *CB–108, 112*; in British Malaya, 64–65; in China, 137; in Japan, 42, 44–45, 48, 84–85, 86; in Korea, 54–55; in South Vietnam, 221, 265, 266; in Taiwan, 106, 146–47
—in India, 212, 424, 467; agricultural production, 338–39; crop loan, 416–18, 337, 419, 448; drought, 529; fertilizer, 447; irrigation, 446; program, 502–03; refugee, 492; small farmer, 509–10; sources of, 448–49, 461
Credit cooperatives, in India, 338, 413–22, 468–70, 505–11; distribution of, 507–08; integrated, 508–10; nature of, 505–06; performance of, 478; sources of, 413, 414–15, 448–49, 461
Credit institutions: in Japan, 42, 76; in South Vietnam, 308
Crop share. *See* Rent
Crops: in Taiwan, seed project for, 140
—in India: buffer, 533, 534; disaster consequences for, 515–16; high-yielding, 410, 414–15; karif, 515–16, 517; loans for, 337, 416–18, 419, 448; rabi, 515–16, 517
CSRE (Crash Scheme for Rural Employment) (India), 501, 502
Cultivation. *See* Agricultural production; Land cultivation acreage

Dandekar, V. M., 466n, 484n
Darling, Malcolm L., 186, 394, 434
Debt, rural: in British Malaya, 64–65; in India, 30–32, 186–87; in Japan, 42, 79, 572, 576–77; Japanese legislation on, 45, 47, 48, 86; in Korea, 54–55; in South Vietnam, 221, 302
Debt Conciliation Act of 1933 (India), 31
Debt relief: in India, 187; in Korea, 55–57
Deccan Agriculturists' Relief Act of 1879 (India), 30–31
Deeds in India, 30–31
Department (or Ministry) for Agrarian Reform (South Vietnam), 247, 272, 273, 310
Department of Agriculture (British Malaya), 64
Department of Agriculture (Japan). See Ministry of Agriculture and Forestry (Japan)
Department of Agriculture (Taiwan), 139
Department of Agriculture of the United Provinces (India), 36
Departmental Examination of Deputy Commissioners and Tehsildars for 1951 (India), 186
Deposit money in China, 123–25
DFS (Dry Farming Scheme) (India), 501–02
Diem, Ngo Dinh, 7, 8, 17, 20, 215, 244; attitude toward, 276, 305; Communist opposition to, 299; Japan and, 239, 243; land reform and, 222, 238, 240, 268–75, 303, 307, 312; land transfer and, 310; local administration and, 241; refugee settlement and, 240–41; rural impressions and, 217; sect conflict with, 241–42, 253–54
Diet (Japan), 83
Disputes, tenant-landlord. *See* Litigation
Domaine Agricole de l'Ouest (South Vietnam), 245, 301
Domaine Gressier (South Vietnam), 256–57
Double cropping in South Vietnam, 266
Drought in India, *CB–125*; conditions of, 514, 517; consequences of, 514, 528, 529–31, 532–35; credit and, 529; employment and, 517, 518–20, 523; expenditures for, 533; food shortages and, 520–22, 559; irrigation for, 517, 518, 533–34; production and, 515, 532, 533, 534; rebellion and, 528–29; refugees and, 527–28; relief for, 518–20, 522–23, 533–34; wages and, 520; water shortages and, 515, 516, 526–27
Dry Farming Scheme (DFS) (India), 501–02

East Pakistan. *See* Refugees in India
ECA (Economic Cooperation Administration), 128, 134, 135, 145
Economic development, *CB–93*
Education: in British Malaya, 63, 64; in Japan, 112–13; in Philippines, 325, 331; in South Vietnam, 221–22. *See also* Extension services in India
Employment in India, 384, 387, 436–38, 464; creation of, 504; double, 443; drought and, 517, 518–20, 523; industrialization and, 539; lack of, 504; prospects for, 438–39; refugee restrictions on, 490–91; scheme for, 501, 502; technology and, 436, 438–39. *See also* Labor

Enforcement Committee for Farm Rent Reduction (Taiwan), 99
Eviction. *See* Tenant eviction
Executive Yuan (China), 99, 114, 118
Expenditures, capital: in India, 174, 201, 501–02; in Japan, 282. *See also* Aid, external
Expenses, agricultural. *See* Rent
Exports, agricultural: by Korea, 573, 577–78; by Netherlands Indies, 60; by Soviet Union, 27; by Taiwan, 573, 577–78
Extension services in India, 36–37; education and, 410; official view of, 407; reform of, 411–13; technology and, 409–10; village-level workers and, 407–09, 412–13

Famine. *See* Food
Famine Commission of 1900 (India), 31
Farm Adjustment War of 1938 (Japan), 47
Farm Debt Adjustment Act (Japan), 45
Farmer-Labor Party (Japan), 81
Farmers' Patriotic Association (Japan), 82
Farmers' Service Society (India), 509–10
Farming. *See* Agricultural production
Fearey, Robert, 5–6, 569
February Revolution of 1917 (Soviet Union), 25
Federated Malay States and Straits Settlements, 65
Fertilizer: in India, 447, 478–79; in Japan, 43, 45, 573, 578; in South Vietnam, 265–66; in Taiwan, 147, 266
Feudalism: in India, 172–73; in South Vietnam, 305; in Soviet Union, 134, 148
—in Japan, 69, 70, 72, 73–74; Meiji Era and, 70n, 80, 284, 285; Taisho Era and, 90; Tokugawa regime and, 73, 284
Fifteenth Party Congress (Soviet Union), 27
Floud Commission (India), 161
FOA (Foreign Operations Administration) (South Vietnam), 240
Fodder shortages in India, 516, 517, 524–26
Food, *CB–13, 19, 21, 23, 56, 57, 138*; projections in India, 537–38, 539. *See also specific crops*
—consumption: in China, 116; in Indonesia, 350
—shortages: in Asia, 353; in India, 31, 516, 520–22, 524–26, 557–58, 559–61; in Indonesia, 349–51; in Korea, 56
—supply: in India, 558, 559–60, 563–64; in Netherlands Indies, 60
Food Corporation (India), 419
Ford Foundation: aid program in Nepal, 293, 315; conferences by, 67, 189, 198; human welfare and, 209–11; India-China border clash and, 324; land reform and, 197, 315, 325, 330–31, 336; report by, 204, 207. *See also* Intensive agricultural district program (IADP)
Foreign aid. *See* Aid, external
Foreign Operations Administration (FOA) (South Vietnam), 240
Foreign policy of United States, 128, 129, 133–34, 242, 278
Foreign relations: India and, 276; South Vietnam and, 242, 276, 278
Formosa. *See* Taiwan
Formosa Provincial Department of Agriculture and Forestry, 139
Formosa Provincial Department of Reconstruction, 139
Formosa Provincial Health Administration and National Institute of Health, 139
Formosa Provincial Veterinary Serum Institute, 139
Fourth International Catholic Congress on Rural Life, 279
French occupation in South Vietnam, 218, 234, 239, 240, 309; Cao Dai and, 242; enterprises of, 244–45, 256–57, 301; land reform and, 257; rent and, 256, 257; revolution and, 186, 199, 395; security and, 228
French Revolution, 186, 199, 395

Gandhi, Indira, 17, 472, 483, 513, 518
Gandhi, Mohandas Karamchand, 393, 431, 485
Gant, George, 289, 290
Government procurement, *CB–128, 130, 135, 137*
Grain. *See specific crops*
Grajdanzev, Andrew J., 72n, 77n
Green Revolution in India, *CB–103, 111, 115, 131*; consequences of, 432, 442, 463–65, 535–36; crop production and, 494–95, 496, 534; farm labor and, 536; income distribution and, 498; landless and, 535–36, 472; landowners and, 535; limitations of, 459–60, 461–62; technology and, 436, 439, 453, 478–79, 536; tenants and, 472, 536
Grist, D. H., 64n, 65n
Gruel kitchens in India, 559–61
Gujarat (India), 337, 380, 391, 420
Gujarat State Cooperative Conference, 420

Harchard Singh Committee (India), 543–44, 547–48
Harijans, 163, 165, 172, 173
Harmony unions (Japan), 82
Haryana (India), 414
Hill, Dr., 292, 293
Hoa Hao: conflict with, 239, 241–42, 254; control by, 230, 236, 237
Hoi, Nguyen Van, 233, 236, 238
Hokkaido (Japan): cultivation rate in, 70, 71–72, 73; land commissions in, 109; land purchase in, 85; lease contracts in, 75; rents in, 76; tenant unions in, 81
Honshu (Japan), 70–71, 73, 81
Huk movement, 326
Human welfare, 67–68; Communist effect on, 204, 205–06, 208–09; economic development and, 208–09; Ford Foundation role in, 209–11; goals of, 204; in India, 208; land reform and, 209–11; magnitude of problem of, 204–06; program identification for, 206–08; technical assistance for, 207, 212–14; Western influence on, 206, 209, 212–13
Hypothec Bank (Japan), 42, 76

IADP. *See* Intensive Agricultural District Program
ICA (International Cooperation Administration), 277, 278
Illiteracy: in China, 137–38; in India, 31, 35, 37–38
Immigrants: in British Malaya, 62–65; in Manchuria, 46; in South Vietnam, 246. *See also* Refugees
Imperial Agricultural Association (Japan), 82
Imperial Council of Agriculture (India), 36
Imports, agricultural: by Indonesia, 350; by Japan, 573, 577–78; by Korea, 55; by Soviet Union, 27
Income: in British Malaya, 63; in Korea, 53–54; in Japan, 43, 77–78
—in India, 170, 450, 451–52, 498; gap in, 440–41; legislation on, 171; refugees and, 491; scales for, 436–38, 520
India, 367, 405–07, 462, 495–99, *CB–10, 21, 50–54, 58, 87, 91, 96–99, 102–08, 110–13, 115–35, 137–40, 142*; *see also* Andhra Pradesh; Bihar; Gujarat; Kashmir; Kerala; Madras; Maharashtra; Punjab; Tamil Nadu; West Bengal; *specific entries*
Indian Immigration Committee (British Malaya), 62, 63
Indian Immigration Fund (British Malaya), 62–63
Indonesia, *CB–14, 80, 90, 92. See also* Bali; Java; Sumatra; *specific entries*
Industrialization: bias, 353–54; in Brazil, 328; in India, 38, 539; in Netherlands Indies, 59–60; in Soviet Union, 23, 25–27, 29
Institutional factors, *CB–103, 105, 139*
Integrated Agricultural Credit Service (India), 509
Intensive Agricultural District Program (IADP) (India), 336–40, 373, 407
Interest rates: in India, 30, 34; in Korea, 54–55; in Japan, 43, 47, 572, 577
International Institute of Agriculture, 36
International Rice Research Institute (Philippines), 325, 331
Investment. *See* Aid, external
Iran, *CB–101, 109*
Irrigation: in China, 138–39, 140; in Philippines, 326; in South Vietnam, 266; in Taiwan, 103–04, 138–39
—in India, 37; credit for, 446; drought and, 517, 518, 533–34; minor, 503–04; project development for, 442, 445–47, 480, 481; rights for, 473–74; tube wells, 446, 447, 476–77, 480, 481
Isobe, Hidetoshi, 40n, 43n, 78n

Jagirdars, 179, 180
Jammu. *See* Kashmir
Jammu and Kashmir Distressed Debtors' Relief Act of 1949, 187
Jammu and Kashmir State Evacuees (Administration of Property) Act (India), 186
Janmi, 175, 176
Japan, *CB–5, 6, 9, 11, 13, 18, 26–34, 36, 44–49, 55–57, 72, 74, 83. See also* Hokkaido; Honshu; Kyushu; *specific entries*
Japan Farmers' Union, 81
Japanese Landowners' Association, 82
Java (Indonesia): agricultural productivity in, 59; food consumption in, 350; immigrant labor in, 62; land availability in, 349; land reform in, 297–98, 340–41, 342, 347, 349; land registration in, 345; land retention in, 346; lease contracts in, 343–44; tenancy status in, 344, 351
JCRR. *See* Joint (U.S. and China) Commission on Rural Reconstruction
Joint (U.S. and China) Commission on Rural Reconstruction (JCRR), 14; China Aid Act

and, 134, 136; irrigation and, 138–39; program costs of, 139, 140, 141–42; program organization of, 139–40; project selection and, 138; results of, 140–41
—in China, 135, 137, 140; credit and, 137; illiteracy and, 137–38; leases and, 123; program goals of, 137–38; rent reduction program and, 127–28, 134, 135
—in Taiwan: programs by, 135, 137–40, 143, 145; rent reduction program and, 95, 108, 134; tenancy evasions and, 101–02
Jute production in India, 444

Kanamdars, 175, 176
Kashmir (India): conditions in, 179–80; conference in, 180; debt in, 186–87; eviction in, 180, 182–83; land commission selection in, 186; land transfer in, 199, 200; landlords in, 180, 183–85, 187–88; rent in, 180, 380; tenancy terms in, 180, 183
—land reform in: benefits of, 188; farmer discussions on, 181–82; implementation of, 179, 196, 200, 210; legislation on, 179–80, 186, 187, 188–89, 196, 210; policies for, 180, 181, 183, 184, 185
Kashmir State Tenancy Act of 1924 (India), 179–80
Kawada, S., 46n, 52n, 75n, 84n, 87n, 88n
Kerala (India), 381, 396, 484, 485
Kerensky, Aleksandr F., 148, 149
Khanh, Nguyen, 235–36, 237, 238
Kharif crops in India, 445, 515–16, 517, 531
Khusro, A. M.: agricultural production and, 407n; credit and, 419n, 424; employment in agriculture and, 384–86; eviction and, 378–79; food shortages and, 495n; income distribution and, 498n; mechanization and, 497n; occupancy rights and, 383; tenancy extent and, 372
Kimura, Magohachiro, 41n, 43n, 76n
Kipling, Rudyard, 130, 208, 404
Koirala, Prime Minister, 17, 292, 294–96
Korea (Chosen), *CB–12. See also specific entries*
Krishna, Raj, 372n, 382n, 387n, 407n
Kulaks, 25–26
Kuomintang Party Conference (China), 117, 118
Kwangsi (China), 121, 128, 129, 130, 137, 140
Kwantung (China), 117, 137, 140
Kyushu (Japan), 73, 81

Labor: in India, 168, 384–86, 450–51, 452.
—in British Malaya: from China, 62, 65; from India, 62–64; from Java, 62
Land, evacuee: in India, 543–48; in South Vietnam, 305
Land, private: in Taiwan, 147
Land, public: in Taiwan, 106; in South Vietnam, 219–20, 224
Land Bank (Philippines), 554, 555
Land bonds: in Philippines, 555; in South Vietnam, 308
Land Bureau (Taiwan), 105
Land ceilings: in Japan, 358; in Nepal, 333; in Taiwan, 358; in West Pakistan, 358–59
—in India, 212, 381–82, 395, 457–58, 472–73, 476, 483–84; guidelines for, 511; importance of, 513; legislation for, 391, 511–12, 540–41; revision of, 512–13
Land Code of 1926 (British Malaya), 61, 62
Land commissions: in China, 125–27; in India, 186; in Indonesia, 346–47; in Japan, 109–12, 113, 152, 363; in South Vietnam, 274–75; in Taiwan under Japanese, 104, 105, 106
Land committees: in India, 337, 338, 379–80, 511; in Indonesia, 348; in Philippines, 553–54; in South Vietnam, 272, 302–03
Land Compensation Committee (India), 184
Land consolidation in India, 35, 374–76, 480
Land councils in South Vietnam, 274, 308
Land cultivation acreage, 70; in Japan, 69–70, 71–72, 73, 75; in South Vietnam, 231; in United States, 366
Land Development Banks (India), 505
Land distribution: in India, 371–73, 381, 477; in Indonesia, 347, 349; in Japan, 71–72; in Korea, 49–50; in South Vietnam, 307, 309; in Taiwan, 96
Land division: in British Malaya, 61–62; in India, 180–81, 183–85, 186
Land expropriation. *See* Land grabbing
Land fragmentation in India, 35–36, 373–74
Land grabbing: in India, 180–81, 482, 483, 484–86, 542–50; in Japan, 575
Land holding: in China, 116; in India, 406–07, 452–53; in Japan, 40, 48, 70–71, 366; in Nepal, 318; in South Vietnam, 220, 226–27, 231, 246, 271–72, 273, 306, 307, 308
Land improvements: in India, 33, 370; in Japan, 69–70, 282; in South Vietnam, 264
Land occupation. *See* French occupation in South Vietnam; Occupation, colonial; Occupation, military

Land ownership, 356, 361; Communism and, 151; in Korea, 50–53; in Netherlands Indies, 60; in Taiwan, 106. *See also* Land ceilings; Landlords
—in India, 156–61, 163, 565; Communism and, 168–69; legislation for, 180, 339, 391–92, 484
—in Japan, 40, 86–89, 281–82, 313; development of, 69; incentive for, 345; protection of, 580
—in South Vietnam, 263–64, 301, 306; public, 219–20, 224; USOM and, 265, 266, 273–74
Land payment terms, 360–61; in Philippines, 554, 555; in South Vietnam, 271, 306–07, 308–09
Land policy: in British Malaya, 62; in India, 197, 202, 211, 338; in South Vietnam, 222, 224–25, 230; in Soviet Union, 294, 300
Land prices: in India, 159, 160–61, 381, 392, 396–97, 543, 545; in Indonesia, 298, 346; in Japan, 87–88, 152, 359, 381; in Philippines, 554; in South Vietnam, 308, 309; in Taiwan, 359
Land purchase, 359–61; in China, 123–24; in India, 156, 191, 391, 396–97; in Japan, 84–86, 110, 113, 152; in Philippines, 361; in South Vietnam, 264, 266, 268; in Taiwan, 103, 107, 144, 147, 360, 397
Land reform, 197, 315, 357–58, *CB–29, 34–38, 44, 46, 47, 50–54, 58, 68, 69, 72, 80, 82, 89, 92, 94–96, 100, 101, 113–15, 119, 121, 122, 127, 129, 136, 141*; in British Malaya, 62; condition of, 355; human welfare and, 209–11; implementation, 361; in Korea, 289; leadership for, 353, 361–62; in Philippines, 325, 330–31, 361, 551–53, 555–56; political implications of, 356, 361; political stability and, 364–65; pressure for, 366; prospects for, 365–66. *See also* Legislation, land reform
—in China, 137, 148, 289; opposition to, 119–21, 134, 135
—in India, 21–22, 289, 336, 361, 565; administration of, 398–400; assessment of, 202–03; assistance for, 201; from below, 393–95, 401, 483; conditions for, 369–70; implementation of, 382–83, 474–75, 538; obstacles to, 188, 371, 402–03, 426–27, 502–05; policy for, 197, 202, 211, 338; political role in, 193, 196, 202, 212; programs for, 403–04, 500–02; project identification and, 503; proposals for, 471–74; recommendations for, 404; results of, 362; review of, 390–93; shortcomings of, 426–27, 431; status of, 211–12; support for, 392–93. *See also specific regions*
—in Indonesia, 348; agreement for, 343–44; nature of, 297–98; proliferation of, 347; results of, 341–44, 347, 349
—in Japan, 5, 6, 167, 355, 395, 579–80; administration of, 363, 400; attitude toward, 356; factors of, 86; limitations of, 92; program for, 152–54, 199, 569–70, 574–78; setting for, 281; success of, 226
—in Nepal, 211, 289–90, 319–20; administration of, 320–21; opposition to, 321; shortcomings of, 332–35
—in South Vietnam, 233, 234–38, 252, 299–300, 307, 312; acceptance of, 305; administration of, 264, 269, 308; attitude toward, 261–64; elections for, 232; French and, 257; implementation of, 247–48, 273, 299, 305; politics and, 222, 224–25, 228, 230; program for, 222–27, 271–75; reaction to, 232. *See also* Diem, Ngo Dinh
—in Taiwan, 134, 151, 289, 355; attitude toward, 356; implementation of, 144, 145, 151
Land Reform Act of 1957 (Nepal), 317–18, 319–20
Land Reform Bill of 1945 (Japan), 79, 86
Land registration in Indonesia, 345–46
Land reservations in British Malaya, 62, 65
Land resumption, landlord: in India, 339, 357, 379–80, 457, 482–83; in South Vietnam, 224–25, 232–33, 250
Land retention, landlord: in Indonesia, 346; in South Vietnam, 271–72, 273, 307, 308
Land Revenue Reforms Committee of Madras of 1950 (Subramanian Committee) (India), 171
Land sale: in India, 562–63; in Japan, 88–89, 110, 113, 152, 281, 286; in Nepal, 319; in South Vietnam, 309–11, 312; in Taiwan, 103
Land survey: in India, 375; in Nepal, 320; in South Vietnam, 273
Land tax, *CB–102, 124*; in India, 171, 179, 377, 480–81; in Japan, 43, 74, 572, 577, 579; in Nepal, 317, 322–23, 333; in South Vietnam, 220–21, 225, 234, 247, 259, 302; in Taiwan, 146
Land tenure. *See* Tenancy
Land transfer: in India, 199, 200, 338, 339, 375–76; in Iran, 199, 334; in Japan, 345, 579–80; in Korea, 361; in South Vietnam, 273, 310; in Soviet Union, 29, 149, 289

Land valuation: in Indonesia, 348–49; in India, 434–36, 449–50; in Japan, 69–70; in Nepal, 322, 323; in Philippines, 553–55; in South Vietnam, 258; in Taiwan, 99, 105–06, 107. *See also* Land prices
Landlessness in India, 34, 383–87, 501, 502, 561–62
Landlords: in China, 117, 119–21, 134, 135; Communism and, 280; in Japan, 41–42, 46, 71–72, 80, 83; in Korea, 50, 52–53; in Nepal, 321; in South Vietnam, 220, 223–25, 232–33, 248, 304, 310; in Soviet Union, 280; in Taiwan, 103. *See also* Compensation; Feudalism; Land resumption; Land retention
—in India, 33, 192–93; abolition of, 180, 183–85, 187–88; absentee, 162, 172; legislation and, 169, 171; opposition by, 402–03; types of, 175, 176, 179, 180, 192. *See also* Zamindari
Latex. *See* Rubber production
Leases: in China, 116, 119–20, 122–23, 125, 126; in Great Britain, 90; in India, 337, 371; in Japan, 74–75, 80, 83, 90; in Taiwan, 97, 99
—contracts for: in China, 119, 121; in India, 370; in Indonesia, 343–44; in Japan, 75, 91–92; in South Vietnam, 247, 248, 262, 305; in Taiwan, 97–98, 99, 103, 357
Lee, Hoon K., 49n, 51n, 55n
Legislation, land reform, 356; in British Malaya, 61–62; in China, 119, 121, 122–23; in Korea, 53; in Nepal, 317–18, 319–20, 321–22; in Philippines, 551; in South Vietnam, 222–23, 302–04, 305, 307–08; in Taiwan, 95, 122–23
—in India, 164, 202–03; British colonial, 376–77; cooperatives and, 387; debt, 30–32, 187; effects of, 181–82, 183, 471–72; estate division, 180–81, 183–85, 186; eviction, 337; income and, 171; lack of, 163, 175–76, 177, 194, 195–96, 337, 396; land ceiling, 511–12, 540–41; land consolidation, 35; land distribution, 381; land expropriation, 180–81; land ownership, 180, 339, 391–92, 484; landlord, 169, 171; as obstacle, 371; recommendations for, 339–40; rent control, 180, 371n, 392, 435; revenue collection, 179; shortcomings of, 210, 340, 371, 378–79, 390, 398; tenancy security, 155, 180, 457, 458–59, 542–43; tenancy status, 33–34, 174, 176–77, 337, 391; zamindari abolition, 190–92, 194, 202, 397, 471. *See also* Punjab Tenants (Security of Tenure) Amendment Act of 1951; Tanjore Tenants and Pannaiyal (Protection) Ordinance of 1952
—in Indonesia: enforcement of, 298, 343, 345, 348; occupancy rights, 344; provisions of, 343, 346; purpose of 341–42; shortcomings of, 297–99, 345, 349; tenancy status, 344–45
—in Japan: farm debt, 45, 47, 86; land commissions and, 111; rent, 79; rice, 44, tenancy, 46, 83–84
Lenin, N., 25, 26, 131, 148–49, 314
Liu, Wen-hui, 122, 128, 135
Linh, Nguyen Ngoc, 247, 248, 257
Litigation: in India, 30–32, 458; in Japan, 41–43, 45–46, 80–83; in Korea, 52–53; in South Vietnam, 305
Livestock. *See* Cattle
Loans. *See* Credit; Credit cooperatives; Credit institutions; Moneylenders
Los Baños Agricultural College (Philippines), 325, 331
Luc, Mayor, 249, 251

MacArthur, Douglas, 17, 148; Communism and, 132, 151–52, 153–54; free enterprise and, 200
—Japanese land reform and, 3, 5–6, 19–20, 287; background of, 150; directive by, 152, 281, 569, 579–80; implementation of, 149, 152–53
McDiarmid, Orville, 487, 491, 492
McNamara, Robert, 17, 474, 550, 557
Madhya Pradesh Land Revenue Code (India), 391–92
Madras (India): agricultural production in, 373; capital expenditures in, 174; class exploitation in, 172–73; conditions in, 162–65, 167–68, 171–72, 173–75, 177; cooperatives in, 34, 415; cultivation expenses in, 166, 167–68, 169–70, 193; debt relief in, 187; eviction in, 165, 166, 170–71, 174–75, 176; land ownership in, 163, 168–69; land reform in, 177–78; land transfer in, 339; landlords in, 162, 169, 172; loans in, 338–39; occupancy rights in, 339; rebellion in, 175, 196; revenue in, 171; sociological neglect in, 406; village-level workers in, 408–09; wages in, 170; water supply in, 172. *See also* Tamil Nadu
—Communism in: crop share and, 169–70; land ownership and, 168–69; land reform and, 165–66, 167–78; rent and, 195

—legislation in, 164; debt, 32; lack of, 163, 175–76, 177, 194, 195–96; recommendations for, 339–40; tenancy status, 174, 176–77, 337
—rent in, 193; Communism and, 195; crop share as, 163, 172; landlord's attitude toward, 174–75; reduction of, 339, 380
—tenancy in, 336, 337; records of, 338–39; security of, 166, 170; system of, 175–76; terms of, 172, 174–75
Madras Agriculturists' Relief Act of 1938 (India), 32
Maharashtra (India): cattle in, 516, 524–26; credit in, 504–05; crop disaster consequences in, 515–16; fodder shortages in, 516, 524–26; land prices in, 392; monsoon in, 515; rents in, 337, 380; wheat production in, 515–16, 517. *See also* Drought
Mahendra, Bir Shah Deva, 17, 315, 316–23, 332–35
Maize in India, 409–10
Malabar Tenancy Act of 1929 (India), 174, 176
Malabar Tenancy Act of 1951 (India), 176–77
Malabar Tenants Improvements Compensation Act (India), 176
Malay Reservation Act of 1913 (British Malaya), 62
Malaya Reservations Committee, 65
Manchuria, *CB–4, 17*
Marcos, Ferdinand, 17, 550, 551, 557
Marginal Farmers and Landless Laborers Scheme (MFAL), (India), 501, 502
Marketing, rural: in India, 416; in Japan, 574, 578; in South Vietnam, 221
Marx, Karl, 131, 287
Mayuram Award of 1948 (India), 165, 168
Mechanization, agricultural: in British Malaya, 64; in India, 384, 479–80, 496–98; in South Vietnam, 245; in Soviet Union, 27, 29
Meiji Era (Japan), 70n, 80, 284, 285
Menon, K. Ramuni, 166, 324
Mensheviks Party (Soviet Union), 148
Mexico, *CB–100*
MFAL (Marginal Farmers and Landless Laborers Scheme) (India), 501, 502
Migration. *See* Immigrants; Refugees
Minhas, R. S., 462, 499, 500
Ministry (or Department) for Agrarian Reform (South Vietnam), 247, 272, 273, 310
Ministry (or Department) of Agriculture and Forestry (Japan): debt and, 42; land commissions and, 111, 112; land ownership and, 40; leases and, 75; living standard and, 77–78; rent and, 41; rice surplus and, 44
Ministry of Finance (Japan), 87
Ministry of Food and Agriculture (India), 202
Mitra, Asok, 16, 495n
Moneylenders: in British Malaya, 64–65; in India, 30–32, 34, 420, 449; in South Vietnam, 221
Monsoon in India, 515, 531
Mortgages: in India, 30–31; in Japan, 74
Moyer, Raymond, 109, 289; Communism and, 206; human welfare and, 213; JCRR and, 136, 137; rent reduction and, 95, 114; U.S. assistance in South Vietnam and, 268
Muafidars, 179
Mukarraries, 179
Mukerjee, Radhakamal, 34, 376n

Nagpur resolution (India), 387
Nasu, Shiroshi: eviction and, 81n; income and, 77n; land prices and, 88n; land utilization and, 40n, 42n, 44, 48n; tenancy leases and, 75n
National Committee for Agrarian Reform (South Vietnam), 272, 302–03
National Cooperative Farming Advisory Board (India), 389
National Council for Agrarian Reform (South Vietnam), 274, 308
National Council of the CPI (Communist Party of India), 485
National Development Council (India), 390, 392
National Rice and Corn Administration (Philippines), 326
National United Front (South Vietnam), 254. *See also* Binh Xuyen; Communism in South Vietnam; Hoa Hao; Viet Minh
Nationalist control in South Vietnam, 228, 246–49
Neale, Walter C., 377n, 393n, 394, 395, 399n
Nehru, Jawaharlal, 132–33, 177, 196, 209, 536
Nepal, 296, *CB–78, 79, 81, 84–86, 89. See also specific entries*
Netherlands Indies. *See* Indonesia; *specific entries*
New Economic Policy (Soviet Union), 25
New York Conference on Land Tenure (1952), 189, 198–201
Noncredit societies in India, 35, 510–11
Norman, E. Herbert, 80n, 285
North Vietnam, 269, 299, 310

Occupancy rights, tenant: in India, 170, 194, 339, 377, 383, 457; in Indonesia, 344; in Nepal, 320; in Taiwan, 357. *See also* Leases
Occupation, colonial: by British in British Malaya, 61–62, 64, 65; by Dutch in Netherlands Indies, 58–60
Occupation, entrepreneurial. *See* French occupation in South Vietnam
Occupation, military: by Japan in Korea, 24, 51, 52, 57–58; by Japan in Taiwan, 104, 105, 106
—by Allies in Japan, 132, 148, 149–50, 151–54; accomplishments of, 93–95, 149–50, 152–54, 287. *See also* MacArthur, Douglas
October Revolution of 1917 (Soviet Union), 25
Okinawa (Japan), 73, 76
Organic Regulations of Chosen Prefectural, District and Island Tenant Committees (Korea), 53
Oriental Development Company (Japan), 52
Overdues in India, 415–16, 506–07

Pakistan. *See* Refugees in India
Panchayats, 538
Pannaiyal, 165, 166, 168, 170–71
Patwari, 156, 157, 397n, 399–400, 546–47
Peasant Proprietors' Agricultural Land Bill of 1932 (Japan), 46
Peasants. *See* Tenants
People's Party (India), 189
Perez, Roderigo, 327, 328
Philippines, 330, 331, *CB–88, 136. See also specific entries*
Planning Board (Nepal), 290, 291, 292
Planning Commission (India), 30; eviction and, 378; first five-year plan and, 377–78; land ceilings and, 382; land policy and, 197, 202, 211; landlord compensation and, 188; Panel on Land Reform and, 377, 378, 379–80; recommendations by, 177, 336, 396–98; report to, 154, 155; role of, 404; tenancy distribution and, 372
Plantations. *See* Agricultural production, large-scale
Planter's Loans Fund (British Malaya), 65
Point IV program, 94, 136
Pope, Cao Dai. *See* Tac, Pham-Cong
Population pressure on land, 366; in India, 38–39, 203, 384–87; in Japan, 40, 68–69, 70–71, 151, 570–71, 579; in South Vietnam, 219; in Taiwan, 95–96
Poverty. *See* Tenant conditions
Presidential Decree No. 27 (Philippines), 551
Prices, agricultural: in Korea for rice, 55; in Taiwan, 143–44, 146
—in India: decline of, 30; for wheat, 433–34
—in Japan: for fertilizer, 45, 573, 578; for rice, 42–44, 76–78, 572–73, 577; for silk, 42, 43, 44–45, 572–73, 577–78
Public Works Administration (Taiwan), 139
Public works program: in India, 501; in Korea, 56–57; in Taiwan, 139
Punjab (India): agricultural productivity in, 432–34; cooperatives in, 34, 35–36; credit in, 421; debt relief in, 187; employment in, 436–39; evacuee land in, 543–48; evictions in, 157–59, 194, 212, 543; Green Revolution in, 432, 436, 439; income gap in, 440–41; land fragmentation in, 373; land grabbing by state in, 542–50; land ownership in, 156–61; land prices in, 159, 160–61, 381, 543, 545; land purchase in, 156; land reform in, 155–61, 392, 441–42, 542–50; land values in, 434–36; mechanization in, 384, 496–97; occupancy rights in, 193; prosperity in, 432–35, 440; rent in, 156, 158, 160, 194–95, 371n, 435–36; technology in, 436, 438–39; tenancy in, 155–59, 160, 208–09, 435–36; wage scales in, 436–38; wheat in, 432–34
—legislation: on debt, 31; effects of, 157–60, 194; enforcement of, 195, 196; remedial measures for, 160–61; on tenancy security, 155–57, 337, 435, 542–43
Punjab Regulation of Accounts Bill of 1930 (India), 31
Punjab Security of Land Tenure Act of 1953 (India), 435, 542–43
Punjab Tenants Security of Tenure Act of 1950 (India), 155, 156, 159
Punjab Tenants Security of Tenure Amendment Act of 1951 (India): enforcement of, 159–60; provisions of, 156; remedies for, 160–61; shortcomings of, 157–59

Rabi crops in India, 445, 515–16, 517, 531
Radio Hanoi, 299, 310
Rainfall in India, 37, 515. *See also* Drought in India
Rajagopahachari, C., 169, 170
Rajasthan (India), 380, 392, 473–74
Ram, Jagjivan, 429–30, 441
Rebellion. *See* Conflict, political

Red Flag Association (India), 167, 195
Records, tenancy: in India, 337, 338–39, 397–98, 482; in South Vietnam, 218, 220, 231
Refugees in India, 487–92, 527–28, 563; land of, 543–48; from West Pakistan, 521
—from East Pakistan, *CB–118*; absorption of, 491; attitude toward, 490; description of, 488–89; dispersal of, 488–89; employment restrictions on, 490–91; living conditions of, 489–90; motivation of, 488; numbers of 487–88; provisions for, 489–90; wages for, 491
Refugees in South Vietnam, 223, 240–41, 246, 305–06
Regulation on the Enforcement of Farm Rent Reduction in Szechwan Province (China), 95, 122–23
Regulation, Registration and Revision of Farm Contracts (China), 119, 121
Reinhardt, Ambassador, 239, 270
Religion: in India, 393; in South Vietnam, *see* Cao Dai; Hoa Hao; Tac, Pham-Cong
Rent, 355; in British Malaya, 62; in China, 95, 114, 117, 118–19, 121–23, 127–28, 134, 135, 141; in Indonesia, 343–44; in Japan, 41, 76, 78, 79, 571–72; in Korea, 52; in Nepal, 318, 319–20, 332–33, 334, 357; in Philippines, 326–27; in United States, 571n
—in India, 370, 373, 391; control of, 337–38, 396; legislation on, 180, 371n, 392, 435; Rent Court and, 339; under zamindari areas, 190–91, 371, 376–77; Communism and, 169–70, 195. *See also specific regions*
—in South Vietnam, 230, 255–56, 258, 262–63, 305; attitude toward, 217; French and, 256, 257; Viet Minh and, 217, 225. *See also* Rent reduction program in South Vietnam
—in Taiwan, 45, 96–97, 98, 99, 100–01, 357; black market in, 101, 102; sugar cane and, 106–07. *See also* Rent reduction program in Taiwan
Rent Court (India), 339
Rent reduction program, 232–33, 312, 356–57; in India, 357, 482; in Japan, 90, 264, 357
—in South Vietnam: effects of, 224, 236–37, 264, 266; landlords and, 232–33; legislation for, 222–23, 302–04
—in Szechwan (China), 114–15, 117–18; assessment of, 127–29; deposit money under, 123–25; evictions under, 122–23, 126; implementation of, 118–26; land commissions under, 125–27; landlords' nonparticipation in, 119–21, 134, 135; lease contracts under, 119–20
—in Taiwan, 95, 357; acceptance of, 105; benefits of, 103, 128, 129, 134, 140; campaign committees and, 104–05; evasions of, 101–02; failings of, 100–08; provisions of, 98–101; results of, 142–47
Reparations from Japan to South Vietnam, 239, 243
Report of the Trustees of the Ford Foundation (1950), 204, 207
Research in India, 36–37, 411
Reserve Bank (India), 413, 414–15
Resettlement. *See* Immigrants; Refugees
Revenue. *See* Land tax
Revolution, agrarian: Communism and, 279–81, 300; in France, 186, 199, 395; in Japan, 47, 279–88; in Soviet Union, 23, 25, 148; in Taiwan, 279–80
Rice: in China, 115–16; in India, 30, 444–45, 534, 558–59; in Indonesia, 349, 350, 356; in Korea, 55, 573, 577–78; in Philippines, 325, 326, 331; in South Vietnam, 244–45, 246, 256, 301. *See also* Rent
—in Japan: crop share of, 41, 76, 78, 571–72; imports of, 573, 577–78; legislation for, 44; prices of, 42, 44, 76–78, 572–73, 577; surplus, 44; yields of, 42, 151, 282–83
—in Taiwan: consumption of, 97; crop share of, 96–97, 99, 100–01, 106–07; exports of, 573, 577–78
Rinderpest, 139
Roxas, Sixto, 327, 331
Royal Commission on Agrarian Reform (Nepal), 319–20, 321
Royal Commission on Agriculture (India), 39; agricultural research and, 36–37; cooperatives and, 35, 36; debt and, 31n; illiteracy and, 38; irrigation and, 37; land fragmentation and, 373
Royal Commission on Taxation and Birta Abolition (Nepal), 323
Rubber production: in British Malaya, 61, 62, 63, 64, 65; in Netherlands Indies, 60; in South Vietnam, 244
Rubber Research Institute of Malaya, 64
Rural Works Program (RWP) (India), 501

Sadjarwo, Minister, 17, 109, 340–52
Samurai, 284
Savings: in Korea, 56; in Japan, 79

SCAP. *See* Supreme Commander for the Allied Powers
Schools. *See* Education
Scissors problem: in Taiwan, 143; in Soviet Union, 26
Seed project in Taiwan, 140
Self-Help Movement (Korea), 56–57
Sen, Ranjit, 498n, 499n
Serednaki, 26
Serfdom. *See* Feudalism
Settlements. *See* Litigation
SFDA. *See* Small Farmer Development Agency
Sharecroppers: in India, 455, 456–57; in Japan, 75
Shen, Tsung-han, 95, 117, 136, 148
Siberia (Soviet Union), 148
Silk, 42, 43, 44–45, 572–73, 577–78
Singh, Srinath, 479, 495n
Sino-Japanese War, 47, 48
Small Farmer Development Agency (SFDA) (India), 466–68, 501, 502–03, 509
Small farmers, *CB–104, 132. See also* Agricultural production, small-scale; Tenants
Socialism. *See* Collectivization
South Keisho Province (Korea), 51, 54
South Korea. *See* Korea; *specific entries*
South Vietnam, 215–16, 265, *CB–60–71, 75, 82*; and Cochin-China, 230; economic condition of, 276, 303, 311–12; journalistic attitude toward, 277. *See also* Central Vietnam; *specific entries*
Soviet Union, *CB– 1, 2, 3, 7, 8. See also specific entries*
Sri Lanka (Ceylon), *CB–25, 141*
Srinagar Board (India), 187
Stalin, J., 25n, 27n, 131, 132
Sterilization in India, 409
Subramanian Committee (Land Revenue Reforms Committee of Madras of 1950), 171
Sugar cane in Taiwan, 106–07, 142–43, 146
Sukarno, President, 17, 298, 299
Sumatra (Indonesia), 59, 340–41, 346, 348–49, 351
Sun, Lien Chuan, 114, 119, 123–24, 125, 126, 127
Supreme Commander for the Allied Powers (SCAP), 6, 94, 95, 579
Swaminathan, M. S., 496n, 536–37, 538
Syndicate Bank (India), 421
Szechwan (China): characteristics of, 114, 115, 130–31; crops in, 115–16; double occupations in, 116; food in, 116; JCRR and, 137, 140; land holding size in, 116; lease contracts in, 116, 122–23; rent in, 95, 114, 118–19, 121–23, 127–28; rice in, 115–16; tenancy in, 115–17

Tac, Pham-Cong, 241, 249; attitude toward Bao Dai, 253–54; attitude toward National United Front, 254; attitude toward President Diem, 253; church of, 251; description of, 251–52; economics of, 252; land reform and, 252; political power struggle and, 253–54; religion of, 252–53
Taccavi loans, 419, 448–49
Taisho Era, 90
Taiwan (Republic of China), *CB–35, 37, 38, 40, 41, 73, 76. See also specific entries*
Taiwan Sugar Corporation, 106, 142, 146
Tam, Nguyen Van, 230, 233, 302–03, 307
Tamil Nadu (India), 473, 483–84
Tanjore Kisan Movement, 169
Tanjore Tenants and Pannaiyal (Protection) Ordinance of 1952 (India), 162; background of, 164, 166; evaluation of, 169–71; exemptions to, 171; farm labor and, 168; landlords and, 169, 171; official attitude and, 166, 176; provisions of, 166, 195; rent reduction and, 371n; shortcomings of, 163; tenants' rights and, 167–68, 170; wages and, 171
Tariff policies in Japan, 572
Taxes. *See* Land tax
Technology in India, 442–43; consequences of, 494–95, 496, 499; employment and, 436, 438–39; extension services and, 409–10; land reform and, 539; production and, 406, 432–34, 495–96; scope of, 441; small farmers and 452–54; social implications of, 462–65. *See also* Industrialization; Mechanization, agricultural
Temporary Farm Debts Liquidation Law of 1938 (Japan), 47
Temporary Prefectural Committee for the Liquidation of Debts (Japan), 48
Tenancy, *CB–5, 27, 28, 37, 84, 91*; in British Malaya, 61–62; human welfare and, 208–09; in Korea, 49–53; in Nepal, 316–17; in Taiwan, 101–02
—in India: distribution of, 372; land improvement compensation and, 370; reform, 377–80; regulation on, 358; restoration of, 458; system, 175–76; terms of, 180, 183; underground, 398
—in Japan, 22–23, 41, 68; extent of, 72–73;

legislation on, 46, 83–84; reduction of, 84–86; reform, 45–46, 47; rise of, 73–74; terms of, 77
Tenancy Act of 1885 (India), 33, 337, 391
Tenancy Act of 1920 (India), 33
Tenant conditions: in China, 115–17; in India, 336–40, 371–72, 454–55, 460–61, 476–78, 498–500; in Indonesia, 351; in Japan, 74–80, 571–72, 574–76, 579; in Korea, 52–53; in Philippines, 326–30; in Nepal, 317, 318–19; in South Vietnam, 245, 249, 250, 266–67, 300–02; in Taiwan, 95–100, 102–03
Tenant eviction: in China, 122–23, 126; in Japan, 42, 47–48, 81, 83–84; in Korea, 52–53
—in India, 182–83, 378–79, 543; landlessness and, 34, 383–87, 501, 502, 561–62; legislation against, 337; reinstatement and, 166, 170–71, 180; replacement and, 165
Tenants: in Korea, 50–51; in South Vietnam, 225, 233–34, 248; in Taiwan, 96–97, 99, 100–05. *See also* Records, tenancy; Sharecroppers; Tenant unions
—in India, 157, 172, 174–75, 191, 391, 435–36; conversion to owners, 380–81; pannaiyals as, 165, 166, 168, 170–71; reform participation by, 363–64, 393–95, 401, 483; underground, 398
—in Japan: expenditures of, 77–78; landlord relations with, 41–42, 46, 80, 83; policies of, 83–86; practices of, 83, 84, 89–91; responsibilities of, 77; unrest of, 80–81
Tenant status: in Great Britain, 74; in Indonesia, 344–45, 351; in Japan, 72, 78, 83–86, 364–65; in South Vietnam, 220, 231, 245–46; in Soviet Union, 23, 25
—in India, 32–34, 155–56, 157–59, 191–92, 202, 208–09, 377, 455, 461; legislation on, 33–34, 174, 176–77, 337, 391; social structure and, 393–95; undertenants and, 33–34. *See also* Tenant eviction
Tenant unions: in Japan, 81–82; in South Vietnam, 255–56, 305
Tenure, insecurity of: in Japan, 75; in Taiwan, 97–98
Tenure, security of, 355, 356–57; in China, 116; in Taiwan, 103. *See also* Occupancy rights
—in India, 160, 202–03, 370, 565; legislation on, 155, 180, 435, 457, 458–59, 542–43; term for, 166, 170
Texan Industries (India), 437
Thailand, *CB–20*
Thapar, Ashok, 410n, 411n, 503n, 519n, 535n
Thorner, Daniel, 377n, 388n, 393n, 425–26, 427n
Threshers. *See* Mechanization, agricultural
Tinker, Hugh, 393n, 394
Tokugawa regime, 73, 284
Tractors. *See* Mechanization, agricultural
Tube wells. *See* Irrigation

Uenoda, Setsuo, 42n, 55n
Ukraine (Soviet Union), 148, 149
Umali, Dean, 327, 330, 331
Undertenants in India, 33–34
United Nations, 293
United Provinces. *See* Uttar Pradesh
United States, 288; external aid of, 134, 136–41, 145, 197, 250–51, 264–67, 268–70, 273–79, 293; foreign policy of, 128, 129, 133–34, 242, 278; Operations Mission in Vietnam, 265, 266, 273–74
Universal Church. *See* Cao Dai; Tac, Pham-Cong
University of the Philippines, 325
Untouchables, 163, 165, 172, 173
Uprisings. *See* Conflict, political
USOM (South Vietnam), 265, 266, 273–74
Usurious Loans Act of 1918 (India), 31
Usurious Loans Amendment Act of 1934 (India), 32
Uttar Pradesh (India), 36; bullocks in, 479; conditions in, 475–76; credit cooperatives in, 478; fertilizer use in, 478–79; Green Revolution in, 476, 478–79; irrigation in, 476–77; land ceilings in, 476; land consolidation in, 480; land distribution in, 477; land policy in, 202, 338; landlords in, 372n, 379; legislation in, 337, 381; mechanization absent in, 479–80; patwari in, 397n, 399; rent in, 190, 194; revenue in, 377, 480–81; tenancy in, 377, 397, 476, 477–78; tube wells in, 476–77, 480, 481; water shortage in, 476, 481; wheat yields in, 479

Viet Minh, 258; attitude toward, 217, 225–26; control by, 218–19, 260; education and, 222; influence of, 227, 233, 234–37, 248–49; land policies of, 222, 224–25, 230; landlords and, 224–25; political conflict with, 228–29; protection against, 235–36; rent and, 217, 225; support for, 259; soldiers of, 260; taxes and, 220, 225, 259; tenants and, 225
Vietnam. *See* Central Vietnam; North Vietnam; South Vietnam; *specific entries*

Village Land Reform Committees (Indonesia), 348
Village-level worker (VLW) in India, 407–09, 412–13

Wada, Hiroo, 286, 287
Wages. *See* Income
War. *See* Conflict, political
Water supply in India, 172; rainfall, 37, 515; shortage, 476, 481, 515, 516, 526–27. *See also* Drought in India; Green Revolution in India: Irrigation
West Bengal (India): agricultural development constraints in, 564–65; cooperatives in, 34; debt relief in, 187; drought in, 559; gruel kitchens in, 559–61; land grab in, 485–86; land ownership in, 565; land sales in, 562–63; landless conditions in, 561–62; landlord land resumption in, 379; legislation in, 33, 484; occupancy rights in, 192; refugees in, 487–92, 563, *CB–118*; rent in, 380; rice hoarding in, 558–59; tenure security measures in, 565–66
—food in: rationing of, 558, 559–60; shortage of, 557–66; smuggling of, 563–64
West Pakistan, 521
Wheat: in India, 432–34, 443–44, 479, 494; in Soviet Union, 26
Working Group on Cooperative Farming (India), 387–88
World Bank, 21–22, 283–84, 367, 462, 475

Yagi, Yoshinosuke, 42n, 45n, 46n, 85n, 87n, 88n
Yoshida, Shigeru, 286, 287

Zamindari, 187n, 362; abolition of, 190–92, 194, 202, 376–77, 397, 471; rent under, 190–91, 371, 376–77
Zamindari Abolition and Land Reform Act (India), 397